Lecture Notes
in Business Information Processing 308

More information about this series at http://www.springer.com/series/7911

Ernest Teniente · Matthias Weidlich (Eds.)

Business Process Management Workshops

BPM 2017 International Workshops
Barcelona, Spain, September 10–11, 2017
Revised Papers

 Springer

Editors
Ernest Teniente 🄳
Department of Service and Information
  System Engineering
Universitat Politècnica de Catalunya
Barcelona
Spain

Matthias Weidlich 🄳
Humboldt-Universität zu Berlin
Berlin, Berlin
Germany

ISSN 1865-1348   ISSN 1865-1356 (electronic)
Lecture Notes in Business Information Processing
ISBN 978-3-319-74029-4   ISBN 978-3-319-74030-0   (eBook)
https://doi.org/10.1007/978-3-319-74030-0

Library of Congress Control Number: 2017963774

Printed on acid-free paper

This Springer imprint is published by Springer Nature
The registered company is Springer International Publishing AG
The registered company address is: Gewerbestrasse 11, 6330 Cham, Switzerland

Preface

Business process management (BPM) is a widely established discipline with a rich body of techniques, methods, and tools. It is also a mature yet highly dynamic field of research, which brings together scholars and industrial researchers from the computer science, information systems, and management fields.

The 15th edition of the International Conference on Business Process Management (BPM 2017) was held in Barcelona, Spain, during September 10–15, 2017. The BPM conference series has a record of attracting innovative research of the highest quality related to all aspects of BPM, including theory, frameworks, methods, techniques, architectures, and empirical findings.

It is a tradition that several workshops precede the main BPM conference on a range of BPM-related topics. The collective goal of these workshops is to promote work-in-progress research that has not yet reached its full maturity, but that has a clear goal and approach as well as promising insights or preliminary results.

In 2017, the following 11 workshops were hosted:

- **BPAI 2017** – First International Workshop on Business Process Innovation with Artificial Intelligence
- **BPI 2017** – 13th International Workshop on Business Process Intelligence
- **BP-Meet-IoT 2017** – First International Workshop on Ubiquitous Business Processes Meeting Internet of Things
- **BPMS2 2017** – 10th Workshop on Social and Human Aspects of Business Process Management
- **CBPM 2017** – First International Workshop on Cognitive Business Process Management
- **CCABPM 2017** – First International Workshop on Cross-Cutting Aspects of Business Process Modeling
- **DeHMiMoP 2017** – 5th International Workshop on Declarative/Decision/Hybrid Mining and Modeling for Business Processes
- **QD-PA 2017** – First International Workshop on Quality Data for Process Analytics
- **REBPM 2017** – Third International Workshop on Interrelations Between Requirements Engineering and Business Process Management
- **SPBP 2017** – First Workshop on Security and Privacy-Enhanced Business Process Management
- **TAProViz-PQ-IWPE 2017** – Joint International BPM 2017 Workshops on Theory and Application of Visualizations and Human-Centric Aspects in Processes (TAProViz 2017), Process Querying (PQ 2017) and Process Engineering (IWPE 2017)

Some of these workshops are well established, having been held for several years. They address long-standing challenges such as how to seamlessly incorporate social and human aspects into process improvement and innovation methods, or how to leverage data analytics techniques for process and decision management. Around half of the workshops, however, were held in their first edition this year. These latter workshops address emerging concerns such as the Internet of Things, artificial intelligence applied to business process innovation, BPM applied to e-government, cognitive business process management, or quality data for process analytics.

We would like to express our sincere gratitude to the individual workshop chairs for their effort to promote, coordinate, and animate their respective workshops and for arranging entertaining, high-quality programs that were well received by the attendees. We are also grateful for the service of the countless reviewers, who supported the workshop chairs and provided valuable feedback to the authors. Several workshops had invited keynote presentations that framed the presented research papers and we would like to thank the keynote speakers for sharing their insights. We would also like to thank Ralf Gerstner and the team at Springer for their support in the publication of this LNBIP volume.

Finally, we are grateful to Josep Carmona and the members of his team for all their efforts in organizing the BPM 2017 conference and the BPM 2017 workshops.

November 2017

Ernest Teniente
Matthias Weidlich

Contents

**13th International Workshop on Business
Process Intelligence (BPI 2017)**

1st International Workshop on Business Processes Meet Internet-of-Things (BP-Meet-IoT 2017)

10th Workshop on Social and Human Aspects of Business Process Management (BPMS2 2017)

**1st International Workshop on Cognitive Business
Process Management (CBPM 2017)**

1st International Workshop on Cross-cutting Aspects of Business Process Modeling (CCABPM 2017)

5th International Workshop on Declarative/Decision/Hybrid Mining and Modeling for Business Processes (DeHMiMoP 2017)

**1st Workshop on Security and Privacy-Enhanced Business Process
Management (SPBP 2017)**

**Joint International BPM 2017 Workshops on Theory
and Application of Visualizations and Human-Centric Aspects
in Processes (TAProViz 2017), Process Querying (PQ 2017),
and Process Engineering (IWPE 2017)**

1st International Workshop on Business Process Innovation with Artificial Intelligence (BPAI 2017)

Introduction to the 1st International Workshop on Business Process Innovation with Artificial Intelligence (BPAI 2017)

Riccardo De Masellis[1], Chiara Di Francescomarino[1], Jana Koehler[2],
Fabrizio Maria Maggi[3], Marco Montali[4], Arik Senderovich[5],
Biplav Srivastava[6], and Heiner Stuckenschmidt[7]

[1] Fondazione Bruno Kessler, Trento, Italy
{r.demasellis,dfmchiara}@fbk.eu
[2] Lucerne University of Applied Sciences and Arts, Lucerne, Switzerland
jana.koehler@hslu.ch
[3] University of Tartu, Tartu, Estonia
f.m.maggi@ut.ee
[4] Free University of Bolzano, Bolzano, Italy
montali@inf.unibz.it
[5] Technion - Israel Institute of Technology, Haifa, Israel
sariks@technion.ac.il
[6] IBM Research, Yorktown Heights, USA
biplavs@us.ibm.com
[7] University of Mannheim, Mannheim, Germany
heiner@informatik.uni-mannheim.de

Abstract. Artificial Intelligence (AI) is receiving high interest from academics, business professionals, and media. It is considered as the next disruptive technology that will significantly impact the workplace and change, innovate, and automate a manifold of business activities. The goal of the workshop was to foster the exchange between AI and Business Process Management (BPM) by taking a closer look at how BPM inspires novel application domains for AI, as well as at how BPM and related fields can benefit from AI solutions. Six full and four short papers were accepted for presentation at the workshop. They stimulated an interesting discussion on potential future synergies between the two disciplines.

Keywords: Artificial intelligence · Machine learning
Business process modelling · Business process mining

1 Aims and Scope

AI is receiving high interest from academics, business professionals, and media, being considered as the next disruptive technology that will impact millions of jobs and change, innovate, and automate a manifold of business activities.

AI examples are:

- Machine Learning and decision-theoretic models for data analysis, e.g., predictive monitoring and customer segmentation;
- Constraint reasoning and related algorithms as the key technology underlying business rule engines;
- Search algorithms for process optimization;
- Intelligent assistants and companions.

These are only a few examples of what AI technology can do to improve and reengineer business processes. Recently, major IT companies have developed cognitive services to make AI technology ready to use for developing applications. Many large companies started projects to plug these services into their processes or to develop their own AI solutions based on these services. At the same time, AI researchers are discussing safety issues and identifying important sources of risks in AI solutions.

The workshop identified many potential sources for synergies between AI and BPM. On the one hand, several AI solutions can be used in the context of BPM, e.g., planning for adapting or composing business processes, machine learning for process mining and analysis, constraint reasoning for process transformation, verification, and compliance checking.

On the other hand, AI will influence the role of humans within a business process and solutions should be developed to address questions such as novel requirements on employee qualification, shared responsibilities between AI and humans, control and impact of automated decision making.

Six full and four short papers were presented at the workshop. The focus of the papers ranged from leveraging Machine Learning approaches for addressing BPM problems to applying planning approaches in BPM scenarios on the one hand, and from analyzing event logs using AI techniques to finding optimal paths in business processes on the other hand.

In particular, Hinkka and colleagues addressed the problem of classifying business process instances based on structural features derived from event logs. Back and colleagues investigated how different measures for entropy could be used to give insights on the complexity of an event log and could act as an indicator of which paradigm (imperative or declarative) should be used for process mining. Comuzzi presented a framework for calculating optimal execution paths in business processes by relying on workflow hypergraph abstraction and using an ant-colony optimisation customised for the hypergraph traversal. Koehler and colleagues showed how AI can support service processes in a variety of ways by proposing three intelligent assistants that support service employees in their complex tasks. Baldoni and colleagues proposed to enrich the definition of business artifact with a normative layer by relying on a multiagent systems approach. De Masellis and colleagues focused on automatically repairing traces with missing information by notably considering not only activities but also data manipulated by them. Wiśniewski and Kluza presented a method of business process composition based on constraint programming. Rietzke and colleagues presented a semantically oriented business process visualization approach developed using a knowledge-based system. Chesani and colleagues leveraged on the abductive declarative language SCIFF for the realization of an event log generator. Finally, Sulis and Di

Leva applied simulation approaches to a real case study of an hospital emergency department.

The importance of the synergy between the AI and BPM fields also came out in the keynote by Andrea Marrella. He showed how automated planning techniques can be leveraged to enable new levels of automation and support for BPM. He discussed several concrete examples of successful application of AI techniques to the different stages of the BPM lifecycle.

The workshop ended with an interesting panel session about the relevance of the synergy between AI and BPM for both fields. Moreover, the discussion covered the future of the BPAI workshop. Everybody agreed to continue this experience at the next year's BPM conference and to take the chance to further foster the interaction between AI and BPM by targeting with the call for paper also other research areas at the intersection between the two fields.

To sum up, the workshop received a significant number of submissions and a good attendance. It clearly showed the strong interest of the BPM community in how the AI and BPM fields can potentially fertilize each other. Of course, many of the questions raised in the call for papers remained unaddressed by this year's submissions, e.g., assessing the business risks of AI technologies, understanding the interplay of humans and robots in business processes, using AI technologies such as AI planning to create dynamic business processes. However, we believe that this only shows how much more can be done in the future in this research direction.

2   Workshop Co-organizers

Riccardo De Masellis	Fondazione Bruno Kessler, Italy
Chiara Di Francescomarino	Fondazione Bruno Kessler, Italy
Jana Koehler	Lucerne University of Applied Sciences and Arts, Switzerland
Fabrizio Maria Maggi	University of Tartu, Estonia
Marco Montali	Free University of Bolzano, Italy
Arik Senderovich	Technion - Israel Institute of Technology, Israel
Biplav Srivastava	IBM Research, USA
Heiner Stuckenschmidt	University of Mannheim, Germany

3   Program Committee

Rama Akkiraju	IBM Almaden Research, USA
Ralph Bergmann	University of Trier, Germany
Andrea Burattin	University of Innsbruck, Austria
Federico Chesani	University of Bologna, Italy
Raffaele Conforti	Queensland University, Australia
Giuseppe De Giacomo	Sapienza University of Rome, Italy
Claudio Di Ciccio	Vienna University of Economics and Business, Austria

Peter Fettke	German Research Center for Artificial Intelligence (DKFI), Germany
Chiara Ghidini	Fondazione Bruno Kessler, Italy
Anna Leontjeva	University of Tartu, Estonia
Henrik Leopold	Vrije Universiteit Amsterdam, Netherlands
Andrea Marrella	Sapienza University of Rome, Italy
Paola Mello	University of Bologna, Italy
Fabio Patrizi	Sapienza University of Rome, Italy
Giulio Petrucci	Fondazione Bruno Kessler, Italy
Andreas Rogge-Solti	Vienna University of Economics and Business, Austria
Ute Schmid	University of Bamberg, Germany
Irene Teinemaa	University of Tartu, Estonia
Stefano Teso	University of Trento, Italy
Ingo Weber	University of New South Wales, Australia

What Automated Planning Can Do for Business Process Management

Andrea Marrella[(✉)]

Sapienza - University of Rome, Rome, Italy
`marrella@diag.uniroma1.it`

Abstract. Business Process Management (BPM) is a central element of today organizations. Despite over the years its main focus has been the support of processes in highly controlled domains, nowadays many domains of interest to the BPM community are characterized by ever-changing requirements, unpredictable environments and increasing amounts of data that influence the execution of process instances. Under such dynamic conditions, BPM systems must increase their level of automation to provide the reactivity and flexibility necessary for process management. On the other hand, the Artificial Intelligence (AI) community has concentrated its efforts on investigating dynamic domains that involve active control of computational entities and physical devices (e.g., robots, software agents, etc.). In this context, Automated Planning, which is one of the oldest areas in AI, is conceived as a model-based approach to synthesize autonomous behaviours in automated way from a model. In this paper, we discuss how automated planning techniques can be leveraged to enable new levels of automation and support for business processing, and we show some concrete examples of their successful application to the different stages of the BPM life cycle.

1 Introduction

Business Process Management (BPM) is a central element of today organizations due to its potential for increase productivity and saving costs. To this aim, BPM research reports on techniques and tools to support the design, enactment and optimization of business processes [1]. Despite over the years the main focus of BPM has been the support of processes in highly controlled domains (e.g., financial and accounting domains), nowadays BPM research is expanding towards new challenging domains (e.g., healthcare [2], smart manufacturing [3], emergency management [4,5], etc.), characterized by ever-changing requirements, unpredictable environments and increasing amounts of data that influence the execution of process instances [6]. Under such dynamic conditions, *BPM is in need of techniques that go beyond hard-coded solutions* that put all the burden on IT professionals, which often lack the needed knowledge to model all possible contingencies at the outset, or this knowledge can become obsolete as process instances are executed and evolve, by making useless their initial effort. Therefore, there are compelling reasons to introduce *intelligent techniques* that

© Springer International Publishing AG 2018
E. Teniente and M. Weidlich (Eds.): BPM 2017 Workshops, LNBIP 308, pp. 7–19, 2018.
https://doi.org/10.1007/978-3-319-74030-0_1

act *autonomously* to provide the reactivity and flexibility necessary for process management [7, 8].

On the other hand, the challenge of building computational entities and physical devices (e.g., robots, software agents, etc.) capable of *autonomous behaviour* under dynamic conditions is at the center of the Artificial Intelligence (AI) research from its origins. At the core of this challenge lies the *action selection problem*, often referred as the *problem of selecting the action to do next*. Traditional hard-coded solutions require to consider every option available at every instant in time based on the current context and pre-scripted plans to compute just one next action. Consequently, they are usually biased and tend to constrain their search in some way. For AI researchers, the question of action selection is: *what is the best way to constrain this search?* To answer this question, the AI community has tackled the action selection problem through two different approaches [9], one based on *learning* and the other based on *modeling*.

In the *learning-based* approach, the controller that prescribes the action to do next is *learned from the experience*. Learning methods, if properly trained on representative datasets, have the greatest promise and potential, as they are able to discover and eventually interpret meaningful patterns for a given task in order to help make more efficient decisions. For example, learning techniques were recently applied in BPM (see [10]) for predicting future states or properties of ongoing executions of a business process. However, a learned solution is usually a "black box", i.e., there is not a clear understanding of how and why it has been chosen. Consequently, the ability to explain why a learned solution has failed and fix a reported quality bug may become a complex task.

Conversely, in the *model-based* approach the controller in charge of action selection is *derived automatically* from a model that expresses the dynamics of the domain of interest, the actions and goal conditions. The key point is that all *models are conceived to be general*, i.e., they are not bound to specific domains or problems. The price for generality is *computational*: The problem of solving a model is computationally intractable in the worst case, even for the simplest models [9].

While we acknowledge that both the *learning* and *model-based* approaches to action selection exhibit different merits and limitations, in this paper we focus on a specific model-based approach called *Automated Planning*. Automated planning is the branch of AI that concerns the automated synthesis of autonomous behaviours (in the form of strategies or action sequences) for specific classes of mathematical models represented in compact form. In recent years, the automated planning community has developed a plethora of *planning systems* (also known as *planners*) that embed very effective (i.e., scale up to large problems) domain-independent heuristics, which has been employed to solve collections of challenging problems from several Computer Science domains.

In this paper, we discuss how automated planning techniques can be leveraged for solving real-world problems in BPM that were previously tackled with hard-coded solutions by enabling new levels of automation and support for business processing and we show some concrete examples of their successful application to the different stages of the BPM life cycle. Specifically, while in Sect. 2 we introduce some preliminary notions on automated planning necessary to understand the rest of the paper, in Sect. 3 we show how instances of some well-known problems from the BPM literature (such as process modeling, process adaptation and conformance checking) can be represented as planning problems for which planners can find a correct solution in a finite amount of time. Finally, in Sect. 4 we conclude the paper by providing a critical discussion about the general applicability of planning techniques in BPM and tracing future work.

2   Automated Planning

Automated planning addresses the problem of generating autonomous behaviours (i.e., *plans*) from a *model* that describes how *actions* work in a *domain* of interest, what is the *initial state* of the world and the *goal* to be achieved. To this aim, automated planning operates on *explicit representations* of states and actions.

The Planning Domain Definition Language (PDDL) [11] is a de-facto standard to formulate a compact representation of a *planning problem* $\mathcal{P}_\mathcal{R} = \langle I, G, \mathcal{P}_\mathcal{D} \rangle$, where I is the description of the initial state of the world, G is the desired goal state, and $\mathcal{P}_\mathcal{D}$ is the planning domain. A planning domain $\mathcal{P}_\mathcal{D}$ is built from a set of *propositions* describing the *state* of the world in which planning takes place (a state is characterized by the set of propositions that are true) and a set of *actions* Ω that can be executed in the domain. An *action schema* $a \in \Omega$ is of the form $a = \langle Par_a, Pre_a, Eff_a, Cost_a \rangle$, where Par_a is the list of *input parameters* for a, Pre_a defines the *preconditions* under which a can be executed, Eff_a specifies the *effects* of a on the state of the world and $Cost_a$ is the *cost* of executing a. Both preconditions and effects are stated in terms of the *propositions* in $\mathcal{P}_\mathcal{D}$, which can be represented through boolean *predicates*. PDDL includes also the ability of defining *domain objects* and of *typing* the parameters that appear in actions and predicates. In a state, only actions whose preconditions are fulfilled can be executed. The values of propositions in $\mathcal{P}_\mathcal{D}$ can change as result of the execution of actions, which, in turn, lead $\mathcal{P}_\mathcal{D}$ to a new state. A *planner* that takes in input a planning problem $\mathcal{P}_\mathcal{R}$ is able to automatically produce a *plan* P, i.e., a controller that specifies which actions are required to transform the initial state I into a state satisfying the goal G.

Example 1. Let us consider a well-known domain in AI: the Blocks World. In Fig. 1 it is shown an instance of this domain where blocks A, B and C are initial arranged on the table. The goal is to re-arrange the blocks so that C is on A and A is on B.

The problem can be easily expressed as a planning problem in PDDL. The variables are the *block locations*: Blocks can be on the table or on top of another block. Two predicates can be used to express (respectively) that a block is *clear* (i.e., with no block on top) or has another block on *top* of it. A single planning action *move* is required to represent the movement of a clear block on top of

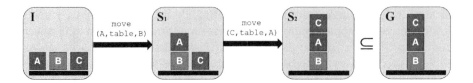

Fig. 1. A plan that solves a simple planning problem from the Blocks World domain.

another block or on the table. Figure 1 shows a possible solution to the problem, which consists of first moving A from the table on top of B (state S_1), and then on moving C from the table on top of A (state S_2). Since S_2 is a state satisfying the goal G, the solution found is a valid *plan*. Furthermore, if we assume that the cost of any *move* action is equal to 1 (i.e., the cost of the plan corresponds to its length), then the plan found is *optimal*, as it contains the minimum number of planning actions to solve the problem. □

There exist several forms of planning models in the AI literature, which result from the application of some orthogonal dimensions [12]: uncertainty in the initial state (fully or partially known) and in the actions dynamics (deterministic or not), the type of feedback (full, partial or no state feedback) and whether uncertainty is represented by sets of states or probability distributions. The simplest form of planning where actions are deterministic, the initial state is known and plans are action sequences computed in advance is called *Classical Planning*. For classical planning, according to [9], the general problem of coming up with a plan is NP-hard, since the number of problem states is exponential in the number of problem variables. For example, in a arbitrary Blocks World problem with n blocks, the number of states is exponential in n, since the states include all the $n!$ possible towers of n blocks plus additional combinations of lower towers.

Despite its complexity, the field of classical planning has experienced spectacular advances (in terms of scalability) in the last 20 years, leading to a variety of concrete planners that are able to feasibly compute plans with thousands of actions over huge search spaces for real-world problems containing hundreds of propositions. Such progresses have been possible because state-of-the-art classical planners employ powerful *heuristic functions* that are automatically derived by the specific problem and allow to *intelligently* drive the search towards the goal. In addition, since the classical approach of solving planning problems can be too restrictive for environments in which information completeness can not be guaranteed, it is often possible to solve non-classical planning problems using classical planners by means of well-defined transformations [13]. A tutorial introduction to planning algorithms and heuristics can be found in [9].

3   Automated Planning for BPM

The planning paradigm (in particular in its classical setting) provides a valuable set of theoretical and practical tools to tackle several challenges addressed by the BPM community and its use may lead to several advantages:

- Planning models are *general*, in the sense that a planner can be fed with the description of any planning problem in PDDL (as defined in Sect. 2) without knowing what the actions and domain stand for, and for any such description it can synthesize a plan achieving the goal. *This means that planners can potentially solve any BPM problem that can be converted into a planning problem in PDDL.*
- Planning models are *human-comprehensible*, as the PDDL language allows to describe the planning domain and problem of interest in a high-level terminology, which is *readily accessible and understandable by IT professionals.*
- The *standardized representation* of a planning model in PDDL allows to *exploit a large repertoire of planners and searching algorithms* with very limited effort.
- Planning models, if encoded with the classical approach, constitute implicit representations of *finite state controllers*, and can be thus *queried by standard verification techniques*, such as Model Checking.
- BPM environments can invoke planners as *external services*. Therefore, *no expertise of the internal working of the planners* is required to build a plan.

A number of research works exist on the use of planning techniques in the context of BPM, covering the various stages of the process life cycle. For the *design-time* phase, existing literature has focused on exploiting planning to automatically generate candidate process models that are able of achieving some business goals starting from a complete or an incomplete description of the process domain. Some research works also exist that use planning to deal with problems for the *run-time* phase, e.g., to adapt running processes to cope with anomalous situations. Finally, for the *diagnosis phase*, the literature reports some works that use planning to perform conformance checking.

In the following sections we discuss in the detail how the use of planning has contributed to tackle the above research challenges from BPM literature.

3.1   Planning for the Automated Generation of Process Models

Process modeling is the first and most important step in the BPM life cycle, which intends to provide a high-level specification of a business process that is independent from implementation and serves as a basis for process automation and verification. Traditional process models are usually well-structured, i.e., they reflect highly repeatable and predictable routine work with low flexibility requirements. All possible options and decisions that can be made during process enactment are statically pre-defined at design-time and embedded in the control-flow of the process.

Challenge. Current BPM technology is generally based on rigid process models making its application difficult in dynamic and possibly evolving domains, where pre-specifying the entire process model is not always possible. This problem can be mitigated through specific approaches to process variability [14], which allow to customize a base process model by implementing specific variants of the process itself. However, this activity is time-consuming and error-prone when more flexible processes are to be modeled, due to their context-dependent nature that make difficult the specification of all the potential tasks interactions and process variants in advance. *The presence of mechanisms that facilitate the design phase of flexible processes by allowing the automated generation of their underlying models is highly desirable* [7].

Application of Planning. The work [15] presents the basic idea behind the use of planning techniques for the *automated generation of a process model*. Process activities can be represented as planning actions together with their preconditions and effects stating what contextual data may constrain the process execution or may be affected after an activity completion. The planning domain is therefore enriched with a set of propositions that characterize the contextual data describing the process domain. Given an initial description of the process domain, the target is to automatically obtain a plan (i.e., a process model and its control-flow) that consists of process activities contextually selected and ordered in a way to satisfy some business goals.

In the research literature, there are four main approaches that use planning on the basis of the general schema outlined above. In [16], the authors exploit a planner for modeling processes in SHAMASH, a knowledge-based system for the design of business processes. The planner, which is fed with a semantic representation of the process knowledge, produces a parallel plan of activities that is translated into SHAMASH and presented graphically to the user. The obtained plan proposes the scheduling of parallel activities that may handle time and resource constraints. Notice that the emphasis here is on supporting processes for which one has complete knowledge.

The work [17] is based on learning activities as planning operators and feeding them to a planner that generates the process model. An interesting result concerns the possibility of producing process models even though the activities may not be accurately described. In such cases, the authors use a best-effort planner that is always able to create a plan, even though the plan may be incorrect. By refining the preconditions and effects of planning actions, the planner will be able to produce several candidate plans, and after a finite number of refinements, the best candidate plan (i.e., the one with the lowest number of unsatisfied preconditions) is translated into a process model.

In the SEMPA approach [18], process actions are semantically described by specifying their input/output parameters with respect to an ontology implemented in OWL. Starting from such a knowledge, planning is used to derive an action state graph (ASG) consisting of those actions whose execution leads to a given goal from a pre-specified initial state. Then, a process model represented as an UML activity diagram is extracted from the ASG by identifying the required control flow for the process. Interestingly, the planning algorithm implemented in SEMPA provides the ability to build the ASG in presence of initial state uncertainty and with different conflicting goals.

The works [19,20] refer to a technique based on partial-order planning algorithms and declarative specifications of process activities in PDDL for synthesizing a library of process templates to be enacted in contextual scenarios. The resulting templates guarantee that concurrent activities of a process template are effectively independent one from another (i.e., they cannot affect the same data) and are reusable in a variety of partially specified contextual environments. A key characteristic of this approach is the role of contextual data acting as a driver for process templates generation.

3.2   Planning for Process Adaptation

Process Adaptation is the ability of a process to react to exceptional circumstances (that may or may not be foreseen) and to adapt/modify its structure accordingly [6]. Exceptions can be either anticipated or unanticipated. An anticipated exception can be planned at design-time and incorporated into the process model, i.e., a (human) process designer can provide an exception handler which is invoked during run-time to cope with the exception. Conversely, unanticipated exceptions generally refer to situations, unplanned at design-time, that may emerge at run-time and can be detected by monitoring discrepancies between the real-world processes and their computerized representation.

Challenge. In many dynamic application domains (e.g., emergency management, smart manufacturing, etc.), the number of possible anticipated exceptions is often too large, and traditional manual implementation of exception handlers at design-time is not feasible for process designers, who have to anticipate all potential problems and ways to overcome them in advance. Furthermore, anticipated exceptions cover only partially relevant situations, as in such scenarios many unanticipated exceptional circumstances may arise during process execution. *The management of processes in dynamic domains requires that BPM environments provide real-time monitoring and automated adaptation features during process execution, in order to achieve the overall objectives of the processes still preserving their structure by minimising any human intervention* [7].

Application of Planning. The first work dealing with this research challenge is [21]. It discusses how planning can be interleaved with process execution to suggest compensation procedures or the re-execution of activities if some anticipated failure arises. The work [22] presents an approach for enabling automated process instance change in case of activity failures occurring at run-time that lead to a process goal violation. The approach relies on a partial-order planner for the generation of a new complete process model that complies with the process goal. The generated model is substituted at run-time to the original process that included the failed task.

The above works use planning to tackle anticipated exceptions or to completely redefine the process model when some activity failure arises. However, in dynamic domains, it would be desirable to adapt a running process by modifying only those parts of the process that need to be changed/adapted and keeps other parts stable, by avoiding to revolutionize the work list of activities assigned to the process participants [7].

In this direction, the SmartPM approach and system [23,24] provides a planning-based mechanism that requires no predefined handler to build on-the-fly the *recovery procedure* required to adapt a running process instance. Specifically, adaptation in SmartPM can be seen as reducing the gap between the *expected reality*, i.e., the (idealized) model of reality that reflects the intended outcome of the task execution, and the *physical reality*, i.e., the real world with the actual values of conditions and outcomes. A recovery procedure is needed during process execution if the two realities are different from each other. A misalignment of the two realities often stems from errors in the tasks outcomes (e.g., incorrect data values) or is the result of exogenous events coming from the environment. If the gap between the expected and physical realities is such that the process instance cannot progress, the SmartPM system invokes an external planner to build a recovery procedure as a plan, which can thereby resolve exceptions that were not designed into the original process. Notice that a similar framework to tackle process adaptation through planning is also adopted in the research works [25–27].

In SmartPM, the problem of automatically synthesize a recovery procedure is encoded as a classical planning problem in PDDL. The planning domain consists of propositions that characterize the contextual data describing the process domain. Planning actions are built from a repository of process activities annotated with preconditions and effects expressed over the process domain. Then, the initial state reflects the physical reality at the time of the failure, while the goal state corresponds to the expected reality. A classical planner fed with such inputs searches for a plan that may turn the physical reality into the expected reality by adapting the faulty process instance.

A similar adaptation strategy is applied in [28], which proposes a goal-driven approach for service-based applications to adapt business processes to run-time context changes. Process models include service annotations describing how services contribute to the intended goal. Contextual properties are modeled as state transition systems capturing possible values and possible evolutions in the case of precondition violations or external events. Process and context changes that prevent goal achievement are managed through an adaptation mechanism based on service composition via planning.

Finally, the work [29] proposes a runtime mechanism that uses dependency scopes for identifying critical parts of the processes whose correct execution depends on some shared variables and intervention processes for solving potential inconsistencies between data. Intervention processes are automatically synthesised through a planner based on CSP techniques. While closely related to SmartPM, this work requires specification of a (domain-dependent) adaptation policy, based on volatile variables and when changes to them become relevant.

3.3   Planning for Conformance Checking

Within the discipline of process mining, conformance checking is the problem of verifying whether the observed behavior stored in an event log is compliant with the process model that encodes how the process is allowed to be executed to ensure that norms and regulations are not violated. The notion of alignment [30] provides a robust approach to conformance checking, which makes it possible to exactly pinpoint the deviations causing nonconformity with a high degree of detail. An alignment between a recorded process execution (log trace) and a process model is a pairwise matching between activities recorded in the log (events) and activities allowed by the model (process activities).

Challenge. In general, a large number of possible alignments exist between a process model and a log trace, since there may exist manifold explanations why a trace is not conforming. It is clear that one is interested in finding the most probable explanation, i.e., one of the alignments with the least expensive deviations (i.e., optimal alignments), according to some function assigning costs to deviations. The existing techniques to compute optimal alignments against procedural [31] and declarative [32] process models provide ad-hoc implementations of the A* algorithm. The fact is that when process models and event logs are of considerable size the existing approaches do not scale efficiently due to their ad-hoc nature and they are unable to accomplish the alignment task. *In the era of Big Data, scalable approaches to process mining are desperately necessary, as also advocated by the IEEE Task Force in Process Mining* [30].

Application of planning. In case of procedural models represented as Petri Nets, the work [33] proposes an approach and a tool to encode the original algorithm for trace alignment [31] as a planning problem in PDDL. Specifically, starting from a Petri net N and an event log L to be aligned, for each log trace $\sigma_L \in L$ it is built *(i)* a planning domain P_D, which encodes the propositions needed to capture the structure of N and to monitor the evolution of its marking, and three classes of planning actions that represent "alignment" moves: synchronous moves (associated with no cost), model moves and log moves; and *(ii)* a planning problem P_R, which includes a number of constants required to properly ground all the domain propositions in P_D; in this case, constants will correspond to the place and transition instances involved in N. Then, the initial state of P_R is defined to capture the exact structure of the specific log trace σ_L of interest and the initial marking of N, and the goal condition is encoded to represent the fact that N is in the final marking and σ_L has been completely analyzed. At this point, for any trace of the event log, an external planner is invoked to synthesize a plan to reach the final goal from the initial state, i.e., a sequence of alignment moves (each of which is a planning action) that establish an optimal alignment between σ_L and N.

Relatively close is the work [34] where authors use planners to recover the missing recording of events in log traces. The concept of missing event recordings is very similar to model moves in [33]. However, in [34] it is assumed that all executions are compliant with the model and, hence, every event that is present in the incomplete log trace is assumed to be correct. In other words, they do not foresee log moves.

In case of declarative process models, where relationships among process activities are implicitly defined through logical constraints expressed in the well-known LTL-f (Linear Temporal Logic on finite traces) formalism, the work [35] leverages on planning techniques to search for optimal alignments. A planning domain is encoded to capture the structure of the finite state automata (augmented with special transitions for adding/removing activities to/from a log trace) corresponding to the individual LTL-f constraints that compose the declarative model. The same can be done for the specific trace to be aligned, which is represented as a simple automaton that consists of a sequence of states. In addition, the definition of specific domain propositions allows to monitor the evolution of any automaton. At this point, the initial state of the planning problem is encoded to capture the exact structure of the trace automaton and of every constraint automaton. This includes the specification of all the existing transitions that connect two different states of the automata. The current state and the accepting states of any trace/constraint automaton are identified as well. Then, the goal condition is defined as the conjunction of the accepting states of the trace automaton and of all the accepting states of the constraint automata. At this point, a planner is invoked with such inputs to synthesize a plan that establishes an optimal alignment between the declarative model and the log trace of interest

Notably, both the works [33,35] report on results of experiments conducted with several planners fed with combinations of real-life and synthetic event logs and processes. The results show that, when process models and event-log traces are of considerable size, their planning-based approach outperforms the existing approaches based on ad-hoc implementations of the A* algorithm [31,32] even by several orders of magnitude, and they are always able to properly complete the alignment task (while the existing approaches run often out of memory).

4   Discussion and Conclusion

We are at the beginning of a profound transformation of BPM due to the recent advances in AI research [8]. In this context, we have shown how Automated Planning can offer a mature paradigm to introduce autonomous behaviour in BPM for tackling complex challenges in a theoretically grounded and domain-independent way. If BPM problems are converted into planning problems, one can seamlessly update to the recent versions of the best performing automated planners, with evident advantages in term of versatility and customization. In addition, planning systems employ search algorithms driven by intelligent heuristics that allow to scale up efficiently to large problems.

On the other hand, although Planning (in particular in its classical setting) embeds properties desirable for BPM, it imposes some restrictions for addressing more expressive problems, including preferences and non deterministic task effects. Furthermore, planning models require that actions are completely specified in term of I/O data elements, preconditions, and effects, and that the execution context can be captured as part of the planning domain. These aspects can frame the scope of applicability of the planning paradigm to BPM.

It is worth to mention that Automated Planning contributed to tackle challenges also from other Computer Science fields, such as Web Service Composition [36,37] and Ubiquitous Computing [38]. As a future work, we aim at developing a rigorous methodology to acquire relevant literature on the use of planning for BPM and derive an common evaluation framework to systematically review and classify the existing methods.

References

1. van der Aalst, W.M.P.: Business process management: a comprehensive survey. ISRN Softw. Eng. **2013**, 37 pages (2013)
2. Lenz, R., Reichert, M.: IT support for healthcare processes - premises, challenges, perspectives. Data Knowl. Eng. **61**(1), 39–58 (2007)
3. Seiger, R., Keller, C., Niebling, F., Schlegel, T.: Modelling complex and flexible processes for smart cyber-physical environments. J. Comput. Sci. **10**, 137–148 (2014)
4. Humayoun, S.R., Catarci, T., de Leoni, M., Marrella, A., Mecella, M., Bortenschlager, M., Steinmann, R.: The WORKPAD user interface and methodology: developing smart and effective mobile applications for emergency operators. In: Stephanidis, C. (ed.) UAHCI 2009. LNCS, vol. 5616, pp. 343–352. Springer, Heidelberg (2009). https://doi.org/10.1007/978-3-642-02713-0_36
5. de Leoni, M., Marrella, A., Russo, A.: Process-aware information systems for emergency management. In: Cezon, M., Wolfsthal, Y. (eds.) ServiceWave 2010. LNCS, vol. 6569, pp. 50–58. Springer, Heidelberg (2011). https://doi.org/10.1007/978-3-642-22760-8_5
6. Reichert, M., Weber, B.: Enabling Flexibility in Process-Aware Information Systems - Challenges, Methods, Technologies. Springer, Heidelberg (2012)
7. Di Ciccio, C., Marrella, A., Russo, A.: Knowledge-intensive processes: characteristics, requirements and analysis of contemporary approaches. J. Data Semant. **4**(1), 29–57 (2015)
8. Hull, R., Motahari Nezhad, H.R.: Rethinking BPM in a cognitive world: transforming how we learn and perform business processes. In: La Rosa, M., Loos, P., Pastor, O. (eds.) BPM 2016. LNCS, vol. 9850, pp. 3–19. Springer, Cham (2016). https://doi.org/10.1007/978-3-319-45348-4_1
9. Geffner, H., Bonet, B.: A Concise Introduction to Models and Methods for Automated Planning. Morgan & Claypool Publishers, San Rafael (2013)
10. Maggi, F.M., Di Francescomarino, C., Dumas, M., Ghidini, C.: Predictive monitoring of business processes. In: Jarke, M., Mylopoulos, J., Quix, C., Rolland, C., Manolopoulos, Y., Mouratidis, H., Horkoff, J. (eds.) CAiSE 2014. LNCS, vol. 8484, pp. 457–472. Springer, Cham (2014). https://doi.org/10.1007/978-3-319-07881-6_31
11. McDermott, D., et al.: PDDL-the planning domain definition language. Technical report DCS TR-1165, Yale Center for Computational Vision and Control (1998)
12. Geffner, H.: Computational models of planning. Wiley Int. Rev. Cogn. Sci. **4**(4) (2013)

13. Geffner, H.: Non-classical planning with a classical planner: the power of transformations. In: Fermé, E., Leite, J. (eds.) JELIA 2014. LNCS, vol. 8761, pp. 33–47. Springer, Cham (2014). https://doi.org/10.1007/978-3-319-11558-0_3

14. La Rosa, M., van der Aalst, W.M., Dumas, M., Milani, F.P.: Business process variability modeling: a survey. ACM Comput. Surv. **50**(1) (2013). Article No. 2

15. Schuschel, H., Weske, M.: Triggering replanning in an integrated workflow planning and enactment system. In: Benczúr, A., Demetrovics, J., Gottlob, G. (eds.) ADBIS 2004. LNCS, vol. 3255, pp. 322–335. Springer, Heidelberg (2004). https://doi.org/10.1007/978-3-540-30204-9_22

16. R-Moreno, M.D., Borrajo, D., Cesta, A., Oddi, A.: Integrating planning and scheduling in workflow domains. Exp. Syst. App. Int. J. **33**(2), 389–406 (2007)

17. Ferreira, H., Ferreira, D.: An integrated life cycle for workflow management based on learning and planning. Int. J. Coop. Inf. Syst. **15**, 485–505 (2006)

18. Henneberger, M., Heinrich, B., Lautenbacher, F., Bauer, B.: Semantic-based planning of process models. In: Multikonferenz Wirtschaftsinformatik (2008)

19. Marrella, A., Lespérance, Y.: Synthesizing a library of process templates through partial-order planning algorithms. In: Nurcan, S., Proper, H.A., Soffer, P., Krogstie, J., Schmidt, R., Halpin, T., Bider, I. (eds.) BPMDS/EMMSAD -2013. LNBIP, vol. 147, pp. 277–291. Springer, Heidelberg (2013). https://doi.org/10.1007/978-3-642-38484-4_20

20. Marrella, A., Lesperance, Y.: A planning approach to the automated synthesis of template-based process models. Serv. Oriented Comput. Appl. **11**, 367–392 (2013)

21. Jarvis, P., et al.: Exploiting AI technologies to realise adaptive workflow systems. In: Proceedings of the AAAI Workshop on Agent-Based Systems in the Business Context (1999)

22. Gajewski, M., Meyer, H., Momotko, M., Schuschel, H., Weske, M.: Dynamic failure recovery of generated workflows. In: DEXA 2005. IEEE Computer Society Press (2005)

23. Marrella, A., Mecella, M., Sardina, S.: SmartPM: an adaptive process management system through situation calculus, IndiGolog, and classical planning. In: KR. AAAI Press (2014)

24. Marrella, A., Mecella, M., Sardina, S.: Intelligent process adaptation in the SmartPM system. ACM Trans. Intell. Syst. Technol. **8**(2) (2016). Article No. 25

25. Marrella, A., Mecella, M.: Continuous planning for solving business process adaptivity. In: Halpin, T., Nurcan, S., Krogstie, J., Soffer, P., Proper, E., Schmidt, R., Bider, I. (eds.) BPMDS/EMMSAD -2011. LNBIP, vol. 81, pp. 118–132. Springer, Heidelberg (2011). https://doi.org/10.1007/978-3-642-21759-3_9

26. Marrella, A., Mecella, M., Russo, A.: Featuring automatic adaptivity through workflow enactment and planning. In: CollaborateCom 2011. IEEE (2011)

27. Marrella, A., Russo, A., Mecella, M.: Planlets: automatically recovering dynamic processes in YAWL. In: Meersman, R., Panetto, H., Dillon, T., Rinderle-Ma, S., Dadam, P., Zhou, X., Pearson, S., Ferscha, A., Bergamaschi, S., Cruz, I.F. (eds.) OTM 2012. LNCS, vol. 7565, pp. 268–286. Springer, Heidelberg (2012). https://doi.org/10.1007/978-3-642-33606-5_17

28. Bucchiarone, A., Pistore, M., Raik, H., Kazhamiakin, R.: Adaptation of service-based business processes by context-aware replanning. In: SOCA 2011. IEEE (2011)

29. van Beest, N.R., Kaldeli, E., Bulanov, P., Wortmann, J.C., Lazovik, A.: Automated runtime repair of business processes. Inf. Syst. **39**, 45–79 (2014)

30. van der Aalst, W.M.P.: Process Mining: Data Science in Action. Springer, Heidelberg (2016)

31. Adriansyah, A., van Dongen, B.F., Zannone, N.: Controlling break-the-glass through alignment. In: SOCIALCOM 2013. IEEE Computer Society (2013)

32. de Leoni, M., Maggi, F.M., van der Aalst, W.M.P.: Aligning event logs and declarative process models for conformance checking. In: Barros, A., Gal, A., Kindler, E. (eds.) BPM 2012. LNCS, vol. 7481, pp. 82–97. Springer, Heidelberg (2012). https://doi.org/10.1007/978-3-642-32885-5_6

33. de Leoni, M., Marrella, A.: Aligning real process executions and prescriptive process models through automated planning. Expert Syst. Appl. **82**, 162–183 (2017)

34. Di Francescomarino, C., Ghidini, C., Tessaris, S., Sandoval, I.V.: Completing workflow traces using action languages. In: Zdravkovic, J., Kirikova, M., Johannesson, P. (eds.) CAiSE 2015. LNCS, vol. 9097, pp. 314–330. Springer, Cham (2015). https://doi.org/10.1007/978-3-319-19069-3_20

35. De Giacomo, G., Maggi, F.M., Marrella, A., Patrizi, F.: On the disruptive effectiveness of automated planning for LTLf-based trace alignment. In: AAAI 2017. AAAI Press (2017)

36. Pistore, M., Traverso, P., Bertoli, P., Marconi, A.: Automated synthesis of composite BPEL4WS web services. In: ICWS 2005. IEEE Computer Society (2005)

37. Pistore, M., Traverso, P., Bertoli, P., Marconi, A.: Automated synthesis of executable web service compositions from BPEL4WS processes. In: WWW 2005. ACM (2005)

38. Georgievski, I., Aiello, M.: Automated planning for ubiquitous computing. ACM Comput. Surv. **49**(4), 1–46 (2016)

Structural Feature Selection for Event Logs

Markku Hinkka[1,2(✉)], Teemu Lehto[1,2], Keijo Heljanko[1,3], and Alexander Jung[1]

[1] Department of Computer Science, School of Science,
Aalto University, Espoo, Finland
{markku.hinkka,keijo.heljanko,alex.jung}@aalto.fi
[2] QPR Software Plc, Helsinki, Finland
teemu.lehto@qpr.com
[3] HIIT Helsinki Institute for Information Technology,
Espoo, Finland

Abstract. We consider the problem of classifying business process instances based on structural features derived from event logs. The main motivation is to provide machine learning based techniques with quick response times for interactive computer assisted root cause analysis. In particular, we create structural features from process mining such as activity and transition occurrence counts, and ordering of activities to be evaluated as potential features for classification. We show that adding such structural features increases the amount of information thus potentially increasing classification accuracy. However, there is an inherent trade-off as using too many features leads to too long run-times for machine learning classification models. One way to improve the machine learning algorithms' run-time is to only select a small number of features by a feature selection algorithm. However, the run-time required by the feature selection algorithm must also be taken into account. Also, the classification accuracy should not suffer too much from the feature selection. The main contributions of this paper are as follows: First, we propose and compare six different feature selection algorithms by means of an experimental setup comparing their classification accuracy and achievable response times. Second, we discuss the potential use of feature selection results for computer assisted root cause analysis as well as the properties of different types of structural features in the context of feature selection.

Keywords: Automatic business process discovery · Process mining
Prediction · Classification · Machine learning · Clustering
Feature selection

1 Introduction

In Process Mining, unstructured *event logs* generated by systems in business processes are used to automatically build real-life process definitions and as-is models behind those event logs. There is a growing need to be able to predict

© Springer International Publishing AG 2018
E. Teniente and M. Weidlich (Eds.): BPM 2017 Workshops, LNBIP 308, pp. 20–35, 2018.
https://doi.org/10.1007/978-3-319-74030-0_2

properties of newly added event log cases, or process instances, based on case data imported earlier into the system. In order to be able to predict properties of the new cases, as much information as possible should be collected that is related to the event log cases and relevant to the properties to be predicted. Based on this information, a model of the system creating the event logs can be created. In our approach, the model creation is performed using machine learning techniques.

One good source of additional case related features is the information stored into the *sequence of activities* visited by cases. This information includes, e.g., number of times an event log case has visited a certain *activity* and the number of times a case has *transitioned* between two specific activities. Features collected in this fashion are often highly dependent on each other. E.g., a patient whose visit to hospital takes long time (outcome) has quite often been in surgery from which he/she has moved into a ward. In this example, we already can easily find three structural features of which any can be used to predict whether the visit lasted long or not: visited surgery, transitioned from surgery to ward, visited ward. However, depending on all the other patients in the data set, it may be that there are no cases where only a subset of these three features occurs, thus making it redundant to have all three features taken into account when building a model for prediction purposes. Thus, one feature could well be enough to give as accurate predictions as having them all.

Another important aspect in Process Mining is that it is often desired to be able to show dependencies between features. Thus, selecting a feature selection algorithm that produces also this information for minimal extra cost is often tempting. For this purpose, the list of the most relevant features and the extent of their contribution should somehow be returned. One example of this kind of root cause analysis technique is influence analysis described in [13].

The primary motivation for this paper is the need to perform classification based on structural features originating from *activity sequences* in event logs as accurately as possible and using a minimum amount of computing resources and maximizing the throughput in order to be able to use the method even in some interactive scenarios. This motivation comes from the need to build a system that can do classification and root cause analysis activities accurately on user configurable phenomena based on huge event logs collected and analyzed, e.g., using *Big Data processing frameworks* and methods such as those discussed in our earlier paper [12]. The response time of this classification system should be good enough to be used as part of a web browser based interactive process mining tool where user wants to perform classifications and expects classification results to be shown within a couple of seconds.

The rest of this paper is structured as follows: Sect. 2 introduces main concepts related to this paper. Section 3 discusses the feature selection concept and gives brief introduction to the methods used in this paper. Section 4 will then present a framework used for comparing performance of the selected feature selection approaches. The results of the tests will be presented in Sect. 5 after which we will discuss related work in Sect. 6. Finally Sect. 7 draws the conclusions from the test results.

2   Problem Setup

The concepts and terminology used throughout this paper follows the principles used commonly in process mining community. For more detailed examples and discussion about *event logs*, *activities*, *cases*, *traces* and other related terminology, see, e.g., the book by van der Aalst [22].

2.1   Structural Features

As opposed to normal case attributes added to cases in event logs, structural feature term in this paper is used for representing properties of activity sequences of cases. Thus, they can be derived directly from actual activity sequences without need to include any additional custom properties. Having a case identifier, activity identifier and order information such as a time stamp for each event occurrence, is enough. In order to simplify the tests and keep requirements for applying the results of this paper to its minimum, we decided to only concentrate on structural features as predictors in this paper. However, in real use cases, the best results are achieved by including also all the available additional case attributes such as duration, age, etc. into features from which the feature selection is performed [14].

There are several different types of structural features to select from. In this paper, we use notations similar to those used in regular expressions [20] combined with notation commonly used for activity sequences [22]. The patterns we focused in this paper are listed in Table 1 with examples of matches when the sequence of activities is illustrated as $\langle S, a, b, b, E \rangle$. As we were interested in minimizing the response time, we decided to consider only the listed patterns because having more complex patterns, such as tandem repeats and maximal repeats [2], would have made the combinations of different predictor types and features to become too big to be able to perform exhaustive tests for and leading to the problem of *curse of dimensionality* [9]. Also, the extraction of all the feature types presented in the literature would have required much more computation time than the selected relatively simple features used in this work.

Table 1. Structural feature types

Pattern/predictor type	Example sequences(s)
Activity	$\langle a \rangle, \langle b \rangle$
Transition/2-grams	$\langle S, a \rangle, \langle a, b \rangle, \langle b, b \rangle, \langle b, E \rangle$
Starter	$\langle S, a \rangle$
Finisher	$\langle b, E \rangle$
Ordering	$\langle a \rangle \rightarrow \langle b \rangle$

For every predictor type listed in the Table 1, there can be several possible implementations. In this paper, we consider structural features of *activity* and *2-gram* predictor types to be such that their values correspond to the number of occurrences of that pattern within each activity sequence. *Starter and finisher* predictor types however are boolean values indicating whether that pattern is valid for an activity sequence. Order feature type is considered to be a boolean value such that it is true only if the first occurrences of both ends of the order relation are in the specified order.

The difference between 2-gram and order pattern is that order allows any number of activities to be between the activities of the ordering relation, whereas in 2-gram, the activities of the relation must be successive in the whole sequence of activities. The importance of predictor types also depends very heavily on the type of the data set and the scenario being predicted.

One more factor to take into account when selecting the actual features is how to handle situations where a feature has more than two different values. E.g., a patient may have visited surgery multiple times while visiting a hospital. In some cases, depending on what we are trying to predict and what kind of prediction models are to be built, it could be better to split these kinds of features into several boolean features. Thus, e.g., we could have a feature for a patient having visited surgery 4 times. However, one has to be careful when to split features like this in order to avoid an explosion in the number of features created for each original feature.

One final step before passing features to the actual model training is to identify and remove any duplicate features that have behaved identically through the whole training set. Some training methods do this automatically, but some do not.

2.2 Classification

Since all the tests performed in this paper are performed on data sets consisting of already completed cases, we are performing *classification* using machine learning prediction algorithms. Classification in machine learning usually consists of two phases: training a model and performing the actual classifications using the trained model. In this paper, we concentrate on building the classification model using *supervised learning methods*, where algorithms are trained using *predictors*, together with their *outcomes*. A core part of the model building is the selection of *features* to be used as predictors. Often the more you have independent features that may have effect in the outcome, the better. As shown in Sect. 2.1, a lot of features can be created directly from the activity sequences of the cases themselves.

Another important factor that has direct effect to the prediction performance of the model is the algorithm that is used for building the model and making the predictions. In this paper, we focus on the feature selection part. However, we need to also validate the performance of the feature selection using a set of algorithms. In order to minimize the skew in the results caused by the validation algorithm itself, we decided to compare the efficiency of the selected features using two different approaches. First, for a given set of selected features, we determined the prediction accuracy obtained by a particular supervised learning method, i.e., the gradient boosting machine (GBM). GBM was selected due to its good reputation [9,18] and performance in both accuracy and response times in our own tests. The second method was to approximate the mutual information score [16] between each of the selected set of features and two different data sets. The first data set consisted of all the available predictors without any feature selection. The second data set consisted only of the outcomes to be predicted.

As we are concentrating on features originating from process mining, at the granularity level of a case, the prediction inputs that are usually used in the field are actually custom case properties such as the customer name or an identifier of the owner of the case. The outcomes that we want to predict are usually values of some custom case properties or some calculated case content dependent values such as durations, some kind of cost of the case or some other metrics measuring the quality of the case. In this paper, we concentrate only in features inherent to the activity sequences inside cases and measure how well certain outcomes can be predicted only based on those features.

The used data set is split into two parts: training and test. Training data set is used for two purposes. First, features are selected from the whole training data set. After this, a model is trained using all the cases in the training data, but only using the selected features as predictors. Finally, once the model has been built, the model is tested against the test data and its performance is estimated using accuracy and mutual information metrics.

3 Feature Selection Methods

The aim of feature selection is to reduce the dimensionality of the structural features constructed from the raw data. Reducing the dimensionality not only reduces the computational complexity of the subsequent prediction methods, it may also lead to an improved prediction accuracy. Indeed, learning algorithms based on a smaller set of features are less prone to overfitting, i.e., the effect of erratic statistical variations of a particular observed dataset is reduced. Finally, feature selection also enhances the interpretability (visualization) of the features and understanding classifications based on them (e.g., if only two numerical features are selected, we can illustrate them by means of a scatterplot).

Initially we also considered testing a couple of feature extraction algorithms. Feature extraction differs from feature selection in that they create new features that will be used instead of the original features. The newly created features try to maximize the variance and expressive power of the features by combining several original features into one new feature. This has a drawback that it hides the original features and makes it harder to understand the properties of the created model. E.g., in root cause analysis, it is often desirable to understand how much the outcome depends of certain features and also to understand which features have an effect to the outcome. Due to this shortcoming, we decided to not include any feature extraction algorithms into this paper.

No additional parallelization techniques were used, thus if the algorithm did not support parallelism out of the box, it was not run in parallel. The following subsections briefly describe the basics of each of the feature selection methods tested in this paper including information on the used R programming language packages and their configurations. We also briefly tested an algorithm based on *Support Vector Machine (SVM)* [1,25] using radial kernel, but decided to leave it out of the paper due to very poor results and extremely long response times, which were order of magnitude slower than with any of the other tested algorithms. The following subsections will briefly describe the details of all the remaining tested algorithms.

Random Selection. The most trivial of all the tested algorithms was a randomized selection where the desired number of features were just randomly selected from all the available features. This method was used as a baseline in order to gain a better understanding on the quality of other used selection algorithms when compared with an algorithm that does not in any way take any properties of the selected features themselves into account. This serves as a baseline selection algorithm. There should not be any algorithm that performs consistently worse than this. In order to alleviate the effect of inherently noisy random selections, median of three separate test runs was used in the experiments. Thus, only the test which yielded the median prediction accuracy was used as the actual result. In the graphs and analysis below, this algorithm is labeled as *Random*.

Feature Clustering. This method is influenced by the idea provided by Covoes et al. [5]. In the algorithm developed for this paper, the training data is first clustered so that every structural feature in the training set constitutes one clustering data point. Each activity sequence in the training data represents one dimension for clustering data points with values equaling the number of times that structural feature occurs within that activity sequence. K-means algorithm is then used to generate K clusters using kmeans R-function which is based on algorithm by Hartigan and Wonget [10]. For each K clusters, the feature having the minimum distance to the mean of that cluster will be selected as the representative for all the features in that cluster. It should be noted also that as a side product of applying this method for feature selection, every selected feature will actually represent all the features within the same cluster. Thus, for every original feature, you have one cluster it belongs into and exactly one feature that is representing that feature in that cluster. This could be useful, e.g., in some root cause analysis scenarios.

Two different versions of this algorithm are tested in this paper. One that first removed all the features having exactly the same occurrence pattern within all the cases thus removing duplicate vectors before the actual clustering step. The other variation of this algorithm does not perform this preprocessing step. The results of different variations being nearly the same except for the processing time, which was clearly faster with the algorithm that first dropped out all the features having exactly the same values for all the cases in the training set. Thus, we decided to limit our tests only to this algorithm variant. In the graphs shown below, this algorithm is labeled as *Cluster*.

Minimum Redundancy Maximum Relevancy. This is a mutual information based approach [6], which uses mutual information as a proxy for computing relevance and redundancy among features. The implementation used in this paper was provided by mRMRe -R package which claims to provide a highly efficient implementation of the mRMR feature selection via parallelization and lazy evaluation of mutual information matrix. We used ensemble method both with solution_count set to 1, which provides results resembling classic mRMR, and also with 5, which does 5 separate runs and combines the results in the end. This time the results were also otherwise quite the same, except the 5-run version provided clearly better mutual information scores. Thus, in the graphs and analysis below, we use only 5 run version labeled as *mRMREns5*.

Least Absolute Shrinkage and Selection Operator (LASSO). This is a regression analysis method that can be used for feature selection [21]. It is related to least squares regression where the solution minimizes the sum of the squares of the errors made. The unique property for this regression technique is the usage of additional regularization that enables discarding irrelevant features and forces usage of simpler models that do not include them. Since the LASSO implementation in *glmnet* R-package in itself did not provide means of sorting features by their importance and since it was not possible to directly adjust the desired number of target features, the actual used algorithm first performed 10 iterations of LASSO algorithm each yielding slightly different results. After this, all the results were collected into a single list with each feature weighted by the number of occurrences of that feature within all the LASSO results. Finally, this list was sorted from the largest height to smallest and the desired number of features were picked from the beginning of this sorted list. Two different variations of this algorithm were tested: one using lambda.1se as the prediction penalty parameter and the other using lambda.min. Due to the results being almost identical in both the cases, we selected the one using lambda.1se prediction penalty parameter. In the results below, this algorithm is labeled as *LASSO1se*.

Markov Blanket. Markov blanket of a variable X is a minimal variable subset conditioned on which all other variables are probabilistically independent of X. [26] For Bayesian networks, Markov blanket of X consists of the union of the following three types of neighbors: the direct parents of X, the direct successors of X, and all direct parents of X's direct successors. Bayesian networks can be inferred from the training data after which Markov blanket for the created network is calculated by selecting the outcome as X. The result is the set of features to select. bnlearn -R package was used to perform Markov Blanket based feature selection. *Hill-Climbing* algorithm is first used to construct a Bayesian network structure out of the training data. After this, Markov Blanket is extracted out of the network for the outcome feature. Finally, out of these results, the desired number of features are selected from the beginning of the returned list, or if the result does not have all the required features, only the returned features are selected. In the results below, this algorithm is labeled as *Blanket*.

Variable Importance. In variable importance based feature selection, some Machine learning algorithm capable of building variable importance information, such as random forest [15], is first performed. After this, the results of the algorithm are used to pick N variables having the greatest effect on the outcome. These N variables are then used as the selected features. Since the performance of variable importance algorithm itself was found to be very poor when using predictor sets having hundreds of features we decided to use a hybrid approach where we first use clustering to remove about 75% of all the features, after which variable importance is calculated for each feature using random forest algorithm and from there, the desired number of the most important target features is picked. In this paper, randomForest -R package is first used to generate a model after which varImp -R function in Caret-package is used to extract the most important features based on the information gathered by the random forest algorithm. This algorithm is labeled as *ClusterImp* in the graphs and analysis below.

Recursive Feature Elimination. Recursive feature elimination [8] starts with estimating the variable importances of all the features in the training data as in the Variable Importance -technique. After this, a smaller subset of the most important features is selected and variable importances are estimated again. This is repeated until the desired feature subset size is reached after which the resulting features can be picked. In this paper, three different variations of this method were tested: a test with only one iteration, another with two iterations and the third one with four iterations. Caret R-package's *rfe* algorithm was used for these tests, with the default method based on random forests. After the initial tests, it was found out that while the accuracy and mutual information of all the cases were very close to each other, the average processing time of the 2-step algorithm was clearly better than the others. Thus, in the graphs and analysis below, we only concentrate on this algorithm labeled as *Rec2S*.

4   Test Setup

Testing was performed on a single system using Microsoft R Open version 3.3, Windows 10 operating system. The used hardware consisted of 3.5 GHz Intel Core i5-6600K CPU with 8 GB of memory. Tests were performed using two publicly available data sets. The first one from Rabobank Group ICT [24] that was also used in BPI Challenge 2014. This data set contained total of 46617 cases of a real-life event log from Rabobank Group ICT company. The average length of a case is nearly 10 events split into 39 different event types. The second used data set is a real-life data set from Dutch academic hospital [23] consisting of 1143 cases. The average length of a case in this data set is slightly over 130 events split into 624 event types.

All the test runs were performed using an R function that ran all the desired test runs in sequence. At the beginning of every test run, random seed was initialized. Thus, the random case samples and other random values generated within the used algorithms behaved the same way in every run, provided that the algorithm used random -methods that support setting the seed using *set.seed* -R function.

In the tests, the training data was first extracted so that 25% of the provided data rows were randomly selected. This training data was first used by the feature selection algorithm to be tested, after which it was used to build the classification *GBM* model for predicting given phenomenon. This classification model was then used for measuring the performance of the feature selection. Mutual information metrics were approximated also at this final phase.

The first run of tests was performed using test data having 4000 cases extracted from the full BPI Challenge 2014 data set. For this first run, all the algorithms were tested so that the number of selected features were 5, 10 and 30. For each of these combinations, 13 different sets of feature patterns were selected. The selected structural feature patterns were different combinations of the following patterns described in Table 1: activity, starter and finisher, 2-grams and ordering. Activity and 2-grams features included occurrence counts, while the others were just boolean values indicating whether the feature occurs at least once in a case. The combinations were created in a way that all the possible combinations of the patterns were tested where activity pattern was present, thus generating 8 different combinations. In addition to these, we also tested 2-grams and ordering separately, as well as having both 2-grams and ordering. Most emphasis was given for activity pattern since it is usually the easiest to extract from event logs and also doesn't yield that many features. Also, we did not want to include any other patterns due to the number of potential new features that would have been needed in order to cover all the possible cases. E.g., adding 3-grams would potentially have generated N^3 additional features where N equals the number of different activities in the training data, which in this case is 39 yielding the maximum of 60000 new features. When using a case sample of 4000 cases, the numbers of features added were as follows: 39 features for activity pattern, 20 features for starter and finisher, 772 features having 2-grams and 1033 features were generated for ordering-pattern. Thus, the total maximum number of features extracted from event logs when all the patterns are included is 1864.

All the tests performed on the first data set were run to predict two different outcomes, which are later in this paper referred to as scenarios. The first scenario was whether the case duration is longer than 7 days. In this case, nearly 37% of all the cases in the small test set had this outcome. This is an example of a prediction that can be trained directly from the event information without any need for additional case or event attributes. The second scenario that was tested is based on additional case-level information provided with the event data: Does the case represent a "request for information" or something else such as an "incident"? In this case, nearly 20% of all the test cases in the small sample had this outcome. For all the tests performed on the same sample size, the actual used cases and their predictors were always the same.

Finally, we compared the results achieved from these earlier tests to tests run on the second Dutch academic hospital data set. In this case, we used all the 1143 cases in the data set in the tests. The number of features added by activity patterns were as follows: 624 features for activity pattern, 36 features for starter and finisher, 4272 features for 2-grams and 79571 features for ordering-pattern.

5   Test Results

We began the actual analysis of our first round of tests by estimating the average classification accuracy of all the tested algorithms for all the tested feature counts and all the tested structural feature patterns using both the test scenarios in 4000 case sample. The results of this analysis are shown in Fig. 1. The first column on the chart labeled *None* shows the accuracy achieved by not performing any feature selection. Based on these results, we can see that the feature selection algorithms ordered in descending order of accuracy are: Recursive, Cluster, Blanket, mRMR, Cluster Importance and LASSO. The rankings of three first algorithms are not changed even with the tested bigger data set size.

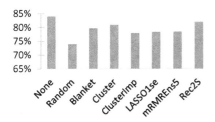

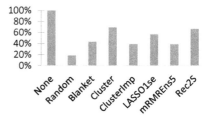

Fig. 1. Average accuracy of all the tested algorithms.

Fig. 2. Average mutual information of all the tested algorithms.

With the same test data, we also measured the average percentage of the mutual information of a data set filtered using feature selection algorithm when compared to the mutual information calculated when the feature selection is not performed at all. This is shown in Fig. 2. The ranking of algorithms in the mutual information case are: Cluster, Recursive, LASSO, Blanket, Cluster Importance and mRMR. Also, in this case, the rankings of three first algorithms are not changed even with the tested bigger data set sizes. It should be noted at this point that identical rankings were obtained in the case the mutual information was calculated between the result of the feature selection algorithm and only the correct outcomes.

After this, we analyzed the response times for all the tested feature selection algorithms with the same test data. This time however, we did not include starter and finisher predictor sets since they would have made the readability of the figure much worse and also would not have provided much additional information due to the small effect they have into the results in the tested scenarios.

As seen in the Fig. 3, the time required to perform feature selection for the tested algorithms and predictor types varied very much. In the worst cases, the difference between the slowest and the fastest algorithm was three orders of magnitudes, with Cluster and mRMR usually performing much faster than all the others.

The largest performance variations within one algorithm were measured using Blanket algorithm which performed in the fastest predictor set case, almost as fast as the two fastest algorithms, but in the slowest case, it performed almost over four orders of magnitudes slower.

Both Clustering and mRMR performed so well in this analysis that they could be incorporated without changes into some interactive process mining systems preferring under ten second response times when the size of the event log used for training is close to 1000 cases. Since we are especially interested in response times, we took Clustering and mRMR algorithms for closer inspection.

First, we analyzed the classification accuracy of both the algorithms separately for both the tested scenarios. Figure 4 shows this information for Cluster algorithm and Fig. 5 for mRMR algorithm.

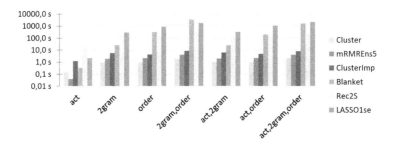

Fig. 3. Average feature selection response time of all the tested algorithms.

From these results, it can be seen that there is a lot of variation between the two tested scenarios. In both the cases, in almost all the predictor sets, predicting case duration produced clearly worse results than in the categorization scenario. It seems that mRMR algorithm was not able to get any additional accuracy into its predictions by including any additional predictor types on top of activity predictors, whereas Cluster algorithm managed in the categorization case to get better accuracy even, e.g., by selecting only *2-gram* or *order* type predictors. Also, in duration scenario, it managed to get better accuracy when adding *order* type predictors in addition to *activities*. Also, in almost all the test cases, having *order* type predictors is more valuable than having *2-gram*

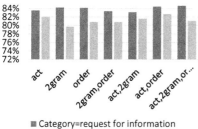

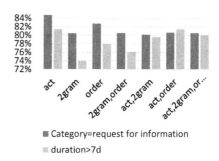

Fig. 4. Average accuracy for cluster algorithm separately for each scenario and predictor types.

Fig. 5. Average accuracy for mRMR algorithm separately for each scenario and predictor types.

type predictors. Having them both is in most cases worse than having *order* only. The number of features has slightly greater effect into the performance of mRMR than for Cluster algorithm. However, for the data set size, the situation is reversed and Cluster is affected clearly more than mRMR. The recommended data size used for feature selection is in our experiments under 10000 cases in order to maintain good response times.

It should be also noted that the time required for building a prediction model with a feature selection algorithm selecting 10 features was only about 1%–3% of the total time required when building the model with all the 1864 tested structural features without any feature selection. When using 3000 cases to build a model, the total measured time difference in the test system was about 250 s. During this time, it would have been possible to run the clustering feature selection several times. Thus, it is clear that having a feature selection performed before model building, at least when GBM is used, is essential when trying to improve the time required for model building.

Finally, we also performed similar tests with the Dutch academic hospital dataset. Tests performed in this dataset indicated that it is absolutely critical to perform some kind of feature selection before training the model since building the model without any selection failed when attempting to use all the 79571 ordering features valid in this event log. From the six tested feature selection finalists, only clustering and clustering with variable importance managed to provide results for all the tested predictor combinations. Out of these two algorithms, clustering with variable importance provided this time slightly more accurate results while both of them still managing to provide relatively good response time of 25–40 s. This result is caused by the fact that clustering itself does not take the outcome into account at all, whereas variable importance based feature selection does. In this case, clustering first filters out most of the features but still tries to maintain as much of the original information content in the predictors as possible. Applying variable importance after this with the specified outcome will filter out all the clusters of original features that do not have anything to do with the outcome.

Thus, in order to fulfill all the requirements, we have for the algorithm, based on the performed tests, the best option from the selected set of algorithms is Cluster based feature selection with only *activity* type predictors with values being the occurrence counts of the *activity* within a *case*. In our experiments, it gave the best overall trade-off in performance considering mutual information, classification accuracy and response time. If a slightly better classification accuracy is required, we recommend adding *order* predictor types to the set of features to pick the actual features from. For event logs having large number of event types and relatively small amount of cases, mixing clustering with variable importance provided better accuracy than performing only clustering.

6 Related Work

In recent years, several papers have been written on the subject of applying data mining and machine learning techniques into predicting outcomes of the business processes. In [3], the authors present a framework that is capable of automatically detecting "signatures" that can be used to discriminate between desired and undesired behavior within traces both seen or unseen. These signatures are essentially combinations of structural features similar to those described in Sect. 2.1. This paper does not in itself specify any automatic feature selection method. Instead, the user is required to specify manually the desired activity sequence patterns, referred to as sequence feature types. After this all the matching features will be used for signature detection. Thus, our research complements the research made in this paper by experimenting with different automatic feature selection methods that could be applied before this signature detection phase in order to reduce the computational cost of signature detection at the cost of some prediction accuracy.

In [17], the authors evaluate the accuracy achieved with three different prediction algorithms using several combinations of structural feature patterns for three different datasets. As result, they find out that just having Activity frequencies often yield, if not the best, then at least almost as good results as the best tested structural feature pattern combination. This finding is visible also in our tests as shown in Figs. 4 and 5.

In [7], the authors present a framework for predicting outcomes of user specified predicates for running cases using clustering based on activity sequence prefixes and classification using attributes associated to events. In [14], the authors have assessed the benefits of including case and event attributes when performing predictions based on sequences of activities. In [19], the authors present a predictive process monitoring framework that is also able to mine unstructured textual information embedded into attributes related to events. In [4], the authors propose a recommendation system that automatically determines the risk that a fault will occur if the input the user is giving to the system will be used to carry on a process instance.

Until now there has not been systematic testing of applying automatic feature selection algorithms after selecting structural feature patterns to use and before building models used for classification. The aim of this feature selection is to minimize the computational cost of the building of classification models. Creating such an approach is crucial for obtaining predictions with interactive response time requirements. This is the primary contribution of this paper.

7   Conclusions

In this paper, we have designed a system for assessing the performance and response times of selected feature selection algorithms specifically tuned into the context of selecting structural features extracted from properties of sequences of activities derived from event logs. Using this system, we tested six feature selection techniques using a total of twelve different algorithms.

Each algorithm was tested using a publicly available real-life Rabobank Group ICT dataset and tuned for two different classification use cases: Predicting whether the duration of a case is longer than seven days, and classifying whether a case is of type request for information. Most of the tests were also run using two different sample sizes out of the full dataset. For sanity checking and benchmarking purposes, we also added test runs without any feature selection and also with randomized feature selection.

We also proposed a rough categorization method for some of the types of structural features that can be extracted from event logs. In this paper, we selected four of these types for closer inspection.

As summary for all the tests and their results, it can be clearly seen that structural features can provide additional means for improving the precision to classifications made for cases in event logs. When the number of selected features is small, the most efficient source of features is activities. Increasing the number of features improves the classification accuracy, but also while doing so, best results are achieved by adding features from other structural feature types such as event type orderings into the set of structural features from which the feature selection is made. However, there is a drawback that having a bigger pool of features to select from makes creating classification models as well as the feature selection slower. As our goal was also to find an algorithm that could perform feature selection and classification with interactive response times using the sample sizes used in this paper, we found out that only one feature selection algorithm of the tested algorithms provided both the speed and accuracy required for the task.

According to the tests, the most consistently well performing algorithm was *Cluster* algorithm we developed for this paper which first used *k-means* algorithm for clustering features into the desired number of clusters by having cases as clustering dimensions, after which the features closest to the center of each cluster were selected as the selected features. This algorithm was not in all the tests as fast as *mRMR*, but it provided feature selections yielding more accurate classifications and it worked without problems with all the tested data sets. Also, it was not as accurate as some other algorithms, such as *Recursive Feature Selection*, in both the scenarios, but it consistently achieved very accurate results in all the tested scenarios within the response time requirements set in this paper. For event logs having very large number of event types, mixing feature clustering with variable importance provided more accurate results than clustering only. For computer assisted root cause analysis, in addition to providing the list of the most important features, *Cluster* algorithm provides a mapping from each of the original structural features to one selected feature that most closely resembles the original feature in the set of selected features.

All the gathered raw information from the performed test runs can be found in the support materials [11].

Acknowledgements. We want to thank QPR Software Plc for funding our research. Financial support of Academy of Finland projects 139402 and 277522 is acknowledged.

References

1. Bennett, K.P., Campbell, C.: Support vector machines: hype or hallelujah? SIGKDD Explor. **2**(2), 1–13 (2000)
2. Jagadeesh Chandra Bose, R.P., van der Aalst, W.M.P.: Abstractions in process mining: a taxonomy of patterns. In: Dayal, U., Eder, J., Koehler, J., Reijers, H.A. (eds.) BPM 2009. LNCS, vol. 5701, pp. 159–175. Springer, Heidelberg (2009). https://doi.org/10.1007/978-3-642-03848-8_12
3. Jagadeesh Chandra Bose, R.P., van der Aalst, W.M.P.: Discovering signature patterns from event logs. In: IEEE Symposium on Computational Intelligence and Data Mining, CIDM 2013, Singapore, 16–19 April 2013, pp. 111–118. IEEE (2013)
4. Conforti, R., de Leoni, M., Rosa, M.L., van der Aalst, W.M.P., ter Hofstede, A.H.M.: A recommendation system for predicting risks across multiple business process instances. Decis. Support Syst. **69**, 1–19 (2015)
5. Covões, T.F., Hruschka, E.R., de Castro, L.N., Santos, Á.M.: A cluster-based feature selection approach. In: Corchado, E., Wu, X., Oja, E., Herrero, Á., Baruque, B. (eds.) HAIS 2009. LNCS, vol. 5572, pp. 169–176. Springer, Heidelberg (2009). https://doi.org/10.1007/978-3-642-02319-4_20
6. Ding, C.H.Q., Peng, H.: Minimum redundancy feature selection from microarray gene expression data. J. Bioinform. Comput. Biol. **3**(2), 185–206 (2005)
7. Francescomarino, C.D., Dumas, M., Maggi, F.M., Teinemaa, I.: Clustering-based predictive process monitoring. CoRR, abs/1506.01428 (2015)
8. Granitto, P.M., Furlanello, C., Biasioli, F., Gasperi, F.: Recursive feature elimination with random forest for PTR-MS analysis of agroindustrial products. Chemom. Intell. Lab. Syst. **83**(2), 83–90 (2006)
9. Han, J., Kamber, M.: Data Mining: Concepts and Techniques. Morgan Kaufmann, San Francisco (2000)
10. Hartigan, J.A., Wong, M.A.: Algorithm AS 136: a k-means clustering algorithm. J. Royal Stat. Soc. Ser. C (Applied Statistics) **28**(1), 100–108 (1979)
11. Hinkka, M.: Support materials for articles (2017). https://github.com/mhinkka/articles. Accessed 13 Mar 2017
12. Hinkka, M., Lehto, T., Heljanko, K.: Assessing big data SQL frameworks for analyzing event logs. In: 24th Euromicro International Conference on Parallel, Distributed, and Network-Based Processing, PDP 2016, Heraklion, Crete, Greece, 17–19 February 2016, pp. 101–108. IEEE Computer Society (2016)

13. Lehto, T., Hinkka, M., Hollmén, J.: Focusing business improvements using process mining based influence analysis. In: La Rosa, M., Loos, P., Pastor, O. (eds.) BPM 2016. LNBIP, vol. 260, pp. 177–192. Springer, Cham (2016). https://doi.org/10.1007/978-3-319-45468-9_11

14. Leontjeva, A., Conforti, R., Di Francescomarino, C., Dumas, M., Maggi, F.M.: Complex symbolic sequence encodings for predictive monitoring of business processes. In: Motahari-Nezhad, H.R., Recker, J., Weidlich, M. (eds.) BPM 2015. LNCS, vol. 9253, pp. 297–313. Springer, Cham (2015). https://doi.org/10.1007/978-3-319-23063-4_21

15. Liaw, A., Wiener, M.: Classification and regression by randomforest. R news **2**(3), 18–22 (2002)

16. Meyer, P.E.: Information-theoretic variable selection and network inference from microarray data. Ph.D. thesis. Université Libre de Bruxelles (2008)

17. Nguyen, H., Dumas, M., La Rosa, M., Maggi, F.M., Suriadi, S.: Mining business process deviance: a quest for accuracy. In: Meersman, R., Panetto, H., Dillon, T., Missikoff, M., Liu, L., Pastor, O., Cuzzocrea, A., Sellis, T. (eds.) OTM 2014. LNCS, vol. 8841, pp. 436–445. Springer, Heidelberg (2014). https://doi.org/10.1007/978-3-662-45563-0_25

18. Ogutu, J.O., Piepho, H.-P., Schulz-Streeck, T.: A comparison of random forests, boosting and support vector machines for genomic selection. In: BMC Proceedings, vol. 5, no. 3, p. S11 (2011)

19. Teinemaa, I., Dumas, M., Maggi, F.M., Di Francescomarino, C.: Predictive business process monitoring with structured and unstructured data. In: La Rosa, M., Loos, P., Pastor, O. (eds.) BPM 2016. LNCS, vol. 9850, pp. 401–417. Springer, Cham (2016). https://doi.org/10.1007/978-3-319-45348-4_23

20. Thompson, K.: Programming techniques: regular expression search algorithm. Commun. ACM **11**(6), 419–422 (1968)

21. Tibshirani, R.: Regression shrinkage and selection via the Lasso. J. Royal Stat. Soc. Ser. B (Methodological) **58**(1), 267–288 (1996)

22. van der Aalst, W.M.P.: Process Mining - Discovery, Conformance and Enhancement of Business Processes. Springer, Berlin (2011)

23. Van Dongen, B.: Real-Life Event Logs - Hospital Log (2011). https://doi.org/10.4121/uuid:d9769f3d-0ab0-4fb8-803b-0d1120ffcf54

24. Van Dongen, B.: BPI Challenge 2014. Rabobank Nederland (2014). http://dx.doi.org/10.4121/uuid:c3e5d162-0cfd-4bb0-bd82-af5268819c35

25. Weston, J., Mukherjee, S., Chapelle, O., Pontil, M., Poggio, T.A., Vapnik, V.: Feature selection for SVMs. In: Leen, T.K., Dietterich, T.G., Tresp, V. (eds.) Advances in Neural Information Processing Systems 13, Papers from Neural Information Processing Systems (NIPS) 2000, Denver, CO, USA, pp. 668–674. MIT Press, Cambridge (2000)

26. Zeng, Y., Luo, J., Lin, S.: Classification using Markov blanket for feature selection. In: The 2009 IEEE International Conference on Granular Computing, GrC 2009, Lushan Mountain, Nanchang, China, 17–19 August 2009, pp. 743–747. IEEE Computer Society (2009)

Towards Intelligent Process Support
for Customer Service Desks: Extracting Problem
Descriptions from Noisy and Multi-lingual Texts

Jana Koehler[1(✉)], Etienne Fux[1], Florian A. Herzog[1], Dario Lötscher[1],
Kai Waelti[1], Roland Imoberdorf[2], and Dirk Budke[2]

[1] School of Information Technology, Lucerne University of Applied Sciences
and Arts, Luzern, Switzerland
{jana.koehler,etienne.fux,florian.herzog,dario.loetscher,
kai.waelti}@hslu.ch
[2] UMB AG, Cham, Switzerland
{roland.imoberdorf,dirk.budke}@umb.ch

Abstract. Customer service is a differentiating capability for companies, but it faces significant challenges due to the growing individualization and connectivity of products, the increasing complexity of knowledge that service employees need to deal with, and steady cost pressure. Artificial intelligence (AI) can support service processes in a variety of ways, however, many projects simply propose replacing employees with chat bots. In contrast to pure automation focusing on customer self-service, we introduce three intelligent assistants that support service employees in their complex tasks: the *scribe*, the *skill manager*, and the *background knowledge worker*.

In this paper, we discuss the technology and architecture underlying the skill manager in more detail. We present the results from an evaluation of commercial cognitive services from IBM and Microsoft on comprehensive real-world data that comprises over 80,000 tickets from a major IT service provider, where problem reports often comprise an email-based conversation in multiple languages. We demonstrate how today's commercially available cognitive services struggle to correctly analyze this data unless they use background ontological knowledge. We further discuss a pattern- and machine-learning based approach that we developed to extract *problem descriptions* from multi-lingual ticket texts, which is key to the successful application of AI-based services.

1 Introduction

Customer service desks are challenged by an increasing complexity and diversity of skills. This challenge is caused by the growing diversity of products that are tailored individually to customer needs as well as internet connectivity that allows users to configure and set up their own product assemblies. Whereas employees in service desks are thus required to perform an increasingly complex work to support customers and solve their problems, their jobs are threatened by outsourcing or automation due to cost pressure.

© Springer International Publishing AG 2018
E. Teniente and M. Weidlich (Eds.): BPM 2017 Workshops, LNBIP 308, pp. 36–52, 2018.
https://doi.org/10.1007/978-3-319-74030-0_3

With the current interest in artificial intelligence (AI), many companies are initiating projects where intelligent chat bots are developed as first contact points for customers. Chat bots can serve as new forms of conversational interfaces for specific types of applications, however, we believe that AI can bring very different and better value to service desks by assisting service-desk employees, see also [9–11,14,19]. Instead of viewing service desks as cost factor and aiming for high automation or outsourcing service desks to external low-cost providers (be they human or artificial), we propose to redirect attention to the tremendous value of the knowledge that is accumulated by service-desk employees.

UMB AG provides IT services (spanning from infrastructure to applications) to several hundred companies in Switzerland. In 2016, over 50,000 service requests arrived at the service desk and a double-digit growth is expected every year as new customers are engaging with the company. The challenges and data available in this service desk are very representative for many swiss and global companies, which will enable us to develop solutions addressing typical service-desk challenges in many industries. We especially look for solutions that are of value to

- New employees entering a service desk: The training and education of young apprentices by companies often requires them to work in the service desk for a certain amount of time, because this gives young professionals detailed and diverse insights into many technologies and customer problems. In the case of UMB, young professionals move on to other business units after approximately two years. This means that a significant fluctuation due to education and career development of service-desk staff takes place and needs to be better supported.
- Enterprise customers: IT service providers often provide service desk infrastructures and offer ticketing solutions for various customer companies. Enhancing the base ticketing systems with intelligent services that allow companies to better understand the problems of their customers helps these companies to improve their engineering and to offer better products and services.

To achieve these goals, we are developing three intelligent assistants, which address key capabilities of a service desk: recording problems, dispatching problems to the best available expert, and finally solving them.

The **Scribe** supports the intelligent recording of a problem. It matches new problems with those it has seen before, executes additional inquiries to clarify a problem and to gather missing information. It precisely records solutions and auto-solves simple problems such as password-resets, which reduces the work load for the team.

The **Skill Manager** has the overview over the skills and competencies of the service-desk team. It can auto-route tickets to the best available expert. The skill map produced by the skill manager helps new employees to learn and grow their skills by quickly finding out who knows what.

The **Background Knowledge Worker** maintains the information sources for the service-desk team and keeps these information sources up to date by analyzing how products and problems evolve. It searches for missing information in internal as well as publicly available knowledge sources and always provides service-desk employees with the information they need to resolve a problem.

The three intelligent assistants provide a number of values:

– Help companies to keep service desks in-house by increasing their business value and reducing process cycle time.
– Offer a better service to customers through a deeper and more individualized understanding of customer problems and needs.
– Gain knowledge that helps to better adapt the engineering of future products.
– Improve the qualification and training of employees within and beyond the service desk.

In the context of service desks, we see tickets of varying complexity that we characterize along three dimensions as shown in Fig. 1: description complexity, knowledge complexity, user "complexity" (user expectations). As for description complexity, problems can be described by a precise simple request such as *"please reset my password"* or they can be exposed in a longer conversation among several users in multiple languages. The knowledge, which is required to solve a problem, can be located in a single narrow domain or can span multiple interconnected domains with intricate dependencies. Furthermore, different types of users require specific response. An end user may just want the problem to be fixed, a business user has perhaps already explored her available resolution options and requires an in-depth investigation, and an expert may want a detailed explanation. In our domain, we are often dealing with tickets resulting from multi-language email conversations among business users and the solution of the described problems requires knowledge from multiple domains (see Fig. 1).

A problem described in a ticket belongs to one or several specific problem categories or topics. Human experts are able to immediately determine the

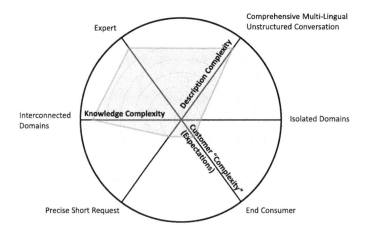

Fig. 1. Three key dimensions that characterize ticket complexity in service desks.

topic(s) from a problem description or they can tell what additional information is required to determine the problem topic in case the description is vague or unclear. Such a deep understanding of problems is key to intelligent support for service desks and a prerequisite for all other insights. Correctly extracting the *topical problem summaries* (TPS) from a complex ticket description is thus essential for the success of our intelligent agents. For example, once the TPS is correctly determined, we can use the set of TPS to compute a *skill map* as the main work product of the skill manager that links TPS with experts, but also shows relationships between the individual TPS. TPS also help the scribe to support the recording of problems, e.g., by relating an initial problem description to a TPS, it can determine missing relevant information. TPS are also used by the background knowledge worker to determine relevant expert knowledge, similar problems, and reusable solutions that can help solving a problem. Being able to compute the TPS is therefore key to intelligent support for the service desk and a major challenge that we need to solve in this project.

In this paper, we report on first results from developing the skill manager and discuss the extraction of problem-related information from unstructured texts as a major prerequisite on the route towards TPS. In Sect. 2, we discuss the architecture of the skill manager and report on our experience in using cognitive services from IBM and Microsoft. In Sect. 3, we present a pattern- and machine-learning based approach to extract TPS-relevant information from a problem description. Section 4 summarizes first lessons learned and concludes the paper. Related work is addressed throughout the paper.

2   The Skill Manager

The skill manager takes a problem description from the ticketing system, extracts the TPS, and links it to the service-desk employee who has solved the problem in the skill map. The ticketing system contains problem descriptions that are auto-generated from automatic monitoring systems and problem descriptions written and submitted by humans. Extracting TPS from automatic systems is straightforward, but extracting TPS from human-authored tickets is a major challenge due to the following observations:

– Ticket descriptions often result from longer email conversations among customers and then extend to involve service-desk employees. Even if customers enter a problem description into a web form, they often copy longer texts from email conversations into the form in order to illustrate the problem background. These email conversations constitute the ticket problem field and frequently, additional comments are added when employees follow up on the ticket to gather more information. We thus need to deal with longer texts, which significantly differ from short and direct questions. These texts contain the common ingredients of emails such as headers, greetings, and signatures that are not contributing directly to the problem description.

– Texts are formulated in a *"technoid"* language where grammar does not matter much, spelling errors occur, and sentences are often incomplete. Furthermore, technical terms are mixed with words in German, English, Italian, French, and local dialects or other languages, which reflects the language diversity of Switzerland and global companies.
– Texts also contain person-related or company-specific information, some users even copy passwords into the ticket. Data is thus sensitive and data protection matters.
– Tickets that involve complex problems are not solved by a single service-desk employee. Typically, such tickets require the contribution of a number of people between whom the ticket is sent back and forth. Often, the employee, who closes the ticket after successful solution, is not the expert who contributed the key knowledge to solving the problem.

Figures 2 and 3 illustrate the challenges on a real-world example from an industrial customer of UMB AG. Names have been replaced by a capital X followed by a number. Numbers are replaced by zeros. The original ticket begins with an email in German where a customer summarizes access problems to a technical system in Finland via a remote control software. The text describes the issues, points to further details in the subsequent email thread, and requests support and a call back, see Fig. 2.

This email is followed by a thread of conversation in German, Finnish, and English emails, which we do not show here as it involves over 200 lines of text. The initial report of the problem in mostly English originates from an office in Finland and can be found at the end of the email conversation shown in Fig. 3. We will discuss the framed and numbered parts of this text fragment further in Sect. 3.

```
Guten Tag
Ich möchte mit X1 auf die X2 in X3 zugreifen.
Unten die Anforderungen von der IT X3.
Ich bitte um Unterstützung beim Beantworten der Fragen und Umsetzung der
Anforderungen.
Bitte rufen sie mich an.
Danke und freundliche Grüsse
X4
---------------------------------------------------------
X AG
X4
mechanical engineer, dipl.Ing.HTL
O & M software engineer
Service
Xtrasse 00, Postfach 000
CH-0000 X
Phone++00(00) 000 00 00
Fax ++00 (00) 000 00 00
mail: ...@....com
web: www....com
```

Fig. 2. Extract from a problem report - initiating text in German.

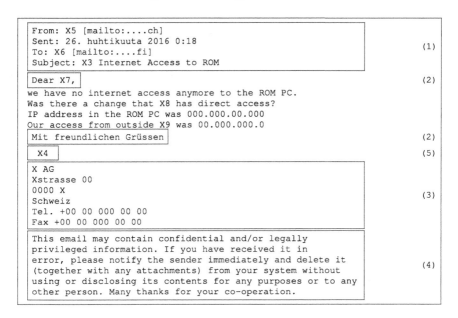

Fig. 3. Extract from a problem report - Final Clarification in English.

Our goal is to automatically compute a TPS that correctly describes the problem, e.g., a failed remote access to a machine. The multi-language conversation requires a dedicated approach to be developed that we discuss subsequently.

2.1   Analyzing Ticket Problem Descriptions with Cognitive Services

A prerequisite for the extraction of a TPS is the determination of key phrases, entities, concepts, and categories that occur in the text. Extracting such information from arbitrary texts is a frequently occurring problem and thus offered as a cognitive service by various commercial vendors such as IBM and Microsoft. Cognitive services wrap algorithms from artificial intelligence into easy to use APIs [7,8,18]. They can potentially facilitate the development of intelligent systems by providing key capabilities, e.g., to deal with unstructured, text-based information. In the following, we summarize our findings from applying cognitive services from IBM and Microsoft to our ticket texts.

From Microsoft, we used the Text Analytics API [15] (TA), which offered sentiment analysis, key phrase extraction, topic detection, and language detection in June 2017. Topic detection is not applicable to our problems as it summarizes topics for collections of multiple documents (at least 100) whereas we want to extract topics from a single ticket. We also experimented with the Entity Linking service (EL), which recognizes and identifies named entities.

From IBM, we used Watson Natural Language Understanding [12] (NLU), which offers sentiment analysis and can extract keywords, entities, concepts, and categories, relations, semantic roles etc. We experimented with the detection of keywords, entities, concepts, and categories. NLU concept detection can identify concepts that are not necessarily directly referenced in the text and uses DBpedia (http://wiki.dbpedia.org/) following an ontology-based information extraction approach. Entity extraction can identify words as belonging to a certain category, e.g., it can determine that New York is a location. The categorization service can categorize content using a five-level classification hierarchy. It relies on a predefined list of categories from DPpedia, which for example contains categories such as arts and entertainment, family and parenting, or education. For our application, the technology and computing category is of interest, which is further refined into subcategories such as hardware, software, programming languages, etc. These subcategories in turn are further refined, for example, the operating systems category is refined into linux, mac os, unix, windows. Concept and category detection was only available for texts in English when we conducted the evaluation.

Sentiment analysis is offered by both vendors and determines the sentiment of the author of a text, i.e., does the text reflect a positive, neutral, or negative mood. We tested these services to find out if the analysis results can be used to adjust the priority of tickets with respect to customer mood. Microsoft expresses sentiment as a number between 0 (negative) and +1 (positive), whereas IBM uses a scale from -1 (negative) to $+1$ (positive). IBM offers an additional emotion analysis, which refines sentiment with respect to detected entities or keywords, which we did not test. Table 1 gives an overview of tested capabilities that are relevant for extracting TPS. Columns 2 and 3 indicate the subset of languages that we are interested in and for which the capability is available: (e)nglish, (g)erman, (i)talian, (f)rench. A letter in braces means that the capability is currently limited or only applicable after providing customized models, which can for example be built with IBM's Watson Knowledge Studio. A dash means that the capability is not available.

We use the example ticket from Figs. 2 and 3 to illustrate typical results that we obtained when applying the services to tickets from our database of over 80,000 tickets from 2015–2017. The problem description of the example is comprised out of 312 lines of text including empty lines. MS TA cannot process this text as it exceeds the document size limit of 10240 bytes. IBM NLU with language set to English processes the text with a sentiment analysis result of slightly positive with score 0.01. Keyword extraction returns problem relevant keywords

Table 1. Overview of relevant cognitive services from IBM and Microsoft.

Capability	IBM NLU	Microsoft MS TA/EL
Sentiment analysis	e, g, i, f	e, g, f
Topic/concept/categorization	e, (f)	-
Entity linking	e, (g, i, f)	e, g, i, f
Keyphrases/ keywords	e, g, i, f	e, g
Language detection	10	>120

such as *X3 Internet Access* (0.95)[1], *ROM* (0.37), *remote access* (0.30), *adapter settings* (0.29), and *remote connection* (0.29), but also many person names and keywords that result from the headers or signatures of the emails such as for example *mailto* (0.81) or *best regards* (0.41). Entity detection mostly returns persons, locations, companies, email and IP addresses as well as job titles such as *software engineer* or *support specialist*. Typing errors can mislead the service, e.g., *RemoteDesktop* is identified as a company. Multi-language usage can also constitute a problem as *mit freundlichen Grüssen* (with best regards) is identified as a person when language is set to English. Concept analysis returns very strong results with only network and routing-related terms such as *IP address* (0.95), *Subnetwork* (0.64) or *Dynamic Host Configuration Protocol* (0.57) and seems to spot the problem source very well. Categories are also strong with *javascript* (0.63), *router* (0.48), and *vpn and remote access* (0.44).

To illustrate how MS TA/EL and IBM NLU compare on the same texts, we also submitted just the text fragments from Figs. 2 and 3 separately and set the language accordingly to German and English. Both services are confused by the email headers and signatures that occur in the raw data. For example, MS EL returns various country, city, and company names contained in headers and signatures as well as other entities that are detected in these text parts. Confidence levels vary significantly with respect to words, but also language. For example, *Fax* is detected with confidence 0.2, in English, but only 0.006 in German. The name *X3* is detected with confidence 0.008, but the entity Switzerland is detected with confidence 0.99. In addition, entities such as *The Internet (0.46)*, *Personal computer (0.90)*, and *IP address (0.02)* are detected in the text from Fig. 3. Interestingly, the term "IT" in Fig. 2 is interpreted as the entity *Italian* (0.16). Microsoft's keyphrase detection splits relevant text phases and returns again many keywords from the signature and headers. The information on the IP address is completely lost. Meaningful parts such as *we have no internet access anymore to the ROM PC* are also completely lost. Language detection returns English as the recognized language for the text. Sentiment analysis rates both texts as very positive with a score above 0.73 and 0.99, respectively.

IBM NLU sentiment analysis rates the individual texts from Figs. 2 and 3 positive with score 0.3 and 0.01, respectively. Submitting a combined text leads to a negative rating with a score of −0.14. Keyword detection is more useful for English than German text fragments, but also returns many recognized names and single words without context. However, words returned with high confidence settle around internet and connection problems. In German, the service returns also parts of numbers, greetings or internet addresses and a few phrases such as *Ich möchte mit (I want with)* or *und Umsetzung der Anforderungen (and implementation of requirements)*. Entity linking is again very strong in detecting names of people and linking them to the entity *person* as well as on detecting locations, job titles (e.g., *software engineer*), and company names. It also discovers the IP addresses contained in the problem description, but also information such as email addresses from headers and signatures. However, it can happen

---

[1] Numbers in braces show returned confidence values.

that an email address is only partially returned without the user name and the user name is identified as a twitter handle. Again, a street name as well as the text *mit freundlichen Grüssen (with best regards)* were recognized as a person. Concept detection in English returns meaningful concepts such as *Internet* and *IP address*, but also information extracted from headers and signatures such as *Engineering* and *Buenos Aires Province* or an *Erdös Woods number*, for which we have no explanation. Category detection in English finds for example *software*, *vpn and remote access*, and *computer*. Concepts and categories are not detected in German and a warning of unsupported text language is returned when applied to the mostly German text from Fig. 2 and the language to be used is set to German.

In summary, we can say that IBM's category and concept detection in English language works quite well on our application domain, even for mixed language texts, whereas returned keywords and sentiment analysis are not really informative enough. Microsoft's entity linking is returning interesting results, but needs properly cleaned problem descriptions. Keyphrase detection returns a mixture of words and word groups. Language detection works well on single-language texts, but returns mostly English on mixed language-texts. Microsoft sentiment analysis also seemed too unreliable.

It becomes obvious that the skill manager needs to clean and extract the problem information from our complex email conversations before any intelligent text analysis can take place. We discuss in detail in Sect. 3 the extractor service that we developed for this purpose. The extractor service returns the *problem description*, see Fig. 4.

Applying the extractor service to the example at hand results in 78 lines of problem-specific text. Applying cognitive services to this text leads to significantly improved results. In particular, Microsoft TA is now applicable as the document size is within the required limits. Keyphrase detection returns a long list of single words that does not appear to be helpful, but word groups of at least two words relate to networking and connection problems. No differentiation of the results based on confidence is provided by the keyphrase service. A sentence like *we have no internet access anymore to the ROM PC* is split into *ROM PC* and *internet access* when invoked on the cleaned full text. Interestingly, when the combined text from both figures is sent to the service, the entire sentence is returned as a single keyphrase. Microsoft's entity linking returns 15 entities, the top five are IP address (1.00), Windows XP (1.00), Microsoft Windows (1.00), Switzerland (0.99), and Domain Name System (0.99). The detected language for the cleaned text with mixed languages is English and sentiment is rated very positive with score 0.97. Interestingly, the sentiment rating of the combined text from the figures is rated neutral with score of 0.5.

The categories returned by IBM NLU on the full cleaned text are *vpn and remote access*, *computer*, and *router*. Sentiment of the full cleaned text is rated negative with score -0.31. Sentiment analysis of the cleaned English text in Fig. 3 is rated very negative with a score of -0.7, whereas the cleaned German text in Fig. 2 is rated as neutral. Keyword detection returns results that differ significantly for each language in particular for the full cleaned text. The occurrence of several languages impacts the usability of results. However, the top rated keywords are quite to the point *IP config* (0.94), *adapter settings* (0.92), *remote connection* (0.91).

2.2 Architecture of the Skill Manager and Current Limitations of Cognitive Services

The experiments with cognitive services showed that indeed these services can contribute significant capabilities to our intelligent assistants, but only when provided with cleaned textual information where the actual problem description is extracted from complex conversational texts. Figure 4 summarizes the current architecture of the skill manager. To achieve the required textual transformation, we developed an extractor service that is run prior to any cognitive services. In the following section, we discuss how the extractor service works and illustrate its effectiveness. Other architectural elements are out of scope of this paper.

We see that results improve after cleaning ticket descriptions, but still important information is lost in keywords or keyphrases and the results differ very much depending on the language of the text. Confidence values, when returned by a service, can sometimes help to further differentiate results, but are neither transparent nor reliable. The specific numbers returned differ significantly and simply setting a threshold value will not work. Mixed language texts clearly constitute a challenge.

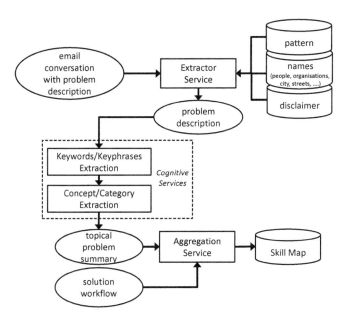

Fig. 4. Architecture of the skill manager.

We decided to completely drop sentiment analysis for the development of our assistants as the results are not reliable enough. We believe that is caused by the difference of our texts from the training data used for these services. However, this cannot be evaluated as no information on the development of cognitive services is available. In particular, we believe that the hidden dependency on unknown training data significantly affects the usability of the services. It has already been discussed in the literature that by slightly shifting patterns in data, classifiers based on machine learning can return completely inaccurate results [16,17]. The phenomenon is known as out-of-range-behavior [1] and we believe that much more research is necessary to precisely specify the range of applicability of cognitive services such that users of these services know what type of information a service can correctly analyze. An extended self-control of cognitive services beyond simple confidence values where services themselves detect data that is outside the trained range and where they also provide an explanation of the results would be even more desirable.

3   The Extractor Service

The goal of the extractor service is to delete text from a description that does not contain problem-relevant information. In our case, this means in particular to delete typical patterns of email-based text fragments that we encounter in our tickets. These text fragments are framed in Fig. 3 to highlight header (1), greetings (2), signature (3), advertisements/disclaimers (4), and names of people (5). Names are handled as part of greetings. In Fig. 3, only the unframed text lines contain problem-relevant information.

Identifying certain types of email fragments has been in the focus of AI researchers for some time and led to a number of approaches on which we can build. Particularly, the approach by Carvalho and Cohen [3,4] turned out to be very applicable to our domain, because it focuses on a line-based identification and uses pattern-based feature annotations, which are fed into various supervised machine-learning algorithms. Approaches using geometric information or zones [5,13] are less applicable as our texts can be arbitrarily structured. Furthermore, machine-learning approaches that exploit regularities in user behavior [6] are not applicable as we cannot expect that problems frequently re-occur for the same user.

Carvalho and Cohen focus on learning signature block lines and reply lines. This comes very close to our goal of removing irrelevant lines from threads of conversation. Their approach treats input text as a sequence of lines. For each line, a set of features is calculated using text patterns and a classifier is learned over this feature space using a supervised machine-learning approach. In the original approach, each feature pattern is applied to the k last lines of an email to detect a signature block. In our case, we apply each feature pattern to each single line contained in our problem description. Furthermore, we split the learning task into four separate subtasks, each addressing the different types of lines that we want to delete: (1) header, (2) signature, (3) greetings, and

(4) disclaimer. For the first three subtasks, we developed an approach that marks each line type using feature patterns. The detection of people names uses a name database built from names occurring in the ticketing system. After line types (1) to (3) have been cleaned, the remaining lines are compared against a disclaimer database. If a line has been seen in at least k different tickets, it is considered a disclaimer and removed. Currently, we set k to 3, which works well for our ticket descriptions.

3.1   Pattern-Based Classification of Text Lines

For each type of text line, we developed a set of patterns taking inspiration from Carvalho and Cohen [4], who introduced patterns such as *line contains URL pattern* or *percentage of punctuation symbols in the line is larger than 20%*. We extended this idea by also assessing the size of the remaining, non-matching text. This allows us to express that a pattern covers at least, e.g., 50% of all characters in a line. In addition, we also apply Carvalho and Cohen's idea to consider a limited context for each line, i.e., we added patterns that check if a previous or following line is empty, only contains punctuation symbols, or matches a header, signature, or greeting pattern. Other patterns that were proposed in [4] such as *mostly capitalized words* work well for English, but do not carry over to languages such as German, French, Czech or Spanish that we need to cover.

Figure 5 shows the regular expressions that we developed to detect greeting patterns in German and English. Similar patterns were defined to detect header keywords, punctuation symbols, telephone numbers, postal codes, job descriptions, organization types, etc. In total, 4 patterns were defined to detect greetings, 7 patterns to detect headers, and 79 patterns to detect signatures as the latter are much more diverse.

We implemented a simple threshold-based extractor using these patterns, which deletes a line whenever a number of patterns at least equal to the threshold matches. Setting the threshold to 3 worked quite well for our application. We used this threshold-based extractor for two purposes: First, to have a baseline of a simple extractor service and second, to facilitate the labeling of training data for a supervised machine-learning approach.

pattern	regular expression
salutation (g)	`(\s|\A)(Sehr gee?hrter?|Herr(en)?|Frau|Damen|Liebe(r|s)?` `|Liebster?|Hallo|Hey|Grezi|Hi|Morgen?` `|Gue?ten? (Tag|Abend|Morgen?|Obig)|Hoi),?.?(\s|\Z)`
closing (g)	`(\s|\A)(Gr(|ue|u)(ss|)e?n?|Alles (Gute|Gueti)|Herzlich|Alles Lieb(e|i)i)|Danke?|Sali` `|Vielen?|Lg|Mfg|Fg|Dame?n?|Herre?n?),?.?(\s|\Z)`
salutation (e)	`(\s|\A)(Dear|Dearest|Hello|Hey|My Dear),?.?(\s|\Z)`
closing (e)	`(\s|\A)(Regards?|Thanks|Cheers|Grateful|Sincerely),?.?(\s|\Z)`

Fig. 5. Text patterns to detect greetings in English and German.

Category	No. of lines	FP	FN	TP	TN	Accuracy (%)	$F_{0.5}$ (%)
header	2516	26	9	2507	17999	99.8	99.1
greeting	3019	109	914	2105	17413	95.02	88.63
signature	7673	7475	130	7543	5393	62.98	55.67

Fig. 6. Performance of threshold-based extractor on sample of 927 tickets (20'541 non-empty lines).

In order to validate the threshold-based extractor and to generate test and training data for the machine learning algorithms, we started with a set of 51,812 tickets from 2016 from which a representative sample that reflects the properties of the population has to be chosen. The population contains tickets from over 70 companies, out of which five companies submit nearly 50% of the tickets. The goal of the sampling was to cover the ticket share of companies. A closer investigation of the population showed that a significant number of tickets are submitted from automatic monitoring systems or contain less than three lines of text. These tickets were removed yielding a population of 27,700 tickets. When conservatively assuming a 99% confidence interval and a 5% sample error, we obtain a sample size of 651 tickets. In the end, we labeled 927 tickets with a total of 33,204 text lines including 12'663 empty lines. 29 tickets contain only problem-relevant information. Headers occur in 845 tickets, signatures in 728, greetings in 843, and disclaimers in 483 tickets. 78 tickets contain lines from one or two categories, 339 tickets contain lines from three categories, and 415 tickets contain all four categories.

The table in Fig. 6 shows the number of lines belonging to the header, greeting, and signature category over the total set of 927 tickets (20'541 non-empty lines) and the absolute numbers for false and true positives/negatives (FP, TP, FN, TN) returned by the extractor. Every line classified into the wrong category is considered an error. We also show values for accuracy and the F-measure. Accuracy gives an impression of how close our extractor service comes to a correct deletion of problem-irrelevant information. We choose the F-measure with $beta = 0.5$ weighing precision four times higher than recall, because high precision is very desirable for the extractor service as this means that only few false positives result from accidentally deleting problem-relevant text lines. In the following, we compare the performance of the threshold-based extractor to the performance of selected machine-learning algorithms.

3.2   Learning Signature Classification

Whereas header and greeting classification work quite well with a simple threshold-based approach, signature detection is clearly insufficient, which motivated us to explore a machine-learning based approach that we describe next.

A key challenge for machine learning is the generation of training and test data. In many cases, the required labeling of data with the correct result has to be done in a time-consuming manual annotation process. The threshold-based extractor helps in significantly reducing the labeling effort. It classifies each line in a given text as belonging to one of the four categories (1) header, (2) signature, (3) greetings, and (4) disclaimer. Lines that do not exceed the threshold are classified as belonging to the problem description. We developed a web-based annotation tool that displays the text from a ticket and marks each classified line by a color linked with the category. A simple and fast-to-use GUI allowed three human annotators to correct or confirm the threshold-based classification of all sample tickets and thereby label our data within one day. The annotation tool logged all inputs made by the human annotators who re-classified each incorrectly classified text line as returned by the threshold-based extractor. These logged corrections were used to automatically determine the error rates of the extractor.

We used the corrected sample tickets as training data to train various classifiers on the problem of signature classification. The feature annotation of a ticket is a 0–1 matrix. For each line in a ticket, a vector of up to 90 features describes which pattern matches a line. Several patterns can contribute to a single feature due to multiple-language support, e.g., the phone number patterns cover the different formats for different countries, but all contribute to the single feature describing whether a phone number occurs in a text line or not. We ran several experiments: First, we were interested how well a classifier can learn to detect signatures based on the 79 signature features. Second, we added the 7 features for headers to allow the classifier to pick up a potential relation between the occurrence of header and signature patterns, which is not available to the simple threshold-based extractor. Finally, we added 4 greetings features which gives a total of 90 features across the three categories.

We used a support vector machine (SVM) and the AdaBoost sequential learning classifier (following results from [4]) from the *scikit-learn* package (http://scikit-learn.org) and a random forest (RF) classifier as the latter is known for good performance, efficient scalability to high-dimensional feature vectors, and is less prone to overfitting [2]. To assess how the model will generalize to an independent data set, cross-validation was used and the data set was split into 80/20% training/test sets. The training set of 741 tickets was used for cross-validation using $k = 5$ to avoid excessively high bias and very high variance. The three classifiers were thus trained 5-times on 4 folds using their default parameters and validated on the remaining fold of 148 tickets every time calculating the validation error.

Figure 7 summarizes the results for each classifier on the test set of 186 tickets containing 6716 empty and non-empty lines. Machine learning clearly outperforms a simple threshold-based approach on the very complex problem of signature classification. From the tested learners, the random forest classifier worked best. Using all the available features yields the best results. The results compare well with [4] who achieved up to 95% accuracy on single English emails even though our texts have a much more complex structure and use multiple languages.

Feature Vector	Learner	FP	FN	TP	TN	Accuracy (%)	$F_{0.5}$ (%)
signature only	SVM	249	696	1344	4427	85.93	79.89
signature only	AdaBoost	218	733	1307	4458	85.84	80.28
signature only	RF	177	453	1587	4499	90.62	87.24
signature + header	SVM	231	670	1370	4445	86.58	81.12
signature + header	AdaBoost	220	686	1354	4456	86.51	81.21
signature + header	RF	175	426	1614	4501	91.05	87.76
all features	SVM	224	671	1369	4452	86.67	81.37
all features	AdaBoost	237	669	1371	4439	86.51	80.91
all features	RF	176	421	1619	4500	**91.11**	**87.80**

Fig. 7. Classifier results for signature line detection on test set of 186 tickets (6716 lines).

Category	No. of lines	FP	FN	TP	TN	Accuracy (%)	$F_{0.5}$ (%)
header	504	4	2	502	6208	99.91	99.29
greeting	604	31	101	503	6081	98.03	91.79
signature	2040	176	421	1619	4500	91.11	87.80

Fig. 8. Final performance of extractor on the test set of 186 tickets (6716 lines).

We also applied the machine-learning approach to the greetings and header classification problem using all features and random forests. Figure 8 summarizes the results for the final version of the extractor service using machine learning on all three text line categories based on random forests with lines labeled by the complete set of 90 features.

3.3 Disclaimer Classification

A pattern-based classification of disclaimers seems impossible as they come in totally different styles and there is no recurring pattern. Nevertheless a disclaimer is not unique - it is repetitively attached to emails of users of the same organization. We thus decided to apply a database approach to detect disclaimers. Every line in a ticket is stored in a database. If a line is seen three times in *different tickets*, the extractor will start marking this line as disclaimer in subsequent tickets. As this is a fairly aggressive approach, we constrained the lines under consideration by their relative positioning in the text. Only lines that occur between the signature and the end of file or that occur between a signature and a subsequent header (remember that we often see forwarded email conversations) are added to the database. Every line in the database is saved with a hash code generated from the original text, which is used to prevent double counting.

Figure 9 shows the performance of the disclaimer detection measured on the whole sample set of 927 tickets. The 281 false positives split into 121 empty lines, 1 header line, 16 greeting lines, and 143 signature lines that are incorrectly

Category	No. of lines	FP	FN	TP	TN	Accuracy (%)	$F_{0.5}$ (%)
disclaimer	1414	281	343	1071	18846	96.69	78.49

Fig. 9. Performance of disclaimer extraction on 927 tickets (20'541 non-empty lines).

labeled as disclaimers and that were not spotted by the before-running header, signature, and greeting detection.

4 Conclusion

We propose to support the complex work of humans in customer service by three different intelligent agents denoted as the skill manager, the scribe, and the background knowledge worker. We discuss the complexity of problem tickets along three dimensions and argue that current chat bot technology is insufficient to help solving problems exhibiting high description, knowledge, and user complexity. We present one of our agents, the skill manager, in more detail and define its capabilities. We discuss the results of an evaluation of commercial cognitive services from IBM and Microsoft to implement the required capabilities and discuss the challenges that arise from our unstructured and multi-lingual text data. We show that out-of-the-box sentiment analysis and other techniques fail on our complex data, but that approaches using ontology-based information extraction work quite well, however, they also need to better understand the problem context to increase accuracy. As a prerequisite to successfully apply ontology-based information extraction, a problem description needs to be extracted out of the raw ticket data, for which we present a pattern- and machine-learning based extractor service that automatically recognizes and removes problem-irrelevant email parts such as headers, greetings, signatures, and disclaimers.

References

1. Amodei, D., Olah, C., Steinhardt, J., Christiano, P., Schulman, J., Mané, D.: Concrete problems in AI safety. arXiv preprint arXiv:1606.06565 (2016)
2. Caruana, R., Karampatziakis, N., Yessenalina, A.: An empirical evaluation of supervised learning in high dimensions. In: 25th International Conference on Machine Learning, pp. 96–103. ACM (2008)
3. Carvalho, V.R., Cohen, W.W.: Ranking users for intelligent message addressing. In: Macdonald, C., Ounis, I., Plachouras, V., Ruthven, I., White, R.W. (eds.) ECIR 2008. LNCS, vol. 4956, pp. 321–333. Springer, Heidelberg (2008). https://doi.org/10.1007/978-3-540-78646-7_30
4. de Carvalho, V.R., Cohen, W.W.: Learning to extract signature and reply lines from email. In: 1st Conference on Email and Anti-Spam (CEAS) (2004)
5. Chen, H., Hu, J., Sproat, R.W.: Integrating geometrical and linguistic analysis for email signature block parsing. ACM Trans. Inf. Syst. (TOIS) **17**(4), 343–366 (1999)

6. Dredze, M., et al.: Intelligent email: aiding users with AI. In: 23rd AAAI Conference, pp. 1524–1527. AAAI Press (2008)
7. Farrell, R., et al.: Symbiotic cognitive computing. AI Mag. **37**(3), 81–93 (2016)
8. Goel, A., et al.: Using Watson for constructing cognitive assistants. Adv. Cognit. Syst. **4**, 1–16 (2016)
9. Gownder, J.P.: The future of jobs, 2025: working side by side with robots. Research report, Forrester (2015)
10. Hui, S.C., Fong, A., Jha, G.: A web-based intelligent fault diagnosis system for customer service support. Eng. Appl. Artif. Intell. **14**(4), 537–548 (2001)
11. Hui, S.C., Jha, G.: Data mining for customer service support. Inf. Manag. **38**(1), 1–13 (2000)
12. IBM: Natural language understanding cognitive service (2017). IBM Bluemix, https://console.ng.bluemix.net/catalog/services/natural-language-understanding
13. Lampert, A., Dale, R., Paris, C.: Segmenting email message text into zones. In: Conference on Empirical Methods in Natural Language Processing, pp. 919–928. Association for Computational Linguistics (2009)
14. Law, Y.F.D., Foong, S.B., Kwan, S.E.J.: An integrated case-based reasoning approach for intelligent help desk fault management. Expert Syst. Appl. **13**(4), 265–274 (1997)
15. Microsoft: Text analytics cognitive service (2017). https://azure.microsoft.com/en-us/services/cognitive-services/text-analytics/
16. Moosavi-Dezfooli, S.M., Fawzi, A., Fawzi, O., Frossard, P.: Universal adversarial perturbations. arXiv preprint arXiv:1610.08401 (2016)
17. Nguyen, A., Yosinski, J., Clune, J.: Deep neural networks are easily fooled: high confidence predictions for unrecognizable images. In: IEEE Conference on Computer Vision and Pattern Recognition, pp. 427–436 (2015)
18. Spohrer, J., Banavar, G.: Cognition as a service: an industry perspective. AI Mag. **36**(4), 71–87 (2015)
19. Wipro: Wipro to transform customer service desk at Nexenta using its artificial intelligence (AI) platform HOLMES (2016). http://www.wipro.com/newsroom/press-releases/Wipro-to-transform-customer-service-desk-at-Nexenta-using-its-artificial-intelligence-platform-HOLMES/. 15 Dec 2016

Towards an Entropy-Based Analysis of Log Variability

Christoffer Olling Back[1]([✉]) [iD], Søren Debois[2] [iD], and Tijs Slaats[1] [iD]

[1] Department of Computer Science, University of Copenhagen,
Emil Holms Kanal 6, 2300 Copenhagen, Denmark
{back,slaats}@di.ku.dk
[2] Department of Computer Science, IT University of Copenhagen,
Rued Langgaards Vej 7, 2300 Copenhagen, Denmark
debois@itu.dk

Abstract. Process mining algorithms can be partitioned by the type of model that they output: *imperative* miners output flow-diagrams showing all possible paths through a process, whereas *declarative* miners output constraints showing the rules governing a process. For processes with great variability, the latter approach tends to provide better results, because using an imperative miner would lead to so-called "spaghetti models" which attempt to show all possible paths and are impossible to read. However, studies have shown that one size does not fit all: many processes contain both structured and unstructured parts and therefore do not fit strictly in one category or the other. This has led to the recent introduction of *hybrid miners*, which aim to combine flow- and constraint-based models to provide the best possible representation of a log. In this paper we focus on a core question underlying the development of hybrid miners: given a log, can we determine *a priori* whether the log is best suited for imperative or declarative mining? We propose using the concept of entropy, commonly used in information theory. We consider different measures for entropy that could be applied and show through experimentation on both synthetic and real-life logs that these entropy measures do indeed give insights into the complexity of the log and can act as an indicator of which mining paradigm should be used.

Keywords: Process mining · Hybrid models · Process variability
Process flexibility · Information theory · Entropy · Knowledge Work

1 Introduction

Two opposing lines of thought can be identified in the literature on process modelling notations. The *imperative* paradigm, including notations such as Petri nets [1] and BPMN [2] focuses on describing the flow of a process and is considered

This work is supported by the Hybrid Business Process Management Technologies project (DFF-6111-00337) funded by the Danish Council for Independent Research.

© Springer International Publishing AG 2018
E. Teniente and M. Weidlich (Eds.): BPM 2017 Workshops, LNBIP 308, pp. 53–70, 2018.
https://doi.org/10.1007/978-3-319-74030-0_4

to be well-suited to structured processes with little variation. The *declarative paradigm*, including notations such as Declare [3], DCR Graphs [4], and GSM [5] focuses on describing the rules of a process and is considered to be well-suited to unstructured processes with large degrees of variation. However, recent studies [6,7] have shown that one size does not fit all: many processes do not fit strictly in one category or the other and instead contain both structured and unstructured parts. This has led to the recent introduction of a *hybrid* paradigm [6,8], which aims to combine the strengths of these two approaches.

Following the introduction of the hybrid modelling paradigm, a number of hybrid mining algorithms have been developed: in [9] the authors use a heuristic approach based on the directly-follows-graph to divide activities between structured and unstructured parts of the model; in [10] the authors take a mixed approach and mine both a declarative and imperative model which are then overlain; and in [11] the authors take a model-based approach, where an imperative model is mined and analysed for pockets of unstructured behaviour, and for these pockets, a declarative alternative is mined.

All these approaches avoid an important research question, first identified in [12]: *can we, based on an a priori analysis of the input log, measure if it is best suited to imperative or declarative mining?* Such a measure:

(i) Would give us greater insights into what type of miner we should use for a log;
(ii) could be combined with existing partitioning techniques [13–16] to determine for each partition if it is more suited for imperative or declarative mining, thereby providing an efficient method to construct a hybrid model; and
(iii) could be used for the development of novel partitioning techniques that specifically aim to separate structured and unstructured behaviour in a log.

In this paper we propose basing such a measure on the notion of *entropy* from the field of information theory. Introduced by Shannon in his seminal 1948 paper [17], entropy measures the information content of a random variable. Intuitively, we can think of entropy as the "degree of surprise" we will experience when obtaining additional information about a system [18].

We propose that the entropy of an event log can serve as a predictor of whether the generating process is best modelled using declarative or imperative models. Highly structured processes should generate more homogeneous (low entropy) traces and more flexible processes should generate more varied (high entropy) traces. While information theoretic tools have been previously applied to predictive modelling [19], our application to discriminating mining techniques is novel.

To find such a measure, we first introduce a number of example logs that we use to illustrate our ideas and concepts in Sect. 2. In Sect. 3 we introduce three entropy measures on event logs: (i) trace entropy measures only the entropy on the level of distinct traces, (ii) prefix entropy measures entropy by taking into account all unique prefixes of the log, and (iii) block entropy measures entropy by considering all unique sub-strings present in the log. In Sect. 4 we report on an implementation of these measures and the results of applying them to both synthetic and real-life logs. We show that block entropy is the most successful measure, but suffers from a high computational complexity which becomes apparent on large logs with long traces. In addition it becomes clear that the current proposed measures are not yet absolute and that both further research and a more detailed evaluation are needed to arrive at such a measure. We discuss how we intend to do so in Sect. 5 and conclude in Sect. 6.

2 Running Example

We will use a running example of three logs to illustrate how we can use entropy to measure the variability of process logs. Recall the definitions of events, traces and logs.

Definition 2.1 (Events, Traces and Logs). *Let Σ be an alphabet of activities. An* event *$e \in \Sigma$ is a* specific occurrence *of an activity. A* trace *$\sigma \in \Sigma^* = \langle e_1, \ldots, e_n \rangle$ is a sequence of events $e_1, \ldots, e_n$, with each $e_i \in \Sigma$. Finally, a* log *is a multiset $[\sigma_1^{w_1}, \ldots, \sigma_n^{w_n}]$ where each $\sigma_i \in \Sigma^*$ and each $w_i \in \mathbb{N}$*

Notice that we have defined a trace as the sequence of activities observed in a particular process instance. A log is a multiset of such traces, representing explicitly the number of process instances exhibiting the particular trace.

Example 2.2. As a running example, consider the three logs L_1, L_2, and L_3 in Fig. 1. L_1 is a very structured log, for which we can easily find a compact imperative model, for example the Petri net shown in Fig. 2. L_2 is the same log, except some traces are now more frequent than others. The last log L_3 is a much less structured, which is more complex to describe with an imperative model, e.g. the Petri net in Fig. 3.

This log can be be more effectively explained by a declarative model, as shown in Fig. 4. The declarative model uses the Declare notation [3] and shows that: (i) a and b can not occur in the same trace, (ii) after an a we always eventually see an h, (iii) we must have seen at least one a before we can see a c, (iv) we must have seen at least one d before we can see a c, (v) we must have seen at least one d before we can see an e, (vi) after an e we always eventually see an f, (vii) we must have seen at least one f before we can see a g, (viii) after an f we will immediately see a g. One should note that in addition to giving a more straightforward view of the process, this model is also much more precise than the Petri net in Fig. 3 (i.e. it allows less behaviour for which there is no evidence in the log).

3 Log Entropy

Entropy is a measure of the information required to represent an outcome of a stochastic variable, intuitively indicating the "degree of surprise" upon learning

L_1
$\langle a, b, c, d, f, g, h \rangle^5$
$\langle a, b, c, e, f, g, h \rangle^5$
$\langle a, b, c, d, f, g \rangle^5$
$\langle a, b, c, e, f, g \rangle^5$
$\langle a, b, b, c, d, f, g, h \rangle^5$
$\langle a, b, b, c, e, f, g, h \rangle^5$
$\langle a, b, b, c, d, f, g \rangle^5$
$\langle a, b, b, c, e, f, g \rangle^5$

L_2
$\langle a, b, c, d, f, g, h \rangle^{15}$
$\langle a, b, c, e, f, g, h \rangle^8$
$\langle a, b, c, d, f, g \rangle^5$
$\langle a, b, c, e, f, g \rangle^2$
$\langle a, b, b, c, d, f, g, h \rangle^3$
$\langle a, b, b, c, e, f, g, h \rangle^4$
$\langle a, b, b, c, d, f, g \rangle^1$
$\langle a, b, b, c, e, f, g \rangle^2$

L_3
$\langle h, a, h, d, c \rangle^5$
$\langle a, d, a, c, a, c, e, h, h, f, g \rangle^5$
$\langle d, e, h, f, g, e, f, g \rangle^5$
$\langle h, b, b, h, h \rangle^5$
$\langle b, h, d, b, e, h, e, d, f, g \rangle^5$
$\langle a, d, a, d, h \rangle^5$
$\langle b, f, g, d \rangle^5$
$\langle f, g, h, f, g, h, h, h \rangle^5$

Fig. 1. Example logs. L_1, L_2 are structured logs, differing only in number of occurrences of complete traces. L_3 is an unstructured log.

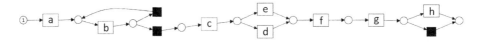

Fig. 2. Petri net for log L_1

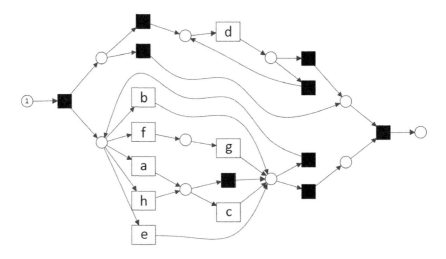

Fig. 3. Petri net for log L_3

a particular outcome [18]. For this paper we focus on Shannon entropy [17], which forms the foundation of the field of information theory.

Given a discrete random variable, X, taking on m possible values with associated probabilities $p_1, p_2, \ldots, p_m$, (Shannon) entropy, denoted H, is given by the expected value of the information content of X:

$$H = -\sum_{i=1}^{m} p_i \log_b p_i \tag{1}$$

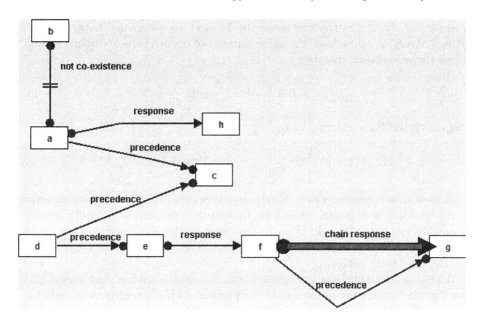

Fig. 4. Declare model for log L_3

Here b corresponds to the choice of coding scheme (i.e. for binary $b = 2$ and for decimal $b = 10$). We shall use the binary logarithm in the sequel.

Shannon justified this choice of measure with the fact that it is (1) continuous w.r.t. p_i (2) monotonically increasing w.r.t. n under uniform distributions and (3) additive under decomposition of choices, i.e., $H(p_1, p_2, p_3) = H(p_1, (p_2 + p_3)) + (p_2 + p_3)H(p_2, p_3)$.

The key question in using entropy as a measure of log complexity is *what would be the random variable implicit in a given log?*

3.1 Trace Entropy

One very simple answer to this question is to take the underlying random variable as ranging over exactly the traces observed in the log, with probabilities exactly the frequencies observed in the log. This idea gives rise to the notion *trace entropy.*

Definition 3.1 (Trace entropy). *Let $L = [\sigma_1^{w_1}, \ldots, \sigma_n^{w_n}]$ be a log. The trace entropy $t(L)$ of L is the entropy of the random variable that takes the value σ_i with the following probability.*

$$p_i = \frac{w_i}{\sum_{i=1}^{m} w_i} \tag{2}$$

Example 3.2. Even though the traces of L_1 and L_3 internally have radically different structure, they have the same number of occurrences of distinct traces, and so the same trace entropy:

$$t(L_1) = t(L_3) = -8 \times \frac{5}{40} \log_2 \frac{5}{40} = 3 \tag{3}$$

Computing the trace entropy of L_2, we find

$$t(L_2) = -\left(\frac{15}{40} \log_2 \frac{15}{40} + \ldots + \frac{2}{40} \log_2 \frac{2}{40} \right) = 2.55 \tag{4}$$

This example demonstrates that trace-entropy is likely *not* a good measure for determining if a model should be modelled imperatively or declaratively: L_1 and L_2 intuitively should mine to the same model, but have distinct trace-entropy. On the other hand, L_3 has much more variable behaviour than L_1, yet has the same trace entropy.

In general, if we are only interested in mining models with perfect fitness [20], then logs that differ only in the number of particular trace occurrences should not mine to different models. We are interested in the number of choices *available* at a particular point in a given trace, not the number of times a particular choice *was made* across all traces. We formalise this observation, using that in this simplistic setting, a "model" is really just a predicate on traces: a *language*.

Definition 3.3 (Language equivalence). *Define logs L, L' to be language equivalent iff they are identical as sets, that is, for each $\sigma^w \in L$, there exists $\sigma^{w'} \in L'$ for some w', and vice versa.*

Lemma 3.4. *Let P be a predicate on traces; lift it to logs pointwise, ignoring multiplicity. Then if logs L, L' are language equivalent, we have $P(L)$ iff $P(L')$.*

Proof. Consider language equivalent logs $L = [\sigma_1^{w_1}, \ldots, \sigma_n^{w_n}]$ and $L' = [\sigma_1^{w_1'}, \ldots, \sigma_n^{w_n'}]$. By definition $P(L)$ iff $\forall i.P(\sigma_i)$ iff $P(L')$.

That is, taking the simultaneously abstract and simplistic view that mining a log L is tantamount to coming up with a predicate P such that $P(L)$, the above Lemma says that a mined model can never be used to distinguish language equivalent logs. Because the output model cannot tell the difference between language equivalent logs, it would be unfortunate for our entropy measure to do so.

Definition 3.5. *An entropy measure is a function from logs to the reals. An entropy measure e respects language equivalence iff for any two language equivalent logs L, L', we have $e(L) = e(L')$.*

Trace entropy is unhelpful, then, because it does not respect language equivalence.

Example 3.6. The logs L_1, L_2 of Example 3.2 are language equivalent. However, they have different trace entropy measures. It follows that trace entropy does not respect language equivalence.

There is on an intuitive level also a second reason that trace entropy is unhelpful: it does not consider the behaviour exhibited *within* the traces. We saw this in Example 3.2, where $t(L_1) = t(L_3)$; that is, trace entropy cannot distinguish internal structure of traces. To consider the full behaviour of a log, we need to determine the entropy on the level of individual events.

3.2   Prefix Entropy

We must find a suitable notion of random variable that "generates" the traces we observe in the log, while at the same time characterises the internal structure of the individual traces.

Recall that a trace is the execution of a single process instance, taking the form of a sequence of events, or activity executions. At each point in a process execution, we will have a prefix of a completed trace. The distribution of these prefixes reflect the structure of the process.

Notation. We write $\langle e_1, \ldots, e_n \rangle$ for a finite string. If s, s' are finite strings, we write $s \sqsubseteq s'$ to indicate that s is a prefix of s'.

Definition 3.7 (Prefix entropy). *Let L be a log. The prefix entropy of L, written $\epsilon(L)$ is defined as the entropy of the random variable which ranges over all prefixes of traces in L, and for each prefix $\langle e_1, \ldots, e_k \rangle \sqsubseteq \sigma$ of a trace σ observed in a log L assigns as its probability the frequency of that prefix among all occurrences of prefixes in L.*

Example 3.8. In the log L_2, the prefix $\langle a, b, c, d \rangle$ occurs in 20 traces; the log contains a total of $15 \times 7 + 8 \times 7 + \ldots + 2 \times 7 = 280$ prefix occurrences, for a probability of $1/14$.

However, this notion of prefix entropy *does not* respect language equivalence, since logs differing only in the number of occurrences of a particular trace also differ in the set of occurrences of prefixes. Intuitively, we are interested in prefixes only as a measure of how much internal structure a log has, not how often various bits of that structure occurs. Hence, we disregard multiplicities of traces, in effect *flattening* the log.

Definition 3.9. *Let f be the function on logs which given a log L produces the corresponding set, i.e.,*

$$f([\sigma_1^{w_1}, \ldots, \sigma_n^{w_n}]) = [\sigma_1^1, \ldots, \sigma_n^1] = \{\sigma_1, \ldots, \sigma_n\} \tag{5}$$

The flattened prefix entropy of L is $\epsilon \circ f(L)$, that is the prefix entropy applied to the flattened log.

Example 3.10. In the log $f(L_2)$, the prefix $\langle a, b, c, d \rangle$ occurs just twice, among a total of only 56 prefix occurrences, for a probability of $1/26$.

We conjecture that transitioning from a log L to a flattened log $f(L)$ does not materially affect prefix entropy; we leave an investigation of exactly which properties are and are not preserved as future work.

Proposition 3.11. *The flattened prefix entropy $\epsilon \circ f$ respects language equivalence.*

Proof. Immediate by definition of ϵ.

Example 3.12. Computing the flattened event entropy of the example logs of Example 3.2, we find:

$$\epsilon \circ f(L_1) = 4.09 = \epsilon \circ f(L_2)$$
$$\epsilon \circ f(L_3) = 5.63$$

While the notion of flattened event entropy seems promising, there is one caveat. Because it is based on prefixes, it fails to appreciate common structure appearing distinct prefixes.

$$\langle a, x, y, z \rangle^5$$
$$\langle a, x, z, y \rangle^5$$
$$\langle b, x, y, z \rangle^5$$
$$\langle b, x, z, y \rangle^5$$
$$\langle c, x, y, z \rangle^5$$
$$\langle c, x, z, y \rangle^5$$
$$\langle d, x, y, z \rangle^5$$
$$\langle d, x, z, y \rangle^5$$
$$\langle e, x, y, z \rangle^5$$
$$\langle e, x, z, y \rangle^5$$

Fig. 5. Log L_4 (highly structured).

Example 3.13. Consider the log L_4 in Fig. 5. This log is highly structured: it always contains exactly 4 activities; the first is a choice between $\{a, b, c, d, e\}$, the second an x, the third and fourth either x, y or y, x. See Fig. 6 for a Petri net admitting this behaviour. However, this log has a trace entropy of $t(L_4) = 4.82$, higher than the apparently *less* structured logs L_1 and L_2.

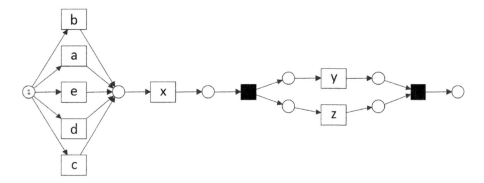

Fig. 6. Petri net for log L_4

3.3   Block Entropy

To address the weaknesses of prefix entropy, we apply ideas from natural language processing [21], where entropy is studied in terms of n-length substrings known as "n-grams".

We consider an individual trace a "word", in which case our log is a multiset of such words, and look at the observed frequencies of arbitrary substrings within the entire log. That is, rather than looking at the frequencies of prefixes, we look at frequencies of substrings.

We shall see that while computationally more expensive, this idea alleviates the problems of prefix entropy and that observed structure is weighted equally, regardless of where it occurs in the log.

Definition 3.14 (*k-block entropy*). *Let L be a log. The k-block entropy of L, written $b_k(L)$ is defined as the entropy of the random variable which ranges over all k-length substrings of traces of L, assigning to each such substring s as probability the frequency of the number of occurrences of that substring among all occurrences of k-length substrings.*

Example 3.15. In the log L_4 in Fig. 5, the 2-block $\langle x, y \rangle$ occurs 5 times; the log contains a total of $10 \times 3 = 30$ occurrences of 2-blocks, for a probability of $1/6$.

Following [21], we compute the k-block entropy $b_k(-)$ directly:

Lemma 3.16. *Let L be a log. The k-block entropy of L is given by*

$$b_k(L) = - \sum_{\langle s_1,\dots,s_k \rangle \in \Sigma^\star} p(\langle s_1, \dots, s_k \rangle) \log p(\langle s_1, \dots, s_k \rangle) \tag{6}$$

Often in the literature on estimating the entropy of natural languages, text corpora are used in which all punctuation has been removed, meaning that sentences are ignored and blocks can cover the end of one sentence and beginning of another. For event logs we want to avoid finding spurious correlations among events at the end of one trace and beginning of another trace, so in our approach we keep a clear separation between traces.

We now define block entropy for all substrings up to the length of the longest trace. That is, instead of restricting the measure to blocks of length k, we include all blocks, from length 1 up to the length of the longest trace, in one entropy measure.

Definition 3.17 (All-block entropy). *Let L be a log. The all-block entropy of L, written $b(L)$, is the entropy of the random variable which ranges over all substrings of traces of L, assigning to each such substring s as probability the ratio of occurrences of that substring over all occurrences of substrings.*

Example 3.18. In the log L_3 in Fig. 1, the substring (2-block) $\langle a, d \rangle$ occurs 3 times, once in the second entry, twice in the sixth. The log contains a total of 248 occurrences of substrings: $\Sigma_1^5 k = 15$ in the first trace, $\Sigma_1^{11} k = 66$ in the second, and so on. Altogether, the probability of $\langle a, d \rangle$ is $3/248$.

As for the prefix entropy, the all-block entropy does not respect language equivalence, but its flattening does.

Proposition 3.19. *The flattened all-block entropy $b \circ f$ respects language equivalence.*

Example 3.20. We give the flattened all-block entropy for the examples L_1 through L_4.

	L_1	L_2	L_3	L_4
$b \circ f(-)$	5.75	5.75	7.04	4.75

Notice how L_3 is still the highest-entropy log, but now L_4 is properly recognised as containing information than does L_1 and L_2.

We conclude this section by noting that while the all-block entropy looks promising, it may be computationally infeasible to apply to large logs. Naively computing the all-block entropy of a log requires, in the worst case, tabulating the frequencies of all substrings seen in that log, an operation that takes polynomial space.

Assume a log has n traces, all of length k. A string of length k contains exactly $k - (i - 1)$ substrings of length i: one starting at each index except for the last $i - 1$ indices, where there is no longer room for a substring of length i.

Thus, by summing over all traces and all substring lengths, we can establish an upper bound on the size of the frequency tables for the substrings of a log:

$$n \times \sum_{i=0}^{k-1} k - i = \mathcal{O}(n \times k^2) \tag{7}$$

So in a concrete case where a log has 20.000 traces of length 100; we would in the worst case need a table of 2×10^{10} substrings. In the next section, we shall in one instance see our naive prototype implementation run out of memory on a somewhat smaller dataset.

4   Implementation and Early Experiments

To test the various measures we implemented a rudimentary ProM [22] plugin with support for computing the trace, prefix and block entropy of a given log. To get an indication of the utility of the entropy measures we applied them to the examples L_1, L_2, L_3, and L_4 of the preceding sections, as well as to a selection of real-life logs. In particular we used the BPI Challenge 2012[1], BPI Challenge 2013 (incidents)[2], hospital[3], sepsis cases[4], and road traffic fines[5] logs.

There is not yet a clear agreement in the literature on which of these logs should be mined imperatively, and which should be mined declaratively. However, it can be observed that the BPI Challenge 2012 log is commonly used as a base-case for declarative mining algorithms and that both the sepsis cases and hospital log result from highly flexible and knowledge-intensive processes within a Dutch hospital. A recent investigation involving the BPI Challenge 2013 (incidents) log seemed to indicate that an imperative approach may be the most successful, but no concrete conclusions were drawn [11].

We ran two sets of experiments: one to contrast the notions of trace, prefix and all-block entropy; and one to investigate more thoroughly the notion of k-block entropy.

4.1   Comparative Measurements

We report measurements of trace, prefix and all-block entropy of the above-mentioned logs in Table 1. The results are promising for the real-life logs we experimented on. In particular, we see that the BPI Challenge 2012 and sepsis cases logs score highly in terms of all-block entropy, whereas the BPI Challenge 2013 log scores somewhat lower, which fits the intuition that the first two are more suited for declarative mining, whereas the latter is more suited for imperative mining.

We were unable to compute the all-block entropy for the hospital log. This log has traces up to length 1800, and thus requires a large amount of memory to store the intermediary table of substring frequencies.

We conclude that (a) the flattened all-block entropy is the most promising measure for indicating whether a log is best mined imperatively or declaratively; and (b) that a computationally more efficient approximation to the all-block entropy is needed.

Furthermore, for using the metric to inform the choice of imperative or declarative miner, we must determine a cut-off entropy value. But as seen in Table 1, prefix and block entropy grow in proportion to the number of unique traces

---

[1] https://data.4tu.nl/repository/uuid:3926db30-f712-4394-aebc-75976070e91f.
[2] https://data.4tu.nl/repository/uuid:a7ce5c55-03a7-4583-b855-98b86e1a2b07.
[3] https://data.4tu.nl/repository/uuid:d9769f3d-0ab0-4fb8-803b-0d1120ffcf54.
[4] https://data.4tu.nl/repository/uuid:915d2bfb-7e84-49ad-a286-dc35f063a460.
[5] https://data.4tu.nl/repository/uuid:270fd440-1057-4fb9-89a9-b699b47990f5.

Table 1. Trace, flattened prefix and flattened all-block entropy measures for select logs.

Log	Event classes	Unique Traces	Shortest Trace	Longest Trace	Entropy Trace	Prefix	All-block
L_1	8	8	6	8	3.00	4.09	5.75
L_2	8	8	6	8	2.55	4.09	5.75
L_3	8	8	4	11	3.00	5.63	7.04
L_4	8	10	4	4	3.32	4.82	4.75
BPI challenge 2012	36	4366	3	175	7.75	12.53	16.01
BPI challenge 2013	13	1511	1	123	6.66	11.32	12.21
Sepsis cases	16	846	3	185	9.34	10.59	14.67
Road traffic fines	11	231	2	20	2.48	6.50	8.73
Hospital log	624	981	1	1814	9.63	DNF	DNF

and event classes. This reflects the second rationale for Shannon's entropy measure, that it be monotonically increasing. This can potentially be addressed by measuring the *entropy rate*, discussed in Sect. 5.2.

4.2 Block Entropy Measures

To understand the flattened block entropy measure in more detail, in particular in the hope of finding an efficient approximation of it, we analyse its constituent parts (1-blocks, 2-blocks etc.) in our selection of logs. The results are visualised in Fig. 7.

We note that when blocks become longer than the longest trace, k-block entropy falls to zero since we are effectively counting the occurrences of impossible events and $lim_{n \to 0} n \log(n) = 0$. This contrasts with a system with one certain outcome in which case we also have $H = 0$ since $1 \log(1) = 0$. We emphasize that this plays no role in our all-block entropy measure since it includes only blocks up to the length of the longest trace.

Note that the entropy of L_3, a declarative process, is never less than those of the more structured logs, L_1, L_2, and L_4. What is otherwise apparent from this figure is that there is no immediately obvious shortcut to the flattened all-block entropy obtainable as any particular k-block size: no single k seems representative of the full measure. This lack of representation is further evident from the number of crossings in the diagram: it would appear that from no single k onwards does the entropy of individual logs maintain their relative positioning. E.g., at $k = 10$, the BPI 2013 log measures more complex than 2012; however, they meet and switch places at $k = 18$.

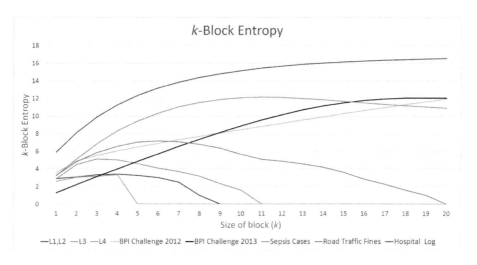

Fig. 7. k-block entropy of flattened logs using different block lengths.

5 Discussion and Future Work

Our experiments show that entropy is a promising indicator of log complexity. However, several questions are left open:

(i) How can we perform a more thorough evaluation of the suitability of our entropy measures?
(ii) Next to flattening the log, should we perform any additional normalisations to arrive at a fair measure of entropy?
(iii) Can we find entropy measures with a reasonable computational complexity so that we can deal with large logs?
(iv) How can we incorporate our approach with clustering techniques? (Both in an effort to find more efficient entropy estimations and use entropy measures to find suitable clusters for hybrid mining).

In this section we shortly discuss these open challenges and provide possible avenues of future research to alleviate them.

5.1 More Thorough Experiments

In the previous section we reflected on the types of models we expected to be most suitable for the real-life logs that we experimented on. These were primarily educated guesses and to do a more thorough evaluation we should perform an analysis of these logs to determine whether they are more suited for imperative or declarative mining.

One way to approach this could be to mine each log with imperative and declarative miners and compare the resulting models according to their precision and simplicity [20]. In addition it would be useful to experiment on a larger set of logs, including both a more comprehensive set of synthetic logs and additional real-life logs.

5.2 Additional Normalisation

Our experiments show that larger logs and more event classes result in a higher entropy measure. It is questionable however whether a larger log by definition always is better suited to declarative modelling, and a measure should not simply proxy for other log attributes. One approach to normalising the entropy measure of different sequences is to use the *entropy rate*, the change in entropy between a block of length k and $k-1$ with increasing k. That is,

$$h = \lim_{k \to \infty} H_k - H_{k-1} \tag{8}$$

Computing this directly, however, requires extremely long sequences, as it breaks down when $d^k \approx K$, where d is the size of the alphabet (events), k is the block length and K the length of the sequence [23]. Fortunately, estimators of h exist, some of which are discussed in Sects. 5.3 and 5.4.

Possible additional normalisations could be based on the number of unique activities in a log, the number of traces, the average length of traces, or the number of events.

5.3 Complexity of Entropy Measures

In certain cases, such as the hospital log, our proposed measure of flattened block entropy is computationally infeasible, at least for our naive implementation. Fortunately, the problem of efficient entropy estimation is well-studied, most notably in physics and natural language processing.

One very efficient approach is based on building prefix trees of non-overlapping blocks. One example of this is the Ziv-Lempel algorithm, which sequentially parses sequences into unique phrases, composed of their previously seen prefix plus a new symbol. In this way a tree structure is built with each phrase defined by a tuple representing a pointer to its prefix and the new symbol. Borrowing an example from [24], the string ABBABBABBBAABABAA would be parsed as:

A	B	BA	BB	AB	BBA	ABA	BAA
(0,A)	(0,B)	(2,A)	(2,B)	(1,B)	(4,A)	(5,A)	(3,A)

With the integers referring to the dictionary reference of earlier encountered prefixes (0 for root prefixes). It can be shown that the compression ratio of the Ziv-Lempel coding scheme converges to the entropy rate, h, of sequences generated by an ergodic process [24].

In [23], the authors found that on very short sequences, block entropy tended to lead to overestimates on low entropy sequences, while they outperformed Ziv-Lempel on high entropy sequences and suggest a two step process which uses a preliminary quick estimate of entropy to inform the choice of proper estimator.

5.4 Clustering of Logs

In determining whether the declarative or imperative modelling paradigm is most appropriate for a given event log, we may want to look more specifically at the similarity *between* traces rather than *within* traces. In other words, an event log consisting of nearly identical, but complex, traces may nonetheless be best modelled using an imperative approach, while a log of simple, but highly varied traces, may be best described by a declarative model. By clustering traces according to some distance metric, we can get an idea of the diversity of an event log by using the distribution of traces across clusters as the probability distribution for calculating Shannon entropy.

Typically, clustering is performed using Euclidean distance metrics, meaning that data must be represented as a d-dimensional vector. Even techniques such as expectation maximisation clustering, that do not rely directly on computing Euclidean distances, do assume that the data is independent and identically distributed. That is, that observed variables are not directly dependent on each other, so $p(a, b|c) = p(a|c)p(b|c)$. This is an issue for sequential data, a well-known problem in natural language processing, where the "bag-of-words" approach nonetheless leads to impressive results, for example in topic modelling and sentiment analysis [25]. In this approach, word order is simply ignored and documents are represented as multisets (a.k.a. bags) of words, which can then be represented as vectors, with word counts comprising the vector elements.

Similar approaches have been used for trace clustering, by representing traces as vectors of event occurrence counts, ignoring event ordering [14,26]. For our purposes, this approach is not adequate: For event logs, event ordering cannot be ignored. The reason for this can clearly be seen from the interpretation of entropy as the amount of information gained upon learning the next symbol in a sequence *given* the preceding sequence. For example, in English the letter q is always followed by u, and so $p(u|q) = 1$ and therefore $h = H_n - H_{n-1} = -p(q, u) \log(u|q) = 0$.

To avoid the loss of ordering information which results from collapsing traces to vectors of event counts, we would need to find ways of estimating entropy which allow us to use non-Euclidean distance metrics, for example string edit distances [16,27,28]. This allows us to distinguish between traces consisting of the same (or similar) events, but in different orderings.

Another issue with techniques like k-means clustering, is that it is often not clear how to choose the optimal number of clusters, k. Previous research in trace clustering has dealt with this in part by using agglomerative hierarchical clustering techniques, which allow one to "zoom in" and "zoom out" on a hierarchy of clusters to find the optimal partitioning [29].

Correlation clustering is a method for grouping a set of objects into optimal clusters without specifying the number of clusters [30,31]. Just as important, correlation clustering is a graph-based approach, meaning that data is defined solely by edge weights between nodes, where edge weight can represent the degree of similarity between nodes. For our current purposes, nodes would represent individual traces and edges the distance measure between them, for example string edit distance.

Another clustering-based estimator that looks promising is Kozachenko and Leonenko's *nearest neighbour entropy estimator* [32], which requires as input only distance measures and some choice of d, the number of dimensions on which the distance metric is defined. This allows us to bypass the actual clustering process, but while it doesn't require the collapsing of traces into vector representations, it does require that we choose a value for d. Using string edit distance, for example, it is not immediately clear what this value should be.

5.5   Noise

Noise is a source of variability, and noisy logs will tend to have a large degree of entropy. The primary challenge is to distinguish between unintentional variability (noise) and intentional variability.

One approach could be to first filter the log for noise using existing techniques, and then measure its entropy afterwards, accepting the risk of accidentally removing interesting behaviour from the log. Alternatively one could assume that the log contains no noise, measure its entropy, mine the log imperatively, declaratively, or hybridly based on the measure, and then analyse the resulting model for unintended flexibility.

6   Conclusion

In this paper we reported on an initial investigation of how entropy can be used as a measure for the complexity of a process log and thereby be used as a basis for determining if a log should be mined and modelled imperatively or declaratively. We investigated three possible entropy measures, each building on the insights gained from the former. We arrived at the notion of block-entropy for process logs and showed through experiments on synthetic and real-life logs that this measure best matches our expectations of log complexity and accordingly which paradigm should be used for mining it. Finally, we proposed 4 distinct paths along which the current work can be extended and we intend to follow-up on these suggestions in future work.

Acknowledgments. We would like to thank both anonymous the reviewers and Jakob Grue Simonsen for valuable and constructive feedback.

References

1. van der Aalst, W.M.P.: The application of petri nets to workflow management. J. Circuits Syst. Comput. **08**, 21–66 (1998)
2. Object Management Group. Business Process Modeling Notation Version 2.0. Technical report, Object Management Group Final Adopted Specification (2011)

3. Pesic, M., Schonenberg, H., van der Aalst, W.M.P.: Declare: full support for loosely-structured processes. In: 2007 EDOC, pp. 287–300 (2007)
4. Debois, S., Hildebrandt, T., Slaats, T.: Safety, liveness and run-time refinement for modular process-aware information systems with dynamic sub processes. In: Bjørner, N., de Boer, F. (eds.) FM 2015. LNCS, vol. 9109, pp. 143–160. Springer, Cham (2015). https://doi.org/10.1007/978-3-319-19249-9_10
5. Hull, R., Damaggio, E., De Masellis, R., Fournier, F., Gupta, M., Heath, F.T., Hobson, S., Linehan, M.H., Maradugu, S., Nigam, A., Noi Sukaviriya, P., Vaculín, R.: Business artifacts with guard-stage-milestone lifecycles: managing artifact interactions with conditions and events. In: 2011 DEBS, pp. 51–62 (2011)
6. Reijers, H.A., Slaats, T., Stahl, C.: Declarative modeling–an academic dream or the future for BPM? In: Daniel, F., Wang, J., Weber, B. (eds.) BPM 2013. LNCS, vol. 8094, pp. 307–322. Springer, Heidelberg (2013). https://doi.org/10.1007/978-3-642-40176-3_26
7. Debois, S., Hildebrandt, T., Marquard, M., Slaats, T.: Hybrid process technologies in the financial sector: the case of BRFkredit. In: vom Brocke, J., Mendling, J. (eds.) Business Process Management Cases. MP, pp. 397–412. Springer, Cham (2018). https://doi.org/10.1007/978-3-319-58307-5_21
8. Slaats, T., Schunselaar, D.M.M., Maggi, F.M., Reijers, H.A.: The Semantics of Hybrid Process Models. In: Debruyne, C., et al. (eds.) OTM 2016. LNCS, vol. 10033. Springer, Cham (2016). https://doi.org/10.1007/978-3-319-48472-3_32
9. Maggi, F.M., Slaats, T., Reijers, H.A.: The automated discovery of hybrid processes. In: Proceedings of 12th International Conference on Business Process Management - BPM 2014, Haifa, Israel, 7-11 September 2014, pp. 392-399 (2014)
10. De Smedt, J., De Weerdt, J., Vanthienen, J.: Fusion miner: process discovery for mixed-paradigm models. Decis. Support Syst. **77**, 123–136 (2015)
11. Schunselaar, D.M.M., Slaats, T., Reijers, H.A., Maggi, F.M., van der Aalst, W.M.P.: Mining hybrid models: a quest for better precision. Unpublished manuscript (2017, Available)
12. Debois, S., Slaats, T.: The analysis of a real life declarative process. In: 2015 CIDM, pp. 1374–1382 (2015)
13. Greco, G., Guzzo, A., Pontieri, L., Sacca, D.: Discovering expressive process models by clustering log traces. IEEE Trans. Knowl. Data Eng. **18**(8), 1010–1027 (2006)
14. Song, M., Günther, C.W., van der Aalst, W.M.P.: Trace clustering in process mining. In: Ardagna, D., Mecella, M., Yang, J. (eds.) BPM 2008. LNBIP, vol. 17, pp. 109–120. Springer, Heidelberg (2009). https://doi.org/10.1007/978-3-642-00328-8_11
15. Makanju, A.A.O., Zincir-Heywood, A.N., Milios, E.E.: Clustering event logs using iterative partitioning. In: Proceedings of the 15th ACM SIGKDD International Conference on Knowledge Discovery and Data Mining, KDD 2009, NY, USA, pp. 1255–1264. ACM, New York (2009)
16. Bose, R.J.C., van der Aalst, W.M.P.: Context aware trace clustering: towards improving process mining results. In: Proceedings of the 2009 SIAM International Conference on Data Mining, pp. 401–412. SIAM (2009)
17. Shannon, C.E.: A mathematical theory of communication. Bell Syst. Tech. J. **27**(3), 379–423 (1948)
18. Bishop, C.M.: Pattern Recognition and Machine Learning. Information Science and Statistics. Springer, New York (2006)
19. Breuker, D., Matzner, M., Delfmann, P., Becker, J.: Comprehensible predictive models for business processes. MIS Q. **40**(4), 1009–1034 (2016)

20. Buijs, J.C.A.M., van Dongen, B.F., van der Aalst, W.M.P.: On the role of fitness, precision, generalization and simplicity in process discovery. In: Meersman, R., Panetto, H., Dillon, T., Rinderle-Ma, S., Dadam, P., Zhou, X., Pearson, S., Ferscha, A., Bergamaschi, S., Cruz, I.F. (eds.) OTM 2012. LNCS, vol. 7565, pp. 305–322. Springer, Heidelberg (2012). https://doi.org/10.1007/978-3-642-33606-5_19
21. Schürmann, T., Grassberger, P.: Entropy estimation of symbol sequences. Chaos: An Interdisciplinary. J. Nonlinear Sci. **6**(3), 414–427 (1996)
22. van der Aalst, W.M.P., van Dongen, B.F., Günther, C.W., Rozinat, A., Verbeek, E., Weijters, T.: ProM: The process mining toolkit. In: BPM (Demos) (2009)
23. Lesne, A., Blanc, J.-L., Pezard, L.: Entropy estimation of very short symbolic sequences. Phys. Rev. E **79**(4), 046208 (2009)
24. Thomas, J.A., Cover, T.M.: Elements of Information Theory. Wiley, Hoboken (2006)
25. Hofmann, T.: Probabilistic latent semantic analysis. In: Proceedings of the Fifteenth Conference on Uncertainty in Artificial Intelligence, pp. 289–296. Morgan Kaufmann Publishers Inc. (1999)
26. Greco, G., Guzzo, A., Pontieri, L., Sacca, D.: Discovering expressive process models by clustering log traces. IEEE Trans. Knowl. Data Eng. **18**(8), 1010–1027 (2006)
27. Delias, P., Doumpos, M., Grigoroudis, E., Matsatsinis, N.: A non-compensatory approach for trace clustering. Int. Trans. Oper. Res. (2017)
28. Ha, Q.-T., Bui, H.-N., Nguyen, T.-T.: A trace clustering solution based on using the distance graph model. In: Nguyen, N.-T., Manolopoulos, Y., Iliadis, L., Trawiński, B. (eds.) ICCCI 2016. LNCS (LNAI), vol. 9875, pp. 313–322. Springer, Cham (2016). https://doi.org/10.1007/978-3-319-45243-2_29
29. Song, M., Yang, H., Siadat, S.H., Pechenizkiy, M.: A comparative study of dimensionality reduction techniques to enhance trace clustering performances. Exper. Syst. with Appl. **40**(9), 3722–3737 (2013)
30. Demaine, E.D., Immorlica, N.: Correlation clustering with partial information. In: Arora, S., Jansen, K., Rolim, J.D.P., Sahai, A. (eds.) APPROX/RANDOM -2003. LNCS, vol. 2764, pp. 1–13. Springer, Heidelberg (2003). https://doi.org/10.1007/978-3-540-45198-3_1
31. Bansal, N., Blum, A., Chawla, S.: Correlation clustering. In: Proceedings of 2002 The 43rd Annual IEEE Symposium on Foundations of Computer Science, pp. 238–247. IEEE (2002)
32. Kozachenko, L.F., Leonenko, N.N.: Sample estimate of the entropy of a random vector. Problemy Peredachi Informatsii **23**(2), 9–16 (1987)

Objective Coordination with Business Artifacts and Social Engagements

Matteo Baldoni$^{(\boxtimes)}$ [ORCID], Cristina Baroglio [ORCID], Federico Capuzzimati [ORCID], and Roberto Micalizio [ORCID]

Università degli Studi di Torino — Dipartimento di Informatica,
c.so Svizzera 185, 10149 Torino, Italy
{matteo.baldoni,cristina.baroglio,federico.capuzzimati,
roberto.micalizio}@unito.it

Abstract. This work studies business artifacts by tackling a limit that we see in the current model, which is: business artifacts are not devised as natural means of coordination in their own right, despite the fact that they have the potential of being natural means of coordination in their own right. Coordination issues are transfered (e.g. by BALSA) to solutions that are already available in the literature on choreography and choreography languages. Instead, we propose to enrich business artifacts with a normative layer that accounts for the social engagements of the parties which interact by using a same business artifact. We explain the advantages, also from a software engineering perspective, and propose an approach that relies on the notion of social commitment.

Keywords: Business artifacts · Normative MAS
Social commitments

1 Introduction

The *artifact-centric* approach [6,10,12] is recently emerging as a viable solution for specifying and deploying business operations by combining both data and processes as first-class citizens. In particular, the notion of *Business Artifact*, initially proposed by Nigam and Caswell [15], opened the way to the development of a data-driven approach to the modeling of business operations. The data-driven approach counterposes a data-centric vision to the activity-centric vision, traditionally used when processes are explicitly modeled in terms of workflows. *Business artifacts* are concrete, identifiable, self-describing chunks of information, the basic building blocks by which business models and operations are described. They are business-relevant objects that are created and evolve as they pass through business operations. A business artifact includes an *information model* of the data, and a *lifecycle model*, the latter capturing the key states through which data, structured according to the information model, evolve together with state transitions. The lifecycle model is used both at run-time to track the evolution of business artifacts, and at design time to distribute

© Springer International Publishing AG 2018
E. Teniente and M. Weidlich (Eds.): BPM 2017 Workshops, LNBIP 308, pp. 71–88, 2018.
https://doi.org/10.1007/978-3-319-74030-0_5

tasks among the actors which operate on a business artifact. The presence of an explicit lifecycle gives business artifacts a semantics that differentiates them from other programming abstractions, like objects, active objects, and artifacts in the sense given to the concept by the A&A meta-model [17]. The lack of autonomy differentiates them from agents.

The work presented in this paper attacks a limit that business artifacts show: they do not provide the programmer with any means for designing and modularizing the *coordination* of those processes which should operate on them. It is a fact that business artifacts encapsulate data which are created and manipulated by many processes. For instance an order is created by a client who interacts with a seller, and is manipulated by the operations that make the transaction between the two proceed. The client, the seller, more in general the processes that operate on a business artifact need to agree on who should do what and sometimes even when to carry out a task of interest. For instance, payment and delivery are up to different parties in the purchase transaction, and they must both occur otherwise the transaction will not reach a happy end. There is a causal relationship between the actions of the merchant and of the client, the latter paying because of the promise (social engagement) of the merchant to deliver the goods.

So, a business artifact, besides having a lifecycle that describes its evolution from creation to some state where it is considered as archivable, is a natural candidate to be a medium of coordination and interaction. However, this latter dimension is not investigated by state-of-art literature on business artifacts.

The proposal we present in this paper aims at overcoming the described lack of business artifacts with a multiagent systems (MAS) approach (along the lines of activity theory, at the basis of A&A, and exploiting commitments to represent social engagements). Specifically, we claim that: (1) services by which it is possible to operate on business artifacts should be encapsulated and organized into *goal-oriented containers*; (2) it is necessary to introduce a *normative layer* to capture the behaviors that are expected of the parties. The paper motivates our research and our proposal, and illustrates it with the help of a simple purchase example.

2   Motivations

In order to understand how business artifacts are currently specified and used, we briefly introduce the BALSA methodology [7] as a significant representative of the current approaches to business artifacts. The BALSA methodology specifies a data-centric declarative model of business operations, and can be summarized in three steps: (1) identify the relevant business artifacts of the problem at hand and their lifecycles, (2) develop a detailed specification of the services (or tasks) that will cause the evolution of the business artifact lifecycles, (3) define a

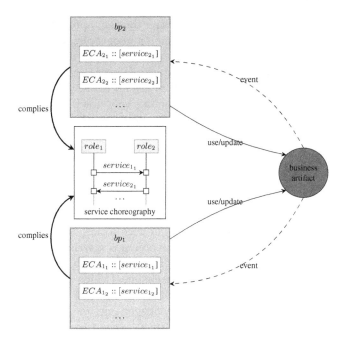

Fig. 1. Synchronized access to a business artifact via a choreography.

number of ECA-rules (Event-Condition-Action) that create associations between services and business artifacts. ECA-rules are the building blocks to define, in a declarative way, processes operating on data.

BALSA (and similar) is extremely interesting, in particular because it introduces a novel perspective on the modeling of business processes. However, for what concerns coordination in presence of more business processes, the methodology simply refers to choreography languages and techniques proposed for service-oriented computing, despite the presence of the business artifacts which could themselves be natural instruments of coordination. The approach is sketched in Fig. 1, where two business processes, bp_1 and bp_2, both defined declaratively in terms of ECA rules, access to the same business artifact. A proper synchronization between the services they invoke is guaranteed by the fact that the two processes comply with a choreography specification.

We deem the absence of an explicit use of the business artifacts to the aim of coordination, which BALSA compensates by referring to the literature and the tools concerning choreographies, as a significant flaw. A choreography induces a form of *subjective coordination* [16]: each business process needs to include within its logic also the interaction logic that refers to the role it plays within the choreography. This lack of separation between the business and the coordination logic has two drawbacks: (1) it makes design and implementation more

complex and reduces the possibility of reusing the same process in different contexts since it depends strictly on a specific choreography, and (2) it demands
for an extensive compliance checking for guaranteeing that the process adheres
to the choreography. In addition, in inherently *destructured* settings (e.g. *cross-
organizational settings*), the involved actors are all peers, each of which has *its
own* business goals, and acts in an autonomous way. Each actor does not know
and does not care about the possible goals of others. They need, however, to
interact to achieve goals they would not be able to achieve alone. This is the
grounding of social engagements and causal relationships.

Interaction is, thus, a critical dimension that needs to be explicitly modeled
to coordinate the usage of shared resources. This poses the question of how to
scale the business artifact model to *coordinate* autonomous entities. We see in
the introduction of a *coordination model within business artifacts* the way to
achieve this goal, and explain what we mean with a simple example.

Let us consider a purchase scenario, involving a merchant and a client. We
claim that in order to coordinate the interaction between the two agents, it is
necessary to add to the plain message exchange (which standard approaches
to business processes envisage as the only means of interaction), one further
abstraction that explicitly represents the engagements each player has towards
the other. We also claim that business artifacts should trace such engagements
and their evolution, in order to enable an effective agent coordination. For example, when offering to sell some goods, the merchant *commits to* the client to ship
the items the client will pay for. Such a commitment is stored by the business
artifact involved in the interaction between the two. Because of his *awareness*
of such a commitment, the client, having paid for the goods, *expects* the ship-
ment to occur. If this does not happen, the commitment progresses into a state
of "violation" and this information, stored in the business artifact, provides a
proof of the merchant's misbehavior. From a different perspective, a client is
enticed to use a business artifact by the merchant's commitment, which makes
explicit the course of interaction the merchant binds to, and creates a right on
the client that such an expected course of action be respected (i.e., my payment
will put an obligation on the merchant to ship the bought items or the merchant
will violate the commitment). On the other hand, the merchant uses commitments inside the business artifact to entice interactions with potential clients –
indeed, the obligation yielded by a commitment is activated only if a client pays
for some goods.

In the example, the commitments that go along with a business artifact
make explicit the behavior the agents are expected to stick to. They also have
a normative flavour, as diverging behaviors will be considered as violations.
This awareness causes agents to take part to an interaction only if they are
fine with the commitments. As such, commitments provide a standard to define
standards of interaction mediated by business artifacts. To realize this vision, we
claim that: (1) services should be encapsulated and organized into *goal-oriented*

containers; (2) it is necessary to introduce a *normative layer*. For what concerns (1), the Agent-Oriented Paradigm is a good candidate. In particular, the Agent and Artifact meta-model (A&A) [17] has already shown how artifacts can be used as environment components that mediate agents' interactions. However, artifacts in the A&A model are radically different from the business artifacts because they *do not come* with an explicit information model for data, and they do not exhibit data lifecycles. Thus, this information cannot be exploited at design time, nor at runtime, to reason about which actions an agent should take. Concerning (2), the normative layer would provide an explicit representation of the business artifacts lifecycles, and of how coordination is expected to occur. Such a representation would allow agents to reason about the use of business artifacts and to create *mutual engagements* for driving their activities. Indeed, we envisage engagements as encoding *causal relations* between the actions of an agent and the goals and actions of another, with a normative power that would allow each agent to have expectations on the behavior of the others. In the purchase example, it is easy to see how the introduction of a norm in form of the commitment *whenever a customer pays, the merchant will ship the goods*, would enhance coordination. The customer now knows that after service *pay*, the merchant will be pushed to consider the service *ship-goods* as one of its next goals because of the social engagement of the merchant with the customer. Should this not happen, the merchant would cause a violation. This provides the customer a guarantee about the achievement of its own goal (or to recoup its losses). An explicit normative layer plays a central role both at the design time, to verify whether all the engagements can converge towards their satisfaction, and at running time to monitor the execution of a system and determine the violation of engagements. In this paper we introduce the notion of normative business artifacts as a means to extend the artifact-centric approach with a normative layer, where engagements and norms are expressed in terms of social commitments [19]. The introduction of a normative layer in the more general setting of business processes is seen as desirable also in [21].

3   Coordination via Normative Business Artifacts

Business artifacts are, by definition, data-aware: they consider data as a first-class primitive that drives the construction of process models [6]. Business artifacts, however, are not an end in themselves: they are business relevant entities that are created, accessed, and manipulated by different services along a business process. We now show how to introduce a normative layer so that business artifacts support coordination.

Destructured business processes call for a modularization of the control flow. Agent-oriented programming [8,25] is conceived exactly for handling multiple and concurrent control flows. Two elements are central in agent-oriented programming: the *agents* and the *environment*. Agents, as abstractions of processes, possess their own control flow, summarized as the cyclic process in which an agent observes the environment (updating its beliefs), deliberates which intentions to achieve, plans how to achieve them, and finally executes the plan [8]. Beliefs concern the environment. Intentions lead to action [25], meaning that if an agent has an intention, then the expectation is that it will make a reasonable attempt to achieve it. In this sense, intentions play a central role in the selection and the execution of actions, which represent the innate capabilities agents have to modify their environment. Among others, (business) artifacts (see A&A-meta model [17]) are privileged elements of an environment. In particular, in contexts where agents cannot achieve their goals on their own, but need to interact with other agents to do so, artifacts provide shared resources that agents will use to mediate their interactions.

We claim that business artifacts should be *norm-aware* in two ways. First, the *lifecycle* of a business artifact should be made explicit by way of norms that specify how data evolve. The agents (i.e., the artifact users), will be able to inspect and reason upon them to decide if and how to operate on an artifact to obtain some result. Second, agents need to coordinate and regulate their interaction *while using* the business artifacts to achieve their goals. Given these two bodies of norms, agents will apply reasoning techniques to plan proper *coordination* that, exploiting the social relationships and possibly without violating any norm, will lead to goal achievement. This is possible because norms enable the creation of *expectations* and *commitments* among agents.

Even though data-awareness and norm-awareness are by and large orthogonal to BDI [25] notions, it is natural to think of agents as BDI agents for a seamless integration of all the aspects of deliberation, including the awareness of data and of their lifecycles. For instance, an agent, that is involved in handling orders, may conclude that, since it has to pick up three items in the warehouse, since each such item is to be packed, since all packagings are performed by a same other agent, and since one of its goals is saving energy, it is preferable to pick them up altogether, and deliver them to the other agent only afterwards, instead of picking and delivering one item at a time. Data-awareness here is awareness that three items of a same kind are requested. Norm-awareness that items are picked because each of them is part of some order, whose lifecycle says that after being picked they will be packed. Again data-awareness allows our agent to know that all parcels are to be made by a same other agent.

Relying on agent-oriented programming is promising also because the agent-based model allows to naturally tackle the issue of coordination by introducing the concept of *norm* [23]. The deliberative cycle of agents is affected by the norms and by the obligations these norms generate as a consequence of the agents' actions. The limit of current agent-based approaches is that they provide no holistic proposal where constitutive norms are used also to specify data operations, and where regulative norms are used to create expectations on the overall evolution of the system (agents behavior and environment evolution).

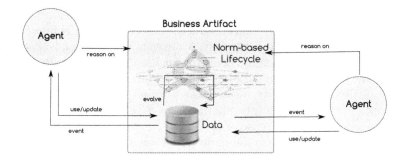

Fig. 2. Environment/Information system based on normative business artifacts.

3.1   Environment/Information Systems Based on Normative Business Artifacts

Figure 2 describes the high-level architecture of the kind of system we propose: (1) involving (normative) business artifacts and agents (with their goals), and (2) holistically norm-aware. Agents interact with each other and with the environment by creating and modifying data which belong to an information system and that are reified by business artifacts. They are goal-driven and capable of coordinating with other agents by creating and exploiting commitments, obligations, permissions, and prohibitions. The conceptual model of the information system is described in terms of the norms that regulate the evolution of data, that is, data lifecycles, capturing how data pass from one state to another as a consequence of actions that are performed by some agent. Moreover, business artifacts will include all those normative elements that regulate the coordination of the agents that interact by way of the artifact. This information is available to the interacting agents in a form that allows agents to reason on it. The agents are aware of the current state (of the lifecycle) of the data, as well as of the obligations, prohibitions, commitments, permissions put on them, and thus they are aware of the tasks expected of them and of their parties. At any time it is possible to check the execution, identifying pending tasks and who is responsible of them, as well as possible violations (e.g. of obligations or commitments), which may activate procedures specifically designed to handle the case.

Comparing our proposal with BALSA (outlined in Fig. 1), we can highlight many advantages of using normative business artifacts, the first and most relevant one residing in the role played by a business artifact: in the BALSA meta-model, a business artifact is just a piece of data, or as pointed out in [15], *the basis for factorization of knowledge* that enables business operations. In our proposal, normative business artifacts become the media through which the interaction among processes (i.e., agents) happens. This idea has a software engineering solid foundation. In his survey on Concurrent Object-Oriented Languages (COOLs), Philippsen [18] highlights the importance of a *locality principle for class correctness*, and advocates that a way to achieve it is to realize a form of *coordination on the side of the callee*. Namely, the coordination is implemented in the class that is accessed concurrently. Moreover, Philippsen advocates that desirable properties of coordination code are *isolation* and *separability*. Isolation means that the code for coordination is isolated from the code that implements class functionality. Separability means that portion of the code for the coordination can be refined while other portions are reused. These two properties promote the modularity and reuse of code.

Normative business artifacts fall entirely in such a coordination model. The coordination is in fact implemented solely inside the artifact itself (i.e. the class called by the interacting parties). Consequently, coordination correctness can be assessed *locally* since all coordination code is part of the artifact implementation. It is no longer required to verify the compliance of agents towards a choreography, but it is sufficient to verify whether agents are capable of using an artifact. This can be done by verifying whether an agent has the capability of playing a role within an interaction, that is, if the agent can invoke (a subset of) role-dependent artifact operations to bring an interaction to conclusion [1].

In contrast to the service choreographies of the BALSA model, normative business artifacts induce a form of *objective coordination* [16], where coordination is addressed outside the interacting agents. Coordination is, thus, explicitly modeled as a first-class element of a MAS. It is responsibility of the coordination designer to identify the objectives of the coordination and the mechanisms through which the space of interactions can be manipulated by the agents. In other words, objective coordination enables a clear separation between the implementation of the business and the coordination logic by explicitly representing the environment where the agents operate. In our proposal, we meet this property by encapsulating a normative layer inside a business artifact. The resulting normative business artifact represents (a portion of) the environment within which the agents can interact by explicitly manipulating their engagements. Thus, the implementation of environment resources (i.e., artifacts) and of the agents can be carried out and verified in *isolation*.

Objective coordination also finds a justification within Activity Theory [14], which postulates that agents control their own behavior from the outside by using and creating "artifacts" through which they interact. In other words, according to activity theory the interaction between two agents is always *mediated* by a further element (e.g., a tool), that manipulated by an agent influences the other. This idea has inspired the A&A meta-model [17] which, alongside agents, introduces the artifact abstraction as a building block of the agents' environment. According to A&A, an artifact is a computational, programmable system resource that can be manipulated by the agents. Thus, an artifact becomes the means through which interaction actually occurs. Notably, A&A relies also on *programmable coordination media*, that can for instance be the *tuple centers* adopted in TuCSoN [24], and derived by the notion of tuple space implemented in LINDA. The existence of a shared dataspace allows agents to coordinate directly on data. Communication becomes *generative* [9], in the sense that agents communicate by generating data in the dataspace, and these data are available to any agent having access to the dataspace. This vision is in contrast to the message passing paradigm, at the basis of choreographies, where communication is only enabled between agents sharing the same channel (typically, one sender and one receiver). In approaches based on tuple centers, *data themselves become a coordination media*.

We see the same potentiality in the business artifacts by Nigam et al. [15], but such a potentiality cannot be exploited within the BALSA methodology for two main reasons: (1) in the BALSA methodology coordination relies on service choreographies, that are based on message passing; and (2) the operations, that make the business artifact lifecycle evolve, are not associated with a precise operational semantics, upon which the coordination can be defined. The normative business artifacts we propose overcome these limitations. In fact, the introduction of a normative layer, based on commitments, allows us to associate operations, that cause the business artifact's progression along its lifecycle, with a semantics given in terms of commitment operations and state changes. Moreover, since commitments have a normative power, they allow the agents to create expectations about the behavior of the other parties, and hence they can be the building blocks upon which coordination can be explicitly defined. Finally, commitments are useful also at design time since they provide a programming interface between agents and their environment, given in terms of those state changes in the environment that are relevant to the agent and that the agent should tackle. This opens the way to agent programming methodologies as the CoSE methodology [2].

4   Building Normative Business Artifacts in JaCaMo+

In this section we explain how the normative business artifact we propose can be implemented by relying on the 2COMM/JaCaMo+ framework [3]. We refer to an implementation where the BDI agents are implemented in the Jason agent programming language, and where agents share artifacts, whose creation and manipulation involves an explicit creation and manipulation of *social commitments* [19]. Social commitments provide the normative layer and enable the coordination of the goal-driven agents.

We exemplify the implementation in the purchase scenario. In this scenario each agent has its own goals: the merchant has the goal of selling goods, while the customer has the goal of getting some goods. We show how they can achieve their goals by using a business artifact as the only means of coordination. To this end we need to present both sides of the interaction: the normative business artifact, on the one side, and how the agents use it, on the other side.

Let us consider the business artifact first. Figure 3 shows the business artifact which represents the transaction, occurring between a merchant and a customer. The specification of this business artifact follows the principles proposed in the BALSA meta-model: an information model specifies the relevant pieces of information (the merchant's and the customer's identifiers, the item sold by the merchant, and the maximum number of pieces that are available). While this information is provided at the time the business artifact is created, three further pieces of information (namely, quotation, quantity and order) are, instead, the result of the operations performed by the agents using the business artifact. These operations appear in the business artifact lifecyle showed here as an automaton: the customer asks the merchant for a quotation of a given quantity of items it wants to buy. Once the quotation is provided, the customer either decides to reject or accept the quotation. In the first case, the business artifact achieves a final state and can be archived. In the second case, a new order number is created to trace the shipping and the payment of the goods. Note that no order upon shipping and payment is imposed. After the payment, the merchant issues the payment receipt, and the business artifact can be archived.

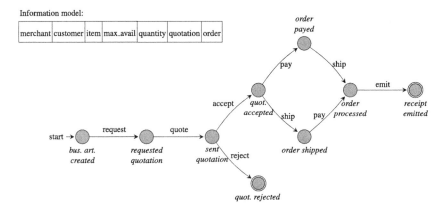

Fig. 3. The business artifact for the purchase scenario.

As explained, such a business artifact is not sufficiently rich to trace the causal relationships captured by social engagements. The customer may have an expectation about the behavior of the merchant, raised by its experience in purchasing things, that the payment of an item will be followed by the merchant giving the item but this expectation has no normative power. We, therefore, extend this artifact with a normative layer that, as anticipated, is expressed in terms of a set of commitments. A social commitment $C(x, y, s, u)$ captures that agent x (debtor) commits to agent y (creditor) to bring about the consequent condition u when the antecedent condition s holds (s and u are conjunctions or disjunctions of events). Only the debtor of a commitment can create it. When s is *true* the commitment is *detached* and turns into an obligation on the debtor. When u is *true* the commitment is *satisfied*. A detached commitment that is canceled or whose consequent becomes *false* is *violated*.

To realize a normative business artifact, thus, it is sufficient to associate each operation on the business artifact with operations (e.g. create, discharge, etc.) on one (or more) commitment(s). It follows that a normative business artifact, besides representing the chunk of information at hand, maintains also the created commitments, that can be inspected by the agents. Specifically agents will be notified of the changes to the business artifact state which include changes occurred to the commitments. Among other events, they will be aware of the detachment of commitments of which they are debtors, and of the satisfaction (violation) of commitments of which they are creditor.

```
commitment ShipGoods merchant to customer
    create quote(quantity, customer)
    detach accept(quotation, quantity, customer)
    discharge ship(customer)
    release reject(price, quantity, customer)

commitment EmitReceipt merchant to customer
    create quote(quantity, customer)
    detach paid(customer)
    discharge emit(customer)
    release reject(price, quantity, customer)

commitment PayForGoods customer to merchant
    create accept(quotation, quantity, customer)
    detach ship(customer)
    discharge paid(customer)
```

Listing 1.1. The normative layer in the purchase scenario

In our example, the normative layer is given by the set of commitments in Listing 1.1. For the sake of readability, the commitments are expressed following the syntax of the Cupid language [11]. The first commitment, *ShipGoods*, is created by the merchant when it executes the *quote* operation. That is, besides giving a value to the *quotation* information, the *quote* operation also commits the merchant towards the customer. Such a commitment is discharged when the customer accept the quotation, and discharged when the merchant ships the bought goods to the customer. Also the second commitment, *EmitReceipt*, is created by the merchant by the execution of the *quote* operation. In this case, the commitment is detached when the customer pays for the goods, and it is discharged when the merchant emits the receipt towards the customer. Both these two commitments are released by the customer when it rejects the quotation provided by the merchant. The last commitment, *PayForGoods*, is created by the customer when it accepts the merchant's quotation. The commitment is detached when the merchant has shipped the goods, and discharged when the customer pays for them.

Now, let us briefly review the resulting *normative* business artifact. As before, the merchant advertises, by creating the artifact, some item to be sold, and specifies the number of available units. An interested customer, by inspecting the artifact, can now see the commitments that the merchant is willing to take towards the customer. That is, the customer can create expectations about the merchant's behavior that have a *normative power*. The customer is, thus, enticed to accept the quotation *because of* the presence of the commitment, as part of the information provided by the business artifact. This, indeed, yields that this action will create an obligation on the merchant to deliver the goods that will make it achieve his goal. On the other hand, the customer will see also the commitments it will take in favor of the merchant, should it join the interaction. So, if the customer starts an interaction by requesting a quotation for a number of units, the merchant will provide such a quotation and, at the same time, it will create a commitment to: (1) ship the goods, provided that the customer accepts the quotation (*ShipGoods*), and (2) emit a receipt upon the payment for the goods (*EmitReceipt*).

```
1  +requestedQuotation ( Quantity ,  Customer_Id )
2       <- quote ( UnitPrice * Quantity ,  Quantity ,  Customer_Id ).
3  +cc ( My_Role_Id ,  Customer_Role_Id ,
4       accept ( Price ,  Quantity ,  Customer_Role_Id ),
5       ship ( Customer_Role_Id ) ,"DETACHED")
6       :  enactment_id ( My_Role_Id )
7       <- ship ( Customer_Role_Id ,  Quantity ).
8  +cc ( My_Role_Id ,  Customer_Role_Id ,  paid ( Customer_Role_Id ),
9                emitReceipt ( Customer_Role_Id ) ,"DETACHED")
10      :  enactment_id ( My_Role_Id )
11      <- emitReceipt ( Customer_Role_Id ).
```

Listing 1.2. Excerpt of the merchant program.

Note how the operations performed by agents on the business artifact make the commitments progress. For instance, the customer's acceptance of the quotation has several effects: (1) on the data side, a new order number is created to trace shipping and payment; (2) the customer commits towards the merchant to pay for the goods once they are delivered (*PayForGoods*); (3) commitment *ShipGoods* is detached, and then the merchant is now asked to ship the goods. As a final comment about the artifact, note how the commitments do not impose any ordering about the payment and shipping. In fact, the customer could pay as soon as it accepts the quotation, assured by the existence of commitment *ShipGoods* that pushes the merchant to actually ship the goods.

A natural way to implement normative business artifacts is to rely on the 2COMM framework [3]. In 2COMM normative business artifacts are reified as commitment-based protocol artifacts, that are provided by the framework as Java classes. Business artifact operations are mapped into protocol actions, whereas the data dimension is captured by the notional social state that protocol artifacts maintain. Indeed, the social state kept by a protocol artifact traces both the data required for the interaction, and the commitments, together with their states, that are created and manipulated along the interaction.

Let us now discuss the other side of the interaction, that is, how the agents use a normative business artifact. To exemplify the agents, we use here the JaCaMo+ framework, consisting in the well-known JaCaMo multi-agent platform enriched with commitment protocols provided by means of the 2COMM framework. Listing 1.2 reports an excerpt of the merchant agent program. In this first plan, the merchant is solicited to act by the reception of a *requestedQuotation* event, that comes from a customer through the business artifact. The body of the plan consists in the execution of *quote*, which sends a quotation to the customer and causes the creation of the merchant's commitments *ShipGoods* and *EmitReceipt*. The second plan captures the detachment of the *ShipGoods* commitment. The detach of the commitment is indeed an event generated by the artifact the merchant is focusing on, and it is the consequence of the accept operation performed by the customer. The body of the plan consists in the *ship* operation. Finally, the third plan captures the detachment of the *EmitReceipt* commitment, also in this case the body of the plan aims at discharging the commitment.

This example shows how the two agents, merchant and customer, can interact with each other by using a business artifact as an interaction medium. The normative layer enables each agent to achieve its own goals by relying on the others. The commitments, in fact, provide an explicit representations of the social engagements existing between the two agents which, behaving so as to accomplish their duties, bring about their goals (and help others to achieve theirs). Without commitments, agents could only rely on the lifecycle of the business artifact, that discloses the possible evolution of the data, but only when the lifecycle is complemented by an explicit representation of the social engagements an agent knows that by performing an action it will create an expectation on the side of the counterpart (and vice versa). In other words, the lifecycle traces the data dimension, and this dimension is orthogonal to the goal/process dimension which is captured by the agents. The normative layer operates as a sort of glue between the dimensions. This is also apparent in the lifecycle in Fig. 3. The two final states model conditions in which the business artifact can be archived independently of the status of the agents using it, that is, independently of whether they have achieved their goals. Looking at the normative layer, however, one can give them a proper semantics in terms of commitment states and goals. One further advantage of adding the normative layer is the possibility of making explicit some of the goals and services agents make available to others. This piece of knowledge can be used by the agents in a practical reasoning step (see [22]) to decide whether to join the artifact.

5   Conclusions

The presented work is strictly related to the problem of interaction in multiagent systems. In these systems, interaction is mainly focused on the modeling of communication patterns (*protocols*), which are concerned with the sequence of messages that can be exchanged between two communicating agents, but disregard the information conveyed by these messages. Recent approaches such as HAPN [26] and BSPL [20] have started to consider also the information dimension. HAPN is formally based on automata where nodes represent states of the interaction and transitions between nodes represent the messages that can be exchanged. Transitions have a complex structure since for each message it is possible to define a guard condition on message sending. A similar approach is BSPL where the information flow is decomposed in a number of "simple protocols", each defining the schema of the messages that can be exchanged together with their parameters. Parameter are decorated as *in* or *out* (meaning it is received or emitted). BSPL provides a formal framework in which it is possible to verify properties such as liveliness and safety of a protocol. Both HAPN and BSPL, however, show some weaknesses in properly handling information. In HAPN, for instance, guards, that enable message sending, may refer to information which is not carried by the message itself, but rather maintained in an external information system, which is not an integral part of the HAPN proposal, and hence the complete verification of an interaction is not actually achievable. BSPL, on the other hand, assumes a distributed view of information. Each participant has its own knowledge base, and the progression of the interaction makes the local knowledge bases evolve. The problem, in this case, is that each participant has just a local view of the information lifecycle. Thus, an agent cannot create expectations about the behaviors of other participants as a consequence of the messages it sends. The approach we propose overcomes these limitations. Business artifacts abstract an information system, and provide the environment in which the agents, which are autonomous loci of control, interact. Both business artifacts and agents are first-class components. The autonomy and flexibility of the agents are preserved and supported; moreover, it is possible to reason both on the evolution of the business artifacts and on the interaction. This work can be extended along three main lines of research. First of all, an explicit normative layer paves the way to formal verification techniques for cross-organizational business processes. In this respect, the notion of *accountability* is rapidly gaining importance since, when more organizations come into play, it is even more important to trace back who is responsible for what. First steps can be found in [4]. Another promising extension is to understand how agents could plan the use of business artifacts for reaching their goals. An initial attempt to use social commitments in planning has been discussed in [5], but business artifacts are yet to be considered. Finally, the standardized lifecycle of commitments can be the key for developing an agent programming methodology, similar to the one discussed in [2]. The idea is to program agents so that they can properly tackle part of the events that are generated in the business artifacts of their interest; specifically, the state transitions that occur to commitments in which they are involved. To conclude, we mention RAW-SYS [13], which enriches the prescriptive process model with data-awareness. Although RAW-SYS looks similar to a (normative) business artifact, the objectives of the two models are quite different. RAW-SYS is essentially a framework for verifying business processes taking into account both the control- and the data-flows. A normative business artifact, instead, aims at coordinating autonomous agents.

Acknowledgements. This work was partially supported by the *Accountable Trustworthy Organizations and Systems (AThOS)* project, funded by Università degli Studi di Torino and Compagnia di San Paolo (CSP 2014). The authors warmly thank the reviewers for their constructive and helpful comments which helped revising the paper.

A   An Example of Enhanced Business Artifact

A.1   merchant.asl

```
1  { include ("jacamoJar/templates/common-cartago.asl") }
2  !sell ("Asus ZenPhone 3", 1000).
3  +!sell (Item, MaxQuantity) : true
4     <-  makeArtifact ("items", "item.Items", [Item, MaxQuantity], ArtId);
5          focus (ArtId); enact ("merchant").
6  +enacted (Id, "merchant", Role_Id) <- +enactment_id (Role_Id).
7  +requestedQuote (Quantity, Customer_Id) <- quote (1000, Quantity, Customer_Id).
8  +cc (My_Role_Id, Customer_Role_Id, AcceptedQuotation, Goods, "DETACHED")
9     :   enactment_id (My_Role_Id)
10     &  jia.getAcceptedQuotationComponentsPrice (AcceptedQuotation, Price)
11     &  jia.getAcceptedQuotationComponentsQuantity (AcceptedQuotation, Quantity)
12     &  jia.getAcceptedQuotationComponentsCustomer (
13          AcceptedQuotation, Customer_Role_Id)
14     <- ship (Customer_Role_Id, Quantity).
15 +cc (My_Role_Id, Customer_Role_Id, Paid, Receipt, "DETACHED")
16     :   enactment_id (My_Role_Id) & .term2string (Term1, Paid)
17     &  Term1 = paid (Customer) & .term2string (Customer, S)
18     &  S == Customer_Role_Id
19     <- emitReceipt (Customer_Role_Id).
```

A.2   customer.asl

```
1  { include ("jacamoJar/templates/common-cartago.asl") }
2  !startRequest.
3  +!startRequest <- lookupArtifact ("items", ArtId);
4         focus (ArtId); enact ("customer").
5  +enacted (Id, "customer", My_Role_Id) <- +enactment_id (My_Role_Id); !buy (10).
6  +!buy (Quantity) <- request (Quantity).
7  +cc (Merchant_Role_Id, My_Role_Id, AcceptedQuotation, Goods, "CONDITIONAL")
8     :   enactment_id (My_Role_Id)
9     &  jia.getAcceptedQuotationComponentsPrice (AcceptedQuotation, Price)
10     &  jia.getAcceptedQuotationComponentsQuantity (AcceptedQuotation, Quantity)
11     &  jia.getAcceptedQuotationComponentsCustomer (AcceptedQuotation, My_Role_Id)
12     <- acceptQuotation (Price, Quantity).
13 +cc (My_Role_Id, Merchant_Role_Id, _, Paid, "DETACHED")
14     :   enactment_id (My_Role_Id)
15     &  .term2string (Term1, Paid) & Term1 = paid (My_Role)
16     <- sendEPO (53530331).
```

A.3 Items.java

```
1  public class Items extends BusinessArtifact {
2    public static String ARTIFACT_TYPE = "Items";
3    public static String MERCHANT_ROLE = "merchant";
4    public static String CUSTOMER_ROLE = "customer";
5    static {    addEnabledRole(MERCHANT_ROLE, Merchant.class );
6               addEnabledRole(CUSTOMER_ROLE, Customer.class );    }
7    public Items() {    super();
8                       socialState = new AutomatedSocialState(this);    }
9    @OPERATION    public void init(String itemName, int maxQuantity)
10   {  defineObsProperty("itemName",    itemName);
11      defineObsProperty("maxQuantity",    maxQuantity);    }
12   @OPERATION    public void quote(int price, int quantity, String customer)
13   {  RoleId merchant = getRoleIdByPlayerName(getOpUserName());
14      RoleId customerId = getRoleIdByRoleName(customer);
15      Commitment c = new Commitment(merchant, customerId,
16        new Fact("acceptedQuotation", price, quantity, customer),
17        new Fact("goods",customer) );
18      createCommitment(c);
19      if (this.socialState.existsFact(new Fact("goods",customer)))
20        satisfyCommitment(c);
21      createCommitment(new Commitment(merchant, customerId,
22        new Fact("paid",customer),
23        new Fact("receipt", customer)));
24      assertFact(new Fact("quotation", price,quantity,customer));    }
25   @OPERATION    public void ship(String customerString, int quantity)
26   {  RoleId merchant = getRoleIdByPlayerName(getOpUserName());
27      RoleId customer = getRoleIdByRoleName(customerString);
28      assertFact(new Fact("goods",customerString));    }
29   @OPERATION    public void emitReceipt(String customerString)
30   {  RoleId merchant = getRoleIdByPlayerName(getOpUserName());
31      RoleId customer = getRoleIdByRoleName(customerString);
32      assertFact(new Fact("receipt", customer.toString()));    }
33   @OPERATION    public void request(int quantity)
34   {  RoleId customer = getRoleIdByPlayerName(getOpUserName());
35      RoleId merchant = getRoleIdByGenericRoleName(MERCHANT_ROLE).get(0);
36      assertFact(new Fact("requestedQuote", quantity, customer.toString()));    }
37   @OPERATION    public void accept(int price, int quantity)
38   {  RoleId customer = getRoleIdByPlayerName(getOpUserName());
39      RoleId merchant = getRoleIdByGenericRoleName(MERCHANT_ROLE).get(0);
40      assertFact(new Fact("acceptedQuotation",
41        price, quantity, customer.toString()));
42      createCommitment(new Commitment(customer, merchant,
43        new Fact("goods",customer.toString()),
44        new Fact("paid",customer.toString())));    }
45   @OPERATION    public void reject(int price)
46   {  RoleId customer = getRoleIdByPlayerName(getOpUserName());
47      RoleId merchant = getRoleIdByGenericRoleName(MERCHANT_ROLE).get(0);
48      assertFact(new Fact("rejectedQuotation", price, customer.toString()));
49      createCommitment(new Commitment(customer, merchant, "goods", "paid"));    }
50   @OPERATION    public void sendEPO(int creditCardNumber)
51   {  RoleId customer = getRoleIdByPlayerName(getOpUserName());
52      RoleId merchant = getRoleIdByGenericRoleName(MERCHANT_ROLE).get(0);
53      assertFact(new Fact("paid", customer.toString()));
54      assertFact(new Fact("EPO", creditCardNumber, customer.toString()));    }
55   public class Merchant extends PARole
56   {  public Merchant(String playerName, IPlayer player)
57      {  super(MERCHANT_ROLE, player);    }  }
58   public class Customer extends PARole
59   {  public Customer(String playerName, IPlayer player)
60      {  super(CUSTOMER_ROLE, player);    }  }
61   public interface MerchantObserver extends ProtocolObserver { }
62   public interface CustomerObserver extends ProtocolObserver { }
63 }
```

References

1. Baldoni, M., Baroglio, C., Capuzzimati, F.: Typing multi-agent systems via commitments. In: Dalpiaz, F., Dix, J., van Riemsdijk, M.B. (eds.) EMAS 2014. LNCS (LNAI), vol. 8758, pp. 388–405. Springer, Cham (2014). https://doi.org/10.1007/978-3-319-14484-9_20

2. Baldoni, M., Baroglio, C., Capuzzimati, F., Micalizio, R.: Empowering agent coordination with social engagement. In: Gavanelli, M., Lamma, E., Riguzzi, F. (eds.) AI*IA 2015. LNCS (LNAI), vol. 9336, pp. 89–101. Springer, Cham (2015). https://doi.org/10.1007/978-3-319-24309-2_7

3. Baldoni, M., Baroglio, C., Capuzzimati, F., Micalizio, R.: Commitment-based agent interaction in JaCaMo+. Fundamenta Informaticae **157**, 1–33 (2018). https://doi.org/10.3233/FI-2018-1600. IOS Press

4. Baldoni, M., Baroglio, C., May, K.M., Micalizio, R., Tedeschi, S.: ADOPT JaCaMo: accountability-driven organization programming technique for JaCaMo. In: An, B., Bazzan, A., Leite, J., Villata, S., van der Torre, L. (eds.) PRIMA 2017. LNCS (LNAI), vol. 10621, pp. 295–312. Springer, Cham (2017). https://doi.org/10.1007/978-3-319-69131-2_18

5. Baldoni, M., Baroglio, C., Micalizio, R.: Social continual planning in open multiagent systems: a first study. In: Chen, Q., Torroni, P., Villata, S., Hsu, J., Omicini, A. (eds.) PRIMA 2015. LNCS (LNAI), vol. 9387, pp. 575–584. Springer, Cham (2015). https://doi.org/10.1007/978-3-319-25524-8_40

6. Bhattacharya, K., Caswell, N.S., Kumaran, S., Nigam, A., Wu, F.Y.: Artifact-centered operational modeling: lessons from customer engagements. IBM Syst. J. **46**(4), 703–721 (2007)

7. Bhattacharya, K., Hull, R., Su, J.: A data-centric design methodology for business processes. In: Handbook of Research on Business Process Modeling, pp. 503–531. IGI Publishing (2009)

8. Bratman, M.E.: What is intention? In: Cohen, P., Morgan, J., Pollack, M. (eds.) Intensions in Communication, pp. 15–31. MIT Press, Cambridge (1990)

9. Busi, N., Ciancarini, P., Gorrieri, R., Zavattaro, G.: Coordination models: a guided tour. In: Omicini, A., Zambonelli, F., Klusch, M., Tolksdorf, R. (eds.) Coordination of Internet Agents, pp. 6–24. Springer, Heidelberg (2001). https://doi.org/10.1007/978-3-662-04401-8_1

10. Calvanese, D., De Giacomo, G., Montali, M.: Foundations of data-aware process analysis: a database theory perspective. In: Proceedings of the 32nd ACM SIGMOD-SIGACT-SIGART Symposium on Principles of Database Systems, PODS, pp. 1–12. ACM (2013)

11. Chopra, A.K., Singh, M.P.: Cupid: commitments in relational algebra. In: Bonet, B., Koenig, S. (eds.) Proceedings of the Twenty-Ninth AAAI Conference on Artificial Intelligence, Austin, Texas, USA, pp. 2052–2059. AAAI Press, 25–30 January 2015

12. Cohn, D., Richard, H.: Business artifacts: a data-centric approach to modeling business operations and processes. IEEE Data Eng. Bull. **32**(3), 3–9 (2009)

13. De Masellis, R., Di Francescomarino, C., Ghidini, C., Montali, M., Tessaris, S.: Add data into business process verification: bridging the gap between theory and practice. In: Proceedings of the Thirty-First AAAI Conference on Artificial Intelligence, San Francisco, California, USA, pp. 1091–1099, 4–9 February 2017

14. Engeström, Y., Miettinen, R., Punamäki, R.-L. (eds.): Perspectives on Activity Theory. Cambridge University Press, Cambridge (1999)

15. Nigam, A., Caswell, N.S.: Business artifacts: an approach to operational specification. IBM Syst. J. **42**(3), 428–445 (2003)
16. Omicini, A., Ossowski, S.: Objective versus subjective coordination in the engineering of agent systems. In: Klusch, M., Bergamaschi, S., Edwards, P., Petta, P. (eds.) Intelligent Information Agents. LNCS (LNAI), vol. 2586, pp. 179–202. Springer, Heidelberg (2003). https://doi.org/10.1007/3-540-36561-3_9
17. Omicini, A., Ricci, A., Viroli, M.: Artifacts in the A&A meta-model for multi-agent systems. Auton. Agents Multi Agent Syst. **17**(3), 432–456 (2008). Special Issue on Foundations, Advanced Topics and Industrial Perspectives of Multi-Agent Systems
18. Philippsen, M.: A survey of concurrent object-oriented languages. Concurr. Pract. Exp. **12**(10), 917–980 (2000)
19. Singh, M.P.: An ontology for commitments in multiagent systems. Artif. Intell. Law **7**(1), 97–113 (1999)
20. Singh, M.P.: Information-driven interaction-oriented programming: BSPL, the blindingly simple protocol language. In: 10th International Conference on Autonomous Agents and Multiagent Systems (AAMAS), pp. 491–498 (2011)
21. Singh, M.P.: NoBPM: supporting interaction-oriented automation via normative specifications of processes (2015). Invited talk, BPM
22. Telang, P.R., Singh, M.P., Yorke-Smith, N.: Relating goal and commitment semantics. In: Dennis, L., Boissier, O., Bordini, R.H. (eds.) ProMAS 2011. LNCS (LNAI), vol. 7217, pp. 22–37. Springer, Heidelberg (2012). https://doi.org/10.1007/978-3-642-31915-0_2
23. Therborn, G.: Back to norms! on the scope and dynamics of norms and normative action. Curr. Sociol. **50**, 863–880 (2002)
24. Viroli, M., Omicini, A., Ricci, A.: Infrastructure for RBAC-MAS: an approach based on agent coordination contexts. Appl. Artif. Intell. **21**(4&5), 443–467 (2007)
25. Wooldridge, M.J.: Introduction to Multiagent Systems, 2nd edn. Wiley, Chichester (2009)
26. Yadav, N., Padgham, L., Winikoff, M.: A tool for defining agent protocols in HAPN: (demonstration). In: Proceedings of the 2015 International Conference on Autonomous Agents and Multiagent Systems, AAMAS, pp. 1935–1936 (2015)

Enhancing Workflow-Nets with Data
for Trace Completion

Riccardo De Masellis[1]([⊠]), Chiara Di Francescomarino[1], Chiara Ghidini[1],
and Sergio Tessaris[2]

[1] FBK-IRST, Bolzano, Italy
{r.demasellis,dfmchiara,ghidini}@fbk.eu
[2] Free University of Bozen-Bolzano, Bolzano, Italy
tessaris@inf.unibz.it

Abstract. The growing adoption of IT-systems for modeling and executing (business) processes or services has thrust the scientific investigation towards techniques and tools which support more complex forms of process analysis. Many of them, such as conformance checking, process alignment, mining and enhancement, rely on *complete* observation of past (tracked and logged) executions. In many real cases, however, the lack of human or IT-support on all the steps of process execution, as well as information hiding and abstraction of model and data, result in incomplete log information of both data and activities. This paper tackles the issue of automatically repairing traces with missing information by notably considering not only activities but also data manipulated by them. Our technique recasts such a problem in a reachability problem and provides an encoding in an action language which allows to virtually use any state-of-the-art planning to return solutions.

1 Introduction

The use of IT systems for supporting business activities has brought to a large diffusion of *process mining* techniques and tools that offer business analysts the possibility to observe the current process execution, identify deviations from the model, perform individual and aggregated analysis on current and past executions.

According to the process mining manifesto, all these techniques and tools can be grouped in three basic types: process discovery, conformance checking and process enhancement (see Fig. 1), and require in input an *event log* and, for conformance checking and enhancement, a *(process) model*. A log, usually described in the IEEE standard XES format[1], is a set of execution traces (or cases) each of which is an ordered sequence of events carrying a payload as a set of attribute-value pairs. Process models instead provide a description of the scenario at hand and can be constructed using one of the available Business Process Modeling Languages, such as BPMN, YAWL and DECLARE.

---

[1] http://www.xes-standard.org/.

E. Teniente and M. Weidlich (Eds.): BPM 2017 Workshops, LNBIP 308, pp. 89–106, 2018.
https://doi.org/10.1007/978-3-319-74030-0_6

Event logs are therefore a crucial ingredient to the accomplishment of process mining. Unfortunately, a number of difficulties may hamper the availability of event logs. Among these are partial event logs, where the execution traces may bring only **partial information** in terms of which process activities have been executed and what data or artefacts they produced.

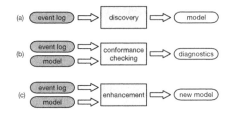

Fig. 1. The three types of process mining.

Thus repairing incomplete execution traces by reconstructing the missing entries becomes an important task to enable process mining in full, as noted in recent works such as [6,14]. While these works deserve a praise for having motivated the importance of trace repair and having provided some basic techniques for reconstructing missing entries using the knowledge captured in process models, they all focus on event logs (and process models) of limited expressiveness. In fact, they all provide techniques for the reconstruction of control flows, thus completely ignoring the data flow component. This is a serious limitation, given the growing efforts to extend business process languages with the capability to model complex data objects along with the fact that considering data in the repair task allows, in general, for reducing the number of possible trace completions, as shown in Sect. 2.2.

In this paper we show how to exploit state-of-the-art planning techniques to deal with the repair of data-aware event logs in the presence of imperative process models. Specifically we will focus on the well established Workflow Nets [16], a particular class of Petri nets that provides the formal foundations of several process models, of the YAWL language and have become one of the standard ways to model and analyze workflows. In particular we provide:

1. a modeling language DAW-net, an extension of the workflow nets with data formalism introduced in [15] so to be able to deal with more expressive data (Sect. 3);
2. a recast of data aware trace repair as a reachability problem in DAW-net (Sect. 4);
3. a sound and complete encoding of reachability in DAW-net in a planning problem so to be able to deal with trace repair using planning (Sect. 5).

The solution of the problem are all and only the repairs of the partial trace compliant with the DAW-net model. The advantage of using automated planning techniques is that we can exploit the underlying logic language to ensure that generated plans conform to the observed traces without resorting to ad hoc algorithms for the specific repair problem. The theoretical investigation presented in this work provides an important step forward towards the exploitation of planning techniques in data-aware processes.

2   Preliminaries

2.1   The Workflow Nets Modeling Language

Petri Nets (PN) is a modeling language for the description of distributed systems that has widely been applied to business processes. The classical PN is a directed bipartite graph with two node types, called *places* and *transitions*, connected via directed arcs. Connections between nodes of the same type are not allowed.

Definition 1 (Petri Net). *A Petri Net is a triple $\langle P, T, F \rangle$ where P is a set of* places; *T is a set of* transitions; *$F \subseteq (P \times T) \cup (T \times P)$ is the flow relation describing the arcs between places and transitions (and between transitions and places).*

The *preset* of a transition t is the set of its input places: $^{\bullet}t = \{p \in P \mid (p,t) \in F\}$. The *postset* of t is the set of its output places: $t^{\bullet} = \{p \in P \mid (t,p) \in F\}$. Definitions of pre- and postsets of places are analogous.

   Places in a PN may contain a discrete number of tokens. Any distribution of tokens over the places, formally represented by a total mapping $M : P \mapsto \mathbb{N}$, represents a configuration of the net called a *marking*.

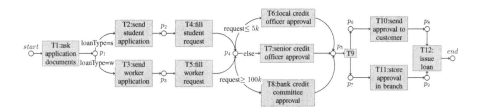

Fig. 2. A process as a Petri Net.

Process tasks are modeled in PNs as transitions while arcs and places constraint their ordering. Figure 2 exemplifies how PNs can be used to model parallel and mutually exclusive choices: sequences *T2;T4-T3;T5* (transitions *T6-T7-T8*) are placed on mutually exclusive paths, while transitions *T10* and *T11* are placed on parallel paths. Finally, *T9* prevents connections between nodes of the same type.

   The expressivity of PNs exceeds, in the general case, what is needed to model business processes, which typically have a well-defined starting (ending) point. This leads to the following definition of a workflow net (WF-net) [16].

Definition 2 (WF-net). *A PN $\langle P, T, F \rangle$ is a WF-net if it has a single source place* start, *a single sink place* end, *and every place and every transition is on a path from* start *to* end, *i.e., for all $n \in P \cup T$, $(start, n) \in F^*$ and $(n, end) \in F^*$, where F^* is the reflexive transitive closure of F.*

A marking in a WF-net represents the *workflow state* of a single case. The semantics of a PN/WF-net, and in particular the notion of *valid firing*, defines how transitions route tokens through the net so that they correspond to a process execution.

Definition 3 (Valid Firing). *A firing of a transition $t \in T$ from M to M' is valid, in symbols $M \xrightarrow{t} M'$, iff*

1. *t is enabled in M, i.e., $\{p \in P \mid M(p) > 0\} \supseteq {}^{\bullet}t$; and*
2. *the marking M' is such that for every $p \in P$:*

$$M'(p) = \begin{cases} M(p) - 1 & \text{if } p \in {}^{\bullet}t \setminus t^{\bullet} \\ M(p) + 1 & \text{if } p \in t^{\bullet} \setminus {}^{\bullet}t \\ M(p) & \text{otherwise} \end{cases}$$

Condition 1. states that a transition is enabled if all its input places contain at least one token; 2. states that when t fires it consumes one token from each of its input places and produces one token in each of its output places.

A *case* of a WF-Net is a sequence of valid firings $M_0 \xrightarrow{t_1} M_1, M_1 \xrightarrow{t_2} M_2, \ldots,$ $M_{k-1} \xrightarrow{t_k} M_k$ where M_0 is the marking indicating that there is a single token in *start*.

Definition 4 (k-safeness). *A marking of a PN is k-safe if the number of tokens in all places is at most k. A PN is k-safe if the initial marking is k-safe and the marking of all cases is k-safe.*

From now on we concentrate on 1-safe nets, which generalize the class of *structured workflows* and are the basis for best practices in process modeling [9]. We also use safeness as a synonym of 1-safeness. It is important to notice that our approach can be seamlessly generalized to other classes of PNs, as long as it is guaranteed that they are k-safe. This reflects the fact that the process control-flow is well-defined (see [8]).

Reachability on Petri Nets. The behavior of a PN can be described as a transition system where states are markings and directed edges represent firings. Intuitively, there is an edge from M_i to M_{i+1} labeled by t_i if $M_i \xrightarrow{t} M_{i+1}$ is a valid firing. Given a "goal" marking M_g, the reachability problem amounts to check if there is a path from the initial marking M_0 to M_g. Reachability on PNs (WF-nets) is of enormous importance in process verification as it allows for checking natural behavioral properties, such as satisfiability and soundness in a natural manner [1].

2.2   Trace Repair

One of the goals of process mining is to capture the as-is processes as accurately as possible: this is done by examining event logs that can be then exploited to perform the tasks in Fig. 1. In many cases, however, event logs are subject to data quality problems, resulting in *incorrect* or *missing* events in the log. In this paper we focus on the latter issue addressing the problem of **repairing execution traces that contain missing entries** (hereafter shortened in trace repair).

The need for trace repair is motivated in depth in [14], where missing entities are described as a frequent cause of low data quality in event logs, especially when the definition of the business processes integrates activities that are not supported by IT systems due either to their nature (e.g. they consist of human interactions) or to the high level of abstraction of the description, detached from the implementation. A further cause of missing events are special activities (such as transition *T9* in Fig. 2) that are introduced in the model to guarantee properties concerning e.g., the structure of the workflow or syntactic constraints, but are never executed in practice.

The starting point of trace repair are *execution traces* and the knowledge captured in *process models*. Consider for instance the model in Fig. 2 and the (partial) execution trace {T3, T7}. By aligning the trace to the model, techniques such as the ones presented in [14] and [6] are able to exploit the events stored in the trace and the control flow specified in the model to reconstruct two possible repairs:

$$\{T1, T3, T5, T7, T9, T10, T11, T12\}$$
$$\{T1, T3, T5, T7, T9, T11, T10, T12\}$$

Consider now a different scenario in which the partial trace reduces to $\{T7\}$. In this case, by using the control flow in Fig. 2 we are not able to reconstruct whether the loan is a student loan or a worker loan. This increases the number of possible repairs and therefore lowers the usefulness of trace repair. Assume now that the event log conforms to the XES standard and stores some observed data attached to $T7$:

$$\{T7[request = 60\mathbf{k}, loan = 50\mathbf{k}]\}$$

If the process model is able to specify how transitions can read and write variables, and furthermore some constraints on how they do it, the scenario changes completely. Indeed, assume that transition $T4$ is empowered with the ability to write the variable *request* with a value smaller or equal than 30**k** (the maximum amount of a student loan). Using this fact, and the fact that the request examined by $T7$ is greater than 30**k**, we can understand that the execution trace has chosen the path of the worker loan. Moreover, if the model specifies that variable *loanType* is written during the execution of $T1$, when the applicant chooses the type of loan she is interested in, we are able to infer that $T1$ sets variable *loanType* to **w**. This example, besides illustrating the idea of trace repair, also motivates why data are important to accomplish this task, and therefore why extending repair techniques beyond the mere control flow is a significant contribution to address data quality problems in event logs.

2.3   The Planning Language $\mathcal{K}$

The main elements of action languages are *fluents* and *actions*. The former represent the state of the system which may change by means of actions. Causation statements describe the possible evolution of the states, and preconditions associated to actions describe which action can be executed according to the current state. A planning problem in $\mathcal{K}$ [7] is specified using a Datalog-like language where fluents and actions are represented by literals (not necessarily ground). The specification includes the list of fluents, actions, initial state and goal conditions; also a set of statements specifies the dynamics of the planning domain using causation rules and executability conditions. The semantics of $\mathcal{K}$ borrows heavily from Answer Set Programming (ASP). In fact, the system enables the reasoning with partial knowledge and provides both weak and strong negation.

A *causation rule* is a statement of the form

> **caused** f **if** $b_1, \ldots, b_k$, **not** b_{k+1}, $\ldots$, **not** b_ℓ
> **after** $a_1, \ldots, a_m$, **not** a_{m+1}, $\ldots$, **not** a_n.

The rule states that f is true in the new state reached by executing (simultaneously) some actions, provided that $a_1, \ldots, a_m$ are known to hold while $a_{m+1}, \ldots, a_n$ are not known to hold in the previous state (some of the a_j might be actions executed on it), and $b_1, \ldots, b_k$ are known to hold while $b_{k+1}, \ldots, b_\ell$ are not known to hold in the new state. Rules without the **after** part are called *static*.

An *executability condition* is a statement of the form

> **executable** a **if** $b_1, \ldots, b_k$, **not** b_{k+1}, $\ldots$, **not** b_ℓ.

Informally, such a condition says that the action a is eligible for execution in a state, if $b_1, \ldots, b_k$ are known to hold while $b_{k+1}, \ldots, b_\ell$ are not known to hold in that state.

Terms in both kind of statements could include variables (starting with capital letter) and the statements must be safe in the usual Datalog meaning w.r.t. the first fluent or action of the statements.

A *planning domain PD* is a pair $\langle D, R \rangle$ where D is a finite set of action and fluent declarations, and R is a finite set of rules, initial state constraints, and executability conditions.

The semantics of the language is provided in terms of a transition system where the states are ASP models (sets of atoms) and actions transform the state according to the rules. A state transition is a tuple $t = \langle s, A, s' \rangle$ where s, s' are states and A is a set of action instances. The transition is said to be legal if the actions are executable in the first state and both states are the minimal ones that satisfy all causation rules. Semantics of plans including default negation is defined by means of a Gelfond-Lifschitz type reduction to a positive planning domain. A sequence of state transitions $\langle s_0, A_1, s_1 \rangle, \ldots, \langle s_{n-1}, A_n, s_n \rangle$, $n \geq 0$, is a trajectory for PD, if s_0 is a legal initial state of PD and all $\langle s_{i-1}, A_i, s_i \rangle$, are legal state transitions of PD.

A *planning problem* is a pair composed of a planning domain PD and a ground goal $g_1, \ldots, g_m$, **not** g_{m+1}, $\ldots$, **not** g_n that has to be satisfied at the end of the execution.

3   Framework

In this section we suitably extend WF-nets to represent data and their evolution as transitions are performed. In order for such an extension to be meaningful, i.e., allowing reasoning on data, it has to provide: (i) a model for representing data; (ii) a way to make decisions on actual data values; and (iii) a mechanism to express modifications to data. We provide (i)–(iii) by enhancing WF-nets with the following elements:

– a set of variables taking values from possibly different domains (provides(i));
– queries on such variables used as transitions preconditions (provides(ii));
– variables updates and deletion in the specification of net transitions (provides(iii)).

Our framework follows the approach of state-of-the-art WF-nets with data [10,15], from which it borrows the above concepts, extending them by allowing reasoning on actual data values as better explained in Sect. 6.

We make use of the WF-net in Fig. 2 extended with data as a running example.

3.1   Data Model

As our focus is on trace repair, we follow the data model of the IEEE XES standard for describing logs, which represents data as a set of variables. Variables take values from specific sets on which a partial order can be defined. As customary, we distinguish between the data model, namely the intensional level, from a specific instance of data, i.e., the extensional level.

Definition 5 (Data model). *A* data model *is a tuple* $\mathcal{D} = (\mathcal{V}, \Delta, dm, ord)$ *where:*

– $\mathcal{V}$ *is a possibly infinite set of variables;*
– $\Delta = \{\Delta_1, \Delta_2, \ldots\}$ *is a possibly infinite set of domains (not necessarily disjoint);*
– $dm : \mathcal{V} \to \Delta$ *is a total and surjective function which associates to each variable* v *its domain* Δ_i;
– ord *is a partial function that, given a domain* Δ_i, *if* $ord(\Delta_i)$ *is defined, then it returns a* partial order *(reflexive, antisymmetric and transitive)* $\leq_{\Delta_i} \subseteq \Delta_i \times \Delta_i$.

A data model for the loan example is $\mathcal{V} = \{loanType, request, loan\}$, $dm(loanType) = \{w, s\}$, $dm(request) = \mathbb{N}$, $dm(loan) = \mathbb{N}$, with $dm(loan)$ and $dm(request)$ being totally ordered by the natural ordering $\leq$ in $\mathbb{N}$.

An actual instance of a data model is a partial function associating values to variables.

Definition 6 (Assignment). *Let* $\mathcal{D} = \langle \mathcal{V}, \Delta, dm, ord \rangle$ *be a data model. An* assignment *for variables in* $\mathcal{V}$ *is a* partial *function* $\eta : \mathcal{V} \to \bigcup_i \Delta_i$ *such that for each* $v \in \mathcal{V}$, *if* $\eta(v)$ *is defined, i.e.,* $v \in img(\eta)$ *where img is the image of* η, *then we have* $\eta(v) \in dm(v)$.

We now define our boolean query language, which notably allows for equality and comparison. As will become clearer in Sect. 3.2, queries are used as *guards*, i.e., preconditions for the execution of transitions.

Definition 7 (Query language - syntax). *Given a data model, the language* $\mathcal{L}(\mathcal{D})$ *is the set of formulas* Φ *inductively defined according to the following grammar:*

$$\Phi \quad ::= \quad true \mid def(v) \mid t_1 = t_2 \mid t_1 \leq t_2 \mid \neg\Phi_1 \mid \Phi_1 \wedge \Phi_2$$

where $v \in \mathcal{V}$ *and* $t_1, t_2 \in \mathcal{V} \cup \bigcup_i \Delta_i$.

Examples of queries of the loan scenarios are $request \leq 5k$ or $loanType = w$. Given a formula Φ and an assignment η, we write $\Phi[\eta]$ for the formula Φ where each occurrence of variable $v \in img(\eta)$ is replaced by $\eta(v)$.

Definition 8 (Query language - semantics). *Given a data model* $\mathcal{D}$, *an assignment* η *and a query* $\Phi \in \mathcal{L}(\mathcal{D})$ *we say that* $\mathcal{D}, \eta$ *satisfies* Φ, *written* $D, \eta \models \Phi$ *inductively on the structure of* Φ *as follows:*

- $\mathcal{D}, \eta \models true$;
- $\mathcal{D}, \eta \models def(v)$ *iff* $v \in img(\eta)$;
- $\mathcal{D}, \eta \models t_1 = t_2$ *iff* $t_1[\eta], t_2[\eta] \notin \mathcal{V}$ *and* $t_1[\eta] \equiv t_2[\eta]$;
- $\mathcal{D}, \eta \models t_1 \leq t_2$ *iff* $t_1[\eta], t_2[\eta] \in \Delta_i$ *for some* i *and* $ord(\Delta_i)$ *is defined and* $t_1[\eta] \leq_{\Delta_i} t_2[\eta]$;
- $\mathcal{D}, \eta \models \neg\Phi$ *iff it is not the case that* $\mathcal{D}, \eta \models \Phi$;
- $\mathcal{D}, \eta \models \Phi_1 \wedge \Phi_2$ *iff* $\mathcal{D}, \eta \models \Phi_1$ *and* $\mathcal{D}, \eta \models \Phi_2$.

Intuitively, def can be used to check if a variable has an associated value or not (recall that assignment η is a partial function); equality has the intended meaning and $t_1 \leq t_2$ evaluates to true iff t_1 and t_2 are values belonging to the same domain Δ_i, such a domain is ordered by a partial order $\leq_{\Delta_i}$ and t_1 is actually less or equal than t_2 according to $\leq_{\Delta_i}$.

3.2 Data-Aware Net

We now combine the data model with a WF-net and formally define how transitions are guarded by queries and how they update/delete data. The result is a Data-AWare net (DAW-net) that incorporates aspects (i)–(iii) described at the beginning of Sect. 3.

Definition 9 (DAW-net). *A DAW-net is a tuple* $\langle \mathcal{D}, \mathcal{N}, wr, gd \rangle$ *where:*

- $\mathcal{N} = \langle P, T, F \rangle$ *is a WF-net;*
- $\mathcal{D} = \langle \mathcal{V}, \Delta, dm, ord \rangle$ *is a data model;*
- $wr : T \mapsto (\mathcal{V}' \mapsto 2^{dm(\mathcal{V})})$, *where* $\mathcal{V}' \subseteq \mathcal{V}$, $dm(\mathcal{V}) = \bigcup_{v \in V} dm(v)$ *and* $wr(t)(v) \subseteq dm(v)$ *for each* $v \in \mathcal{V}'$, *is a function that associates each transition to a (partial) function mapping variables to a finite subset of their domain.*
- $gd : T \mapsto \mathcal{L}(\mathcal{D})$ *is a function that associates a guard to each transition.*

Function gd associates a guard, namely a query, to each transition. The intuitive semantics is that a transition t can fire if its guard $gd(t)$ evaluates to true (given the current assignment of values to data). Examples are $gd(T6) = request \leq 5k$ and $gd(T8) = \neg(request \leq 99999)$. Function wr is instead used to express how a transition t modifies data: after the firing of t, each variable $v \in \mathcal{V}'$ can take any value among a specific finite subset of $dm(v)$. We have three different cases:

- $\emptyset \subset wr(t)(v) \subseteq dm(v)$: t nondeterministically assigns a value from $wr(t)(v)$ to v;
- $wr(t)(v) = \emptyset$: t deletes the value of v (hence making v undefined);
- $v \notin dom(wr(t))$: value of v is not modified by t.

Notice that by allowing $wr(t)(v) \subseteq dm(v)$ in the first bullet above we enable the specification of restrictions for specific tasks. E.g., $wr(T4) : \{request\} \mapsto \{0 \ldots 30k\}$ says that $T4$ writes the *request* variable and intuitively that students can request a maximum loan of 30k, while $wr(T5) : \{request\} \mapsto \{0 \ldots 500k\}$ says that workers can request up to 500k.

The intuitive semantics of gd and wr is formalized next. We start from the definition of DAW-net state, which includes both the state of the WF-net, namely its marking, and the state of data, namely the assignment. We then extend the notions of state transition and valid firing.

Definition 10 (DAW-net state). *A state of a DAW-net $\langle \mathcal{D}, \mathcal{N}, wr, gd \rangle$ is a pair (M, η) where M is a marking for $\langle P, T, F \rangle$ and η is an assignment for $\mathcal{D}$.*

Definition 11 (DAW-net Valid Firing). *Given a DAW-net $\langle \mathcal{D}, \mathcal{N}, wr, gd \rangle$, a firing of a transition $t \in T$ is a valid firing from (M, η) to (M', η'), written as $(M, \eta) \xrightarrow{t} (M', \eta')$, iff conditions 1. and 2. of Def. 3 holds for M and M', i.e., it is a WF-Net valid firing, and*

1. $\mathcal{D}, \eta \models gd(t)$,
2. *assignment η' is such that, if $WR = \{v \mid wr(t)(v) \neq \emptyset\}$, $DEL = \{v \mid wr(t)(v) = \emptyset\}$:*
   - *its domain $dom(\eta') = dom(\eta) \cup WR \setminus DEL$;*
   - *for each $v \in dom(\eta')$:*

$$\eta'(v) = \begin{cases} d \in wr(t)(v) & if\ v \in WR \\ \eta(v) & otherwise. \end{cases}$$

Condition 1. and 2. extend the notion of valid firing of WF-nets imposing additional pre- and postconditions on data, i.e., preconditions on η and postconditions on η'. Specifically, 1. says that for a transition t to be fired its guard $gd(t)$ must be satisfied by the current assignment η. Condition 2. constrains the new state of data: the domain of η' is defined as the union of the domain of η with variables that are written (WR), minus the set of variables that must be deleted (DEL). Variables in $dom(\eta')$ can indeed be grouped in three sets depending on the effects of t: (i) OLD $= dom(\eta) \setminus$ WR: variables whose value is unchanged after t; (ii)

NEW $= $ WR $\setminus dom(\eta)$: variables that were undefined but have a value after t; and (iii) OVERWR $=$ WR $\cap dom(\eta)$: variables that did have a value and are updated with a new one after t. The final part of condition 2. says that each variable in NEW $\cup$ OVERWR takes a value in $\mathsf{wr}(t)(v)$, while variables in OLD maintain the old value $\eta(v)$.

A *case* of a DAW-net is defined as a case of a WF-net, with the only difference that the assignment η_0 of the initial state (M_0, η_0) is empty, i.e., $dom(\eta_0) = \emptyset$.

4   Trace Repair as Reachability

In this section we provide the intuition behind our technique for solving the trace repair problem via reachability. Full details and proofs are contained in [5].

A *trace* is a sequence of observed *events*, each with a payload including the transition it refers to and its effects on the data, i.e., the variables updated by its execution. Intuitively, a DAW-net case is *compliant* w.r.t. a trace if it contains all the occurrences of the transitions observed in the trace (with the corresponding variable updates) in the right order.

As a first step, we assume without loss of generality that DAW-net models start with a special transition $start_t$ and terminate with a special transition end_t. Every process can be reduced to such a structure as informally illustrated in the left hand side of Fig. 3 by arrows labeled with (1). Note that this change would not modify the behavior of the net: any sequence of firing valid for the original net can be extended by the firing of the additional transitions and vice versa.

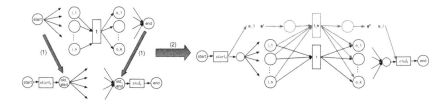

Fig. 3. Outline of the trace "injection"

Next, we illustrate the main idea behind our approach by means of the right hand side of Fig. 3: we consider the observed events as transitions (in red) and we suitably "inject" them in the original DAW-net. By doing so, we obtain a new model where, intuitively, tokens are forced to activate the red transitions of DAW-net, when events are observed in the trace. When, instead, there is no red counterpart, i.e., there is missing information in the trace, the tokens move in the black part of the model. The objective is then to perform reachability for the final marking (i.e., to have one token in the *end* place and all other places empty) over such a new model in order to obtain all and only the possible repairs for the partial trace.

More precisely, for each event e with a payload including transition t and some effect on variables we introduce a new transition t_e in the model such that:

- t_e is placed in parallel with the original transition t;
- t_e includes an additional input place connected to the preceding event and an additional output place which connects it to the next event;
- $\mathsf{gd}(t_e) = \mathsf{gd}(t)$ and
- $\mathsf{wr}(t_e)$ specifies exactly the variables and the corresponding values updated by the event, i.e. if the event set the value of v to d, then $\mathsf{wr}(t_e)(v) = \{d\}$; if the event deletes the variable v, then $\mathsf{wr}(t_e)(v) = \emptyset$.

Given a trace τ and a DAW-net W, it is easy to see that the resulting *trace workflow* (indicated as W^τ) is a strict extension of W (only new nodes are introduced) and, since all newly introduced nodes are in a path connecting the start and sink places, it is a DAW-net, whenever the original one is a DAW-net net.

We now prove the soundness and completeness of the approach by showing that: (1) all cases of W^τ are compliant with τ; (2) each case of W^τ is also a case of W and (3) if there is a case of W compliant with τ, then that is also a case for W^τ.

Property (1) is ensured by construction. For (2) and (3) we need to relate cases from W^τ to the original DAW-net W. We indeed introduce a projection function Π_τ that maps elements from cases of the enriched DAW-net to cases of elements from the original DAW-net. Essentially, Π_τ maps newly introduced transitions t_e to the corresponding transitions in event e, i.e., t, and also projects away the new places in the markings. Given that the structure of W^τ is essentially the same as that of W with additional copies of transitions that are already in W, it is not surprising that any case for W^τ can be replayed on W by mapping the new transitions t_e into the original ones t, as shown by the following:

Lemma 1. *If C is a case of W^τ then $\Pi_\tau(C)$ is a case of W.*

This lemma proves that whenever we find a case on W^τ, then it is an example of a case on W that is compliant with τ, i.e., (2). However, to reduce the original problem to reachability on DAW-net, we need to prove that *all* the W cases compliant with τ can be replayed on W^τ, that is, (3). In order to do that, we can build a case for W^τ starting from the compliant case for W, by substituting the occurrences of firings corresponding to events in τ with the newly introduced transitions. The above results pave the way to the following:

Theorem 1. *Let W be a DAW-net and $\tau = (e_1, \ldots, e_n)$ a trace; then W^τ characterises all and only the cases of W compatible with τ. That is*

$\Rightarrow$ *if C is a case of W^τ containing t_{e_n} then $\Pi_\tau(C)$ is compatible with τ; and*
$\Leftarrow$ *if C is a case of W compatible with τ, then there is a case C' of W^τ s.t. $\Pi_\tau(C') = C$.*

Theorem 1 provides the main result of this section and is the basis for the reduction of the trace repair for W and τ to the reachability problem for W^τ. In fact, by enumerating all the cases of W^τ reaching the final marking (i.e. a token in *end*) we can provide all possible repairs for the partial observed trace. Moreover, the transformation generating W^τ is preserving the safeness properties of the original workflow:

Lemma 2. *Let W be a DAW-net and τ a trace of W. If W is k-safe then W^τ is k-safe as well.*

This is essential to guarantee the decidability of the reasoning techniques described in the next section.

5 Reachability as a Planning Problem

In this section we exploit the similarity between workflows and planning domains in order to describe the evolution of a DAW-net by means of a planning language. Once the original workflow behaviour has been encoded into an equivalent planning domain, we can use the automatic derivation of plans with specific properties to solve the reachability problem. In our approach we introduce a new action for each transition (to ease the description we will use the same names) and represent the status of the workflow – marking and variable assignments – by means of fluents. Although their representation as dynamic rules is conceptually similar we will separate the description of the encoding by considering first the behavioural part (the WF-net) and then the encoding of data (variable assignments and guards).

5.1 Encoding DAW-net Behaviour

Since we focus on 1-safe WF-nets the representation of markings is simplified by the fact that each place can either contain 1 token or no tokens at all. This information can be represented introducing a propositional fluent for each place, true iff the corresponding place holds a token. Let us consider $\langle P, T, F \rangle$ the *safe WF-net* component of a DAW-net system. The declaration part of the planning domain will include:

- a fluent declaration p for each place $p \in P$;
- an action declaration t for each task $t \in T$.

Since each transition can be fired[2] only if each input place contains a token, then the corresponding action can be executed when place fluents are true: for each task $t \in T$, given $\{i_1^t, \ldots, i_n^t\} = {}^\bullet t$, we include the executability condition:

> **executable** t **if** $i_1^t, \ldots, i_n^t$.

As valid firings are sequential, namely only one transition can be fired at each step, we disable concurrency in the planning domain introducing the following rule for each pair of tasks $t_1, t_2 \in T$[3]

> **caused** false **after** t_1, t_2.

---

[2] Guards will be introduced in the next section.

[3] For efficiency reasons we can relax this constraint by disabling concurrency only for transitions sharing places or updating the same variables. This would provide shorter plans.

Transitions transfer tokens from input to output places. Thus the corresponding actions must clear the input places and set the output places to true. This is enforced by including

caused $-i_1^t$ **after** t. ... **caused** $-i_n^t$ **after** t.
caused o_1^t **after** t. ... **caused** o_k^t **after** t.

for each task $t \in T$ and $\{i_1^t, \dots, i_n^t\} = \mathbf{^{\bullet}} t \setminus t^{\bullet}$, $\{o_1^t, \dots, o_k^t\} = t^{\bullet}$. Finally, place fluents should be inertial since they preserve their value unless modified by an action. This is enforced by adding for each $p \in P$

caused p **if not** $-p$ **after** p.

Planning problem. Besides the domain described above, a planning problem includes an initial state, and a goal. In the initial state the only place with a token is the source:

initially: *start*.

The formulation of the goal depends on the actual instance of the reachability problem we need to solve. The goal corresponding to the state in which the only place with a token is *end* is written as:

goal: *end*, **not** p_1, ..., **not** p_k?

where $\{p_1, \dots, p_k\} = P \setminus \{end\}$.

5.2   Encoding Data

For each variable $v \in \mathcal{V}$ we introduce a fluent unary predicate var$_v$ holding the value of that variable. Clearly, var$_v$ predicates must be functional and have no positive instantiation for undefined variables.

We also introduce auxiliary fluents to facilitate the writing of the rules. Fluent def$_v$ indicates whether the v variable is *not* undefined – it is used both in tests and to enforce models where the variable is assigned/unassigned. The fluent chng$_v$ is used to inhibit inertia for the variable v when its value is updated because of the execution of an action.

DAW-net includes the specification of the set of values that each transition can write on a variable. This information is static, therefore it is included in the background knowledge by means of a set of unary predicates dom$_{v,t}$ as a set of facts:

dom$_{v,t}$(e).

for each $v \in \mathcal{V}$, $t \in T$, and $e \in \mathrm{wr}(t)(v)$.

Constraints on variables. For each variable $v \in \mathcal{V}$:

– we impose functionality
    caused false **if** var$_v$(X), var$_v$(Y), X != Y.
– we force its value to propagate to the next state unless it is modified by an action (chng$_v$)
    caused var$_v$(X) **if not** $-$var$_v$(X), **not** chng$_v$
            after var$_v$(X).
– the defined fluent is the projection of the argument

caused def$_v$ **if** var$_v$(X).

Variable updates. The value of a variable is updated by means of causation rules that depend on the transition t that operates on the variable, and depends on the value of wr(t). For each v in the domain of wr(t):

- wr(t)(v) $= \emptyset$: delete (undefine) a variable v
   caused false **if** def$_v$ **after** t.
   caused chng$_v$ **after** t.
- wr(t)(v) $\subseteq$ dm(v): set v with a value nondeterministically chosen among a set of elements from its domain
   caused var$_v$(V) **if** dom$_{v,t}$(V), **not** $-$var$_v$(V) **after** t.
   caused $-$var$_v$(V) **if** dom$_{v,t}$(V), **not** var$_v$(V) **after** t.
   caused false **if not** def$_v$ **after** t.
   caused chng$_v$ **after** t.

If wr(t)(v) contains a single element d, then the assignment is deterministic and the first three rules above can be substituted with[4]
   caused var$_v$(d) **after** t.

Guards. To each subformula φ of transition guards is associated a fluent grd$_\varphi$ that is true when the corresponding formula is satisfied. To simplify the notation, for any transition t, we will use grd$_t$ to indicate the fluent grd$_{\text{gd}(t)}$. Executability of transitions is conditioned to the satisfiability of their guards; instead of modifying the executability rule including the grd$_t$ among the preconditions, we use a constraint rule preventing executions of the action whenever its guard is not satisfied:
   caused false **after** t, **not** grd$_t$.

Translation of atoms (ξ) is defined in terms of var$_v$ predicates. For instance $\xi(v = w)$ corresponds to var$_v$(V), var$_w$(W), V $==$ W. That is $\xi(v, T) = $ var$_t$(T) for $t \in \mathcal{V}$, and $\xi(d, T) = $ var$_t$T $== d$ for $d \in \bigcup_i \Delta_i$. For each subformula φ of transition guards a static rule is included to "define" the fluent grd$_\varphi$:

$true$: **caused** grd$_\varphi$ **if true** .
def(v) : **caused** grd$_\varphi$ **if** def$_v$.
$t_1 = t_2$: **caused** grd$_\varphi$ **if** $\xi(t_1, $T1$)$, $\xi(t_2, $T2$)$, T1 $==$ T2 .
$t_1 \leq t_2$: **caused** grd$_\varphi$ **if** $\xi(t_1, $T1$)$, $\xi(t_2, $T2$)$, ord(T1,T2) .
$\neg\varphi_1$: **caused** grd$_\varphi$ **if not** grd$_{\varphi_1}$.
$\varphi_1 \wedge \ldots \wedge \varphi_n$: **caused** grd$_\varphi$ **if** grd$_{\varphi_1}, \ldots,$ grd$_{\varphi_n}$.

5.3 Correctness and Completeness

We provide a sketch of the correctness and completeness of the encoding. Proofs can be found in [5].

---

[4] The deterministic version is a specific case of the non-deterministic ones and equivalent in the case that there is a single dom$_{v,t}$(d) fact.

Planning states include all the information to reconstruct the original DAW-net states. In fact, we can define a function $\Phi(\cdot)$ mapping consistent planning states into DAW-net states as following: $\Phi(s) = (M, \eta)$ with

$$\forall p \in P, \; M(p) = \begin{cases} 1 \text{ if } p \in s \\ 0 \text{ otherwise} \end{cases} \qquad \eta = \{(v, d) \mid \mathrm{var}_v(d) \in s\}$$

$\Phi(s)$ is well defined because s it cannot be the case that $\{\mathrm{var}_v(d), \mathrm{var}_v(d')\} \subseteq s$ with $d \neq d'$, otherwise the static rule

caused false **if** $\mathrm{var}_v(\mathsf{X})$, $\mathrm{var}_v(\mathsf{Y})$, $\mathsf{X} \mathrel{!=} \mathsf{Y}$.

would not be satisfied. Moreover, 1-safeness implies that we can restrict to markings with range in $\{0, 1\}$. By looking at the static rules we can observe that those defining the predicates def_v and grd_t are stratified. Therefore their truth assignment depends only on the extension of $\mathrm{var}_v(\cdot)$ predicates. This implies that grd_t fluents are satisfied iff the variables assignment satisfies the corresponding guard $\mathrm{gd}(t)$. Based on these observations, the correctness of the encoding is relatively straightforward since we need to show that a legal transition in the planning domain can be mapped to a valid firing. This is proved by inspecting the dynamic rules.

Lemma 3 (Correctness). *Let W be a DAW-net and $\Omega(W)$ the corresponding planning problem. If $\langle s, \{t\}, s' \rangle$ is a legal transition in $\Omega(W)$, then $\Phi(s) \xrightarrow{t} \Phi(s')$ is a valid firing of W.*

The proof of completeness is more complex because – given a valid firing – we need to build a new planning state and show that it is minimal w.r.t. the transition. Since the starting state s of $\langle s, \{t\}, s' \rangle$ does not require minimality we just need to show its existence, while s' must be carefully defined on the basis of the rules in the planning domain.

Lemma 4 (Completeness). *Let W be a DAW-net, $\Omega(W)$ the corresponding planning problem and $(M, \eta) \xrightarrow{t} (M', \eta')$ be a valid firing of W. Then for each consistent state s s.t. $\Phi(s) = M$ there is a consistent state s' s.t. $\Phi(s') = M'$ and $\langle s, \{t\}, s' \rangle$ is a legal transition in $\Omega(W)$.*

Lemmata 3 and 4 provide the basis for the inductive proof of the following theorem:

Theorem 2. *Let W be a safe WF-net and $\Omega(PN)$ the corresponding planning problem. Let (M_0, η_0) be the initial state of W – i.e. with a single token in the source and no assignments – and s_0 the planning state satisfying the initial condition.*

$(\Rightarrow)$ *For any case $\zeta : (M_0, \eta_0) \xrightarrow{t_1} (M_1, \eta_1) \ldots (M_{n-1}, \eta_{n-1}) \xrightarrow{t_n} (M_n, \eta_n)$ in W there is a trajectory $\eta : \langle s_0, \{t_1\}, s_1 \rangle, \ldots, \langle s_{n-1}, \{t_n\}, s_n \rangle$ in $\Omega(W)$ such that $(M_i, \eta_i) = \Phi(s_i)$ for each $i \in \{0 \ldots n\}$ and viceversa.*

$(\Leftarrow)$ *For each trajectory $\eta : \langle s_0, \{t_1\}, s_1 \rangle, \ldots, \langle s_{n-1}, \{t_n\}, s_n \rangle$ in $\Omega(W)$, the sequence of firings $\zeta : \Phi(s_0) \xrightarrow{t_1} \Phi(s_1) \ldots \Phi(s_{n-1}) \xrightarrow{t_n} \Phi(s_n)$ is a case of W.*

Theorem 2 above enables the exploitation of planning techniques to solve the reachability problem in DAW-net. Indeed, to verify whether the final marking is reachable it is sufficient to encode it as a condition for the final state and verify the existence of a trajectory terminating in a state where the condition is satisfied. Decidability of the planning problem is guaranteed by the fact that domains are effectively finite, as in Definition 9 the wr functions range over a finite subset of the domain and by the fact that the planner takes as input the maximum length of the plan to be returned (note that this allows for dealing with loops).

6   Related Work and Conclusions

The key role of data in the context of business processes has been recently recognized.

A number of variants of PNs have been enriched so as to make tokens able to carry data and transitions aware of the data, as in the case of Workflow nets enriched with data [10,15], the model adopted by the business process community. In [15] Workflow Net transitions are enriched with information about data (e.g., a variable *request*) and about how it is used by the activity (for reading or writing purposes). Nevertheless, these nets do not consider data values (e.g., in the example of Sect. 2.2 we would not be aware of the values of the variable *request* that *T4* is enabled to write). They only allow for the identification of whether the value of the data element is **defined** or **undefined**, thus limiting the reasoning capabilities that can be provided on top of them. For instance, in the example of Sect. 2.2, we would not be able to discriminate between the worker and the student loan for the trace in Sect. (2.2), as we would only be aware that *request* is **defined** after *T4*.

The problem of incomplete traces has been investigated in a number of works of trace alignment in the field of process mining, where it still represents one of the challenges. Several works have addressed the problem of aligning event logs and procedural models, without [2] and with [10,11] data. All these works, however, explore the search space of possible moves in order to find the best one aligning the log and the model. Differently from them, we assume that the model is correct and we focus on the repair of incomplete execution traces. Moreover, we exploit state-of-the-art planning techniques to reason on control and data flow rather than solving an optimisation problem.

We can overall divide the approaches facing the problem of reconstructing flows of model activities given a partial set of information in two groups: quantitative and qualitative. The former rely on the availability of a probabilistic model of execution and knowledge. For example, in [14], the authors exploit stochastic PNs and Bayesian Networks to recover missing information (activities and their durations). The latter stand on the idea of describing "possible outcomes" regardless of likelihood; hence, knowledge about the world will consist of equally likely "alternative worlds" given the available observations in time, as in this work. For example, in [3] the same issue of reconstructing missing information has been tackled by reformulating it in terms of a Satisfiability(SAT) problem rather than as a planning problem.

Planning techniques have already been used in the context of business processes, e.g., for the construction and adaptation of autonomous process models [13]. In [4] automated planning techniques have been applied for aligning execution traces and declarative models. As in this work, in [6], planning techniques have been used for addressing the problem of incomplete execution traces with respect to procedural models.

However, differently from the two approaches above, this work uses for the first time planning techniques to target the problem of completing incomplete execution traces with respect to a procedural model that also takes into account data and the value they can assume.

Despite this work mainly focuses on the problem of trace completion, the proposed automated planning approach can easily exploit reachability for model satisfiability and trace compliance and furthermore can be easily extended also for aligning data-aware procedural models and execution traces. Moreover, the presented encoding in the planning language $\mathcal{K}$, can be directly adapted to other action languages with an expressiveness comparable to $\mathcal{C}$ [12]. In the future, we would like to explore these extensions and implement the proposed approach and its variants in a prototype.

Acknowledgments. This research has been partially carried out within the Euregio IPN12 KAOS, which is funded by the "European Region Tyrol-South Tyrol-Trentino" (EGTC) under the first call for basic research projects.

References

1. Aalst, W.M.P.: Verification of workflow nets. In: Azéma, P., Balbo, G. (eds.) ICATPN 1997. LNCS, vol. 1248, pp. 407–426. Springer, Heidelberg (1997). https://doi.org/10.1007/3-540-63139-9_48

2. Adriansyah, A., van Dongen, B.F., van der Aalst, W.: Conformance checking using cost-based fitness analysis. In: Proceedings of the 2011 IEEE 15th International Enterprise Distributed Object Computing Conference (EDOC 2011), pp. 55–64. IEEE Computer Society (2011)

3. Bertoli, P., Di Francescomarino, C., Dragoni, M., Ghidini, C.: Reasoning-based techniques for dealing with incomplete business process execution traces. In: Baldoni, M., Baroglio, C., Boella, G., Micalizio, R. (eds.) AI*IA 2013. LNCS (LNAI), vol. 8249, pp. 469–480. Springer, Cham (2013). https://doi.org/10.1007/978-3-319-03524-6_40

4. De Giacomo, G., Maggi, F.M., Marrella, A., Sardiña, S.: Computing trace alignment against declarative process models through planning. In: Proceedings of the 26th International Conference on Automated Planning and Scheduling, pp. 367–375. AAAI Press (2016)

5. De Masellis, R., Di Francescomarino, C., Ghidini, C., Tessaris, S.: Enhancing workflow-nets with data for trace completion (2017). https://arxiv.org/abs/1706.00356

6. Di Francescomarino, C., Ghidini, C., Tessaris, S., Sandoval, I.V.: Completing workflow traces using action languages. In: Zdravkovic, J., Kirikova, M., Johannesson, P. (eds.) CAiSE 2015. LNCS, vol. 9097, pp. 314–330. Springer, Cham (2015). https://doi.org/10.1007/978-3-319-19069-3_20
7. Eiter, T., Faber, W., Leone, N., Pfeifer, G., Polleres, A.: A logic programming approach to knowledge-state planning, II: the DLVK system. Art. Intell. **144**(1–2), 157–211 (2003)
8. van Hee, K., Sidorova, N., Voorhoeve, M.: Soundness and separability of workflow nets in the stepwise refinement approach. In: van der Aalst, W.M.P., Best, E. (eds.) ICATPN 2003. LNCS, vol. 2679, pp. 337–356. Springer, Heidelberg (2003). https://doi.org/10.1007/3-540-44919-1_22
9. Kiepuszewski, B., ter Hofstede, A.H.M., Bussler, C.J.: On structured workflow modelling. In: Bubenko, J., Krogstie, J., Pastor, O., Pernici, B., Rolland, C., Sølvberg, A. (eds.) Seminal Contributions to Information Systems Engineering. Springer, Heidelberg (2013)
10. de Leoni, M., van der Aalst, W.: Data-aware process mining: discovering decisions in processes using alignments. In: Proceedings of the 28th ACM Symposium on Applied Computing (SAC 2013), pp. 1454–1461. ACM (2013)
11. de Leoni, M., van der Aalst, W.M.P., van Dongen, B.F.: Data- and resource-aware conformance checking of business processes. In: Abramowicz, W., Kriksciuniene, D., Sakalauskas, V. (eds.) BIS 2012. LNBIP, vol. 117, pp. 48–59. Springer, Heidelberg (2012). https://doi.org/10.1007/978-3-642-30359-3_5
12. Lifschitz, V.: Action languages, answer sets and planning. In: Apt, K.R., Marek, V.W., Truszczynski, M., Warren, D.S. (eds.) The Logic Programming Paradigm: A 25-Year Perspective, pp. 357–373. Springer, Heidelberg (1999)
13. Marrella, A., Russo, A., Mecella, M.: Planlets: automatically recovering dynamic processes in YAWL. In: Meersman, R., Panetto, H., Dillon, T., Rinderle-Ma, S., Dadam, P., Zhou, X., Pearson, S., Ferscha, A., Bergamaschi, S., Cruz, I.F. (eds.) OTM 2012. LNCS, vol. 7565, pp. 268–286. Springer, Heidelberg (2012). https://doi.org/10.1007/978-3-642-33606-5_17
14. Rogge-Solti, A., Mans, R.S., van der Aalst, W.M.P., Weske, M.: Improving documentation by repairing event logs. In: Grabis, J., Kirikova, M., Zdravkovic, J., Stirna, J. (eds.) PoEM 2013. LNBIP, vol. 165, pp. 129–144. Springer, Heidelberg (2013). https://doi.org/10.1007/978-3-642-41641-5_10
15. Sidorova, N., Stahl, C., Trčka, N.: Soundness verification for conceptual workflow nets with data. Inf. Syst. **36**(7), 1026–1043 (2011)
16. van der Aalst, W., van Hee, K., ter Hofstede, A., Sidorova, N., Verbeek, H., Voorhoeve, M., Wynn, M.: Soundness of workflow nets: classification, decidability, and analysis. Formal Aspects of Comput. **23**(3), 333–363 (2010)

Optimal Paths in Business Processes: Framework and Applications

Marco Comuzzi[✉]

Ulsan National Institute of Science and Technology (UNIST), 50 UNIST-gil,
Ulju-gun, Ulsan 44919, Republic of Korea
mcomuzzi@unist.ac.kr

Abstract. We present an innovative framework for calculating optimal
execution paths in business processes using the abstraction of workflow hypergraphs. We assume that information about the utility associated with the execution of activities in a process is available. Using
the workflow hypergraph abstraction, finding a utility maximising path
in a process becomes a generalised shortest hyperpath problem, which
is NP-hard. We propose a solution that uses ant-colony optimisation
customised to the case of hypergraph traversal. We discuss three possible applications of the proposed framework: process navigation, process
simulation, and process analysis. We also present a brief performance
evaluation of our solution and an example application.

Keywords: Process simulation · Process navigation
Process analysis · Optimal path · Workflow hypergraph

1 Introduction

Business processes are at the heart of modern organisations and are increasingly supported by so-called Process-Aware Information Systems (PAIS), that
is, software systems supporting different phases of the typical business process
management lifecycle, such as process modelling, enactment, and monitoring [7].
PAIS enable collecting large amount of data about business process executions.
These data are normally available in the form of process *event logs*. Event logs
enable fine-grained observation and analysis of business processes to improve
processes and guide future execution.

Typical ways to exploit data from previous executions, possibly combined
with *a priori* knowledge about processes, are recommendation and predictive
monitoring. Approaches to predictive monitoring and recommendation in the
literature tend to either rely solely on past execution [19], e.g., suggesting activities executed more often in past cases under similar circumstances, consider
alternative non-procedural process modelling techniques, such as declarative processes [2], or rely on simple definitions of utility involving only time and cost
performance [5,18].

© Springer International Publishing AG 2018
E. Teniente and M. Weidlich (Eds.): BPM 2017 Workshops, LNBIP 308, pp. 107–123, 2018.
https://doi.org/10.1007/978-3-319-74030-0_7

Utility in a procedural workflow, however, is a multi-attribute concept which synthesises in a single scalar value the importance for the user of several dimensions, such as cost, quality, availability, or reliability of activities in a workflows. Utility functions can be defined by different users or user classes based on different preferences, contexts of executions, or other runtime features. This approach is adopted, for instance, by workflow optimisation in service-oriented and cloud computing, where a priori knowledge and data about previous service invocations and deployments are exploited to calculate the utility of candidate services to fulfill specific functionality in a workflow. Utility information is then used to plan in real time the best possible workflow to satisfy a user request [11, 15].

In this paper, we consider procedural process models and we assume that information about the utility associated with the execution of individual activities in a process is available. We then develop a framework to determine optimal, i.e., utility maximising, paths to complete the execution of a business process from a start to an end state in the process. Extending the framework to calculate optimal paths from any given state of execution of a process to a subsequent intermediate state is straightforward.

For achieving the objective delineated above, this paper introduces the abstraction of workflow hypergraphs. Workflow hypergraphs are a different way of representing process models that is particularly useful to tackle combinatorial problems involving utility and paths in a process. Note that the workflow hypergraphs introduced in this paper are a full-fledged procedural process modeling technique equivalent to WF-nets. In this regard, they differ substantially from the process hypergraphs introduced in [16] in the context of flexible process graphs.

Using workflow hypergraphs, utility can be seen as a weight associated to nodes, i.e., activities, in a hypergraph, and the problem of determining optimal paths becomes the problem of determining shortest paths in a hypergraph. While shortest paths can be calculated in polynomial time in the case of graphs, determining shortest paths in hypergraphs is a NP-hard problem [8]. We propose a solution based on ant-colony optimisation, i.e., a swarm intelligence meta-heuristic particularly suitable for optimisation on graph structures [6].

To show the flexibility of our framework, we discuss three applications of it in three different phases of the business process management lifecycle: process navigation, process simulation, and process analysis. Optimal paths naturally can be used to support *process navigation*, that is, suggesting to process users the next steps to complete optimally a process. As far as process simulation is concerned, the ant-colony optimisation solution is extended to support simulated *what if* scenarios, e.g., determining optimal paths for cases with specific attributes or for different properties of decision points in a process. Finally, determining optimal paths enables ex-post process analysis in an innovative way, e.g., determining whether cases have followed the optimal path or assessing which decision points in a process have deviated from the optimal path.

The paper is organised as follows. Related work is reviewed in Sect. 2. Section 3 introduces workflow hypergraphs. Section 4 presents the ant-colony optimisation solution to determine optimal paths. Application scenarios are discussed in depth in Sect. 5, whereas an evaluation with real event logs is provided in Sect. 6. Conclusions are finally drawn in Sect. 7.

2   Related Work

The literature about service composition in both service-oriented and cloud computing has flourished in the last 10 years [11,15]. The idea of utility associated with executions of activities in a process is directly derived from this literature. In service composition, in fact, utility is associated with the individual services selected to fulfill a given workflow required by a user. A workflow in service composition is fixed by a user's request and optimisation focuses on selecting an optimal set of services to execute a workflow. On the contrary, in our approach, activities in a process are fixed, but there are possible multiple ways, i.e., paths, to bring a process to an end state.

In the context of process recommendation, in the early work of Schonenberg et al. [19] a technique is presented to recommend activities that minimise the current instance completion time based on the similarity of the current case with previous cases. In the work of Haisjackl et al. [10], recommendation is also based on the most similar previously executed process instance. Barba et al. [2] describe a recommendation system that can optimally select and schedule remaining activities in a declarative process based on constraint-based optimisation. Vanderfeesten et al. [24] discuss a framework based on product-based workflow modelling that suggests the steps to optimally produce a workflow product. The problem is reduced to a Markov decision process and optimisation considers a combination of cost and time required to produce intermediate artefacts necessary for completing a workflow product. In the context of case-based systems, Lakshmanan et al. [13] propose a recommendation system suggesting a sequence of steps to complete cases based on previous case history, minimising the likelihood of occurrence of undesired outcomes in the current case. Ghattas et al. [9] have proposed a semi-automated process recommendation system that exploits experience from previous executions and adopts the notion of process goals.

The broader research area of process predictive monitoring is also tightly linked to the issue of process operational support. One typical issue in this context is time prediction, i.e., predicting when a given case will complete. van der Aalst et al. [22] propose a technique to predict completion time of activities using transition systems annotated with time-related information extracted from previous process executions. Similarly, Rogge-Solti and Weske [18] propose a time prediction framework which uses stochastic Petri nets. Prediction may also focus on process constraint violations or process faults. A comprehensive framework for prediction of different types of constraints in declarative process models is provided in [5]. In the context of procedural process models, Conforti et al. [4] propose a framework to reduce the likelihood of process faults, which also prompts users the risk level of possible choices to be made during process execution.

As far as hypergraph-based modelling of business processes is concerned, Polyvyanyy and Weske [16] have proposed to use (undirected) hypergraphs to model flexible processes. Multiple nodes connected by one edge in a hypergraph, in this case, signify multiple options that users can choose from to complete the execution of a process. Product-based workflows [24] are also a special class of hypergraphs, referred to as B-hypergraphs or AND/OR graphs. Comuzzi et al. [3] have proposed an ant-colony based optimisation algorithms for paths in product-based workflows. This algorithm, however, can only deal with B-Hypergraphs. In this paper, we use directed hypergraphs to model procedural process models and propose an ant-colony based optimisation algorithm that serves the more general case of BF-Hypergraphs.

3   Preliminaries

We consider business processes modelled as workflow nets (WF-net), which are a restricted class of Petri nets (see Fig. 1(a) for a sample WF-net).

Definition 1 (Petri net [23]). *A PN is a triple (P, T, F) such that:*

- *P is a finite set of places;*
- *T is a finite set of transitions, with $P \cap T = \emptyset$*
- *$F \subseteq (P \times T) \cup (T \times P)$ is a set of arcs (flow relation)*

A place p in a Petri net is an *input place* (*output place*) of a transition t if and only if there exists a directed arc from p to t (from t to p). We use $\bullet p$ ($p \bullet$) to denote the set of transitions sharing p as input (output).

Definition 2 (Strongly connected PN [23]). *A PN is strongly connected iff, for every pair of nodes (i.e., places or transitions) x and y, there exists a directed path leading from x to y.*

The notion of strong connectivity of a PN can be used to define WF-nets.

Definition 3 (WF-net [23]). *A Petri net $PN = (P, T, F)$ is a workflow net if and only if:*

1. *PN has a* source *place i such that $\bullet i = \emptyset$ and a* sink *place o such that $o \bullet = \emptyset$*
2. *If a dummy transition $\hat{t}$ such that $\bullet \hat{t} = o$ and $\hat{t} \bullet = i$ is added to PN, then the resulting Petri net is strongly connected.*

WF-nets can be used to represent process models using transitions as *activities* (or *tasks*) and places as *conditions* along the execution of a process. Note that, in Definition 3, condition 1 imposes that a WF-net has one source (or start) and one sink (or end) activity, whereas condition 2 avoids dangling tasks or conditions.

Directed hypergraphs are a generalisation of directed graphs allowing each directed hyperedge to have multiple source (tail) vertices and multiple destination (head) vertices.

Definition 4 (Directed hypergraph [8]**).** *A directed hypergraph $\mathcal{H}$ is a pair $\langle V, E \rangle$ such that:*

- *V is a finite set of nodes (or vertices);*
- *$E \subseteq 2^V \times 2^V$ is a finite set of hyperedges*
- *For every $e = (T(e), H(e)) \in E$, $T(e) \neq \emptyset$, $H(e) \neq \emptyset$, and $T(e) \cap H(e) = \emptyset$.*

For a hyperedge $e \in E$, we denote $T(e)$ as the *tail* of e and $H(e)$ as the *head* of e.

For a node $v \in V$, we define its *forward star* (*backward star*) as the set of edges in $\mathcal{H}$ having v in their tail (head), that is $FS_{\mathcal{H}}(v) = \{e \in E : v \in T(e)\}$ $(BS_{\mathcal{H}}(v) = \{e \in E : v \in H(e)\})$. For example, in the hypergraph of Fig. 1(b), $FS(\text{C}) = \{(\{C\}, \{E, D\})\}$ and $BS(\text{H}) = \{(\{G\}, \{H\}), (\{F\}, \{H\})\}$.

For a set of nodes $\bar{V} \subseteq V$, we define their *predecessors* (*successors*) as the set of edges in $\mathcal{H}$ having $\bar{V}$ as their tail (head), that is $prec(\bar{V}) = \{e \in E : \bar{V} = H(e)\}$ $(succ(\bar{V}) = \{e \in E : \bar{V} = T(e)\})$. For example, in Fig. 1(b), $prec(\{E, D\}) = \{(\{C\}, \{E, D\})\}$.

Given a hypergraph $\mathcal{H}$, we define as its *symmetric image* the hypergraph $\bar{\mathcal{H}}$, where $\bar{\mathcal{H}} = \{\langle E, V \rangle : \langle V, E \rangle = \mathcal{H}\}$.

We define workflow hypergraphs (WF-hypergraph) as a restricted class of hypergraphs that can be used to model business processes. As an example, Fig. 1(b) shows the WF-Hypergraph equivalent to the WF-net of Fig. 1(a).

Definition 5 (WF-Hypergraph). *A directed hypergraph $\mathcal{H} = \langle V, E \rangle$ is a workflow hypergraph if and only if:*

1. *$\mathcal{H}$ has a source node i and a source node o such that $BS_{\mathcal{H}}(i) = \emptyset$ and $FS_{\mathcal{H}}(o) = \emptyset$;*
2. *For all nodes v in $\mathcal{H}$, $v \neq i, o$, $BS_{\mathcal{H}}(v) \neq \emptyset$ and $FS_{\mathcal{H}}(v) \neq \emptyset$.*

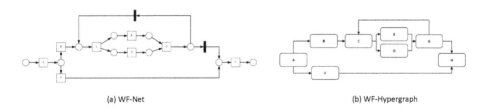

(a) WF-Net   (b) WF-Hypergraph

Fig. 1. Translating a WF-net process model into a WF-hypergraph

The translation from WF-nets to WF-hypergraphs is straightforward and its formal characterisation is omitted in this paper for the sake of conciseness (see Fig. 1 for an example). Note that parallelism in WF-hypergraphs is modelled by edges with more than one node in their heads/tails, whereas choice is modelled by nodes with multiple edges in their forward and backward stars.

Different notions of connectivity in hypergraphs can be defined, which lead to different definitions of hyperpaths from a node s to a node t in a hypergraph $\mathcal{H}$.

Definition 6 (Simple path in hypergraph [20]). *A simple path $P_{s,t}$ from s to t in $\mathcal{H}$ is a sequence $(v_1, e_1, \ldots, v_k, e_k, v_{k+1})$ of nodes v_i and distinct hyperedges e_j such that $s = v_1$, $t = v_{k+1}$, and, for every $1 \leq i \leq k$, $v_i \in T(e_i)$ and $v_{i+1} \in H(e_i)$. If, in addition, $t \in T(e_1)$, then $P_{s,t}$ is a simple cycle.*

Definition 7 (B-Hyperpath [20]). *A B-hyperpath $B_{s,t}$ from s to t in $\mathcal{H}$ is defined using the notion of B-connection to a vertex s, which is defined recursively as follows: (i) a vertex s is B-connected to itself and (ii) if there is a hyperedge e such that all nodes in $T(e)$ are B-connected to s, then every node in $H(e)$ is B-connected to s. A B-hyperpath $B_{s,t}$ is a minimal (with respect to deletion of nodes and hyperedges) subhypergrah of $\mathcal{H}$ such that t is B-connected to s.*

Definition 8 (F-Hyperpath [20]). *A F-hyperpath $F_{s,t}$ from s to t in $\mathcal{H}$ is a subhypergarph of $\mathcal{H}$ such that $\bar{B}_{s,t}$ is a B-hyperpath from t to s in $\bar{\mathcal{H}}$.*

Definition 9 (BF-Hyperpath [20]). *A BF-hyperpath $BF_{s,t}$ from s to t in $\mathcal{H}$ is a subhypergraph that is both a B-hyperpath and an F-hyperpath from s to t.*

Without loss of generality, in the remainder of this paper we consider block-structured business process models [17]. In the WF-Hypergraph representation of a block-structured process model, any path connecting the source node i to the sink node o is a BF-Hyperpath. For instance, the WF-Hypergraph of Fig. 1(b) contains two BF-paths between nodes A and H, i.e., $p_1 = \{A, F, H\}$ and $p_2 = \{A, B, C, D, E, G, H\}$. Hence, in the remainder of this paper we consider only BF-Hyperpaths and we refer to them simply as *paths*. When cycles are present, the definition of path provided is ambiguous, since any path involving an arbitrary number of walks through the cycle is also a path. However, as we clarify later, since cycles cannot be considered in optimal paths, we consider only paths that do not involve cycles. Note that, using a typical process mining terminology, there could be different execution traces in which an optimal path can be executed. In this regard, the traces (A, B, C, D, E, G, H) and (A, B, D, C, E, G, H) represent possible execution traces of the path p_2 of Fig. 1(b), since nodes C and D can be executed in any order.

WF-hypergraphs are syntactically simpler than WF-nets, since they involve only two types of elements, i.e., nodes and hyperedges, compared to places, arcs, and transitions appearing in WF-nets. Given a business process, its WF-Hypergraph representation is also more compact than the corresponding WF-net one, i.e., it involves a lower number of graphical elements. Most importantly, WF-hypergraphs are to be preferred in those applications for which notable results from the extensive literature on hypergraphs can be reused. WF-Hypergraphs are not intended to substitute other commonly used business process model notations in applications where process models have to be graphically laid out. They should rather be adopted as intermediate representations only in specific use cases, such as the one considered in this paper.

4   Characterising and Calculating Optimal Paths

We define optimality of a path based on a notion of utility associated with the execution of activities in a path. The utility of a path $P_{i,o}^{\mathcal{H}}$ between a source node i and a sink node o in a WF-Hypergraph $\mathcal{H}$ is, in general, a function of the utility associated to the set of R nodes $V_P = \{v_1, \ldots, v_R\}$ belonging to $P_{i,o}^{\mathcal{H}}$, that is $U\left(P_{i,o}^{\mathcal{H}}\right) = f\left(V_P\right)$.

We propose in this paper an approach that works with any utility function $U\left(P_{i,o}^{\mathcal{H}}\right)$. In our evaluation of Sect. 6, we consider 4 utility dimensions, i.e., *cost*, *time*, *availability*, and *quality*, which capture particular types of process performance. Note that, in order to be combined in a global utility function, the partial utility functions of individual dimensions should be comparable, e.g., normalised within the same range.

Cost. It captures the cost of executing activities along a path. Costs are summative and can be normalised in respect of the cost associated with the most expensive activity in a process. Cost has also a negative relation with utility, that is, the higher the cost incurred, the lower the utility of a path. Based on these assumptions, given the cost $cost(v_r)$ of a node v_r, the utility of a path in respect of the cost dimension is $f^{cost}\left(P_{i,o}^{\mathcal{H}}\right) = \sum_{r=1}^{R}\left(1 - \frac{cost(v_r)}{\max_{v \in \mathcal{H}} cost(v)}\right)$.

Time. It captures the execution time of activities in a path. Similarly to cost, time has a negative relation with utility. Execution time of a path can be calculated from the execution time $time(v_r)$ of individual activities v_r using standard aggregation patterns as long as the probabilities of choosing particular paths in XOR splits is known a priori [12]. These standard aggregation patterns are captured by the function $cycle\_time$ in the definition of the partial utility function for the time dimension, that is $f^{time}\left(P_{i,o}^{\mathcal{H}}\right) = 1 - \frac{cycle\_time(v_1, \ldots, v_R)}{\max_{P \subset \mathcal{H}} cycle\_time(P)}$.

Note that, in this formula, path execution time is normalised using the worst case scenario, i.e., the execution time of the worst performing path in a process. Note that this requires the enumeration of all possible paths to reach an end state, which may be inefficient. Alternatively, a reference worst case scenario of process execution in the past can be considered for normalising path execution time.

Availability. Availability models the probability that the resources, e.g., human resources and/or raw material, required for execution of an activity are available when an activity is enabled. Availability is aggregated on a path using the product of availability associated to the activities in a path, that is $f^{avail}\left(P_{i,o}^{\mathcal{H}}\right) = \prod_{r=1}^{R} avail(v_r)$.

Quality. Quality models the quality of the output of an activity, which varies depending on the type of activity involved. In a service activity involving an end customer, quality has to be intended as the quality of service perceived by customers. If the output of an activity is an artefact, e.g., a document, then quality is defined as the quality of the artefact. Finally, in a manufacturing activity, quality refers to standard dimensions such as defect rate. Quality in a path is aggregated using the min function, that is, the quality of the output of a list of activities cannot be higher than the minimum quality of one of these outputs [1], that is $f^{qual}\left(P_{i,o}^{\mathcal{H}}\right) = \min_{r\in[1,R]} qual(v_r)$.

Note that utility values of activities in a process may be known a priori, e.g., a manufacturing company may have detailed knowledge of the costs of each step in a complex production process, or may be reconstructed from available data, e.g., from start and end timestamps of activities. Note also that by construction $0 \leq f^x\left(P_{i,o}^{\mathcal{H}}\right) \leq 1$, for $x = \{cost, time, avail, qual\}$. This allows to create global utility functions combining the partial utility functions defined above using arbitrary patterns. In our evaluation, we consider a global utility function built using a simple weighted sum.

After having introduced a notion of utility of a path, our problem becomes the one of searching for a path between a source and a sink node in a WF-Hypergraph with maximum utility. This is equivalent to a generalised shortest path problem, that is, calculating a shortest path using an arbitrary distance function between nodes, simply considering the inverse of utility as distance function. More specifically, in the literature about shortest path in hypergraphs (and graphs), utility, i.e., weights, is usually an attribute of edges, rather than nodes. Our problem can be reduced to a traditional hypergraph shortest path problem with weighted hyperedges by (i) assigning the utility of a node to all the hyperedges in that node's backward star and (ii) creating a dummy starting node connected to the original source node and assigning to the connecting hyperedge the utility of the source node.

Calculating shortest hypergraph (B-, F-, or BF-)paths is a NP-hard problem in the case of generic utility functions [8]. In this paper we provide a heuristic solution based on the Ant-Colony Optimisation (ACO) [6] meta-heuristic specifically tailored to the case of BF-hyperpath hypergraph traversal. ACO mimics the behaviour of ants while search for food. Initially, ants walk randomly until they find a source of food. Then, they will return to the nest leaving a *pheromone* trail for fellow ants to follow. The strength of this trail is inversely proportional to the length of the path. Over time, the shorter paths are the ones with a stronger trail that will be marched over more frequently. The ants will eventually converge marching on the shortest path from the nest to the source of food. In our scenario, edges in a WF-Hypergraph are initialised with the same level of pheromone. Ants wander randomly and at a XOR-split node choose the edge to follow using a uniform probability distribution proportional to the level of pheromone deposited on each hyperedge. The utility of a path represents the pheromone that ants deposit on the path that brought them to the an end state of the process. To avoid committing to initial sub-optimal solutions, pheromone levels are not updated by individual ants, but after a *colony* of ants have completed their walks.

Algorithm 1 shows the main ant colony procedure. Each colony comprises a number of ants and each ant walks the WF-Hypergraph $\mathcal{H}$ in search of a path

```
1  $\mathcal{BF}_{opt}, \mathcal{BF} \leftarrow \emptyset$ ;
2  foreach colony do
3  |    $\mathcal{K} \leftarrow \mathcal{H}$ ;
4  |    foreach ant do
5  |    |    $\mathcal{BF} \leftarrow$ aco_search($\mathcal{H}$) ;
6  |    |    if $U(\mathcal{BF}) \geq U(\mathcal{BF}_{opt})$ then $\mathcal{BF}_{opt} \leftarrow \mathcal{BF}$; ant_phero_update($\mathcal{K},\mathcal{BF}$);
7  |    end
8  |    colony_phero_update($\mathcal{H},\mathcal{K}$) ;
9  end
10 return $BF_{opt}$
```

Algorithm 1. Ant colony main procedure

from the source node i to the sink node o. The current optimal path is stored by the variable $\mathcal{BF}_{opt}$. If the utility of a discovered path is higher than the one of the current optimal path walked by an ant, then the optimal path is updated (line 6). Each ant in a colony updates the pheromone level on a local copy $\mathcal{K}$ of the WF-Hypergraph $\mathcal{H}$ (line 6), while, as mentioned above, the pheromone in $\mathcal{H}$ is updated once per colony (line 8).

The function ψ controlling the pheromone update from an iteration, i.e., an ant, it to the next iteration $it + 1$ in the procedure ant_phero_update() is shown in Eq. 1. At each iteration, the pheromone level on an edge is updated using the utility value of the optimal path discovered $\mathcal{BF}$ and through the evaporation of the existing pheromone level. Note that pheromone on an edge e *evaporates* of a fraction τ, with $0 \leq \tau \leq 1$, at each iteration it. Note also that all edges in $\mathcal{H}$ must be initialised to the same level of pheromone at the start of the procedure. The procedure colony_phero_update() (line 8) simply copies the pheromone levels of the copy $\mathcal{K}$ into the WF-hypergraph $\mathcal{H}$.

$$\psi\left(e\right)^{it+1} = \left(1 - \tau\right)\psi\left(e\right)^{it} + U\left(\mathcal{BF}\right), \tag{1}$$

The procedure aco_search() represents the *walk* of an individual ant from the source node i to the sink node o and it is built on the following logic:

1. When visiting a node v, the ant chooses the next edge to follow according to a uniform probability distribution proportional to the level of pheromone currently on each available edge, that is, each edge in the successors $succ(v)$ of node v;
2. If $|H(e)| = 1$ then the ant simply moves to the node in $H(e)$;
3. If $|H(e)| > 1$, then the ant is *spawned*, that is, one ant is created to visit each node in $H(e)$. Note that this occurs in the case of parallel blocks in the process.
4. When an ant reaches a node v such that $prec(v) > 1$, then the ant will wait to synchronise with other ants visiting the nodes in $T(v)$. That is, node v is

visited by a single ant only after all the nodes in $T(v)$ have been visited by fellow ants;

5. If an ant reaches the sink node o, and there are no other ants waiting to be synchronised still walking, then the procedure terminates.

Shortest paths in a hypergraph (or graph) are computable only if cycles, when present, have non-decreasing length (i.e., non-increasing utility in the general case). This assumption fits well the scenario of WF-hypergraph optimisation, since any cycle in a process model is usually introduced to model rework, which clearly decreases utility. Ants in our framework are equipped with the ability to escape cycles. When traversing a graph, an ant detects to have traversed a cycle each time a node is visited a second time during a walk. In traversing hypergraphs, however, recognising cycles is more challenging, since nodes may be visited multiple times because of the ant *spawning* behaviour in the case of parallel blocks in a process. A cycle is detected only when an ant visits a node a second time and there are no further spawned ants waiting to complete their traversal on a parallel block. In this case, an ant retraces its steps back by navigating the backward star of nodes. While retracing its step, in each node an ant checks whether an edge leading to node(s) not visited yet in the current traversal can be found. If such an edge is found, then an ant visits its head, thus *escaping* from the cycle.

5   Applications and Framework Functionality

On top of the main path optimisation functionality introduced in the previous sections, a set of additional functionalities of our framework is determined by three application scenarios, i.e., process navigation, process simulation, process analysis.

Process Navigation. Process navigation was introduced in the BPM literature in 2009 [21], to transpose the metaphor of GPS-supported vehicle navigation into the case of business process execution. The idea in process navigation is to build systems providing process users with suggestions about what steps to take to terminate the process optimally. While vehicle navigation relies on the abstraction of destinations as nodes in a graph, for reasons that we discussed already in Sect. 3, business processes cannot be modelled as simple graphs, but rather, hypergraphs. The framework proposed in this paper naturally fits the scenario of process navigation. The path suggested by a process navigation application, in fact, should be the one maximising some sort of utility for the process user, such as minimising time, cost, or a combination thereof. Our framework can be used to recalculate the optimal path after the execution of each activity in a process, so as to suggest to process users the optimal way to terminate a process.

Process Simulation. Besides optimisation, ant-colony is also a flexible framework that enables simulation of complex systems [6]. In the specific case of optimal paths in a business process, our ant-colony optimisation procedure has been extended to enable simulation of *what-if* scenarios. In this case, instead of being guided entirely by pheromone, the behaviour of ants is partially controlled by additional simulation features. Two types of features are considered:

1. *Case-based attributes*: a decision point in a WF-Hypergraph is modelled as a node v (activity) for which $|FS(v)| > 1$, e.g., node A in Fig. 1(b). The path chosen at specific decision points in a process is often determined by case-based attributes, i.e., features characterising a case. This can range from customer type to other characteristics, such as the time of the day in which a case started, e.g., the type of customer or a case starting time determine the outcome chosen at specific decision points during process execution; This feature is implemented in our framework by considering *smart* ants, which are characterised by a set of case-based attributes and that know, for each activity in a case, when to follow the pheromone trail and when to base their choice on other attributes;

2. *Decision point-specific attributes*: the outcome at a decision point can be controlled, deterministically or stochastically, by the value of one or more variables internal to the process or external. While a deterministic outcome can be embedded directly in the process model as a condition on the outgoing branches of a xor-split gateway, stochastic relations between process variables and process execution can be captured by a simulation framework, e.g., an outcome at a decision point is chosen with 90% probability if the exchange rate between KRW and EUR (that is, a variable external to the process) is above a certain threshold. This feature is implemented by extending our framework with additional global simulation variables, which can be updated by the process execution or by invoking external services at specific points. Ants consider these variables as additional criteria guiding their choice during their path exploration.

Process Analysis. This application can be seen as a special case of conformance analysis whereby conformance is checked against a particular ideal path in a process, i.e., the optimal one discovered by our framework, rather than an entire process model. In particular, the framework implements the following two use cases:

1. *Evaluating whether a case has followed the optimal path.* This can be achieved by enumerating the possible traces that conform to the optimal path and determining whether a case adheres to one of these traces;
2. *Evaluating decisions taken in a case.* For cases that did not follow the optimal path, it is possible to analyse which decisions have deviated from the optimal path. Decision points in the optimal path can be enumerated. Then, each decision point in a case is analysed to determine whether the decision taken conforms to the optimal path or not.

The first use case allows separating *optimal* and *non-optimal* cases. This can be used as a criterion to divide-and-conquer an event log and to facilitate process analysis by isolating common features of cases in the same class, e.g., comparing resources involved in optimal and non-optimal cases. The first use case is also a prerequisite step enabling the second use case, in which cases can be analysed to understand why they have deviated from the optimal path.

6   Evaluation and Example

As an evaluation of our framework, we first present a brief performance analysis of the ant-colony optimisation algorithm. Then, we discuss an example application.

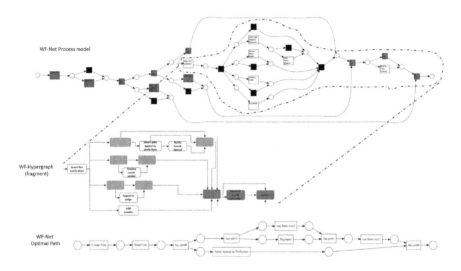

Fig. 2. WF-net with highlighted optimal path and corresponding WF-Hypergraph (event log *Road Fine Management*) (Color figure online)

As far as performance evaluation is concerned, the framework[1] has been scripted in Python 3 and uses the `halp`[2] library for hypergraph implementation; experiments run on a Intel i5 machine equipped with 8GB memory. Experiments run on WF-net process models mined using the inductive miner [14] available in ProM 6.5 from widely used event logs in the process mining community[3]. Utility information has been randomly generated, since it is not available in the event logs. Figure 2 shows an example of the output of our framework: a WF-net process model in which the transitions corresponding to activities in the discovered optimal path are highlighted in red, a WF-net model of the discovered optimal path only, and the corresponding WF-hypergraph with highlighted optimal path.

---

[1] Code available at: https://github.com/emettelatripla/opsupport.

[2] http://murali-group.github.io/halp/.

[3] Available at: https://data.4tu.nl/repository/collection:event_logs.

Table 1. Performance evaluation

Event log description			Simulation parameters		Results	
Id	Act	DP	Ants	Colonies	Utility (%)	Runtime (s)
Purchasing	28	3	10	2	100	0.11
			10	2	100	0.20
Road Fine Management	26	8	10	2	97.2	0.27
			10	4	98.1	0.54
			10	6	100	0.78
			10	8	100	1.01
Loan application	43	7	10	2	91.2	0.31
			10	6	98.3	0.87
			10	12	100	1.87
			10	13	100	2.00
Hospital	351	18	10	3	90.5	1.44
			10	5	93.7	1.65
			15	7	97.1	2.26
			15	9	100	4.16

Table 1 shows the results of our performance evaluation. It focuses on (i) the utility of the path discovered by the optimisation algorithm, expressed as a fraction (percentage) of the utility of the optimal path calculated using a brute force approach (*Uti*), and (ii) the runtime in seconds (*Runtime*). Because of the random initialisation of utility, performance measures are the average values obtained from 10 replications. The complexity of a process mined from an event log is expressed by the number of activities (Act) and the number of decision point (*DP*) in the process model. The remaining columns *Col* and *Ants* show the number of colonies and the number of ants per colony adopted in an experiment, respectively. As far as utility is concerned, above a certain number of ant colonies the algorithm converges to the optimal solution, i.e., with 100% utility. This threshold increases with the complexity of the process model, that is, a larger population of ants is required to optimise more complex process models. The runtime remains limited and above two seconds only for the more complex *spaghetti* hospital process model.

In discussing an example application, our aim is to demonstrate the potential of the framework proposed in this paper to provide additional insights into processes in different phases of the process lifecycle. For brevity, in this paper we concentrate on the third application, i.e., process analysis, leaving a more thorough evaluation of the other two applications to other publications. We consider the event log *Purchasing* of a purchasing process built from ERP data. Information about the utility associated with the execution of individual activities has been generated randomly, since it is not contained in the original event log. Figure 3 shows the process model discovered from the event log using the

Table 2. Analysis of the *Purchasing* event logs

N. traces in event log	609
N. activities in optimal path	17
N. possible optimal traces	72
Optimal traces in event logs	88 (14.4%)
Deviations from optimal path	234 cases at *Create Purchase Requisition*
	238 cases at *Analyze Request for Quotation*

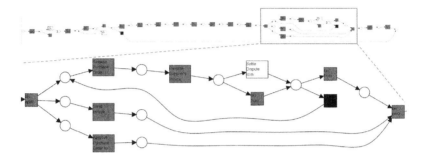

Fig. 3. Process fragment with highlighted optimal path (Color figure online)

inductive miner in Prom 6.5, in which activities belonging to the optimal path calculated using our framework have been highlighted in red. Table 2 reports some descriptive statistics related to the analysis of this event log.

The *Purchasing* event log contains 609 traces. The discovered optimal path contains 17 activities (note that some highlighted transitions in Fig. 3 are introduced by the inductive mining process, but do not represent process activities) and yields 72 possible traces, i.e., 72 possible ways in which a process can be executed from a start to an end state achieving maximum utility. The variability of optimal traces is determined by the way in which the parallel block zoomed in Fig. 3 is executed. Before the transition *tau split0* and after the transition *tau join0*, in fact, the process involves only activities in sequence. The optimal traces are obtained by enumerating all the possible ways in which activities in the parallel block can be executed.

Among the total 609 traces in the event log, only 88 (14.4 %) have followed the optimal path, i.e., they coincide with one of the optimal traces. The event log involves some traces that did not terminate, i.e., they did not reach the sink transition in the WF-net model, and several traces that completed, but that did not follow the optimal path.

The mined process contains three decisions points, that is, after the activities *Create Purchase Requisition*, *Analyze Request for Quotation*, *Release Supplier's Invoice*. The two former activities appear in the first part of the process, i.e., before the transition *tau join0*, the latter is part of the zoomed in parallel block of Fig. 3. We refer to these activities as the *antecedent* of a decision point.

Our framework allows (i) to check whether an antecedent has been reached by a process trace during its execution and (ii) to check whether the *consequent*, i.e., the next activity after the decision point, conforms to the optimal path. In the 521 cases that do not follow the optimal path, process execution deviates at the first two decision points (234 times for Create Purchase Requisition and 238 times for Analyze Request for Quotation) more often than at the last one (80 times for Release Supplier's Invoice). Note that many traces do not even reach the last antecedent, since it appears towards the end of the process. Note also that the sum of deviations from the optimal path $(234 + 238 + 80 = 552)$ is not equal to the number of non optimal traces (521). This because the last antecedent appears in a loop (see Fig. 3) and, therefore, it is encountered more than once in a number of cases.

To summarise, our framework allows to separate the cases that followed the optimal path from the ones that did not, which enables further analysis to compare specific features of these two groups of cases. Moreover, it allows to detect that a large part of the non optimal cases deviate from the optimal case at one of the first two decision points in the process.

7   Conclusions

This paper has presented a framework for determining optimal paths in business process executions based on the abstraction of workflow hypergraphs. Three applications of the proposed framework have also been discussed, namely process navigation, simulation, and analysis.

As future work, we are extending the features of ant-colony optimisation to implement finer grained options for process simulation, possibly adopting the DMN (Decision Modeling and Notation) standard. We are also developing a more generic framework to define domain specific utility functions combining domain experts a priori knowledge with information about process outcomes that could be extracted from event logs. Finally, we are also working on integrating our framework with actual PAIS. This latter development is particularly relevant for the process navigation scenario, in which a usable interface for process users is fundamental to achieve user acceptance.

References

1. Ardagna, D., Pernici, B.: Adaptive service composition in flexible processes. IEEE Trans. Softw. Eng. **33**(6), 369–384 (2007)
2. Barba, I., Weber, B., Del Valle, C., Jiménez-Ramírez, A.: User recommendations for the optimized execution of business processes. Data Knowl. Eng. **86**, 61–84 (2013)
3. Comuzzi, M., Vanderfeesten, I.T.P., Wang, T.: Optimized cross-organizational business process monitoring: design and enactment. Inf. Sci. **244**, 107–118 (2013)
4. Conforti, R., de Leoni, M., La Rosa, M., van der Aalst, W.M., ter Hofstede, A.H.: A recommendation system for predicting risks across multiple business process instances. Decis. Support Syst. **69**, 1–19 (2015)

5. Di Francescomarino, C., Dumas, M., Federici, M., Ghidini, C., Maggi, F.M., Rizzi, W.: Predictive business process monitoring framework with hyperparameter optimization. In: Nurcan, S., Soffer, P., Bajec, M., Eder, J. (eds.) CAiSE 2016. LNCS, vol. 9694, pp. 361–376. Springer, Cham (2016). https://doi.org/10.1007/978-3-319-39696-5_22

6. Dorigo, M., Birattari, M., Stutzle, T.: Ant colony optimization. IEEE Comput. Intell. Mag. **1**(4), 28–39 (2006)

7. Dumas, M., La Rosa, M., Mendling, J., Reijers, H.A., et al.: Fundamentals of Business Process Management, vol. 1. Springer, Heidelberg (2013). https://doi.org/10.1007/978-3-642-33143-5

8. Gallo, G., Longo, G., Pallottino, S., Nguyen, S.: Directed hypergraphs and applications. Discret. Appl. Math. **42**(2–3), 177–201 (1993)

9. Ghattas, J., Soffer, P., Peleg, M.: Improving business process decision making based on past experience. Decis. Support Syst. **59**, 93–107 (2014)

10. Haisjackl, C., Weber, B.: User assistance during process execution - an experimental evaluation of recommendation strategies. In: zur Muehlen, M., Su, J. (eds.) BPM 2010. LNBIP, vol. 66, pp. 134–145. Springer, Heidelberg (2011). https://doi.org/10.1007/978-3-642-20511-8_12

11. Jula, A., Sundararajan, E., Othman, Z.: Cloud computing service composition: a systematic literature review. Expert Syst. Appl. **41**(8), 3809–3824 (2014)

12. Laguna, M., Marklund, J.: Business Process Modeling, Simulation and Design. CRC Press, Boca Raton (2013)

13. Lakshmanan, G.T., Shamsi, D., Doganata, Y.N., Unuvar, M., Khalaf, R.: A Markov prediction model for data-driven semi-structured business processes. Knowl. Inf. Syst. **42**(1), 97–126 (2015)

14. Leemans, S.J.J., Fahland, D., van der Aalst, W.M.P.: Discovering block-structured process models from event logs containing infrequent behaviour. In: Lohmann, N., Song, M., Wohed, P. (eds.) BPM 2013. LNBIP, vol. 171, pp. 66–78. Springer, Cham (2014). https://doi.org/10.1007/978-3-319-06257-0_6

15. Lemos, A.L., Daniel, F., Benatallah, B.: Web service composition: a survey of techniques and tools. ACM Comput. Surv. (CSUR) **48**(3), 33 (2016)

16. Polyvyanyy, A., Weske, M.: Hypergraph-based modeling of ad-hoc business processes. In: Ardagna, D., Mecella, M., Yang, J. (eds.) BPM 2008. LNBIP, vol. 17, pp. 278–289. Springer, Heidelberg (2009). https://doi.org/10.1007/978-3-642-00328-8_27

17. Polyvyanyy, A., Smirnov, S., Weske, M.: Business process model abstraction. In: vom Brocke, J., Rosemann, M. (eds.) Handbook on Business Process Management 1. IHIS, pp. 147–165. Springer, Heidelberg (2015). https://doi.org/10.1007/978-3-642-45100-3_7

18. Rogge-Solti, A., Weske, M.: Prediction of business process durations using non-Markovian stochastic Petri nets. Inf. Syst. **54**, 1–14 (2015)

19. Schonenberg, H., Weber, B., van Dongen, B., van der Aalst, W.: Supporting flexible processes through recommendations based on history. In: Dumas, M., Reichert, M., Shan, M.-C. (eds.) BPM 2008. LNCS, vol. 5240, pp. 51–66. Springer, Heidelberg (2008). https://doi.org/10.1007/978-3-540-85758-7_7

20. Thakur, M., Tripathi, R.: Linear connectivity problems in directed hypergraphs. Theor. Comput. Sci. **410**, 2592–2618 (2009)

21. Aalst, W.M.P.: TomTom for business process management (TomTom4BPM). In: van Eck, P., Gordijn, J., Wieringa, R. (eds.) CAiSE 2009. LNCS, vol. 5565, pp. 2–5. Springer, Heidelberg (2009). https://doi.org/10.1007/978-3-642-02144-2_2

22. Van der Aalst, W.M., Schonenberg, M.H., Song, M.: Time prediction based on process mining. Inf. Syst. **36**(2), 450–475 (2011)
23. van der Aalst, W.M.P., van Hee, K.M., ter Hofstede, A.H.M., Sidorova, N., Verbeek, H.M.W., Voorhoeve, M., Wynn, M.T.: Soundness of workflow nets: classification, decidability, and analysis. Form. Asp. Comput. **23**(3), 333–363 (2010)
24. Vanderfeesten, I., Reijers, H.A., van der Aalst, W.M.: Product-based workflow support. Inf. Syst. **36**(2), 517–535 (2011)

An Agent-Based Model of a Business Process: The Use Case of a Hospital Emergency Department

Emilio Sulis$^{(\boxtimes)}$ and Antonio Di Leva

Computer Science Department, University of Torino, Turin, Italy
{sulis,dileva}@di.unito.it

Abstract. An application of Artificial Intelligence is computational simulation which reproduces the behavior of a system, such as an organization. Simulations provide benefits into business process management, also by combining scenarios and what-if analysis. This study explores the adoption of agent-based modeling technique, in addition to traditional discrete event simulations. The focus is on a real case study of an hospital emergency department. Following the construction of a new hospital, managers are interested in simulating the actual flows in the new configuration before the moving. In our model, patients and operators are agents, acting due to simple behavioral rules in the environment. The different activities are placed on the map of the department, to provide immediate understanding of bottlenecks and queues. While first results were validated from managers, next steps include the comparison of resulting flows between the new and the old department. Logistics analysis includes the time for moving agents between different wards.

Keywords: Health-care processes
Agent based modeling and simulation · Emergency Department

1 Introduction

In Artificial Intelligence, several efforts were done to model complex and changing environments. Three main types of computer-based simulations were recently developed by considering different modeling techniques: system dynamics, discrete event simulations, and agent based modeling.

This study investigates an application of simulations in the framework of Business Process Management (BPM) [1]. The discipline includes modeling to facilitate organizations in describe, analyze, test, and optimize business processes. Through the comparison of actual and simulated process indicators, computer-based decision support systems [2] provide effective and efficient performance analysis. Planning, management and decision-making would greatly benefit from the analysis of the outcomes of different scenarios. We performed an agent-based simulation with NetLogo, which is considered "by far the most professional platform in its appearance and documentation" [3]. Simulated results

© Springer International Publishing AG 2018
E. Teniente and M. Weidlich (Eds.): BPM 2017 Workshops, LNBIP 308, pp. 124–132, 2018.
https://doi.org/10.1007/978-3-319-74030-0_8

allow to detect inefficiencies, bottlenecks, constraints, and risks, as well as to estimate the performance of the system when process modifications have to be applied (new strategies, an increase of the workload, changes in workforce etc.).

The use case is a public hospital Emergency Department (ED) in Italy. We focus on public sector, as in many European Countries it is increasingly required to provide better public services at lower costs [4]. Moreover, EDs are one of the more complex areas in health-care systems, facing several challenges such as long wait times, medical risk, lack of resources. Our interest here concerns the exploration of Agent-Based Modeling (ABM), setting up the ED model to perform different experiments and analyze early results.

Hospital managers are interested in understanding the functioning of new ED, which is actually under construction. Mapping old flows into the new configuration would provide benefits to business organisation. Patients and operators are agents acting due to behavioral rules. The different activities are placed on the current map of the department, to provide immediate understanding of bottlenecks and queues. While first results were validated from managers, next steps include the comparison of flows between the new and the old ED, with a early logistics analysis, i.e. time needed to move patients and operators between different wards. In the following we introduce a review of related works in Sect. 2, while Sect. 3 describes data from our case study and the methodological framework. Section 4 gives an insight of the model and early results. Finally, future works conclude the paper in Sect. 5.

2   Related Works

In the framework of business process modeling [5], most studies focused on design, control, and analysis of operational tasks involving humans, documents, applications [6]. Generally speaking, simulation techniques in the field of BPM have not yet been developed as they deserve [7]. Nevertheless, some applications of computer-based simulations to business process modeling regards both industrial re-engineering [8,9], as well as the public sector by modeling public policies [10], services [11], public administration processes [12], contact centers [13,14], political decision-making [15], care processes in the medical field [16]. Scenario analysis has been applied to explore different options for restructuring an existing process [17] before any change is effectively made. Although agent-based computing represents a relatively new research sector, several applications in different fields were already performed also in BPM [18]. A review of related works in the area of ABM in health sector can be found in [19]. The case of ED was implemented by agent-based simulations with respect to evacuation [20], optimization scheduling [21], disease propagation analysis [22]. The above mentioned studies largely adopted NetLogo, which easily allows to investigate different scenarios by changing configurations. In addition, the possibility to address logistics of ED seem promising. Some existing studies already focused on this task in health-care [23], typically treating Lean techniques [24].

3   The Use Case: Data and Methodology

Our case study refers to an hospital ED located in the immediate surrounding of
Turin, one of the largest Italian city. Managers were interested in understanding
the flows in the new ED, before the moving scheduled in winter 2017.

3.1   Dataset

We considered real data of the ED in 2016, when patients involved were 46,497.
The daily access follows an hourly distribution of arrivals with two peaks, one
in the morning and one in the afternoon (Fig. 1).

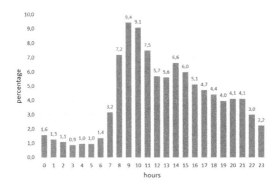

Fig. 1. The arrival of patients in the ED on average per hours in 2016

A four-level scale of Emergency Severity Index defines the urgency of patients
in Italy, usually named by colors ranging from red (R) which are the more urgent
cases to yellow (Y), green (G), and white (W) (Table 1).

Table 1. Distribution of patients by ESI level in 2016

	R	Y	G	W	Tot
Number of patients	314	5,292	39,561	1,330	*46,497*
Percentage values	0.7	11.3	85.1	2.9	*100*

3.2   Methodology

In the framework of BPM, our analysis starts with the implementation of the
actual situation of organization processes (As-Is). Simulations results will suggest
different scenarios for re-engineering the restructured processes (To-Be).

Simulations. In this context, we performed a simulation to describe events,
activities and decisions concerning As-Is, with the actual business processes.

In particular, we based on previous works dealing with the traditional modeling technique Discrete Event Simulation (DES) of the same organization [25]. In this case, we explored a different approach for the same case study, focusing on ABM to create agents acting (or moving) in ED.

NetLogo. We adopt the open-source programmable modeling environment Net-Logo [26], which is a popular modeling software, suitable for this kind of simulations. The program manages thousands of agents (so called turtles) operating independently in a landscape made of static agents, building the background of the simulation (patches). This makes it possible to distinguish different areas of the environment corresponding to the activities of the ED, in which agents act. Figure 2 describes the ED map and the colored version of patches.

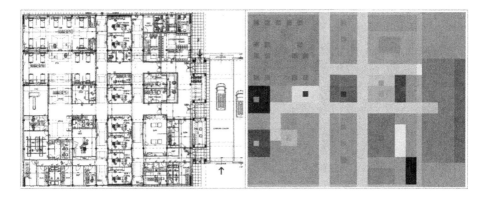

Fig. 2. ED map (left) and the corresponding NetLogo colored version (right) (Color figure online)

Agents. The two main types of individual agents in an ED are patients and operators. We model them as turtles having several attributes which determine the flows toward different paths or activities. Main attributes for operators discriminate if they are waiting for the next activity or busy in working with a patient. Two patients' attributes concern the level of priority (ESI level) as well as the arrival (see Table 2) which influence the paths towards different ED areas.

Activities. All the activities are modeled as tasks which involve the corresponding agents (patients and operators) detailed in Sect. 4. As NetLogo is well suited for modeling systems developing over time, each activity has a certain duration, obtained by a preliminary study on the basis of real values. For instance, Table 3 describes the post-triage waiting time of patients, which largely depends on both ESI level and hour of arrival: in central hours of the day (12–16) urgent cases have a low waiting-time (7.1 min), while not urgent cases have to wait on average about three hours (174.4 min).

Key Performance Indicators. The performance of the model is actually evaluated on the basis of two business process indicators:

Table 2. Main attributes of agents (Patients or Operators)

Agents	Variable	Description	Cases
	op-move?	Reach the activity	T/F
O	op-in-wl?	Waiting	T/F
	busy?	Working	T/F
	ESI	Level of priority	red/yellow/green/white
	arrival	Arrive in ED	ambulance/walk-in
P	pat-move?	Reach the activity/patch	T/F
	served?	Served	T/F
	pat-in-waiting-list?	Waiting	T/F

Table 3. Average duration of post-triage waiting time depending on the hour of arrival (6 time slots) and ESI level (White, Green, Yellow, Red), in 2016

	W	G	Y	R
1–4	21.3	18.3	9.1	3.0
5–8	79.6	30.5	14.3	5.3
9–12	112.8	84.9	16.6	7.0
13–16	174.4	145.5	22.8	7.1
17–20	166.8	146.6	19.1	5.5
21–24	75.9	75.9	19.9	14.0

- *Door-to-Doctor-Time (DTDT)*: The number of minutes a patient takes from the arrival in the ED to see a doctor.
- *Lenght-of-Stay (LoS)*: The number of minutes from patient arrival to the exit from the ED.

Figure 3 describes two monitors of the simulation detailing the results of KPIs as well as the number of patients in waiting-room.

Fig. 3. Monitors related to waiting lists and KPIs

4   Initial Settings and Early Results

This section describes the initial settings concerning agents and the environment.

Agents: patients and operators. As previously discussed, we set the creation of patients by referring to the average values of the previous year (2016). Operators involved in the ED are three doctors (PHY), two triage nurses (TRN), two ED nurses (EDN) and three healthcare worker (OSS), as in the actual organisation. By modifying a form it would be possible to change these parameters: Fig. 4 shows the initial configuration of the simulation.

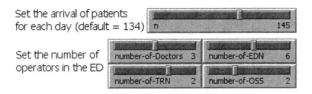

Fig. 4. NetLogo sliders for the creation of individual agents

Patches: ED activities. The activities correspond to our ED areas. Table 4 describes their colors in NetLogo map, patients (pat) or operators (op) involved, duration (by triangular distributions, seconds).

Table 4. Table of activities, patch colors, agents involved, and durations (seconds)

Activity	Color	Agents	Duration
Registration	Blue	Pat, EDN	Tri(100,140,120)
Triage room	Light green	Pat, TRN	Tri(300,900,600)
Visit room	Orange	Pat, PHY	Tri(940,400,1140)
Radiology	Purple	Pat, EDN	Tri(650,100,700)
Shock-Room	Red	Pat, EDN, PHY (2)	Tri(1550,500,1800)
Observation	Orange	Pat, EDN, OSS	Tri(3600,32400,18000)
Waiting room	Dark green	Pat, OSS	*undefined*

Managing agents: rules of behavior. The behavior of individual agents (operators and patients) follows simple rules. If agents are not involved in an activity, they are managed by two First Input First Output queues. Urgent patients (red) follow a different path going immediately in shock-room. Yellow cases have the priority over the two other cases. Once an activity (corresponding to patch agents in NetLogo) are free, they call the first agent. We are planning to adopt more complicated features by adding realistic parameters (as some breaks for operators, queue scheduling management, and so on). For instance, abandons of patients who left the ED without being seen can be modeled depending on both the urgency level and the number of patients in waiting rooms.

Early Results. First results of the ABM[1] give us an idea of how patients move in ED areas. Managers confirm the initial suggestions coming from simulation: the blue area, corresponding to Short Intensive Observation area, is often over-crowded, as well as the long queue in waiting room (see Fig. 5).

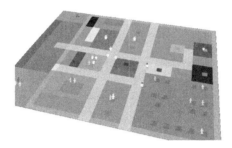

Fig. 5. Simulation output after two weeks: blue area and waiting room are overcrowded

5   Conclusions

This paper discusses the adoption of an agent-based simulation to the use case of an hospital Emergency Department. Initial parameters are derived from 2016 data. Individual agents of our simulation are dynamically created, as well as the environment in which they move is based on the real map. We are implementing the simulation closely with ED managers, before the moving in the new ED, scheduled for winter 2017. In future works we are planning to improve ABM simulation by adding more realistic features to extend our set of KPIs. In addition, we plan to compare DES and ABM results.

Acknowledgments. We are grateful to Dr. Adriana Boccuzzi and Dr. Franco Ripa of the San Luigi Gonzaga hospital of Orbassano (Italy). The City of Health and Science of Turin has funded the business processes analysis working group, including the research work of Emilio Sulis.

References

1. Dumas, M., Rosa, M.L., Mendling, J., Reijers, H.A.: Fundamentals of Business Process Management. Springer, Heidelberg (2013). https://doi.org/10.1007/978-3-642-33143-5
2. Shang, S., Seddon, P.B.: Assessing and managing the benefits of enterprise systems: the business manager's perspective. Inf. Syst. J. **12**(4), 271–299 (2002)

---

[1] See OpenABM archive at the address: https://www.openabm.org/model/5832.

3. Railsback, S.F., Lytinen, S.L., Jackson, S.K.: Agent-based simulation platforms: review and development recommendations. Simulation **82**(9), 609–623 (2006)
4. OECD: OECD science, technology and industry outlook 2012, OECD Publishing (2012)
5. Williams, S.: Business process modeling improves administrative control. Automation **44**, 50 (1967)
6. van der Aalst, W.M.P., ter Hofstede, A.H.M., Weske, M.: Business process management: a survey. In: van der Aalst, W.M.P., Weske, M. (eds.) BPM 2003. LNCS, vol. 2678, pp. 1–12. Springer, Heidelberg (2003). https://doi.org/10.1007/3-540-44895-0_1
7. Van der Aalst, W.M.P., Nakatumba, J., Rozinat, A., Russell, N.: Business process simulation. In: Brocke, J., Rosemann, M. (eds.) Handbook on BPM 1. International Handbooks on Information Systems, pp. 313–338. Springer, Heidelberg (2010). https://doi.org/10.1007/978-3-642-00416-2_15
8. Scheer, A.-W., Nüttgens, M.: ARIS architecture and reference models for business process management. In: van der Aalst, W., Desel, J., Oberweis, A. (eds.) Business Process Management. LNCS, vol. 1806, pp. 376–389. Springer, Heidelberg (2000). https://doi.org/10.1007/3-540-45594-9_24
9. Reijers, H.A. (ed.): Design and Control of Workflow Processes. LNCS, vol. 2617. Springer, Heidelberg (2003). https://doi.org/10.1007/3-540-36615-6
10. Lempert, R.: Agent-based modeling as organizational and public policy simulators. Proc. Natl. Acad. Sci. **99**(suppl.3), 7195–7196 (2002)
11. Gulledge Jr., T.R., Sommer, R.A.: Business process management: public sector implications. Bus. Process Manag. J. **8**(4), 364–376 (2002)
12. Kovacic, A., Pecek, B.: Use of simulation in a public administration process. Simulation **83**(12), 851–861 (2007)
13. Vinai, M., Sulis, E.: Health and social public information office simulation (2015). https://www.openabm.org/model/4778/version/2
14. Di Leva, A., Sulis, E., Vinai, M.: Business process analysis and simulation: the contact center of a public health and social information office. Intell. Inf. Manag. **9**(05), 189 (2017)
15. Rouchier, J., Thoyer, S.: Votes and lobbying in the european decision-making process: application to the european regulation on GMO release. J. Artif. Soc. Soc. Simul. **9**(3), 1 (2006). http://jasss.soc.surrey.ac.uk/9/3/1/citation.html
16. Di Leva, A., Sulis, E.: Process analysis for a hospital emergency department. Int. J. Econ. Manag. Syst. **2**(1), 34–41 (2017)
17. Lam, K., Lau, R.: A simulation approach to restructuring call centers. Bus. Process Manag. J. **10**(4), 481–494 (2004)
18. Jennings, N.R., Norman, T.J., Faratin, P., O'Brien, P., Odgers, B.: Autonomous agents for business process management. Appl. Artif. Intell. **14**(2), 145–189 (2000)
19. Friesen, M.R., McLeod, R.D.: A survey of agent-based modeling of hospital environments. IEEE Access **2**, 227–233 (2014)
20. Ren, C., Yang, C., Jin, S.: Agent-based modeling and simulation on emergency evacuation. In: Zhou, J. (ed.) Complex 2009. LNICST, vol. 5, pp. 1451–1461. Springer, Heidelberg (2009). https://doi.org/10.1007/978-3-642-02469-6_25
21. Cabrera, E., Taboada, M., Iglesias, M.L., Epelde, F., Luque, E.: Simulation optimization for healthcare emergency departments. Procedia Comput. Sci. **9**, 1464–1473 (2012)
22. Jaramillo, C., Taboada, M., Epelde, F., Rexachs, D., Luque, E.: Agent based model and simulation of mrsa transmission in emergency departments. Procedia Comput. Sci. **51**, 443–452 (2015)

23. Landry, S., Philippe, R.: How logistics can service healthcare. Supply Chain Forum Int. J. **5**(2), 24–30 (2004)
24. Chan, H., Lo, S., Lee, L., Lo, W., Yu, W., Wu, Y., Ho, S., Yeung, R., Chan, J.: Lean techniques for the improvement of patients flow in emergency department. World J. Emerg. Med. **5**(1), 24 (2014)
25. Di Leva, A., Sulis, E.: A business process methodology to investigate organization management: a hospital case study. WSEAS Trans. Bus. Econ. **14**, 100–109 (2017)
26. Wilensky, U.: Netlogo, Technical report, Center for Connected Learning and Computer-Based Modeling, Northwestern University, Evanston (1999)

Constraint-Based Composition of Business Process Models

Piotr Wiśniewski[(✉)], Krzysztof Kluza, Mateusz Ślażyński, and Antoni Ligęza

AGH University of Science and Technology, al. Mickiewicza 30,
30-059 Krakow, Poland
{wpiotr,kluza,mslaz,ligeza}@agh.edu.pl
http://www.agh.edu.pl

Abstract. Process models help organizations to visualize and optimize their activities, and achieve their business goals in a more efficient way. Modeling a business process requires exact information about possible execution sequences of the activities as well as process modeling notation knowledge. We present a method of business process composition based on the constraint programming technique. Taking task specifications as the input, our solution can generate a workflow log which can be used to discover the model using any process mining technique.

Keywords: Business processes · BPMN · Automated planning
Constraint programming · Process mining
Business process composition

1 Introduction

As business processes are subject to constant changes, the possibility to effectively redesign a process model becomes a valuable feature [1]. This action usually requires the cooperation of domain experts with process designers and impacts the sequence flow [2]. To create the improved model, it is favorable to use the process composition technique, which can be defined as combining different models into a large one [1]. These component models can be represented either by subprocesses or primitive tasks, each representing a single activity.

The method we propose allows a domain expert to compose a business process model based on the simple task specification including data entities present within the process, gathered from various participants of the process (see the approach outline in Fig. 1). An unordered list of activities, with their inputs and outputs, and the initial and goal state are required to generate a process model. The algorithm verifies the satisfaction of the set of predefined constraints which ensure the correct process flow and generates all admissible execution sequences of the activities. These sequences, also known as workflow traces, are then used as an input to the process mining algorithm which produces a final BPMN model. The information about the sequence flows of the process does

© Springer International Publishing AG 2018
E. Teniente and M. Weidlich (Eds.): BPM 2017 Workshops, LNBIP 308, pp. 133–141, 2018.
https://doi.org/10.1007/978-3-319-74030-0_9

Fig. 1. Overview of the business process composition approach.

not have to be known before using the proposed method. Thus, the constraint-based composition is a suitable way to perform fast reconstructions of business processes in case of requirement modification.

The rest of the paper is organized as follows. Section 2 presents the state of the art solutions. Section 3 describes our process composition method. The evaluation of the solution is given in Sect. 4. The paper is concluded in Sect. 5.

2   Related Works

Business process automated planning aims to support business analysts and modelers in the design or rebuild phase. The graph-based construction of exclusive choices [3] optimizes the model by removing unnecessary redundancies in alternative flows. Processes can be also planned with respect to their context, such as changing environmental conditions [4]. This can help modelers in time-efficient process construction. For example, Havur et al. [5] propose resource allocation planning in business processes using timed Petri Nets and Answer Set Programming technique.

Constraint satisfaction in processes can be used to align real workflow traces, with artifacts or missing relevant events, with the predefined declarative model specification [6]. Declarative models are also used to generate optimal enactment plans for Business Process Execution [7]. Constraints can also be extracted from a BPMN model [8] and then be used as an input to planning systems.

Our work refers also to service composition, which can be described as combining existing functionalities into a new workflow [9]. In this case, the heuristic search methods are used to generate a list of admissible service invocations. Composition of BPMN models disassembled into several subprocesses was presented in [10]. In this approach business rules are used to build a complete model in real time. Another solution is dedicated to the integration of inter-organizational business workflows [11]. Component processes are verified in terms of compatibility and if the result of this comparison is positive, an adapter is generated when needed. Synthetic process models can be also generated using pre-defined Relevant Process Fragments (RPF) stored in a database [12].

3   Constraint-Based Process Composition

Constraint Programming is a paradigm for solving combinatorial search problems by exploring and reducing the search space using predefined constraints over variables. Our method uses a set of constraints to generate a complete workflow log of the process based on previously identified tasks and their specifications. The generated log, covering all the admissible workflow traces, is then used to compose a process model in the BPMN notation.

3.1   Task Identification

Since a business process is mainly defined as a partially ordered set of activities, tasks constitute the crucial part of the process. Thus, in the first step of process modeling, it is necessary to identify the main elements that form the workflow [13]. Each task has some prerequisites and produces an effect of its execution. Knowing this information, it is possible to define input and output variables for the tasks. For further processing, we introduce a notion of data entity as the additional process specification used by our method.

3.2   Workflow Trace Discovery

The created process composition algorithm can be applied when process tasks along with their inputs and outputs were defined. Therefore, prerequisites for this type of process modeling can be described as a set of tasks $\mathbb{T}$ and a set of data entities Δ.

Let n be the number of all identified tasks and m the number of all data entities in the analysed process. The presence of data entities can be represented as a certain state of the process, which changes after a task is executed. A state is therefore an m-dimensional binary vector where an element is equal to 1 if the corresponding data entity is present and 0 otherwise. In the next step, it is necessary to define the task conditions and task effects matrices of size (n, m). Their form is described by Matrix 1:

$$\begin{bmatrix} x_{11} & x_{12} & x_{13} & \dots & x_{1m} \\ x_{21} & x_{22} & x_{23} & \dots & x_{2m} \\ \vdots & \vdots & \vdots & \ddots & \vdots \\ x_{n1} & x_{n2} & x_{n3} & \dots & x_{nm} \end{bmatrix} \tag{1}$$

where: $x_{ij} \in \{-1, 0, 1\}$, $i = 1 \dots n$, $j = 1 \dots m$.

The task conditions matrix M_{TC} determines which data entities are required to be present in order to execute each task. The value of the element x_{ij} is equal to:

- -1, if the presence of j-th entity is not relevant for the execution of i-th task,
- 0, if j-th entity is forbidden for the execution of i-th task,
- 1, if j-th entity is required for the execution of i-th task.

The task effects matrix M_{TE} specifies what are the effects of each task. The matrix elements x_{ij} admit one of three values:

- -1, if j-th entity stays unchanged after the execution of i-th task,
- 0, if j-th entity is removed by i-th task,
- 1, if j-th entity is added by the execution of i-th task.

Another two mathematical structures that are required for process modeling are two m-element vectors which represent starting and goal states. The initial state of the process needs to be explicitly defined and thus the initial state vector s_0 can have values equal to 0, if i-th state is not present before the process execution and 1 otherwise. On the other hand, elements of the final state vector s_T can be equal to -1, as presence of some data entities may not be relevant for the process execution.

If all the tasks and data entities were defined as well as values of M_{TC}, M_{TE} matrices and s_0, s_T vectors were set, it is possible to run the algorithm for process composition which produces a workflow log W as its result: $W = \{\sigma_1, \sigma_2, \dots \sigma_L\}$, where σ is a workflow trace defined as an ordered sequence of tasks, L is the number of generated workflow traces and K is the number of non-idle tasks in a trace. To differentiate the order of tasks executed in a workflow trace from the identified tasks, it was denoted by upper indices: $\sigma = (\tau^{(1)}, \tau^{(2)}, \dots \tau^{(K)})$.

A state vector s satisfies the requirement if it matches all non-negative elements in the corresponding row in M_{TC} matrix. The state satisfiability predicate for the execution of i-th task is shown in Formula 2.

$$sat(s, TC(\tau_i)) \iff \forall j = 1 \dots m : s_j = TC(\tau_i)_j \lor TC(\tau_i)_j = -1 \qquad (2)$$

where $TC(\tau_i)$ is the condition vector for i-th task and as a consequence, the i-th row in M_{TC} matrix.

In order to clarify the notation the symbols $s_{input}(\tau)$ and $s_{output}(\tau)$ are used to describe the input and output state of task τ respectively. The resulting workflow trace is configured by satisfying the following constraints:

1. Tasks should not repeat: $\forall i, j = 1 \dots K : i \neq j \implies \tau^{(i)} \neq \tau^{(j)}$.
2. The maximum length of the process flow is equal to n: $K \leq n$.
3. In order to make the number of tasks and states equal, an additional empty task is added at the end of the process: $\tau^{(K+1)} = \emptyset$.
4. The input state of the first executed task should be equal to s_0: $s_{input}(\tau^{(1)}) = s_0$.
5. The last task of the execution sequence should not be preceded by any idle task and all tasks after the last one should be idle:
   $(\tau^{(i)} \neq \emptyset \iff i \leq K) \quad \land \quad (\tau^{(i)} = \emptyset \iff K < i \leq n)$
6. The process should end when the final state is achieved and the output of the last task should satisfy the goal states: $sat(s_{output}(\tau^{(K)}), s_T)$.
7. Idle (empty) tasks should not change the states: $s_{input}(\emptyset) = s_{output}(\emptyset)$.
8. Every task can be only executed when the current state satisfies its input conditions: $\forall \tau \in \mathbb{T} : \tau \in \sigma \iff sat(s_{input}(\tau), TC(\tau))$.
9. Every state has to be changed according to the executed task:

$$\forall j = 1 \ldots m : s_{output_j}(\tau) = \begin{cases} s_{input}(\tau), & \text{when } TC_j(\tau) = -1 \\ TE_j(\tau), & \text{otherwise} \end{cases},$$

where $TE(\tau)$ is the effect vector defined for the task and the corresponding row in M_{TE}.

The proposed process composition model was created in the MiniZinc constraint programming language [14], where the following data were used:

- Inputs: unordered lists of tasks and data entities, matrices M_{TC} and M_{TE}, initial state vector s_0, final state vector s_T.
- Decision variables: last task index, process workflow, process states.
- Constraints: as defined by the list above.

The method can be executed using Gecode solver [15] with the goal to satisfy the predefined constraints. In order to obtain the complete workflow log, the option to print all solutions should be set and solution compression should be turned off. Such a workflow log is the input to the next step – process mining.

3.3   Process Mining

The workflow log W obtained after solving the constraint satisfaction problem is used as an input to the process mining algorithm. The workflow log needs to be converted from the ordered list of tasks into an event log. Next, based on such an event log any process mining algorithm can be run. For our approach, we chose one of the most popular process mining algorithms used in the ProM Lite software [16] – the Heuristic Miner (with the default parameters). In the following section, we provide a simple step by step example presenting our method.

4   Evaluation

Let us analyze the method presented in Sect. 3 on the real-life example of a bank account opening process. A sample BPMN model of such case is shown in Fig. 2. The verification of our algorithm consisted in rebuilding the source model having its tasks and identified data entities as inputs.

In the process presented in Fig. 2, 19 different data entities were defined and each task was assigned input and output data entities. For the analyzed process, a table containing a list of tasks was prepared (the selected part of which is presented in Table 1).

In the next step, it is necessary to define task condition and task effects matrices, which were described in Sect. 3.2. Using the example data from Table 1 both matrices are of size $(3, 4)$ as there are 3 identified tasks $\mathbb{T} = \{e, o, f\}$ and 4 different data entities Δ in this part of the process δ_1 for AccountDetails, δ_2 for AccountID, δ_3 for DepositChecked, and δ_4 for DepositSet.

In this case, the matrices M_{TC} and M_{TE} are described by Formula 3. The orders of their elements correspond to the alphabetical order of task ids and data entities, respectively.

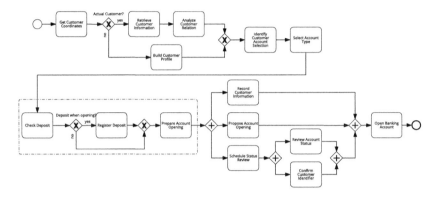

Fig. 2. An example of a process model of a bank account opening.

Table 1. An excerpt of a list of identified tasks and data entities.

Task Id	Task name	Task inputs	Task outputs
e	Check deposit	AccountID	DepositChecked
f	Prepare account opening	AccountID, DepositChecked, (DepositSet)	AccountDetails
o	Register deposit	DepositChecked	DepositSet

$$M_{TC} = \begin{bmatrix} 0 & 1 & 0 & 0 \\ 0 & 1 & 1 & -1 \\ 0 & 1 & 1 & 0 \end{bmatrix} \quad M_{TE} = \begin{bmatrix} -1 & -1 & 1 & -1 \\ 1 & -1 & -1 & -1 \\ -1 & -1 & -1 & 1 \end{bmatrix} \quad (3)$$

Task o (Register Deposit) is bypassed by an alternative flow in the original model (Fig. 2). Therefore, it is not executed in every process instance and it is coherent with the created specification. Regarding the output data entity from task o (DepositSet), the only activity that can be executed in the next step is task f where this input is optional (element $(2, 4)$ of matrix M_{TC} is equal to -1).

At the same time, it is possible to set values of vectors $s_0 = \begin{bmatrix} 0 & 1 & 0 & 0 \end{bmatrix}$ and $s_T = \begin{bmatrix} 1 & -1 & -1 & -1 \end{bmatrix}$ which determine the initial and final state of the process. Before the execution only AccountID is needed to run the process, and the only required output value is AccountDetails data entity; other values are not relevant.

It can easily be observed that in the trivial example presented in Table 1 only two workflow traces are possible. Thus, $W = \{(e, f), (e, o, f)\}$ is the complete workflow log. Using one of the process mining techniques, this would mostly generate a simple output BPMN model, depicted in Fig. 3.

After running the algorithm for the complete list of tasks (created based on the model shown in Fig. 2), 160 different traces were generated. After transforming the log to the required form and running the process mining algorithm in ProM environment the output model was generated. It is shown in Fig. 4.

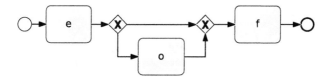

Fig. 3. A sample model for the workflow.

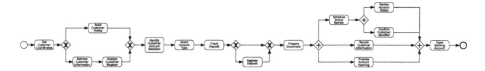

Fig. 4. Generated BPMN model.

The number of tasks and the flow of the generated result (Fig. 4) is the same as in the source model presented in Fig. 2. All the gateways and branches were accurately reproduced.

As illustrated by the example, the proposed solution extends the existing approaches by simplifying input information needed to execute the algorithm and providing workflow logs which may be used to discover a complete business process model. Table 2 presents a summary of selected workflow planning techniques in comparison to the proposed approach.

Table 2. Comparison of the existing process planning approaches.

Authors	Heinrich and Schön [4]	van Beest et al. [17]	Rodriguez [7]	Quin et al. [18]	*Our approach*
Year	2015	2014	2011	2009	2017
Input	Planning graph	Intention Spec	Declarative model	Activity Functions	Task Spec.
Method	Graph search	AI Planner[a]	CSP solving	Graph search	CSP solving
Output	Context-aware model	Event logs	Enactment plan	Grid workflow	BPMN Model[b]
Alternative flow	◑	●	●	●	●
Parallel Flow	○	◑	●	●	●

[a]In [17], the exact planner was not specified.
[b]The final model is generated using a process mining algorithm, as described in Sect. 3.3.

5   Conclusions

We provide an approach to business process composition, represented as a constraint satisfaction problem. The method uses an unordered list of tasks accompanied by data entities to produce all the admissible execution sequences which satisfy the set of provided constraints. This ensures the correctness of the model. As a consequence, a workflow log is used to generate a complete BPMN model using process mining technique. Our algorithm supports parallel and alternative control flows as well as the definition of the initial and goal states of the process.

The advantage of our approach is a simple and user-friendly form of data required to compose a business process model. Generating all the admissible workflow logs using the constraint programming technique may also lead to a further optimization of the process, regarding its complexity and robustness. Therefore, such solution can be valuable for domain experts who intend to design or improve business processes within their organizations.

In future work, we will extend the support for further BPMN constructs like swimlanes, loops, and events. We will also include the resource availability in the planning algorithm which better corresponds to actual process execution.

References

1. van der Aalst, W.M.P.: Business process management: a comprehensive survey. ISRN Softw. Eng. **2013**, 1–37 (2013)
2. Ghattas, J., Soffer, P., Peleg, M.: Improving business process decision making based on past experience. Decis. Support Syst. **59**, 93–107 (2014)
3. Heinrich, B., Schön, D.: Automated planning of process models: the construction of simple merges. In: European Conference on Information Systems (ECIS) (2016)
4. Heinrich, B., Schön, D.: Automated planning of context-aware process models. In: Becker, J., vom Brocke, J., de Marco, M. (eds.) ECIS (2015)
5. Havur, G., Cabanillas, C., Mendling, J., Polleres, A.: Automated resource allocation in business processes with answer set programming. In: Reichert, M., Reijers, H.A. (eds.) BPM 2015. LNBIP, vol. 256, pp. 191–203. Springer, Cham (2016). https://doi.org/10.1007/978-3-319-42887-1_16
6. De Giacomo, G., Maggi, F., Marella, A., Sardina, S.: Computing trace alignment against declarative process models through planning. In: Proceedings of the International Conference on Automated Planning and Scheduling, pp. 367–375 (2016)
7. Barba Rodriguez, I.: Constraint-based planning and scheduling techniques for the optimized management of business processes. Ph.d. Universidad de Sevilla (2011)
8. Schneeweis, D.: Constraint-based scheduling and planing in medical facilities with BPMN. In: The 19th International Conference on Principles and Practice of Constraint Programming. Doctoral Program, pp. 115–120 (2013)
9. Meyer, H., Weske, M.: Automated service composition using heuristic search. In: Dustdar, S., Fiadeiro, J.L., Sheth, A.P. (eds.) BPM 2006. LNCS, vol. 4102, pp. 81–96. Springer, Heidelberg (2006). https://doi.org/10.1007/11841760_7
10. Graml, T., Bracht, R., Spies, M.: Patterns of business rules to enable agile business processes. In: 11th IEEE International Enterprise Distributed Object Computing Conference (EDOC 2007), pp. 365–365 (2007)

11. Kumar, A., Shan, Z.: Algorithms based on pattern analysis for verification and adapter creation for business process composition. In: Meersman, R., Tari, Z. (eds.) OTM 2008. LNCS, vol. 5331, pp. 120–138. Springer, Heidelberg (2008). https://doi.org/10.1007/978-3-540-88871-0_11

12. Skouradaki, M., Andrikopoulos, V., Leymann, F.: Representative BPMN 2.0 process models generation from recurring structures. In: Proceedings of the 23rd IEEE International Conference on Web Services, pp. 468–475. IEEE, June 2016

13. Kluza, K., Wiśniewski, P.: Spreadsheet-based business process modeling. In: 2016 Federated Conference on Computer Science and Information Systems (FedCSIS), pp. 1355–1358. IEEE (2016)

14. Nethercote, N., Stuckey, P.J., Becket, R., Brand, S., Duck, G.J., Tack, G.: MiniZinc: towards a standard CP modelling language. In: Bessière, C. (ed.) CP 2007. LNCS, vol. 4741, pp. 529–543. Springer, Heidelberg (2007). https://doi.org/10.1007/978-3-540-74970-7_38

15. Schulte, C., Stuckey, P.J.: Efficient constraint propagation engines. ACM Trans. Program. Lang. Syst. **31**(1), 2:1–2:43 (2008)

16. van der Aalst, W.M., van Dongen, B.F., Günther, C.W., Rozinat, A., Verbeek, E., Weijters, T.: ProM: the process mining toolkit. BPM (Demos) **489**(31), 2 (2009)

17. van Beest, N.R., Russell, N., ter Hofstede, A.H., Lazovik, A.: Achieving intention-centric BPM through automated planning. In: 2014 IEEE 7th International Conference on Service-Oriented Computing and Applications, pp. 191–198 (2014)

18. Qin, J., Fahringer, T., Prodan, R.: A novel graph based approach for automatic composition of high quality grid workflows. In: Proceedings of the 18th ACM International Symposium on High Performance Distributed Computing, HPDC 2009, pp. 167–176. ACM (2009)

Semantically-Oriented Business Process Visualization for a Data and Constraint-Based Workflow Approach

Eric Rietzke[1]([✉]), Ralph Bergmann[2], and Norbert Kuhn[1]

[1] University of Applied Science Trier, Trier, Germany
e.rietzke@umwelt-campus.de
[2] University of Trier, Trier, Germany

Abstract. This paper introduces a novel approach which unifies a data-centric and a constraint-based workflow principle to support the requirements of knowledge intensive business processes. By the integration of a knowledge-based system, process definition and execution relevant data coincide on an ontology-based semantic net. The data, mainly driving the process, can be delivered by different sources or can be the result of an inference step by the underlying ontology. In such a case, AI technology plays an active role during the process execution and result in a division of labor with human actors. This paper presents a concept for a semantically-oriented process visualization for the introduced unified approach.

1 Introduction

Today, business process visualization (BPV) approaches are developed for an activity-centric (usually imperative) workflow principle where the activities are directly related to each other and form a control-flow or constraints define some rules for their execution. However, with view to the demand of knowledge intensive processes for flexibility at design- and run-time [6], new concepts based on data-centric approaches are subject of investigation and have formed an active field of research [1,3] over the last decade. These approaches have in common that the activities are no longer directly related to one another. Instead, the activities are bound to the data elements which are required to perform an activity (input) or which are the result of it (output).

Such data-centric approaches come along with characteristics, fitting very well to an ontology-based data management and form the preconditions for further intelligent process contributions. In this way, the process data can be used to create new information by simple inference mechanisms, exploiting the accessible knowledge. Moreover, the semantic description of the relations between data and activities can be utilized for a sophisticated process visualization.

In the following we present a BPV approach based on a knowledge-based system. Section 2 presents the foundations while Sect. 3 introduces a new unified

E. Teniente and M. Weidlich (Eds.): BPM 2017 Workshops, LNBIP 308, pp. 142–150, 2018.
https://doi.org/10.1007/978-3-319-74030-0_10

approach for business process modeling and execution by an example. The new concept of a semantically oriented BPV is presented in Sect. 4. We conclude our paper by giving an outlook on our future research in Sect. 5.

2   Foundations

In the following, we briefly summarize relevant previous research.

2.1   Business Process Modelling and Execution

Business Process Model and Notation (BPMN) is up-to-date the de facto standard for designing and describing business processes world-wide. In the center of this approach reigns a control-flow coordination of process-steps (activities). A less restrictive, but still activity-centric perspective is supported by constraint-based approaches [11], which allow flexibility in a scalable manner. Alternatively, there are several approaches with the intention to gain flexibility based on the control-flow principles [12]. Despite of the consideration of data-flow in such processes, the data is just integrated in a kind of an afterthought [5]. Opposed to this, knowledge intensive processes are usually barely structured and the execution is driven by user decisions and business data [6]. Activity-centric approaches are not sufficient to achieve knowledge intensive business goals [1,9] which is why several new approaches were brought up, putting the data into the center for the process design and process execution.

The existing workflow principles can be differentiated regarding the rationale for selecting activities for execution. Under a control-flow, the activities are chosen for the execution primarily by the connected ancestors, while a constraint-based model selects the activities by considering a set of restrictions. Both principles put the activity into the center of the view. This changes with data-centric approaches, where activities are executable as soon as the necessary information is available and the expected outcome on a new information is still required for further process activities. This is expressed by the taxonomy of workflow principles, shown in Fig. 1.

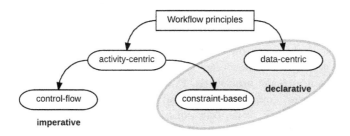

Fig. 1. Workflow principles

The data-centric and the constraint-based concepts have in common that in both cases the relations between objects are described, while the execution order is deduced on the fly. The nature of both declarative principles is their inherent flexibility. Nonetheless there are major differences. With a data-centric approach, possibilities for the execution of activities are described, while constraints define restrictions between activities. In a nutshell, data-centric principle defines the GO's, while constraints are beneficial to describe the NO-GO's. In this paper, we argue that both declarative principles can be combined to build a unified approach which will be introduced in Sect. 3.

2.2 Visualization

The importance of visualization within all fields with a human-computer interaction is well established [4] and is subject to research in many segments. This paper considers mainly the field of business process visualization [2,8].

The most established toolset to model business processes (BPMN) comes along with a detailed graphical notation definition. Graphs are built during the modeling phase and are usually used directly, when it comes to a process visualization later on. It represents the perspective of a process designer, which in general does not fit to the demands and needs of a process controller or an actor during the process execution. Additionally, since the modeling procedure is done to create a process template, the temporal situation of a process instance is usually just reflected by a state presentation and has no structural impact on the graph. An example of structural changes according to the process-state was introduced in Proviado [2]. Further research can be named [8] and has introduced interesting tools and methods to go beyond the static process graph created by the process designer. However, all have in common that they are based on the control-flow oriented approaches like BPMN. A data perspective is available only as an add-on to the dominating activity-centric principles.

3   A Unified Approach

As described in Sect. 2.1, the data-centric and the constraint-based approaches follow the same declarative paradigm. In this work, we argue that both principles can be combined to a unified approach. We expect that this will offer a seamless scalability regarding flexibility and strictness, from an unstructured process task-list up to a narrow restrictive process model. The approach would also serve the demands of knowledge intensive processes by integrating data as a first-class citizen into the process [1]. Finally, we see the possibilities to assure the existing enterprise compliance rules by explicit restrictions based on additional activity constraints. With our work in the SEMAFLEX[1] [7] project, we have already considered this by combining flexible workflow management and knowledge-based document management.

[1] SEMAFLEX is funded by Stiftung Rheinland-Pfalz für Innovation, grant no. 1158.

With the unified approach, we do not only allow data to influence a process instance but to directly control its execution. Like in similar approaches [10] the demand for information within a process specifies the set of activities which can yield this information. As a result, all such activities whose preconditions are satisfied may be candidates for execution. From the data-centric perspective the preconditions are the availability of the required input information. However, through the combination with the constraint-based principle an additional layer for preconditions becomes possible. Such constraints may define direct dependencies between activities and guarantee a predefined execution order.

Nonetheless, data has an essential influence on the process execution and can be delivered by manual activities, system activities, external sources like documents [7] and can be inferred based on the ontology. This mechanism will be explained by the following short example.

3.1   Example

Figure 2 shows a segment of an order process which delivers the information, whether a customer should be served on account or whether an upfront payment is required. The most important part in this example is the required information (Upfront payment), which is represented by a tristate value (true, false, unknown). All data elements are shown as a circle while the activities are presented in rectangles. Activities can be distinguished between manual activities (human symbol) and service activities (cogwheel). Edges between data and activities define the direction of the data flow. The edges between activities represent additional constraints. The transfer of an activity result can be obligated (solid lines) or optional (dashed lines).

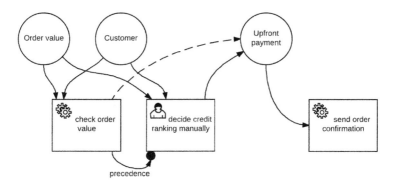

Fig. 2. Segment of an order process

The process is mainly data-driven as the activities depend on the input and the output data. This means that the activities are executable if the potential output data (Upfront payment) is still required and as soon as the necessary input data (Order value, Customer) is available. Since two activities can provide the required output, an additional precedence constraint [11] is defined. This constraint is used to assure that the decision regarding the payment condition is made in a controlled and predefined order, following the enterprise compliance rules. In the following, three different cases for this process segment are discussed and present the basic mechanisms of the introduced unified approach.

Case 1: Enterprise Knowledge. In the first case, we assume that the required information *Upfront payment* is available right from the beginning. The information might be delivered from outside by a document. Alternatively, the customer might be known through previous order-processes and his payments were always according the payment conditions. Thus, the customer is credit-worthy, which is stored in a global knowledge store. Now, the information *Upfront payment* could be deduced by the integrated knowledge base, utilizing the underlying ontology which combines the process data with a global knowledge store. This way and without an explicit activity, the results, produced by the inference mechanisms, have an impact on the further execution of the process. Once the information *Upfront payment* is available, none of the two activities need to be executed.

Case 2: Check Order Value. In the case that the customer is new and thus no information about his credit-worthiness exists in the global knowledge store, the two activities are potentially executable from the data-perspective. Because of the additional constraint, only the activity *check order value* can be executed. As a system-activity, this can be performed without further user interaction. We assume that an upfront payment is required for any unknown customer if the *Order value* is above a certain threshold. In this case, the information *Upfront payment* is true and all other activities become obsolete again. If the *Order value* is below the level, the activity is successfully executed without delivering an answer for *Upfront payment* and the next activity can take over.

Case 3: Manual Decision. This case happens if there is a new customer without knowledge about his credit-worthiness in the global knowledge store, who places an order above a certain *order value*. Now, all preconditions including the precedence constraint are fulfilled for a manual credit decision. A process actor, who is allowed to perform this activity, is asked for a decision, which can be *true* or *false*. However, since this is the last possibility to get an answer, this activity is obligated to deliver a result (represented by the solid line) in case of its execution.

The example has presented the data-oriented enactment of activities of the unified approach. In the execution, it follows the information-state rather than a predefined control-flow and offers a high degree of flexibility. To supervise the flexibility, constraints are used to guard predefined compliance rules.

The briefly presented unified approach combines the two declarative paradigms with a knowledge-based system. Besides the possibilities for an artificial process contribution, the ontology will be utilized for the following BPV approach.

4   A Semantically-Oriented Business Process Visualization

Some studies have introduced adaptation techniques [2] like reduction and aggregation steps to adapt a process graph for control-flow oriented approaches like BPMN. However, these adaptation techniques cannot be easily transferred to the described unified approach because of its declarative and data-oriented principle. In general, the proposed unified approach requires a process visualization which allows to explore and identify the expected possibilities and allows the different user-groups (like process designer, process controller, process actor) to understand the process in its definition and its behavior during the execution.

Additionally, the integrated knowledge base offers new opportunities for a process view. An adapted process visualization can be generated for the specific purpose of each user-group. Considering the possibilities for deduced information, humans have the need for a comprehensible presentation of the artificial process contribution. Since such new information is deduced based on the ontology, the source can be determined and the explanation can be visualized.

4.1   Visualization Concept

The intended visualization concept focuses on three presentation relevant factors. The *user perspective* considers the role of the user and the assigned activities. The *process-state* is the result of the state of each process element and its current influence to the process execution. Both factors are inherent to the situation and cannot be controlled by the user directly. However, the user can express the *personal focus* by an interaction with the BPV. These three factors build the foundation of our visualization concept and define the individual points of interest together. These points of interest (explained in the next section) and the knowledge base will be utilized to generate an adapted process view.

To a large extent, this concept is similar to existing approaches [2]. The new aspect in this concept is the underlying data-centric paradigm as well as the knowledge base, which will be used for the view adaptation.

4.2   Points of Interest

A point of interest (POI) represents a process element which has some particular importance for the process visualization. Since the unified approach follows the data-oriented principles, a process element can either be an activity or a data-element. The three introduced factors (*user perspective, process-state, personal focus*) will be applied to the process elements to identify the elements with an elevated importance. Thus, we get a list of POIs which filter the most important data and activity elements for the following process presentation.

The relevance of each further element depends on the individual meaning related to each POI. This meaning is reflected by the integrated knowledge base and the ontology can be utilized to calculate the relevance for each element. In a nutshell, the closer an element is related to a POI and the more meaningful it is for the execution or understanding of the POI, the higher is its relevance. Such a relevance value can be used for each element for the process adaptation.

4.3   View Adaptation

The view adaptation is the central transformation step of the view generator. With different techniques like reduction and aggregation, single elements up to process segments can be transformed and thus can be presented with a variable granularity. The most relevant elements can be presented in the highest level of detail while the less important elements can be presented in a more abstract view with a lower granularity. In the following, different adaptation methods are introduced, which are partially already examined [2] under the control-flow paradigm. These methods will be transferred to the introduced unified approach and take further advantage of the integrated knowledge base.

Reduction: One possibility to lower the granularity of a segment is the reduction step where single elements are taken out of the view. In existing BPV approaches often system activities where reduced with the expectation that they are less important than user activities. With the introduced approach, the knowledge base is used to calculate the relevance in semantic relation to the POIs.

Assuming that a specific data-element (D1) has just enough relevance for an unchanged presentation, the producing activity of this information might be not important enough and thus maybe reduced from the presentation. Unless the user is not clicking on D1 and thus increasing the relevance of this element by adding his personal focus, the producing activity remains invisible in the BPV.

Aggregation: Another option to reduce granularity is to aggregate a group of activities or data elements to a single representative element. Unlike a reduction step where the elements are completely removed, the representing symbol is still visible and the user can expand it to the fine granularity any time. In existing BPV approaches, the aggregation step is used to combine activities which follow a narrow route within the control-flow and do not split to further activities. Under this new approach, elements can be grouped and aggregated by a similar meaning. For example, data-elements with the same type of relation to a common activity can be replaced by a single element, which substitutes the data-elements.

Assuming that several data-elements (D1, D2, D3) were required for an already executed activity (A1), these data-elements share the same type of relation to the activity (A1). By replacing the data-elements with a single element (D1-3) the granularity of this process segment can be reduced.

Expansion: The most obvious way to adapt a view is lowering the granularity of process segments. However, with view to the underlying ontology new information can also be deduced by inference steps. Referring to the example in Sect. 3.1, the credit-worthy might be stored in a global knowledge store. This source of information is not represented in the process as an explicit activity, nonetheless it has some impact to the process execution. With the method of expansion, the process view can be expanded by further elements which are defined in the ontology and which are not an explicit part of the process definition.

Assuming that data-element (D1) is deduced from existing data (D2) by utilizing the knowledge-base, the process view can expand D1 with a further data-element D2 by representing the relation through an additional edge.

5   Conclusion

In this paper, we have shown that the data-centric as well as the constraint-based approaches follow the same declarative paradigm and can be combined to a unified approach. Through the integration of a knowledge based system, artificial intelligence methods should be able to participate directly in a process execution. This allows a division of labor between humans and an AI system. A new concept for business process visualization was presented to support a common understanding and a comprehensive presentation of the unified approach. The presented adaptation of the process view is taking advantage of the integrated knowledge base and a semantically-oriented process visualization is conceivable. In our ongoing work, we will implement the described approaches in a prototype to prove the presented concepts.

References

1. Bhattacharya, K., Hull, R., Su, J.: A data-centric design methodology for business processes. In: Cardoso, J., van der Aalst, W. (eds.) Handbook of Research on Business Process Modeling, pp. 503–531. IGI Global Publishing, Hershey (2009)
2. Bobrik, R., Reichert, M., Bauer, T.: View-based process visualization. In: Alonso, G., Dadam, P., Rosemann, M. (eds.) BPM 2007. LNCS, vol. 4714, pp. 88–95. Springer, Heidelberg (2007). https://doi.org/10.1007/978-3-540-75183-0_7
3. Bruno, G.: A data-flow language for business process models. Procedia Technol. **16**, 128–137 (2014)
4. Card, S., Mackinlay, J.D., Shneiderman, B.: Readings in Information Visualization: Using Vision to Think. Morgan Kaufmann, San Francisco (1999)
5. Cohn, D., Hull, R.: Business artifacts: a data-centric approach to modeling business operations and processes. IEEE Data Eng. Bull. **32**(3), 3–9 (2009)
6. Di Ciccio, C., Marrella, A., Russo, A.: Knowledge-intensive processes: characteristics, requirements and analysis of contemporary approaches. J. Data Semant. **4**(1), 29–57 (2015)
7. Grumbach, L., Rietzke, E., Schwinn, M., Bergmann, R., Kuhn, N.: SEMAFLEX-semantic integration of flexible workflow and document management. In: CEUR Workshop Proceedings: LWDA, Potsdam, vol. 1670, pp. 43–50 (2016)
8. Kolb, J., Reichert, M.: A flexible approach for abstracting and personalizing large business process models. Appl. Comput. Rev. **13**(1), 6–17 (2013)
9. Kuenzle, V., Reichert, M.: PHILharmonicFlows: towards a framework for object-aware process management. J. Softw. Maint. Evol. **23**(4), 205–244 (2011)
10. Meyer, A., Pufahl, L., Fahland, D., Weske, M.: Modeling and enacting complex data dependencies in business processes. In: Daniel, F., Wang, J., Weber, B. (eds.) BPM 2013. LNCS, vol. 8094, pp. 171–186. Springer, Heidelberg (2013). https://doi.org/10.1007/978-3-642-40176-3_14

11. Pesic, M., Schonenberg, H., van der Aalst, W.M.P.: DECLARE: full support for loosely-structured processes. In: Proceedings - IEEE International Enterprise Distributed Object Computing Workshop, EDOC, pp. 287–298 (2007)
12. Sauer, T., Minor, M., Bergmann, R.: Inverse workflows for supporting agile business process management. In: Proceedings 6th Conference Knowledge Management, vol. 182, pp. 194–203 (2011)

Abduction for Generating Synthetic Traces

Federico Chesani$^{(\boxtimes)}$ ⓘ, Anna Ciampolini ⓘ, Daniela Loreti ⓘ,
and Paola Mello ⓘ

Department of Computer Science and Engineering,
University of Bologna, viale Risorgimento 2, 40136 Bologna, Italy
{federico.chesani,anna.ciampolini,daniela.loreti,paola.mello}@unibo.it

Abstract. In this paper we report our preliminary experience on the design of a generator of synthetic logs. Sometimes real logs might not be available, or their quality might not be good enough: synthetic logs instead can be generated with all the desired features and characteristics. Our tool takes as input a structured workflow model, encoded in the abductive declarative language SCIFF, and provides as output a log containing positive traces, i.e. traces deemed as conformant w.r.t. the model. Distinctive features of our approach are the capability of generating trace templates as well as grounded traces, the possibility of taking into account user-specified constraints on data and timestamps, and the capability of generating traces starting from a user-specified partial trace. Although our tool is still in its preliminary version, we have successfully exploited it to generate synthetic logs of different dimension, thus proving the viability of our approach.

Keywords: Structured workflows · Synthetic log generation
Abduction

1 Introduction

In the Business Process Management (BPM) research community, an important research field is related to Process Mining [1], and in particular to the tasks of learning a model from a log, and of verifying the conformance of a log w.r.t. a model. Roughly speaking, the learning task takes as input a log containing the observed (logged) execution of a process, being the log made of *traces* or *instances* of the process execution, and return a model of the process described in some formal language. The conformance task compares a log towards a model, establishing *how much* and *how well* the model and the log fit each other.

Both the tasks start from a log. However, in a number of real contexts such logs might not be available: for example, a newly designed process might have a model but it might miss any execution log (due to its novelty). Another concern is about the log *quality*, since it can heavily affect both the mining and the conformance checking results. Also, logs are often looked for to compare different mining and conformance checking approaches: clearly, a benchmark suite containing logs with different characteristics is desirable when dealing with novel

© Springer International Publishing AG 2018
E. Teniente and M. Weidlich (Eds.): BPM 2017 Workshops, LNBIP 308, pp. 151–159, 2018.
https://doi.org/10.1007/978-3-319-74030-0_11

approaches to mining and conformance. An answer to the mentioned issues has been provided by generators of synthetic logs, software tools that take as input a model and synthetically generate logs with the desired features. They differentiate each other for the model language they support, and for the features that the generated logs will exhibit [5, 6, 10, 12, 13, 17–20].

In previous works [7, 9] we exploited abduction to complete partial logs and traces, in order to support a notion of *weak compliance* of incomplete traces w.r.t. a process model. The focus was on determining if incomplete logs (traces with missing events, or only partially specified events) might be considered conformant to a process model, and under which conditions. To that end, we exploited an abductive logic programming framework, and in particular the hypotheses-making reasoning capabilities typical of abduction.

In this work we further investigate the link between abduction and process mining, by exploiting the same abductive framework to generate synthetic logs with a number of custom-selectable properties. Among them, the user can select to get trace templates (i.e., with data field not specified), or grounded traces. Users can also select the minimum and/or the maximum length of traces (i.e., the minimum/maximum number of events in a trace), as well as the minimum/maximum number of traces for each control-flow path allowed by the process model. User-defined time- and data-constraints can be added as well, so as to introduce a bias in the synthetic log. Finally, users can provide a partial, incomplete trace and generate a log whose traces complete the partial one given in input.

As a test case, we consider the temporal workflow model presented in [15]. The chosen example is quite meaningful for the presence of temporal constraints, thus providing us the opportunity of testing our solution even in presence of data and time constraints. Starting from the chosen test case, we generated synthetic logs of different dimensions (from 200 KB up to 25 GB), that we plan to use for testing a distributed compliant monitoring architecture based on a Map-Reduce approach [16].

2   Background

In previous works [7, 9] we tackled the problem of evaluating the conformance of a log with incomplete information with respect to a model. We considered as incompleteness dimensions (i) the absence in the log of traces forecasted by the model; (ii) traces with missing events (w.r.t. what prescribed by the model); and (iii) events with only partial information, for example missing the timestamp, some data, or even the activity name itself.

The main idea was then to exploit *abduction*, a well known AI technique, to generate hypotheses about the missing information. Abduction [14] is a general, logic-based technique for making hypotheses, originally thought as a mechanism for providing explanations given some observations. Usually, the hypothesis-making process is not completely unbound: users provide some rules/constraints that guide the hypothesis-making process. Such rules/constraints capture the domain knowledge, so as for example to forbid two contradicting hypotheses. In the BPM setting, we encoded the process model in terms of such rules: hypotheses (made to complete a trace) were generated on the base of what described in the model, interpreting it as a sort of mandatory description of the preferred/expected traces.

From the practical viewpoint, we resorted to exploit the SCIFF Abductive Logic Programming Framework [4], that provides a language based on logic programming for expressing the domain knowledge, together with a proof procedure supporting the abduction reasoning process. Procedural models are represented in terms of SCIFF's *Integrity Constraints* (IC), forward rules of the form *body* → *head*. Roughly speaking, when *body* becomes true, also *head* must be true. The *body* contains conjunctions of special terms with functor **H**, with functor **E** or with functor **ABD**, while the *head* is made up of disjunctions of conjunctions of terms with functor **E** or with functor **ABD**. A term $\mathbf{H}(EvDesc, T)$ stands for the happening/the observation of an event described by *EvDesc* at timestamp T. A term $\mathbf{E}(EvDesc, T_e)$ indicates instead that an event described by *EvDesc* is expected to be observed at the timestamp T_e. Finally, a term $\mathbf{ABD}(EvDesc, T_a)$ is used to indicate that the happening of an event *EvDesc* at time T_a can be hypothesised.

Let us consider a simple model where (the execution of) an activity a must be followed by (the execution of) an activity b, and b must be followed by c. The SCIFF formalisation of such sequence is given by the following ICs:

$$\mathbf{H}(a, T_a) \rightarrow \mathbf{E}(b, T_b) \wedge T_b > T_a \tag{1}$$
$$\mathbf{H}(b, T_b) \rightarrow \mathbf{E}(c, T_c) \wedge T_c > T_b \tag{2}$$

It is worthy to notice that timestamps are subjected to dis-equalities (e.g., $T_b > T_a$), so as to capture the proper ordering imposed by the sequence. Thus, Eq. 1 states that whenever the execution of activity a is observed, then also b is expected to be observed, at a later time.

Incompleteness of a trace can be simply addressed by introducing in the *head* of integrity constraints also the possibility of making hypotheses (i.e., *abducing*) about the happening of events. Let us suppose that activity b is never logged down (because, for example, of a human mistake). A model allowing the absence of b in the traces would be the following:

$$\mathbf{H}(a, T_a) \rightarrow \mathbf{E}(b, T_b) \wedge T_b > T_a \quad \vee \quad \mathbf{ABD}(b, T_b) \wedge T_b > T_a \tag{3}$$
$$\mathbf{H}(b, T_b) \rightarrow \mathbf{E}(c, T_c) \wedge T_c > T_b \tag{4}$$
$$\mathbf{ABD}(b, T_b) \rightarrow \mathbf{E}(c, T_c) \wedge T_c > T_b \tag{5}$$

Equation 3 again captures the consequences of observing the happening of activity a: however, either b is expected later, or the happening of b is hypothesised. Equation 5 has been added: if the happening of b is hypothesised, then the happening of c is expected as well.

3   Generation of a Synthetic Log

A synthetic log generator takes as input a model, expressed in some formal language, and provides as output a log of traces. Several tools and approaches are available for generating synthetic logs (see Sect. 4 for a brief comparison). We identified a number of features and desiderata that, to the best of our knowledge, are not yet covered by a unified approach. In particular:

1. Synthetic logs should contain *positive* traces, i.e. traces deemed as fully conformant with the provided model. If needed, also *negative* traces should be included as well, with both the possibilities of labelling them, or leave them undistinguished from positive traces (for benchmarking purposes).
2. Users should be able to choose between trace templates (where data field and timestamps are left unbound, and variables are put as placeholders), and completely grounded traces (where all the data field and timestamps are filled up with values allowed by the model).
3. Users should be able to specify the minimum/maximum length of each trace, intended as the minimum/maximum number of events included in each trace composing the log.
4. Structured processes implicitly bring the idea of a finite number of different finite flow-paths. Users should be able to specify the maximum number of traces for each possible flow-path allowed by the model.
5. Users should be able to provide additional constraints (w.r.t. the model) on data, as well as on timing of the events, so as to provide some bias to the generated log.
6. Users should be able to provide some trace template (for example, an incomplete/partial trace) and ask for a log whose traces are all instances of the given input trace, so as to support *operational decision support* as discussed in [20].

Our synthetic log generator accomplishes the features above through a three-steps procedure: after properly representing the model in a suitable language (the

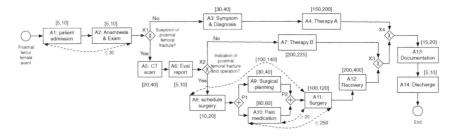

Fig. 1. A workflow for the hospital treatment of femur, taken from [15]. Intervals placed nearby the activities indicate the minimum and maximum temporal duration of each activity. Dashed arrows indicate minimum and maximum time difference between the start/end of an activity and the start/end of another one.

SCIFF language, step (1)), we exploit the SCIFF Proof Procedure to generate all the trace templates (step (2)). Finally, trace templates are grounded with values respecting the constraints imposed by the model (step (3)).

As a test case, let us consider a workflow extended with temporal deadlines as the one proposed in [15], and shown in Fig. 1. The workflow captures the hospital treatment of a femure fracture, with temporal constraints about the duration of each activity, as well as temporal constraints among the start/end of different activities.

3.1   Model Representation Through SCIFF

We do not report here a complete translation of a structured workflow into a SCIFF model: the interested reader can refer to [8] for a detailed introduction. We will point out here that, with respect to our previous works [7,9], here the context implicitly imposes that all the activities are *non-observable*.

To better provide the idea, let us consider activity A2 and the choice point X1 (xor split) in Fig. 1: the model prescribe on one path the execution of activity A3, and on another path A5. The choice point would be represented as:

$$\mathbf{ABD}(a2\_end, T_2) \rightarrow \mathbf{ABD}(a3\_start, T_3) \wedge T_3 > T_2$$
$$\vee \mathbf{ABD}(a5\_start, T_5) \wedge T_5 > T_2 \qquad (6)$$

Similarly, the implicit constraint that links the start of an activity with its end can be mapped as follow:

$$\mathbf{ABD}(a2\_start, T_{2s}) \rightarrow \mathbf{ABD}(a2\_end, T_{2e}) \wedge T_{2e} \geq T_{2s} + 5 \wedge T_{2e} \leq T_{2s} + 10 \qquad (7)$$

Notice however that, in the case of Eq. 7, there are also two constraints ($T_{2e} \geq T_{2s} + 5$ and $T_{2e} \leq T_{2s} + 10$) as the consequence of the minimum and maximum allowed duration of activity A2.

3.2   Generating Trace Templates

Once a model has been properly represented through the SCIFF language, the SCIFF proof procedure can be exploited to generate trace templates. The starting point of each trace is always the special event start, representing the entry point for the workflow. To fulfil the model constraints, the SCIFF will hypothesise the happening of the event a1_start, then the happening of the event a1_end, and so forth.

The SCIFF proof procedure will compute the so called *abductive answer*, i.e. it will provide the set of all the hypotheses (abducibles) that are needed to fulfil the constraints. For example, it will provide the following trace template:

$$T_1 \equiv \{ \mathbf{ABD}(start, T_{start}),$$
$$\mathbf{ABD}(a1\_start, T_{a1\_start}), \ \mathbf{ABD}(a1\_end, T_{a1\_end}),$$
$$\mathbf{ABD}(a2\_start, T_{a2\_start}), \ \mathbf{ABD}(a2\_end, T_{a2\_end}),$$
$$\mathbf{ABD}(a3\_start, T_{a3\_start}), \ \mathbf{ABD}(a3\_end, T_{a3\_end}),$$
$$\mathbf{ABD}(a4\_start, T_{a4\_start}), \ \mathbf{ABD}(a4\_end, T_{a4\_end}), \qquad (8)$$
$$\mathbf{ABD}(a13\_start, T_{a13\_start}), \ \mathbf{ABD}(a13\_end, T_{a13\_end}),$$
$$\mathbf{ABD}(a14\_start, T_{a14\_start}), \ \mathbf{ABD}(a14\_end, T_{a14\_end}),$$
$$\mathbf{ABD}(stop, T_{stop}) \}$$

Timestamps are still shown as variables, even if they are constrained by the model: for example, the model prescribes that $T_{a1\_start} + 5 \leq T_{a1\_end} \leq T_{a1\_start} + 10$. The SCIFF framework integrates Constraint Logic Programming (CLP) techniques, and keeps track of all these constraints (not reported in Eq. 8). SCIFF can be queried also for further abductive answers: the proof procedure will explore all the available alternatives (choice points X1 and X2 in Fig. 1), thus producing all the other trace templates.

3.3   Generating Grounded Traces from Templates

Trace template T_1 does not contain ground values: in particular, all the timestamps are still in the form of a placeholder variable. Generating ground traces is trivial: it suffices to perform the *labeling* step of the CLP solver, in order to get all the possible ground values for all the variables.

Possibly, the grounding step might generate a number of infinite ground traces: for example, activity A1 can simply start at any time point, thus allowing for infinite traces. We overcome this possible issue by imposing two conditions: first of all, the time structure underlying our approach is *discrete*. This means that given any time interval, there is always a finite number of allowed time points belonging to the interval.

The second condition is on the input parameters provided to our model: we require that either the user provides a maximum time for the final activity (namely, the stop activity), or we require the user to specify the maximum number of traces to be generated. In both cases, the SCIFF proof procedure will terminate.

4   Related Work

A number of works [5,6,10,12,13,17–20] focuses on the *Business Process Simulation*, and support log generation through the specification of procedural business process models plus additional information like, e.g., mean execution time of activities, probability of choices, etc. Moreover, a number of approaches support also the simulation of limited resources, their allocation, queues, etc.

Being based on procedural models, these approaches show some limitations when the process is characterized by high variability and allows for many alternative execution paths. In these cases, declarative process models are proven to perform better. In fact, CPN Tools [19] has presented an extension of the traditional procedural-based modeler to graphically add Declare constraints. The work by Di Ciccio et al. [10] present a synthetic log generator that allow the user to define the process model by listing a number of declarative constraints expressed in Declare language. More recently, another log generator based on a declarative approach has been presented in [2,3]. The SCIFF approach can inherently give support to the definition of both declarative as well as procedural models and naturally exploit abduction to enable the generation.

5   Conclusions and Future Works

In this preliminary report, we sketch a novel approach for generating synthetic logs starting from a model. Distinctive features of our solution are the possibility of generating trace templates as well as completely grounded traces, together with the possibility of specifying further user constraints on data and time.

We are currently investigating the possibility of generating also *negative traces*, i.e. traces that are not conformant with the given model. Since the model is described in terms of a set of (a conjunction of) SCIFF Integrity Constraints, a trace would be *negative* if it violates one or more ICs. Starting from such observation, we could think of taking the initial model and "negate" one or more ICs. These new models would be used to feed the proof procedure, thus generating traces that would violate the initial process model. The "negation" of an IC would consist on modifying its *head*, for example by non generating an event, or by generating it with a timestamp that violates the temporal constraints.

Confronting with the existing literature on business process simulation, our approach lacks also a number of features that we plan to add in the near future. Namely, the injection of random errors into positive traces so as to simulate *noise*, useful for benchmarking purposes for mining algorithms, as discussed in [11]. Resources, their management and their allocation is not taken into consideration, although it could be done by adding ad-hoc declarative constraints. Data constraints are currently supported only by directly modifying the SCIFF constraints, while a more high-level support is required to meet non-technical users' needs. Finally, the generation of traces and their grounding does not take into account any probability information: some traces might be more probable (frequent) than others, as well as some data values and activity durations might be more frequent than others.

We plan to make the tool available to the BPM and AI communities: currently, we successfully exploited it to generate logs of various dimensions. For example, the generation of a 200 Kb log containing 600 traces (200 traces for each flow-path allowed by the model in Fig. 1) required few seconds on a medium PC architecture, while a log of 25 Gb took roughly a couple of hours.

References

1. van der Aalst, W., et al.: Process Mining Manifesto. In: Daniel, F., Barkaoui, K., Dustdar, S. (eds.) BPM 2011. LNBIP, vol. 99, pp. 169–194. Springer, Heidelberg (2012). https://doi.org/10.1007/978-3-642-28108-2_19
2. Ackermann, L., Schönig, S.: MuDePs: Multi-perspective Declarative Process Simulation. In: BPM Demo Track 2016. CEUR, vol. 1789, pp. 12–16. CEUR-WS.org (2016). http://ceur-ws.org/Vol-1789/bpm-demo-2016-paper3.pdf
3. Ackermann, L., Schönig, S., Jablonski, S.: Simulation of multi-perspective declarative process models. In: Dumas, M., Fantinato, M. (eds.) BPM 2016. LNBIP, vol. 281, pp. 61–73. Springer, Cham (2017). https://doi.org/10.1007/978-3-319-58457-7_5
4. Alberti, M., Chesani, F., Gavanelli, M., Lamma, E., Mello, P., Torroni, P.: Verifiable agent interaction in abductive logic programming: the SCIFF framework. ACM Trans. Comput. Log. 9(4), 29:1–29:43 (2008). https://doi.org/10.1145/1380572.1380578
5. van den Broucke, S.: Advances in Process Mining: Artificial Negative Events and Other Techniques. Ph.D. thesis, Katholieke Universiteit Leuven, Belgium (2014). https://lirias.kuleuven.be/handle/123456789/459143
6. Burattin, A., Sperduti, A.: PLG: A Framework for the Generation of Business Process Models and Their Execution Logs. In: zur Muehlen, M., Su, J. (eds.) BPM 2010. LNBIP, vol. 66, pp. 214–219. Springer, Heidelberg (2011). https://doi.org/10.1007/978-3-642-20511-8_20
7. Chesani, F., De Masellis, R., Di Francescomarino, C., Ghidini, C., Mello, P., Montali, M., Tessaris, S.: Abducing compliance of incomplete event logs. In: Adorni, G., Cagnoni, S., Gori, M., Maratea, M. (eds.) AI*IA 2016. LNCS (LNAI), vol. 10037, pp. 208–222. Springer, Cham (2016). https://doi.org/10.1007/978-3-319-49130-1_16
8. Chesani, F., De Masellis, R., Francescomarino, C.D., Ghidini, C., Mello, P., Montali, M., Tessaris, S.: Abducing compliance of incomplete event logs. CoRR abs/1606.05446 (2016). http://arxiv.org/abs/1606.05446
9. Chesani, F., De Masellis, R., Francescomarino, C.D., Ghidini, C., Mello, P., Montali, M., Tessaris, S.: Abducing workflow traces: a general framework to manage incompleteness in business processes. In: ECAI 2016. Frontiers in Artificial Intelligence and Applications, vol. 285, pp. 1734–1735. IOS Press (2016). https://doi.org/10.3233/978-1-61499-672-9-1734
10. Di Ciccio, C., Bernardi, M.L., Cimitile, M., Maggi, F.M.: Generating event logs through the simulation of declare models. In: Barjis, J., Pergl, R., Babkin, E. (eds.) EOMAS 2015. LNBIP, vol. 231, pp. 20–36. Springer, Cham (2015). https://doi.org/10.1007/978-3-319-24626-0_2
11. Di Ciccio, C., Mecella, M., Mendling, J.: The effect of noise on mined declarative constraints. In: Ceravolo, P., Accorsi, R., Cudre-Mauroux, P. (eds.) SIMPDA 2013. LNBIP, vol. 203, pp. 1–24. Springer, Heidelberg (2015). https://doi.org/10.1007/978-3-662-46436-6_1
12. Garcia-Banuelos, L., Dumas, M.: Towards an open and extensible business process simulation engine. In: CPN Workshop 2009 (2009)
13. van Hee, K.M., Liu, Z.: Generating benchmarks by random stepwise refinement of Petri Nets. In: Proceedings of PETRI NETS 2010, CEUR, vol. 827, pp. 403–417. CEUR-WS.org (2010), http://ceur-ws.org/Vol-827/31_KeesHee_article.pdf

14. Kakas, A.C., Kowalski, R.A., Toni, F.: Abductive logic programming. J. Log. Comput. **2**(6), 719–770 (1992). https://doi.org/10.1093/logcom/2.6.719

15. Kumar, A., Sabbella, S.R., Barton, R.R.: Managing controlled violation of temporal process constraints. In: Motahari-Nezhad, H.R., Recker, J., Weidlich, M. (eds.) BPM 2015. LNCS, vol. 9253, pp. 280–296. Springer, Cham (2015). https://doi.org/10.1007/978-3-319-23063-4_20

16. Loreti, D., Chesani, F., Ciampolini, A., Mello, P.: Distributed compliance monitoring of business processes over MapReduce architectures. In: ICPE 2017, pp. 79–84. ACM (2017). https://doi.org/10.1145/3053600.3053616

17. Medeiros, A.K.A.D., De Medeiros, A.K.A., Günther, C.W.: Process mining: using CPN tools to create test logs for mining algorithms. In: Proceedings of the 6th works on practical use of coloured Petri Nets and the CPN tools, pp. 177–190 (2005)

18. Stocker, T., Accorsi, R.: Secsy: A security-oriented tool for synthesizing process event logs. In: Proceedings of the BPM Demo Sessions 2014. CEUR, vol. 1295, p. 71. CEUR-WS.org (2014). http://ceur-ws.org/Vol-1295/paper13.pdf

19. Westergaard, M., Slaats, T.: CPN tools 4: A process modeling tool combining declarative and imperative paradigms. In: Proceedings of the BPM Demo sessions 2013, CEUR, vol. 1021. CEUR-WS.org (2013). http://ceur-ws.org/Vol-1021/paper_3.pdf

20. Wynn, M.T., Dumas, M., Fidge, C.J., ter Hofstede, A.H.M., van der Aalst, W.M.P.: Business process simulation for operational decision support. In: ter Hofstede, A., Benatallah, B., Paik, H.-Y. (eds.) BPM 2007. LNCS, vol. 4928, pp. 66–77. Springer, Heidelberg (2008). https://doi.org/10.1007/978-3-540-78238-4_8

13th International Workshop
on Business Process Intelligence
(BPI 2017)

Introduction to the 13th International Workshop on Business Process Intelligence (BPI 2017)

Boudewijn van Dongen[1], Jochen De Weerdt[2], Andrea Burattin[3], and Jan Claes[4]

[1] Eindhoven University of Technology, Eindhoven, The Netherlands
[2] KU Leuven, Leuven, Belgium
[3] Technical University of Denmark, Kongens Lyngby, Denmark
[4] Ghent University, Ghent, Belgium

1 Aims and Scope

Business Process Intelligence (BPI) is a growing area both in industry and academia. BPI refers to the application of data- and process-mining techniques to the field of Business Process Management. In practice, BPI is embodied in tools for managing process execution by offering several features such as analysis, prediction, monitoring, control, and optimization.

The main goal of this workshop is to promote the use and development of new techniques to support the analysis of business processes based on run-time data about the past executions of such processes. We aim at bringing together practitioners and researchers from different communities, e.g. Business Process Management, Information Systems, Database Systems, Business Administration, Software Engineering, Artificial Intelligence, and Data Mining, who share an interest in the analysis and optimization of business processes and process-aware information systems. The workshop aims at discussing the current state of ongoing research and sharing practical experiences, exchanging ideas and setting up future research directions that better respond to real needs. In a nutshell, it serves as a forum for shaping the BPI area.

The 13th edition of this workshop attracted 16 international submissions. Each paper was reviewed by at least three members of the Program Committee. From these submissions, the top eight were accepted as full papers for presentation at the workshop.

The papers presented at the workshop provide a mix of novel research ideas, evaluations of existing process mining techniques, as well as new tool support. *Burattin and Carmona* propose a framework for online conformance checking. *Deeva, De Smedt, De Koninck and De Weerdt* compared process mining and sequence mining techniques for dropout prediction in MOOCs. *Korneef, Solti, Leopold and Reijers* propose a probabilistic approach towards identifying most probable alignments. *Sanchez-Charles, Carmona, Muntés-Mulero and Solé* investigate the use of word embedding for reducing the amount of labels in an event log by combining events with semantically similar names. *Seeliger, Stein and Mühlhäuser* present a novel approach which provides suggestions for redesigning business processes by using discovered as-is process models from event logs and apply motif-based graph adaptation. *Fani Sani, Van Zelst and van der Aalst* address the problem of complex and incomprehensible discovered process models with a general purpose filtering method that exploits observed conditional probabilities between sequences of activities. *Syamsiyah, van Dongen and van der Aalst* focus on recurrent process mining, i.e. the application of process discovery to systems from which data can be extracted near real time, by keeping an intermediate structure persistent in the database thus reducing the time to rediscover process models from the same data source. Finally, *Rehse, Fettke and Loos* analyse the influence of unobserved behaviour on the quality of discovered process models.

This year, the BPI workshop is also co-located with the second Process-Discovery Contest, organized by Josep Carmona, Massimiliano de Leoni, Benoit Depaire and Toon Jouck. With the patronage of the IEEE Task Force on Process Mining, the contest aims to assess tools and techniques that discover business process models from event logs. Compared with the 2016 edition, this year the contest aims to ensure that the models provide business values for process owners. Another change compared to 2016, is the introduction of trace completeness. Five out of the 10 event logs marked are characterized by containing 20% of incomplete traces. Those traces are incomplete in the sense that they are missing the last events. This is very common in reality because event logs are usually extracted from information systems in which a certain number of process executions are still being carried on. The objective is to compare the effectiveness of techniques to discover process models that provide a proper balance between "overfitting" and "underfitting". For the purpose of the contest, 10 "reference" models were created. For each process model, a perfectly-fitting training event log was generated. These training logs were used by the contestants to discover 10 process models. Contestants were allowed to use any technique or combination of techniques. The winner is the contestant that could discover process models that are the closest to the original process models. To assess this, a classification perspective was used. For every process model, an undisclosed "reference" test log was created containing 20 traces, of which 10 positive traces (traces recording behavior compliant with the "reference" model) and 10 negative (the trace recording behavior not compliant with the "reference" model). The winner is the group that discovers the models with the highest accuracy, namely which contains the largest number of true positive and the lowest number of true negative traces, within the "reference" model. As an example, a false positive is a trace that is complaint with the discovered model but, in fact, is not complaint with the "reference" model. The winner was the team of P. Dixit and H. Garcia Caballero, who generated accurate models (98.5%) that were considered as highly understandable by the jury.

As with previous editions of the workshop, we hope that the reader will find this selection of papers useful to keep track of the latest advances in the BPI area, and we are looking forward to keep bringing new advances in future editions of the BPI workshop.

2 Workshop Co-organizers

Boudewijn van Dongen	Eindhoven University of Technology, The Netherlands
Jochen De Weerdt	KU Leuven, Belgium
Andrea Burattin	Technical University of Denmark, Denmark
Jan Claes	Ghent University, Belgium

3 Program Committee

Joos Buijs	Eindhoven University of Technology, The Netherlands
Josep Carmona	Universitat Politècnica Catalunya, Spain
Raffaele Conforti	Queensland University of Technology, Australia
Johannes De Smedt	The University of Edinburgh, UK
Benoit Depaire	Universiteit Hasselt, Belgium
Claudio Di Ciccio	Vienna University of Economics and Business, Austria
Chiara Di Francescomarino	Fondazione Bruno Kessler – IRST, Italy
Marlon Dumas	University of Tartu, Estonia
Diogo R. Ferreira	IST, University of Lisbon, Portugal
Gianluigi Greco	University of Calabria, Italy
Daniela Grigori	Laboratoire LAMSADE, University Paris-Dauphine, France
Mieke Jans	Universiteit Hasselt, Belgium
Gert Janssenswillen	Universiteit Hasselt, Belgium
Anna Kalenkova	Higher School of Economics, Russia
Michael Leyer	University of Rostock, Germany
Fabrizio Maggi	University of Tartu, Estonia
Jan Mendling	Wirtschaftsuniversität Wien, Austria
Steven Mertens	Ghent University, Belgium
Jorge Munoz-Gama	Pontificia Universidad Católica de Chile, Chile
Viara Popova	University of Tartu, Estonia
Manfred Reichert	University of Ulm, Germany
Pnina Soffer	University of Haifa, Israel
Andreas Rogge-Solti	Vienna University of Economics and Business, Austria
Suriadi Suriadi	Queensland University of Technology, Australia
Toon Jouck	Universiteit Hasselt, Belgium
Seppe vanden Broucke	KU Leuven, Belgium
Eric Verbeek	Eindhoven University of Technology, The Netherlands
Matthias Weidlich	Humboldt-Universität zu Berlin, Germany
Hans Weigand	Tilburg University, The Netherlands

A Framework for Online Conformance Checking

Andrea Burattin[1,2(✉)] and Josep Carmona[3]

[1] Technical University of Denmark, Kongens Lyngby, Denmark
andbur@dtu.dk
[2] University of Innsbruck, Innsbruck, Austria
[3] Universitat Politècnica de Catalunya, Barcelona, Spain
jcarmona@cs.upc.edu

Abstract. Conformance checking – a branch of process mining – focuses on establishing to what extent actual executions of a process are in line with the expected behavior of a reference model. Current conformance checking techniques only allow for *a-posteriori* analysis: the amount of (non-)conformant behavior is quantified after the completion of the process instance. In this paper we propose a framework for *online conformance checking*: not only do we quantify (non-)conformant behavior as the execution is running, we also restrict the computation to constant time complexity per event analyzed, thus enabling the online analysis of a *stream* of events. The framework is instantiated with ideas coming from the theory of regions, and state similarity. An implementation is available in ProM and promising results have been obtained.

Keywords: Online process mining · Conformance checking
Event stream

1 Introduction

Process mining [1] represents an important research and industrial topic comprising the analysis of data regarding business processes in order to extract knowledge. Within process mining, different problems are typically identified and, in this paper, we focus on *conformance checking*. Conformance checking techniques, given as input a reference process model and an execution trace, compute the extent to which the executed actions conform the given model. Since most information systems allow for a certain amount of flexibility and deviations, conformance checking represents an extremely valuable tool.

All techniques available nowadays require a complete trace in order to calculate their conformance. From a business point of view, however, this represents an important limitation: if the trace is already finished, the countermeasures needed to fix the deviation can be implemented at a very late stage (i.e., when the process instance is already completed). In this paper, we drop such requirement and present a technique capable of computing the conformance for *running* process instances. Therefore, if a deviation from the reference behavior is observed, the

© Springer International Publishing AG 2018
E. Teniente and M. Weidlich (Eds.): BPM 2017 Workshops, LNBIP 308, pp. 165–177, 2018.
https://doi.org/10.1007/978-3-319-74030-0_12

system notices immediately the problem and allows for an immediate response. When these errors are accumulating, the "seriousness" of the process instance is raised, thus providing stronger alerts to the process administrator. In this paper, we focus on imperative process models, such as Petri nets.

In the context of this paper, with the term *online* we refer to the type of input of our technique: we assume to have an *event stream* which, basically, is a *data stream* of events. According to [4,5,15], a data stream consists of an unbounded sequence of data items which are generated at very high throughput. To cope with such data streams, in the literature, typically the following assumptions are made: *(i)* each item is assumed to contain just a small and fixed number of attributes; *(ii)* algorithms processing data streams should be able to process an infinite amount of data, without exceeding memory limits; *(iii)* the amount of memory available to an algorithm is considered finite, and typically much smaller than the data observed in a reasonable span of time; *(iv)* there is a small upper bound on the time allowed to process an item, e.g. algorithms have to scale linearly with the number of processed items: typically the algorithms work with one pass of the data; *(v)* stream "concepts" (i.e. models generating data) are assumed to be stationary or evolving. The literature reports several algorithms for the analysis of data stream [4,13,14]. However, typically these works cope with different problems, such as classification, frequency counting, time series analysis, and changes diagnosis (concept drift detection).

In the remainder of the paper, Sect. 2 presents the state of the art. Sect. 3 describes the technical details of our proposal, while Sect. 4 presents the implementation and performance results. Sect. 5 concludes the paper.

2   Related Work

In [23], authors compared an event log with a Petri net to compute *fitness* and *appropriateness* measures. A different family of approaches relies on *alignments* [3]: in [2] the idea is to "align" a given trace with the most similar one that can be generated by the given model. Optimized versions have also been proposed [6,20,21]. All these techniques, however, cannot be applied in online settings since they require a complete trace. The main contribution of this paper is to drop this requirement, as we can compute the conformance during the execution of the process, and not *a-posteriori*.

Online process mining has also been investigated, but just concerning the control-flow discovery problem: algorithms generating Petri nets [7,12,24] as well as Declare models [10,11,17] have been proposed. These approaches, however, are not capable of checking the conformance. Mixed approaches to discovery and guarantee conformance values with one pass over data are available [16] but these still require a finite log with complete traces.

A related field of research is operational support. In this case, the system is capable of providing contextual information for running process instances, e.g. [18,19] for the Declare context. However, as soon as a deviation is observed, corresponding Declare constraint are marked as permanently violated. In conformance checking, instead, the behavior might come back to a normal state.

3   Proposed Approach

Given a process model, we want to analyze an event stream, in order to detect (and notify to the process analyst) running cases that are deviating from the behavior prescribed by the model. We assume our input process model is represented as a Petri net [22] and we are going to leverage the notion of region theory and states similarity to achieve our goal.

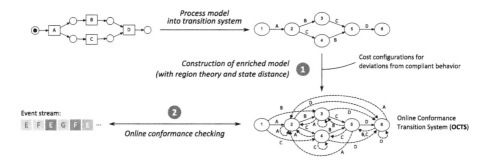

Fig. 1. Approach representation. Circled numbers represent the involved steps.

The strictness of the rules governing the online scenario is playing a fundamental role in this case. In particular, most recent approaches for conformance checking on Petri nets are based on finding the *optimal alignment* between an observed trace and the *closest* possible trace allowed by the model. By moving into the online scenario, we know in advance the impossibility to achieve the same goal (i.e., find the optimal alignment). This is due to the impracticability of backtracking operations while analyzing an event of the stream. To avoid these operations, we devised a two-steps approach as depicted in Fig. 1: we first (i.e., offline, before the analysis) embed all computations in an augmented model, and then we perform the online analysis on such augmented model. The model extension has to provide information on how to deal with any uncompliant scenarios and these "wrong behaviors" are associated with costs larger than 0. Then, with such a model, it is possible to process the stream by analyzing one event at a time in constant time and, therefore, fulfilling the requirements for online algorithms. The online analysis consists of *replaying* the events of each trace of the stream on the model and accumulating the costs associated with each execution. All running process instances with costs larger than 0 are deviating from the reference model. Additionally, the larger the cost the more problematic is the process instance.

In online scenarios, it is not always relevant to mimic the concept of alignments. For example, in very critical environments, we might want to immediately raise alarms when deviations take place. In such cases, we need to extend the model with *sink states* collecting all deviations. We might also associate different costs to different sink connections, thus providing fine-grained alarm levels.

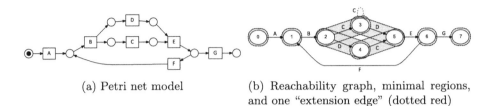

(a) Petri net model   (b) Reachability graph, minimal regions, and one "extension edge" (dotted red)

Fig. 2. A simple Petri net and the corresponding reachability graph. (Color figure online)

In this paper, however, we would like to leverage some alignments concepts. In order to precompute sub-optimal solutions that are similar to optimal alignments, we will use both region theory and states similarity. For example, considering the model depicted in Fig. 2a, and corresponding reachability graph in Fig. 2b we can highlight the corresponding 8 minimal regions (in dashed line). Informally, a region (e.g., the set of states $\{2, 3\}$ in Fig. 2b), denotes a set of states for which arcs have an homogeneous relation. Each region corresponds to a place in the Petri net (e.g., the aforementioned region corresponds to place between transition B and D in Fig. 2a. Let's now assume the following trace $\langle A, B, C, C \rangle$. After executing $\langle A, B, C \rangle$ the trace reaches state 3. From this state, however, a new execution of C (which represents a deviation from reference behavior) should remain in state 3 since such state belongs to a region where C does not change the behavior of the model. Therefore, we extended our transition system with such self loop (highlighted in dotted red). Consider now a partial execution $\langle A, C \rangle$. After executing A we have uncompliant behavior and we would like to synchronize the execution with the model again. To do that, we should connect state 1 with a state which has a transition labeled C entering (i.e., 3 or 5). In this case, it is important to check the "similarity" of the assumptions of two target states. In this paper, we analyze these cases by checking the activities that lead to the possible states and considering the most likely configuration.

Concerning the computation of regions, please note this is a very expensive operation. However, the transition system we use as input is actually generated from a Petri net. And, in this case, each place of the Petri net defines a region. Therefore, given the set of reachable states and a place, all those states where the place has a token define the corresponding region.

The rest of the paper uses these preliminary definitions:

Definition 1 (Sequence). *Given the first n positive natural numbers $\mathbb{N}_n^+ = \{1, 2, \ldots, n\}$ and a target set A, a sequence σ is a function $\sigma : \mathbb{N}_n^+ \to A$. We say that σ maps indexes to the corresponding elements of A. For simplicity, in this text, we refer a sequence using its string interpretation: $\sigma = \langle a_1, a_2, \ldots, a_n \rangle$, where $a_i = \sigma(i)$ and $a_i \in A$.*

We assume typical operators are available over sequences and behave as expected. For example, given a sequence $\sigma = \langle a_1, a_2, \ldots, a_n \rangle$ and an element a, we have $a \in \sigma$ (reads "element a is contained in σ") if $\exists i \in \mathbb{N}_n^+$ such that $a_i = a$. In the context of this work, we refer to a trace as a sequence of events. Formally:

Definition 2 (Trace). *Given a set of activities A (e.g., the tasks of a process model), a trace T of length n is a sequence $T : \mathbb{N}_n^+ \to A$. Activities are grouped in the same trace when they are part of the same process instance.*

Please note that, in the online context, though we assume an infinite number of traces, the length of each of them is typically assumed finite.

Definition 3 (Event Stream). *Given the event universe $\mathcal{E} = A \times \mathcal{C}$, where A is the set of activities and $\mathcal{C}$ is the set of possible case ids, an event stream Ψ is an infinite sequence $\Psi : \mathbb{N}^+ \to \mathcal{E}$.*

3.1   Construction of the Enriched Model

The input of our approach is a Petri net:

Definition 4 (Petri net). *A Petri net N is a tuple $N = (P, T, F)$ where P is a set of places; T is a non-empty, finite, set of transitions, such that $P \cup T = \emptyset$; and $F \subseteq (P \times T) \cup (T \times P)$ is a flow relation.*

Given a Petri net $N = (P, T, F)$, a *marking* of N is a function $M : P \to \mathbb{N}_0$ mapping each places to the number of tokens it contains. The set of all possible markings is denoted with $\mathbb{M}$.

Definition 5 (Petri net system). *A Petri net system PS is a tuple $PS = (P, T, F, m_0)$ where $N = (P, T, F)$ represents a given a Petri net and $m_0 \in \mathbb{M}$ is the initial marking of N.*

As explained before, in order to compute in advance all possible configurations we may have to deal with, it is necessary to construct a behavioral model describing the different configurations and their interactions. To represent such model, we use a labeled transition system:

Definition 6 (Labeled transition system). *A labeled transition system TS is a tuple $TS = (S, \Sigma, \delta)$ where S is a finite set of states; Σ is a finite alphabet; and $\delta \subseteq S \times \Sigma \times S$ is a state transition relation.*

Converting a Petri net into a labeled transition system is a well-studied operation and, in particular, it results in the construction of the so-called *reachability graph*. The reachability graph is finite only if the starting Petri net is bounded, but algorithms which deal with unbounded cases to create a *coverability graph*[1] have been proposed. The idea is to map reachable markings of the Petri net system into states of the transition system (i.e., S represents the subset of $\mathbb{M}$ which is reachable). State transitions, in turn, connect different reachable markings and are labeled according to the Petri net transition leading to the target state. In order to compute online conformance checking, however, we need to extend the transition system definition in order to deal with initial state and costs. Therefore, we define:

---

[1] The *coverability graph*, actually, does not represent a good transition system for conformance purposes, as it allows for more behavior with respect to the original Petri nets. Therefore, in this paper, we assume that the given Petri net is bounded. This assumption is typically fulfilled in many real-world applications.

Definition 7 (Extended transition system). *Given a transition system $T = (S, \Sigma, \delta)$ we define an extended transition system as $T_{ext} = (S, \Sigma, \delta, w, s_0)$ where $w : \delta \to \mathbb{N}_0$ is a cost function which associates transitions to a cost, and $s_0 \in S$ is an initial state.*

Moreover, given a Petri net system $PS = (P, T, F, m_0)$, and the corresponding transition system $TS = (S, \Sigma, \delta)$ (e.g., the reachability graph) we can construct its extended transition system by considering all states, all transitions and all labels of the transition systems and setting all initial weights to 0 (i.e., $w = \{(d, 0) \mid d \in \delta\}$) and associating s_i to the state corresponding to m_0. On top of an (extended) transition system, it is possible to define the concept of *region*.

Definition 8. *Given a transition system $TS = (S, \Sigma, \delta)$, let $S' \subseteq S$ be a subset of states and $\sigma \in \Sigma$ be a letter of the alphabet. We define:*

$$\text{nocross}(\sigma, S') \equiv \exists (s_1, \sigma, s_2) \in \delta : s_1 \in S' \Leftrightarrow s_2 \in S'$$
$$\text{enter}(\sigma, S') \equiv \exists (s_1, \sigma, s_2) \in \delta : s_1 \notin S' \wedge s_2 \in S'$$
$$\text{exit}(\sigma, S') \equiv \exists (s_1, \sigma, s_2) \in \delta : s_1 \in S' \wedge s_2 \notin S'$$

Based on these conditions, we can define a region:

Definition 9 (Region). *Given a transition system $TS = (S, \Sigma, \delta)$, a set of states $R \subseteq S$ is called* region *if, for all $\sigma \in \Sigma$, both these conditions are fulfilled:*

- $\text{enter}(\sigma, R) \Rightarrow \neg\text{nocross}(\sigma, R) \wedge \neg\text{exit}(\sigma, R)$;
- $\text{exit}(\sigma, R) \Rightarrow \neg\text{nocross}(\sigma, R) \wedge \neg\text{enter}(\sigma, R)$.

Let R and R' be regions of TS. R is minimal *if there is no R' such that $R' \subset R$.*

Informally, a region is a subset of states where all transitions with the same label share the same enter/exit relationship. Algorithms for the identification of minimal regions have been proposed in the literature.

An extended transition system of a Petri net allows for the replay just of traces that conform the process and, to be general enough, we have to consider deviations. To do so, we add additional transitions to the system, associating them with costs larger than 0. In the end, the transition system has to allow the execution of all possible events from any given state. We call such transition system an *online conformance transition system* (OCTS):

Definition 10 (Online Conformance Transition System (OCTS)). *An extended transition system $T_{ext} = (S, \Sigma, \delta, w, s_i)$ is an OCTS if: $\forall s \in S, \forall \sigma \in \Sigma$ we have $|\delta(s, \sigma)| = 1$.*

An OCTS is always deterministic since, for each state, it has exactly one transition for each label in our alphabet of possible transitions. An OCTS can be used to *replay* traces:

Definition 11 (Replay). *Given an OCTS $O = (S, \Sigma, \delta, w, s_i)$ and a (partial) trace $t = \langle t_1, \ldots, t_n \rangle$, a replay of trace t in O is $R_t^O = \langle \delta_1, \ldots, \delta_n \rangle$ such that for all $(s_i^s, l_i, s_i^t) \in R_t^O$ we have $l_i = t_i$ (i.e., the transitions of the replay correspond to the activities of the trace) and $\forall j \in \{1, \ldots, n-1\} s_j^t = s_{j+1}^s$ (i.e., the target state of each transition corresponds to the source state of the following one). The cost of the replay R_t^O is: $cost(R_t^O) = \sum_{d \in R_t^O} w(d)$.*

Since OCTSs are deterministic, given an OCTS O and a trace t, the replay R_t^O is unique. Necessary additional properties of an OCTS O are: *(i)* given a conformant trace t, then $cost(R_t^O) = 0$; *(ii)* given a non-conformant trace t, then $cost(R_t^O) > 0$.

The way a transition system of a Petri net is extended into an OCTS represents how we are dealing with deviations. Different ways of dealing with deviations might be implemented in this framework and, in the rest of this section, we suggest one. Let's, for example, consider the process in Fig. 2a with its corresponding reachability graph (cf. Fig. 2b) and the scenario in which the replay reached state 1 (i.e., the trace $\langle A \rangle$ was observed). At this point, if the stream contains activity C, there are three choices: *(i)* to ignore C, or to execute C by assuming to go into state *(ii)* 3 or *(iii)* 5. In this case, we prefer to execute C to state 3 and this is due to the "contextual" information: considering the past histories of the traces leading to the two states, 3 is more similar to 1 with respect to 5. In some other situations, instead, we'll stay in the same state even though we did observe an unexpected activity. This option (i.e., what we indicate with option *(i)*) is considered if in any of the regions the current state belongs to there is a transition that is labeled as the observed activity and that does not cross the border of the region. The rationale is that the occurrence of activities that do not cross regions are not affecting the local state of the underlying system, and therefore it can be assumed that whilst an activity was observed, the system remains in the current state.

A formal representation of the enrichment approach is reported in Algorithm 1. It takes as input an extended transition system (e.g., created starting from the reachability graph), and 3 cost parameters. The procedure is structured in 3 parts: the first (line 1) is in charge of setting the costs to 0 (i.e., correct behavior) for the transitions already in place. The second part (lines 2–5) adds to each state the possibility to replay any activity not belonging to the process alphabet (i.e., $* \setminus \Sigma$). The third part of the algorithm (lines 6–19) is in charge of extending the set of transitions to cope with deviations. The algorithm needs to process each state (line 6) in order to allow the execution of any event (line 8, which filter those not already in place). At this stage there are two options. The first case deals with transitions that are not crossing the regions the current state belongs to (lines 7 and 9–12). In this case, we just add a self-loop (line 10) and set the proper cost (line 11). If there's no transition with the same label in the region, then we first select the candidate states (line 13, those with and incoming transition labeled as the activity we're dealing with) and then we pick the candidate maximizing the cosine similarity of the vector representation of

---

Algorithm 1. Enrich an extended transition system into an OCTS

Input: $T = (S, \Sigma, \delta, w, s_i)$: an extended transition system
        c_s: cost of skipping the activity
        c_j: cost of jumping to the next synchronous move
        c_u: cost of activities not in the alphabet

    ▷ Set initial costs to 0 for all conformant transitions
1 **foreach** $d \in \delta$ **do** $w(d) \leftarrow 0$

    ▷ Add self loops for activities not in the alphabet
2 **foreach** $s \in S$ **do**
3      $\delta \leftarrow \delta \cup (s, * \setminus \Sigma, s)$
4      $w \leftarrow w \cup ((s, * \setminus \Sigma, s), c_u)$
5 **end**

    ▷ Add transitions to deal with deviations
6 **foreach** $s \in S$ **do**
      ▷ Construct the set of states sharing a region with s
7      $R \leftarrow \bigcup_{S' \in Regions(T)} \{s' \in S \mid s \in S' \wedge s' \in S'\}$
      ▷ Consider all possible following activities except those allowed by the model
8   **foreach** $e \in \Sigma \setminus \{e' \mid (s, e', s') \in \delta\}$ **do**
        ▷ Check if at least 1 transition labeled e connects 2 states in the regions of s
9   **if** $e \in \{e' \mid (s^s, e', s^t) \in \delta \wedge s^s \in R \wedge s^t \in R\}$ **then**
          ▷ Option 1: Skipping the activity
10          $\delta \leftarrow \delta \cup (s, e, s)$ ▷ Add a self loop when e is observed
11          $w \leftarrow w \cup ((s, e, s), c_s)$ ▷ Add proper cost for the self loop
12   **else**
          ▷ Option 2: Align to synchronous move
13          $C \leftarrow \{s^t \mid (s^s, e, s^t) \in \delta\}$ ▷ Set of candidate states
          ▷ State maximizing cosine similarity with s. Details about vec on the text
14          $s_{goal} \leftarrow \text{argmax}_{c \in C} {vec(c, e) \cdot vec(s, e)}/{\|vec(c, e)\| \| vec(s, e)\|}$
15          $\delta \leftarrow \delta \cup (s, e, s_{goal})$
16          $w \leftarrow w \cup ((s, e, s_{goal}), c_j)$
17   **end**
18   **end**
19 **end**

---

the candidate with the current state (line 14). Corresponding edges and costs (lines 15 and 16) are added to create the OCTS.

The vectorial representation of a state s, considering the target activity execution e, indicated as $vec(s, e)$, consists of an n-dimensional vector, where $n = |\Sigma|$ (i.e., the size of the alphabet, which is the number of different transition labels in the original Petri net) and where components correspond to letters of our alphabet. Each v_i, referring to letter Σ_i, is valued as follow:

- if $\Sigma_i = e$ (i.e., if the letter refers to the target activity), then the value is 0;
- if there exists a path from the start state to the current state s, containing label Σ_i, then the value is 1;
- in all other cases, the value is 0.

Considering again the example reported in Fig. 2b, given state 1 and activity C, we have the following representations:

$$
\begin{array}{c}
A\ B\ C\ D\ E\ F\ G \\
vec(1,\ C) = [1\ 0\ 0\ 0\ 0\ 0\ 0]
\end{array}
\qquad
\begin{array}{c}
A\ B\ C\ D\ E\ F\ G \\
vec(3,\ C) = [1\ 1\ 0\ 0\ 0\ 0\ 0] \\
vec(5,\ C) = [1\ 1\ 0\ 1\ 0\ 0\ 0]
\end{array}
$$

The cosine similarity between $vec(1, C)$ and $vec(3, C)$ is 0.71, whereas the similarity between $vec(1, C)$ and $vec(5, C)$ is 0.58. For this reason, state 3

Algorithm 2. Online conformance checking

Input: $O = (S, \Sigma, \delta, w, s_i)$: the OCTS of the reference model
$\quad\quad\quad \Psi$: event stream
$\quad\quad\quad m$: maximum number of parallel instances

1 $M_O : C \rightarrow S \times \mathbb{N}^+ \times \mathbb{N}^+$ ▷ Hash map which, given a case id, returns a tuple with a pointer to the current state in O, the cost of the process instance so far, and the time of last update

2 **forever do**

3 $\quad$ $(a, c) \leftarrow observe(\Psi)$ ▷ Obtain a new event from stream Ψ

4 $\quad$ **if** $analyze((a, c))$ **then**

$\quad\quad$ ▷ Obtain the replayer status for the given process instance

5 $\quad\quad$ $(state, cost, time) \leftarrow M_O(c)$

6 $\quad\quad$ **if** $(state, cost, time) = \bot$ **then**

$\quad\quad\quad$ ▷ There is no replayer associated, a new one is needed

7 $\quad\quad\quad$ $M_O(c) \leftarrow (state, cost, time) \leftarrow (s_i, 0, now)$

8 $\quad\quad$ **end**

$\quad\quad$ ▷ Fetch the transition to follow (OCTSs are deterministic: only one transition is labeled a)

9 $\quad\quad$ $(state, a, new\text{-}state) \in \{(d_s, \sigma, d_t) \in \delta \mid d_s = s_i, \sigma = a\}$

10 $\quad\quad$ $M_O(c) \leftarrow (new\text{-}state, cost + w((state, a, new\text{-}state)), now)$ ▷ Replay

$\quad\quad$ ▷ Cleanup the map, removing old elements

11 $\quad\quad$ **if** $|M_O| > m$ **then**

12 $\quad\quad\quad$ $R \leftarrow \text{argmin}_{(c,s,c,t) \in M_O} \{t\}$ ▷ Get oldest elements in M_O

13 $\quad\quad\quad$ Remove R from M_O

14 $\quad\quad$ **end**

15 $\quad$ **end**

16 **end**

---

is preferred. Therefore, the algorithm will introduce the edge with label C connecting state 1 and state 3[2].

Mapping the construction of an OCTS to alignments, please note that edges with cost 0 correspond to synchronous moves. Self loops with cost larger than 0 correspond to log moves. Those edges with cost larger than 0 that are not self loops, correspond to model moves on silent actions (i.e., there's a change in the model state) followed by a synchronous move. By using such mapping an alignment can be provided for the latest events observed. Please note, however, this alignment might be sub-optimal.

3.2 Online Conformance

Given an OCTS, the actual online conformance procedure is reported in details in Algorithm 2. Specifically, the algorithm expects an OCTS and an infinite event stream as input, as well as the maximum number of process instances that we expect to have in parallel. Then, the algorithm constructs a hash map M_O (line 1) which, given the case id of an observed event, returns a tuple containing, for that specific instance, the state of the OCTS reached so far, the deviations cost until now, and a numerical representation of the last update (e.g., the Unix timestamp). The algorithm, then, begins the actual online procedure by repeating forever (line 2) the main loop which consists of the observation of a new event from the stream (line 3), a decision whether the event has to be analyzed or not (line 4) and the actual analysis (lines 5–14).

---

[2] The OCTS of the system reported in Fig. 2b is available at https://andrea.burattin. net/public-files/online-conformance/octs.pdf.

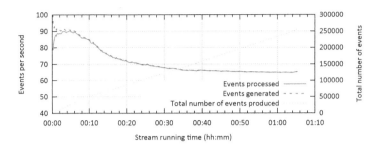

Fig. 3. Performance evaluation: events generated versus the events processed.

The analysis starts obtaining the state of the current process instance from M_O (line 5). If the map does not contain information (e.g., because this is the first event with this case id ever observed) then a new process instance is assumed (line 7). After that, the algorithm computes which is the next state reachable with the lowest cost (line 9) and updates the status of the current process instance with the new state, new total cost, and last update time (line 10). The last operation performed by the algorithm is a cleanup: this is necessary to keep the memory bounded and consists in dropping all states referring to the old executions (lines 12–13). This is necessary to keep the memory bounded and to avoid that an infinite stream causes an infinite memory usage.

Please note that the operations not requiring constant time are the selection of the next state (line 9) and the removal of old elements from M_O (line 12). Since OCTS is a deterministic transition system, the former has complexity linear on the alphabet size (i.e., the number of transitions in the original Petri net), which is assumed to be constant over time. The latter has linear complexity on the size of M_O which, again, is constant over time. Therefore, the theoretical computational complexity of the algorithm makes it a viable solution for online applications.

4   Implementation and Results

The entire approach has been implemented[3] in the process mining toolkit ProM. As for other online plugins, the current implementation connects to a stream source that emits events (via a TCP connection) and processes each of them independently. This approach allows a strong decoupling between the source of the events and the actual online process mining tool.

In order to assess the feasibility of our approach, we simulated the stream of a process model comprising 26 tasks, and 20 gateways using PLG2 [8][4]. We have been able to simulate an unlimited event stream, with events referring to

---

[3] The implementation is available in the **StreamConformance** package of ProM: https://svn.win.tue.nl/repos/prom/Packages/StreamConformance/.

[4] The BPMN model is available at https://andrea.burattin.net/public-files/online-conformance/model.pdf.

the given model. Specifically, we configured PLG to generate up to 90 events per second. Then, we tested the capabilities of our implementation, by running the conformance checker for about 1 h and 10 min. As plotted in Fig. 3, after this period of time, 256110 events were generated. The generator, however, was not capable of keeping the given pace (we used a standard office laptop machine), and the system stabilized in simulating about 65 events per second. As the chart reports, all generated events were processed in time by our prototype. This demonstrates the actual feasibility of the approach, even in prototypical implementations.

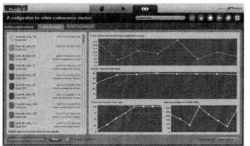
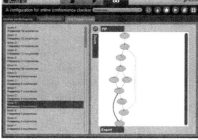

(a) Main dashboard of the plugin   (b) Routing of frequent errors

Fig. 4. Screenshots of the ProM plugin implementing the approach.

The ProM plugin implemented comprises a "dashboard" (cf. Fig. 4a) which shows, on its left-hand side, the running process instances (color-coded by severity and sorted by update time or by severity). The right-hand side of the dashboard contains system charts with number of errors every 5 s, number of processed events per second, number of traces in memory, and total memory consumption. The second component (cf. Fig. 4b) reports the frequent deviations on the behavioral model. More information on the implementation in [9].

5   Conclusion and Future Work

This paper presents the first approach to compute conformance checking for online data streams. The fundamental advantage of this technique, with respect to previous off-line approaches, is the ability to check deviations from the reference behavior in real-time, i.e., immediately after they occurred. This way, possible corrections can be immediately enacted. The input of the presented technique is a Petri net, which is converted into a transition system. Such transition system is decorated with additional arcs in order to allow for deviations. Non-zero costs are associated with transitions representing deviations. Behavioral properties are employed to detect the target state of deviating transitions. The whole approach has been implemented in ProM.

We plan to continue the work presented on this paper by improving the conversion from Petri net to transition system using more conformance-oriented techniques as well as by exploiting contextual information (e.g., data associated with states).

Acknowledgements. We would like to thank Jorge Munoz-Gama for discussing early stage ideas of the approach. This work was partially funded by the Spanish Ministry for Economy and Competitiveness (MINECO) and the EU (FEDER funds) under grant COMMAS (TIN2013-46181-C2-1-R).

References

1. van der Aalst, W.M.: Process Mining: Discovery Conformance and Enhancement of Business Processes. Springer, Heidelberg (2011). https://doi.org/10.1007/978-3-642-19345-3
2. van der Aalst, W.M., Adriansyah, A., van Dongen, B.: Replaying history on process models for conformance checking and performance analysis. Wiley Interdisc. Rev. Data Min. Knowl. Discov. **2**(2), 182–192 (2012)
3. Adriansyah, A.: Aligning observed and modeled behavior. Ph. D. thesis, Technische Universiteit Eindhoven (2014)
4. Aggarwal, C.C.: Data Streams: Models and Algorithms, Advances in Database Systems, vol. 31. Springer, Boston, MA (2007). https://doi.org/10.1007/978-0-387-47534-9
5. Bifet, A., Holmes, G., Kirkby, R., Pfahringer, B.: MOA: massive online analysis learning examples. J. Mach. Learn. Res. **11**, 1601–1604 (2010)
6. vanden Broucke, S.K.L.M., Munoz-Gama, J., Carmona, J., Baesens, B., Vanthienen, J.: Event-based real-time decomposed conformance analysis. In: Meersman, R., Panetto, H., Dillon, T., Missikoff, M., Liu, L., Pastor, O., Cuzzocrea, A., Sellis, T. (eds.) OTM 2014. LNCS, vol. 8841, pp. 345–363. Springer, Heidelberg (2014). https://doi.org/10.1007/978-3-662-45563-0_20
7. Burattin, A.: Process Mining Techniques in Business Environments: Theoretical Aspects, Algorithms, Techniques and Open Challenges in Process Mining. LNBIP, vol. 207. Springer, Cham (2015). https://doi.org/10.1007/978-3-319-17482-2
8. Burattin, A.: PLG2 : Multiperspective process randomization with online and offline simulations. In: Proceedings of the BPM Demo Track. CEUR-WS.org (2016)
9. Burattin, A.: Online conformance checking for petri nets and event streams. In: Online Proceedings of BPM Demo Track. CEUR-WS.org (2017)
10. Burattin, A., Cimitile, M., Maggi, F.M., Sperduti, A.: Online discovery of declarative process models from event streams. IEEE TSC **8**(6), 833–846 (2015)
11. Burattin, A., Maggi, F.M., Cimitile, M.: Lights, camera, action! business process movies for online process discovery. In: Proceedings of TAProViz (2014)
12. Burattin, A., Sperduti, A., van der Aalst, W.M.: Control-flow discovery from event streams. In: Proceedings of IEEE CEC, pp. 2420–2427. IEEE (2014)
13. Gaber, M.M., Zaslavsky, A., Krishnaswamy, S.: Mining data streams: a review. ACM Sigmod Rec. **34**(2), 18–26 (2005)
14. Gama, J.: Knowledge Discovery from Data Streams. Chapman & Hall/CRC, Boca Raton (2010)
15. Golab, L., Özsu, M.T.: Issues in data stream management. ACM SIGMOD Rec. **32**(2), 5–14 (2003)

16. Leemans, S.J.J., Fahland, D., van der Aalst, W.M.P.: Scalable process discovery and conformance checking. Software & Systems Modeling, pp. 1–33. Springer, Heidelberg (2016). https://doi.org/10.1007/s10270-016-0545-x
17. Maggi, F.M., Burattin, A., Cimitile, M., Sperduti, A.: Online process discovery to detect concept drifts in LTL-based declarative process models. In: Meersman, R., Panetto, H., Dillon, T., Eder, J., Bellahsene, Z., Ritter, N., De Leenheer, P., Dou, D. (eds.) OTM 2013. LNCS, vol. 8185, pp. 94–111. Springer, Heidelberg (2013). https://doi.org/10.1007/978-3-642-41030-7_7
18. Maggi, F.M., Montali, M., van der Aalst, W.M.: An operational decision support framework for monitoring business constraints. In: Proceedings of FASE (2012)
19. Maggi, F.M., Montali, M., Westergaard, M., van der Aalst, W.M.P.: Monitoring business constraints with linear temporal logic: an approach based on colored automata. In: Rinderle-Ma, S., Toumani, F., Wolf, K. (eds.) BPM 2011. LNCS, vol. 6896, pp. 132–147. Springer, Heidelberg (2011). https://doi.org/10.1007/978-3-642-23059-2_13
20. Munoz-Gama, J.:: Conformance Checking and Diagnosis in Process Mining - Comparing Observed and Modeled Processes. LNBIP. Springer, Cham (2016). https://doi.org/10.1007/978-3-319-49451-7
21. Munoz-Gama, J., Carmona, J., van der Aalst, W.M.P.: Conformance checking in the large: partitioning and topology. In: Daniel, F., Wang, J., Weber, B. (eds.) BPM 2013. LNCS, vol. 8094, pp. 130–145. Springer, Heidelberg (2013). https://doi.org/10.1007/978-3-642-40176-3_11
22. Murata, T.: Petri nets: properties, analysis and applications. Proc. IEEE **77**(4), 541–580 (1989)
23. Rozinat, A., van der Aalst, W.M.: Conformance checking of processes based on monitoring real behavior. Inf. Syst. **33**(1), 64–95 (2008)
24. van Zelst, S.J., van Dongen, B.F., van der Aalst, W.M.P.: Event stream-based process discovery using abstract representations. Knowl. Inf. Syst. **53**, 1–29 (2017)

Recurrent Process Mining with Live Event Data

Alifah Syamsiyah$^{(\boxtimes)}$, Boudewijn F. van Dongen, and Wil M. P. van der Aalst

Eindhoven University of Technology, Eindhoven, The Netherlands
{A.Syamsiyah,B.F.v.Dongen,W.M.P.v.d.Aalst}@tue.nl

Abstract. In organizations, process mining activities are typically performed in a recurrent fashion, e.g. once a week, an event log is extracted from the information systems and a process mining tool is used to analyze the process' characteristics. Typically, process mining tools import the data from a file-based source in a pre-processing step, followed by an actual process discovery step over the pre-processed data in order to present results to the analyst. As the amount of event data grows over time, these tools take more and more time to do pre-processing and all this time, the business analyst has to wait for the tool to finish. In this paper, we consider the problem of recurrent process discovery in live environments, i.e. in environments where event data can be extracted from information systems near real time. We present a method that preprocesses each event when it is being generated, so that the business analyst has the pre-processed data at his/her disposal when starting the analysis. To this end, we define a notion of intermediate structure between the underlying data and the layer where the actual mining is performed. This intermediate structure is kept in a persistent storage and is kept live under updates. Using a state of the art process mining technique, we show the feasibility of our approach. Our work is implemented in the process mining tool ProM using a relational database system as our persistent storage. Experiments are presented on real-life event data to compare the performance of the proposed approach with the state of the art.

Keywords: Recurrent process mining · Live event data
Incremental process discovery

1 Introduction

Process mining is a discipline where the aim is to improve an organization's processes given the information from the so called *event logs* [16]. Process mining techniques have been successfully demonstrated in various case studies such as health care, insurance, and finance [5,10,12,23]. In many of these cases, a one-time study was performed, but in practice, process mining is typically a recurring activity, i.e. an activity that is performed on a routine basis.

As an illustration, suppose that each manager of an insurance company has the obligation to report her/his work to a director once in each month to see

© Springer International Publishing AG 2018
E. Teniente and M. Weidlich (Eds.): BPM 2017 Workshops, LNBIP 308, pp. 178–190, 2018.
https://doi.org/10.1007/978-3-319-74030-0_13

the company's progress since the beginning of a year. Typical analysis in such regular report incorporates the observations from previous month, last three months, or last year. This type of reporting requires an analyst to repeatedly produce analysis results from data that grows over time.

Existing process mining tools are not tailored towards such recurrent analyses. Instead, they require the analysts to export the event data from a running system, import it to the mining tool which pre-processes the data during importing and then use the tool on the pre-processed data. As the amount of data grows, the import and pre-processing phase takes longer and longer causing the analyst to waste time.

However, most information systems nowadays record real-time event data about business processes during their executions. This enables analysis techniques to import and pre-process the data when it "arrives". By shifting the pre-processing time an analyst is able to do process mining on the pre-processed data instantly.

The idea of this live scenario is sketched in Fig. 1. In the live scenario, we introduce a persistent storage for various structures that are kept as "an intermediate structure" by process mining algorithms. We then show how such intermediate structure can be updated without the need for full recomputation every time an event arrives. Using the intermediate structure of the state-of-the-art process mining technique, we show the feasibility of our approach in the general process mining setting and using experiments on real-life data, we show the added time-benefit for the analyst.

This paper is organized as follows. In Sect. 2 we introduce recurrent process mining and we focus on a traditional technique in this setting. In Sect. 3, we show how recurrent process mining can be performed on live event data and

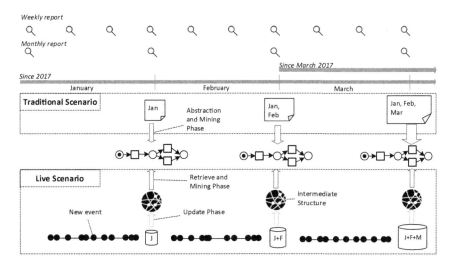

Fig. 1. Traditional vs live scenario in process discovery

we prove that the technique of Sect. 2 can be made incremental. We show the improvements of our work using real-life data in Sect. 4. Then, in Sect. 5 discusses some related work. Finally, we conclude the paper in Sect. 6.

2   Recurrent Process Mining

The challenge in process mining is to gain insights into an operational system in which activities are executed. The execution of these activities is reflected through events, i.e. events are occurrences of activities. Furthermore, in process mining, activities are being executed in the context of cases (also referred to as process instances) at a particular point in time.

Definition 1 (Events, cases, and activities). *Let A be the universe of activities, let C be the universe of cases, and let Υ be the universe of events. We define $\#_{act} : \Upsilon \to A$ a function labeling each event with an activity. We define $\#_{cs} : \Upsilon \to C$ a function labeling each event with a case. We define $\#_{tm} : \Upsilon \to \mathbb{R}$ a function labeling each event with a timestamp.*

In process mining, the general assumption is that event data is provided in the form of an event log. Such an event log is basically a collection of events occurring in a specific time period, in which the events are grouped by cases sequentialized based on their time of occurrence.

Definition 2 (Event log and trace). *Let $E \subseteq \Upsilon$ be a collection of events and $t_s, t_e \in \mathbb{R}$ two timestamps with $t_s < t_e$ relating to the start and the end of the collection period.*
A trace $\sigma \in E^*$ is a sequence of events such that the same event occurs only once in σ, i.e. $|\sigma| = |\{e \in \sigma\}|$. Furthermore, each event in a trace refers to the same case $c \in C$, i.e. $\forall_{e \in \sigma} \#_{cs}(e) = c$ and we assume all events within the given time period are included, i.e. $\forall_{e \in \Upsilon}(\#_{cs}(e) = c \wedge t_s \leq \#_{tm}(e) \leq t_e) \implies e \in \sigma$.
An event log $L \in \wp(E^*)^1$ is a set of traces.

Note that the time-period over which an event log is collected plays an important role in this paper. In most process mining literature, this time period is neglected and the event log is assumed to span eternity. However, in practice, analysis always consider event data pertaining to a limited period of time. Therefore, in this paper, we explicitly consider the following process mining scenarios (as depicted in Fig. 1):

- Analysts perform process mining tasks on a recurrent schedule at regular points, e.g. once a week about the last week, or once a month about the last month, or
- Analysts perform process mining on the data since a pre-determined point in time, e.g. since 2017, or since March 2017.

---

¹ $\wp(E^*)$ denotes a powerset of sequences, i.e. $L \subseteq E^*$.

2.1 Traditional Recurrent Process Mining

To execute the two process mining scenarios, a multitude of process mining techniques is available. However, all have two things in common, namely (1) that the starting point for analysis is an event log and (2) that the analysis is performed in three phases, namely: *loading*, *abstraction*, and *mining*.

While the details may differ, all process mining techniques build an *intermediate structure* in memory during the abstraction phase. Typically, the time needed to execute this phase is linear in the number of events in the event log. This intermediate structure (of which the size is generally polynomial in the number of activities in the log, but not related to the number of events anymore) is then used to perform the actual mining task. In this paper, we consider the state-of-the-art process discovery algorithm, namely the Inductive Miner [6,7], which is known to be flexible, to have formal guarantees, and to be scalable. For the Inductive Miner, the intermediate structure is the so-called *direct succession relation* (Definition 3) which counts the frequencies of direct successions.

Definition 3 (Direct succession [16]**).** *Let L be an event log over $E \subseteq \Upsilon$. The direct succession relation $>_L : \mathcal{A} \times \mathcal{A} \to \mathbb{N}$ counts the number of times activity a is directly followed by activity b in some cases in L as follows:*

$$>_L (a,b) = \Sigma_{\sigma \in L} \; \Sigma_{i=1}^{|\sigma|-1} \begin{cases} 1, & \text{if } \#_{act}(\sigma(i)) = a \; \wedge \; \#_{act}(\sigma(i+1)) = b \\ 0, & \text{otherwise.} \end{cases}$$

To illustrate how the Inductive Miner algorithm uses the direct succession relation described above, we refer to Table 1. Here, we show (by example), how the mining technique uses the relation as an intermediate structure to come to a result. In order to apply this traditional technique in a recurrent setting, the loading of the data, the abstraction phase, and the mining phase have to be repeated. When over time the event data grows, the time to execute the three phases also grows, hence performing the recurrent mining task considering one year of data will take 52 times longer than considering one week of data.

In Table 2 we show an example of the Inductive Miner applied to a real-life dataset of the BPI Challenge 2017 [17] which contains data of a full year. We record the times to perform the three phases of importing, abstraction, and mining on this dataset after the first month, at the middle of the year, and at the end of the year. It is clear that indeed the importing and abstraction times grow considerably, while the actual mining phase is orders of magnitude faster.

Table 1. Examples of the use of the intermediate structure in Inductive Miner

×	An exclusive-choice cut of L is a cut $(\times, A_1, ..., A_n)$ such that $\forall i, j \in \{1, ..., n\} \; \forall a \in A_i \; \forall b \in A_j \; i \neq j \Rightarrow \not>_L (a,b)$
→	A sequence cut of L is a cut $(\rightarrow, A_1, ..., A_n)$ such that $\forall i, j \in \{1, ..., n\} \; \forall a \in A_i \; \forall b \in A_j \; i < j \Rightarrow (>_L^+ (a,b) \wedge \not>_L^+ (b,a))$

Table 2. Process mining times (in seconds) in the traditional setting on data of the
BPI Challenge 2017 [17]

Week	Inductive Miner		
	Loading	Abstraction	Mining
5	0.8520	2.8531	0.0254
26	3.6319	30.9528	0.0257
52	9.5854	93.5118	0.0291

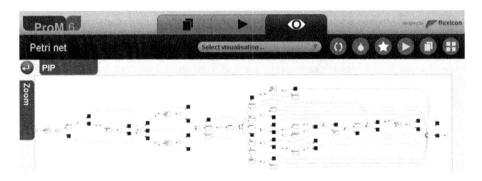

Fig. 2. Screenshot of ProM showing the result of the Inductive Miner on the BPI
Challenge 2017 data considering the first 5 weeks of data.

Figure 2 shows the result of the Inductive Miner after the first 5 weeks of
data, i.e. if an analyst has produced this picture on January 29th 2016, it would
have taken 3.7305 s ($= 0.8520 + 2.8531 + 0.0254$) in total to load the event log,
build the abstraction, and do the mining in order to produce this picture using
the Inductive Miner in ProM.

In Sect. 3, we present a method to store the intermediate structure in a
persistent storage and to keep it live under updates, i.e. every time an event is
generated, the intermediate structure is updated (in constant time). This way,
when a process mining task is performed, the time spent by the analyst is limited
to the time to retrieve the intermediate structure and to do the actual mining.

3   Recurrent Process Mining with Live Event Data

It is easy to see that given an event log L, the relation in Definition 3 can
be computed during a single linear pass over the event log, visiting each event
exactly once. In this section, we present a live process mining technique based
on the Inductive Miner which does not require the analyst to repeat the import
and analysis phases every time a process mining task is performed.

In order to enable live process mining, we use a persistent event storage,
called DB-XES [13], which uses a relational database to store event data. The
full structure of this relational database is beyond the scope of this paper, but

what is important is that given a trace, it is possible to quickly retrieve the last event recorded for that trace.

Definition 4 (Last event in a trace). *Let $E \subseteq \Upsilon$ be a collection of events and let $c \in C$ be a case. The function $\lambda : C \rightarrow E \cup \{\bot\}$ is a function that returns the last event in E belonging to case c, i.e.*

$$\lambda(c) = \begin{cases} \bot, & \text{if } \forall_{e \in E} \#_{cs}(e) \neq c, \\ e \in E, & \text{if } \#_{cs}(e) = c \wedge \not\exists e' \in E(\#_{cs}(e') = c \wedge \#_{tm}(e') > \#_{tm}(e)). \end{cases}$$

Using DB-XES as a persistent storage and making use of the ability to query for the last event in a trace, we present the Incremental Inductive Miner in Sect. 3.1.

3.1 Incremental Inductive Miner

The Inductive Miner uses only the frequency of direct successions between activities as input as defined in Definition 3. Therefore, to enable an incremental version of the Inductive Miner, we present an update strategy that, given the relation $>_L$ for some log L and a new event e, we can derive the relation $>_{L'}$ where L' is the log L with the additional event e.

Theorem 1 (Updating relation $>$ is possible). *Let $E \subseteq \Upsilon$ be a set of events and L a log over E. Let $e \in \Upsilon \setminus E$ be a fresh event to be added such that for all $e' \in E$ holds $\#_{ts}(e') < \#_{ts}(e)$ and let $E' = E \cup \{e\}$ be the new set of events with L' the corresponding log over E'. We know that for all $a, b \in \mathcal{A}$ holds that:*

$$>_{L'} (a, b) = >_L (a, b) + \begin{cases} 0 & \text{if } \lambda(\#_{cs}(e)) = \bot, \\ 1 & \text{if } \#_{act}(\lambda(\#_{cs}(e))) = a \wedge \#_{act}(e) = b, \\ 0 & \text{otherwise.} \end{cases}$$

Proof. Let $c = \#_{cs}(e) \in C$ be the case to which the fresh event belongs.

If for all $e' \in E$ holds that $\#_{cs}(e') \neq c$, then this is the first event in case c and we know that $L' = L \cup \langle e \rangle$. Hence relation $>$ does not change from L to L' as indicated by case 1.

If there exists $e' = \lambda(\#_{cs}(e)) \in E$ with $\#_{cs}(e') = c$, then we know that there is a trace $\sigma_c \in L$. Furthermore, we know that $L' = (L \setminus \{\sigma_c\}) \cup \{\sigma_c \cdot \langle e \rangle\}$, i.e. event e gets added to trace σ_c of L. This implies that $>_{L'} (\#_{act}(e'), \#_{act}(e)) = >_L (\#_{act}(e'), \#_{act}(e)) + 1$ as indicated by case 2.

In all other cases, the number of direct successions of two activities is not affected. ■

Using the simple update procedure indicated in Theorem 1 the Incremental Inductive Miner allows for recurrent process mining under live updates. In Sect. 4 we show how the effect of keeping relation $>$ lives on the total time needed to perform the recurrent process mining task.

4 Experimental Results

We implemented the algorithm presented as a ProM[2] plug-in called *Database-Incremental Inductive Miner (DIIM)*[3]. DIIM is designed for recurrent process discovery based on live event data. It uses DB-XES as the back-end storage and the Inductive Miner as the mining algorithm. In this section, we show the experimental results of applying DIIM in a live scenario and we compare it to traditional Inductive Miner.

For the experiment, we used a real dataset from BPI Challenge 2017 [17]. This dataset pertains to the loan applications of a company from January 2016 until February 2017. In total, there are 1,202,267 events and 26 different activities which pertain to 31,509 loan applications.

In this experiment, we looked into some weekly reports where we were interested to see process models of the collective data since 2016. The last working day, i.e. Friday, was chosen as the day when we performed the process discovery to have a progress report for that week. In live scenario, we assumed that each event was inserted to the DB-XES database precisely at the time stated in the timestamp attribute of the event log. Then, the DB-XES system immediately processed each new event data as it arrived using triggers in the relational database, implementing the update procedures detailed in Sect. 3, thus keeping the relations live under updates. In traditional scenario, we split the dataset into several logs such that each log contained data for one week. For the *n*-th report, we combined the log from the first week until the *n*-th week, loaded it into ProM, and discovered a process model.

Figure 3 shows the experimental results of recurrent process discovery using DIIM and the Inductive Miner. The x-axis represents the *n*-th week, while the y-axis represents the time spent by user (in seconds) to discover process models. The blue dots are the experiment using DIIM which includes the total times to insert new events, update the intermediate structure, retrieve the values from the

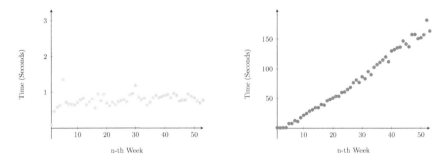

Fig. 3. The comparison of recurrent process discovery using DIIM (left) vs traditional Inductive Miner (right) (Color figure online)

---

[2] See http://www.processmining.org and http://www.promtools.org.

[3] https://svn.win.tue.nl/repos/prom/Packages/DatabaseInductiveMiner/Trunk/.

DB-XES, and mine the process models, while the red dots are the experiment using traditional Inductive Miner which includes the time to load the XES event logs, build the intermediate structure, and mine the process models.

As shown in Fig. 3, our incremental technique is much faster, even when considering the time needed to insert events in the relational database, a process that is typically executed in real time and without the business analyst being present. More important however, is the trendlines of both approaches.

As expected, the time to perform the process mining task in the traditional setting is growing linear in the size of the event data (the arrival rate of events in this dataset is approximately constant during the entire year). This is due to the fact that the first two phases of loading the data and doing the abstraction into the intermediate structure scales linearly in the number of events, whereas the mining scales in the number of activities. The latter is considerably smaller than the former in most practical cases as well as in this example. Our incremental algorithms are more stable over time as the update time only depends on the number of *newly inserted* events and both the retrieval and mining times depend on the number of activities rather than the number of events.

The variations in the recorded values of the DIIM are therefore explained by the number of inserted events in a day. The higher the number of newly inserted events, the longer it takes to do the update in the relational database system of the intermediate structure. However, the total update time remains limited to around 1.4 s per day.

In order to see the average time for doing an update for a single event, we normalized the total update time with the number of events inserted in each day as shown in Fig. 4. The x-axis represents the n-th day, while the y-axis represents the update time per event. As shown from the Fig. 4, the average time to conduct an update for a single event stabilizes around 0.000545 s, i.e. the database system can handle around 1800 events per second and this includes the insertion of the actual event data in the underlying DB-XES tables.

To validate the fact that the update time scales linearly in the number of activities, we refer to Fig. 5. For this experiment, we used a different dataset with 31 different activities and eleven thousands of events, provided to us by Xerox Services, India. The x-axis represents the total number of different activities which has been inserted to the database, while the y-axis represents the time in seconds to update an event. From the figure, it is clear that the update time indeed depends linearly on the number of activities.

It is important to realize that the results of the process mining techniques in both the traditional and the live setting are not different, i.e. the process models are identical. Figure 6 shows a screenshot of a process model in ProM, produced by the Incremental Inductive Miner considering all the data in the original file. In the traditional setting, it would have taken an analyst 103.1263 s to load the event log, build the abstraction and do the mining in order to get this picture on December 30th 2016. Due to the availability of the intermediate structure in the database, it would take the analyst only 0.0392 s to produce the same result using the Incremental Inductive Miner.

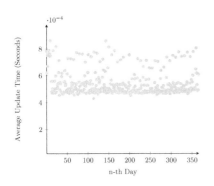
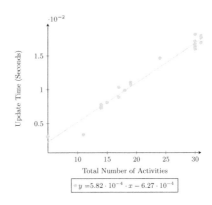

Fig. 4. Average update time per event

Fig. 5. The influence of number of activities to update time

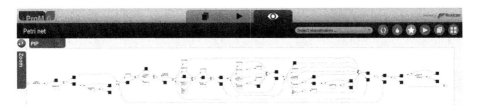

Fig. 6. Screenshot of ProM showing the result of the Inductive Miner on the BPI Challenge 2017 data considering all data.

5   Related Work

For a detailed explanation about process mining, we refer to [16]. Here we primarily focus on the work related to recurrent process discovery in process mining and its applications on live event data.

In the current setting of process discovery, event data from a file-based system is imported to a process mining tool. This technique potentially creates redundancy of data reloading in environments which necessitate some repetitions in the discovery. Therefore, some researches have been looking to the area of databases, Hadoop, and other ways to store event data in a persistent manner.

A study in [22] examined a tool called XESame. To access the data in XESame, one needs to materialize the data by selecting and matching it with XES [4] elements. It does not provide a direct access to the database. A more advanced technique using ontology was proposed in [1,2]. In this work, data can be accessed on-demand using query unfolding and rewriting techniques, called ontology-based data access. However, performance issues make this approach unsuitable for large event logs.

Relational XES, or RXES, was introduced in [18]. The RXES schema was designed with a focus on XES standard. Using experiments with real life data, it was shown that RXES typically uses less memory compared to the file-based OpenXES and MapDB XESLite implementations [9]. As an improved version of RXES, DB-XES was introduced in [13] to enable process mining in the large. In [14] DB-XES basic schema is extended to allow instant social network mining, especially the handover of work networks.

Process mining not only covers procedural process discovery, but also declarative process discovery. The work in [11] deals with declarative process discovery using SQL databases. Building on top of RXES, the authors introduce a mining approach that directly works on relational event data by querying the log with conventional SQL. Queries can be customised and cover process perspective beyond the control flow, e.g., organisational aspects. However, none of these techniques handles live event data, the focus is often on static data that has been imported in a database.

Treating event data as a dynamic sequence of events has been explored in [19, 20]. This work presented single-node stream extension of process mining tool ProM which enables researchers and business users to also apply process mining on streaming-data. The applicability of the framework on the cooperative aspects of process mining was demonstrated in [21]. Here the authors define the notion of case administration to store for each case the (partial) sequence of (activity, resource)-pairs seen so far.

Another online process discovery technique based on streaming technology was proposed in [8]. This work presented a novel framework for the discovery of LTL-based declarative process models from streaming event data. The framework continuously updates a set of valid business constraints based on the events occurring in the event stream. Moreover, it gives the user meaningful information about the most significant concept drifts occurring during process execution.

However, from the real-time techniques using streaming event data that we have seen so far, none of them deals with recurrent process discovery. Intermediate results are not stored after presenting the results. Therefore, all data (old and new) needs to be reloaded and processed each time a new analysis is needed. Building on from that concern, this work explores both in the ability to process live event data and to handle recurrent questions in process discovery.

6   Conclusion

Process mining aims to give insights into processes by discovering process models which adequately reflect the behavior seen in an event log. One of the challenges in process mining is the recurrent nature of mining tasks which, using traditional tools and techniques, require analysts to import and pre-process larger and larger datasets.

In this paper we focused on recurrent process discovery on live event data. Using the Inductive Miner technique as an example, we show how we can reduce the time needed for an analyst to do process mining by storing intermediate structure in a persistent storage. Then we show how to keep this intermediate structure alive under insertion of new events.

Using a concrete implementation, we use the relational database called DB-XES to store the event data and the intermediate structure. In this relational database, we added triggers to update the intermediate structure with the insertion of each event and we implemented the Incremental Inductive Miner as a version of the existing technique which is able to use this persistent intermediate structure directly.

We tested the performance of the proposed approach and compared it to the traditional techniques using real-life datasets. We show that loading and mining time of the traditional approach grows linearly as the event data grows. In contrast, our incremental implementation shows constant times for updating (per event) and the retrieval and mining times are independent of the size of the underlying data.

The core ideas in the paper are not limited to control flow. They are, for example, trivially extended to store intermediate structures keeping track of average times between activities or for social networks. Moreover, they are not restricted towards procedural process discovery. In [15] we show how the work extends into declarative process discovery, particularly using MINERful [3] as the discovery technique. We introduce a so-called controller function which we keep live under updates. Then we show that, using the controller function, we can keep all MINERful relations live under updates.

A more fundamental challenge for future work is the updating of the intermediate structures in batches of events, rather than for each event separately. Furthermore, we aim to enable these techniques to keep these structures live under removal of past events.

References

1. Calvanese, D., Kalayci, T.E., Montali, M., Tinella, S.: Ontology-based data access for extracting event logs from legacy data: the onprom tool and methodology. In: Abramowicz, W. (ed.) BIS 2017. LNBIP, vol. 288, pp. 220–236. Springer, Cham (2017). https://doi.org/10.1007/978-3-319-59336-4_16
2. Calvanese, D., Montali, M., Syamsiyah, A., van der Aalst, W.M.P.: Ontology-driven extraction of event logs from relational databases. In: Reichert, M., Reijers, H.A. (eds.) BPM 2015. LNBIP, vol. 256, pp. 140–153. Springer, Cham (2016). https://doi.org/10.1007/978-3-319-42887-1_12
3. Di Ciccio, C., Mecella, M.: Mining constraints for artful processes. In: Abramowicz, W., Kriksciuniene, D., Sakalauskas, V. (eds.) BIS 2012. LNBIP, vol. 117, pp. 11–23. Springer, Heidelberg (2012). https://doi.org/10.1007/978-3-642-30359-3_2
4. Günther, C.W.: XES Standard Definition (2014). www.xes-standard.org
5. Jans, M.J., Alles, M., Vasarhelyi, M.A.: Process Mining of Event Logs in Auditing: Opportunities and Challenges (2010). SSRN 2488737

6. Leemans, S.J.J., Fahland, D., van der Aalst, W.M.P.: Discovering block-structured process models from event logs - a constructive approach. In: Colom, J.-M., Desel, J. (eds.) PETRI NETS 2013. LNCS, vol. 7927, pp. 311–329. Springer, Heidelberg (2013). https://doi.org/10.1007/978-3-642-38697-8_17

7. Leemans, S.J.J., Fahland, D., van der Aalst, W.M.P.: Discovering block-structured process models from event logs containing infrequent behaviour. In: Lohmann, N., Song, M., Wohed, P. (eds.) BPM 2013. LNBIP, vol. 171, pp. 66–78. Springer, Cham (2014). https://doi.org/10.1007/978-3-319-06257-0_6

8. Maggi, F.M., Burattin, A., Cimitile, M., Sperduti, A.: Online process discovery to detect concept drifts in LTL-based declarative process models. In: Meersman, R., Panetto, H., Dillon, T., Eder, J., Bellahsene, Z., Ritter, N., De Leenheer, P., Dou, D. (eds.) OTM 2013. LNCS, vol. 8185, pp. 94–111. Springer, Heidelberg (2013). https://doi.org/10.1007/978-3-642-41030-7_7

9. Mannhardt, F.: XESLite managing large XES event logs in ProM. BPM Center Report BPM-16-04 (2016)

10. Rojas, E., Munoz-Gama, J., Sepúlveda, M., Capurro, D.: Process mining in healthcare: a literature review. J. Biomed. Inform. **61**, 224–236 (2016)

11. Schönig, S., Rogge-Solti, A., Cabanillas, C., Jablonski, S., Mendling, J.: Efficient and customisable declarative process mining with SQL. In: Nurcan, S., Soffer, P., Bajec, M., Eder, J. (eds.) CAiSE 2016. LNCS, vol. 9694, pp. 290–305. Springer, Cham (2016). https://doi.org/10.1007/978-3-319-39696-5_18

12. Suriadi, S., Wynn, M.T., Ouyang, C., ter Hofstede, A.H.M., van Dijk, N.J.: Understanding process behaviours in a large insurance company in australia: a case study. In: Salinesi, C., Norrie, M.C., Pastor, Ó. (eds.) CAiSE 2013. LNCS, vol. 7908, pp. 449–464. Springer, Heidelberg (2013). https://doi.org/10.1007/978-3-642-38709-8_29

13. Syamsiyah, A., van Dongen, B.F., van der Aalst, W.M.P.: DB-XES: enabling process mining in the large. In: SIMPDA 2016, pp. 63–77 (2016)

14. Syamsiyah, A., van Dongen, B.F., van der Aalst, W.M.P.: Discovering social networks instantly: moving process mining computations to the database and data entry time. In: Reinhartz-Berger, I., Gulden, J., Nurcan, S., Guédria, W., Bera, P. (eds.) BPMDS/EMMSAD -2017. LNBIP, vol. 287, pp. 51–67. Springer, Cham (2017). https://doi.org/10.1007/978-3-319-59466-8_4

15. Syamsiyah, A., van Dongen, B.F., van der Aalst, W.M.P.: Recurrent process mining on procedural and declarative approaches. BPM Center Report BPM-17-03 (2017)

16. van der Aalst, W.M.P.: Process Mining: Data Science in Action. Springer, Heidelberg (2016)

17. van Dongen, B.F.: BPI Challenge 2017 (2017)

18. van Dongen, B.F., Shabani, S.: Relational XES: data management for process mining. In: CAiSE 2015, pp. 169–176 (2015)

19. van Zelst, S.J., Burattin, A., van Dongen, B.F., Verbeek, H.M.W.: Data streams in ProM 6: a single-node architecture. In: BPM Demo Session 2014, p. 81 (2014)

20. van Zelst, S.J., van Dongen, B.F., van der Aalst, W.M.P.: Know what you stream: generating event streams from CPN models in ProM 6. In: BPM Demo Session 2015, pp. 85–89 (2015)

21. van Zelst, S.J., van Dongen, B.F., van der Aalst, W.M.P.: Online discovery of cooperative structures in business processes. In: Debruyne, C., et al. (eds.) OTM 2016. LNCS, vol. 10033, pp. 210–228. Springer, Cham (2016). https://doi.org/10.1007/978-3-319-48472-3_12

22. Verbeek, H.M.W., Buijs, J.C.A.M., van Dongen, B.F., van der Aalst, W.M.P.: XES, XESame, and ProM 6. In: Soffer, P., Proper, E. (eds.) CAiSE Forum 2010. LNBIP, vol. 72, pp. 60–75. Springer, Heidelberg (2011). https://doi.org/10.1007/978-3-642-17722-4_5
23. Zhou, Z., Wang, Y., Li, L.: Process mining based modeling and analysis of workflows in clinical care - a case study in a Chicago Outpatient Clinic. In: ICNSC 2014, pp. 590–595 (2014)

Reducing Event Variability in Logs by Clustering of Word Embeddings

David Sánchez-Charles[1][(✉)], Josep Carmona[2], Victor Muntés-Mulero[1], and Marc Solé[1]

[1] CA Strategic Research, CA Technologies, Barcelona, Spain
{David.Sanchez,Victor.Muntes,Marc.SoleSimo}@ca.com
[2] Universitat Politècnica de Catalunya, Barcelona, Spain
jcarmona@cs.upc.edu

Abstract. Several business-to-business and business-to-consumer services are provided as a human-to-human conversation in which the provider representative guides the conversation towards its resolution based on her experience, following internal guidelines. Several attempts to automatize these services are becoming popular, but they are currently limited to procedures and objectives set during design step. Process discovery techniques could provide the necessary mechanisms to monitor event logs derived from textual conversations and expand the capabilities of conversational bots. Still, variability of textual messages hinders the utility of process discovery techniques by producing non-understandable unstructured process models. In this paper, we propose the usage of word embedding for combining events that have a semantically similar name.

Keywords: Unstructured processes · Process discovery
Machine learning · Word embedding

1 Introduction

Recent trends in Natural Language Processing and Machine Learning allow machines to understand and answer simple queries in the form of free text. Lots of textual data is still generated for describing actions performed during the execution of procedures or services in which the human interaction is a key component. For instance, software development teams textually describe changes on source code, customer support channels record conversations with customer and the actions performed by the support engineers. Although it is well known in the industry that improving the efficiency of such services and procedures lead to more efficient businesses[1], the Business Process Management arena is behind on applying the most recent developments on Natural Language Processing and Machine Learning.

---

[1] For instance, a faster customer support channel leads to lower customer churn rates. https://www.salesforce.com/blog/2017/03/effective-strategies-to-reduce-customer-churn.html.

© Springer International Publishing AG 2018
E. Teniente and M. Weidlich (Eds.): BPM 2017 Workshops, LNBIP 308, pp. 191–203, 2018.
https://doi.org/10.1007/978-3-319-74030-0_14

Process Modelling and Discovery have the potential for advancing towards the creation of models of human-to-human, or human-to-computer, interactions, in which events would represent the invent of a human action or message. Nevertheless, one of the most frequent assumptions in the literature of business process modelling is that events are well defined (i.e. a set of activities fixed during design time). This assumption is no longer applicable in this context, as events are manually defined by humans and, hence, high variability on the event space is expected.

In this paper, we investigate the problem of event name variability for process discovery and propose an approach to resolve this problem through event log preprocessing. In particular, we introduce an approach for clustering event names based on novel similarity metrics between textual data [14] and, afterwards, create a new refined log by projecting events to the discovered clusters. In Sect. 2, we describe the problem and a general overview of a solution. Then, in Sect. 3, we explain the details of our solution which is later validated during Sect. 4. Related work and a discussion on the work presented are provided in Sects. 5 and 6, respectively.

2   Log Abstraction via Event Variability Reduction

Due to the inherent freedom of language and communication, a textual description of a human activity may never be reused for two executions. Analysis of human-described events is hindered by such variability. In this section, we define an event log abstraction consisting as an event projection that generalizes sets of messages. Its main objective is to reduce the number of distinct events and increase the ratio of shared events among conversations. Later in Sect. 3, we explain how this generalization can consider the semantic similitude between event names.

> **Support Engineer**: How can we help you?
> **Customer**: My installation of Product A is not working properly,
> the system prompts me an error code TEST001.
> **Support Engineer**: Please, try the fix provided in KB00001.

Fig. 1. Example of a conversation between a Customer and a Support Engineer. Each interaction in the conversation is considered as a text-based event with an event name consisting of the factual textual message. In this example, the trace consists of three events.

We define[2] the alphabet *MSG* of all the possible messages that may be interchanged in a conversation. We will assume that this alphabet is formed by words, sentences and paragraphs, albeit other non-textual messages may be

---

[2] For the sake of simplicity, the definitions and examples of the paper are tailored to the context of conversations between humans and, possibly, computers. In spite of this, the theory of the paper can be applied to general event logs as defined in [18].

interchanged. A **text-based event** is any instance of an element of the alpha-bet *MSG*, and a **conversation** is a trace of text-based events. I.e. a non-empty sequence of instances of elements in *MSG*. Finally, an **event log** is a collection of conversations of a conversational-based human-to-human service. Figure 1 depicts an example of a typical conversation considered in this paper.

Definition 1. *Given an alphabet A, a log L with the set of traces defined in A and a surjective mapping α between the alphabet A and an alphabet B, the* **Event Variability Reduction** *(EVR) is the projected event log*

$$L' = \{\langle \alpha(e_0), \alpha(e_1), \ldots, \alpha(e_n) \rangle \mid \langle e_0, e_1, \ldots, e_n \rangle \in L\}$$

Generally speaking, the EVR replaces all repetitions of an event e by its image $\alpha(e)$. Due to the surjection of the mapping α, the size of B is smaller than the original alphabet A and, hence, we are performing a reduction of the event space. Although performing such abstraction reduces the information contained in the event log, it enables practitioners to compare different traces. In fact, the major benefit of the EVR technique is when coupled with other techniques. In Sect. 4, we will evaluate the combination of EVR with process discovery and sequence classification techniques.

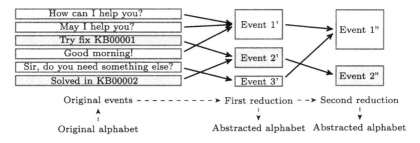

Fig. 2. Graphical representation of two executions of an EVR method over 6 fictional textual messages. The color of the box depicts the final abstract event.

Figure 2 depicts a graphical example of the application of EVR techniques to the alphabet of an event log L, which is compromised by 6 distinct messages in a conversation. A first run of the EVR technique reduces the number of distinct messages to 3, and generated an abstracted alphabet. The resulting event log after the first EVR contains the projected events as specified by the arrows. Notice that the two first events in the abstracted alphabet are now referring to the same abstract *Event 1'* instead of *How can I help you?* and *May I help you?*. A second run of the EVR method simplifies even more the event space to only 2 distinct events.

3   Approach

The utility of the EVR is based on the quality of the event set mapping that we consider. In general, it is expected that if two events e_1 and e_2 are projected into the same event e, then both events have a property in common that is not necessarily shared with events not projected to e. Assuming that the name of the events is a good representative of the real action performed by a human, we propose to group together those events that have similar event names. We will use a novel technique, explained in Sect. 3.1, for measuring the similarity of two event names. Such similarity compares the semantics of words and sentences instead of the exact repetition of words as in traditional bag-of-words techniques.

Figure 3 summarizes our methodology. First, we retrieve all event names from the log and we compute the embeddings of all the words contained in the event name, as specified in Sect. 3.1. These word embeddings allow us to compute a word similarity, that is later averaged for measuring a similarity metric between event names such that text-based events with similar semantics are very similar according to this metric. Finally, we consider a clustering technique for discovering group of events and we use this as the event set partition of the embedding-based EVR.

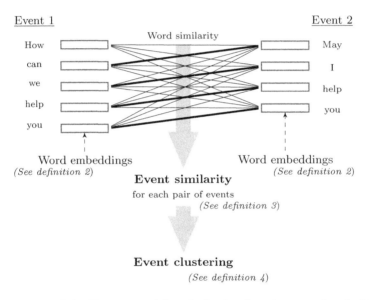

Fig. 3. Overview of the Event Variability Reduction based on word embeddings. A clustering technique utilizes an embedding-based text similarity for creating groups of events.

3.1 Word Embeddings

An embedding is a function that generates representations of objects difficult to analyze into a well-known space, such as a vector space, allowing further analysis. In this paper, we focus on word embeddings.

Definition 2 (Bengio et al. [5]). *A word feature vector or **word embedding** is a function that converts words into points in a vector space. Word embeddings are usually injective functions (i.e. two words do not share the same word embedding), and highlight not-so-evident features of words. Hence, one usually says that word embeddings are an alternative **representation** of words.*

Word2Vec [14] is a word embedding in which words are mapped into a fixed-length vector space, such that the cosine similarity of the embedding of two words is a good estimator of its semantic similarity. The major benefit of using Word2Vec is that the training method does not need to manually build or validate complex taxonomies, but it learns by extracting the *meaning* of a word by considering its adjacent words in a set of sentences. Typically, a large textual corpora such as Wikipedia is considered, but it could also leverage information from knowledge-specific documentation. Moreover, accuracy with respect to unsupervised count-based techniques [4] positions Word2Vec as the perfect candidate for measuring similarity of textual data.

Authors in [11] extended the results obtained by the Word2Vec technique in order to compute a similarity between short messages. Their approach measures the pairwise similarity between words of the two sentences, averaging by the inverse document frequency[3] of words. Although it is out of the scope of this paper, other embedded-based similarity metrics could also be considered [13].

Definition 3 (Tom Kenter and Maarten de Rijke [11]). *Given two event names E_1 and E_2 (with E_1 shorter shorter than E_2), its embedded-based similarity is*

$$Sim(E_1, E_2) = \sum_{w_1 \in E_1} idf(w_1) \cdot \frac{(c+1) \cdot \max_{w_2 \in E_2} Sim(w_1, w_2)}{\max_{w_2 \in E_2} Sim(w_1, w_2) + c \cdot \left(1 + b - b \cdot \frac{|E_2|}{average\ length}\right)},$$

where w_i is a word of the sentence E_i, $Sim(w_1, w_2)$ is the cosine similarity between the Word2Vec embeddings of w_1 and w_2 and b, c are two regularization constants[4].

3.2 Event Rediscovery via Document Embedding Clustering

In the previous subsection, we have defined a similarity between event names based on a word embedding known as Word2Vec. We propose to use a clustering technique based on this similarity metric in order to retrieve groups of similar event names to discover a set of abstract activities that will be consumed by an EVR.

---

[3] We follow the classical definition $idf(w) = \log \frac{Number\ of\ documents}{Occurrences\ of\ w}$.

[4] During the evaluation of this approach, we set c to 1.2 and b to 0.75 as proposed by [11].

Definition 4. *Given a log L, with a set of distinct events E, and a parti-tion of the event set $\{E_i\}_{i \in I}$*[5] *obtained by running a clustering technique on E with the embedding-based similarity metric, we define an* **Embedding-based Event Variability Reduction** *as the EVR defined over the mapping α such that $\alpha(e) = i \in I$ such that $e \in E_i$.*

After applying an embedding-based EVR, the newly discovered event log has reduced the variability of event names by discarding the wording used and, instead, focusing on the semantic of the words. Depending on the clustering technique used and parameters, one could obtain event logs with different levels of granularity. Notice that the event identifier of the newly discovered event log is a set of events in the original event log, the end-user may need to check the set of events to understand the abstracted event log.

4   Evaluation

To evaluate the approach presented in this paper, we chose a dataset in which *event names* hold information about an unknown activity, there are some guide-lines on how those *events* should be named and positioned in the trace (i.e. the system process) and results can be easily interpreted. With such dataset, we will perform a preliminary process analysis that would have been impossible with-out the use of the EVR. Afterwards, we apply the techniques to an industrial dataset comprising textual conversations between technical support engineers and customers. Prior to the technique described in the paper, the lack of struc-ture in textual messages did not allow for a mechanism to monitor the evolution of conversations.

4.1   Structure of Documents in Wikipedia

Mass collaboration projects usually rely on guidelines to palliate the variability of human outcomes, and sub-communities are created to ensure better coherence. Wikipedia is a great example of such a complex collaboration project, with over 200 guidelines[6], ranging from behavioral recommendations on a discussion to naming rules. We hypothesize that the Table of Contents of Wikipedia articles follows some of these guidelines. We will evaluate how the embedding-based EVR helps process discovery techniques in discovering such guideline.

Discovery of the Structural Process Model. We retrieved over 800 articles from Wikipedia[7], selected from the list of featured articles[8] of the *Media, Litera-ture and Theater, Music biographies, Media biographies, History biographies* and

---

[5] i.e. a finite collection of sets $\{E_i\}_{i \in I}$ such that $\cup_{i \in I} E_i = E$ and $E_i \cap E_j = \emptyset$ for any $i \neq j$.

[6] https://en.wikipedia.org/wiki/Wikipedia:List_of_guidelines.

[7] 8th August 2016. The dataset is publicly available on data.4tu.nl [16].

[8] https://en.wikipedia.org/wiki/Wikipedia:Featured_articles.

Video gaming categories. From the list of articles we extracted the structure of the document, i.e. sections and subsections of the text.

Directly applying process discovery techniques over such event logs generates the *flower model*[9] in all cases, primarily due to the high ratio of distinct events over events seen in the log. In fact, several events are only seen once and, hence, patterns in the structure of the document are difficult to find. Besides, comparison of traces between categories is almost impossible. To overcome this challenge, we applied the embedded-based EVR technique for discovering a set of 50 abstract events shared among all the Wikipedia articles. After discovering such abstract events, we see a complete different picture allowing us to further analyze this dataset.

Table 1. 5 randomly chosen section titles from the first 6 out of 50 discovered clusters from the Wikipedia dataset.

Cluster 1	Cluster 2	Cluster 3
In culture	Travels	Aftermath of centralia
Philosophy	First voyages	History of the manuscripts
Re-discovery	Privateering expedition	With the five
Mythology	Journey	The poems
Historiography	Trans-pacific voyage	In the media
Cluster 4	Cluster 5	Cluster 6
Writing and publishing	Return to United States	Lineups
Writing career	Return to France	Collaborations
Writing style	Return to Missouri	Recording
Writing history	Return to Canada	Discography
Writing	Return to Japan	Collaborators

Table 1 shows six activities discovered by applying the embedded-based EVR. One may check that Cluster 4 trivially refers to sections involving *writing*, and Cluster 5 combines sections containing *return to*. Nevertheless, other combinations are less trivial such as the sections included in Cluster 2. Unfortunately, some groups are not as accurate as the aforementioned. For instance, the first cluster seems to contain topics related to *philosophy*, *religion* and *history*. The three topics are certainly related, but a better granularity on such cluster might be necessary.

The embedded-based EVR replaces all the listed Wikipedia titles in Table 1 with the assigned Cluster number. Therefore, one may take the set of titles as the new *event name*. This may hinder the understandability of the discovered abstract activities, as the practitioner needs to take a look into the clustered items, as we have done in the above paragraph, to have an understanding of their relation.

---

[9] The flower model is a model that allows any possible behavior.

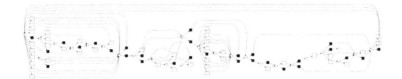

Fig. 4. Petri net discovered from articles in the Music biography category.

Continuing the log analysis, we run a process discovery method on each of the logs. In particular, we run the Inductive Miner[10]. An example of a process model discovered after applying the EVR is depicted in Fig. 4. Most of them have a small subprocess with a flower-like behavior, and an in-depth analysis of the traces and abstract events highlighted several section and subsections with similar names that may happen in any ordering. Nevertheless, in general, a more detailed pattern has been found in this dataset thanks to the event abstraction.

Table 2. Table summarizing quality of the discovered process models. For each row, fitness and precision of a process model is measured with respect to all the logs.

		History biography	Media biography	Music biography	Literature	Video games	Media
History biography	F	**0.94**	0.79	0.72	0.79	0.65	0.68
	P	0.37	0.37	0.37	0.39	0.41	0.43
Media biography	F	0.73	**0.91**	0.77	0.69	0.59	0.61
	P	0.47	0.42	0.46	0.49	0.50	0.52
Music biography	F	0.88	0.92	**0.97**	0.86	0.69	0.72
	P	0.35	0.34	0.34	0.36	0.38	0.39
Literature	F	0.71	0.66	0.66	**0.81**	0.75	0.80
	P	0.41	0.39	0.45	0.48	0.46	0.48
Video games	F	0.44	0.49	0.55	0.64	**0.87**	0.73
	P	0.66	0.64	0.65	0.64	0.55	0.65
Media	F	0.74	0.75	0.75	0.83	**0.97**	0.93
	P	0.43	0.42	0.44	0.42	0.43	0.46

Comparison of the Structure on Wikipedia Articles. The process model depicted in Fig. 4 is a first approach to find the underlying guideline for writing articles in their respective categories. Table 2 depicts alignment-based fitness [2] and precision [1] of the discovered process models with respect to all the abstract event logs, and serves as a mechanism to compare the underlying guidelines between the categories. One may notice that the three *biography* process models have high fitness with all the biography logs, but it is significantly lower with

---

[10] We run the infrequent version of the Inductive Miner, with default parameters, on ProM 6.5.1.

respect to *Video gaming* and *Media* categories. The contrary also holds, as fitness of the *Video gaming* and *Media* process models is significantly lower with respect to the three biography logs. On the other hand, the *Literature* category fits fairly well in both groups.

These results were expected, as all *biography* articles should have a similar structure (although talking about different types of artists or personalities) and *videogames* are nowadays produced as popular films and series (as they appear in the *media* category). On the contrary, *literature* articles usually talk about the authors and historical context of the book, and also about its plot (which is very common in *videogames* and *media* articles).

4.2 Application of the Event Variability Reduction to Trace Monitoring of Human-Driven Processes

Some textual documents such as Tickets in Support systems, or live support chats, evolve over time. For example, tickets consist of a sequence of messages exchanged between a customer and one or more support engineers. The first messages usually provide a first description of the problem. Nevertheless, the content may evolve throughout the chain of messages and derive to other topic as the root-cause of the issue is being discovered. When the conversation between the customer and the support team ends, the ticket is usually enriched with extra information about the conversation outcomes such as the product causing the issue, type of fix needed, solution proposed, time to complete the issue, or satisfaction of the client with the support team. If this final information is known during the conversation, a support engineer would be able to better guide the conversation.

It is very important for the industry to predict the level of satisfaction of a customer with respect to the service offered. Customer escalation is the formal mechanism that customers have to warn support engineers that the resolution of an issue is not as fast and smooth as they expected. In fact, the number of escalations is used as a Key Performance Indicator (KPI) for measuring the quality of support teams, and it is clearly an indicator of customer dissatisfaction and churning.

We applied the EVR technique in an industrial dataset provided by CA Technologies. This dataset contains the messages interchanged between support engineers and customers during 2015, as well as all customer escalations during the same period. We applied the EVR technique for discovering models specifics to escalated support cases and non-escalated cases, with the objective of building a predictor for future cases. Figure 5 depicts the fuzzy models [19] of both categories. Notice the structural difference between the two process models. The escalated process model is more unstructured than the non-escalated process models. This might indicate that the support engineers need a broader exploration phase of such cases.

We trained a pair of Hidden Markov Models [7] for building a predictor capable of classifying non-complete traces. At the initialization step, we used the process models in the two figures as the structure of the hidden states. Then,

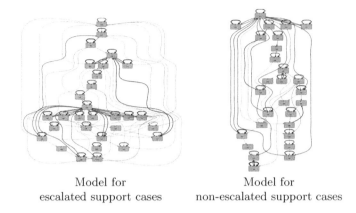

Model for   Model for
escalated support cases   non-escalated support cases

Fig. 5. Fuzzy Nets obtained from the support cases after applying the embedded-based EVR. Escalated cases are more unstructured than non-escalated cases, indicating that either support engineers need a broader exploration of the issue or that customers felt that the case was not properly handled.

probabilities were tuned during the training of the model for maximizing the accuracy of the classifier. Despite the structural difference provided by the fuzzy models and the existence of differential, albeit not-so-frequent, small sequential patterns, roughly 10% of accuracy was achieved when detecting escalated cases[11]. The high imbalance between the escalated and non-escalated datasets may have caused the low ratio of detected escalations. Besides, it also indicates that escalations are not primarily caused by a global property of the conversation, but other external features must be taken into account. Nevertheless, it has been acknowledged by the company as a useful tool as it reduces the number of cases that need to be closely monitored for having an impact on customer satisfaction.

5   Related Work

Different approaches exist in the literature for the problem of discovery and management of process models with a large set of supported activities. For instance, the Fuzzy Miner [19] allows practitioners to choose a level of abstraction for the discovered process model, and the algorithm automatically merges different events into a single, and more abstract, cluster of events. Nevertheless, these approaches are either based on the *directly-follows* relation [8,19], a temporal correlation [9] or satisfying a particular known pattern [17] (or initially unknown patterns [6]). Our approach is not based on the fact that events may have a sequential or temporal relation between them, but that the event name similarity indicates how similar are two events.

---

[11] Results are consistent with respect to a 20%-out cross-validation.

The art of modelling human conversations, known as Speech Acts or Dialog Acts, is a well known challenge in Computer Science [15]. Nevertheless, to the best of our knowledge, current applications of Speech Acts follow a top-down approach in which textual messages are classified based on a list of known Intents[12]. As an example, Authors in [10] combine Intent detection for abstracting Events and then measures the likelihood of changing from one Intent to another.

Bag-of-words techniques, i.e. using the frequency of each word in a document, have been largely used for comparing two texts and discover a list of topics in a set of documents. Process Matching [12] has considered these techniques for matching activities of different known processes, and [3] consumes activity names, and their descriptions, to map events to activities of a known process model. None of them considers event name abstraction for the discovery of the process model. Besides, the approach presented in this paper enables the comparison of word semantics instead of considering exact word matches.

6   Discussion

In this paper, we have developed a method for reducing variability of event names by grouping them according to their similarity. Recent developments on Natural Language Processing allowed us to compute this similarity based on semantic information, instead of traditional bag-of-words techniques or creating a complex ontology.

We applied this technique on a dataset compromising the structure of articles in Wikipedia. Initially, it was impossible to find any common structure because section names were almost never repeated in the dataset. Nevertheless, after applying EVR, common process discovery technique already discovered some patterns on the data and enable us to compare two different articles. Although this use case is very simplistic, the results validate the methodology and motivates further research in this direction. We have also used this technique for analyzing the structure of support cases, and we arrived to the conclusion that there is a weak relation between customer satisfaction and the content of the sequence of messages interchanged between support engineers and customers.

This paper focus on the event name similarity of two events, but it does not consider the frequency of the events nor its role in the trace. Further research should consider how this information can be leveraged to discover better abstract events.

Acknowledgements. This work is funded by Secretaria de Universitats i Recerca of Generalitat de Catalunya, under the Industrial Doctorate Program 2013DI062, and the Spanish Ministry for Economy and Competitiveness, the European Union (FEDER funds) under grant COMMAS (Ref. TIN2013-46181-C2-1-R).

---

[12] ISO 224617-2 defines 57 generic communicative functions, that one may enrich or refine depending with domain knowledge.

References

1. Adriansyah, A., Munoz-Gama, J., Carmona, J., Dongen, B.F., Aalst, W.M.: Measuring precision of modeled behavior. Inf. Syst. E-bus. Manag. **13**(1), 37–67 (2015)
2. Adriansyah, A., van Dongen, B.F., van der Aalst, W.M.P.: Conformance checking using cost-based fitness analysis. In: Proceedings of the 2011 IEEE 15th International Enterprise Distributed Object Computing Conference, EDOC 2011, Washington, DC, USA, pp. 55–64. IEEE Computer Society (2011)
3. Baier, T., Mendling, J., Weske, M.: Bridging abstraction layers in process mining. Inf. Syst. **46**, 123–139 (2014)
4. Baroni, M., Dinu, G., Kruszewski, G.: Don't count, predict! A systematic comparison of context-counting vs. context-predicting semantic vectors. In: Proceedings of Association for Computational Linguistics (ACL), vol. 1 (2014)
5. Bengio, Y., Ducharme, R., Vincent, P., Jauvin, C.: A neural probabilistic language model. J. Mach. Learn. Res. **3**(Feb), 1137–1155 (2003)
6. Jagadeesh Chandra Bose, R.P., van der Aalst, W.M.P.: Abstractions in process mining: a taxonomy of patterns. In: Dayal, U., Eder, J., Koehler, J., Reijers, H.A. (eds.) BPM 2009. LNCS, vol. 5701, pp. 159–175. Springer, Heidelberg (2009). https://doi.org/10.1007/978-3-642-03848-8_12
7. Da Silva, G.A., Ferreira, D.R.: Applying hidden Markov models to process mining. Sistemas e Tecnologias de Informação. AISTI/FEUP/UPF (2009)
8. Günther, C.W., Rozinat, A., van der Aalst, W.M.P.: Activity mining by global trace segmentation. In: Rinderle-Ma, S., Sadiq, S., Leymann, F. (eds.) BPM 2009. LNBIP, vol. 43, pp. 128–139. Springer, Heidelberg (2010). https://doi.org/10.1007/978-3-642-12186-9_13
9. Günther, C.W., van der Aalst W.M.P.: Mining activity clusters from low-level event logs. Beta, Research School for Operations Management and Logistics (2006)
10. He, Z., Liu, X., Lv, P., Wu, J.: Hidden softmax sequence model for dialogue structure analysis. In: Proceedings of the 54th Annual Meeting of the Association for Computational Linguistics (2016)
11. Kenter, T., de Rijke, M.: Short text similarity with word embeddings. In: Proceedings of the 24th ACM International on Conference on Information and Knowledge Management, CIKM 2015, pp. 1411–1420. ACM, New York (2015)
12. Klinkmüller, C., Weber, I., Mendling, J., Leopold, H., Ludwig, A.: Increasing recall of process model matching by improved activity label matching. In: Daniel, F., Wang, J., Weber, B. (eds.) BPM 2013. LNCS, vol. 8094, pp. 211–218. Springer, Heidelberg (2013). https://doi.org/10.1007/978-3-642-40176-3_17
13. Le, Q.V., Mikolov, T.: Distributed representations of sentences and documents. In: Proceedings of the 31th International Conference on Machine Learning, ICML 2014, 21–26 June 2014, Beijing, China, pp. 1188–1196 (2014)
14. Mikolov, T., Sutskever, I., Chen, K., Corrado, G., Dean, J.: Distributed representations of words and phrases and their compositionality. CoRR, abs/1310.4546 (2013)
15. Morelli, R.A., Bronzino, J.D., Goethe, J.W.: A computational speech-act model of human-computer conversations. In: Proceedings of the 1991 IEEE Seventeenth Annual Northeast Bioengineering Conference, pp. 263–264. IEEE (1991)
16. Sanchez-Charles, D.: Title and subtitles of wikipedia articles (2017). https://doi.org/10.4121/uuid:61fb9665-40ab-4b70-8214-767c521cc950
17. Tax, N., Sidorova, N., Haakma, R., van der Aalst, W.M.P.: Event abstraction for process mining using supervised learning techniques. CoRR, abs/1606.07283 (2016)

18. van der Aalst, W.M.P.: Process Mining - Discovery Conformance and Enhancement of Business Processes. Springer, Berlin (2011)
19. van der Aalst, W.M.P., Günther, C.W.: Finding structure in unstructured processes: the case for process mining. In: ACSD, pp. 3–12. IEEE Computer Society (2007)

Automatic Root Cause Identification Using Most Probable Alignments

Marie Koorneef[1], Andreas Solti[2], Henrik Leopold[1(✉)], and Hajo A. Reijers[1,3]

[1] Department of Computer Sciences, Vrije Universiteit Amsterdam,
Amsterdam, The Netherlands
h.leopold@vu.nl
[2] Institute for Information Business, Vienna University of Economics and Business,
Vienna, Austria
[3] Department of Mathematics and Computer Science,
Eindhoven University of Technology, Eindhoven, The Netherlands

Abstract. In many organizational contexts, it is important that behavior conforms to the intended behavior as specified by process models. Non-conforming behavior can be detected by aligning process actions in the event log to the process model. A probable alignment indicates the most likely root cause for non-conforming behavior. Unfortunately, available techniques do not always return the most probable alignment and, therefore, also not the most probable root cause. Recognizing this limitation, this paper introduces a method for computing the *most probable alignment*. The core idea of our approach is to use the history of an event log to assign probabilities to the occurrences of activities and the transitions between them. A theoretical evaluation demonstrates that our approach improves upon existing work.

Keywords: Conformance checking · Root cause analysis
Most probable alignments

1 Introduction

In many organizations, it is important that employees execute the tasks of processes in conformance with certain rules. For example, employees of a bank must check the credit history of a customer *before* granting a loan and call center agents must verify the identity of a caller *before* providing support. Possible implications of violating such rules might be severe. So-called *conformance checking* tools are able to automatically check whether the recorded process actions in an event log match the intended behavior as specified by a process model [1,12,14]. In this way, these tools intend to help organizations to automatically monitor the level of conformance and to identify problems.

However, automatically checking the conformance between event logs and process models is a complex task. The key challenge in this context is to create an *alignment* between the event log and the process model in order to explain the

ⓒ Springer International Publishing AG 2018
E. Teniente and M. Weidlich (Eds.): BPM 2017 Workshops, LNBIP 308, pp. 204–215, 2018.
https://doi.org/10.1007/978-3-319-74030-0_15

behavior as captured in the log using the process model. An approach commonly used to obtain such an alignment is the *cost-based approach*. This approach aims to find the alignment with the least expensive deviations using a manually defined cost function [4]. One property of this cost-based approach is that it often leads to several alignments with the same costs. In practice, this means that an organization is provided with a set of possible root causes for non-conforming behavior, instead of with the most probable one. Recognizing this limitation, Alizadeh et al. [5,6] introduced the notion of a *probable alignment*, which focuses on follow relationships in the fitted event log. However, their approach does not always return the most probable alignment and, therefore, also not the most probable root cause for non-conforming behavior.

In this paper, we propose a different technique for computing *probable alignments*. Our technique builds on the probabilities from the event log to compute the *most probable alignment*. More specifically, within the event log it assigns a probability to the occurrence of an activity; and in the transition system of the Petri net it assigns a probability to a transition. On this basis, we provide a principled approach that solves the most probable alignment problem in conformance checking.

The remainder of this paper is organized as follows. Section 2 discusses the background of our work. Section 3 introduces the preliminary concepts that we use. Section 4 explains our technique for computing probable alignments. Section 5 presents a theoretical evaluation of our technique. Section 6 discusses related work before Sect. 7 concludes the paper and discusses future work.

2   Background

In this section, we discuss the background of our work. Section 2.1 introduces the running example we use throughout this paper. Section 2.2 explains the details of the cost-based approach. Section 2.3 then discusses the shortcomings of current techniques for computing probable alignments.

2.1   Running Example

To illustrate the problem of finding the most probable alignment, consider the simple process model from Fig. 1, which captures the intended behavior of an organization as a Petri net. It defines that conforming behavior consists of a sequence of A, at least one B, followed by a choice between C and D. Thus, possible conforming traces to this model are: $\langle A,B,C \rangle$, $\langle A,B,B,C \rangle$, ..., $\langle A,B,D \rangle$, $\langle A,B,B,D \rangle$, and so on.

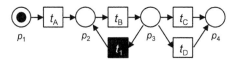

Fig. 1. Petri net capturing the intended behavior

Suppose the trace $\mathtt{tr}_1 = \langle A,B,D,C\rangle$ is observed, which is generated by an information system that tracks the actual behavior. This trace does not conform to the behavior specified by the process model from Fig. 1, since the process model defines a choice between C and D. The remainder of this section illustrates existing alignment techniques and their shortcomings based on this example.

2.2   Cost-Based Alignment

The cost-based alignment technique aligns each observed trace to a path in the process model; thereby it identifies missing events, additional events or incorrect ordering of events. Having modeled the trace $\mathtt{tr}_1$ as a sequential Petri net, the *Petri net product* in Fig. 2 models all possible movements by taking the product of the trace and process model. The Petri net of the trace represents log moves (additional events) and the Petri net of the process model represents model moves (missing events). Synchronous moves are created by pairing each activity in the trace to a transition in the model that corresponds to the same activity.

The transition system of the Petri net product is created, such that states are reachable markings and transitions are log, model and synchronous moves of an activity type. To find an optimal alignment, costs are assigned manually to the move types (i.e. log, model and synchronous move). The A^*-algorithm can then be used to find the shortest path from initial state to end state in the state space of the product net, which corresponds to a weighted transition system [1,3,4,9].

For the trace $\langle A,B,D,C\rangle$, assuming equal costs for any deviation, the cost-based alignment finds two optimal alignments with the least expensive deviations (i.e. the alignments with the lowest costs), as displayed in Fig. 3. The $\gg$ symbol represents no progress in the replay on the respective side, e.g. the step for D in the alignment in Fig. 3a is a log move.

Note that the costs of a move type can be individually set for each activity. However, it is not possible to make the costs conditional on e.g. other activities in the sequence. This implies that move types of an activity are treated as independent from each other.

2.3   Shortcomings

Both optimal alignments in Fig. 3 have an equal cost of one log move, so either transition D or C is in excess. Without further knowledge, the choice between these alignments is arbitrary. This means that it is not guaranteed that the most probable root cause is taken.

Suppose in the event log $\langle A,B,C\rangle$ is observed 80 times and $\langle A,B,D\rangle$ is observed 20 times. Then $\langle A,B,C\rangle$ is the more likely alignment, so alignment 1 (see Fig. 3a) gets a higher probability than alignment 2 (see Fig. 3b).

The notion of a *probable alignment* as introduced by Alizadeh et al. [5] provides a solution by considering the event log history. They favor synchronous moves over log and model moves and they calculate a log move using the *never eventually follow* relation conditioned on a conforming partial trace. For example, the probability that D never eventually follows after $\langle A,B\rangle$ (i.e. the log

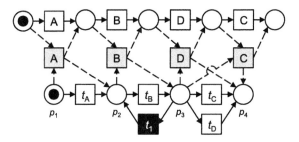

Fig. 2. Petri net product of the trace $\mathbf{tr}_1 = \langle A,B,D,C \rangle$ on top and the process model in Fig. 1 on the bottom. Synchronous execution is marked as shaded transitions in the middle.

trace	A	B	D	C
model	A	B	$\gg$	C
	t_A	t_B		t_C

(a) Optimal alignment 1

trace	A	B	D	C
model	A	B	D	$\gg$
	t_A	t_B	t_D	

(b) Optimal alignment 2

Fig. 3. Cost-based optimal alignments between the trace $\mathbf{tr}_1 = \langle A,B,D,C \rangle$ and the process model from Fig. 1

move probability of D given $\langle A,B \rangle$) is 80%. The log move probability of C given $\langle A,B,D \rangle$ is 100%. Maximizing the probable moves, the technique results in $\langle A,B,D \rangle$ (see Fig. 3b) as the most probable alignment, even though C occurred more frequently in the event log history.

In line with the intuition that C is more probable to occur, our approach identifies alignment 1 as the most probable alignment.

3   Preliminaries

This section gives a definition for *event log*, *Petri net* and *marking*.

An *event log* captures information about a running process, where each case (or instance) is represented with a trace of events that correspond to activities in the process. Formally, an event log is defined as follows.

Definition 1 (Event Log). *Let $\mathcal{A}$ be the set of activities, and $\mathcal{E}$ denote the set of events. The activity function $\alpha_L : \mathcal{E} \rightarrow \mathcal{A}$ assigns an activity to each event. A trace $\mathbf{tr}_E$ is a sequence of events, i.e., $\mathbf{tr}_E \in \mathcal{E}^*$. We use α_L to project sequences of events to activity traces $\mathbf{tr}_A \in \mathcal{A}^*$. Thus, an event log $\mathcal{L}$ is a multiset of activity traces.*

Table 1 captures the event log history used in the remainder of this paper. This history event log contains all events observed in the past and is non-empty. The corresponding process model is represented as a *Petri net* in Fig. 1. Note that the last two traces in Table 1 do not conform with the process model. Formally, a Petri net is defined as follows.

Table 1. Event log history

trace	frequency
$\langle$A,B,C$\rangle$	40
$\langle$A,B,D$\rangle$	10
$\langle$A,B,B,C$\rangle$	80
$\langle$A,B,B,D$\rangle$	20
$\langle$A,B,B,B,C$\rangle$	40
$\langle$A,B,B,B,D$\rangle$	10
$\langle$A,B,B,B,B,B,B,C$\rangle$	1
$\langle$A,B,B,B,B$\rangle$	1
$\langle$A,C,B$\rangle$	1

Definition 2 (Petri Net, Marking). *A Petri net* over a set of activities $\mathcal{A}$ is *a tuple* $(P, \mathcal{T}, F, \alpha_1, m_i, m_f)$ *where* P *and* $\mathcal{T}$ *are sets of places and transitions, respectively.* $F : (P \times \mathcal{T}) \cup (\mathcal{T} \times P) \to \mathbb{N}$ *is a flow relation between places and transitions; and* $\alpha_1 : \mathcal{T} \nrightarrow \mathcal{A}$ *is a partial function mapping transitions to activities. A state of a Petri net is determined by the* marking $(\mathcal{M} : P \to \mathbb{N})$ *of a net, which specifies the number of tokens on the places.* m_i *and* m_f *are the initial marking and the final marking, respectively.*

In addition, a transition $t \in \mathcal{T}$ is *invisible* if $t \notin Dom(\alpha_1)$. A single activity can be represented by multiple transitions (i.e. *duplicate* transitions). Finally, we only consider *easy sound* Petri nets, that is, Petri nets of which the final state is reachable from its initial state [2].

4   Method

In this section, we introduce our method for computing the most probable alignment. Section 4.1 first introduces the probabilistic model. Section 4.2 then explains the computation of the most probable alignment.

4.1   Probabilistic Model

To derive the most probable alignment, we assess the probabilities of individual moves of one alignment. To this end, this section explains the probability calculation for a log-, model- and synchronous move of an activity.

Probability of an Activity in the Event Log. Given the event log history in Table 1, an activity X is a discrete random variable with an outcome in the set of activities $\mathcal{A} = \{A, B, C, D\}$. Let us extend the set of activities to allow for unobserved activities $\mathcal{B}$, for example $\mathcal{B} = \{\star\}$. The probability of seeing outcome i is:

$$P(X = i) = \theta_i \; \forall \, i \in \mathcal{A} \cup \mathcal{B}, \tag{1}$$

where $\boldsymbol{\theta} = (\theta_A, \theta_B, \theta_C, \theta_D, \theta_\star)$, $0 \leq \theta_i \leq 1$ and $\sum_{i \in \mathcal{A} \cup \mathcal{B}} \theta_i = 1$. So each (un)observed activity gets a (positive) probability of occurrence.

We assume that the random variables in the event log $X_1, \ldots, X_n$ are a *random sample* of size n from the population. This means that all random variables are independent and identically distributed (hereafter: iid). The sample is denoted by $\boldsymbol{X} = (X_1, \ldots, X_n)$ and is categorically distributed with parameter $\boldsymbol{\theta}$. We estimate $\boldsymbol{\theta}$ based on our event log history (see Table 1).

Let $\hat{\boldsymbol{\theta}}$ be the estimate of the true probability $\boldsymbol{\theta}$. Given our observations $\boldsymbol{X} = \boldsymbol{x}$, $\hat{\boldsymbol{\theta}}$ equals the sample mean:

$$\hat{\theta}_i = E(|X = i|) = \frac{|X = i|}{\sum_{j \in \mathcal{A} \cup \mathcal{B}} |X = j|} = \frac{|X = i|}{n} \; \forall \, i \in \mathcal{A} \cup \mathcal{B}, \tag{2}$$

So for our event log history in Table 1, the probability of seeing activity A ($\hat{\theta}_A$) is $\frac{203}{817} = 24.9\%$, because Table 1 contains 203 A's of 817 activities in total. We do not observe activity $\star$, so $\hat{\theta}_\star = 0$.

To avoid zero probabilities, we can use Bayesian theory. In the Bayesian framework, the prior embodies a belief about the distribution of $\boldsymbol{\theta}$. Let parameter $\boldsymbol{\theta}$ be treated as a random variable: we define a prior and posterior density distribution of parameter $\boldsymbol{\theta}$. The posterior distribution is a combination of our prior belief and what we observe in the data. In determining the posterior distribution, the effect of the prior distribution decreases when the sample size increases [10]. For example, assuming each outcome in the event log is equally probable, the prior assigns $\frac{1}{5}$ to each outcome {A, B, C, D, $\star$}. Example 1 below illustrates the application of the prior[1].

Probability of a Log Move. To calculate the probability of a log move, we want to remove the least likely fitting activity from the event log to align the remainder with the model. So the least frequent activity in an event log should be the most probable log move. To calculate the log move probability, we define the log move of an activity X^L as a monotone decreasing function of X ($X \mapsto h(X) = X^L$) [8], such that the estimated probability of a log move equals:

$$\hat{\theta}_i^L = E(|X^L = i|) = \frac{1 - \hat{\theta}_i}{k - 1} \; \forall \, i \in \mathcal{A} \cup \mathcal{B}, \tag{3}$$

where k is the number of possible outcomes in $\mathcal{A} \cup \mathcal{B}$, $0 \leq \hat{\theta}_i^L \leq 1$, and $\sum_{i \in \mathcal{A} \cup \mathcal{B}} \hat{\theta}_i^L = 1$. It follows that log moves X^L are iid categorically($\hat{\boldsymbol{\theta}}^L$) distributed.

---

[1] A conjugate prior means that the prior is from the same family of distributions as the posterior. The conjugate prior for the categorical distribution is the Dirichlet distribution [7].

i	$\lvert X=i \rvert$		i	$\lvert X=i \rvert$		i	$\hat{\theta}_i$		i	$\hat{\theta}_i^L$	
A	1		A	203+1		A	$\frac{204}{822}$		A	$\frac{618}{822\cdot4}$	≈ 0.19
B	1		B	412+1		B	$\frac{413}{822}$		B	$\frac{409}{822\cdot4}$	≈ 0.12
C	1		C	162+1		C	$\frac{163}{822}$		C	$\frac{659}{822\cdot4}$	≈ 0.20
D	1		D	40+1		D	$\frac{41}{822}$		D	$\frac{781}{822\cdot4}$	≈ 0.24
$\star$	1		$\star$	1		$\star$	$\frac{1}{822}$		$\star$	$\frac{821}{822\cdot4}$	≈ 0.25
	5			817+5			1			1	
(a) Assign prior			(b) Add observations			(c) Calculate $\hat{\boldsymbol{\theta}}$			(d) Calculate $\hat{\boldsymbol{\theta}}^L$		

Fig. 4. Estimation of the log move probability using Table 1

Example 1. Figure 4 illustrates the estimation of the log move probability $\hat{\boldsymbol{\theta}}^L$. A prior is assigned to each activity $i \in \{A, B, C, D, \star\}$ in Fig. 4a, i.e. pseudo-observations are assigned to each activity. In Fig. 4b the observations of Table 1 are added to the prior. The probability of an activity $\hat{\boldsymbol{\theta}}$ is calculated with Eq. (2). The probability of a log move of an activity $\hat{\boldsymbol{\theta}}^L$ is calculated with Eq. (3), where $k = 5$, since i can take 5 different outcomes. Figure 4c and d display the results. For example, the probability of a log move D equals 24%.

Probability of a Model Move. The probability of a model move is equal to the probability of a transition given a marking in a Petri net[2]. Let marking M be the state of a Petri net. Given the Petri net in Fig. 1, a marking M is a discrete random variable with outcomes in the set of markings $\mathcal{M} = \{\{p_1\}, \{p_2\}, \{p_3\}, \{p_4\}\}$. A Petri net can be represented by a Markov Chain where states are markings, this means independence of firing history [13].

Given the Petri net in Fig. 1, a transition T is a discrete random variable with outcomes in the set of transitions $\mathcal{T} = \{t_A, t_B, t_C, t_D, t_1\}$. A transition conditioned on marking $M = m$ (hereafter: $T \mid m$) is also a discrete random variable. The probability of seeing outcome i given $M = m$ is:

$$P(T = i \mid M = m) = \phi_{i\mid m} \; \forall \, i \in \mathcal{T}, \tag{4}$$

where $0 \le \phi_{i\mid m} \le 1$ and $\sum_{i\in\mathcal{T}} \phi_{i\mid m} = 1$. $T \mid m$ is categorically distributed with parameter $\boldsymbol{\phi}_m = (\phi_{t_A\mid m}, \ldots, \phi_{t_1\mid m})$.

We estimate the true probability $\boldsymbol{\phi}_m$ by $\hat{\boldsymbol{\phi}}_m$ for all $m \in \mathcal{M}$. Given $\boldsymbol{T} = \boldsymbol{t} \mid \boldsymbol{M} = \boldsymbol{m}$, $\hat{\boldsymbol{\phi}}_m$ equals the sample mean:

$$\hat{\phi}_{i\mid m} = E(\lvert t = i \mid m \rvert) = \frac{\lvert t = i \mid m \rvert}{\sum_{j\in\mathcal{T}} \lvert t = j \mid m \rvert} \; \forall \, t \in \mathcal{T} \tag{5}$$

Moreover, we add a prior to embody our belief of the distribution of $\hat{\boldsymbol{\phi}}_m$ for all $m \in \mathcal{M}$. The most frequent transition is the most probable model move,

---

[2] The probability of a transition given a marking uses the same logic as in Stochastic Petri Nets [13].

i.e. when an event is missing (a model move) the most likely activity is added. Model moves are independent, because markings M are memoryless.

Example 2. Figure 5 illustrates the estimation of the model move probability $\hat{\phi}_m$. The states (markings) and the transitions of the Petri net are represented in a transition system in Fig. 5a. First, a prior is assigned to the transitions in Fig. 5b. Second, given the optimal *cost-based alignments*, the frequencies are calculated as the sum of synchronous and model moves for a transition. In Fig. 5c these frequencies are added to the prior. Third, this sum is used to estimate the probabilities with Eq. (5) in Fig. 5d.

We use the cost-based alignment to calculate the observed random sample $\boldsymbol{T} = \boldsymbol{t} \mid \boldsymbol{M} = \boldsymbol{m}$. For this, we consider all optimal cost-based alignments with minimal number of model moves of invisible transitions. The cost-based alignment handles duplicate and invisible transitions in the model. Next to this, it non-deterministically chooses one alignment from all optimal alignments. In calculating our model move probabilities, we assume that all optimal alignments are equally likely, e.g. for $\mathtt{tr}_1$ in Fig. 3 this implies that both alignments have a 50% likelihood (frequency $= \frac{1}{2}$).

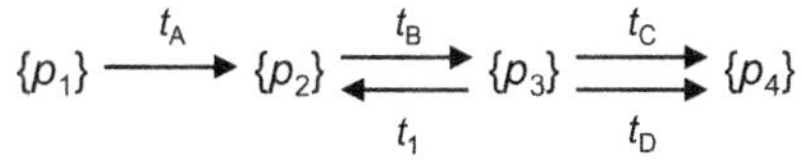

(a) The transition system of the process model in Fig. 1

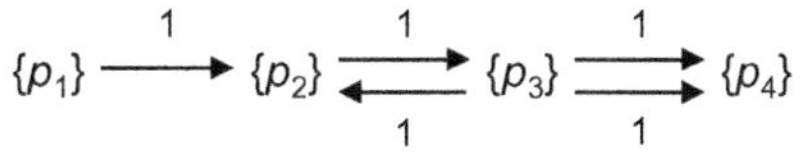

(b) The transition system with assigned priors

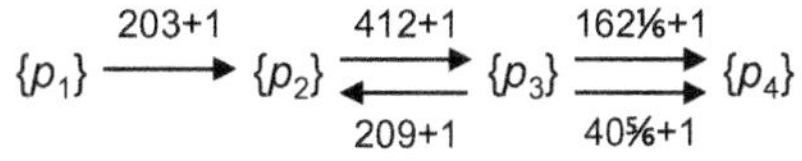

(c) The transition system with assigned priors and aligned frequencies of synchronous and model moves

(d) The model move probabilities given a marking $\hat{\phi}_m$

Fig. 5. Estimation of the model move probability using the transition system of the process model in Fig. 1 and the optimal cost-based alignments

Probability of a Synchronous Move. In the Petri net product of Fig. 2 we can perform either a synchronous move of A, or a model move t_A plus a log move of A. In line with existing techniques, we prefer a synchronous move. Let ψ_i denote the probability of a synchronous move of activity $i \in \alpha_2(\mathcal{T})$, where $\alpha_2 : \mathcal{T} \nrightarrow \mathcal{A} \cup \mathcal{B}$ is a partial function mapping transitions to all activities $\mathcal{A} \cup \mathcal{B}$. We assume that the probability of the synchronous move is the maximum of the log and model move probability, so:

$$\psi_i = \max\{\theta_i^L, \phi_{i|m}\} \ \forall \ i \in \alpha_2(\mathcal{T}) \tag{6}$$

That is, the equation $\psi_i > \theta_i^L \phi_{i|m}$ is satisfied, which ensures that a synchronous move is preferred over a separate log plus model move.

Example 3. Given Figs. 4 and 5 and Eq. (6), the probability of a synchronous move of A equals $\psi_A = \max\{\theta_A^L, \phi_{t_A|\{p_1\}}\} = \max\{0.19, 1\} = 1$.

4.2 Most Probable Alignment

Consider the transition system of the Petri net product in Fig. 2, where we assign probabilities to the arcs. By the independence assumptions, the probabilities of successive moves (a path) can be multiplied. The most probable path is the path with the maximum probability.

A $\log(\cdot)$ transformation enables us to transform the product of probabilities to a sum of log probabilities. To find the most probable path, we search for the maximum sum. Maximizing the sum of $\log(\cdot)$ is equivalent to minimizing the sum of $-\log(\cdot)$. Hence, the $-\log(\cdot)$ transformation allows us to use the A^*-algorithm to find the shortest path in the transition system of the Petri net product of log and model. The shortest path with transformed probabilities then corresponds to the most probable alignment.

Example 4. Figure 3 displays two alignments for the trace $\mathtt{tr}_1$. The probability for alignment 1 in Fig. 3a equals 9.3%:

$$\psi_A \psi_B \theta_D^L \psi_C = \max\{0.19, 1\} \times \max\{0.12, 1\} \times 0.24 \times \max\{0.20, 0.39\}$$
$$= 0.24 \times 0.39$$

Similarly, the probability for alignment 2 in Fig. 3b equals 4.8%. Hence the most probable alignment for trace $\mathtt{tr}_1$ is alignment 1 (see Fig. 3a).

5   Theoretical Evaluation

This section theoretically evaluates the performance of our approach.

Given the process model in Fig. 1 and the event log in Table 1 we calculate the most probable alignment for four traces: $\mathtt{tr}_1 = \langle A,B,D,C\rangle$, $\mathtt{tr}_2 = \langle A,B,C,D\rangle$, $\mathtt{tr}_3 = \langle A,B,B,B\rangle$ and $\mathtt{tr}_4 = \langle A,A,B,C\rangle$. Two techniques are applied; the technique from Alizadeh et al. [5] (see Fig. 6)[3] and our technique (see Fig. 7). In this section, we show that our technique improves upon the shortcomings of [5].

First, the order of transitions in the trace (log moves) determines the outcome (see Fig. 6a and c and Sect. 2). Given that the number of observations of C is larger than the number of observations of D in Table 1, a synchronous move of C is more probable than a synchronous move of D independent of the order, as is shown in Fig. 7a and c.

---

[3] To calculate probabilities Alizadeh et al. [5] use the fitted event log ($\mathcal{L}_{fit}$) of Table 1 (first 6 rows). Process executions are mapped onto a state, for which a state representation function is used: either a sequence, multiset or set abstraction. As a cost function $g = 1 + \log\left(\frac{1}{\theta}\right)$ is used.

Second, the model behavior is restricted to behavior in the fitting part of the log $\mathcal{L}_{\text{fit}}$ (see Fig. 6b). $\mathcal{L}_{\text{fit}}$ contains traces with 1,2,3 and 7 repetitions of activity B. For trace $\mathbf{tr_3}$, Alizadeh et al. [5] find the outcome $\langle$A,B,B,B,C$\rangle$ if the state representation function is a sequence or multiset abstraction[4]. We argue that it is more likely that the 4th B is a synchronous move instead of a log move, given the Markov property of the process model. We thus get the outcome $\langle$A,B,B,B,B,C$\rangle$, as is shown in Fig. 7b.

Third - in line with [4] - the location of a synchronous move does not matter in a repetition of log moves, where one synchronous move is possible according to the model. So we find two alignments in Fig. 7d. In contrast, using [5] the synchronous move is assigned to the first activity of the repetition (see Fig. 6d).

Beyond addressing the shortcomings of [5], our approach to determine the most probable alignment is unaffected by noise. This is another strength. We consider two extreme variants of noise: (1) An activity is not in the model, but is in the log. (2) An activity is not in the log, but is in the model. The prior

trace	A	B	D	C
model	A	B	D	≫
	t_A	t_B	t_D	

(a) $\mathbf{tr_1} = \langle$A,B,D,C$\rangle$

trace	A	B	≫	B	≫	B	B	≫
model	A	B	τ	B	τ	B	≫	C
	t_A	t_B	t_1	t_B	t_1	t_B		t_C

(b) $\mathbf{tr_3} = \langle$A,B,B,B,B$\rangle$ and sequence or multiset abstraction

trace	A	B	C	D
model	A	B	C	≫
	t_A	t_B	t_C	

(c) $\mathbf{tr_2} = \langle$A,B,C,D$\rangle$

trace	A	A	B	C
model	A	≫	B	C
	t_A		t_B	t_C

(d) $\mathbf{tr_4} = \langle$A,A,B,C$\rangle$

Fig. 6. The probable alignments for traces $\mathbf{tr_1}, \mathbf{tr_2}, \mathbf{tr_3}$ and $\mathbf{tr_4}$ with cost function $g = 1 + \log\left(\frac{1}{\theta}\right)$ according to [5]

trace	A	B	D	C
model	A	B	≫	C
	t_A	t_B		t_C

(a) $\mathbf{tr_1} = \langle$A,B,D,C$\rangle$

trace	A	B	≫	B	≫	B	≫	B	≫
model	A	B	τ	B	τ	B	τ	B	C
	t_A	t_B	t_1	t_B	t_1	t_B	t_1	t_B	t_C

(b) $\mathbf{tr_3} = \langle$A,B,B,B,B$\rangle$

trace	A	B	C	D
model	A	B	C	≫
	t_A	t_B	t_C	

(c) $\mathbf{tr_2} = \langle$A,B,C,D$\rangle$

trace	A	A	B	C		trace	A	A	B	C
model	A	≫	B	C	and	model	≫	A	B	C
	t_A		t_B	t_C				t_A	t_B	t_C

(d) $\mathbf{tr_4} = \langle$A,A,B,C$\rangle$

Fig. 7. The most probable alignments for traces $\mathbf{tr_1}, \mathbf{tr_2}, \mathbf{tr_3}$ and $\mathbf{tr_4}$ using our technique

---

[4] Alizadeh et al. [5] find the outcome $\langle$A,B,B,B,B,C$\rangle$ if the state representation function is a set abstraction. Our technique obtains the same result (see Fig. 7b).

assigns a positive probability to the model and log moves (and therefore the synchronous move) of that activity. Neither of these has an effect on the probability estimation[5].

We always prefer synchronous moves, even if frequencies in the event log are missing or not balanced. Suppose in the event log of the running example (see Fig. 1) activity C is observed 999 times and D is observed once; and we observe a trace $\mathtt{tr_5} = \langle A,B,D \rangle$. A synchronous move of D is preferred over a log move of D plus a model move of C (due to Eq. (6)).

6   Related Work

The cost-based alignment technique is used to calculate the fitness conformance metric. Fitness measures how well the process model captures the observed behavior as recorded in an event log.

In this paper we focus on the control-flow perspective. Some approaches extend alignment-based techniques to support conformance checking based on multiple perspectives (e.g. control-flow, data, resource and time). de leoni and Aalst [11] build alignments by first considering the control-flow and, second, refine the computed alignments using other perspectives. In contrast, Felix et al. [12] balance the deviations with respect to all perspectives (not prioritizing control-flow). Both approaches use a (customizable) cost function and do not consider a cost function based on probabilities. Alizadeh et al. [6] consider multiple perspectives to calculate probable alignments, but their technique has the same shortcoming as [5] (mentioned in Sects. 2.3 and 5).

7   Conclusion

In this paper, we address the problem of computing the *most probable alignment* in the context of conformance checking. The core idea of our approach is to use the history of the event log to assign probabilities to the occurrence of activities and to the transitions between them. We apply Bayesian theory to avoid zero probabilities. The theoretical evaluation demonstrates that our approach improves upon existing work by Alizadeh et al. [5]. Moreover, it is unaffected by noise.

In future work, we plan to extend this current work with an empirical evaluation. To this end, we intend to implement the presented technique in the context of the ProM Framework. Further, we plan to relax the independence assumption of activities in the event log to take into account the correlation between activities, e.g. long distance dependencies. Finally, we intend to extend our approach to the multi-perspective scenario, in which we would take into account attribute values.

Acknowledgement. We thank Massimiliano de Leoni for validating our understanding of [5].

---

[5] Note for both variants $\mathcal{L}_{\text{fit}} = \emptyset$, so the technique in [5] does not work.

References

1. van der Aalst, W.M.P., Adriansyah, A., van Dongen, B.F.: Replaying history on process models for conformance checking and performance analysis. Wiley Interdisc. Rev. Data Min. Knowl. Discov. **2**, 182–192 (2012)
2. van der Aalst, W.M.P., van Hee, K.M., ter Hofstede, A.H., Sidorova, N., Verbeek, H.M.W., Voorhoeve, M., Wynn, M.T.: Soundness of workflow nets: classification, decidability, and analysis. Formal Aspects Comput. **23**, 333–363 (2011)
3. Adriansyah, A., van Dongen, B.F., van der Aalst, W.P.: Conformance checking using cost-based fitness analysis. In: 2011 15th IEEE International on EDOC, pp. 55–64 (2011)
4. Adriansyah, A., van Dongen, B.F., van der Aalst, W.M.P.: Memory-efficient alignment of observed and modeled behavior. In BPM Center Report, 03–03 (2013)
5. Alizadeh, M., de Leoni, M., Zannone, N.: History-based construction of log-process alignments for conformance checking: discovering what really went wrong. In: SIM-PDA, pp. 1–15 (2014)
6. Alizadeh, M., de Leoni, M., Zannone, N.: Constructing probable explanations of nonconformity: A data-aware and history-based approach. In: 2015 IEEE Symposium Series on Computational Intelligence. IEEE (2015)
7. Barber, D.: Bayesian Reasoning and Machine Learning. Cambridge University Press, Cambridge (2012)
8. Casella, G., Berger, R.L.: Statistical Inferences. Duxbury, Pacific Grove, CA (2002)
9. Dechter, R., Pearl, J.: Generalized best-first search strategies and the optimality of A. J. ACM **32**, 505–536 (1985)
10. Greenberg, E.: Introduction to Bayesian Econometrics. Cambridge University Press, Cambridge (2012)
11. de Leoni, M., van der Aalst, W.M.P.: Aligning event logs and process models for multi-perspective conformance checking: an approach based on integer linear programming. In: Daniel, F., Wang, J., Weber, B. (eds.) BPM 2013. LNCS, vol. 8094, pp. 113–129. Springer, Heidelberg (2013). https://doi.org/10.1007/978-3-642-40176-3_10
12. Felix, M., de Leoni, M., Reijers, H.A., van der Aalst, W.M.P.: Balanced multi-perspective checking of process conformance. Computing **98**, 407–437 (2016)
13. Molloy, M.K.: Performance analysis using stochastic Petri nets. IEEE Trans. Comput. **31**, 913–917 (1982)
14. Munoz-Gama, J., Carmona, J., van der Aalst, W.M.P.: Single-entry single-exit decomposed conformance checking. Inf. Syst. **46**, 102–122 (2014)

Improving Process Discovery Results by Filtering Outliers Using Conditional Behavioural Probabilities

Mohammadreza Fani Sani$^{(\boxtimes)}$, Sebastiaan J. van Zelst,
and Wil M. P. van der Aalst

Department of Mathematics and Computer Science,
Eindhoven University of Technology,
P.O. Box 513, 5600 MB Eindhoven, The Netherlands
{m.fani.sani,s.j.v.zelst,w.m.p.v.d.aalst}@tue.nl

Abstract. Process discovery, one of the key challenges in process mining, aims at discovering process models from process execution data stored in event logs. Most discovery algorithms assume that all data in an event log conform to correct execution of the process, and hence, incorporate all behaviour in their resulting process model. However, in real event logs, noise and irrelevant infrequent behaviour are often present. Incorporating such behaviour results in complex, incomprehensible process models concealing the correct and/or relevant behaviour of the underlying process. In this paper, we propose a novel general purpose filtering method that exploits observed conditional probabilities between sequences of activities. The method has been implemented in both the ProM toolkit and the RapidProM framework. We evaluate our approach using real and synthetic event data. The results show that the proposed method accurately removes irrelevant behaviour and, indeed, improves process discovery results.

Keywords: Process mining · Process discovery · Noise filtering
Outlier detection

1 Introduction

Process mining is a research discipline that positioned at the intersection of data driven methods like machine learning and data mining and Business Process Modeling (BPM) [1]. There are three types of process mining; *process discovery*, *conformance checking* and *process enhancement*. Process discovery aims at discovering process models from event logs. Conformance checking aims at assessing to what degree a process model and event log conform to one another in terms of behaviour. Finally, process enhancement aims at improving process model quality by enriching them with information gained from the event log.

Within process mining/process identification projects, process discovery is often used to quickly get insights regarding the process under study [1]. A business process analyst simply applies a process discovery algorithm on the

© Springer International Publishing AG 2018
E. Teniente and M. Weidlich (Eds.): BPM 2017 Workshops, LNBIP 308, pp. 216–229, 2018.
https://doi.org/10.1007/978-3-319-74030-0_16

extracted event log and analyzes its result. Most process discovery algorithms assume that event logs represent accurate behaviour. So, they are designed to incorporate all of the event log's behaviour in their resulting process model as much as possible.

Real event logs contain both *noise* and *infrequent behaviour* [2]. In general, noise refers to behaviour that does not conform to the process specification and/or its correct execution. Examples of noise are, amongst others, incomplete logging of process behaviour, duplicated logging of events and faulty execution of the process. Infrequent behaviour relates to behaviour that is may be supposed to happen, yet, in very exceptional cases of the process. For example, additional checks may be required when a loan request exceeds $10.000.000. Incorporating either noise and/or infrequent behaviour results in complex, incomprehensible process models concealing the correct and/or relevant behaviour of the under-lying process. As such, when using process discovery for the purpose of process identification, we are often unable to gain any actionable knowledge by applying process discovery algorithms directly.

In this paper, we focus on improving process discovery results by applying general purpose event log filtering, i.e. filtering the event log prior to apply-ing any arbitrary process discovery algorithm. Distinguishing between noise and infrequent behaviour is a challenging task and is outside the scope of this paper. Hence, we consider both noise and infrequent behaviour as *outliers* and aim at identifying and removing such outliers from event logs. We propose a generic filtering approach based on conditional probabilities between *sequences of activi-ties*. The approach identifies whether certain activities are likely to happen based on a number of its preceding activities. Using the ProM (http://promtools.org) [3] based extension of RapidMiner (http://rapidminer.com), i.e. RapidProM [4], we study the effectiveness of our approach, using synthetic and real event data. The results of our experiments show that our approach adequately identifies and removes outliers, and, as a consequence increases the overall quality of process discovery results. Additionally we show that our filtering method outperforms other general purpose filtering techniques in the process mining domain.

The remainder of this paper is structured as follows. Section 2 motivates the need for general purpose event log filtering methods. In Section 3, we discuss related work and after that, in Sect. 4, we explain our proposed method. Details of the evaluation and corresponding results are given in Sect. 5. Finally, Sect. 6 concludes the paper and presents some future work in this domain.

2   Motivation

An interpretable process model helps business process analysts to understand what is going on in an event data. However, often process discovery algorithms return results that are complicated and not understandable, because of outliers within the event logs used. In Fig. 1 we show how the application of filtering greatly reduces the complexity in a *real event log*, i.e. the event log of the *Business Process Intelligence Challenge 2012 (BPIC 2012)*. Figure 1a shows a process

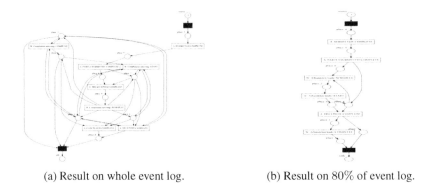

(a) Result on whole event log.   (b) Result on 80% of event log.

Fig. 1. Process models discovered by applying the ILP Miner [5] on the BPIC 2012 log.

model discovered using the ILP Miner of [5][1] for this event log, whereas Fig. 1b shows the result of applying the same process discovery algorithm on 80% of the most frequent original behaviour.

In process mining, two quality measures are defined for measuring the behavioural quality of process models, i.e. fitness and precision [6]. Fitness computes how much behaviour in the event log is also described by the process model. On the other hand, precision measures the amount of behaviour described by the model that is also present in the event log. The fitness values of Figs. 1a and 1b are 0.57 and 0.46 whereas their precision values are 0.68 and 1.0 respectively. This means that the model in Fig. 1a describes more behaviour that is also presented in the event log, however, in order to do this it is greatly underfitting. Hence, it allows for much more behaviour compared to the model in Fig. 1b. As a consequence, the model in Fig. 1a is more complex and ambiguous. However, arbitrarily removing behaviour based on frequency is too ad-hoc and does not work when there is a lot of variety present within an event log. Therefore, we need more advanced filtering methods that take into account and exploit the actual behaviour described by the event log. Clearly, by incorporating all possible behaviour, the model in Fig. 1a is overly complex and conceals the dominant/main-stream behaviour of the underlying process. The process model in Fig. 1b, on the other hand, is much simpler while still covering 80% of the observed behaviour allowed by the other model. At the same time, the model describes less behaviour compared to the model in Fig. 1a, and is therefore less under-fitting. Thus, by removing 20% of behaviour, we obtain a simpler model that still accurately describes the underlying process.

3   Related Work

In recent years, many process discovery algorithms have been proposed [7–12]. The first algorithms where designed to incorporate all behaviour in the event

---

[1] HybridILPMiner package in ProM.

log [7,11,12]. However, later more algorithms have more recently been extended to be able to handle outliers [13,14]. However, these extended filtering techniques are tailored towards the internal working of the corresponding algorithm and hence do not work as a general purpose filtering technique. Other algorithms [8,10], are designed to cope with noisy and infrequent behaviour, however, these algorithms do not result in process models with a clear semantics. Most of commercial process mining tools using these algorithms and their filtering are based on just frequency of activities and their direct relations. Moreover, the filters are relatively ad-hoc and require significant user input.

In this paper, we propose to *separate concerns*, and thus develop a novel, general purpose filtering technique that pre-processes event logs. In such way, any process discovery algorithm is able to benefit from effective identification and removal of outlier behaviour. In the remainder of this section, we focus on techniques developed for general purpose filtering in the process mining domain.

Separating outliers from event logs and focusing just on them rather than all behaviour also has been studied [15], however, a detailed treatment of such work is outside the scope of this paper.

The vast majority of process mining research has an accompanying implementation in the process mining toolkit ProM. Most work on general purpose event log filtering concerns ad-hoc filtering implementations within ProM. Many of these implementations are useful when we aim at using specific subsets of traces/events of an event log instead of the whole event log. In Table 1, the main filtering plugins are listed, accompanied by a brief description of their applications and methods. All plugins take an event log as an input and return a *filtered*

Table 1. Overview of filtering plugins in ProM

Plug-in	Applications	Main method
Filter log using simple heuristics	Helpful for removing traces and activities based on frequency of events or the presence of certain start/end events	Frequency/position of events
Filter log on event/trace attributes	Useful when we want to just keep events/traces with specific attribute values	Attribute values
Dotted chart	Allows us to visually select specific traces in event logs (usually base on a time frame)	Time window
Transition systems miner	Helpful to project traces/events on specific transitions and/or states	Frequency of transitions
Filter log using prefix-closed language	Allows us to remove events from traces	Rule based

event log as an output. Moreover, they need some form of domain knowledge to work properly. In addition, typically the user needs to set one or more (complex) settings. However, they do not support generic outlier detection, i.e. in cases where we possess no or little domain knowledge.

Surprisingly, little research has been done in the field of general purpose filtering. In [16] a graph-based outlier detection method is proposed to detect inaccurate data in an event log. In [17] a method is proposed that detects non-fitting behaviour based on a given reference model and then repair event log. As we want to improve process discovery results and, in general, we do not have a reference model, this method is not useful for general purpose filtering. In [18] the authors propose to provide training traces to the PRISM algorithm [19] which returns rules for detecting outliers. However, in real event logs, providing a set of training traces that covers all possible outliers is impractical. Also, in this method, deciding about level of filtering is impossible.

The most relevant research in the area of general purpose log filtering is the work in [20]. The authors propose to construct an Anomaly Free Automaton (AFA) based on the whole event log and a given threshold. Subsequently, all events that do not fit the AFA are removed from the filtered event log. Filtering event logs using AFA indeed allows us to detect and remove noisy and/or infrequent behaviour. However, the technique does not allow us to detect all types of outliers like incomplete traces, i.e. traces that fit the AFA perfectly yet do not terminate properly. Incorporation of such behaviour can still lead to infeasible process discovery results.

4   Filtering with Conditional Behavioural Probabilities

As indicated in Sect. 3, the most filtering approaches are not suitable for process discovery because they need additional information like reference model or a set of outlier traces. Furthermore, the AFA filter, which is the most suitable general purpose event log filter, has trouble identifying irrelevant infrequent behaviour. Therefore, we present a general purpose filtering method that is able to deal with all types of outliers. The main purpose of the filter is to identify the likelihood of the occurrence of an activity, based on its surrounding behaviour, e.g. how likely is it that activity a follows the sequence of activities $\langle b, c \rangle$. To detect such likelihood it uses the conditional probability of activity occurrences, given a sequence of activities. As we just consider a sample of behaviour in the underlying process, i.e. event log, all computed probabilities are estimation of behaviour that are really happened. Prior to presenting the filtering method, we present some basic notations used throughout the paper.

4.1   Basic Notation and Definitions

Given a set X, a multiset M over X is a function $M \colon X \to \mathbb{N}_{\geq 0}$, i.e. it allows certain elements of X to appear multiple times. We write a multiset as $M = [e_1^{k_1}, e_2^{k_2}, ..., e_n^{k_n}]$, where for $1 \leq i \leq n$ we have $M(e_i) = k_i$ with $k_i \in \mathbb{N}$. If $k_i = 1$, we omit its superscript, and if for some $e \in X$ we have $M(e) = 0$, we omit it from the multiset notation. Also, $M = [\,]$ is an empty multi set if $\forall e \in X$, $M(e) = 0$. We let $\overline{M} = \{e \in X \mid M(e) > 0\}$, i.e. $\overline{M} \subseteq X$. The set of all possible multisets over a set X is written as $\mathcal{M}$.

Let $\mathcal{A}$ denote the set of all possible activities and let $\mathcal{A}^*$ denote the set of all finite sequences over $\mathcal{A}$. A finite sequence σ of length n over $\mathcal{A}$ is a function $\sigma \colon \{1, 2, ..., n\} \to \mathcal{A}$, alternatively written as $\sigma = \langle a_1, a_2, ..., a_n \rangle$ where $a_i = \sigma(i)$ for $1 \leq i \leq n$. The empty sequence is written as ϵ. Concatenation of sequences σ and σ' is written as $\sigma \cdot \sigma'$. We let $hd \colon \mathcal{A}^* \times \mathbb{N}_{\geq 0} \nrightarrow \mathcal{A}^*$ with, given some $\sigma \in \mathcal{A}^*$ and $k \leq |\sigma|$, $hd(\sigma, k) = \langle a_1, a_2, .., a_k \rangle$, i.e., the sequence of the first k elements of σ. Note that $hd(\sigma, 0) = \epsilon$. Symmetrically $tl \colon \mathcal{A}^* \times \mathbb{N}_{\geq 0} \nrightarrow \mathcal{A}^*$ is defined as $tl(\sigma, k) = \langle a_{n-k+1}, a_{n-k+2}, ..., a_n \rangle$, i.e., the sequence of the last k elements of σ. Again, $tl(\sigma, 0) = \epsilon$. Finally, sequence $\sigma' = \langle a'_1, a'_2, ..., a'_k \rangle$ is a subsequence of sequence σ if and only if we are able to write σ as $\sigma_1 \cdot \langle a'_1, a'_2, ..., a'_k \rangle \cdot \sigma_2$, where both σ_1 and σ_2 are allowed to be ϵ, i.e. σ is a subsequence of itself.

Event logs describe sequences of executed business process activities, typically in context of some case, e.g. a customer or some order-id. The execution of an activity in context of a case is referred to as an *event*. A sequence of events is referred to as a *trace*. Thus, it is possible that multiple traces describe the same sequence of activities, yet, each trace itself contains different events. An example event log, adopted from [1], is presented in Table 2.

Table 2. Fragment of a fictional event log (each line corresponds to an event).

Case-id	Activity	Resource	Time-stamp
...	...	...	...
1	register request (a)	Sara	2017-04-08:08.10
1	examine thoroughly (b)	Ali	2017-04-08:09.17
2	register request (a)	Sara	2017-04-08:10.14
2	check ticket (d)	William	2017-04-08:10.23
1	check ticket (d)	William	2017-04-08:10.53
2	examine causally (b)	Ava	2017-04-08:11.13
1	reject request (h)	Ava	2017-04-08:13.05
...	...	...	...

Consider all activities related to *Case-id 1*. Sara *registers a request*, after which Ali *examines it thoroughly*. William *checks the ticket* after which Ava *examine causally* and *reject the request*. The execution of an *activity* in context of a business process is referred to as an *event*. A sequence of events, e.g. the sequence of events related to case *1*, is referred to as a *trace*.

Definition 1 (Trace, Variant, Event Log). *Let $\mathcal{A}$ be a set of activities. An event log is a multiset of sequences over $\mathcal{A}$, i.e. $L \in \mathcal{M}(\mathcal{A}^*)$. $\sigma \in \mathcal{A}^*$ is a trace in L and $\sigma \in \overline{L}$ is a variant.*

Observe that each $\sigma \in \overline{L}$ describes a *trace-variant* whereas $L(\sigma)$ describes how many traces of the form σ are present.

Definition 2 (Subsequence Frequency). *Let L be an event log over a set of activities $\mathcal{A}$ and let $\sigma' \in \mathcal{A}^*$. The subsequence frequency of σ' w.r.t L, written as $freq(\sigma', L)$, denotes the number of times σ' occurs as a subsequence of any trace present in L.*

Given a simple example event log $L = [\langle a, b, c, d \rangle^5, \langle a, c, b, d \rangle^3]$, we have $freq(\langle a \rangle, L) = freq(\epsilon, L) = 8$, $freq(\langle a, b \rangle, L) = 5$, etc.

Definition 3 (Conditional Occurrence Probability). *Let L be an event log over a set of activities $\mathcal{A}$ and let σ' be a subsequence. Given some $a \in \mathcal{A}$, the conditional probability of occurrence of activity a, given σ' and L, i.e. $COP(a, \sigma', L)$ is defined as:*

$$COP(a, \sigma', L) = \begin{cases} \frac{freq(\sigma' \cdot \langle a \rangle, L)}{freq(\sigma', L)} & \text{if } freq(\sigma', L) \neq 0 \\ 0 & \text{otherwise} \end{cases}$$

Clearly, the value of any $COP(a, \sigma, L)$ is a real number in $[0, 1]$. A high value of $COP(a, \sigma, L)$ implies that after the occurrence of σ', it is very probable that activity a occurs. For example, $COP(a, \sigma, L) = 1$ implies that if σ occurs, a always happens directly after it. Based on the previously used simple event log, we have $COP(b, \langle a \rangle, L) = \frac{5}{8}$.

4.2   Outlier Detection

We aim to exploit conditional probabilities present within event logs for the purpose of filtering event logs. Conceptually, after a given subsequence, activities that have a particularly low COP-value are unlikely to have happened and therefore their occurrence may be seen as outlier. However, to account for dependencies between activities and previously occurred activities at larger distances, we compute COP-values for subsequences of increasing length.

In our proposed method, for each $i \in \{1, 2, ..., k\}$ we construct a COP-Matrix. Assume there are a total of m unique subsequences with length $1 \leq l \leq k$ in an event log. A COP-Matrix $\mathbf{A}_{COP}^l$ for length l is simply an $m \times |\mathcal{A}|$-matrix, where $\mathbf{A}_{COP}^l(\sigma', a) = COP(a, \sigma', L)$.

We additionally compute conditional probabilities for *start* and *end* subsequences relatively. We let $\mathbf{A}_S^l$ denote a matrix describing the occurrence probability matrix of all subseqeunces $\sigma' = hd(\sigma)$ with $|\sigma'| = l$ for $\sigma \in L$. We are able to compute such probability by dividing the number of traces that start with σ' over the total number of traces in the log. Similarly we define $\mathbf{A}_E^l$ denote a matrix describing the conditional probability matrix of all subseqeunces $\sigma' = tl(\sigma)$ with $|\sigma'| = l$ for $\sigma \in L$ that is equal to $a = \epsilon$ in $\mathbf{A}_{COP}^l$. By doing so, we be able to handle outliers which occur in the start and the end parts of trace.

Given our different COP-Matrix, and a user-defined threshold κ, we identify each entry $\mathbf{A}^l(\sigma', a) < \kappa$ as an outlier. The pseudo-code of detecting outliers is present in Algorithm 1. In this fashion, it is possible to detect outliers that occur in start, middle or end part of traces. There are two ways to handle detected outliers. We are able to simply remove the corresponding event from the trace, i.e. *event-level filtering*, or, remove the trace as a whole, i.e. *trace-level filtering*. However, removing an improbable event in a trace may make the trace to have more outlier behaviour. Hence, we just focus on *trace-level filtering*.

Algorithm 1. Outlier Detection Algorithm

> **procedure** OUTLIERDTECTION(L, k, κ)
> *Computing Probabilities*:
>   **for** $(l \le k)$ **do**
>   Build $\mathbf{A}_{COP}^l$, $\mathbf{A}_E^l$ and $\mathbf{A}_S^l$
> $FilteredEventLog \leftarrow EmptyEventLog$
> *Filtering*:
>   **for** (each Trace σ in the L) **do**
> $Outlier \leftarrow false$
>   **for** $(l = 1 : k)$ **do**
>   **for** (each subsequence σ' with length l and its following activity a) **do**
>   Find corresponding $COP(a, \sigma', L)$ in $\mathbf{A}_{COP}^l$, $\mathbf{A}_E^l$ and $\mathbf{A}_S^l$
>   **if** $(\kappa > COP(a, \sigma', L))$ **then**
> $Outlier \leftarrow true$
>   **if** $(Outlier = false)$ **then**
> $FilteredEventLog \leftarrow Add(FilteredEventLog, \sigma)$
>   **return** $FilteredEventLog$

With increasing value of k (maximum length of subsequences), the complexity of the filtering method increases. The number of different strings we can generate over $\mathcal{A}$ with length k is $(|\mathcal{A}|)^k$ and total possible subsequences for some k: $\sum_{i=1}^{k} \mathcal{A}^i$ where $|\mathcal{A}|$ is the number of activities in the L. However, there is no need to compute $COPs$ of all possible subsequences. For subsequences with length $k + 1$, it is sufficient to just consider $\sigma'.\langle a \rangle$ in level k that $\kappa \le COP(a, \sigma', L)$. For example, if at $k = 1$ $COP(c, \langle b \rangle) \le \kappa$, there is no need to consider $\langle b, c \rangle$ as a subsequent at $k = 2$. Even the $COP(a, \langle b, c \rangle)$ be higher than κ.

4.3   Implementation

To be able to combine the proposed filtering method with any process discovery algorithm, we implemented the *Matrix Filter* plugin (*MF*) in the ProM framework[2]. The plugin takes an event log as input and outputs a filtered event log. Furthermore, the user is able to specify threshold κ and whether event-level or trace-level filtering needs to be applied. The maximum subsequence length to be considered also needs to be specified.

In addition, to apply our proposed method on various event logs with different filtering thresholds and applying different process discovery algorithms with different parameters, we ported the *Matrix Filter* (*MF*) plugin to RapidProM. RapidProM is an extension of RapidMiner that combines scientific workflows [21] with a range of (ProM-based) process mining algorithms.

---

[2] MatrixFilter plugin svn.win.tue.nl/repos/prom/Packages/LogFiltering.

5   Evaluation

To evaluate the usefulness of filtering outliers using our method, we have conducted several experiments using both synthetic and real event data. The purpose of these experiments is to answer the following questions:

1. Does *MF* help process discovery algorithms to return more precise models?
2. How does the performance of *MF* compare to *AFA* filtering method?

To evaluate discovered process models, we use fitness and precision (introduced in Page 2). There is a trade off between these measures [22]. Sometimes, putting aside little behaviour cause a few decrease in fitness value, however more increasing in the precision value. To make a balance between fitness and precision, we use the F-Measures metric that combines fitness and precision: $\frac{2 \times Precision \times Fitness}{Precision + Fitness}$. Also, filtering time and process model discovery time in milliseconds have been measured. Note that in all experiments, filtered event logs are only used in the process discovery part. Computing the F-Measure for all process models is done using the corresponding raw, unfiltered event logs. Furthermore, we only consider subsequences with length k in $[0, 2]$.

In the first experiment we investigate the effect of changing the κ in the *MF* threshold on the F-Measure w.r.t. different process discovery algorithms. We use the Inductive Miner [12] (IM) and the ILP Miner (ILP) [11]. Additionally we assess the interaction between our filtering technique and integrated filtering within the Inductive Miner, i.e. we use the IMi variant [13] with noise thresholds 0.1 and 0.3. We apply these algorithms and filtering methods on the BPIC2012 log. The results for this experiment are shown in Fig. 2.

In this figure, each line corresponds to a discovery algorithm. The x-axis represents the threshold level of *MF*, the y-axis represents the corresponding F-Measure. Hence, for each technique, the data point $x = 0$ corresponds to not applying behavioural conditional probability filtering. We thus aim at finding out whether there exist regions of threshold values for which we are able to increase the F-measure when applying filtering. The F-measure of *IM* on this event log without using *MF* is 0.45. However, using the proposed filter increases the F-Measure of the discovered model to 0.80. Even for *IMi*, which uses an embedded filter, the *MF* increases the F-Measure from 0.69 and 0.7 to 0.81. As the ILP miner is more sensitive to outliers, *MF* helps more and its enhancement for this algorithm is higher. However, with increasing the threshold of *MF*, all the traces in the filtered event log are removed and the fitness and F-Measure of the discovered model will equal to 0. The best result, i.e. an F-measure of 0.81, is achieved by IMi with threshold 0.1 and *MF* threshold of 0.09.

To illustrate the effect of filtering on the discovered process models, in Fig. 3, we apply *IMi* with 11 internal thresholds ranging from 0.0 to 0.5 on the raw BPIC2012 and the filtered event log using *MF* with threshold 0.09. Here, each circle or square correspond to fitness and precision values related to one discovered model. A circle is related to applying *MF*, whereas squares relate to using the raw event log. As the results show, *MF* usually causes a little decrease in fitness value, yet yields an increase in precision value. The average of F-Measures

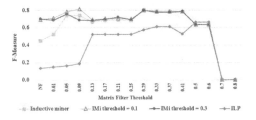

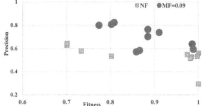

Fig. 2. Applying process discovery algorithms on the BPIC2012 log with different *MF* thresholds.

Fig. 3. Comparing process models discovered by 11 noise thresholds on the BPIC 2012 log with/without filtering.

when applying no filtering is 0.66 versus 0.77 in case of *MF* (with threshold 0.09). Thus, Figs. 2 and 3 indicate that *MF* improve process discovery results, i.e. the process models have an overall higher F-Measure.

In next experiment, using the BPIC2012 and BPIC2017 event log we additionally assess what the maximal obtainable level of F-measure is for different process discovery algorithms, using different levels of internal filtering. We computed F-measures based on the unfiltered event log, and, maximized the F-measure result for both *MF* and *AFA*. With a workflow in the RapidMiner, for both filtering methods we filtered the event log using 40 different thresholds. The results are presented in Fig. 4. This figure shows *MF* allows us to discover process models with higher F-Measures.

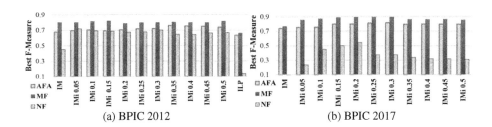

(a) BPIC 2012   (b) BPIC 2017

Fig. 4. Effect of filtering on best F-Measure of discovered models.

In Fig. 5, we compare the average required time of applying the process discovery algorithms with/without filtering methods. In this figure, the *y*-axis represents the time in milliseconds with logarithmic scale. According to this figure, filtering methods reduce the required time for discovering process models, because there are fewer traces in the filtered event logs. Although, in *AF* the discovery time reduction is higher, the filtering time for this method is much higher than *MF* method. Therefore, generally *MF* is faster than *AF*.

In the last experiment, to evaluate the ability of our proposed filtering method in detecting traces that contain outliers and corresponding effects on quality of

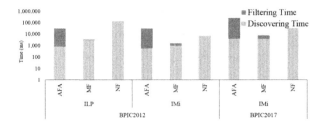

Fig. 5. Average of required time for process discovery with/without filtering.

process discovery algorithms, we use three synthetic event logs; *a12f0n*, *a22f0n* and *a32f0n*. These event logs are artificially added by 0, 10, 20 and 50 percent of different types of noise [2]. The last two characters of event log indicate the percentage of noise added to it, for example, *a22f0n20* correspond to *a22f0n* that contains 20% noise. Here, noisy event logs are used for process discovery and the original synthetic event logs (which contain no noise) use for computing F-Measure. Similar to the experiment in Fig. 3, *IMi* algorithm with 11 various

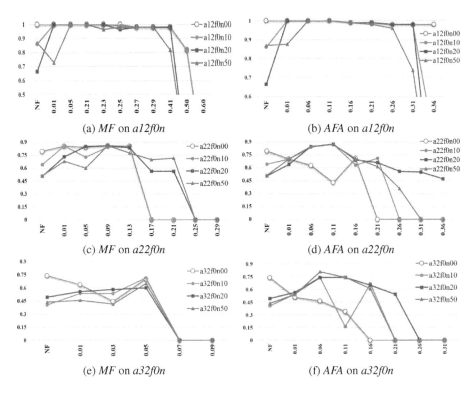

Fig. 6. Effect of filtering thresholds on F-Measures of synthetic event logs. *y*-axises are indicating values of best F-Measure and *x*-axises are showing the filtering thresholds.

internal noise thresholds has been used, but we show results of best F-Measure. The results of this experiment is presented in Fig. 6. According to this figure, F-Measures of models improve when applying filtering methods. This improvement is much more substantial for event logs that contain more percentage of noise.

For *a12f0n* event log which has the simplest structure among these event logs, both methods lead to similar results. However, for *a22f0n*, applying *MF* results in better F-Measures. Finally, in *a32f0n* that corresponds to the most complex model with lots of parallelism, *AFA* performs better than *MF*. This can be explained by the fact that when a lot of paralelism is present, the conditional probability of non-outlier behaviour is low as well, i.e. parallelism implies a lot of variety in behaviour. In this situations it seems using short subsequences (e.g. $k = 1$) or applying smaller κ value are better choices for *MF*.

These experiments indicate that the proposed filtering method is useful for process discovery algorithms to have models with higher F-Measure and it reduces their required discovery time. This way tends to outperform state-of-the-art process mining filtering techniques.

6   Conclusion

Process discovery is used to extract process models from event logs. However, real event logs contain noise and infrequent behaviour that make to discovery process model from the whole event log problematic. Separating these outliers from event logs is beneficial for process discovery techniques and helps to improve process discovery results.

To address this problem, we propose a filtering method that takes an event log as an input and returns a filtered event log based on a given threshold. It usesn the conditional probability of occurrence of an activity after sequence of activities. If this probability is lower than the given threshold, the activity is considered as an outlier.

To evaluate the proposed filtering method we developed a plugin in the `ProM` platform and also offer it through `RapidProM`. As presented, we have applied this method on real event logs, and several process discovery algorithms. Additionally, we use the proposed method on three synthetic event logs. The results indicate that the proposed approach is able to help process discovery algorithms to discover models that better balance between different behavioural quality measures. Furthermore, using these experiments we show that our filtering method outperforms related state-of-the-art process mining filtering techniques.

We plan to evaluate the effect of using different values of k, i.e. length of subsequences. Also, other metrics besides F-Measure like simplicity, generalization and structuredness could be analyzed. we want to apply event-level filtering and also assess different ways of using κ. The other approach in this domain will be providing event-level filtering that we ignore it in this paper.

References

1. van der Aalst, W.M.P.: Process Mining - Data Science in Action, 2nd edn. Springer, Heidelberg (2016)
2. Maruster, L., Weijters, A.J.M.M., van der Aalst, W.M.P., van den Bosch, A.: A rule-based approach for process discovery: dealing with noise and imbalance in process logs. Data Min. Knowl. Discov. **13**(1), 67–87 (2006)
3. Van der Aalst, W.M., van Dongen, B.F., Günther, C.W., Rozinat, A., Verbeek, E., Weijters, T.: Prom: The process mining toolkit. BPM (Demos) 489(31) (2009)
4. van der Aalst, W.M.P., Bolt, A., van Zelst, S.J.: RapidProM: Mine your processes and not just your data. CoRR abs/1703.03740 (2017)
5. van Zelst, S., van Dongen, B., van der Aalst, W., Verbeek, H.: Discovering Relaxed Sound Workflow Nets using Integer Linear Programming. arXiv preprint arXiv:1703.06733 (2017)
6. Buijs, J.C.A.M., van Dongen, B.F., van der Aalst, W.M.P.: On the role of fitness, precision, generalization and simplicity in process discovery. In: Meersman, R., Panetto, H., Dillon, T., Rinderle-Ma, S., Dadam, P., Zhou, X., Pearson, S., Ferscha, A., Bergamaschi, S., Cruz, I.F. (eds.) OTM 2012. LNCS, vol. 7565, pp. 305–322. Springer, Heidelberg (2012). https://doi.org/10.1007/978-3-642-33606-5_19
7. van der Aalst, W., Weijters, T., Maruster, L.: Workflow mining: discovering process models from event logs. IEEE Trans. Knowl. Data Eng. **16**(9), 1128–1142 (2004)
8. Weijters, A.J.M.M., Ribeiro, J.T.S.: Flexible Heuristics Miner (FHM). In: IEEE Symposium on Computational Intelligence and Data Mining (CIDM). IEEE (2011)
9. van der Aalst, W.M.P., Rubin, V., Verbeek, H.M.W., van Dongen, B.F., Kindler, E., Günther, C.W.: Process mining: a two-step approach to balance between underfitting and overfitting. Softw. Syst. Model. **9**(1), 87–111 (2008)
10. Günther, C.W., van der Aalst, W.M.P.: Fuzzy mining – adaptive process simplification based on multi-perspective metrics. In: Alonso, G., Dadam, P., Rosemann, M. (eds.) BPM 2007. LNCS, vol. 4714, pp. 328–343. Springer, Heidelberg (2007). https://doi.org/10.1007/978-3-540-75183-0_24
11. van der Werf, J.M.E.M., Dongen van Dongen, B.F., Hurkens, C.A.J., Serebrenik, A.: Process discovery using integer linear programming. Fundam. Inform. **94**(3–4), 387–412 (2009)
12. Leemans, S.J.J., Fahland, D., van der Aalst, W.M.P.: Discovering block-structured process models from event logs - a constructive approach. In: Colom, J.-M., Desel, J. (eds.) PETRI NETS 2013. LNCS, vol. 7927, pp. 311–329. Springer, Heidelberg (2013). https://doi.org/10.1007/978-3-642-38697-8_17
13. Leemans, S.J.J., Fahland, D., van der Aalst, W.M.P.: Discovering block-structured process models from event logs containing infrequent behaviour. In: Lohmann, N., Song, M., Wohed, P. (eds.) BPM 2013. LNBIP, vol. 171, pp. 66–78. Springer, Cham (2014). https://doi.org/10.1007/978-3-319-06257-0_6
14. van Zelst, S.J., van Dongen, B.F., van der Aalst, W.M.P.: Avoiding over-fitting in ILP-based process discovery. In: Motahari-Nezhad, H.R., Recker, J., Weidlich, M. (eds.) BPM 2015. LNCS, vol. 9253, pp. 163–171. Springer, Cham (2015). https://doi.org/10.1007/978-3-319-23063-4_10
15. Yang, W., Hwang, S.: A process-mining framework for the detection of healthcare fraud and abuse. Expert Syst. Appl. **31**(1), 56–68 (2006)

16. Ghionna, L., Greco, G., Guzzo, A., Pontieri, L.: Outlier detection techniques for process mining applications. In: An, A., Matwin, S., Raś, Z.W., Ślęzak, D. (eds.) ISMIS 2008. LNCS (LNAI), vol. 4994, pp. 150–159. Springer, Heidelberg (2008). https://doi.org/10.1007/978-3-540-68123-6_17

17. Wang, J., Song, S., Lin, X., Zhu, X., Pei, J.: Cleaning structured event logs: a graph repair approach. In: IEEE 31st International Conference on Data Engineering, ICDE, pp. 30–41 (2015)

18. Cheng, H.J., Kumar, A.: Process mining on noisy logs –can log sanitization help to improve performance? Decis. Support Syst. **79**, 138–149 (2015)

19. Cendrowska, J.: PRISM: An algorithm for inducing modular rules. Int. J. Man Mach. Stud. **27**(4), 349–370 (1987)

20. Conforti, R., La Rosa, M., ter Hofstede, A.H.M.: Filtering out infrequent behavior from business process event logs. IEEE Trans. Knowl. Data Eng. **29**(2), 300–314 (2017)

21. Bolt, A., de Leoni, M., van der Aalst, W.M.P.: Scientific workflows for process mining: building blocks, scenarios, and implementation. Int. J. Softw. Tools Technol. Transfer. **18**(6), 607–628 (2016). https://doi.org/10.1007/s10009-015-0399-5

22. Weerdt, J.D., Backer, M.D., Vanthienen, J., Baesens, B.: A robust F-measure for evaluating discovered process models. In: Proceedings of the CIDM, pp. 148–155 (2011)

Can We Find Better Process Models?
Process Model Improvement Using
Motif-Based Graph Adaptation

Alexander Seeliger[✉], Michael Stein, and Max Mühlhäuser

Telecooperation Lab, Technische Universität Darmstadt, Darmstadt, Germany
{seeliger,stein,max}@tk.tu-darmstadt.de

Abstract. In today's organizations efficient and reliable business processes have a high influence on success. Organizations spend high effort
in analyzing processes to stay in front of the competition. However, in
practice it is a huge challenge to find better processes based on process
mining results due to the high complexity of the underlying model. This
paper presents a novel approach which provides suggestions for redesigning business processes by using discovered as-is process models from event
logs and apply motif-based graph adaptation. Motifs are graph patterns
of small size, building the core blocks of graphs. Our approach uses the
LoMbA algorithm, which takes a desired motif frequency distribution
and adjusts the model to fit that distribution under the consideration of
side constraints. The paper presents the underlying concepts, discusses
how the motif distribution can be selected and shows the applicability
using real-life event logs. Our results show that motif-based graph adaptation adjusts process graphs towards defined improvement goals.

Keywords: Business process optimization · Graph adaptation
Business process analytics · Data mining · Tool support

1 Introduction

The efficiency and the reliability of business processes have a high influence on
organizations' success. High effort is spent in analyzing processes to stay in front
of the competition. With process mining, organizations get valuable insights
into how their business processes are really executed, by using available event
logs recorded by information systems. These discovered process models help to
identify bottlenecks, compliance violations, and other process problems, aiming
to support organizations to improve their business processes. Analyzing processes
using process mining helps to understand the actual use of information systems
in organizations, but process mining does not necessarily provide improvement
advice. It turns out that improving process models automatically is a challenging
problem because the system must be able to provide automatic adjustments for
existing models by providing better alternatives using extracted knowledge [1].

© Springer International Publishing AG 2018
E. Teniente and M. Weidlich (Eds.): BPM 2017 Workshops, LNBIP 308, pp. 230–242, 2018.
https://doi.org/10.1007/978-3-319-74030-0_17

With respect to business process model improvement, there are basically two goals that must be fulfilled (see Fig. 1): On the one hand, any process improvement is focused on a specific goal (*improvement goal*) that should be fulfilled. For example, organizations may want to resolve detected process problems, reduce process duration times, increase the throughput or reduce the complexity. On the other hand, provided suggestions for improvement must guarantee that the process is still executable and achieves the goal that was originally intended (*business goal*). Both the improvement goal and the business goal must be satisfied to get useful improvement suggestions for process models. However, the improvement and the business goal are often conflicting with each other. Thus, it is a challenging task to balance both goals.

This paper presents an approach that provides suggestions for process model improvement with respect to defined goals. Our approach is inspired by the systematic adaptation of communication network topologies based on motifs [8,9,15]. Motifs are graph patterns of small size [10]. Various studies, e.g., [10,14], have investigated the frequency distribution of motifs, the so-called motif signature, for different kinds of graphs. These studies consistently observe that the motif signature of a graph strongly correlates with important structural metrics of the graph. Apparently, motifs are simple building blocks of complex graphs [10]. The bespoke communication network approaches [8,9,15] build upon this observed correlation in the following way: In a first step, a target motif signature is extracted from a network topology that performs well with respect to selected metrics. In a second step, topologies that perform badly with respect to these metrics will be adapted such that the topology approximates the discovered target motif signature.

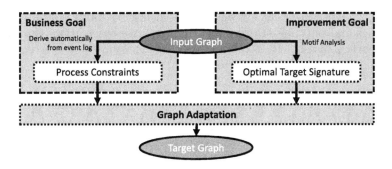

Fig. 1. Overview of our business process motif-based graph adaptation approach.

We are the first to transfer the idea of motif-based graph adaptation to the domain of business process models. A process model can also be represented as a graph with a set of nodes and a set of edges. The motif signature of a process graph represents the core structure of the process, which also reflects various aspects, such as the complexity and standardization in processes. The basic idea is to find an optimal motif signature that can be used to adapt non-optimized process models by removing or adding edges to the process graph. If we can approximate the process model to an optimal target signature, the resulting process model will also have desired characteristics of the optimal model. We use the LoMbA (Local Motif-based Adaptation) algorithm [15], which modifies the graph to approximate the target signature to improve process graphs. LoMbA permits the specification of constraints to retain certain aspects, e.g., graph connectivity. We use these constraints to make sure that the resulting process model is still executable and follows the original process model to achieve the desired goal. The advantage of our approach is that we can adapt process graphs with respect to different improvement goals by selecting different target signatures. These signatures can also be reused for other process graphs of different domains.

In this paper, we provide three main contributions: (1) In Sect. 2 we show how target signatures can be derived from existing process models to optimize business processes for certain goals. (2) We present how to use motif-based graph adaptation to improve business processes in Sect. 3.2. (3) In Sect. 4 we conduct an experimental evaluation based on real-life event logs. The results show that motif-based graph adaptation actually improves process models. Finally, we conclude the paper by outlining open research problems.

2   Finding Optimal Target Motif Signature

Before any motif-based graph adaptation algorithm can be executed, an optimal target signature must be found reflecting a very good process. For better understanding how the motif signature looks like for different process models, we investigated 5 different event logs and examined the frequency of each 3 node motif (Fig. 2). For each event log we used the heuristics miner [16] in the standard settings and the "noise-free log" setting (positive observations threshold = 100, dependency threshold = 100, best to relative threshold = 0.0) to extract 10 process models (heuristics nets) in total.

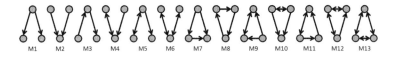

Fig. 2. All directed ($M1$-$M13$) motifs with 3 nodes

A motif signature is defined as a vector $M = (M_1, ..., M_l)$ which contains the relative frequency of a motif M_i in the examined motif space (here: 3 node motifs). Figure 3 shows an aggregated view on the motif signature over all investigated process models using standard and "noise-free log" setting. We can see that $M1$ to $M3$ are the most dominant motifs in process models, whereas the frequencies of motifs $M4$ to $M13$ are less dominant for the standard setting.

This distribution is not surprising because processes follow a specific forward path through the graph. A process has some starting and end events which are connected with some other nodes in between. $M1$ and $M2$ reflect branches in a process whereas $M3$ is a simple forward transition between events. If a process model contains events that are executed in turn, we see the occurrence of $M4 - M6$. $M7$ and $M8$ occur if the process model contains loops and $M9 - M13$ are combinations. A slightly different histogram can be seen for the "noise-free log" setting where $M1 - M6$ are the most dominating motifs. With a higher threshold we also mine lower frequent behaviors which results in more edges in the model (see Fig. 3). In conclusion we can say that motifs can characterize the structure of the process as they are the core building blocks for processes.

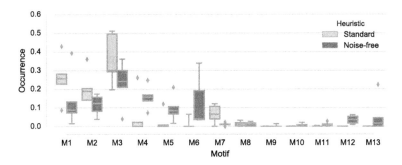

Fig. 3. 3 node motif frequency over the investigated process models with two different heuristic miner settings.

Besides the relative frequency of motifs we searched for Pearson correlations between the motifs and process model properties. The different motifs have different correlations to the number of edges, events, the graph density and the clustering coefficient ($\frac{3 \cdot \text{number of triangles}}{\text{number of connected triplets of vertices}}$). Although the sample of 10 models may be small, the analysis gives a quick overview how the different process properties are linked to motifs, enabling to determine a possible target signature to optimize process models. Within our analysis, we found the

	M1	M2	M3	M4	M5	M6	M7	M8	M9	M10	M11	M12	M13
Edge Correl	-0.21	-0.25	-0.75	0.07	-0.09	0.77	0.12	-0.34	0.87	0.66	-0.08	0.57	0.92
Node Correl	0.21	0.20	-0.81	0.03	-0.21	0.29	0.41	-0.24	0.64	0.51	-0.19	0.16	0.46
Density Correl	-0.39	-0.40	-0.60	0.07	0.03	0.86	-0.04	-0.33	0.51	0.83	0.10	0.67	0.95
Clustering Coefficient	-0.31	-0.26	-0.71	0.07	-0.02	0.77	0.16	-0.31	0.89	0.68	0.01	0.64	0.89

Fig. 4. Pearson correlations between motif and graph property. Correlations > 0.55 are significant with $p < 0.05$; correlations > 0.6 with $p < 0.01$.

following correlations: $M3$ has a significant negative correlation on the number of edges, number of nodes, graph density and clustering coefficient. $M6$, $M9$, $M10$, $M12$, $M13$ have a significant positive correlation on the number of edges (see Fig. 4). This is not surprising because these motifs have a high connectivity between the nodes themselves which leads to a higher graph density in larger graphs.

The observations depicted in Fig. 4 show the correlation details for each motif. If we would like to reduce the complexity of a process graph, we should decrease the occurrence of motifs which have a positive correlation with the graph density. Analogue to this observation, other analysis can be performed with different optimization goals. Another approach would be to use an optimized model and calculate the signature that can then be used as the target signature for improvement. This paper will focus on the reduction of complexity of process models. However, our approach is not limited in this respect.

3   Motif-Based Process Graph Adaptation

In the previous section we have seen how an optimal target motif signature can be found. Our approach assumes the existence of such a target signature to adapt the given process model. In practice an analyst can use reference process models to gather an optimal target motif signature or try different target signatures and analyze their behavior to the process graph. Next we will describe how our graph adaptation algorithm tries to improve process graphs and how optimization constraints can be defined with DECLARE in order to retain the original business goal of the process.

3.1   Basic Definitions

The original local motif-based adaptation approach (LoMbA) was first introduced for the adaptation of network topologies by Stein et. al. in [15]. In this paper we will transfer this algorithm to the domain of business process model improvement, thus we will first introduce some basic definitions and formalize the graph adaptation problem in the business process domain:

Definition 1 (Process Graph). *Let $G = (V, E)$ be a simple graph, where nodes $v \in V$ are the event classes that can occur in a process and edges $e \in E$ between nodes represent the possible transitions between events.*

Definition 2 (Motifs). *A motif M is a small subgraph pattern of small size, typically 3 or 4 nodes (see Fig. 2). We denote M as a finite set of motifs of the same node size.*

As motifs are the building blocks of graphs we can count their relative frequency in a process graph:

Definition 3 (Motif Signature). *The occurrence of M in G is represented as a motif signature $s(G)$ which is a real-valued vector that stores the relative frequency of motif M_i.*

For comparing multiple graphs regarding their motif signature, we can define a function which returns the distance between two motif signatures:

Definition 4 (*Motif Distance*). *Furthermore, distance(x, y) of motif signatures x and y is the Euclidean distance between x and y.*

In this paper we aim to improve process graphs by modifying the process graphs' edges. One way of gathering real process graphs is to extract event logs from information systems. Event logs contain the actual use of the process involved information systems and provide the ideal input for process improvement approaches. A process mining discovery algorithm such as the heuristics miner [16] can be used to extract a process model. Due to the fact that we cannot modify the process graph arbitrarily, we need to specify a function that checks if the proposed change is allowed. In comparison to the function defined in [15] we define a function that checks various business constraints (see Sect. 3.3).

Definition 5 (*Process Constraint Functions*). *Let $f(G) \rightarrow \{true, false\}$ be a function that receives the original graph G and returns a boolean value. f returns true if the graph G fulfills given process constraints and false if not.*

3.2 Algorithm

The goal of the graph adaptation algorithm is to modify the input graph G such that a target signature t is approximated under the process constraint function f. The algorithm periodically iterates through all nodes $v \in V$ of the graph G (see Algorithm 1). The algorithm is divided into two steps: (1) the algorithm searches for a proper set of modification operations and (2) it selects an operation that fulfills the process constraint function and checks if the modification has approximated G to the target signature t.

The following modifications for each node $v \in V$ are sufficient to modify a graph in our domain: *Remove-edge, Add-edge, Move-edge* operation. Due to the large amount of possible operations, the algorithm reduces the search space by calculating the *edge operation indicator* (lines 2–3). It ranks the generated modification operations by using a simple heuristic that decides which graph operations are more appropriate to approximate G to the target signature:

$$EOI = \frac{\sum (t_i - s_i(G)) \cdot |E(M_i)|}{l}$$

t_i is the relative frequency of motif M_i in the target signature, $s_i(G)$ returns the relative frequency of motif M_i in the input graph G, and $|E(M_i)|$ returns the number of edges in motif M_i. l is the number of inspected motifs. EOI calculates the average weighted ratio between the amount of edges in the target signature and the current graph signature. A value larger than 0 indicates that more edges need to be added to the graph, thus the algorithm will prefer the *Add-edge* operator. If *EOI* is smaller than 0, the algorithm will prefer the *Remove-edge* operator. A value near 0 indicates that neither edges should be removed nor added, thus the algorithm will prefer the *Move-edge* operator.

Algorithm 1. Motif-based Process Graph Adaptation Algorithm [15]

---

1 its ← {Remove (G, v), Add (G, v), Move (G, v)};

2 EOI ← GetEdgeIndicator (G, t);
3 itOrder ← GetIteratorOrder (its, EOI, eoiThreshold);

4 **for** $it: itOrder$ **do**
5   foundValid ← false, maxSteps ← ∞, doneSteps ← 0;
6   currentDistance ← Distance (s(G), t);
7   **while** $op ← it.next()$ and doneSteps ¡ maxSteps **do**
8   doneSteps ← doneSteps + 1;
9          $G' ←$ CreateCandidate (G, op);
10   **if** $f(G')$ **then**
11   foundValid ← true;
12   **if** Distance $(s(G'), t)$ ¡ currentDistance **then**
13                 $G ← G'$;
14   **if** maxSteps $= ∞$ **then**
15   maxSteps ← doneSteps;
16   doneSteps ← 0, currentDistance ← Distance (s(G), t);
17   **end**
18   **end**
19   **end**
20   **end**
21   **if** foundValid **then** break;
22 **end**

In the second step of the algorithm, a graph operation is selected for $v \in V$ and a candidate graph G' with the applied operation is generated. Now the algorithm checks the process constraints. If f is violated then the operation is discarded. Otherwise, the motif signature distance between G' and t is calculated. If the motif signature of G' is closer to the target signature than G, then v modifies G. In the original algorithm a sampled graph is used instead of the complete graph to reduce the computational complexity of counting motifs [17]. As our graphs tend to be smaller than network topologies, we can operate on the complete graph. If the algorithm has found a valid graph operation for v that fulfills f, it is likely that other operations are valid within the given operation iterator. Thus the algorithm will check another *doneSteps* operations from the iterator. If none of the operations could be applied to G due to the violations of f, the algorithm will expand the search space with additional iterators.

The complexity of the tackled graph problem is NP-hard [15]. Due to the much smaller process graphs compared with network topology graphs, we observed a feasible average runtime of 130 s over 5 rounds in our experiments.

3.3 Specification of Constraints Using Declare Models

In order to retain the feasibility of the optimized process model, our algorithm allows the specification of a process constraint function that must be fulfilled after each graph modification operation. We use DECLARE [12] models to specify the process constraint function that the graph adaptation algorithm will respect. DECLARE is an LTL-based declarative process modeling language which allows the specification of process models by the instantiation of templates [2]. For instance, the *response* template specifies that a certain event must be executed if another specific event has been executed before. Each time our algorithm generates possible modifications to the process model, it checks if all given constraints are still fulfilled after the application of proposed modifications using automata described in [5]. If so, the algorithm will apply the modification otherwise it will reject the change (see Sect. 3.2). This allows us to restrict the algorithm to certain modifications such that the original process model is still followed in the improved one.

Constraints can be either specified by hand using the graphical representation of DECLARE models [2] or by declarative process discovery using historic process executions [6]. We decided to use declarative process discovery from historic executions gathered in event logs to automatically generate constraints for process models. By specifying a confidence level we can determine which quality the constraints should have in order to force them to be fulfilled in the resulting improved model. Still, it is possible to edit the constraints, disable specific constraint types or add custom ones. The advantage of the automatic discovery is that hidden constraints will be uncovered from the historic executions, reducing the manual constraint definition time and leading to more relevant constraints.

4 Experimental Evaluation

In this experimental evaluation we show the general applicability of motif-based graph adaptation in the domain of business process model improvement.

4.1 Setup

We tested our approach with 5 real-life event logs (see Table 1) and compared if the improvement goal was reached. For each event log we generated two process models, one with the standard setting and one with the "noise-free" setting (marked with a star *), using the heuristics miner.

Table 1. Characteristics of the used real-life event logs in our evaluation.

#	Event log	Instances	Variants	Events	Events/Case	Constraints
I	BPI Challenge'12	13 087	7 179	36	20.04	214
II	Large	651 709	30460	35	5.95	122
III	Small	873	101	45	7.77	26
IV	Midsize	90 536	1630	30	9.06	56
V	Environmental	1 434	381	27	5.98	49

The improvement goal for all process models is to reduce the complexity of the model. Based on the observations in Sect. 2, we select three different process independent target signatures: (1) $M1, M2, M3 = 33.3\%$, (2) $M1 = 20\%, M2 = 20\%, M3 = 60\%$ and (3) $M1 = 22.8\%, M2 = 22.8\%, M3 = 34.7\%, M8 = 19.0\%$. For retaining the original business goal, we automatically gather the constraints by generating DELARE rules using MINERful (see Sect. 3.3) and pick the all rules with a confidence of at least 0.95.

4.2   Results and Discussion

Table 2 shows the results of the process graph improvement using the three different target signatures after 5 rounds. Target signatures (1) and (2) actually decreased the number of edges, the density and the clustering coefficient, achieving our improvement goal. Only target signature (3) did not work as well as the other two, resulting in an increase of the clustering coefficient (also see Fig. 5). Most graph operations were made to the "noise-free setting" process models. Here more edges could be removed which is not surprising because these models in general have more edges. The graph constraints are the same for both models thus these additional edges were now removed by our improvement approach.

Table 2. Evaluation result before and after motif-based graph adaptation.

		I	I*	II	II*	III	III*	IV	IV*	V	V*
Initial Edges		39	54	106	244	29	37	73	61	40	47
Initial Density		0,071	0,098	0,089	0,205	0,069	0,088	0,084	0,070	0,057	0,067
Initial Cl.Coeff.		0,059	0,100	0,120	0,331	0,065	0,072	0,110	0,062	0,033	0,097
(1)	Δ Edges	−1	−19	−20	−152	0	−7	0	−12	−3	−16
	Δ Cl.Coeff	−0,059	−0,100	−0,140	−0,325	−0,065	−0,056	−0,110	−0,056	−0,021	−0,097
(2)	Δ Edges	0	−19	−38	−168	0	−8	−23	−17	−3	−18
	Δ Cl.Coeff	−0,059	−0,100	−0,140	−0,331	−0,065	−0,072	−0,110	−0,062	−0,033	−0,097
(3)	Δ Edges	1	−11	0	−138	0	0	0	0	−3	−10
	Δ Cl.Coeff	0,133	0,092	0,023	−0,178	0,122	0,113	0,076	0,130	0,151	0,086

We run the algorithm with a maximum of 5 rounds. Figure 6 shows the distance to the target signatures over the rounds. We can see that the algorithm approximates the graph towards the target signature. For target signature (1) we were able to reach the signature for all datasets within 4 rounds, whereas we did not reach it for signature (3). The largest number of applied operations were already made in the first round, resulting in a quick converge towards the target signature. In addition, we can see that the size of the input graph influenced how fast the algorithm converges to the target.

We also investigated how many graph operations (*Add-edge*, *Remove-edge*, *Move-edge*) were performed over the number of rounds (see Fig. 7). After round 2 the number of added and removed edges is almost equal for all datasets indicating that only *Move-edge* operations are performed because the target signature is

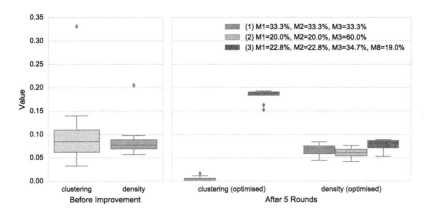

Fig. 5. Normed number of edge operations performed for target signature (1) and (3).

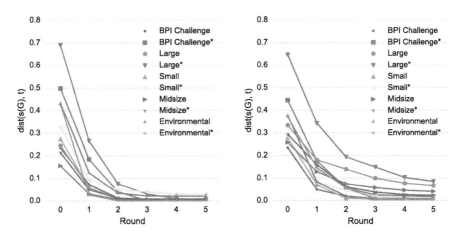

Fig. 6. Distance to target signatures (1) and (3) over the number of rounds.

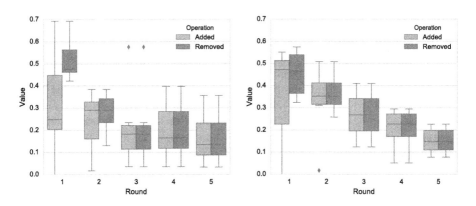

Fig. 7. Normed number of edge operations performed for target signature (1) and (3).

almost reached. Here it would make sense to stop the algorithm early until the distance to the target signature is close enough because this would retain more structures of the original process.

5   Related Work

According to the BPM survey [1] from 2013 the improvement of existing process models using process mining is not much researched. Currently there exists no tool, neither from research nor from industry, that provides automatic guidance for redesigning processes, although there is the need for proper tools [3].

In [13] best practices for redesigning business processes are presented. The authors evaluate different approaches and provide an overview of the most common redesign operations. Niedermann et al. [11] propose a semi-automatic process optimization platform which matches new processes to existing processes using similarity metrics (e.g. syntactic, linguistic and context). Best practice optimization patterns that consist of a detection and an application component (based on [13]) are matched and applied to the original process.

Process model improvement can also be indirectly achieved using conformance checking methods such as presented in [4]. [7] propose the use of reference models (ITIL) that usually best practices to improve existing process models. The authors present an approach which overcomes the problems of different levels of detail, partial views and overemphasis of the order. Another approach was presented in [18] which calculate various process performance indicators. By matching and clustering PPIs together with other process models from different organizations, the presented system can make suggestions and recommendations for performance improvement.

6   Conclusion

This paper first showed that motif analysis can also be applied to business processes to find relations between the core structure and other process properties. We have found correlations between specific motifs and process improvement goals. We further adapted the motif-based graph adaptation algorithm to modify process graphs in order to improve them using an optimized motif target signature based on the observed motif histogram. We also showed how to automatically define required process constraints to retain the feasibility. Finally, in our experimental evaluation we presented the applicability of our approach for real event logs. The results show that process models were improved based on a given target signature to achieve the improvement goal.

In future work, we will investigate how target signatures can be defined for specific business goals and how they can be derived from existing process models or reference models. Another aspect that we will address is the heuristic that ranks the possible graph operations. Currently, the heuristic is just target signature focused but it should also be targeting the improvement goal. Lastly, we will expand our approach to use larger motifs with more complex graph patterns.

Acknowledgement. This project (HA project no. 522/17-04) is funded in the framework of Hessen ModellProjekte, financed with funds of LOEWE-Landes-Offensive zur Entwicklung Wissenschaftlich-ökonomischer Exzellenz, Förderlinie 3: KMU-Verbundvorhaben, by the LOEWE initiative (Hessen, Germany) within the NICER project [III L 5-518/81.004] and by the German Research Foundation (DFG) as part of project A1 within the Collaborative Research Center (CRC) 1053 – MAKI.

References

1. van der Aalst, W.M.P.: Business process management : a comprehensive survey. ISRN Soft. Eng. **2013**, 1–37 (2013)
2. Burattin, A., Maggi, F.M., van der Aalst, W.M.P., Sperduti, A.: Techniques for a posteriori analysis of declarative processes. In: EDOC, pp. 41–50 (2012)
3. Cater-Steel, A., Tan, W.G., Toleman, M.: Challenge of adopting multiple process improvement frameworks. In: ECIS, pp. 1–12 (2006)
4. de Leoni, M., van der Aalst, W.M.P.: Aligning event logs and process models for multi-perspective conformance checking: an approach based on integer linear programming. In: Daniel, F., Wang, J., Weber, B. (eds.) BPM 2013. LNCS, vol. 8094, pp. 113–129. Springer, Heidelberg (2013). https://doi.org/10.1007/978-3-642-40176-3_10
5. Di Ciccio, C., Maggi, F.M., Montali, M., Mendling, J.: Resolving inconsistencies and redundancies in declarative process models. Inf. Syst. **64**, 425–446 (2017)
6. Di Ciccio, C., Schouten, M.H.M., de Leoni, M., Mendling, J.: Declarative process discovery with MINERful in ProM. In: CEUR Workshop Proceedings, vol. 1418, pp. 60–64 (2015)
7. Gerke, K., Tamm, G.: Continuous quality improvement of IT processes based on reference models and process mining. In: AMCIS (2009)
8. Krumov, L., Fretter, C., Müller-Hannemann, M., Weihe, K., Hütt, M.T.: Motifs in co-authorship networks and their relation to the impact of scientific publications. Eur. Phys. J. B **84**(4), 535–540 (2011)
9. Krumov, L., Schweizer, I., Bradler, D., Strufe, T.: Leveraging network motifs for the adaptation of structured peer-to-peer-networks. In: GLOBECOM, pp. 1–5 (2010)
10. Milo, R., Shen-Orr, S., Itzkovitz, S., Kashtan, N.: Network motif: simple building blocks of complex networks. Science **298**(5594), 298 (2002)
11. Niedermann, F., Radeschütz, S., Mitschang, B.: Design-time process optimization through optimization patterns and process model matching. In: CEC, pp. 48–55 (2010)
12. Pesic, M., Schonenberg, H., Van Der Aalst, W.M.P.: DECLARE: full support for loosely-structured processes. In: EDOC, pp. 287–298 (2007)
13. Reijers, H.A., Liman Mansar, S.: Best practices in business process redesign: an overview and qualitative evaluation of successful redesign heuristics. Omega **33**(4), 283–306 (2005)
14. Schreiber, F., Schwöbbermeyer, H.: Motifs in biological networks. In: Statistical and Evolutionary Analysis of Biological Networks, pp. 45–64 (2010)
15. Stein, M., Weihe, K., Wilberg, A., Kluge, R., Klomp, J.M., Schnee, M., Wang, L., Mühlhäuser, M.: Distributed graph-based topology adaptation using motif signatures. In: ACM-SIAM Meeting on Algorithm Engineering & Experiments (2017)
16. Weijters, A.J.M.M., van der Aalst, W.M.P., Medeiros, A.K.A.D.: Process Mining with the HeuristicsMiner Algorithm. BETA Working Paper Series 166, pp. 1–34 (2006)

17. Wernicke, S.: Efficient detection of network motifs. In: IEEE/ACM Transactions on Computational Biology and Bioinformatics. vol. 3, pp. 347–359 (2006)
18. Yilmaz, O., Karagoz, P.: Generating performance improvement suggestions by using cross-organizational process mining. In: SIMPA, vol. 6, pp. 3–17 (2015)

Dropout Prediction in MOOCs: A Comparison Between Process and Sequence Mining

Galina Deeva[1]([✉]), Johannes De Smedt[2], Pieter De Koninck[1],
and Jochen De Weerdt[1]

[1] Department of Decision Sciences and Information Management,
Faculty of Economics and Business, KU Leuven, Leuven, Belgium
`galina.deeva@kuleuven.be`
[2] Management Science and Business Economics Group, Business School,
University of Edinburgh, Edinburgh, Scotland

Abstract. Recently, Massive Open Online Courses (MOOCs) have experienced rapid development. However, one of the major issues of online education is the high dropout rates of participants. Many studies have attempted to explore this issue, using quantitative and qualitative methods for student attrition analysis. Nevertheless, there is a lack of studies which (1) predict the actual moment of dropout, providing opportunities to enhance MOOCs' student retention by offering timely interventions; and (2) compare the performance of such predicting algorithms. In this paper, we aim to predict student drop out in MOOCs using process and sequence mining techniques, and provide a comparative analysis of these techniques. We perform a case study based on the data from KU Leuven online course "Trends in e-Psychology", available on the edX platform. The results reveal, that while process mining is better capable to perform descriptive analysis, sequence mining techniques provide better features for predictive purposes.

Keywords: Dropout prediction · Process mining
Sequence classification · Massive Open Online Course
Educational data mining

1 Introduction

The important growth in success of Massive Open Online Courses (MOOCs) over the last five years, makes that these learning platforms are becoming a key focus for academic research. Given that students can freely join and leave, MOOCs are usually characterized by high dropout rates. Given the scale of these courses, personalization of engagement initiatives must be automated, thus data driven. In this paper, the goal is to investigate whether the first step, namely the identification of disengagement or dropout prediction can be performed based on so-called behavioural MOOC data, i.e. the logging of actions that learners undertake while interacting with the MOOC platform. More specifically, the key

© Springer International Publishing AG 2018
E. Teniente and M. Weidlich (Eds.): BPM 2017 Workshops, LNBIP 308, pp. 243–255, 2018.
https://doi.org/10.1007/978-3-319-74030-0_18

contribution of this paper is an investigation of the usefulness of both process mining and sequence mining techniques for dropout prediction.

The paper is structured as follows. In Sect. 2, the related work on analyzing MOOC data is discussed. Next, in Sect. 3 the approaches used for the analysis are reviewed and subsequently applied and discussed in Sect. 4. Section 5 concludes the paper and sets out the lines for future work.

2   Related Work

Over the past few years, MOOCs have provided educational researchers with data on a nearly unprecedented scale, which has led to a vast surge in research in the field of learning analytics. Given that students can join and leave MOOCs freely, some studies have been addressing questions regarding the reasons for students enrolment and attrition and the ways to increase the level of student engagement in a course from a qualitative perspecive [1–3].

From a quantitative perspective, researchers have investigated social network analytics [4], regression and random-forest classification [5], and hidden markov models [6] for analyzing predictors of dropout. In [7], a prediction model was based on student actions, measuring activity frequency and absenteeism. Process mining techniques have also been applied for analyzing action-level MOOC data [8]. Regarding the application of sequence mining techniques, sequence classification techniques were successfully applied in [9], in order to identify and compare segments of students' productive and unproductive learning behaviors. Furthermore, extraction of sequential patterns in an e-learning context was used, for instance in [9,10].

All these applications suggest that both process and sequence mining have a great potential to provide powerful tools in the field of online education. However, observe that patterns of regular event sequences are only conceptually interesting if they are sufficiently explained in theoretical terms. Moving from exploration straight to normative guidelines without having a real explanatory account is risky and hard to justify, as discussed in [11].

3   Behavioral Analysis of Dropout in MOOCs Using Process Mining and Sequence Mining

In this section, an overview of both process and sequence mining techniques that have potential to be applied to behavioral MOOC data for dropout prediction is described. However, firstly, a more accurate description of the typical input data and preprocessing is discussed.

3.1   Behavioral Data in MOOCs

Given the online-based format and nature of MOOCs, it is possible to track students' activities following the individual clicks they make on the course webpages. Most contemporary MOOC platforms are capable of writing event records to log files as users interact with the features of the platform. The data generated in this way can provide insights into when and how students watch videos, interact on forums, solve quizzes questions, etc. In this work we restrict ourselves to behavioral data and don't take into account demographic (age, gender and degree), geographical (country of origin, native language) or psychological (goal striving, social belonging and growth mindset) factors.

Obtaining an Event Log. As students participate in MOOCs activities, the system collects data about their content interactions. The data is logged at the level of events and timestamped to the millisecond. These events represent fine-grained actions such as closing a web page, clicking a button, opening a new window, or tagging a selected text. Oftentimes, the data is available in structured files like JSON, however the transformation of such files to the typical event log format used in process mining requires adequate preprocessing. An example of such an obtained event log, after preprocessing the raw data, and similar to the data used in this paper, is shown in Table 1.

Table 1. An example of an event log obtained after preprocessing the native JSON file used for storage in the edX platform.

Timestamp	Case	$AG1$	$AG2$	$AG3$
2015-11-20 13:25:36.844	User1	enrolled	enrolled	enrolled
2015-11-21 01:31:58.687	User2	page_close	page_close/FAQ	Page_close/FAQ/C5S5
2015-11-22 20:23:23.042	User2	load_video	load_video/Intro	load_video/Intro/C1S1
2015-11-22 20:28:34.492	User2	pause_video	pause_video/Intro	pause_video /Intro/C1S1
2015-11-22 20:29:00.230	User3	forum_post	forum_post/e-Cigarette	forum_post/e-Cigarette/C2S1
2015-11-22 20:30:48.177	User3	problem_save	problem_save/Intro	problem_save/Intro/C2S3

Activity Granularity. One particular difficulty when dealing with MOOC-based behavioral data, relates to activity granularity. Logging of actions is often at a rather technical level, similar to raw web analytics data. Clicking on a link or a button, press the play button of a movie element, etc., are rather meaningless without information about the actual content element to which the action was applied. As such, appending action information to content element information can be expected to enhance the outcomes of analytics techniques. In this work, we look at three different levels of granularity of event types, denoted in Table 1 as $AG1$, $AG2$ and $AG3$. In this case, $AG1$ level indicates the activity or action, which is conducted by a student, while $AG2$ and $AG3$ provide information about a module and a chapter of this module, in which an action took place.

Window-Based Approach. Another key issue when analysing MOOC data for dropout analysis, deals with how actionable derived models and patterns are. Most importantly, it is claimed in this paper that an appropriate window-based approach is necessary that guarantees that no data leakage is present (e.g. information about a period after dropout is assessed, is taken into account for students who don't dropout). Furthermore, it is more valuable to predict whether a specific student is going to drop out as early as possible during the course. Hence, we take into account the data only for the first 3 weeks of the course. Using the original event log, we construct 3 data sets, which include events happening before the end of Week 1, Week 2 and Week 3. Each data set only considers students, who have not dropped out before the cut-off date, since we are only interested in students which will drop out in the future, thus allowing to provide timely interventions in order to retain these students.

3.2 Process Mining

Process mining offers a comprehensive set of tools, techniques and algorithms to perform log data analysis and support process improvements [12]. The techniques are usually classified along three main perspectives, *process discovery*, *conformance checking* and *process enhancement*.

In the context of MOOCs, process mining can provide information about how the learners interact with the MOOC platform and analyse the underlying process. The following techniques are considered most worthwhile for analyzing behavioural MOOC data with the intention to discriminate between dropout/non-dropout.

Process Discovery. First of all, plain simple process discovery can be used to investigate whether differences exist in terms of how students who drop out participate in the learning activities compared to students who don't drop out. Given the low-level nature of the source data and the unstructured nature of the process, it can, however, be expected that interpretation of discovered process models could be difficult.

Trace Clustering Explanation Techniques. Trace clustering is the task of finding similar process traces and grouping them accordingly [13]. Trace clustering can be applied to event logs containing behavioural MOOC data, however, given the problem at hand, i.e. dropout prediction, we already have clusters: dropout vs. non-dropout. Therefore, trace clustering explanation methods could be worthwhile.

In [14], a technique called SECPI is proposed that can accurately explain why traces are included in a certain cluster. As such, applied to the context of this paper, SECPI can learn individual rules, based on control-flow characteristics derived from the event sequences of a single student, which explains why that particular student is part of the dropout or non-dropout cluster. An example explanation would be "Student Jane Doe would have been predicted to drop out if she had not visited the student forum and posted a new thread".

Predictive Process Mining. The use of process mining for predictive analytics has seen a vast surge over recent years [15]. Researchers have focused

on a number of applications, such as predicting the remaining execution time
[16–19], the next event to execute [19,20], the tendency towards violation in
timed contexts [21], or the violation of predicates [22].

Predictive process mining techniques could be tailored towards predicting
whether an (artificial) dropout event will occur in a certain timeframe (as next
event prediction is not a realistic choice in this case). However, as far as we
know, predictive process mining techniques are not available off-the-shelve. Furthermore, the challenge to predict the occurrence of a particular event in a certain window is more complex as compared to existing predictive process mining
applications.

3.3 Sequence Mining

Within the sequence mining field, a wide range of techniques to extract knowledge from sequential data has been proposed. Given the nature of our data and
our goal to detect dropout, there are two key application scenarios which can be
envisioned: sequential pattern mining and sequence classification.

Descriptive Analysis of Sequential Data: Sequential Pattern Mining.
Sequential pattern mining is a process of discovering a set of frequent subsequences among a set of sequences in a given database, such that the occurrence
of such subsequences exceeds a defined minimum support. The *support* of a
sequential pattern is the percentage of sequences containing this pattern [23].
By investigating frequent patterns from event sequences of dropout and non-
dropout sequences, a descriptive comparative analysis can be performed along
the lines of process discovery, however in this case, in a less visually appealing
manner. Given the large number of items (i.e. types of events) in our data, it
can also be expected that performing such a comparison based on patterns is far
from easy.

Sequence Classification. Given that sequences have labels (as is the case
here), sequence mining techniques that are specifically developed to find discriminative patterns between such labeled sequences [24] seem to be even more
worthwhile for dropout prediction. The sequence classification task can be formulated as labelling new sequences based on the information gained in the training stage [25]. Some studies combine pattern mining techniques and classification, for instance, pattern based sequence classifier [26] or classification based
on association rules [27]. Such combined algorithms can provide useful results
and also give some comprehensible information regarding the characteristics of
the dataset. Sequence classification can be either direct, i.e., the sequences are
retained according to their predictive quality, or indirect, i.e., frequent sequences
are generated and used as features for prediction. The majority of approaches
adheres to the second principle. An overview of techniques that are used for
sequence classification can be found in [28]. Sequence classification has the advantage that these techniques should provide more accurate and insightful differences between dropout/non-dropout and in a much more direct way, as compared
to descriptive sequential pattern mining.

4   Experimental Evaluation

This section illustrates the actual application of a selection of process and sequence mining techniques to one particular MOOC data set. Using both qualitative and quantitative methods, a comparative analysis is made. However, first, we would like to introduce the used data in a bit more detail.

4.1   Overview of the Case Study edX MOOC

In this paper, we analyze the KU Leuven MOOC "ePSYx Trends in e-Psychology", hosted on the edX platform. This online course ran for 5 weeks from 17 of November 2015 until 22 of December 2015, with the participation of 4,511 students. From these 4,511 participants, we only take into account students who conducted more than 10 actions, leaving us with 1,394 students, which corresponds to 299,225 events in total.

edX keeps track of all students activity details useful for our analysis. We extract and preprocess students' behavioral data into a sequence of events. After obtaining the event log as shown in Table 1, we perform descriptive analysis, extracting three groups of students: (1) successful students, which obtained certificate, (2) dropped out students, which actively participated in the beginning of the course, but stopped participation afterwards, and (3) undefined category of students, which cannot be readily classified in either group.

Table 2. Descriptive statistics about the nine datasets used. #Events is the total number of events, #Event types is total number of different types of events/activities, #Cases is the number of students, and Dropout %.

Activity granularity	Statistic	Week1	Week1+2	Week1+2+3
AG_1	#Events	38755	75591	105153
	#Event types	28	27	29
	#Cases	465	383	334
	Dropout %	73	64	62
AG_2	#Events	38755	75591	105153
	#Event types	65	83	110
	#Cases	465	383	334
	Dropout %	73	64	62
AG_3	#Events	38755	75591	105153
	#Event types	206	339	491
	#Cases	465	383	334
	Dropout %	73	64	62

In line with our explanations in Sect. 3.2, we perform analysis on different levels of activity granularity and we apply a window-based approach. More specifically, in our experiments, we look at three different time windows (Week1,

Week1 + 2, and Week $1 + 2 + 3$) and three different levels of activity granularity (denoted as AG_i), thus resulting in nine different subsets of data. Detailed statistics of the nine datasets can be found in Table 2.

4.2   Results

In this section, the results of applying both process mining and sequence mining approaches are presented.

Process Mining: Process Discovery. In order to analyze the data from a process discovery perspective, we perform analysis using the Disco tool[1] and the Inductive Miner (IM) plug-in in ProM. The results reveal that it is challenging to build process models that allow to clearly distinguish between the two groups of students, especially for lower level of event abstraction in AG_2 and AG_3. The higher the level of granularity (i.e. the more detailed), the less understandable the discovered models become both for IM and Disco. To make it possible to compare such "spagetti-like" process models, we vary the number of activities and paths shown in Disco and ProM to a very low level, which goes at the cost of losing some information regarding behavioural patterns. Figures 1 and 2 provide an example of such process models for granularity level AG2 for a subset of activities happened before the end of Week 1.

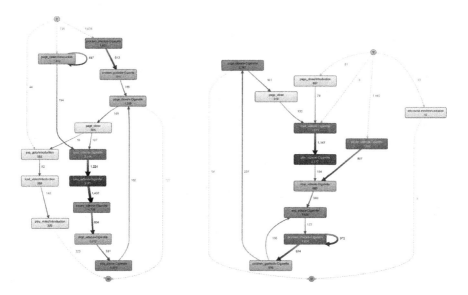

Fig. 1. Process models built in Disco for dropout (left) and non-dropout (right) cases. Parameters: Week 1 and level of granularity AG_2

---

[1] http://fluxicon.com/disco.

Additionally, the process discovery tools allow to compare models visually, while, as will be clear from the rest of this section, other techniques are able to provide formal and detailed verification of the results.

Process Mining: Trace Clustering Explanation Methods. Next, we consider dropout and non-dropout students as two trace clusters, and explain the differences between the two clusters with SECPI. To validate the model, we apply 10-fold cross-validation. In Table 3, the results of this technique on the dataset are presented. The classification techniques underlying the SECPI-technique are shown to lead to excellent results in terms of accuracy, and the explanations on instance-level are typically rather short, especially at a higher hierarchical abstraction level. All configurations can explain certain instances with a single rule, but Table 3 contains the results averaged over all instances in the dataset. The application of SECPI on the most fine-grained datasets of week 2 and 3 did not yield any results, since the number of events and event types in those datasets is simply too large. Observe that SECPI only provides rules at an instance level, and therefore, can only be used to explain which control-flow aspects had an influence on dropout. In general, SECPI would not be useful for dropout prediction, unless the individual rules are aggregated, and general dropout patterns can be identified.

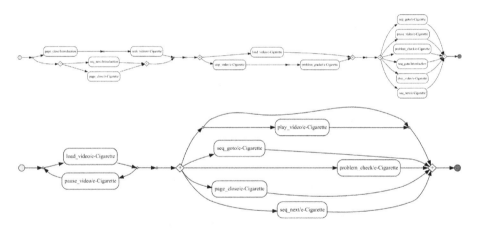

Fig. 2. Process models built by Inductive Miner in ProM for dropout (top) and non-dropout (bottom) cases. Parameters: Week 1 and level of granularity AG_2

Sequence Mining: Sequence Classification. Two sequence classification approaches are applied to the MOOC event log. We use Random Forests as a classifier and, similarly to SECPI case, perform 10-fold cross-validation. Both techniques generate sequence constraints in the form of $\alpha_1, \alpha_2 \implies \alpha_3$, to express that α_3 appears after α_1 and α_2. cSPADE [29] is used to generate sequences over all traces at once, while SCIP [25] generates sequence constraints

Table 3. Accuracy (Acc.) and average explanation length (EL) of applying the SECPI-technique on the MOOC-dataset, using the *SomeTimesDirectlyFollows*-attribute, on the 9 datasets defined according to Activity Granularity and time windows and weeks.

Activity granularity	Metric	Week1	Week1+2	Week1+2+3
AG_1	Acc.	87.96%	91.39%	93.41%
	EL	2.6	3.2	4.1
AG_2	Acc.	91.4%	94.5%	98.2%
	EL	2.6	3.2	4.1
AG_3	Acc.	94.8%	DNF	DNF
	EL	4.8	DNF	DNF

for the labels separately. They both apply support, and SCIP is operated with an interestingness level of 0.05. Only sequences of length two are generated, as the computational effort rises significantly when longer ones are mined for. However, it was shown that longer lengths do not significantly improve classification performance [25]. The support was used in a range between 0.35 and 1, and the accuracy as well as the number of constraints generated displayed in Figs. 3 and 4. For SCIP, the number of constraints generated for either class (dropout/non-dropout) is shown.

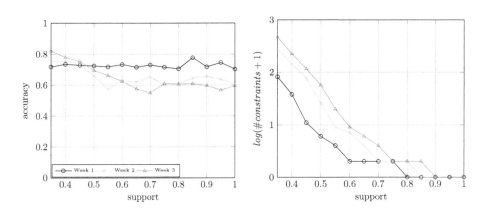

Fig. 3. Overview of the performance of cSPADE.

As can be learned from the results, majority class prediction is capable of providing 60–70% accuracy when no constraints are found (this is true for both techniques at a support level between 0.9 and 1.0). cSPADE is not capable of vastly improving on these results, except for Week 2 and 3 predictions, where eventually an accuracy of 80% is achieved at a 0.4 support level. SCIP is able to vastly improve on this result, achieving near perfect accuracy as of a support level of 0.8 for week 2 and 3 predictions, and a level of 0.35 for week 1 predictions.

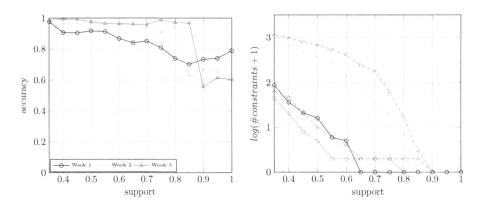

Fig. 4. Overview of the performance of SCIP. The number of constraints generated for the dropouts is indicated with a lighter shade and different marker.

A trade-off exists between the number of constraints generated and the support level, however, high accuracy is already achieved at a support for 0.8 and only between 10 and 100 constraints per class. Notice the great discrepancy between dropouts and non-dropouts in the number of constraints that are generated. Since the non-dropouts generate more events and follow a more streamlined process to obtain the degree, this observation is not unexpected.

To illustrate what the patterns look like, consider the output for weeks 2 and 3 in Tables 4 and 5 for 80% support. Clearly, the difference in the amount of sequences generated becomes apparent.

Table 4. Sequences of length 2 mined with SCIP with support 80% for week 2.

	Activity 1	Activity 2
D	page_close/Introduction/C1S1	
ND	page_close/e-Cigarette/C2S2	page_close/e-Cigarette/C2S4
	page_close/e-Cigarette/C2S4	
	problem_graded/e-Cigarette/C2S2	
	problem_check/e-Cigarette/C2S2	problem_check/e-Cigarette/C2S2
	problem_check/e-Cigarette/C2S2	
	problem_check/e-Cigarette/C2S2	problem_graded/e-Cigarette/C2S2

Discussion. The results reveal that the analytical models are better capable of processing such unstructured data, containing information about learning behaviour, especially for predictive purposes. While process discovery techniques provide useful tools for descriptive purposes, they are hardly capable to express the magnitude of the difference between different types of learners. Regarding the quantitative comparison between clustering explanation techniques and sequence classification, it should be observed that accuracy levels are comparable, however, SECPI only works on an instance level, while cSPADE and SCIP provide a result that holds at the group level and can actually be put into configuration in an actual dropout prediction operationalization.

Table 5. Sequences of length 2 mined with SCIP with support 80% for week 3.

	Activity 1	Activity 2
D	page_close/Introduction/C1S1	
ND	problem_check/e-Cigarette/C2S2	page_close/e-Cigarette/C2S2
	problem_graded/e-Cigarette/C2S2	page_close/e-Cigarette/C2S2
	load_video/e-Cigarette/C2S2	load_video/e-Cigarette/C2S2
	problem_check/e-Cigarette/C2S2	problem_graded/e-Cigarette/C2S2
	load_video/e-Cigarette/C2S2	problem_check/e-Cigarette/C2S2
	page_close/e-Cigarette/C2S2	page_close/e-Cigarette/C2S4
	problem_graded/e-Cigarette/C2S2	problem_graded/e-Cigarette/C2S2
	page_close	
	load_video/e-Cigarette/C2S2	page_close/e-Cigarette/C2S2
	problem_graded/e-Cigarette/C2S2	problem_check/e-Cigarette/C2S2
	page_close/e-Cigarette/C2S4	
	page_close/e-Mental	
	load_video/e-Cigarette/C2S2	problem_graded/e-Cigarette/C2S2
	problem_check/e-Cigarette/C2S2	problem_check/e-Cigarette/C2S2
	problem_check/e-Cigarette/C2S2	
	problem_graded/e-Cigarette/C2S2	

5   Conclusion and Future Work

In this paper, we performed a qualitative and quantitative comparison of process and sequence mining techniques for the purpose of dropout prediction in MOOCs, based on event data gathered from the edX platform. Given the recent growth in popularity of MOOCs, the issue of high students attrition rates should be investigated carefully and given the magnitude of such courses, automated dropout prediction and engagement intervention technique development is considered a worthwhile avenue for research. While it is also possible to analyze student dropout in MOOCs using other types of student data, we aim to exploit behavioral data to be able to explore learning process of MOOC participants and predict whether a student is going to drop out from the course. Our results reveal that, while process mining provides useful tools for descriptive analysis, sequence mining is better capable to deal with the rather unstructured data produced by learning processes and provides better techniques for predictive purposes.

In future work, we would like to enlarge our experimental setup to more and different MOOCs. In addition, it would be worthwhile to investigate the effect of combining different types of data for dropout prediction. Finally, real-life experiments with e.g. a sequence classification-based dropout preventing intervention mechanisms should reveal the capabilities of behavior-based prediction models.

References

1. Milligan, C., Margaryan, A., Littlejohn, A.: Patterns of engagement in massive open online courses. J. Online Learn. Technol. **9**(2), 149–159 (2013)
2. Zheng, S., Rosson, M.B., Shih, P.C., Carroll, J.M.: Understanding student motivation, behaviors and perceptions in MOOCs. In: Proceedings of the 18th ACM Conference on Computer Supported Cooperative Work & Social Computing, pp. 1882–1895. ACM (2015)
3. Eriksson, T., Adawi, T., Stöhr, C.: "Time is the bottleneck": a qualitative study exploring why learners drop out of MOOCs. J. Comput. High. Educ. **29**(1), 133–146 (2017)
4. Yang, D., Sinha, T., Adamson, D., Rosé, C.P.: Turn on, tune in, drop out: anticipating student dropouts in massive open online courses. In: Proceedings of the 2013 NIPS Data-driven Education Workshop. vol. 11, p. 14 (2013)
5. Adamopoulos, P.: What makes a great MOOC? An interdisciplinary analysis of student retention in online courses (2013)
6. Balakrishnan, G., Coetzee, D.: Predicting student retention in massive open online courses using hidden markov models. Electrical Engineering and Computer Sciences, University of California at Berkeley (2013)
7. Halawa, S., Greene, D., Mitchell, J.: Dropout prediction in MOOCs using learner activity features. Experiences Best Pract. Around MOOCs **7**, 3–12 (2014)
8. Mukala, P., Buijs, J., Van Der Aalst, W.: Exploring students' learning behaviour in moocs using process mining techniques. Technical report, Eindhoven University of Technology, BPM Center Report BPM-15-10, BPMcenter.org (2015)
9. Kinnebrew, J.S., Loretz, K.M., Biswas, G.: A contextualized, differential sequence mining method to derive students' learning behavior patterns. JEDM-J. Educ. Data Min. **5**(1), 190–219 (2013)
10. Luan, J.: Data mining and its applications in higher education. New Dir. Inst. Res. **2002**(113), 17–36 (2002)
11. Reimann, P., Markauskaite, L., Bannert, M.: e-Research and learning theory: what do sequence and process mining methods contribute? Br. J. Educ. Technol. **45**(3), 528–540 (2014)
12. van der Aalst, W.M.P.: Process Mining - Data Science in Action. Springer, Heidelberg (2016). https://doi.org/10.1007/978-3-662-49851-4
13. De Weerdt, J., vanden Broucke, S.K.L.M., Vanthienen, J., Baesens, B.: Active trace clustering for improved process discovery. IEEE Trans. Knowl. Data Eng. **25**(12), 2708–2720 (2013)
14. De Koninck, P., De Weerdt, J., vanden Broucke, S.K.L.M.: Explaining clusterings of process instances. Data Min. Knowl. Discov. **31**(3), 774–808 (2017)
15. de Leoni, M., van der Aalst, W.M.P., Dees, M.: A general process mining framework for correlating, predicting and clustering dynamic behavior based on event logs. Inf. Syst. **56**, 235–257 (2016)

16. van der Aalst, W.M.P., Schonenberg, M.H., Song, M.: Time prediction based on process mining. Inf. Syst. **36**(2), 450–475 (2011)
17. Rogge-Solti, A., Weske, M.: Prediction of remaining service execution time using stochastic petri nets with arbitrary firing delays. In: Basu, S., Pautasso, C., Zhang, L., Fu, X. (eds.) ICSOC 2013. LNCS, vol. 8274, pp. 389–403. Springer, Heidelberg (2013). https://doi.org/10.1007/978-3-642-45005-1_27
18. Polato, M., Sperduti, A., Burattin, A., de Leoni, M.: Data-aware remaining time prediction of business process instances. In: IJCNN, pp. 816–823. IEEE (2014)
19. Polato, M., Sperduti, A., Burattin, A., de Leoni, M.: Time and activity sequence prediction of business process instances. CoRR abs/1602.07566 (2016)
20. Breuker, D., Matzner, M., Delfmann, P., Becker, J.: Comprehensible predictive models for business processes. MIS Q. **40**(4), 1009–1034 (2016)
21. Westergaard, M., Maggi, F.M.: Looking into the future. Using timed automata to provide a priori advice about timed declarative process models. In: Meersman, R., Panetto, H., Dillon, T., Rinderle-Ma, S., Dadam, P., Zhou, X., Pearson, S., Ferscha, A., Bergamaschi, S., Cruz, I.F. (eds.) OTM 2012. LNCS, vol. 7565, pp. 250–267. Springer, Heidelberg (2012). https://doi.org/10.1007/978-3-642-33606-5_16
22. Francescomarino, C.D., Dumas, M., Maggi, F.M., Teinemaa, I.: Clustering-based predictive process monitoring. arXiv preprint arXiv:1506.01428 (2015)
23. Srikant, R., Agrawal, R.: Mining sequential patterns: generalizations and performance improvements. In: Apers, P., Bouzeghoub, M., Gardarin, G. (eds.) EDBT 1996. LNCS, vol. 1057, pp. 1–17. Springer, Heidelberg (1996). https://doi.org/10.1007/BFb0014140
24. Cheng, H., Yan, X., Han, J., Hsu, C.W.: Discriminative frequent pattern analysis for effective classification. In: IEEE 23rd International Conference on Data Engineering 2007, ICDE 2007, pp. 716–725. IEEE (2007)
25. Zhou, C., Cule, B., Goethals, B.: Pattern based sequence classification. IEEE Trans. Knowl. Data Eng. **28**(5), 1285–1298 (2016)
26. Lesh, N., Zaki, M.J., Oglhara, M.: Scalable feature mining for sequential data. IEEE Intell. Syst. Appl. **15**(2), 48–56 (2000)
27. Wang, J., Karypis, G.: Harmony: efficiently mining the best rules for classification. In: Proceedings of the 2005 SIAM International Conference on Data Mining, SIAM, pp. 205–216 (2005)
28. Egho, E., Gay, D., Boullé, M., Voisine, N., Clérot, F.: A parameter-free approach for mining robust sequential classification rules. In: ICDM, IEEE Computer Society, pp. 745–750 (2015)
29. Zaki, M.J.: SPADE: an efficient algorithm for mining frequent sequences. Mach. Learn. **42**(1/2), 31–60 (2001)

Process Mining and the Black Swan: An Empirical Analysis of the Influence of Unobserved Behavior on the Quality of Mined Process Models

Jana-Rebecca Rehse$^{(\boxtimes)}$, Peter Fettke, and Peter Loos

Institute for Information Systems (IWi) at the German Center
for Artificial Intelligence (DFKI GmbH), Saarland University,
Campus D3 2, 66123 Saarbruecken, Germany
{Jana-Rebecca.Rehse,Peter.Fettke,Peter.Loos}@iwi.dfki.de

Abstract. In this paper, we present the epistomological problem of induction, illustrated by the metaphor of the black swan, and its relevance for Process Mining. The quality of mined models is typically measured in terms of four dimensions, namely fitness, precision, simplicity, and generalization. Both precision and generalization rely on the definition of "unobserved behavior", i.e. traces not contained in the log. This paper is intended to analyze the influence of unobserved behavior, the potential black swan, has on the quality of mined models. We conduct an empirical analysis to investigate the relation between a system, its observed and unobserved behavior and the mined models. The results show that the unobserved behavior, mainly determined by the nature of the unknown system, can have a significant impact on the quality assessment of mined models, hence eliciting the need to explicate and discuss the assumptions underlying the notions of unobserved behavior in more depth.

Keywords: Process mining · Process discovery · Evaluation metrics
Process model quality

1 Introduction

The goal of process discovery is to describe the behavior of an information system in form of a business process model, based on the observed behavior of said information system, represented as an event log [1]. Traditionally, the quality of process discovery algorithms is assessed in terms of four dimensions [1,2]. *Fitness* measures how much of the observed behavior is captured by the model. *Simplicity* measures the complexity of the model. *Generalization* measures how well the model explains unobserved behavior. *Precision* measures how much unobserved behavior exists in the model.

Several metrics exist for each dimension, quantifying the quality of a model with respect to a given log. Process discovery is targeted towards achieving a

© Springer International Publishing AG 2018
E. Teniente and M. Weidlich (Eds.): BPM 2017 Workshops, LNBIP 308, pp. 256–268, 2018.
https://doi.org/10.1007/978-3-319-74030-0_19

satisfactory balance among them [2]. A model should be able to reproduce the behavior in the log, resulting in a high fitness, while being as simple as possible. Also, a model should be precise, i.e. not allowing for behavior greatly different from the log, but also generalizing, i.e. not only allowing for the behavior seen in the log.

Current quality measures typically evaluate the discovered model against the given log. However, an event log is just a limited sample of the behavior of an unknown process system [3]. Since the system is unknown, its behavior is usually divided into "observed behavior", i.e. behavior contained in the given log, and "unobserved behavior", i.e. behavior that is *not* contained in the given log. Both notions are necessary for the quality dimensions above, however, since the unobserved behavior is by definition unknown, it is epistemologically problematic to make universal statements about it without explicating further assumptions. This relates to the epistemological problem of induction, illustrated by the metaphor of the black swan.

The objective of this paper is to demonstrate the relevance the problem of induction has for Process Mining by means of an experiment, analyzing the potential impact unobserved behavior has on the measured quality of mined models. Therefore, we introduce the metaphor of the black swan in Sect. 2. Definitions and preliminary considerations are given in Sect. 3. In Sect. 4, we present our experiment and discuss its impact in Sect. 5. We report on Related Work in Sect. 6, before concluding the paper in Sect. 7.

2   The Black Swan - A Metaphor

In logic, an inference allows the derivation of new statements, called conclusions, from existing ones, called premises. We differentiate deductive inferences from inductive ones. A deductive inference concludes a singular statement from a universal one, whereas an inductive inferences concludes a universal statement from one or several singular ones. From a philosophical point of view, inductive inferences are highly problematic. No matter how many singular statements are considered, any conclusion drawn in this manner may turn out to be false [4].

The "Black Swan" is a metaphor introduced by philosopher Karl Popper to illustrate the difficulties arising with inductive inferencing. In his *Logic of Scientific Discovery*, he states that "no matter how many instances of white swans we may have observed, this does not justify the conclusion that all swans are white" [4]. Independent from the observed number of white swans, the next swan can always be black. The question under which circumstances inductive inferences are justified is what Popper calls the "problem of induction" [4].

This problem also applies to empirical sciences, i.e. those disciplines that try to formulate theories and hypotheses based on experience and observation [4]. Process mining in general and process discovery in particular can be considered empirical science disciplines. Their objective is to gather empirical knowledge in the form of an event log and use it to draw conclusions about the nature of the unknown system underlying this event log. Discovering a model may be interpreted as stating a hypothesis about the design of said system. Assessing its quality then corresponds to testing the validity of the hypothesis.

Given this interpretation of process discovery, each observed trace corresponds to a singular statement, while the discovered model is a universal statement intended to conform with as many of the singular statements as possible. The unobserved behavior is the potential Black Swan. No matter which behavior has been observed so far, neither can it be justified that this is all the observable behavior, nor that the unobserved behavior will follow the same patterns.

So far, this aspect is mainly implicitly addressed by the four dimensions used for assessing the quality of discovered models. In their informal definition above, both precision and generalization rely on the notion of "unobserved behavior". Addressing generalization is particularly complicated, since typically no assumptions can be made as to the nature of the unobserved behavior. However, such assumptions are necessary to make justifiable statements regarding a model's ability to generalize, since different assumptions may lead to completely different understandings of the unobserved behavior and a high generalization depends on correctly assessing this behavior.

3   On Logs, Models, and Systems

3.1   Definitions

Our experiments rely on several definitions and considerations. According to [3], we define traces, logs, models, and systems as follows.

Definition 1 (Traces [3]). *Let $\mathcal{A}$ be the activity alphabet. A trace $\sigma \in \mathcal{A}$ is a finite sequence of activities, where $\mathcal{A}^*$ the set of all finite sequences over $\mathcal{A}$, i.e. the universe of traces.*

Definition 2 (Event Log [3]). *Let $\mathcal{A}^*$ be the universe of traces. An event log $L \in \mathbb{B}(\mathcal{A}^*)$ is a multiset of traces, where $\mathbb{B}(\mathcal{A}^*)$ is the set of all multisets of $\mathcal{A}^*$.*

Definition 3 (Model and System [3]). *Let $\mathcal{A}^*$ be the universe of traces. A model $M \in \mathbb{P}(\mathcal{A}^*)$ and a system $S \in \mathbb{P}(\mathcal{A}^*)$ are subsets of the universe of traces, where $\mathbb{P}(\mathcal{A}^*)$ is the power set of $\mathcal{A}^*$.*

3.2   Observed and Unobserved Behavior

Definition 4 (Process Discovery). *Let L be an event log. A process discovery technique is a function $D : \mathbb{B}(\mathcal{A}^*) \to \mathbb{P}(\mathcal{A}^*)$ which assigns a discovered model $M \in \mathbb{P}(\mathcal{A}^*)$ to a given event log $L \in \mathbb{B}(\mathcal{A}^*)$.*

In a typical process mining setting, the system is unknown. The objective of process discovery is to find a model, which is as close to the system as possible. The input for this task is a log, i.e. a multiset of traces that is generated by the system. From an epistemological point of view, the log can be seen as empirical evidence, gathered by means of observing the running system. This means that the system behavior S is divided into two subsets, determined by the point of observation t^*. All system behavior before t^* is potentially observable behavior. This behavior again consist of two subsets. Observed behavior is the system behavior that was witnessed by the observer, i.e. the behavior contained in the log. Unobserved behavior is the system behavior that was not observed, i.e. behavior that happened before t^*, but that is not included in the log. All system behavior after t^* lies in the future and is by definition unobserved. These relations are illustrated in Fig. 1.

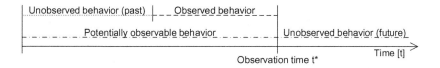

Fig. 1. Temporal characterization of observed behavior

3.3   Perspectives on Measuring Model Quality

Fitness, precision, and generalization measures all relate a model M to a log L and an underlying system S. According to the above definitions, S, M, L are all subsets of the same trace universe $\mathcal{A}^*$. Their relations are shown as a Venn diagram in Fig. 2 [5]. Focusing on different interjections in this diagram leads to three different perspectives on measuring the quality of a mined model, each with a different focus and objective.

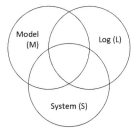

Fig. 2. Perspectives on process behavior [5]

The *Model-Log Perspective* relates the model with the log without considering the system. As systems are typically unknown when discovering a process, this perspective is usually chosen when assessing the quality of discovery approaches [3]. The *Model-System Perspective* inspects the relation between the model and the system, i.e. the model's ability to replicate the unknown system behavior. This is especially relevant when measuring generalization, as this dimension measures a model's ability to reproduce behavior outside of the considered log. The *Log-System Perspective*, relating the log to the generating system, is often not considered in Process Mining, because it does not directly relate to the process model [5]. However, this perspective plays an important role in process discovery, as the relationship between log and system has a direct influence on the ability of a mined model to replicate the system.

4   Demonstrating the Influence of Unobserved Behavior

4.1   Experimental Outline and Goals

The objective of this paper is to analyze the influence unobserved behavior can have on the quality assessment of discovered models. Therefore, we perform an experiment comparing the Log-System Perspective with the Model-Log Perspective. Normally, the logs used for process discovery are neither complete nor accurate. Hence, dealing with the influence of unobserved behavior is a common occurrence in process discovery. Depending on the size of the log and the frequency distribution among the traces, the discovered model may neglect or wrongly emphasize certain system behavior.

To analyze this effect, eight systems over the same activity universe are chosen and different logs are generated for each of them. Each log L is then randomly divided into 10 samples of equal size, $L_0, ..., L_9$. For each sample L_i, we discover a model M_i. These sample-based models are compared against the system S, considering the complete log L. To estimate whether M_i is a good approximation of S, we measure the differences in fitness, precision, and generalization:

$$\Delta F(S, M_i, L) = F(S, L) - F(M_i, L) \tag{1}$$

$$\Delta P(S, M_i, L) = P(S, L) - P(M_i, L) \tag{2}$$

$$\Delta G(S, M_i, L) = G(S, L) - G(M_i, L) \tag{3}$$

If $\Delta F, \Delta P, \Delta G$ are positive, the model M_i underestimates the respective value of S. If they are negative, M_i overestimates S. The smaller the values are, the better model M_i estimates S and the smaller is the influence of unobserved behavior. Please note that we do not evaluate the fourth dimension, simplicity, as it does not directly relate system and log.

4.2   Set-Up

The experiment was designed according to the evaluation framework for comparing process mining algorithms described in [6]. The CoBeFra Benchmarking Framework for Conformance Checking [7] was used for all evaluations.

1. Choose systems: Eight systems, representing different semantics over the same activity universe and shown in Fig. 3 [1], are the basis of our experiment. The models were specifically designed to illustrate the impact of fitness, precision, and generalization, which is why we adopted them here.
2. Generate logs and log samples: For each system, we generate nine logs, varying in size (100, 200, 500 traces) and frequency distribution (uniform, binomial, poisson distribution). Each system is separately used as input for the log generator [8]. Wherever possible, a complete log is generated, otherwise a limit is set on the maximum number of loop iterations. If the possible behavior exceeds the intended log size, traces are picked randomly. Hence, each system

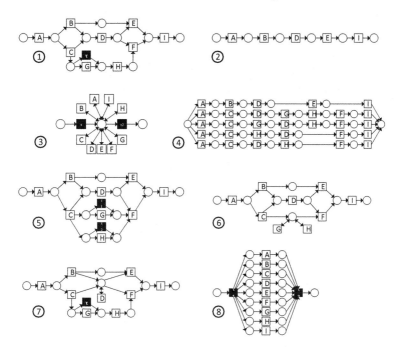

Fig. 3. Systems used in the experiments [1]

log contains only behavior that can be executed on that system, but some behavior is contained in multiple logs. Frequencies are assigned by computing a pre-defined distribution for the intended log size and assigning them randomly. Finally, each generated log is randomly divided into 10 samples of equal size, resulting in 792 different logs.

3. Discover models: Models are discovered for each sample. To mitigate side effects of individual algorithms, three different miners are used (Heuristics Miner [9], Inductive Miner [10], ILP [2,11]), resulting in 2,376 total models.

4. Evaluate systems and models: The discovered models are evaluated by comparing the differences in fitness, precision, and generalization between the models and the systems, with regard to the complete log. To mitigate side effects of individual quality metrics, multiple metrics are used. Fitness is measured using Alignment-based Fitness [12], Negative event recall [13], Token-based Fitness [14]. Precision measures are Alignment-based Precision [12], Best-align-etc Precision [15], Negative event Precision [13], One-align-etc Precision [15]. Alignment-based Generalization [12], Negative event Generalization [13] are used to assess generalization.

4.3 Results

The experiment results are shown in Figs. 4, 5 and 6. Figure 4 shows in the top left corner the distribution of ΔF, ΔP, ΔG over all systems. The remaining plots

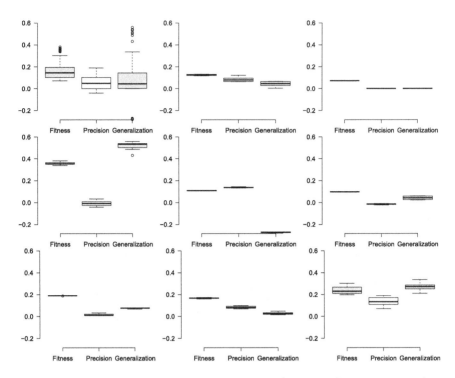

Fig. 4. Distribution of $\Delta F, \Delta P, \Delta G$ for all systems (upper left) and system 1 (upper center) to system 8 (lower right)

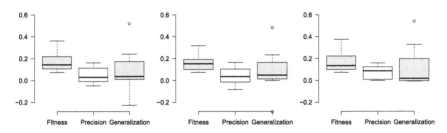

Fig. 5. Distribution of $\Delta F, \Delta P, \Delta G$ for Binomial (left), Poisson (center), Uniform (right) frequency distribution

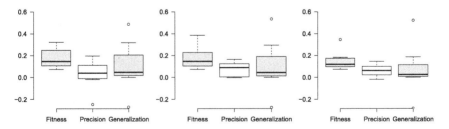

Fig. 6. Distribution of $\Delta F, \Delta P, \Delta G$ for log size 100 (left), 200 (center), 500 (right)

refer to each system separately, from S1 in the top center to S8 in the bottom right, going from left to right and top to bottom. The distributions of $\Delta F, \Delta P, \Delta G$ for the different frequency distributions are shown in Fig. 5 and Fig. 6 shows how $\Delta F, \Delta P, \Delta G$ are distributed for the different log sizes.

Inspecting the distribution of $\Delta F, \Delta P, \Delta G$ across all systems, the values span a large range, but are fairly similar within a system. This range is therefore caused by the differences between the systems. This also explains the outliers, particularly for the generalization dimension. The system is the main discriminator regarding $\Delta F, \Delta P, \Delta G$, as neither the log sizes nor the frequency distributions exhibit large deviations. Size 500 covering a slightly smaller range is due to the larger samples providing a better estimation of system behavior. The different frequency distributions in the logs have virtually no effect on the measures. This is remarkable, as the frequency assignment of the binomial and poisson distributions caused many traces originally contained in the log to be left out. Apparently, the remaining traces are frequent enough to represent typical process behavior in each sample. The uniform distribution with the most traces does not yield better quality measures, probably caused by the fairly high amount of singular traces, which could be considered as noise.

As shown in Fig. 5, the distribution of $\Delta F, \Delta P, \Delta G$ differs by system characteristics. All values are considerably different from 0, which would indicate a perfect system approximation. The only exception is S2, but since S2 contains only one trace, this is not surprising. In general, we see that the more behavior a system allows, the larger $\Delta F, \Delta P, \Delta G$ are. Very generalizing systems (S3 and S8) cause a high discrepancy in both fitness and generalization, as none of the logs can be considered a representative sample of system behavior. S4 does not generalize at all, causing a high overestimation by the discovered models. Even for small systems like S1 or S5, the values are fairly high, indicating a considerable deviation between the discovered model and the underlying system.

5   Analyzing the Impact of Unobserved Behavior

The experiment in Sect. 4 demonstrates that unobserved behavior can influence the quality of discovered models. Thus, the problem of induction, as described in Sect. 2 also applies to process discovery as an empirical research discipline. This does not mean that process discovery as a whole is meaningless. Its practical applicability and usefulness, as demonstrated by numerous projects and contributions, is beyond all question. However, with this paper, we want to establish the awareness among researchers and practitioners that not only does this problem exist, but it also has an impact on their everyday work.

As the previous sections have shown, a model's quality mainly depends on the (unknown) system character. Hence, assumptions on the unobserved are necessary for correctly interpreting a quality measure. In the following, we explicate two possible assumptions, relating the unobserved (past) behavior with the unobservable (future) behavior. The first dimension concerns model continuity. We can assume a system to be *continuous*, i.e. expect the unobserved behavior of the future to repeat the unobserved behavior of the past, or *non-continuous*, i.e. expect the future behavior to differ from the past. Second, we distinguish between *descriptive* models intended to represent the past and *prescriptive* models intended to draw conclusions about the future. Combining these two assumptions leads to four different scenarios, as outlined in Fig. 7.

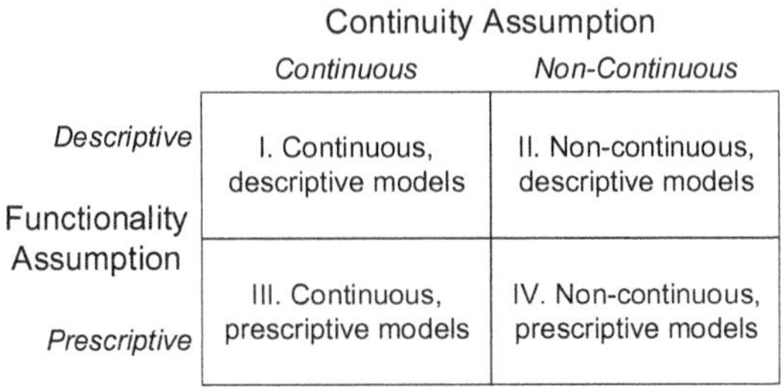

	Continuity Assumption	
	Continuous	*Non-Continuous*
Descriptive	I. Continuous, descriptive models	II. Non-continuous, descriptive models
Prescriptive	III. Continuous, prescriptive models	IV. Non-continuous, prescriptive models

Functionality Assumption

Fig. 7. Four different models types for quality assessment

Models of **Type I** are assumed to be continuous and descriptive. The future behavior is not expected to deviate from the past, while the model is primarily used to describe the observed behavior. These models are retrospective, focusing solely on the known past. Assessing their quality should be focused on high fitness and precision values. Generalization can be neglected, as the unobserved behavior is not considered relevant to the model quality. In such a case, an enumerating model such as S4 is seen as ideal.

Models of **Type II** are assumed to be non-continuous and descriptive. While the model is focused on representing the past, the future behavior is expected to contain new patterns. This leads to a contradiction; should the unobserved behavior be considered in the model or not? If it is considered, the model cannot claim to be strictly descriptive anymore, as it includes behavior from an assumed future. However, if it is not considered, the model deliberately excludes behavior that is assumed to be possible, purposely reducing the model quality.

Models of **Type III** are assumed to be continuous and prescriptive. The model will be used to reason about a future, which is not expected to considerably deviate from the past. Observed behavior equates to unobserved, making it easy to reason about the latter based on the former. With regard to quality assessment, high fitness and precision and medium generalization values should be pursued to achieve a balanced model with a limited degree of generalization.

Models of **Type IV** are assumed to be non-continuous and prescriptive. The future behavior, which the model reasons about, will contain unseen patterns. This implies drawing conclusions about unknown model behavior without any additional knowledge. We must expect to see all different kinds of behavior, eliciting a need for maximum generalization. Fitness and precision are less relevant, as the unobserved behavior can be expected to be large.

All four types are generally possible, however, the continuity assumptions necessitates strong evidence from the system context in order to be valid. Continuity might be given in a highly automated process, but the more degrees of freedom a process allows, the less realistic this is. In any application of process discovery, explicating these assumptions will simplify correctly interpreting the resulting model.

6 Related Work

With many technical issues in process discovery now solved, new discovery techniques supposed to outperform existing ones rely on commonly agreed and scientifically sound instruments for comparing their performance. A first attempt towards the development of comparative frameworks has recently been suggested [7]. It includes a variety of previously suggested evaluation metrics, such as the ones used in this paper.

The need for an explicit differentiation between system, log, and model in process discovery is originally discussed in [5]. The author considers the system as either a concrete information system implementation, or, more likely, as the context of a process, i.e. its organization, rules, economy, etc. Such a definition is rather intangible and does not allow for measuring concrete process behavior. The author introduces and shortly discusses the notion of an "unknown system", but neglects the Log-System perspective. To analyze the system in a more formal way, our paper follows [3], defining a system as a set of traces.

The strict distinction between system, log, and model in process discovery has not yet been widely adopted. In [16], the author states that "both process discovery and conformance checking aim to tell something about the unknown real process rather than the example traces in the event log". As pointed out in [3], this suggests that "the one and only goal of process discovery would be to represent the true underlying process". Typically used quality metrics, however, do not take on the Log-System perspective, but evaluate the discovered model with regard to the event log. Little empirical work exists on the impact of this assumption. Rogge-Solte et al. illustrate the difficulties of the Model-Log perspective in their Generalized Conformance Checking Framework [17]. Janssenswillen et al. provide a first analysis of the Log-System perspective by measuring the effect that noise and incompleteness of event logs may have on discovered models [3]. However, their main objective is to analyze the dependability of certain fitness and precision metrics in terms of bias and ranking, whereas our contribution is focused on the influence the unobserved behavior has on the generic quality assessment.

7 Discussion and Conclusion

In this paper, we present the well-known epistemological problem of induction [4] and discuss its consequences for the field of Process Mining. By means of an empirical analysis, we demonstrate that the quality of discovered models can be significantly influenced by unobserved system behavior. These findings elicit a need for Process Miners to not only explicate the assumptions they make on the nature of their underlying system, but also to develop these assumptions on a more detailed and context-driven level. They are closely related to the choice of log as well as the intended objective of the task at hand. Altogether, these factors can have a significant impact on the success of any Process Mining project.

Our work is not intended as a full analysis or solution. Instead, we want to raise awareness on how existing epistemological discussions may relate to Process Mining and spark a discussion on how they influence the day-to-day work of researchers and professionals. Our experiment is intended to demonstrate potential influences under realistic circumstances. Therefore, we need to make several assumptions to its design. The systems we use for our experiment are small and artificial. They are able to demonstrate the potential effects of unobserved behavior, however, in how far the results are generalizable remains to be seen. Also, although we try to mitigate side effects of individual miners or quality measures by working with average values, they cannot be completely eliminated.

We also do not consider log completeness, but focus on size and frequency distribution instead. This is done for matters of comparability; the effect of unobserved behavior is influenced by the amount of unobserved and unobservable behavior, which depends on the system itself. Simplicity was not included in our experiment, since it does not directly relate a model to a log. Another shortcoming is that we do not consider the presence of noise in neither our experiment nor the framework in Sect. 5. Noise can play an important role, as it may influence both the discovered model and its quality assessment. We are aware of these influences, however, we decided to focus on the system perspective here.

This contribution can be the starting point for a number of future activities. The influence of log completeness and noise should be investigated, using larger and more realistic systems. The framework in Sect. 5 should be elaborated and enriched with additional dimensions. Its validity could also be investigated by means of a larger case study, where we want to demonstrate the differences between the model types. There is also a need to deepen the epistemological discussion, and also take log choices and goals of Process Mining into account.

One can also argue that the metaphor of the Black Swan (and thus the necessity to falsify a hypothesis) is of little significance to the reality of Process Mining, as it can also be interpreted as positivist research, eliciting a need to verify a theory based on empirical data. These differences, although subtle, are highly relevant when it comes to the interpretation of scientific results, including Process Mining as a discipline. As this is ultimately a philosophical and epistemological discussion, we are not favoring one interpretation over the other. However, we want to encourage researchers and professionals to think about the implications of this discussion on their everyday work.

References

1. van Dongen, B.F., Carmona, J., Chatain, T.: A unified approach for measuring precision and generalization based on anti-alignments. In: La Rosa, M., Loos, P., Pastor, O. (eds.) BPM 2016. LNCS, vol. 9850, pp. 39–56. Springer, Cham (2016). https://doi.org/10.1007/978-3-319-45348-4_3

2. Buijs, J., van Dongen, B., van der Aalst, W.M.P.: Quality dimensions in process discovery: the importance of fitness, precision, generalization and simplicity. Int. J. Coop. Inf. Syst. **23**(01), 1440001 (2014)

3. Janssenswillen, G., Jouck, T., Creemers, M., Depaire, B.: Measuring the quality of models with respect to the underlying system: an empirical study. In: La Rosa, M., Loos, P., Pastor, O. (eds.) BPM 2016. LNCS, vol. 9850, pp. 73–89. Springer, Cham (2016). https://doi.org/10.1007/978-3-319-45348-4_5

4. Popper, K.: The Logic of Scientific Discovery. Routledge, London (2005)

5. Buijs, J.: Flexible evolutionary algorithms for mining structured process models. Ph.D. thesis, Technische Universiteit Eindhoven (2014)

6. Weber, P., Bordbar, B., Tiňo, P., Majeed, B.: A framework for comparing process mining algorithms. In: GCC Conference and Exhibition (GCC), pp. 625–628. IEEE (2011)

7. vanden Broucke, S., De Weerdt, J., Vanthienen, J., Baesens, B.: A comprehensive benchmarking framework (coBeFra) for conformance analysis between procedural process models and event logs in prom. In: Proceedings of the IEEE Symposium on Computational Intelligence and Data Mining (CIDM 2013), pp. 254–261. IEEE (2013)

8. vanden Broucke, S., De Weerdt, J., Vanthienen, J., Baesens, B.: An improved process event log artificial negative event generator. Feb research report, KU Leuven - Faculty of Economics and Business, Leuven, Belgium (2012)

9. Weijters, A., van Der Aalst, W.M.P., De Medeiros, A.: Process mining with the heuristics miner-algorithm. Technical Report 166, TU Eindhoven (2006)

10. Leemans, S.J.J., Fahland, D., van der Aalst, W.M.P.: Discovering block-structured process models from event logs containing infrequent behaviour. In: Lohmann, N., Song, M., Wohed, P. (eds.) BPM 2013. LNBIP, vol. 171, pp. 66–78. Springer, Cham (2014). https://doi.org/10.1007/978-3-319-06257-0_6

11. van der Werf, J.M.E.M., van Dongen, B.F., Hurkens, C.A.J., Serebrenik, A.: Process discovery using integer linear programming. In: van Hee, K.M., Valk, R. (eds.) PETRI NETS 2008. LNCS, vol. 5062, pp. 368–387. Springer, Heidelberg (2008). https://doi.org/10.1007/978-3-540-68746-7_24

12. van der Aalst, W.M.P., Adriansyah, A., van Dongen, B.: Replaying history on process models for conformance checking and performance analysis. Wiley Interdisc. Rev. Data Min. Knowl. Discov. **2**(2), 182–192 (2012)

13. vanden Broucke, S., De Weerdt, J., Vanthienen, J., Baesens, B.: Determining process model precision and generalization with weighted artificial negative events. IEEE Trans. Knowl. Data Eng. **26**(8), 1877–1889 (2014)

14. Rozinat, A., van der Aalst, W.M.P.: Conformance checking of processes based on monitoring real behavior. Inf. Syst. **33**(1), 64–95 (2008)

15. Adriansyah, A., Munoz-Gama, J., Carmona, J., van Dongen, B.F., van der Aalst, W.M.P.: Alignment based precision checking. In: La Rosa, M., Soffer, P. (eds.) BPM 2012. LNBIP, vol. 132, pp. 137–149. Springer, Heidelberg (2013). https://doi.org/10.1007/978-3-642-36285-9_15

16. van der Aalst, W.M.P: Mediating between modeled and observed behavior: the quest for the "right" process. In: IEEE International Conference on Research Challenges in Information Science (RCIS 2013), pp. 31–43 (2013)
17. Rogge-Solti, A., Senderovich, A., Weidlich, M., Mendling, J., Gal, A.: In log and model we trust? A generalized conformance checking framework. In: La Rosa, M., Loos, P., Pastor, O. (eds.) BPM 2016. LNCS, vol. 9850, pp. 179–196. Springer, Cham (2016). https://doi.org/10.1007/978-3-319-45348-4_11

1st International Workshop on Business Processes Meet Internet-of-Things (BP-Meet-IoT 2017)

1st International Workshop on Business Processes Meet Internet-of-Things (BP-Meet-IoT 2017)

Agnes Koschmider[1], Massimo Mecella[2], Estefania Serral Asensio[3],
and Victoria Torres[4]

[1] Karlsruhe Institute of Technology, Karlsruhe, Germany
agnes.koschmider@kit.edu
[2] Sapienza Università di Roma, Rome, Italy
mecella@diag.uniroma1.it
[3] Katholieke Universiteit Leuven, Leuven, Belgium
estefania.serralasensio@kuleuven.be
[4] Universitat Politecnica de Valencia, Valencia, Spain
vtorres@pros.upv.es

Business Process Management (BPM) is the body of methods, techniques and tools to manage the processes of an organization (i.e., the chains of events, activities, and decisions performed in a coordinated manner) in order to achieve (business) goals. Ubiquitous Computing (Ubicomp) enables computing to appear anytime and everywhere, becoming "invisibly" embedded in physical objects to sense and respond to their surrounding environment. This embedding allows building a bridge between the digital and the physical worlds. Ubicomp, heavily based nowadays on Internet-of-Things (IoT) technology, is revolutionizing many areas, including real processes in cyber-physical domains. But the two areas, which are undoubtedly strictly related and with reciprocal influences, have not yet meet significantly in the research and practice.

In the traditional sense, processes represent a specific ordering of activities across time and place to serve a business goal. So far, the predominant paradigm to design (business) processes has been based on the Model-Enact paradigm, i.e., the process has been depicted as a (graphical) process model that then could be executed by a Business Process Management System (BPMS). This largely follows a top-down approach. With the emergence of IoT, the Discover-Predict paradigm becomes an alternative one. The Discover-Predict paradigm is characterized as a bottom-up approach where data is generated from physical devices sensing their environment and producing events. These events are then correlated to detect (complex) higher-level events and to discover patterns. Respective higher-level events or patterns can then be used as input for process mining algorithms. An IoT-aware Discover-Predict paradigm has plenty of potentials. However, also several challenges must be meet.

Research is necessary in addressing questions such as:

- How does the central role of communication in IoT fit with the control-flow centric view of most BPM approaches?
- How to exploit IoT within BPM and vice versa?

- How to bridge the abstraction gap between low-level (sensor data) and high-level events?
- How to integrate complex-event processing technologies into BPM?

BP-MEET-IoT 2017, the 1st International Workshop on Business Processes Meet Internet-of-Things, has been held on September 11, 2017 in Barcelona, Spain, in conjunction with the 15th International Conference on Business Process Management (BPM 2017), and consisted of four papers, a keynote talk and a panel. The objective of the workshop has been to attract novel research at the intersection of the above mentioned areas, by bringing together practitioners and researchers from both communities that are interested in making IoT-based ubiquitous business processes a reality.

Those four accepted papers, namely *(i)* "Technology-Enhanced Process Elicitation of Worker Activities in Manufacturing" by Knoch et al., *(ii)* "Discovering process models of activities of daily living from sensors" by Cameranesi et al., *(iii)* "An Habit is a Process: a BPM-based Approach for Smart Spaces" by Sora et al., and *(iv)* "From BPM to IoT" by Cherrier et al., out of several submissions, exactly provide a good overview of the above mentioned topics. Each paper was reviewed by three members of the internationally renowned Program Committee.

The program was completed by a keynote talk by Barbara Weber (Technical University of Denmark) on "BPM and IoT: a Dream Team?", and by a closing panel on "BP meet IoT : new wine in old bottles?", moderated by Massimo Mecella (Sapienza Università di Roma, Italy), and having as panelist Barbara Weber (Technical University of Denmark), Massimiliano de Leoni (Eindhoven University of Technology, The Netherlands), Marcello La Rosa (Queensland University of Technology, Australia), Victoria Torres (Universitat Politecnica de Valencia, Spain), and Estefania Serral Asensio (Katholieke Universiteit Leuven, Belgium).

In addition, the organizers of the workshop, together with the growing up BPM-IoT community, announced the recently published manifesto titled "The Internet-of-Things Meets Business Process Management: Mutual Benefits and Challenges", cf. https://arxiv.org/abs/1709.03628.

The organizers would like to thank all the Program Committee members for their valuable work in selecting the papers, and the BPM 2017 organizing committee for the support in this successful event.

Workshop Chairs

Agnes Koschmider	Karlsruhe Institute of Technology, Germany
Massimo Mecella	Sapienza Università di Roma, Italy
Estefania Serral Asensio	Katholieke Universiteit Leuven, Belgium
Victoria Torres	Universitat Politecnica de Valencia, Spain

Program Committee

Adrian Mos	Xerox, France
Alaaeddine Yousfi	Hasso-Plattner-Institut at the University of Potsdam, Germany
Andrea Delgado	INCO, Universidad de la República, Uruguay
Andrea Marrella	Sapienza Università di Roma, Italy
Andreas Oberweis	Karlsruhe Institute of Technology, Germany
Anne Monceaux	Airbus Group Innovations, France
Antonio Ruiz-Cortés	University of Seville, Spain
Armando Walter Colombo	University of Applied Sciences Emden/Leer, Schneider Electric, Germany
Avigdor Gal	Technion – Israel Institute of Technology, Israel
Barbara Weber	Technical University of Denmark, Denmark
Bart Baesens	KU Leuven
Christian Janiesch	University of Wurzburg, Germany
Ferry Pramudianto	North Carolina State University, USA
Francisco Ruiz	University of Castilla-La Mancha, Spain
Gero Decker	Signavio GmbH, Germany
Hajo Reijers	Eindhoven University of Technology, the Netherlands
Jan Mendling	Vienna University of Economics and Business, Austria
Jianwen Su	University of California at Santa Barbara, USA
Liang Zhang	Fudan University, China
Manfred Reichert	University of Ulm, Germany
Marta Indulska	University of Queensland, Australia
Mathias Weske	Hasso-Plattner-Institut at the University of Potsdam, Germany
Matthias Weidlich	Imperial College London, UK
Pnina Soffer	University of Haifa, Israel
Selmin Nurcan	Universite Paris 1-Pantheon-Sorbonne, France
Sylvain Cherrier	University Marne-la-Vallée, France
Udo Kannengießer	eneon IT-solutions GmbH, Austria
Vicente Pelechano	Universidad Politecnica de Valencia, Spain

Technology-Enhanced Process Elicitation of Worker Activities in Manufacturing

Sönke Knoch[(✉)], Shreeraman Ponpathirkoottam, Peter Fettke, and Peter Loos

German Research Center for Artificial Intelligence (DFKI), Saarland University,
Stuhlsatzenhausweg 3, 66123 Saarbrücken, Germany
Soenke.Knoch@iwi.dfki.de

Abstract. The analysis of manufacturing processes through process mining requires meaningful log data. Regarding worker activities, this data is either sparse or costly to gather. The primary objective of this paper is the implementation and evaluation of a system that detects, monitors and logs such worker activities and generates meaningful event logs. The system is light-weight regarding its setup and convenient for instrumenting assembly workstations in job shop manufacturing for temporary observations. In a study, twelve participants assembled two different product variants in a laboratory setting. The sensor events were compared to video annotations. The optical detection of grasping material by RGB cameras delivered a Median F-score of 0.83. The RGB+D depth camera delivered only a Median F-score of 0.56 due to occlusion. The implemented activity detection proofs the concept of process elicitation and prepares process mining. In future studies we will optimize the sensor setting and focus on anomaly detection.

Keywords: Process elicitation · Activity recognition · Manufacturing

1 Introduction

Workers retain flexibility in semi-automated assembly systems becoming more and more complex. Especially during unforeseen occurrences or incidents a large degree of flexibility during assembly is required, e.g., for small and medium-sized companies [5]. Production planners need accurate information about the assembly process to make realistic assumptions during the plant design or in the operational phase balancing work load between production lines. The manual work in manufacturing processes is analyzed using predetermined motion time systems, such as Methods-Time Measurement (MTM), cf. mtm-international.org, and REFA, cf. refa.de. In MTM, a person documents all motions in assembly tasks under different plant settings and looks up the standard time for relevant motions in the MTM catalogue. In REFA, organization-specific catalogues are created timing each motion with a stop watch.

Process mining from Business Process Management (BPM) bridges the gap between data and process science [1]. Process mining is grounded on meaningful log data. Especially regarding manual activities in BPM this data is sparse.

E. Teniente and M. Weidlich (Eds.): BPM 2017 Workshops, LNBIP 308, pp. 273–284, 2018.
https://doi.org/10.1007/978-3-319-74030-0_20

Thus, we suggest a process elicitation system that tracks and pre-processes manual activities during assembly processes at workstations in job shop manufacturing. Today, the elicitation of worker motions in manufacturing requires an expensive manual procedure. Activity recognition in combination with existing sensors has the potential to increase efficiency and effectiveness during this process by delivering detailed information about manual activities. Sensor data is pre-processed and filtered by applying Complex Event Processing (CEP), efficiently handling huge amounts of heterogeneous data. Operative work steps are integrated into process models in compliance with the standards of formal modeling and enabling process monitoring during execution time. This new knowledge helps to understand how the work plans, process models defining assembly processes created during the product's industrial engineering, are executed on the shop floor. It delivers input for process discovery, conformance checking and enhancement to foster the detection of anomalies and to uncover optimization capabilities.

Following the design-science research approach, a light-weight artifact was developed addressing the problem of process elicitation in manual assembly facing a heterogeneous sensor setting, which has not been addressed by current work that is mainly focused on log data of existing software systems. The design's efficacy was evaluated, which proofs that an easy and fast instrumentation of common assembly workstations can be applied to enable temporary monitoring tasks connecting the business process to an Internet-of-Things in Industrie 4.0 factories. To integrate the suggested sensors with existing software, machines and sensors, an event-driven architecture couples state-of-the-art modules loosely keeping the system extremely flexible. Raw events of multiple sensors are aggregated and combined in event rules to detect patterns using CEP. The system was set up in a laboratory setting and enables activity tracking during the assembly of a realistic artificial product by fusing information from four sensor applications together with events from a worker guidance system and mapping them to basic motions to break down the underlying process activity. In an experiment with 12 participants the optical detection of grasping material was analyzed. Methods and metrics from the activity recognition domain were applied to measure the artifact's performance proving the concept of sensor-driven process elicitation.

2   Problem Statement

In process engines, manual activities are usually black boxes documenting human tasks which are beyond the process engine's reach. A system that couples sensor data with such activities would enable *process monitoring* and the following three types of *process mining*. To position them, we use the *BPM life cycle* by [10] who introduced a new and comprehensive life cycle concept partitioned into the phases (1) strategy development, (2) definition/modeling, (3) implementation, (4) execution, (5) monitoring/controlling, and (6) optimization/improvement.

Process discovery can be applied when no a-priori information is available, typically during work plan creation in the definition and modeling phase (2).

Production engineers and workers test best practices to assemble a new product variant to create the work plan. An instrumented prototypical assembly workstation, cf. Sect. 3, can deliver event logs during that test phase and supports the discovery of detailed work plans/assembly processes. In manufacturing, discovery would be applied during the elicitation of the process while conducting observations, measurements and workshops with workers, team leaders and engineers. Since the main high-level process is already known during the execution phase (4), discovery could reveal hidden sub-processes.

Conformance checking compares an existing work plan/process model against the generated event logs (5). It addresses the detection of anomalies and deviations by matching reality against existing models. In manufacturing, the detection of assembly faults is one of the major applications. Assembly faults lead to longer process execution times compared to the scheduled times or reduce the number of accurate pieces compared to the output that was planned.

Process enhancement can be used to improve a work plan/process model (6) using knowledge contained in event logs. If an optimization potential (best practices or anti pattern) is mined, the work plan is adapted. For example, the order or duration of assembly tasks is adjusted. In the worst case, a complete redesign of the process model could become necessary.

To enable these three types of process mining the generation of meaningful log data is essential. Thus, a system is needed that tracks progress of process instances consisting of activity instances that occur during a concrete case. The case id is equivalent to the product id that is read through automatic identification and data capture. If no case id is available activities associated to resources such as assembly workstations, tools, and materials are tracked. During the product assembly a set of `Activities` is performed which are associated with certain `Resources` observed by `Sensors`. Activities can be composed of multiple sub-activities, e.g., start/stop and have additional properties, e.g., duration or performed left/right handed. The sensing of worker activities, its domain specific requirements and the filtering of events with convenient rules enhances the potential of process mining in manufacturing benefitting from new Internet-of-Things infrastructures in Industrie 4.0 factories.

3   Solution Within a Laboratory Setting

Within the BPM life cycle, our system is intended to run during execution (4) as a monitoring application (5), when work plans are executed in assembly lines connecting multiple assembly workstations on the shop floor. Additionally, it can be applied in early prototypes during the design of assembly workstations and work plans (2). During the monitoring of a work plan execution, the assembly of a concrete product, the process instance, is observed. The outcome delivers input for the design, optimization and improvement (6) of assembly workstations, work plans and assembly processes in further life cycle iterations.

The suggested activity detection and event pre-processing is an event log generating software system in process mining, cf. Fig. 1. It observes and senses

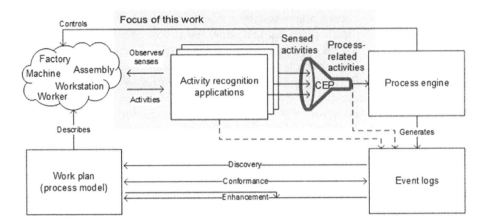

Fig. 1. Activity recognition in the context of process mining (extension of [1]).

worker activities filtered by a CEP engine and generates new event logs. These events can be integrated in business processes executed in an engine running processes defined a-priori and formally described in a standardized representation such as the Business Process Model and Notation (BPMN).

3.1 Requirements and Setup

For the artifact, as stated in [11], a *fast setup* and the use of *light hardware* was one aspect to enable a fast instrumentation of the assembly workstation for temporary tracking. In addition, the worker's *degree of freedom* during work is not restricted, which excludes heavy sensor technology and on-body motion sensors. To address medium-sized and small companies, the equipment *costs should be limited* acquiring and operating such a system. Finally, only tracking technology *appropriate for manufacturing* environments is considered.

Implementing the artifact, these basic aspects were addressed by using affordable and light sensor hardware. These sensors (see below) do not afford any direct contact with the worker (not restricting the degree of freedom) and work appropriately for manufacturing (usually stable light conditions). Additionally, a flexible event-based architecture was chosen, which allows a flexible integration and removal of sensor components depending on the desired tracking case. Regarding the software that was developed to configure the sensor applications a fast setup to run the optical tracking components even by non-experts is supported. A simple user interface allows the drawing, labeling and configuring of activity zones to observe areas in the view of the respective camera, cf. [11].

The assembly workstation is made from carton prototyping material and instrumented with low-cost sensors from the consumer electronics domain: 3 RGB (top: Logitech C920 HD Pro, bottom: 2x Creative Live! Cam Voice), 1 RGB+Depth (Microsoft Kinect 2), 1 infrared (Leap Motion) cameras and

2 ultrasonic (GHI HC-SR04) sensors (cf. Fig. 2). The event-bus was realized using MQTT. All events are logged to a database for process elicitation and mining.

In the current setting two product variants can be assembled: The bill of material of variant "BG" and "BCD" consists of seven and six different materials, see table in Fig. 2. Three materials are variant-specific resulting in a total number of ten materials available in small load carriers (SLC) at the assembly workstation. Both products consist of two 3D-printed cases filled with three interconnected printed circuit boards fixed with screws while the number of screws varies per variant ("BG": 8, "BCD": 4). The only tool available during assembly is a manual torsional screw driver. To assemble a product, the workpiece holder in front of the worker is fed with the variant-specific Top_Casing part (Task 1). Next, the Mainboard is inserted into the Top_Casing (2) and fixed with two Small_Screws (3). Afterwards, the Application_Board is inserted (4) and the Connecting_Board that has to be connected to Mainboard and Application_Board (5). Once that is finished, the Application_Board is fixed with two Small_Screws (6). Finally, the Bottom_Casing is fastened together with the Top_Casing with clips (7) or four Big_Screws in case of variant BG (7+8). The products are artificial, meaning that they have no purpose or function.

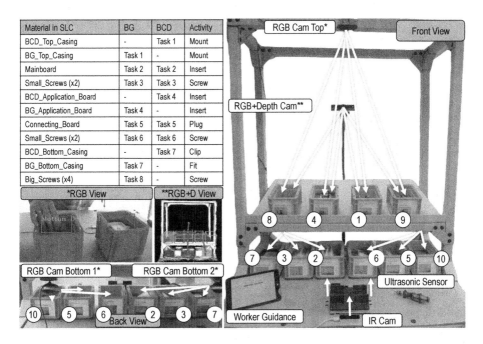

Material in SLC	BG	BCD	Activity
BCD_Top_Casing	-	Task 1	Mount
BG_Top_Casing	Task 1	-	Mount
Mainboard	Task 2	Task 2	Insert
Small_Screws (x2)	Task 3	Task 3	Screw
BCD_Application_Board	-	Task 4	Insert
BG_Application_Board	Task 4	-	Insert
Connecting_Board	Task 5	Task 5	Plug
Small_Screws (x2)	Task 6	Task 6	Screw
BCD_Bottom_Casing	-	Task 7	Clip
BG_Bottom_Casing	Task 7	-	Fit
Big_Screws (x4)	Task 8	-	Screw

Fig. 2. The instrumented assembly workstation made from carton prototyping material in front and back view. Arrows indicate the sensor perspectives. Camera perspectives are shown on the mid left. The numbers from 1 to 10 refer to the IDs of Table 1. On the top left, the assembly tasks of both product variants are provided.

3.2   Work Step Events: Composition, Format and Pattern Detection

Work Step Composition. A work step is split up into five fundamental motions that were adapted from the MTM used to analyze the performance of manual operations. MTM is based on empirically gathered data aggregated in time catalogues that focus on activities which are 100% influenceable. The motion cycle covers the five motions *reach* (move an empty hand to a thing), *grasp* (bring a thing under control), *move* (move a thing by hand), *position* (fit things into each other), and *release* (intended loose control of a thing).

Message Format. The event generating sensor applications are identified by topics that clients can subscribe to communicating over a publish/subscribe protocol, such as MQTT. Topics are simple text strings hierarchically structured with forward slashes, e.g., `Sensor/Resource/Activity`. The `Sensor` contains the name of the data delivering sensor application, e.g., RGB+D, and the `Resource` denotes the observed area at the assembly workstation containing a tool or material, e.g., screws. The `Activity` indicates the actual action, e.g., In or Out, if someone reached in or out of the SLC with screws. This allows a device independent subscription to sensor events using the single- or multi-level wildcard character $(+/\#)$, e.g., *subscribeTo(`#/Screws`)*. The event's payload is en-/decoded to/from an internal format within the processing engine.

Event Pattern. Two types of rules were applied to match event patterns: patterns consisting of bounding events and patterns consisting of multiple events from the same type. For the first, a rule waits for the start and end event limiting, e.g. the grasping of the material screws. It checks if two events `e1` and `e2` arrive in a successive order without any event `e3` between and within a time window of carefully defined `T` seconds (20 in our current setting). The exclusion of `e3`, having `e3.topic=e2.topic`, ensures accurate detection when the worker grasps twice successively in the same SLC within `T`. The positioning of a material in the assembly area represents the second type of patterns, since its boundaries are not strict. An aggregation query processes a series of events and fires every time an additional event from the same type occurs. Once a pattern has been detected, each query calculates the time difference between relevant events and sends a high-level event about the detected operation to the event bus.

3.3   Process Elicitation Through Activity Detection

Four different sensor applications were implemented. The *Local Activity Detection* with RGB web cameras detects activities within activity zones defined by the user. The competing *Local Activity Detection and Skeleton Tracking* with RGB+D camera tracks the skeleton and activities in three-dimensional rectangular areas wrapping SLCs on a digital representation of the assembly workstation. It delivers information about the body posture, location and the hand that is used. The *Hand Detection* with IR camera senses whether a hand entered or left an a-priori defined area. The *Motion Detection* recognizes motions based on an ultrasonic distance measurement, similar to a light barrier. In the following, we will focus on the Local Activity modules optically sensing material grasps.

Local Activity Detection with RGB Cameras. Activity detection can be achieved by performing background subtraction and foreground extraction using RGB web cameras. Therefore, the cameras are arranged so that their field of view observes the SLCs. Usually, there is minimal/no motion in the observation regions other than motion of hands during the assembly process. However, the system is highly sensitive to changing background intensity and effects of unavoidable artifacts like shadows and repeated adding/removing of objects (material). The system should be robust against such effects and able to adapt continuously and learn the presence or absence of objects in the scene as a change in the background and not indicate activities in these regions. Here we adopt a time tested method involving fast adapting mixture of Gaussians by [14]. The values that each pixel can take is modelled as a mixture of Gaussians. An advantage of the method is that it chooses the number of Gaussians for each pixel automatically and independently of other pixels not wasting memory as in the case of a pre-defined number of Gaussians. For each pixel, weights are assigned to the Gaussians modelling it. In a particular pixel, if the weights are concentrated in a few Gaussians, a stable state is achieved thereby indicating inactivity. If the weights are distributed over a lot of Gaussians this indicates activity in the specific pixel. The parameters of the algorithm are adjusted in a way that the Gaussians modelling the background adapt quickly enough to ignore the added/removed tool but slowly enough to detect the movement of the hand in and out of the activity regions. The activity regions are rectangles defined by the user. If more than half the pixels in this region has been identified as foreground by the algorithm, an activity is detected.

Local Activity Detection with RGB+D Cameras. The Kinect sensor performs skeleton tracking as well as activity detection by forming a two and a half dimensional (2.5D) model of the assembly workstation. The 2.5D model is obtained by using the Kinect fusion functionality, which is a part of the Kinect SDK. The model is shown from the camera's perspective. Since a user's perspective is more intuitive, during setup Kinect is positioned on the user side in front of the workstation and moved and rotated through 180° around the workstation to its final position. This process is the geometric transformation between the two diagonally opposite positions of the Kinect sensor (in front of the assembly workstation and behind). Once the final position has been fixed, the vectors normal to the worktop surface and two mutually orthogonal vectors along the worktop surface are determined using three user-defined points on the surface in the model. It is assumed that the Kinect's X axis and the table surface are in parallel. Kinect's skeletal tracking provides a total of 25 joints which delivers the necessary data for visualizing and analyzing motions of the worker. The hand joints are used for testing the overlap with activity regions. To allow tolerance to joint position accuracy as given by the Kinect, a sphere of a 25 cm radius (identified empirically) surrounds the hand position. The activity in an activity region (cuboid in 3D space) is triggered when the sphere around a given hand joint overlaps with its cuboid, see Kinect view in Fig. 2. The condition for overlap: the projection of the vector joining the centre of the sphere and the centre of the cuboid along any of the three cuboid dimensions is less than the sum of the radius of the sphere and half of the respective dimension.

4   Evaluation

In the following, we provide insights into a first study analyzing the performance of the optical sensor applications using RGB and RGB+D cameras. The events generated by these applications enable the use of CEP and the elicitation of operative work steps mapped to activities in formal process models.

4.1   Experimental Setup and Data Analysis

The suggested system was evaluated with 12 participants consisting of students and campus staff. To evaluate the system, an artificial but realistic chain of assembly tasks had to be solved, cf. Sect. 3.1. Each participant had to assemble four artificial products, two of each variant. The participants were split into two counter-balanced groups. One group started with the variant "BG", the other with "BCD". The four products were assembled in an alternating order. The instructions about the assembly process were provided on a simple worker guidance system (WGS), a web page shown on an eight inch tablet. Text, images and videos were used to explain the different assembly steps. Difficulties in understanding were answered by a supervisor who was present during the whole experiment. In addition, the WGS provided temporal boundaries for relevant sensor events by delivering an event each time a new assembly step was started. Therefore, the user had to confirm each step which is common practice in industrial manufacturing. Regarding the assembly, apart from a correctly assembled product, no limitations were given which lead to a large degree of freedom making the recognition of activities more realistic while exacerbating it.

To measure the system's precision, the ground truth (GT) was captured filming the assembly process with the consent of the participant. Three persons manually annotated all videos with a given set of tags. One tag represented one GT activity and consisted of the name of the observed material in the SLC and the start and stop time taken from the video. The timestamps of video and sensor events had to be synchronized to allow a comparison of the GT against the system's event log. Each system event between two instructions from the WGS was matched to a corresponding GT entry based on the material referenced in event and GT. This lead to a bipartite graph with two independent sets of vertices (GT entries and system events) such that every edge connects a vertex in the one set (GT) to one in the other (system). To find the best matching system event for a given GT entry, the optimum matching had to be calculated. This represents the assignment problem of finding the minimum weight matching in a weighted bipartite graph. The weight of an edge connecting a vertex representing a GT entry and a vertex representing a system event is the time distance. After construction, we received a simple undirected weighted bipartite graph. JGraphT, cf. jgrapht.org, was used to compute the maximum weight matching in $O(V|E|^2)$, where E is the set of edges and V the set of vertices.

4.2   Results

Table 1 shows the results of the experiment in numbers. For each SLC carrying a certain material the results of the three RGB cameras observing all SLCs and the results of the RGB+D camera observing the four upper SLCs true positives (TP), false negatives (FN) and false positives (FP) in absolute numbers such as precision, recall and the F-score are provided. In the last column the number of events clustered when the material was equal within a time distance of T reducing the number of insertions (FP) are listed. Anomalies (percentage provided in brackets under the respective SLC/material name) represent a deviation from the instructions, such as picking into the wrong SLC or moving the hand over several SLCs while searching the correct material, e.g., when they look alike.

The RGB cam application delivered satisfying results with an F-score higher than 0.8 for seven SLCs and higher than 0.9 for two. Only the observation of the SLC containing the material "BCD_Bottom_Casing" was with F = 0.68 low. Having a look at the physical setup during the experiment as shown in Fig. 2, it can be seen that the angle of the camera to the SLC is adverse.

Table 1. Results of activity detection per SLC containing a certain material.

ID	SLC/material (anomalies)	Sensor	TP	FN	FP	Precision	Recall	F-score	Clustered
1	BCD_Application_Board (50%)	RGB	32	4	10	0.76	0.89	0.82	-
		RGB+D	28	8	40	0.41	0.78	0.54	7
2	BCD_Bottom_Casing (37%)	RGB	26	9	16	0.62	0.74	0.68	-
3	BCD_Top_Casing (39%)	RGB	32	1	6	0.84	0.97	0.90	-
4	BG_Application_Board (54%)	RGB	34	7	2	0.94	0.83	0.88	-
		RGB+D	26	15	23	0.53	0.63	0.58	7
5	BG_Bottom_Casing (35%)	RGB	24	2	9	0.72	0.92	0.81	-
6	BG_Top_Casing (29%)	RGB	27	1	10	0.73	0.96	0.83	-
7	Big_Screws (14%)	RGB	131	40	24	0.85	0.77	0.80	-
8	Connecting_Board (20%)	RGB	46	4	10	0.82	0.92	0.87	-
		RGB+D	30	20	39	0.44	0.6	0.50	8
9	Mainbaord (15%)	RGB	54	0	26	0.68	1.00	0.81	-
		RGB+D	45	9	14	0.76	0.83	0.80	27
10	Small_Scres (12%)	RGB	65	2	8	0.89	0.97	0.93	-

Another restriction evolved by the chosen setup is the shock resistance. Physical shocks caused by participants by bumping against a material box or the whole assembly workstation led to FPs. A similar effect was discovered by arm/hand shadows generating activities in adjacent SLCs. Although both did not occur very often, it can be addressed in industrial settings by using a solid assembly workstation, putting SLCs on tracks and illuminating the workplace from top. Finally, the precision can be improved by using industrial cameras where the area of interest is observed with a fixed focus setting delivering a higher resolution.

Analyzing the results of the RGB+D camera, it was discovered that the optical sensor delivered a large amount of insertions (FP). Compared to the web cam which generates exactly two events per activity (hand in and hand out), a strong flickering was discovered regarding the event generation of the RGB+D sensor. Thus, we clustered sensor events by removing all events with a time difference of T compared to a matching event of the same type (RGB+D sensor and material) from the set of insertions (FP). We set $T = D(G^{v,a})$ where D is the duration of the ath ground truth action in the vth video. Nevertheless, it can be seen that the F-score of the RGB+D sensor is low. One reason is the high number of anomalies for the SLCs "BCD_" (50%) and "BG_Application_Board" (54%). The other reason is the position of the RGB+D sensor behind the assembly workstation. Large parts of the worker's body are covered by the workstation which leads to a less accurate skeleton tracking that we are using to estimate the hand position. This fact is generating a lot of flickering which is the reason for a high number of insertions (FP) and failures (FN) during activity detection.

5   Related Work

In the pervasive computing community Funk et al. [7] suggest a cognitive assistance system that aims to increase efficiency and assistance in manufacturing processes based on motion detection with an RGB+D camera. A permanent and calibration intensive integration of the sensor is necessary to provide feedback in contrast to our approach that aims to provide a temporary and light-weight sensor setup. Instead of using additional instrumentation, Bader et al. [4] use existing sensors (RFIDs) to provide assistance on a display. RFID could be integrated in our system through CEP but is alone insufficient to provide details on worker activities. Quint et al. [13] suggest a system architecture for assistance in manual tasks with the aim to combine components, such as visualization techniques, interaction modalities and sensor technologies. Sensor events are matched from RGB and RGB+D cameras to states in state machines. Compared to our approach the emphasis lies on the information model and the architecture more than on studying the artifact's accuracy and application in BPM.

In the augmented reality domain several systems realize activity tracking to provide assistance during assembly tasks in the psychomotor phase of a work-flow. Henderson and Feiner's prototype [8] provides several forms of assistance visualizing arrows, labels, highlighting effects and motion paths using markers attached to all tracked objects. In contrast, Peterson et al. [12] focus on the technical realization of markerless spatiotemporal tracking with uncalibrated cameras to automate the creation of augmented reality video manuals from a single first-person view video example. Both papers feature a strong emphasis on augmented reality and do not touch the BPM world. Nevertheless, the suggested technologies provide the potential of analyzing worker activities.

Data fusion, which is necessary to integrate heterogeneous sensor information and application events, matches the concept of CEP and is supported by our even-driven architecture, similar to Bruns et al. [3], who integrate CEP and finite state machines fusing sensor data to determine the actual states of ambulance vehicles and not the overall process. The interconnection between BPM and CEP was introduced as event-driven BPM by von Ammon et al. [2] and demonstrated, e.g., by Estruch et al. [6] who suggest an approach for CEP with BPMN 2.0 in the manufacturing domain. They focus on the modelling aspects in a top-down manner and leave out the sensor-level. Not tackling the manufacturing domain, Herzberg et al. [9] provide a framework to correlate specific process instances by enriching events already recorded with context data resulting in process events and show its feasibility applying process mining in the logistics domain. They focus on the event correlation and leave out the data elicitation.

Summarized, work on activity tracking is mainly focused on assistance using existing or additional sensor technology. Work interconnecting BPM and CEP is sparse in the manufacturing domain. The technology-enhanced process elicitation suggested here is related to both worlds and aims to deliver detailed information about worker activities, which has not been addressed so far.

6   Conclusion

In this work, an artifact was implemented and evaluated that supports process elicitation tasks in manual assembly systems. The system was set up in a laboratory setting and showed a satisfying performance in activity tracking during the assembly of an artificial product regarding the RGB camera application to proof the concept of process elicitation. The competing RGB+D camera application did not satisfy the performance criteria and will in future only be applied to deliver additional information on the worker (e.g., left or right handed). The experiment showed that the number of false negatives and false positives of the RGB activity detection still needs to be reduced. The evaluation of relevant factors was influenced by the high degree of freedom during product assembly and the laboratory setting within the experiment. Especially carton prototyping allows the construction of a testing environment very fast, but has the drawback of being shock sensitive. Similarly, consumer electronics are low in price but less stable regarding the result. In future, we plan additional experiments with more participants in multiple scenarios controlling and analyzing these side effects to increase the precision. It will be combined with the detection of material parts completing the picture of tracking manual assembly tasks. To support process enhancement and operational support, we will examine the anomalies detected during the study and follow the question how the manufacturing process and its model can be reflected and improved as a basis for a cyber-physical BPM.

Acknowledgments. This research was funded in part by the German Federal Ministry of Education and Research (BMBF) under grant number 01IS16022E (project BaSys4.0). The responsibility for this publication lies with the authors.

References

1. van der Aalst, W.M.P.: Process Mining: Data Science in Action, pp. 3–23. Springer, Heidelberg (2016). https://doi.org/10.1007/978-3-662-49851-4_1
2. von Ammon, R., Ertlmaier, T., Etzion, O., Kofman, A., Paulus, T.: Integrating complex events for collaborating and dynamically changing business processes. In: Dan, A., Gittler, F., Toumani, F. (eds.) ICSOC/ServiceWave -2009. LNCS, vol. 6275, pp. 370–384. Springer, Heidelberg (2010). https://doi.org/10.1007/978-3-642-16132-2_35
3. Bader, S.,et al.: Tracking assembly processes and providing assistance in smart factories. In: 6th International Conference on Agents and Articial Intelligence, ICAART, vol. 1, pp. 161–168. SCITEPRESS, Science and Technology Publications, Lda, Angers, France (2014)
4. Bruns, R., et al.: Using complex event processing to support data fusion for ambulance coordination. In: 17th International Conference on Information Fusion, FUSION, pp. 1–7, July 2014
5. Dencker, K., et al.: Proactive assembly systems - realising the potential of human collaboration with automation. Ann. Rev. Control **33**(2), 230–237 (2009)
6. Estruch, A., Heredia Álvaro, J.A.: Event-driven manufacturing process management approach. In: Barros, A., Gal, A., Kindler, E. (eds.) BPM 2012. LNCS, vol. 7481, pp. 120–133. Springer, Heidelberg (2012). https://doi.org/10.1007/978-3-642-32885-5_9
7. Funk, M., et al.: Cognitive assistance in the workplace. Pervasive Comput. **14**(3), 53–55 (2015)
8. Henderson, S.J., et al.: Augmented reality in the psychomotor phase of a procedural task. In: 10th International Symposium on Mixed and Augmented Reality, ISMAR, pp. 191–200. IEEE, Oct 2011
9. Herzberg, N., et al.: An event processing platform for business process management. In: 17th International Enterprise Distributed Object Computing Conference, EDOC, pp. 107–116. IEEE, September 2013
10. Houy, C., et al.: Empirical research in business process management - analysis of an emerging field of research. Bus. Process Manag. J. **16**(4), 619–661 (2010)
11. Knoch, S., et al.: Automatic capturing and analysis of manual manufacturing processes with minimal setup effort. In: International Joint Conference on Pervasive and Ubiquitous Computing, UbiComp, pp. 305–308. ACM, Heidelberg, Germany, September 2016
12. Petersen, N., et al.: Real-time modeling and tracking manual workflows from first-person vision. In: International Symposium on Mixed and Augmented Reality, ISMAR, pp. 117–124. IEEE, October 2013
13. Quint, F., et al.: A system architecture for assistance in manual tasks. In: Intelligent Environments, IE. vol. 21, pp. 43–52, Ambient Intelligence and Smart Environments. IOS Press (2016)
14. Zivkovic, Z., et al.: Efficient adaptive density estimation per image pixel for the task of background subtraction. Pattern Recogn. Lett. **27**(7), 773–780 (2006)

Discovering Process Models of Activities of Daily Living from Sensors

Marco Cameranesi, Claudia Diamantini, and Domenico Potena[⊠]

Dipartimento di Ingegneria dell'Informazione,
Università Politecnica delle Marche,
via Brecce Bianche, 60131 Ancona, Italy
{m.cameranesi,c.diamantini,d.potena}@univpm.it

Abstract. In recent years, more and more effort was put in the design and development of smart environments, which are aimed at improving the life quality of people, providing users with advanced services supporting them during their daily activities. In order to implement these services, smart environments are equipped with several sensors that continuously monitor the activities performed by a user. Sensor data are activation sequences and could be seen as the execution of a process representing daily user behaviors and performed activities. In this paper we propose a methodology, which exploit Process Mining techniques to discover both the *daily behavior model* and *macro activities model*s. The former represents the "standard" behavior of the user in the form of a process model. The latter is a set of process models representing the flow of sensors activations when given tasks or macro activities are performed. A real-world case study is introduced to empirically show the efficacy of the proposed methodology.

Keywords: Ambient assisted living · Activities of daily living
Process mining · Process discovery · Sensor data mining

1 Introduction

During the last years several research efforts have been done in the field of Ambient Assisted Living (AAL) technologies. Several kinds of different solutions have been proposed aimed at improving the life quality of people, supporting their activities during their daily life. Nowadays, the development of new technologies have made available several kinds of devices, which can be easily deployed to set-up a *smart environment* [11]. With this term, we refer to a generic environment (e.g. hospital, house, schools or other places) where the pervasive/ubiquitous computing paradigm [2] is being applied in order to provide advanced services to users which interact with the surrounding environment, e.g., health monitoring, home security, energy saving, light management, and so forth.

In order to implement these advanced services, a smart environment needs several kinds of actuators and, in particular, sensors that continuously monitor the environment and hence collect huge amount of data. In addition to

© Springer International Publishing AG 2018
E. Teniente and M. Weidlich (Eds.): BPM 2017 Workshops, LNBIP 308, pp. 285–297, 2018.
https://doi.org/10.1007/978-3-319-74030-0_21

be exploited to deploy the aforementioned services, sensors data represent the activities performed by a user each day. Hence, these data could be profitably analyzed to discover models representing daily user behaviors and performed activities; in the Literature, these goals are known as discovering of Activity of Daily Living (ADL) [9]. It is noteworthy that each task performed by a user is a sequence of activities. For instance, when a user performs a cooking task in a house, she moves towards the light switch to turn it on, opens the fridge and after a while also switches on the oven to warm the food, and so on until the user leaves the kitchen. Each of these activities triggers even several sensors (e.g. motion sensors, doors, and so on). Hence, the sensor activation sequence could be seen as the execution (i.e., a trace) of a process.

In this paper we propose a methodology, which exploits Process Mining techniques to discover the behavior of a user. In particular, we refer to Process Discovery algorithms, which are aimed at finding the process model representing a set of given traces, to discover two kinds of models: *daily behavior model* and *macro activity model*. The former represents the "standard" behavior of the monitored user in the form of a process model, where activities are daily tasks or *macro activities*; e.g. cooking, reading, watching TV, and so on. The latter is a process model representing the flow of sensors activations when a given macro activity is performed. Both these models could be used to enable several smart services, e.g., allowing the home automation manager to choose different scenarios on the basis of both habits of the user and sensors that are actually triggered; or identification of anomalous behaviors which could be due to an accident or other type of emergencies. The novelty of the paper is twofold: on the one hand, event logs produced by sensors are characterized in terms of macro and micro activities; on the other hand, the use of Process Mining techniques to represent daily living activities (i.e., macro activities) as a process where activities are sensor activations (i.e., micro activities).

We experimentally show the efficacy of proposed methodology by referring to a real-world case study from the CASAS project of the Washington State University, where a user and her pet have been monitored in a house equipped with different types of sensors.

The rest of the paper is organized as follows. Section 2 is devoted to present the case study, which will be used as illustrative example throughout the paper. In Sect. 3, we provide a detailed description of the proposed methodology and show experimental results. Section 4 provides an overview of the related work. Finally, Sect. 5 draws some conclusions and presents future works.

2   Case Study

This section is devoted to describe a case study we will use as running example throughout the paper. In particular, we will refer to the *Milan* dataset, which is part of data collected by the CASAS project of the Washington State University[1]. The selected dataset describes behaviors of a user living with her pet for

---

[1] http://ailab.wsu.edu/casas/datasets.html.

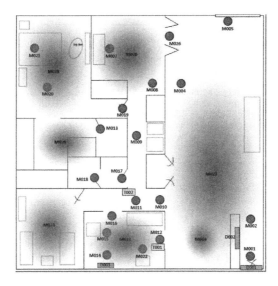

Fig. 1. The map of the Milan house and position of equipped sensors.

58 days in a house, which has been equipped with some different types of sensors (Fig. 1 shows the map of the house and sensors positioning), as follows:

- **Motion sensors** which track the user moving in the various rooms of the house. These sensors should be placed at least one per room, in order to understand the user movements during her daily life activities. In our example, each motion sensor triggers an ON state when the user is nearby. In Fig. 1, these sensors are denoted with the prefix M followed by an identifying number (e.g., M002);
- **Open/close sensors** that are able to detect the status of a door. They are installed on the external doors of the house, allowing us to know when someone is leaving or coming back home. These kinds of sensors are denoted with the letter D (e.g., D003);
- **Environmental sensors.** the house is also equipped with some temperature sensors, which are represented with the prefix T (e.g., T002).

Sensors have collected data about the daily activities of the user, in the form of an event log, where each event represents a measurement by a sensor and is described by the record ⟨*timestamp, sensor, value*⟩:

- *timestamp*, which identifies when an event is recorded, namely the time when the sensor is triggered;
- *sensor*: the unique identifier of the sensor;
- *value*, which is the measurement of the sensor; e.g., ON/OFF state, a temperature value, OPEN/CLOSE state.

timestamp	sensor	value	macro activity
...	...	...	...
2009-10-16 08:45:38.000076	M016	ON	Kitchen_Activity begin
2009-10-16 08:45:39.000094	M003	ON	NULL
2009-10-16 08:45:40.000041	M015	ON	NULL
2009-10-16 08:45:40.000090	M023	ON	NULL
...	...	...	...
2009-10-16 08:58:52.000004	M016	ON	Kitchen_Activity end
2009-10-16 08:58:52.000053	M023	OFF	NULL
...	...	...	...
2009-10-16 08:59:02.000054	M019	ON	Chores begin
...	...	...	...
2009-10-16 09:14:47.000090	M006	ON	Chores end
...	...	...	...

Fig. 2. An excerpt of the input dataset. The record ⟨*timestamp, sensor, value*⟩ is automatically acquired by sensors; the feature *macro activity* is manually added.

Noteworthy that, beside the pet sometimes guests come to visit the main user. Since the guest and the pet interact with the surrounding environment, triggering home sensors, they represent a source of noise of the case study.

During the project, the user has been asked to annotate the activity she has been executing, e.g. cooking, watching TV, sleeping and so on. Each of these activity, hereafter be called *macro activity*, could be represented as a sequence of low level activities that can be measured by sensors (i.e., *micro activities*). For instance, when cooking, the user moves towards the kitchen light switch to turn it on, moves to fridge and after a while also switches on the oven to warm the food, and so on until the user leaves the kitchen. Each of these activities triggers even several sensors. In this simple example the macro activity cooking starts with the user entering the kitchen and ends when she leaves the room. The user must manually specify when she starts and ends the performed macro activity. From the case study description, we cannot say whether the user records parallel activities (e.g., taking medicines while watching TV). Indeed, from a preliminary analysis of data, it seems that macro activities are stored in strict sequence; except rare cases that can be considered noise (maybe due to annotation errors). Hence, we assume the user annotates only the main macro activities.

The macro activities in the dataset are: Bed To Toilet, Chores, Desk Activity, Dining Room Activity, Morning Meds, Evening Meds, Guest Bathroom, Kitchen Activity, Leave Home, Meditate, Master Bathroom, Read, Master Bedroom Activity, Watch TV, Sleep. In Fig. 2 a short excerpt of the dataset is shown as example. The dataset is characterized by non-uniform daily distribution of macro activities, as shown in Fig. 3. The average number of macro activities performed by the user in a day is around 41 with a standard deviation of 13.7.

3   Methodology

In this work, we use macro activities to characterize the daily behavior of the user, while micro activities will be used to model a macro activity. In particular, since user activities can be seen as instances of a process describing the

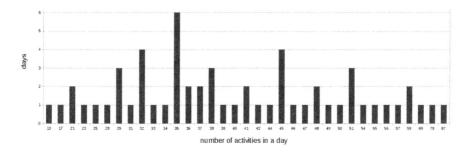

Fig. 3. Daily distribution of macro activities.

daily behavior, Process Mining (PM) techniques will be adopted to analyze the sequence of macro activities in a day as well as micro activities occurring during each macro activity. To this end, in the next Subsection we discuss pre-processing operations, which are needed to handle possible annotation errors and to properly set up the data for applying specific PM techniques, which are presented in Subsect. 3.2.

3.1   Pre-processing

As first step, we fixed inconsistencies due to the manual annotation of macro activities. Indeed, macro activities that by their nature are sequential sometimes are reported in the log as parallel, namely macro activity B has started before the end of macro activity A. This is the case, for instance, of `Bed To Toilet` and `Sleep` because the user can not do any activity while she is sleeping. We detected these activities and added artificial events in the log to fix the issue: the artificial end of A is added just before the reported begin of B and the artificial begin of A is added just after the reported end of B.

In order to analyze macro activities, the first step is to remove from the dataset unnecessary data, namely sensor id and value. Now, in order to apply PM techniques we need to identify the begin and end of the instance of the process in the log, namely the trace. Since we are interested in daily activities, the issue becomes to identify the timestamp when a day starts/ends. Since a fixed time could fall within a macro activity (e.g. `Sleep`), hence splitting it into two days, we have chosen to set the end of a day dynamically, when the user went to sleep. However, sometimes she went to sleep after 12pm, hence in a new day. Furthermore, sometimes she wakes up during the night to go to the toilet, even several times in a night; hence, we have several sleeping activities in a day. For these reasons, we defined the trace related to a given day as the period between the time when the user goes to sleep for the first time after 9pm and the first sleeping activity after 9pm of the next day.

An excerpt of the *macro activity event log* obtained at the end of the preprocessing is reported in Fig. 4. A day is composed by several rows, each representing the begin and the end of a given activity. Several instances of the same activity could occur in the same day.

day	activity	timestamp begin	timestamp end
...	...	...	...
1	kitchen_activity	2009-10-16 08:45:38.000076	2009-10-16 08:58:52.000004
1	chores	2009-10-16 08:59:02.000054	2009-10-16 09:14:47.000090
...	...	...	...

Fig. 4. An excerpt of the macro activity event log.

As concerns micro activities, the input dataset has been divided into several parts, an event log for each macro activity. The structure of each *micro activity event log* is similar to the one reported in Fig. 2, where the *macro activity* field has been deleted and an identifier representing the different instances (i.e. traces) of the macro activity has been added. In order to reduce noise only motion sensors have been considered.

It is noteworthy that traces in the same micro activity event log could have different activities in the first and last position. This is due to both diverse ways (i.e., process variants) a macro activity is executed (e.g., Kitchen Activity could start triggering the sensor closed to one of the two kitchen doors, i.e., M012 and M016) and the noise introduced by the user during the manual annotation of macro activities. In order to have unique start and end points for any instance of the process, for each trace of both event logs, artificial events corresponding to START and END activities have been added as first and last event respectively. This is a widespread practice aimed at improving the soundness of Petri net resulting by a Process Discovery algorithm [10].

3.2   Process Discovery

Among the PM techniques, here we focus on process discovery algorithms which are aimed at extracting process models starting from event logs. In particular, among of the most important discovery techniques there are: Alpha Miner [1], Fuzzy Miner [7], Heuristic Miner [17], and infrequent Inductive Miner [10].

The Alpha Miner algorithm has been developed to extract a process model from "clean" conditions, namely log without noise. It works well with structured processes (known in Literature as *lasagna-like* processes), whereas it fails in discovering a valid model for highly variable processes (known as *spaghetti-like* processes) returning overgeneralizing "flower" models, which are not useful knowledge as any behavior is represented (see [4]). As we can notice in Fig. 3, the dataset we are dealing with is characterized by a high standard deviation value; this suggests that we are dealing with a log representing a highly variable process. Hence, we should use Fuzzy, Heuristic or infrequent Inductive Miner. Fuzzy Miner filters and groups infrequent behaviors and returns a high level (i.e., abstract) representation of the mined model. The outcome of the algorithm is a scheme involving just the most relevant activities, displayed as single activities or aggregated in clusters, and their precedence relations. Fuzzy Miner is not able to represent split and join constructs, i.e. the obtained model does not have an executable semantics. Heuristic Miner algorithm can be viewed as an extension of the Alpha Miner. This latter finds all causal relations among activities in the log, whereas the Heuristic Miner takes into account the frequency of relations and applies some heuristics to determine relevant sequence and parallelism relations. Finally, infrequent Inductive Miner returns a model by iteratively refining a process tree, where each node represents either an activity of the process (leaf node), or an operator (branch node). As for Heuristic, also this algorithm filters infrequent behaviors and the output can be represented by a Petri net. However, we like to note that infrequent Inductive Miner guarantees to always return a sound model, i.e. a model where all process steps can be executed and the final marking of the Petri net is always reachable. From what has been said, we chosen to adopt the infrequent Inductive Miner Algorithm [10], and in the experiments we used the implementation provided within ProM[2], a well-known framework which implements several PM algorithms through a modular plugin system. In the next experiments we set only the noise threshold, which range from 0 to 1, where 0 means that the infrequent paths are not filtered at all.

In next Sections, we present results of two sets of experiments: the discovering of the daily behavior model, representing the monitored user which performs daily tasks, and the discovering of macro activities models, representing the flow of micro activities when a macro activity is performed.

One of the most common measures used to evaluate the quality of the discovered model is the *fitness*, which evaluates the ability of the model to replay traces in the log. A fitness value closed to 1 means that each trace in the log can be associated with an execution path specified by the process model, lower values represent a model that is not able to represent some traces or pieces of them. In order to compute the fitness, conformance checking algorithms [1] are used to compare the log with the model.

3.3   Daily Behavior Model

In this Section we present experiments aimed at discovering the daily behavior model of the user, using the infrequent Inductive Miner [10] algorithm. As input, the *macro activity event log* (see Subect. 3.3) has been used.

As a first experiment, we have characterized each macro activity as two events: the beginning of the activity and the end. All the Petri nets resulting from such experiments have been quite difficult to understand, independently by the filtering settings of the algorithm. For this reason, we have decided to remove any END events from the event log; resulting the log as in Table 1.

In Table 2 is shown the fitness obtained by using the infrequent Inductive Miner with different values of the threshold parameter. Lower threshold values returns a model where almost all causal relations are considered, hence the model has high fitness but is not simple; i.e., it is represented through a big amount of transitions, places and arcs so it might be difficult to understand. On the other hand, high threshold values leads to a very simple, quite trivial, model which does not bring any contribution to the knowledge. In our experiment, a

---

[2] http://www.promtools.org.

Table 1. An excerpt of macro activity event log without END of activities.

Day 1:	[Sleep, 2009-10-16 21:03:28.034]
	[Bed_to_Toilet, 2009-10-16T21:29:11.043]
	...
Day 2:	[Sleep, 2009-10-16 03:58:44.068]
	[Morning_Meds, 2009-10-16 08:42:01.077]
	[Watch_TV, 2009-10-16 08:43:59.024]
	...

Table 2. The fitness obtained on macro activity event log by using the infrequent Inductive Miner with different values of the threshold parameter.

Threshold	0.00	0.05	0.10	0.15	0.20	0.25	0.30	0.35	0.40	0.50	0.60
Fitness (%)	100.0	99.4	98.2	99.4	91.4	86.7	82.4	66.2	51.5	44.7	44.3

good compromise is obtained setting the threshold to 0.20; the fitness of the discovered model is 91.4% and the model is quite simple (Fig. 5).

In order to give an example, Fig. 6 focuses on the first part of the model, corresponding to the way in which the monitored user begins her typical day. From that excerpt, we can notice that sometimes during the night the user stops her sleeping for going to the toilet, and then comes back sleeping. We see also that after waking up the user usually takes her morning medicines. Then several scenarios are possible, which allow us to derive useful insights. For instance, Kitchen Activity and Master Bathroom usually are not in sequence, while Master Bedroom Activity and Dining Room Activity usually occur after Meditate and/or Chores.

3.4 Macro Activity Model

This Section is devoted to present experiments for discovering macro activities models, representing the flow of micro activities (i.e., sensors) triggered when a macro activity is performed. In order to give an example here we focus on the Watch TV macro activity. In the whole micro activity event log, there are 114 traces describing Watch TV, which are composed of 207.8 micro activities on average, with a standard deviation of 153.91. In order to reduce the noise (mainly generated by a pet and some guests that sometimes visit the monitored user), we only consider motion sensors closed to the room where the TV is; such sensors are: M004, M007, M008, M009, M013, M019 and M026 (see Fig. 1)[3].

We have executed the infrequent Inductive Miner algorithm five times, varying the threshold parameter from 0.1 to 0.5. In Table 3 we report the fitness

---

[3] The event log is available at http://kdmg.dii.univpm.it/?q=content/watch-tv-event-log-bp-meet-iot.

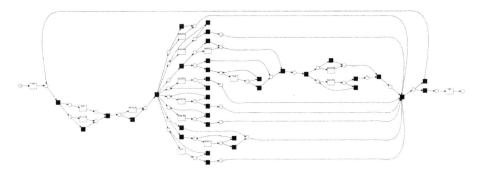

Fig. 5. Petri net of the daily behavior process model obtained with a threshold value of 0.20 of the Infrequent Inductive Miner.

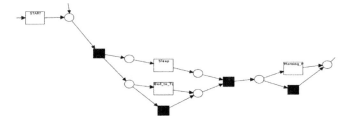

Fig. 6. An excerpt of the daily behavior process model.

Table 3. Fitness values for the Watch TV macro activity model, computed using both all sensors and the sensors closest to the TV room.

Threshold	Fitness (all sensors)	Fitness (room sensors)
0.1	72.15%	97.68%
0.2	73.16%	99.36%
0.3	68.44%	91.88%
0.4	65.47%	88.15%
0.5	17.54%	24.67%

values obtained with the resulting models both on reduced event log and on the whole Watch TV macro activity event log. As we can notice from the results, the best model is obtained with a threshold value of 0.20; as expected the fitness of the model on the whole log is lower, however the chosen threshold returns the best relative fitness. The Petri net of such a model is shown in Fig. 7.

In order to evaluate the quality of the resulting model, we have also computed the fitness of the model in Fig. 7 with respect to event logs of other macro activities, as obtained at the end of preprocessing step (see Subsect. 3.1). In order to be a good model for the Watch TV macro activity, the fitness of the discovered model on the whole Watch TV log is expected to be a significant

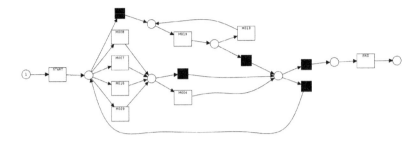

Fig. 7. Watch TV macro activity model.

Table 4. Fitness of Watch TV macro activity model with respect to different logs.

Macro activity	Watch TV	Read	Kitchen activity	Meditate
Fitness	73.16%	7.07%	7.89%	9.84%

value and higher than the same metrics computed using the same model on other logs. In particular, in addition to Watch TV we have calculated the fitness with respect to Read, Kitchen Activity and Meditate macro activities. Results are reported in Table 4. As expected, the model shows a good fitness on Watch TV macro activity event log and very low values on other macro activities event logs. The values decrease significantly when other logs are filtered using sensors closest to the TV room. Hence, the discovered model brings useful knowledge to represent the chosen macro activity, and only it.

4   Related Work

In the Ambient Assisted Living (AAL) field, some researches in last years have focused on recognizing user's activities of daily living (ADL), starting from raw data collected through a set of sensors. AAL community classifies these researches into two groups: Activities of Daily Living and Instrumental Activities of Daily Living (IADLs) [9]. The main difference is that in IADLs there is an interaction between the user and electronic devices (e.g. phone, home appliances), whereas there is not in ADLs. Our work is mainly focused on ADL group. In [16] the authors propose an approach which is focused to predict a macro activity from values collected by simply and cheap sensors. The prediction is performed by using a Naive Bayes classifier. The results achieved by this work demonstrates that ubiquitous, simple sensor devices can be used to recognize activities of daily living. However, classification models are not able to describe the user behavior, but only to recognize that a trace is an execution of a given activity. In [12] an unsupervised approach is introduced for discovering interesting patterns of ADLs. The proposed approach is able to identify a pattern regardless of the time intervals between activating a sensor and the next one. Noteworthy that the approach we propose is time independent too, since timestamps are only used to identify an ordering among activities.

Recently some researches have proposed the use of PM techniques to recognize user's ADLs [5,6,8,14,15]. A formalization based on an ontology is introduced in [8] to represent user's activities and to define a process model representing standard/expected user's behavior; then PM and conformance checking techniques are used to discover a model from sensor data and to compare it with the expected behavior. In [15], the use of PM has been proposed to accomplish several tasks. In particular, they share with our work the analysis of user's behavior, defined as a process model of (macro) activities; where an activity has been defined by the user through smartphone and smartwatch. We like to note that the use of personal and wearable devices provide more accurate data to track user behavior, avoiding the noise due to the activation of multiple sensors or the presence of another user or pet. In [6], authors have proposed a system to analyze the behavior of health staff in a surgical area of an hospital. First an indoor location systems is used to collect data, then each datum is assigned to a (macro) activity, finally PM techniques are used to discover the model representing the staff behavior. In the same mainstream, [5,14] propose pre-processing approaches to map sensor data to event logs based on domain knowledge; in particular [5] deals with the problem to identify habits of multiple users living in the same house. Differing from our work, all these researches do not deal with the discovery of macro activity models from sensor data.

5   Conclusions

In this paper we have proposed a methodology, which exploit Process Mining techniques to discover both the daily behavior model of an user and macro activities models. The former describes the way a user performs a sequence of tasks, or macro activities (e.g., cooking, reading and watching TV), in her environment. The latter is a process model representing the flow of sensors activations when a given macro activity is performed. The proposed methodology starts with a pre-processing stage, which is aimed at handling possible errors and subsequently recognizing the trace representing the execution of daily activities. Then, Process Discovery techniques, and in particular the infrequent Inductive Mining algorithm, have been used for extracting the two kinds of models. As concerns the macro activity model, a subset of sensors has been chosen to reduce the complexity of the model without decreasing the quality of the discovered model, measured in term of fitness. We experimentally show the efficacy of proposed methodology by referring to a real-world case study from the CASAS project of the Washington State University, where a user and her pet have been monitored for 58 days in a house equipped with different types of sensors.

We plan to apply the proposed methodology to other datasets from CASAS as well as from HicMO [13], a project funded by Marche region which involved several enterprises and universities to establish a laboratory for active and healthy aging. In order to emphasize frequent behaviors, we plan also to adopt sub-graph mining technique, like the one described in [3].

References

1. van der Aalst, W.M.P.: Process Mining: Discovery Conformance and Enhancement of Business Processes, 1st edn. Springer, Heidelberg (2011)
2. Atzori, L., Iera, A., Morabito, G.: The internet of things: a survey. Comput. Netw. **54**(15), 2787–2805 (2010)
3. Diamantini, C., Genga, L., Potena, D., Storti, E.: Pattern discovery from innovation processes. In: 2013 International Conference on Collaboration Technologies and Systems, pp. 457–464, May 2013
4. Diamantini, C., Genga, L., Potena, D., van der Aalst, W.M.P.: Building instance graphs for highly variable processes. Expert Syst. Appl. **59**, 101–118 (2016)
5. Dimaggio, M., Leotta, F., Mecella, M., Sora, D.: Process-based habit mining: experiments and techniques. In: 2016 International IEEE Conferences on Ubiquitous Intelligence & Computing, Advanced and Trusted Computing, Scalable Computing and Communications, Cloud and Big Data Computing, Internet of People, and Smart World Congress, pp. 145–152, 18–21 July 2016
6. Fernández-Llatas, C., Lizondo, A., Sanchez, E.M., Benedí, J., Traver, V.: Process mining methodology for health process tracking using real-time indoor location systems. Sensors **15**(12), 29821–29840 (2015)
7. Günther, C.W., van der Aalst, W.M.P.: Fuzzy mining – adaptive process simplification based on multi-perspective metrics. In: Alonso, G., Dadam, P., Rosemann, M. (eds.) BPM 2007. LNCS, vol. 4714, pp. 328–343. Springer, Heidelberg (2007). https://doi.org/10.1007/978-3-540-75183-0_24
8. Jaroucheh, Z., Liu, X., Smith, S.: Recognize contextual situation in pervasive environments using process mining techniques. J. Ambient Intell. Hum. Comput. **2**, 53–69 (2011)
9. Katz, S.: Assessing self-maintenance: activities of daily living, mobility, and instrumental activities of daily living. J. Am. Geriatr. Soc. **31**(12), 721–7 (1983)
10. Leemans, S.J.J., Fahland, D., van der Aalst, W.M.P.: Discovering block-structured process models from event logs containing infrequent behaviour. In: Lohmann, N., Song, M., Wohed, P. (eds.) BPM 2013. LNBIP, vol. 171, pp. 66–78. Springer, Cham (2014). https://doi.org/10.1007/978-3-319-06257-0_6
11. Rashidi, P., Mihailidis, A.: A survey on ambient-assisted living tools for older adults. IEEE J. Biomed. Health Inform. **17**(3), 579–590 (2013)
12. Rashidi, P., Cook, D.J.: Mining and monitoring patterns of daily routines for assisted living in real world settings. In: Proceedings of the 1st ACM International Health Informatics Symposium, pp. 336–345. ACM, New York (2010)
13. Rossi, L., Belli, A., Santis, A.D., Diamantini, C., Frontoni, E., Gambi, E., Palma, L., Pernini, L., Pierleoni, P., Potena, D., Raffaeli, L., Spinsante, S., Zingaretti, P., Cacciagrano, D., Corradini, F., Culmone, R., Angelis, F.D., Merelli, E., Re, B.: Interoperability issues among smart home technological frameworks. In: International Conference on Mechatronic and Embedded Systems and Applications, pp. 1–7, 27 September 2014
14. Senderovich, A., Rogge-Solti, A., Gal, A., Mendling, J., Mandelbaum, A.: The ROAD from sensor data to process instances via interaction mining. In: Nurcan, S., Soffer, P., Bajec, M., Eder, J. (eds.) CAiSE 2016. LNCS, vol. 9694, pp. 257–273. Springer, Cham (2016). https://doi.org/10.1007/978-3-319-39696-5_16
15. Sztyler, T., Völker, J., Carmona, J., Meier, O., Stuckenschmidt, H.: Discovery of personal processes from labeled sensor data - an application of process mining to personalized health care. In: Proceedings of the International Workshop on Algorithms and Theories for the Analysis of Event Data, pp. 31–46 (2015)

16. Tapia, E.M., Intille, S.S., Larson, K.: Activity recognition in the home using simple and ubiquitous sensors. In: Ferscha, A., Mattern, F. (eds.) Pervasive 2004. LNCS, vol. 3001, pp. 158–175. Springer, Heidelberg (2004). https://doi.org/10.1007/978-3-540-24646-6_10
17. Weijters, A., van der Aalst, W., de Medeiros, A.: Process mining with the heuristics miner-algorithm. Technische Universiteit Eindhoven, Technical report WP 166 (2006)

An Habit Is a Process: A BPM-Based Approach for Smart Spaces

Daniele Sora, Francesco Leotta, and Massimo Mecella[✉]

Dipartimento di Ingegneria Informatica,
Automatica e Gestionale Antonio Ruberti,
Sapienza Università di Roma, Rome, Italy
{sora,leotta,mecella}@diag.uniroma1.it

Abstract. Among the most important decisions to be taken in modeling human habits in smart spaces there is the choice of the technique to be adopted: models can be expressed by using a multitude of formalisms, all with differently proven effectiveness. However, a crucial aspect, often underestimated in its importance, is the readability of the model: it influences the possibility of validating the model itself by human experts. Possible solutions for the readability issue are offered by Business Process Modeling techniques, designed for process analysis: to apply process automation and mining techniques on a version of the sensor log preprocessed in order to translate raw sensor measurements into human actions. The paper also presents some hints of how the proposed method can be employed to automatically extract models to be reused for ambient intelligence, analysing the challenges in this research field.

1 Introduction

The aim of a *smart space* is providing people with automatic or semi-automatic services realizing the concept of *ambient intelligence* (AmI). The input for these intelligent services is represented by a *sensor log*, which is a sequence of measurement values acquired from sensors deployed across the monitored space. Many approaches have been proposed in the literature to automatically analyze sensor logs at runtime to understand the current context and to make decisions on the basis of user preferences and habits. All of these solutions are based on *models* that relate the output of the sensors during a (potentially very short) temporal window, to a specific contextual information that can be then employed to act or reason on the state of the environment.

Business Process Management (BPM) and specifically process mining techniques aims at mining a model from process logs; the most of these techniques

Results in this paper have been obtained with an academic license of Disco freely provided by Fluxicon. The work of Daniele Sora has been partly supported by the Lazio regional project *SAPERI & Co* (FILAS-RU-2014-1113), the work of Francesco Leotta has been partly supported by the Lazio regional project *Sapientia* (FILAS-RU-2014-1186), all the authors have been also partly supported by the Italian projects *NEPTIS*, *SM&ST* and *RoMA*.

E. Teniente and M. Weidlich (Eds.): BPM 2017 Workshops, LNBIP 308, pp. 298–309, 2018.
https://doi.org/10.1007/978-3-319-74030-0_22

are designed for dealing with quite structured business processes. The ambition of our research is to provide an answer to the question "Are the human habits structured enough to be modeled with BPM techniques?". Probably yes, but the adaptation procedure is challenging and not straightforward.

In the literature, smart space activities' models have been either manually defined (*specification-based methods*) or obtained through machine learning techniques (*learning-based methods*). In the former case, models are usually based on logic-based formalisms, relatively easy to read and validate (once the formalism is known to the reader), but their creation requires a great effort in terms of expert time. In the latter case, the model is automatically learned from a training set (whose labeling cost may vary according to the proposed solution), but employed formalism are usually taken from statistics, making them less immediate to understand. Another difference between the two approaches is that whereas specification-based methods use human actions as main modeling elements, learning-based ones directly refer to sensor measurements, thus loosing the focus on human actions and making even more difficult to visually inspect and validate produced models. On the other hand, taking as input raw sensor measurements usually makes learning-based methods easier to apply in a practical context; whereas, in the vast majority of cases, specification-based methods do not face the problem of translating sensor measurements into actions.

Our research is based on the intuition that human habits can be considered as *personal processes*. Under this particular perspective, BPM techniques could be successfully exploited to represent human habits through readable models. In fact, one of the most relevant problems of data-driven approaches in smart space modeling, is the lack of model readability. This implies a poor user interaction with the model itself, which makes difficult some operations as, for instance, the user model validation. Moreover, as argued in [16], applying methods originally taken from the area of *business process mining* [1] to human habits may represent a compromise between specification-based and learning-based methods, provided that the gap between raw sensor measurements and human actions can be filled in by performing a log preprocessing step. Such a log preprocessing step may consist of simple inferences on data or complex machine learning algorithms.

On the line of this argument, in this paper we will discuss how to transform raw movement measurements into actions, adapting sensor logs for application of BPM techniques, and we will present an example of using fuzzy mining to build an habit model, providing intuitions of how human readable models can be reused for compliance checking and proactive actions in smart spaces.

The paper is organized as it follows. Section 2 recaps relevant literature. Section 3 introduces the preliminary background for this work. Then Sect. 4 outlines how it is possible to preprocess the sensor logs in order to derive an event log ready for successive process mining, which is finally presented in Sect. 5 adopting fuzzy mining. Concluding remarks are presented in Sect. 6.

2 Related Work

The literature about representing models of human habits is wide. In this section, we will focus on those formalisms that are human readable, thus being easy to validate by a human expert or by the final user (once the formalism is known).

Initial approaches to the development of context-aware systems able to recognize situations were based on *predicate logic*. Loke [18] introduced a PROLOG extension called LogicCAP, and different reasoning techniques with situations including selecting the best action to perform in a certain situation, understanding what situation a certain entity is in (or the most likely) and defining relationships between situations. The same author [19] later defined a formal methodology to compose different context-aware pervasive systems.

Ontologies have increasingly gained attention as a generic, formal and explicit way to "capture and specify the domain knowledge with its intrinsic semantics through consensual terminology and formal axioms and constraints" [24]. They provide a formal way to represent sensor data, context, and situations into well-structured terminologies, which make them understandable, shareable, and reusable by both humans and machines. A considerable amount of knowledge engineering effort is expected in constructing the knowledge base, while the inference is well supported by mature algorithms and rule engines. Among relevant systems based on ontologies, we can cite CoBrA [9] (based on OWL), Gaia [20] (based on DAML + OIL) and SOCAM [13] (also called CONON, based on OWL). Another example of using ontologies in identifying situations is given by [22] (later evolved in [14,23]). Instead of using ontologies to infer activities, they use ontologies to validate the result inferred from statistical techniques.

Augusto [4] introduces the concept of Active DataBase (ADB) composed by Event-Condition-Action (ECA) rules. An ECA rule basically has the form "ON *event* IF *condition* THEN *action*", where conditions can take into account time. These have been extended in [3] with more complex temporal operators (`ANDlater` and `ANDsim`) and by adding uncertainty management. The APUBS [5] system extends previous works by mining ECA rules instead of relying on a specification-based approach.

The first attempts to apply techniques taken from the BPM area [10] were the usage of workflow specifications to anticipate user actions. A workflow is composed by a set of tasks related by qualitative and/or quantitative time relationships. Authors in [12] present a survey of techniques for *temporal calculus* (i.e., Allen's Temporal Logic and Point Algebra) and *spatial calculus* aiming at decision making. The SPUBS system [6,7] automatically retrieve these workflows from sensor data. The method basically consists of the following steps:

1. *Identifying frequent sets*: during this step, thoroughly inspected in [7], frequent item-sets are mined by employing the seminal APriori algorithm [2]. A database is obtained by segmenting the dataset on a per time-slot basis. Frequent item-sets are than refined in order to include frequent events discarded by APriori.
2. *Identifying topology*: this step has been deeply influenced by previous works on workflow mining. Once the frequent item-set has been obtained during

the previous step, a weighted directed graph is constructed where weights are computed as the number of times vertices of the corresponding edges come one after the other in the sensor log. As a first refinement, events that happen multiple time are replicated in such a way to reflect the context when they happen. Then events without any ordering relation are merged into single events.

3. *Identifying time relations*: at this step, for each edge in the graph created at the previous step, a quantitative/qualitative time relation is mined.
4. *Identifying conditions:* during the last step, conditions are mined for those edges that happen only if certain conditions of the environment (e.g., temperature, day of the week) are satisfied.

In this paper, we decided to apply a specific technique named *fuzzy mining* that, as we will explain in following, extract models that are more suitable than workflows employed in SPUBS to describe processes (like human habits) that are flexible (less structured) by nature.

3   Discussing Sensor and Event Logs

A smart space produces, at runtime, a sensor log containing raw measurements from available sensors. Given a set S of sensors, a sensor log $\mathcal{S}$ is an ordered sequence of measurements of the kind $\langle ts, s, v \rangle$ where ts is the timestamp of the measurement, $s \in S$ is the source sensor and v the measured value, belonging to a domain D_s that is a vectorial space (in the vast majority of cases with a single dimension) where any component can be either nominal (categorical) or numeric (quantitative). Measurements can be produced by a sensor on a periodic base (e.g., temperature) or whenever a particular event is detected (e.g., a door opening).

Many solutions instead expect as input an *event log*. Given a set $E = \{e_1, \ldots, e_n\}$ of event types, an event log $\mathcal{E}$ is a sequence of pairs $\langle e, ts \rangle$ where $e \in E$ and ts is the occurrence time of the event e. Events are usually obtained by filtering and aggregating measurements from the sensor log $\mathcal{S}$; thus the granularity of $\mathcal{E}$, in general, is coarser than the one observed in $\mathcal{S}$ and can potentially lead to loss of information.

Authors in the field of smart spaces use, sometimes as synonyms, a variety of terms to refer to the state of the environment and the tasks humans perform in it. In this paper, we will use the following terminology:

Context: the state of the environment including the human inhabitants. The state of a human inhabitant includes the action he/she is performing.
Action: atomic interaction with the environment or a part of it (e.g., a device). Recognizing actions can more or less difficult depending on the sensors installed. Certain methods only focus on actions. In some cases, methods to recognize activities and habits skip the action recognition phase, only relying on the raw measurements in the sensor log.

Activity: a sequence of actions (one in the extreme case) or sensor measurements/events with a final goal. In some cases an action can be an activity itself (e.g., ironing). Activities can be collaborative, including actions by multiple users, and can interleave one each other. The granularity of considered activities cannot be precisely specified. According to the given approach, tidying up a room can be an activity whereas others approaches may generically consider tidying up the entire house as an activity. In any case, some approaches may hierarchically define activities, where an activity is a combination of sub-activities.

Habit: a sequence or interleaving of activities that happen in specific contextual conditions (e.g., what the user does each morning between 08:00 am and 10:00 am).

In this paper, we aim at showing how habits can be modeled by using business process modeling technique, cf. [16]. In BPM, a business process is a collection of related events, activities, and decisions that involve a number of actors and resources and that collectively lead to an outcome that is of value for an organization or a user. The process logic is explicitly described in terms of a process schema (i.e., the model), and a specific execution of a process is named process instance, or also case. The progress of a process instance produces a trace of execution, which may be stored in an event log and can be used for process mining [1], e.g., discovering a process model from the event log or checking the compliance of the log with the model.

In order to apply a process discovery technique to a smart space, the sensor log must be turned into an event log, where the granularity chosen for the aggregation should be the same one of tasks in the process model. Additionally, the log must be segmented into traces, i.e., repetitions of the same process schema (corresponding to process cases).

Notably, as pointed out also in [1], depending on the process mining technique used, the challenge is to extract event logs from the data sources; when merging and extracting data, both syntax and semantics play an important role. Moreover, depending on the questions one seeks to answer, different views on the available data are needed. *Scoping* is of the utmost importance: the problem is not the syntactical conversion but the selection of suitable data. Depending on the questions and viewpoint chosen, different event logs may be extracted from the same data set. In [1], the five most important challenges encountered when attempting to apply process mining techniques to heterogeneous data sources (as sensor logs) are *(i)* correlation, proper management of *(ii)* timestamps and *(iii)* snapshots, the *(iv)* scoping and the identification of the proper *(v)* granularity. A notable contribution of this paper is devising an approach for solving the above challenges in the context of smart spaces.

4   Environment, Datasets and Log Preprocessing

Collecting and labeling large data sets from sensorized environments is a key step in order to deal with smart spaces. The data used in this work have been collected exploiting a specific toolkit provided by the Smart Home in a Box

(SHiB) project of Washington State University through the CASAS laboratory [21]. It is a large-scale smart environment data and tool infrastructure, allowing researchers to monitor behavioral patterns over months or years by retrieving longitudinal data. Each kit consists of about one hundred elements divided in one server, relays, adhesives, batteries, door open/close, temperature sensors and motion/lighting detectors. Each toolkit produces a measurements' log, that can be considered for the model production. Some already prepared datasets are freely available on CASAS project site[1]. It offers heterogeneous datasets, either unlabeled and labeled, single and multi users. Some of our experiments have been performed over the Kyoto and Aruba, two of the datasets presented above: the Kyoto dataset consists of single user traces from a total of 20 different subjects performing 5 different activities. The Aruba dataset consists instead of 2 years of life of a single person labeling ten different activities. However, we also installed our own kit at a residential environment, to obtain a dataset extracted from a real context.

In this work, we assume that only discrete output sensors are available. In particular, among the sensors provided with the CASAS datasets, we only employ Presence InfraRed (PIR) sensors. A PIR sensor is usually installed on the ceiling and triggers whenever an object moves inside his activation area. We also assume that the granularity of available PIR sensors is enough detailed for the type of habits we want to extract, e.g., if the sensor on top of the basin triggers we are only interested in the fact that someone is using the basin without getting into the details of the action s/he is performing (e.g., washing hands, washing face, cleaning teeth).

In [8], a granularity problem is introduced: there exists a clear gap between the granularity of sensor logs and that of traces traditionally employed for process mining. In other words, there is no one-to-one correspondence between sensor measurements and actions performed by people. There are several techniques to bridge this gap, usually adapted from the machine learning field. Here we outline our proposal based on trajectory analysis, the reader interested to all details can refer to [17].

Log preprocessing. We need to translate the sensor log $\mathcal{S}$ into an event log $\mathcal{E}$. If we denote with D_s the domain of the sensor $s \in S$, the transformation between $\mathcal{S}$ and $\mathcal{E}$ will be driven by a function $f(\langle ts, s, v \rangle) = \langle ts, f_s(v) \rangle$ where $f_s : D_s \mapsto E$ is a function that converts the values from D_s in events of E.

The first step is filtering the useless measurements. In particular, as PIR sensors resets automatically after a fixed period, only activation measurements are considered. At this point, the log is composed by time irregular activation records, in the form $\langle ts, s, ON \rangle$: given two consecutive activation records $\mathcal{A}$ and $\mathcal{B}$ corresponding to different timestamps, respectively $ts_{\mathcal{A}}$ and $ts_{\mathcal{B}}$ (in milliseconds), we can have two different cases, $ts_{\mathcal{B}} - ts_{\mathcal{A}} > 1000\,\text{ms}$ and $ts_{\mathcal{B}} - ts_{\mathcal{A}} < 1000\,\text{ms}$. Our approach requires equidistant records, so we want to impose $ts_{\mathcal{B}} - ts_{\mathcal{A}} = 1\,\text{s}$. The interval chosen warranties a proper granularity for maintaining information

---

[1] cfr. http://ailab.wsu.edu/casas/datasets/.

about the activity contained and at the same time cutting off sensors' dirty triggers. The two approaches adopted to obtain that are:

$ts_\mathcal{B} - ts_\mathcal{A} > \mathbf{1000\,ms}$: $\mathcal{A}$ is replicated in $\tilde{\mathcal{A}}$, with $ts_{\tilde{\mathcal{A}}} = ts_\mathcal{A} + 1000$. This operation is iterated until $ts_{\tilde{\mathcal{A}}} < ts_\mathcal{B}$

$ts_\mathcal{B} - ts_\mathcal{A} < \mathbf{1000\,ms}$: in this case $\mathcal{B}$ is removed and if there is no record until time stamp $ts_\mathcal{A} + 1000$, then $\mathcal{B}$ is inserted with $ts_\mathcal{B} = \mathcal{A} + 1000$, otherwise $\mathcal{B}$ is considered an outlier and is simply skipped.

Segmentation through trajectory analysis. The log segmentation is a core task, since a common prerequisite of process mining techniques is to have a trace log explicitly segmented in traces (process instances). This assumption is clearly not met by sensor logs as labeling is generally an expensive task to be performed by humans. We adapted a portion of the TRACLUS algorithm [15] for performing segmentation. It is a trajectory analysis algorithm, focused on sub-trajectory automatic definition and their clustering, originally devised for describing hurricanes' trajectories. The sub-trajectory detection system finds the points where the behavior of a trajectory changes rapidly. The separation of sub-trajectories can be seen as an unsupervised segmentation technique. In our work, the segmentation module takes as input the preprocessed log containing the sampled and uniformed measurements (see previous step), and from this one, we obtain several sub-logs, i.e., the segments that have to be classified.

Classification of segments. The separation in segments is a preliminary step for building a model of the habits. The following step is to reduce the log's granularity to improve readability for users and efficiency for miners. To obtain that, we exploited the segments extracted in the previous step: we create a new log composed by a sequence of classification results (over segments). In other terms, each log's portion corresponding to a segment is substituted with its classification.

The classification technique adopted is a novel one specifically developed for this scenario [17]. In general terms, a classification problem consists in assigning to an element of a domain, a value (a label) taken from a set of other values (co-domain). So given a domain of elements Δ and a set of labels Γ, a classification function f can be described as $\gamma = f(\delta)$ where $\gamma \in \Gamma$ and $\delta \in \Delta$. In our context Δ is the set of all the segments extracted by the TRACLUS algorithm in the previous step, and the set Γ of labels consists of *stay*, *area*, and *movement* (i.e., $\Gamma = \{\mathbf{STAY}, \mathbf{AREA}, \mathbf{MOVEMENT}\}$). The classes were identified by visual inspection, by employing a specific feature of our visual analysis tool [17]. Intuitively, the three classes characterize what happens in the segment: *(i)* **MOVEMENT** if the sub-trajectory contains a "fast" movement from a sensor to another one, *(ii)* **AREA** if the action involves many sensor triggers in the very same specific area, and *(iii)* **STAY** if the user stays hold under a given sensor "long" enough.

5   Habits Identification Through Fuzzy Mining

The previous classification results are used for obtaining the log reduction. Entire portions of the log corresponding to segments, are replaced by a single event consisting of:

$$day_{id} \quad ts_s \quad ts_e \quad \langle \gamma \quad \psi \rangle$$

where day_{id} is a constant integer value that identifies the day, ts_s and ts_e are respectively the starting and ending timestamps, $\gamma \in \Gamma = \{movement, area, stay\}$ and $\psi \in \Psi$. Ψ is the *set of locations* and consists of all the positions, rooms and appliances covered by at least one sensor. In other words, everything is covered by a sensor is included into Ψ. The mapping between locations ψ and sensors' names (or set of sensors) is made manually.

Generally, the **AREA** label is assigned to a room, and many sensors are related to the room; it is an indication that the inhabitant moves withing the room area, e.g., $< AREA \quad Living >$. Conversely, the **STAY** label is related to a specific point or element, e.g., $< STAY \quad kitchen.table >$). When the label is **MOVEMENT**, the ψ value is not assigned: for our aims, if the inhabitant is moving from a source a to a destination b, the sequence between a and b and the movement duration are interesting, not the followed path. This new log, consisting of entries structured as described above, is the event log $\mathcal{E}$), and is fed into a fuzzy miner.

If no filtering is applied, the fuzzy miner produces a weighted graph (with weighted nodes) that is difficult to analyse, and in which single habits cannot be distinguished one each other, and not frequent edges cause readability lacks. In order to identify habits, we therefore need to use the tuning parameters of fuzzy miner. At this granularity level, the fuzzy output, properly filtered, is strongly readable and an user level analysis is feasible also for non technical users.

The application of the previous steps over the specific `Aruba` dataset, as an example, provides the fuzzy model shown in Fig. 1. A very important step is to identify the habits (or the activities) from this general model.

To obtain that, our approach recurs to the segmentation procedure described above: all the instances of each activity (i.e. *meal preparation*) are processed either with TRACLUS algorithm and Fuzzy Miner. This procedure's output is a graph, model of the activity. This step is repeated for each activity, to have a model for them all. Successively these graphs are matched to subgraphs of the complete graph to identify activities. In particular subgraphs corresponding to the same nodes of the activity's graph are considered for the comparison. This procedure allows us to obtain two important goals:

– single activity identification inside the complete graph (i.e. Fig. 2)
– to make an evaluation of the segmentation algorithm goodness

The second point is not immediate. The positive evaluation is due to the creation of comparable segments classified in the same way either for the complete log and for the log divided in actions. Intuitively this fact shows the approach's

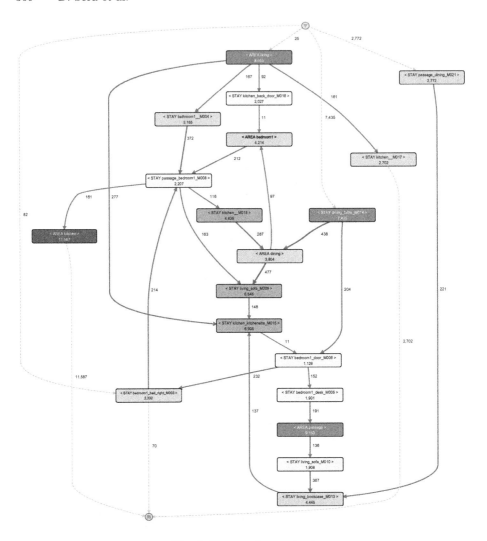

Fig. 1. Fuzzy miner output

capability of isolating single activities in a complete log. This property guarantees the possibility of obtaining a log granularity reduction that makes the process mining algorithms effective also in a smart space context. Examples of activity identifications are shown in Fig. 2. Subgraphs corresponding to the activities (in the example *Eating* and *Meal Preparation*) are highlighted in red.

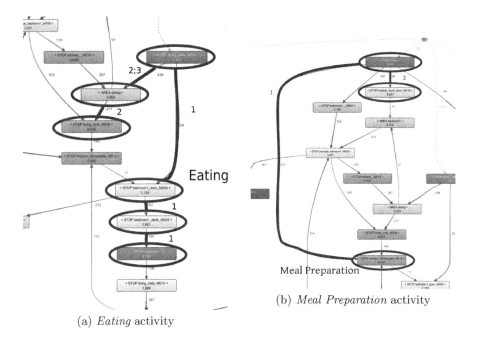

(a) *Eating* activity

(b) *Meal Preparation* activity

Fig. 2. Extracted sub-sequences (examples)

6   Conclusions

Our work is based on the intuition that human habits can be modeled as personal processes. Under this particular perspective, BPM techniques could be successfully exploited to represent human habits through readable models. In fact one of the most relevant problems of data-driven approaches in smart spaces, is the lack of model readability. This implies a poor user interaction with the model itself that makes difficult some operations, as for instance the user model validation.

In this paper, we have outlined about the potential of BPM techniques in modeling a smart space. In particular, we proposed to employ an established visualization and analysis methodology, namely fuzzy mining, for analysis of human habits. To this aim, the sensor log produced by sensors deployed across a smart space must be turned into an event log consisting of human actions. The proposed technique is validated for the case of CASAS datasets and SHiB – Smart Home in a Box – toolbox, where vast majority of sensors are represented by PIR sensors, whose activation sequence is segmented using TRACLUS.

Finally a few considerations about the achieved results and how they open to the possibility of future works. The proposed solution is focused on a purely analytical phase of human habits. Process mining techniques need to take as input a set of cases of the same process, i.e., the same human habit in our scenario. So far, we analyzed the graph obtained by the fuzzy miner by considering as case an entire day, and we detected in it the sub-graphs obtained by considering as cases the repetitions of single human habits by employing a manually labeled dataset. The models obtained in the second case are of course more precise and represent the gold standard, but labeling a dataset is an expensive task. As readability represents the ultimate and most relevant goal of our research, we are currently working on two solutions to this problem. The first solution is a subdivision of the obtained graph into habit sub-graphs by analyzing cluster of nodes identified by looking at edge weights. Another possible approach is to apply data mining in order to automatically segment sub-traces of human actions from the processed logs. Inspiring similar approaches can be found in [11] for what concerns the BPM literature and in [21] in the literature specific to smart spaces.

Models of human habits have an utility that goes beyond the mere analysis. A way to recognize their occurrence at runtime must be devised. At this point the models can be used either for conformance checking (e.g., identifying possible misalignments with the foreseen behavior) or to make the smart space work as a proactive agent. In the second case, controllable services provided by the smart house (e.g., automatic shutters) should be part of the model.

References

1. van der Aalst, W.M.: Process Mining: Data Science in Action. Springer, Heidelberg (2016). https://doi.org/10.1007/978-3-662-49851-4
2. Agrawal, R., Srikant, R., et al.: Fast algorithms for mining association rules. In: Proceedings of 20th International Conference on Very Large Data Bases (VLDB), vol. 1215, pp. 487–499 (1994)
3. Augusto, J.C., Liu, J., McCullagh, P., Wang, H., Yang, J.B.: Management of uncertainty and spatio-temporal aspects for monitoring and diagnosis in a smart home. Int. J. Comput. Intell. Syst. 1(4), 361–378 (2008)
4. Augusto, J.C., Nugent, C.D.: The use of temporal reasoning and management of complex events in smart homes. In: ECAI 2014
5. Aztiria, A., Augusto, J.C., Basagoiti, R., Izaguirre, A., Cook, D.J.: Discovering frequent user-environment interactions in intelligent environments. Pers. Ubiquit. Comput. 16(1), 91–103 (2012)
6. Aztiria, A., Izaguirre, A., Basagoiti, R., Augusto, J.C., Cook, D.J.: Automatic modeling of frequent user behaviours in intelligent environments. In: 2010 6th International Conference on Intelligent Environments (IE), pp. 7–12. IEEE (2010)
7. Aztiria, A., Izaguirre, A., Basagoiti, R., Augusto, J.C., Cook, D.J.: Discovering frequent sets of actions in intelligent environments. In: 2009 5th International Conference on Intelligent Environments (IE), pp. 153–160. IEEE (2009)
8. Baier, T., Mendling, J.: Bridging abstraction layers in process mining by automated matching of events and activities. In: Daniel, F., Wang, J., Weber, B. (eds.) BPM 2013. LNCS, vol. 8094, pp. 17–32. Springer, Heidelberg (2013). https://doi.org/10.1007/978-3-642-40176-3_4
9. Chen, H., Finin, T., Joshi, A.: An ontology for context-aware pervasive computing environments. Knowl. Eng. Rev. 18(03), 197–207 (2003)
10. Dumas, M., La Rosa, M., Mendling, J., Reijers, H.A., et al.: Fundamentals of Business Process Management. Springer, Heidelberg (2013). https://doi.org/10.1007/978-3-642-33143-5

11. García-Bañuelos, L., Dumas, M., La Rosa, M., De Weerdt, J., Ekanayake, C.C.: Controlled automated discovery of collections of business process models. Inf. Syst. **46**, 85–101 (2014)

12. Gottfried, B., Guesgen, H.W., Hübner, S.: Spatiotemporal reasoning for smart homes. In: Augusto, J.C., Nugent, C.D. (eds.) Designing Smart Homes. LNCS (LNAI), vol. 4008, pp. 16–34. Springer, Heidelberg (2006). https://doi.org/10.1007/11788485_2

13. Gu, T., Pung, H.K., Zhang, D.Q.: A service-oriented middleware for building context-aware services. J. Netw. Comput. Appl. **28**(1), 1–18 (2005)

14. Helaoui, R., Riboni, D., Stuckenschmidt, H.: A probabilistic ontological framework for the recognition of multilevel human activities. In: Proceedings of the 2013 ACM International Joint Conference on Pervasive and Ubiquitous Computing, pp. 345–354. ACM (2013)

15. Lee, J.G., Han, J., Whang, K.Y.: Trajectory clustering: a partition-and-group framework. In: Proceedings of the 2007 ACM SIGMOD International Conference on Management of Data, pp. 593–604 (2007)

16. Leotta, F., Mecella, M., Mendling, J.: Applying process mining to smart spaces: perspectives and research challenges. In: Persson, A., Stirna, J. (eds.) CAiSE 2015. LNBIP, vol. 215, pp. 298–304. Springer, Cham (2015). https://doi.org/10.1007/978-3-319-19243-7_28

17. Leotta, F., Mecella, M., Spinelli, G., Sora, D.: Pipelining user trajectory analysis and visual process maps for habit mining. In: 14th Intl IEEE Conference on Ubiquitous Intelligence & Computing (UIC), 4–8 August 2017. (to appear)

18. Loke, S.W.: Logic programming for context-aware pervasive computing: language support, characterizing situations, and integration with the web. In: Proceedings of the 2004 IEEE/WIC/ACM International Conference on Web Intelligence, pp. 44–50 (2004)

19. Loke, S.W.: Incremental awareness and compositionality: a design philosophy for context-aware pervasive systems. Pervasive Mob. Comput. **6**(2), 239–253 (2010)

20. Ranganathan, A., McGrath, R., Campbell, R., Mickunas, M.: Use of ontologies in a pervasive computing environment. Knowl. Eng. Rev. **18**(03), 209–220 (2003)

21. Rashidi, P., Cook, D.J.: COM: a method for mining and monitoring human activity patterns in home-based health monitoring systems. ACM Trans. Intell. Syst. Technol. (TIST) **4**(4), 64 (2013)

22. Riboni, D., Bettini, C.: Context-aware activity recognition through a combination of ontological and statistical reasoning. In: 2009 International IEEE Conference Ubiquitous Intelligence & Computing (UIC), pp. 39–53 (2009)

23. Riboni, D., Sztyler, T., Civitarese, G., Stuckenschmidt, H.: Unsupervised recognition of interleaved activities of daily living through ontological and probabilistic reasoning. In: Proceedings of the 2016 ACM International Joint Conference on Pervasive and Ubiquitous Computing, pp. 1–12. ACM (2016)

24. Ye, J., Coyle, L., Dobson, S., Nixon, P.: Ontology-based models in pervasive computing systems. Knowl. Eng. Rev. **22**(4), 315–347 (2007)

From BPM to IoT

Sylvain Cherrier[1](✉) and Varun Deshpande[2](✉)

[1] Université Paris-Est, Laboratoire d'Informatique Gaspard Monge
(CNRS: UMR8049), Champs-sur-Marne, France
`cherrier@u-pem.fr`
[2] Université Paris-Est Marne-la-Vallée, Champs-sur-Marne, France
`vdeshpan@etud.u-pem.fr`

Abstract. IoT's presence is increasingly felt. There are already more connected devices to the internet than total human population and sales are starting to rise. As there is extensive research ongoing in order to propose a global architecture to build IoT applications, the domain will rise as soon as a common solution will be widespread. Further, we would need to integrate IoT applications into the legacy data processing solutions. Business Process Model, for example, is a common concept used to build and compose software. Could the Internet of Things become an active shareholder in this approach? The Internet of Things, unlike BPM (which is more static), has specific constraints and peculiar organisations based on its dynamicity and heterogeneity. We propose, in this paper, a gateway to adapt our own IoT platform to the needs of the BPM approach and we also try to analyse the difficulties that may arise therein.

Keywords: BPM · IoT · BPM in IoT

1 Introduction

BPM is made for building a workflow of services to be provided. BPM describes the interactions and the sequence of request/responses to/from services. The Internet of Things has an extremely broad definition. In principality, it's interconnected network of things but if viewed objectively in pursuit of this paper, can also be viewed as a heterogeneous network of sensors and actuators where data is collected by sensor(s) and the triggering of action(s) is done by actuator(s).

There are similarities between the two when we visualise the requests/ responses of objects as services provided by these objects. But there are differences as well: In BPM, services at stack are unique and rich (offering complex interactions and detailed responses). There are very distinct, and stable in time. On the contrary, in the IoT, data gathered can indistinctly come from only one sensor or from multiple sensors (average temp, consumption, etc.). Objects' sensing or acting can change, they may be unreachable or running out of energy, their responses may be intermittent, they may be replaced by a different one giving the same data but under another format, etc.

E. Teniente and M. Weidlich (Eds.): BPM 2017 Workshops, LNBIP 308, pp. 310–318, 2018.
https://doi.org/10.1007/978-3-319-74030-0_23

In a nutshell, there is a conflict between the stability, unicity, meaningfulness of the services at work composed in BPM as opposed to the unreliable and changing environment IoT offers. But with the emergence of the IoT as part of the IT, there is an increasing need to interconnect the two worlds and make them interact. We propose in this paper the description and the intended functioning of a conceptual software bridge. This software bridge aims to be a connecting link between the stability needed by BPM and the dynamic IoT environment.

This paper is organised as follows: Sect. 2 presents related works and background for our solution. Section 3 describes the differences in architecture, as this paper presents a conceptual middle-ware intended to serve as a bridge between IoT services and a BPM application. Section 4 gives an examples through an use cases. Finally, concluding remarks end this paper in Sect. 5.

2 State of Art

Business process management (BPM) is a field in operations management that focuses on improving corporate performance by managing and optimizing a company's business processes [1]. It can therefore be described as a "process optimization" process. It is argued that BPM enables organizations to be more efficient, more effective and more capable of change than a functionally focused, traditional hierarchical management approach. These processes can impact the cost and revenue generation of an organization.

The objective of BPM is to facilitate the management of the services offered by an organization. It aims to describes the enterprise activity as a set of processes, and to help the invocation of these processes, from within the organization or from its partners (clients or providers) [2]. The modeling of the Business Process can be done using tools [3], describing the coordination of services, the procedures constraints and the work-flow that involve the different tasks, input and output at stack.

Coming to the Internet of things (IoT), it is the inter-networking of physical devices (also referred to as "connected devices" and "smart devices"), buildings, vehicles and other items embedded with electronics, software, sensors, actuators, and network connectivity which enable these objects to collect and exchange data [4].

The key foundation underlying a global IoT network is its true incorporation of heterogeneity. This, on one hand, brings tremendous possibilities with improved solutions, on other hand, makes the system more difficult to handle. A fully inclusive approach for connected objects is not only difficult for its flawless implementation but also subject of scrutiny in connected world's domain.

The essence and success of IoT lies in its "Remote Robustness" approach. This approach constraints the energy resources available at disposal. Energy for operation of devices is very limited because objects must remain autonomous (some of them are not wired to any energy source). There are others constraints as well: devices are not always physically accessible (for example, a metal sensor in the pavement of parking lot) so battery can't be replaced, the network protocol used to offer the connectivity is sometimes dramatically thrifty, etc.

To cope with this energy issue [5], IoT uses application and network protocols that are very different (and often not compatible) from those used in the Internet of Data. Small payload, small throughput, sleep mode, etc. are used to reduce energy consumption [6]. Smaller payloads and throughput severely limits the use of available bandwidth (which is already reduced due to increased sleep time). This, in turn forces one to use efficient and innovative network protocols.

The purpose of IoT application is mainly to access the data gathered by devices or using the ability to act on the physical world. In order to facilitate object's integration with a well-known programming paradigm, Erik Wilde had proposed to give access to data and actions provided by objects through a REST API [7]. This Resource Oriented Approach is described by Guinard et al. in [8,9]. Using such approach, the writing of applications using IoT devices is simplified because SOA, ROA and REST are mastered by programmers. It also gives an opportunity to integrate IoT solutions in a more global vision of IT Enterprise services [10].

Recently, 6LowPAN, [11] a version of IPv6 for IoT devices, has been standardized by the IETF. Opening a direct access to devices to the IP world, a specific REST protocol has been proposed by the same author: CoAP [12]. A more global access to IoT devices called LwM2M [13], secure, with configuration settings and over-the-air firmware update is proposed on the top of CoAP and 6LowPAN. If the proposal is widely adopted by the industry, it will provide a common way to access, use, configure and update any kind of devices. This unification of a standard access to a huge variety of devices is also the purpose of architectural solutions such as LWM2M [13], oneM2M [14], or IoTA and SENSEI [15] for example.

Proposing an Architecture for IoT applications [16] is a very active domain in which the academic research and the industry is trying to cope with the issues raised by the IoT, mainly in their wireless part. Providing a network access to everyday life objects adds new opportunities, but is subject to many constraints [17].

An architectural solution provides a complete structure aiming to interact with objects. It solves not only the access to services provided by each object, but also their description, the requests (i.e., which object can answer a query, which objects implement this service), security, authentication [16].

Still, the IoT world and the Internet of Data are different in their characteristics, in response time, unicity of provider, reliability of the network, persistence of the service provider, throughput, payload, etc. One of the IoT's architectural approach is based on data: sensors measure the physical world, data is sent to a central point with powerful processing capabilities. This popular architectural approach is provided by solutions such as Thread, OneM2M, IoTA, LwM2M in addition to closed source manufacturer solutions. As none of them became the market leader, the integration of the constantly moving IoT resources in a stable BPM may encounter compatibility issues, forcing company to be linked to an unique provider, or facing the impossibility to integrate new devices as they are not taken into account because of their IoT architectures.

Besides this data-centric architectural approach, IoT proposes multiple architectural approaches viz., *Three-layer, Middle-ware, SOA, Five-layer* [17]. In these architectural approaches, IoT offers services to be used by applications. Kannengiesser describes one architectural solution [18] to integrate IoT in a Business Process Management: *"Smart devices and processes at all levels in the industrial control hierarchy need to interact"*. Following the same line, we propose in this paper a reflection about the difference between the solidity of BPM and the agility of the IoT, and how it could be solved.

3   Differences in Architecture

IoT architectures provide solutions to use objects, with identification, localization, data adaptation to the needs. They must react to various issues and multiple changes. First, IoT devices are heterogeneous. They can dramatically differ in terms of network protocols, operating systems and software used to solve the same problem. As industries give tools to access these devices, the users may encounters issues to make them interact (same manufacturer, same protocol), or simply gather data under the same format (from temperature to motion, accelerometer value, sensitivity, etc.).

Another issue is the number of objects that are at stake for one defined purpose. For example, the temperature value gathered for a part of a building may be the average of data gathered from an important number of devices. The reaction to this aggregated value can be to switch on a set of air conditioners. This specificity of a unique value/service which is gathered/triggered from multiple sources is different from the common BPM approach in which a service/event is unique in its format and source. This rigidity of implementation in BPM works well when situation in question is non-dynamic/non-evolving in a very short period of time. But in IoT's context, such rigidity almost act as a disadvantage. Moreover, the multiplicity leads to another major difference as some elements can fail and can be replaced by new ones, that may differ in terms of API(s), services, and format or a mix of them. As a conclusion, the direct use of Objects of the IoT by BPM seems difficult to achieve. The mediation of specialized IoT middle-ware can introduce a loosely coupled link between the two domains, providing a easier integration.

Last issue is related to network bandwidth and throughput. These characteristic are fluctuating and often limited compared to usual values encountered in wired networks. Because of their specific constraints (especially energy), wireless networks used by sensors and actuators are slow and not reliable. In order to prevent energy loss, the number of messages sent (depending on the direction) can be limited in time. For example, long range protocols (such as Lora or Sigfox) used by embedded devices can send data more often than they receive (only few messages per day). Thus data-centric approach may not be the best solution to integrate IoT in BPM as its effectiveness can be challenged.

On the other side, BPM is an approach that aims to build a centralized control flow that require services provided by other entities for its use. This work-flow is characterized by events, activities, flows and gateways. In BPM, an event is the input or an output, and represents something that happens. On the contrary, an activity is a task that must be done; it is an action that is under responsibility. Flows (or connections) represent the sequence i.e., the order in which the activities are done. Gateway show the different decisions that can be taken, and their impacts on the work-flow.

To be usable by BPM, we propose a software bridge, that on one side provides the stability and meaningfulness required by BPM, and on the other side the agility needed to drive an IoT set of Objects (see Fig. 1). This software bridge can be seen as a link between the IoT network, which is dynamic and the BPM architecture, which is strongly defined. This software bridge is a conceptual model which is underway to its physical and practical implementation.

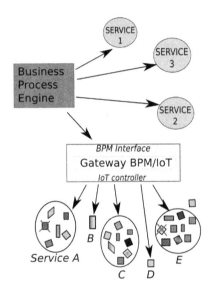

Fig. 1. The proposed conceptual gateway runs a software bridge that offers on one side, a BPM Interface to interact as a standard service used by BPM (the Business Process Engine), while on the other side, is able to interact with multiple IoT services. IoT *Service A*, *C* and *E* are build from multiple devices. *Service B* and *D* are offered by a single device. Devices are provided by different industries. One device of *Service A* is not running. In *Service E*, one device is broken, and another one is out of reach.

To realise this, the proposed software gateway provides the BPM side with an invariable API to access data and services provided by the IoT platform. On the BPM side, the gateway along with software bridge, gives a stable point of view over the IoT devices, hiding their heterogeneity, the difficulty in accessing data in terms of response time, throughput, and failure. The gateway is responsible for transferring events detected inside the IoT network as an input to the BPM. The Gateway also provides services that can be invoked from a BPM process in order to trigger actions on the physical world as a response to activity in BPM.

On the IoT side, the gateway is aware of the various technologies used, their different APIs, and the data format. It gathers data, computes them if the BPM side needs a unique value made of multiple sources, store values to provide quick response time and provide a stable way to actuate devices even if they change with time. Data is treated by objects themselves, and reactions are triggered by actuator(s) as described by the IoT programmer. This computed data, changed into events, can be transmitted to the BPM via gateway. Orders from the BPM can be transformed in an incoming event in the IoT, triggering actuation and reaction inside Objects.

To materialise the conceptual BPM-IoT gateway, we plan to use our ingeniously developed BeC^3 [19] framework. BeC^3 is a framework that shows all the devices as generic ones inside the framework. Then, it is possible to remotely program them, using pre-written set of commands, in order to build a distributed applications. In BeC^3 applications (we call them *compositions*) and Objects act/react depending their environment and others Objects. Programming an object to compute data and send results/orders to other objects provides an event-centric approach that is useful to interact with BPM architecture.

The software bridge running on the gateway is written following the design patterns *Adapter* (a programming Design pattern proposed by the Gang of Four [21]) which interacts with other IoT Objects handled in BeC^3. This Adapter gives an unique representation of actions that are available through IoT objects (such as switch on a device, set a value on a device, display a message, measure a physical value, etc.). It can also collect events (as we use an event-centric approach) from sensors.

4   Use Cases

IoT can be used in various environment(s) and for different purpose(s), such as home automation, smart cities, electronic health or green monitoring. Smart building is a branch of IoT: The supervision of different physical values (water consumption, electricity, temperature) measured by sensors is used in IoT applications to build automation, gather data, offer new service(s) to inhabitants and save energy.

At first, as IoT provides sensors specialized in very specific measure, we can imagine a scenario using activity sensors as a first link between BPM and IoT. For example, sensors can detect presence of people in rooms, floors, lavatories, and actuators can control lights, door locks, temperature, connectivity, depending on the activity detected. Human presence detected at a given floor, in certain rooms can trigger the call to a service, inform security dashboard, activate a work-flow, etc. In another example, the opened door and electrical consumption of the computer reveals the activity of a given staff member. Inside the IoT architecture, this computed event can order a reconfiguration of the user's room (temperature, lights, connectivity). On the contrary (no electrical consumption and door locked), lights are switched off, heaters hibernated to lower energy consumption by the IoT.

These interactions inside the IoT can be used to trigger processes in the BPM, to interact with a procedure or coordination of services provided by the IT. E.g. the BPM receives the information (the member staff is absent) so the PBX is configured to store incoming calls in the member's voice box, or route them to another member, generating voice box message mails. At the IT level, through the BPM, the cleaning service is informed that the room is empty and has to be cleaned. Beside, any work-flow involving orders from this member is aware that no action will be taken for a while. The cleaning person receives a notification (this is IoT work) and the central service is notified (BPM side) that the cleaning service is running in the room.

Thus, IoT events become part of the BPM processes involving the IT management of employees, thereby, monitoring and logging the person's activity in the Human Resource Management's IT Application which can be included as a constraint that drives the IT procedural treatment of the service activity in the Services Management statistics, in the building maintenance organisation, etc.

Events detected by the IoT, treated inside the IoT or even inferred from multiple internal measure(s) can be a source of a more conventional procedure of the BPM. On the other side, the BPM can trigger action in the real world through the IoT, and be a source of information for an IoT application, e.g. office supplies such as papers in printers/photocopiers, even trash bins (empty or full) can be input for the IoT. At the local stage, the IoT can stop the printer/copier, redirect the printing order to another printer automatically while at the global stage (BPM), an alarm is sent, an automatic re-servicing is (see Fig. 2), and the work-flow organization is rescheduled accordingly.

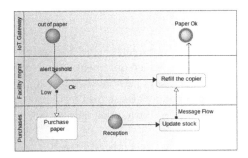

Fig. 2. This figure shows a BPMN workflow. The IoT application has detected an empty paper-tray in a printer, and therefore, reconfigured the default printer for users. The IoT gateway sends an event to the BPM application in order to trigger the paper refill workflow. This work-flow may purchase paper, if the stock is low.

Already existent processes handled by BPM can be improved by the use of the IoT and its connection to events generated by the physical world. For instance, the cleaning service can be more efficient inside the building. If the thrash bins are equipped with sensors sensing if they are full (smart trash is a successful example of IoT impact [20]), the human cleaning service can be reorganized based on real needs and not on a pre-established sequence that may not always fit the needs. Event(s) sent by the IoT smart trash application is/are used as input(s) in the BPM in order to trigger the human activity and to start a specific associated process at the BPM level. The real time spend for cleaning is measured, treated as an input in IT applications, costs are affected based on a real measure. As a result, the building is cleaner while the internal service is more efficiently requested, and costs are more precisely evaluated. The inputs of the BPM work-flow in charge of setting the cleaning service schedule reacts to events generated by the IoT, while outputs of the BPM workflow can become events taken into account inside the IoT application for other needs. For example, the automatic closing of the trash bins to avoid its usage, and an automatic alert sent on the cleaning person's smart-phone.

This example can be extended to sensors that are in charge of the specific activity in a company, for example on assembly lines, inside warehouses, in malls, shopping centers, in offices, etc. Physical measures are treated inside the IoT, and linked to BPM processes as new inputs. These inputs can be useful to create new workflow that add more information to the global perception of the activity. A link between IoT and BPM also gives the ability to offer an output to BPM, that triggers a physical response to the workflow handled in BPM.

5   Conclusion

In this paper, we have presented an example of the possible interactions between the IT usual approach of Business Process Management and the upcoming domain of the Internet of Things. As the IoT is an extension of the Internet to the physical world, its inclusion within the standard paradigm of the computerized management of a company is imperative. We describe a software gateway between the IoT event-centric applications and the BPM work-flow. The gateway offers a proposed bridge to make the two worlds communicate, while it hides their specificity, such as multiplicity of sensor, non-reliability of IoT networks, dynamic changes of the set of devices, etc. On one side, it offers BPM a stable view of services provided by the IoT, providing physical inputs and outputs while on the other side, it take into account the dynamicity of the IoT environment, the heterogeneity of devices at stack, and the specificity of the IoT protocols.

In future, we will explore the effect of this gateway in terms of services discovery and fault-tolerance.

References

1. Panagacos, T.: The ultimate guide to business process management: everything you need to know and how to apply it to your organization (2012). https://books. google.fr/books?id=AyCQMQEACAAJ
2. Danylevych, O., Karastoyanova, D., Leymann, F.: Service networks modelling: an SOA & BPM standpoint. J. UCS **16**(13), 1668–1693 (2010)
3. Mendling, J., Recker, J.C., Wolf, J.: Collaboration features in current BPM tools. EMISA Forum **32**(1), 48–65 (2012). Gesellschaft für Informatik eV

4. Brown, E.: Who needs the Internet of Things? (2016). https://www.linux.com/news/who-needs-internet-things

5. Atzori, L., Iera, A., Morabito, G.: The Internet of Things: a survey. Comput. Netw. **54**(15), 2787–2805 (2010). http://www.sciencedirect.com/science/article/pii/S1389128610001568

6. Anastasi, G., Conti, M., Di Francesco, M., Passarella, A.: Energy conservation in wireless sensor networks: a survey. Ad Hoc Netw. **7**(3), 537–568 (2009)

7. Wilde, E.: Putting things to rest, School of Information (2007)

8. Guinard, D., Trifa, V., Mattern, F., Wilde, E.: From the Internet of Things to the web of things: resource-oriented architecture and best practices. In: Uckelmann, D., Harrison, M., Michahelles, F. (eds.) Architecting the Internet of Things, pp. 97–129. Springer, Heidelberg (2011). https://doi.org/10.1007/978-3-642-19157-2_5

9. Guinard, D., Trifa, V., Wilde, E.: A resource oriented architecture for the web of things. In: Proceedings of IoT (2010)

10. Spiess, P., Karnouskos, S., Guinard, D., Savio, D., Baecker, O., Souza, L., Trifa, V.: SOA-based integration of the Internet of Things in enterprise services. In: IEEE International Conference on Web Services, ICWS 2009, pp. 968–975. IEEE (2009)

11. Shelby, Z., Bormann, C.: 6LoWPAN: The Wireless Embedded Internet. Wiley, Chichester (2010)

12. Shelby, Z.: Embedded web services. IEEE Wirel. Commun. **17**(6), 52–57 (2010)

13. Rao, S., Chendanda, D., Deshpande, C., Lakkundi, V.: Implementing LWM2M in constrained IoT devices. In: 2015 IEEE Conference on Wireless Sensors (ICWiSe), pp. 52–57. IEEE (2015)

14. Datta, S.K., Gyrard, A., Bonnet, C., Boudaoud, K.: oneM2M architecture based user centric IoT application development. In: 2015 3rd International Conference on Future Internet of Things and Cloud (FiCloud), pp. 100–107. IEEE (2015)

15. Krco, S., Pokric, B., Carrez, F.: Designing IoT architecture(s): a European perspective. In: 2014 IEEE World Forum on Internet of Things (WF-IoT), pp. 79–84. IEEE (2014)

16. Gubbi, J., Buyya, R., Marusic, S., Palaniswami, M.: Internet of Things (IoT): a vision, architectural elements, and future directions. Futur. Gener. Comput. Syst. **29**(7), 1645–1660 (2013)

17. Al-Fuqaha, A., Guizani, M., Mohammadi, M., Aledhari, M., Ayyash, M.: Internet of Things: a survey on enabling technologies, protocols, and applications. IEEE Commun. Surv. Tutor. **17**(4), 2347–2376 (2015)

18. Kannengiesser, U., Neubauer, M., Heininger, R.: Subject-oriented BPM as the glue for integrating enterprise processes in smart factories. In: Ciuciu, I., et al. (eds.) OTM 2015 Workshops. LNCS, vol. 9416, pp. 77–86. Springer, Cham (2015). https://doi.org/10.1007/978-3-319-26138-6_11

19. Cherrier, S., Salhi, I., Ghamri-Doudane, Y., Lohier, S., Valembois, P.: BeC^3: Behaviour crowd centric composition for IoT applications. Mob. Netw. Appl. **19**, 1–15 (2013). https://doi.org/10.1007/s11036-013-0481-8

20. Glouche, Y., Couderc, P.: A smart waste management with self-describing objects. In: Leister, W., Jeung, H., Koskelainen, P. (eds.) The Second International Conference on Smart Systems, Devices and Technologies (SMART 2013), Rome, Italy. IARIA, June 2013. https://hal.inria.fr/hal-00924270

21. Gamma, E., Helm, R., Johnson, R., Vlisside, J.: Design Patterns. Addison-Wesley Professional Computing Series

10th Workshop on Social and Human Aspects of Business Process Management (BPMS2 2017)

Introduction to the BPMS2 Workshop 2017

Rainer Schmidt[1] and Selmin Nurcan[2]

[1] Faculty of Computer Science and Mathematics, Munich University of Applied Sciences, Lothstrasse 64, 80335 Munich, Germany
Rainer.Schmidt@hm.edu
[2] Sorbonne School of Management, CRI, University Paris 1 Panthéon-Sorbonne, Paris, France

1 Introduction

Social software [1] [2] is a new paradigm that is spreading quickly in society, organizations, and economics. It enables social business that has created a multitude of success stories. More and more enterprises use social software to improve their business processes and create new business models. Social software is used both in internal and external business processes. Using social software, the communication with the customer is increasingly bi-directional. E.g., companies integrate customers into product development to capture ideas for new products and features. Social software also creates new possibilities to enhance internal business processes by improving the exchange of knowledge and information, to speed up decisions, etc. Social software is based on four principles: weak ties, social production, egalitarianism and mutual service provisioning.

Up to now, the interaction of social and human aspects with business processes has not been investigated in depth. Therefore, the objective of the workshop is to explore how social software interacts with business process management, how business process management has to change to comply with weak ties, social production, egalitarianism and mutual service, and how business processes may profit from these principles.

The workshop discussed the three topics below.

1. **Social Business Process Management (SBPM)**, i.e., the use of social software to support one or multiple phases of the business process lifecycle
2. **Social Business: Social software supporting business processes**
3. **Human Aspects of Business Process Management**

Based on the successful BPMS2 series of workshops since 2008, the goal of the 10th BPMS2 workshop is to promote the integration of business process management with social software and to enlarge the community pursuing the theme.

During the workshop, eight teams presented the results of their research:

In his keynote on the impact of Social Software on Modelling, Michael Möhring gave an insight on new research. He showed that Social Software approaches can reduce the modeling effort and cost in business process modeling projects.

In their paper "Lightweight Process Support with Spreadsheet-Driven Processes: A Case Study in the Finance Domain" the authors Michael Stach, Rüdiger Pryss,

Maximilian Schnitzlein, Tim Mohring, Martin Jurisch and Manfred Reichert showed how the advantages of traditional process management technology could be combined with the ones of spreadsheets. They also showed how spreadsheet-driven processes could improve collaborative work.

How Adaptive Case Management concepts are adapted to support human-intensive process is investigated in the Paper of Ioannis Routis, Mara Nikolaidou and Dimosthenis Anagnostopoulos with the title "Using CMMN to model Social Processes". The authors provide an extension to the Adaptive Case Management meta-model and overall approach is demonstrated in a case study.

Johannes Tenschert, Jana-Rebecca Rehse, Peter Fettke and Richard Lenz investigated interactions in the IT Infrastructure Library (ITIL) in their paper with the title "Speech Acts in Actual Processes: Evaluation of Interfaces and Triggers in ITIL". The authors also demonstrated the importance, prevalence, and diversity of interactions in triggers. They also showed that further abstraction of interactions could improve the reusability of process patterns and identified the great potential of speech act theory for process management.

In their paper titled "SLA-based Management of Human-based Services in Business Processes for Socio-Technical Systems" Mirela Riveni, Tien Dung Nguyen and Schahram Dustdar investigated Service Level Agreements for social computing, including non-functional parameters for human-based services. The authors also investigated Service-Level Agreements at run-time.

How to trigger and influence business processes using smart edge devices is shown in the paper "Using Smart Edge Devices to Integrate Consumers into Digitized Processes: the Case of Amazon Dash-Button" from Michael Möhring, Barbara Keller, Rainer Schmidt, Lara Pietzsch, Leila Karich, Carolin Berhalter and Karsten Kilian. The authors explored the potential value provided by smart edge devices and how they can extend the reach of business processes.

Based on previous work, Maryam Razavian, Irene Vanderfeesten and Oktay Turetken discussed how cognitive biases might negatively influence the design phase of business processes in their paper "Towards a Solution Space for BPM Issues based on Debiasing Techniques". Their research can be used to decrease or dodge the cognitive biases leading to BPM issues.

In their paper with the title "A Framework for Improving User Engagement in Social BPM", the authors Vanisha Gokaldas and Mohammad Ehson Rangiha present a user engagement framework for social BPM. It syndicates theories from psychology and computer science disciplines designed to the organizational, social software and task level.

By collecting and classifying a variety of articles, the paper titled "A Systematic Literature Review of the Use of Social Media for Business Process Management" from Jana Prodanova and Amy Van Looy explores the evolution of social media implementations within the BPM discipline. The authors also point out future research and strategic implication.

We wish to thank all the people who submitted papers to BPMS2 2017 for having shared their work with us, the many participants creating fruitful discussion, as well as the members of the BPMS2 2017 Program Committee, who made a remarkable effort in reviewing the submissions. We also thank the organizers of BPM 2017 for their help with the organization of the event.

2   Program Committee

Renata Araujo	Department of Applied Informatics, UNIRIO
Jan Bosch	Chalmers University of Technology
Marco Brambilla	Politecnico di Milano
Lars Brehm	Munich University of Applied Sciences
Claudia Cappelli	UNIRIO
Monique Janneck	Fachhochschule Lübeck
Ralf Klamma	RWTH Aachen University
Sai Peck Lee	University of Malaya
Michael Möhring	Munich University of Applied Services
Selmin Nurcan	Université de Paris 1 Panthéon - Sorbonne
Andreas Oberweis	Karlsruhe Institute of Technology (KIT)
Michael Rosemann	Queensland University of Technology
Gustavo Rossi	LIFIA-F. Informatica, UNLP
Flavia Santoro	NP2Tec/UNIRIO
Rainer Schmidt	Munich University of Applied Sciences
Pnina Soffer	University of Haifa
Frank Termer	Bitkom e.V.
Irene Vanderfesten	Eindhoven University of Technology

3   References

1. Schmidt, R., Nurcan, S.: BPM and social software. In: Ardagna, D., Mecella, M., Yang, J., Aalst, W., Mylopoulos, J., Rosemann, M., Shaw, M.J., and Szyperski, C. (eds.) Business Process Management Workshops, pp. 649–658. Springer, Berlin (2009)
2. Bruno, G., Dengler, F., Jennings, B., Khalaf, R., Nurcan, S., Prilla, M., Sarini, M., Schmidt, R., Silva, R.: Key challenges for enabling agile BPM with social software. J. Softw. Maint. Evol.: Res. Pract. 23, 297–326 (2011)

Lightweight Process Support with Spreadsheet-Driven Processes: A Case Study in the Finance Domain

Michael Stach[1]([✉]), Rüdiger Pryss[1], Maximilian Schnitzlein[1], Tim Mohring[1], Martin Jurisch[2], and Manfred Reichert[1]

[1] Institute of Databases and Information Systems, Ulm University, Ulm, Germany
{michael.stach,ruediger.pryss,maximilian.schnitzlein,tim.mohring,
manfred.reichert}@uni-ulm.de
[2] AristaFlow GmbH, 89231 Neu-Ulm, Germany
martin.jurisch@aristaflow.com
http://www.uni-ulm.de/dbis
http://www.aristaflow.com

Abstract. The use of process management technology constitutes a salient factor for a multitude of business domains as it particularly addresses the flexibility demands of the digital enterprise. Still, spreadsheet applications are more likely to be used in many scenarios in which process management technology appears to be a more appropriate solution. Especially in the context of human-centric and knowledge-intensive processes, spreadsheets are widely used, even if more business-tailored applications exist. For example, financial service providers, like banks or insurers, prefer spreadsheet applications for accomplishing their daily business. However, this kind of usage reveals drawbacks when working collaboratively based on the same spreadsheet document. To remedy these drawbacks, we suggest the use of spreadsheet-driven processes, which shall combine the advantages of traditional process management technology with the ones of spreadsheets. Using a sophisticated scenario from the financial domain, this paper shows how spreadsheet-driven processes improve collaborative work, as required in the context of business processes, significantly. Moreover, a proof-of-concept prototype is presented to evaluate the approach in practice. Altogether, first results indicate that spreadsheet-driven processes may be a promising technical solution for everyday business involving human resources.

Keywords: Human-centric · Spreadsheet-driven process
Financial field

1 Introduction

During the last decade, the support of human-centric and knowledge-intensive processes has gained increasing attention [11]. In general, corresponding processes face unplanned and complex situations, and their enactments depends on

© Springer International Publishing AG 2018
E. Teniente and M. Weidlich (Eds.): BPM 2017 Workshops, LNBIP 308, pp. 323–334, 2018.
https://doi.org/10.1007/978-3-319-74030-0_24

the expertises and skills of the stakeholders involved. In particular, an unmanageable set of factors makes it impossible for process stakeholders to always foresee all details of the tasks they shall perform and coordinate with other stakeholders. Particularly, these processes comprise mutual phases of planning work and performing it. Usually, in these phases, spreadsheet documents are used to support the respective planning and working tasks. In addition, spreadsheet documents are used to coordinate the tasks of the collaborating stakeholders. Human-centric, knowledge-intensive processes of this kind can be found in domains like, for example, healthcare or finance [5].

Due to regulatory changes in the financial field [12], recently, for many service providers (e.g., banks), the support of human-centric, knowledge-intensive processes is increasingly demanded; e.g., the risk reporting procedure provides a scenario in which recent regulatory changes have generated high demands for any IT solution. More precisely, the management of human resources is particularly challenging. For example, many decisions concerned with the evaluation of risks must be accomplished by more than two experts. To coordinate such decision making process with the help of IT solutions, in turn, often requires a costly adaptation of existing software systems.

In a case study that we conducted in a bank, we learned that risk reporting procedures are still mainly managed based on spreadsheet documents. Regarding the considered bank in particular, and financial service providers in general, the use of spreadsheet documents poses three major advantages:

- Spreadsheet documents are easy to manage, intuitive, and well established.
- Spreadsheet documents can be easily extended.
- Spreadsheet documents provide powerful off-the-shelf features (e.g., calculation functions, chart-based data visualization).

The large-scale use of spreadsheet documents, however, reveals many drawbacks. For example, it is a challenging endeavor for all collaborating stakeholders to work on the same spreadsheet document concurrently without generating unintended side effects (e.g., a corrupt document state). More specifically, avoiding undesired side effects must be accomplished manually, which, in turn, is an error-prone task. To remedy this drawback, while still using the well-known spreadsheet metaphor, we propose spreadsheet-driven processes combining spreadsheet with process management technology. Conceptionally, spreadsheet-driven processes aim to preserve the aforementioned advantages of spreadsheet documents on one hand, while avoiding unintended side effects in the context of collaborative work on one and the same spreadsheet document on the other. Technically, spreadsheet-driven processes are enabled by enhancing existing process management technology in two respects. First, we tightly integrate spreadsheet applications with a robust process engine. Second, we adjust the data handling concept used by common process engines in order to integrate spreadsheet documents for handling data.

In the aforementioned case study, we presented the concept of spreadsheet-driven processes to the stakeholders involved in the risk reporting procedure. As our basic ideas were rated very positively, we realized a proof-of-concept prototype enabling spreadsheet-driven processes based on the *AristaFlow Business Process Management Suite* [2]. This paper reports on three aspects. First, we present the case study, and second, we introduce the concept of spreadsheet-driven processes, and third, we give insights into the proof-of-concept prototype.

The remainder of the paper is structured as follows: Sect. 2 introduces the risk management procedure, the reporting procedure as well as the risk-bearing capacity procedure in detail. Note that this procedure constitutes a fundamental part of the overall risk management procedure. Section 3 then introduces the concept of spreadsheet-driven processes. The proof-of-concept prototype is presented in Sect. 4. Finally, Sect. 5 discusses related work and Sect. 6 concludes the paper with a summary and outlook.

2   Risk Management Procedure

This section presents the overall risk management procedure in general as well as its risk reporting part in particular. Note that the risk-bearing capacity procedure this work is exemplarily focusing on constitutes one crucial part of the risk reporting. Figure 1 illustrates the risk management procedure. In a nutshell, it consists of two parts. The first one is risk inventory, which comprise the strategical tasks, whereas the second part is risk management, which comprises the operational tasks. In the context of risk inventory, in turn, risks (e.g., deficits) are identified and rated, while in the context of risk management they will be handled. To be more precise, the risk inventory part consists of activities *risk detection* and *risk rating*. In turn, *risk measurement*, *risk reporting*, *risk controlling*, and *risk monitoring* constitute the activities of the risk management procedure. At first glance, the risk management procedure seems to be of poor complexity. In practice, however, its handling is challenging as the involved actors need to be coordinated. For example, for almost all activities to be performed, a binding of duties becomes necessary. Thereby, a binding of duty entailment constraint expresses that two independent persons shall approve an activity.

Fig. 1. Risk management procedure

2.1   Risk Reporting Procedure

To give an idea of the complexity of risk management [10], Fig. 2 illustrates major parts of the risk reporting procedure. First, overall risk reporting is accomplished quarterly. Second, it comprises five sub-procedures: *management summary, risk metrics, risk-bearing capacity, single risks*, and *issue management*. Ideally, the five sub-procedures can be executed in parallel in order to minimize overall execution time. After completing all sub-procedures, a second phase of the risk reporting procedure, consisting of seven activities, needs to be coordinated. This second phase starts with the assembly of a preliminary version of the risk report ①. For this purpose, documents created by the mentioned sub-procedures are collected and compiled to a single document ②, i.e., the risk report. Using the compiled report, a consultation with the risk controller takes place ③. As an outcome of this consultation, the risk controller decides whether or not adjustments to the risk report become necessary. If adjustments are required, activity *include additions* is performed ④. If no or no more adjustments are needed, the risk report must be discussed with a responsible board member ⑤. Again, it needs to be decided whether additional adjustments are required ⑥. After the risk report is approved by the responsible board member, it is presented to the executive board ⑦. The resulting risk report is then dispatched to a specified list of addressees ⑧. Note that Fig. 2 only reflects a simplified version of the risk reporting procedure. In daily business, there are many other interdependencies to be considered. Nevertheless, the presented risk reporting procedure provides a proper basis for discussing the risk-bearing capacity procedure.

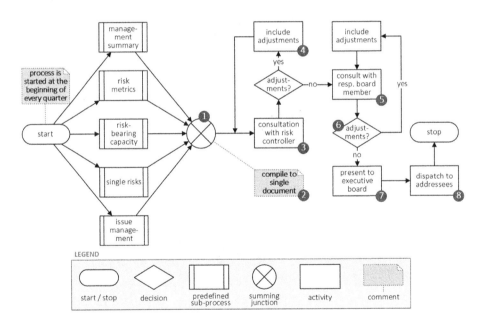

Fig. 2. Risk reporting procedure (in Flowchart Notation)

2.2 Risk-Bearing Capacity Procedure

In the context of the aforementioned case study, we learned that the risk-bearing capacity procedure (RBCP) [6] constitutes the one of the five risk reporting sub-procedures for which the coordination efforts between the involved actors might be significantly decreased when providing a proper IT support. Based on Fig. 3, we illustrate the specific aspects of the RBCP that require sophisticated IT support not existing so far:

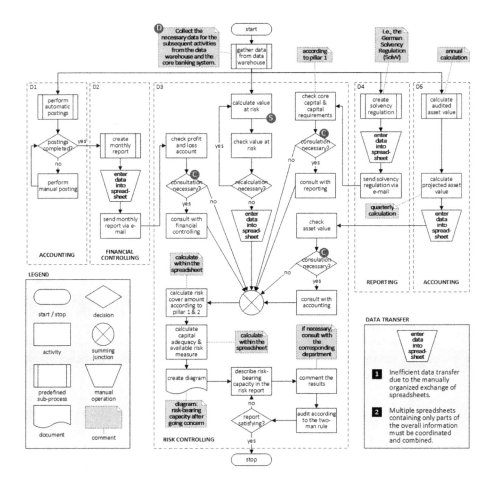

Fig. 3. Risk-bearing capacity procedure (in Flowchart Notation)

- In total, the activities from five different departments (cf. Fig. 3, D1-D5) need to be coordinated, i.e., *accounting, financial controlling, risk controlling, reporting*, and *accounting of leasing company*. Note that the involvement of the leasing company requires the consideration of different organizational scopes.
- RBCP comprises challenging data collection tasks (cf. Fig. 3, (D)) involving numerous heterogeneous IT systems. Several of these tasks, in turn, require the use of simulation tools (cf. Fig. 3, (S)). As a drawback, currently, simulation results need to be manually transferred to the relevant spreadsheet.

– Several RBCP parts show variations in practice. In Fig. 3, they are represented
  as sub-procedure activities (e.g., *create monthly report*).
– Data, exchanged between departments, needs to be approved (cf. Fig. 3, Ⓒ),
  necessitating costly consultation activities.
– In the current practice, spreadsheets are used to coordinate data exchange
  activities between the stakeholders involved in the RBCP. In turn, this results
  in spreadsheet-related drawbacks SRD1-SRD5:

  SRD1: Spreadsheet management (i.e., to send, receive, store spreadsheets)
  is manually organized through e-mails, file sharing, or phone calls.

  SRD2: A sophisticated rights management for the use of spreadsheet documents is difficult to realize.

  SRD3: Manual spreadsheet management hampers fine-grained rights management that allows distinguishing rights regarding the different parts of
  a spreadsheet document. In practice, such a fine-grained rights management would be highly welcome in the RBCP context.

  SRD4: Though spreadsheets constitute the primary data handling instrument, manual data transfer between the spreadsheets and other systems
  (e.g., the simulation tool used for activity *calculate value at risk*) is often
  required.

  SRD5: Recall that the risk reporting procedure, which includes RBCP as
  sub-procedure, is executed quarterly (cf. Fig. 2). In practice, the order of
  the execution activities and the used spreadsheets (i.e., how many spreadsheets are actually used) of the risk reporting procedure vary quarterly.
  Therefore, users want to operate with the same and only one spreadsheet document. To cope with multi-user scenarios in the context of single spreadsheets, a feature must be provided that enables a parallel use
  in this context. Such feature, however, cannot be easily realized using
  off-the-shelf spreadsheet applications.

3   Towards Spreadsheet-Driven Processes

Section 2 has confirmed that the RBCP requires considerable coordination efforts
and involves many stakeholders. Furthermore, RBCP activities are mostly executed in the same sequence each quarter of the year. Considering these two issues,
the use of flexible process management technology for the RBCP is promising [14]. Even the use of contemporary process management technology for the
support of risk management procedures in general, and the RBCP in particular,
however, has revealed several drawbacks as reported by the stakeholders during our case study. To obtain deeper insights into these drawbacks, we consider
Fig. 4. Assume the following scenario: The RBCP shall be executed by a process
engine [15]. Then, additional information is required to transform the RBCP
to an executable process. For example, data elements need to be defined and
assigned to the activities reading or writing them. In current practice, lots of
data from different sources need to be handled. Consider the *data warehouse*
rectangle shown in Fig. 4. It illustrates that the different sources to be handled

lead to a massive definition of data elements and their assignments to activities. Furthermore, for each created data element, the rights of the involved stakeholders, who may operate on these data elements, need to be defined. On one hand, the use of process management technology allows for a technically sound representation of the RBCP. On the other, the maintenance of the aforementioned required data elements is still complex. Taking this into account, we propose to combine advanced process management technology with spreadsheets (cf. Fig. 4). This integration follows three goals:

1. It addresses the discussed spreadsheet-related drawbacks of the current processing of the RBCP (cf. Sect. 2.2). Concerning *SRD1*, for example, the spreadsheet management is automated.
2. It copes with the data handling drawbacks discussed in this section, i.e., it addresses the scenario depicted in Fig. 4, ①.
3. It allows users to continue working in the spreadsheet environment.

Considering these goals, the approach applies the following concept:

- Data entry as well as data retrieval are performed using spreadsheets.
- The spreadsheet management (i.e., to send, receive, store spreadsheets) is governed through a well-defined process as specified for the RBCP.
- Data handling of any supported business process is completely managed through one and the same spreadsheet.

From a technical perspective, the following adaptations are required when integrating existing process management technology with spreadsheets:

1. All activities reading or writing data are assigned to a spreadsheet document (cf. Fig. 4, ②). Note that this requires an enhancement of the modeling component as the structure of the spreadsheet must be obtained automatically.
2. The process engine executing a process needs to be adjusted for the use of spreadsheets as well. In this context, three further adaptations are required:
   (a) The process engine needs to be tightly integrated with a spreadsheet application (e.g., *Microsoft Excel*). For example, the process engine needs to be able to automatically run the spreadsheet application.
   (b) The process engine needs to be able to use spreadsheets for data handling as illustrated in Fig. 4, ②.
   (c) The process engine should ensure that the spreadsheet (1) shall be automatically provided solely to the assigned editor (i.e., the user processing the spreadsheet). Further, it needs to guarantee that the spreadsheet (2) never runs into a corrupt state (e.g., through a multi-user access). Finally, (3) a sophisticated rights management (e.g., cell protection) needs to be implemented.

When following such an approach, several advantages emerge:

1. An editor has read access to the entire spreadsheet while performing any activity. In particular, every editor can obtain all data already entered by other editors.

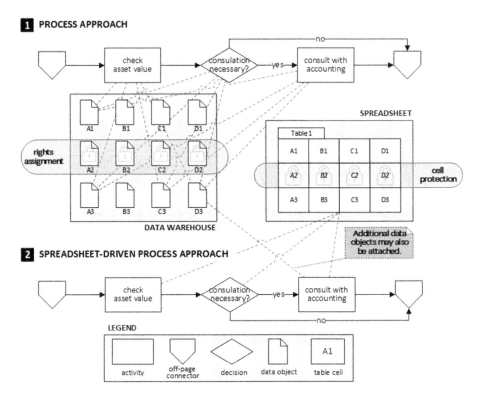

Fig. 4. Data elements in spreadsheet-driven processes (in Flowchart Notation)

2. When processing an assigned activity, an editor may implicitly obtain the current process status (e.g., through data that has not been entered yet, but still has to be provided).
3. When processing an assigned activity, the spreadsheet application may comprise additional and useful features not provided by the process engine. For example, the spreadsheet application can be used to visualize data based on the supported charts.

In summary, we aim at adopting modern process management technology to enable Scenario 2 (cf. Fig. 4) as replacement of Scenario 1 (cf. Fig. 4). We denote processes executed according to Scenario 2 as *Spreadsheet-Driven Processes*. The mentioned case study revealed that the latter will foster RBCP support in particular and risk reporting in general.

4   Proof-of-Concept Prototype

This section illustrates the proof-of-concept prototype we implemented in order to enable the Spreadsheet-Driven Process as such as the one supporting the RBCP. The prototype was realized with *Spreadsheet Router*, a software toolset recently released by the AristaFlow Inc. The toolset, in turn, is based on the AristaFlow Business Process Management Suite [2]. Due to the lack of space, only selected features of the proof-of-concept prototype are presented. Figures 5 and 6 show selected screens of *Spreadsheet Router*. Figure 5 illustrates a screen during the creation of the spreadsheet-driven process supporting the RBCP, while Fig. 6 illustrates a screen when executing the created process. Specifically, Fig. 5 shows one of the three configuration steps required to create the spreadsheet-driven process for the RBCP. To be more precise, the screen shows:

① **A structured activity list:** Based on this list, process activities may be created. In addition, within this screen one may obtain the entire process view of the modeled activities (see bottom of Fig. 5).

② **Spreadsheet area:** Using this area, spreadsheet parts (i.e., so-called spreadsheet areas) of the document can be assigned to the process activities.

③ **Resource assignment:** Using this area, it is specified which editors get the right (e.g., specifying roles) to perform a particular process activity; i.e., getting the right to access an area of the spreadsheet in the context of the specified process activity.

Figure 6 depicts a screen during the execution of the RBCP process. In particular, it shows a form for the activity *check core capital & capital requirements* (cf. Fig. 6, ①, ②). Currently, this activity was processed by an editor through the use of the spreadsheet application. If the editor wants to again change already processed data, he or she must start the spreadsheet application again with the

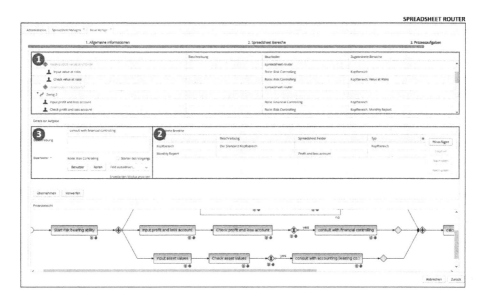

Fig. 5. Spreadsheet-driven process creation

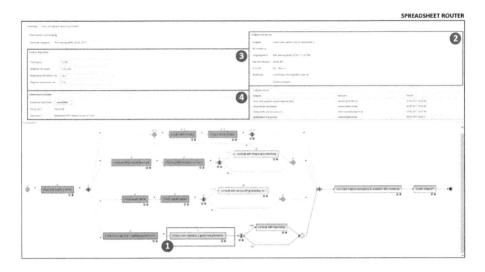

Fig. 6. Spreadsheet-driven process execution

button shown in Fig. 6 at position ④. Note that in the presented case the spreadsheet area consists of four table cells (cf. Fig. 6, ③). Furthermore, *Spreadsheet Router* provides a process-oriented view, e.g., to monitor the current execution status of the process (see the bottom of Fig. 6).

Overall, we realized a proof-of-concept prototype with the *Spreadsheet Router* that supports all features presented in Sect. 3. When using the prototype in practice, the editors demanded another feature that allows assigning process activities to a presentation tool (e.g., *Microsoft PowerPoint*). Such feature would enable them, for example, to easily create presentations for the management board. Altogether, the presented approach offers proper IT support of the risk-bearing capacity procedure within a bank. Currently, we realize other parts of the risk reporting procedure as spreadsheet-driven processes. Finally, a limitation of the prototype is discussed. Therefore, recall the discussed case study (cf. Sect. 2) in which we revealed five spreadsheet-related drawbacks (SRD1-SRD5). The realized prototype provides features that address SRD1-SRD4. A feature that addresses SRD5, in turn, is currently not provided. Therefore, we work on this issue for the second version of the prototype.

5   Related Work

Three categories of related work need to be discussed in the context of this paper. **First**, there are approaches that apply process management technology to support activities for financial service providers. For example, [16] discusses performance issues of business processes in the context of ITIL processes, whereas [8] deals with the re-engineering of business processes in an international bank. Furthermore, there exist approaches that refer to financial scenarios in order to identify the characteristics of the underlying business processes [7]. Overall, the discussed approaches neither focus on human-centric, knowledge-intensive processes in the financial domain nor on the support of risk reporting procedures.

Second, approaches have to be considered that deal with office automation on one hand and spreadsheet-driven approaches on the other. Concerning the first category, such approaches already exist for quite some time. For example, in [3,13] office automation approaches and their characteristics have been introduced. In particular, these approaches focus on the dynamic changes of office environments [4]. Furthermore, these approaches also discuss the need of workflows in the context of office automation and work distribution [9]. However, a tight integration of process management with spreadsheet technology is not presented in more detail by these approaches. Concerning the second category, many approaches exist in the context of information systems that apply spreadsheet-driven approaches to support the paradigm of end-user programming [1]. However, the use of spreadsheets as an advanced concept to perform process activities as we do is not considered so far.

Third, there are commercial tools that support spreadsheet-driven processes. Currently, the best match to this category constitutes the *Microsoft Sharepoint Portal Server*, which allows defining a workflow and assigning spreadsheet documents to its activities. The goal of such workflows is to easily exchange an document to coordinate procedures that particularly require the processing of numerical data. However, parts of an spreadsheet cannot be directly assigned to process activities with the standard features of the Sharepoint Portal Server.

6   Summary and Outlook

The paper presented an approach for the support of human-centric, knowledge-intensive business processes in the field of financial services. Regarding these processes, a variety of spreadsheet documents is used to cope with tasks and their respective processes in everyday business. However, existing IT solutions lack the combined support of spreadsheet technology with contemporary process management systems as required in the context of the present scenario. We propose a novel approach that exploits the benefits of using spreadsheets to perform process activities. Thereby, we illustrated the advantages of this approach in the context of risk reporting procedures. The latter are crucial for many financial service providers as regulatory requirements have been recently changed, but no proper IT solutions exist. We further showed how the approach can be applied to the risk-bearing capacity procedure as a fundamental pillar of risk reporting within a bank. Furthermore, we used the proof-of-concept prototype, which was realized on top of the *Spreadsheet Router* to execute the risk-bearing capacity procedure in terms of a spreadsheet-driven process. We consider the latter processes as crucial when targeting at the support of human-centric, knowledge-intensive business processes in the field of financial services. Altogether, with its spreadsheet-driven approach, the paper shows that the combination of process management with spreadsheet technology is promising.

References

1. Burnett, M., et al.: End-user software engineering with assertions in the spreadsheet paradigm. In: 25th International Conference on Software Engineering, pp. 93–103. IEEE (2003)
2. Dadam, P., et al.: From ADEPT to AristaFlow BPM suite: a research vision has become reality. In: Rinderle-Ma, S., Sadiq, S., Leymann, F. (eds.) BPM 2009. LNBIP, vol. 43, pp. 529–531. Springer, Heidelberg (2010). https://doi.org/10.1007/978-3-642-12186-9_50
3. Ellis, C., Nutt, G.: Office information systems and computer science. ACM Comput. Surv. **12**(1), 27–60 (1980)
4. Ellis, C., et al.: Dynamic change within workflow systems. In: Proceedings of Conference on Organizational Computing Systems, pp. 10–21. ACM (1995)
5. Eppler, M., et al.: Improving knowledge intensive processes through an enterprise knowledge medium. In: Meckel, M., Schmid, B.F. (eds.) Kommunikationsmanagement im Wandel, pp. 371–389. Springer, Wiesbaden (2008)
6. Garleanu, N., Pedersen, L.: Liquidity and risk management. Technical report, National Bureau of Economic Research (2007)
7. Korherr, B., List, B.: Extending the EPC and the BPMN with business process goals and performance measures. In: ICEIS, vol. 3, pp. 287–294 (2007)
8. Küng, P., Hagen, C.: The fruits of business process management: an experience report from a Swiss bank. Bus. Process Manag. J. **13**(4), 477–487 (2007)
9. Lapourtre, C., Rolf, G.: Distributed office automation system with specific task assignment among workstations. US Patent 5,136,708 (1992)
10. Mittnik, S.: Solvency ii calibrations: Where curiosity meets spuriosity. Center for Quantitative Risk Analysis (CEQURA), Department of Statistics, University of Munich, Munich (2011)
11. Mundbrod, N., Reichert, M.: Process-aware task management support for knowledge-intensive business processes: findings, challenges, requirements. In: 3rd International Workshop on Adaptive Case Management and Other Non-workflow Approaches to BPM. IEEE Computer Society Press (2014)
12. Nicolas, M., Firzli, J.: A critique of the basel committee on banking supervision. Revue Analyse Financière, 10 November 2011
13. Olson, M., Lucas Jr., H.: The impact of office automation on the organization: some implications for research and practice. Commun. ACM **25**(11), 838–847 (1982)
14. Reichert, M., Weber, B.: Enabling Flexibility in Process-Aware Information Systems: Challenges, Methods, Technologies. Springer, Heidelberg (2012)
15. Schobel, J., Pryss, R., Wipp, W., Schickler, M., Reichert, M.: A mobile service engine enabling complex data collection applications. In: Sheng, Q.Z., Stroulia, E., Tata, S., Bhiri, S. (eds.) ICSOC 2016. LNCS, vol. 9936, pp. 626–633. Springer, Cham (2016). https://doi.org/10.1007/978-3-319-46295-0_42
16. Spremic, M., et al.: IT and business process performance management: case study of ITIL implementation in finance service industry. In: Proceedings of the 30th International Conference on Information Technology Interfaces, pp. 243–250. IEEE (2008)

Using CMMN to Model Social Processes

Ioannis Routis[(✉)], Mara Nikolaidou, and Dimosthenis Anagnostopoulos

Harokopio University, Eleftheriou Venizelou 70, Athens, Greece
`i.routis@hua.gr`

Abstract. Adaptive Case Management is an alternative approach to support human-intensive processes. It may be served by Case Management Modeling Notation (CMMN) language for modeling purposes. In this paper, ACM concepts are adopted to support human-intensive processes executed in social environments, referred to as social processes. As a first step, the usage of CMMN language is explored to model them. Corresponding CMMN social processes models could be executable within a social network platform. For this purpose, ACM meta-model is extended to incorporate execution properties within a social environment. To demonstrate the potential of the proposed enhancement an example of a social process is used as a case study.

Keywords: Adaptive Case Management
Case Management Modeling Notation · Social BPM
Human-centric processes · Collaborative processes · Executable models

1 Introduction

Adaptive Case Management is a data-driven methodology for business processes targeting human-centric processes and considers data creation and data exchange as the center of its designing philosophy, rather than the process itself [23]. This methodology uses the created case data that may lie even in case actors email inboxes [19], as useful knowledge about the case, in order to support the knowledge workers [20] in decision making. Thus, ACM is supportive and human-centered as its main initiative lies to facilitating the involved participants as well as agile and adaptive as a methodology, especially because it provides an organization with the ability to change according to its needs, a feature that is essential in continuously changing human-centric business domains [26]. However, dependence on changing data and other (possibly unknown) circumstances makes automated support for case management processes extremely challenging [24].

On the other hand, social platforms promote the collaboration between actors and may provide a powerful environment for the execution of human-intensive processes [7], since they provide the means for exchanging messages, create and share data, as well as, keeping track of all the activities belonging in an interaction stream [4]. ACM could be considered to support collaboration between

© Springer International Publishing AG 2018
E. Teniente and M. Weidlich (Eds.): BPM 2017 Workshops, LNBIP 308, pp. 335–347, 2018.
https://doi.org/10.1007/978-3-319-74030-0_25

knowledge workers within a social network platform, as proposed in [17]. However, is it possible to model human-intensive processes executed in a social environment, referred to as social processes, using standard corresponding notation languages? Case Management Modeling Notation (CMMN) proposed by OMG is designed to serve ACM, in a similar manner as BPMN serves BPM for modeling purposes [14].

The goal of this paper is to examine whether CMMN could be effectively used to model social processes. Furthermore, small extensions of corresponding ACM meta-model are proposed to enable CMMN social process models to become executable in a social network platform.

In Sect. 2 of this paper, related work is presented. Section 3 projects the Adaptive Case Management extension through the addition of social characteristics into its core meta-model. Additionally, a social process, namely ReWeee electronic product exchange process, is used as an ACM example. Section 4 presents how the extended meta-model for ACM can be expressed through, utilizing the Case Management notation for the design of ReWeee process. The final Section refers to the conclusion that can be drawn including any added value created from this research work, as well as some future challenges set by the authors.

2 Related Work

The theory of ACM seems to have a continuously growing reputation and competitiveness as far as the business process modeling approaches is concerned, a fact that justifies its growing establishment in research. Thus, it is important for ACM the fact that CMMN has to be widely adapted.

For that reason the CMMN standard defines a common meta-model and notation for modeling and graphically expressing a Case, as well as an interchange format for exchanging Case models among different tools [21]. This specification is intended to capture the common elements that Case management products use, while also taking into account current research contributions on Case management [21]. Based upon this specification for Case Management, several research attempts were made, casting their focus either upon the analysis of the Case Management Specification [6,9,10,12,15], or upon extending and improving this specification [5,25]. However, it still seems to lack in execution effectiveness, especially as far as automation in run-time phase is concerned [8,16,24].

This later argument arises from the fact that different ACM modeling tools had been developed to support this theory, in addition to ACM platforms that had been designed, implementing the core philosophy of this approach. To be more specific, several ACM-empowered platforms and modeling tools exist, such as Cognoscenti, an experimental system for exploring different approaches to supporting of complex, unpredictable work patterns [27], a collaboration platform for ACM as designed and presented in [11], a wiki-based ACMS named Darwin Wiki [13] which empowers knowledge workers with limited modeling capabilities and introduces inference opportunities for structured attributes of a case [28] and ISIS Papyrus [1]. Additionally, the fact that Oracle with Oracle BPM Suite for ACM [2] and IBM with Case Manager tool have implemented suites that support ACM methodology enhances the argument that it is an established approach.

However, the above mentioned platforms and tools cannot be considered as standardized and compliant with the theory of Case Management. The main reason behind this assumption lies to the fact that there is no common view for these tools and platforms and there is not a common graphical language that these tools and platforms can be based upon in order to support adaptive process design and modeling. Thus, what is required is a more standardized concept for ACM and Case Management theory in general.

To this end, the authors of this paper propose an extension to Case Management, by infusing social characteristics into the methodology of ACM, inspired from Casebook [17], which combines the main Case Management elements with social features. To be more specific, Casebook was chosen as it embraces social and collaboration technology, analytics, and intelligence to advance the state of the art in case management from systems of record to a system of engagement for knowledge workers [17]. Moreover, Casebook's main capabilities include case planning, execution and assistance, measuring and learning, and a community-curated case catalog, alongside with the fact that it enables collaboration among case workers to facilitate case resolution in a social environment [17]. Thus, what is inspiring about Casebook it the fact that it implements all the essential ingredients of effective execution of human-centric models.

In this paper, the applicability of the CMMN standard upon social-driven, human-centric processes is examined, as well as the way in which CMMN as a modeling standard can be valuable in modeling a social process. This attempt consists of two stages. The first one includes the infusion of ACM with minor social enhancements in order to enable social processes support, while in second stage what is examined is whether the execution of ACM models was facilitated or not. A benefit from this convergence is the fact that collaboration, which is characterized as a vital ACM feature ([26] in [11]), as it allows actors with suitable expertise to advance the case or provide information which can help to advance the case [25], can be enhanced in a social and knowledge intensive environment. Additionally, what else can be emphasized through this is the event feature. That way, efficient automation levels can be established for a case as an event can express in a more abstract manner the conditions under which an action should be initiated [3].

This way, a social-driven approach of Adaptive Case Management is proposed. It is expected to be implemented within the context of Social Adaptive Case Management, a notion that already exists in research [18, 22]. This alteration of ACM, supported by CMMN, seems effective in modeling human-intensive processes especially, when the executable CMMN models are to be automated and executed within social environments.

3   Extending ACM Meta-Model with Social Features - A Case Study

Is ACM the proper paradigm to model social processes? The authors wondered on this issue when tackling with the ReWeee collaboration process. This collaboration process lies within the framework of the LIFE ReWeee Project, and the scope of the ReWeee web-based collaborative platform, that has as main goal to facilitate and promote Electrical and Electronic Equipment exchange and donation among households or households and public/private bodies. Its success lies within the social communication between volunteers and their collaboration in order to achieve the best possible result.

3.1   ReWeee Social Process as an ACM Example

ReWeee social process allows volunteers to exchange electronic devices through a social network platform, namely the ReWeee platform. Below, a use case diagram of the ReWeee process is presented (Fig. 1). In this, a generic view about both the user categorization is projected and a summary of the primary actions that each type of user can take during the process.

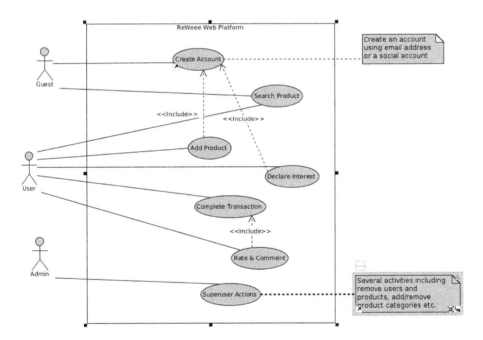

Fig. 1. The ReWeee process use case.

As far as the categorization of actors is concerned, there are three types of users. These are guest users, registered users and administrators, that differentiate themselves as far as their granted permissions upon the use of the platform is concerned.

More specifically, when any unregistered user visits the web platform for the first time, he gets prompted to register into it, by creating a user account. This account can be created either as usual, namely, by signing up via an email and a password or by signing up via a social network account, as well as giving to the platform the necessary permissions for using personal user data. After a successful registration, the, from now on, registered platform user, is able to submit an advertisement donating or exchanging an item, to declare interest for an existing product and propose an offer to acquire it, as well as to communicate with any other user who owns a desirable electric device.

Moreover, a registered user is not only able to search a product based on some conditions, namely, filters like item categories, item state, donating-user region, but also to either suggest changed regarding the item's category for which he is searching, or even to comment in any advertisement that he had made use of. That way, either the platform administrators or the appropriate users will be notified for either the category change proposal or the commenting in an advertisement.

Finally, registered users have a profile in which they are able to be notified for any recycling actions taken via a news-feed as well as being informed for general topics regarding recycling and its benefits. Within each user's profile, a calendar exists via which a user can be informed for any recycling events taking place.

3.2   ACM Meta-Model Extended with Social Features

In this subsection an extended meta-model for ACM is provided as it is projected below (Fig. 2). Based upon the CMMN standard the authors kept ACM main features unchanged, namely, the "Case" and "Task" elements. These Case management features represent the main characteristics of a human-centric environment as it includes cases that need to be handled properly, while it additionally includes, the appropriate human tasks assigned to the involved actors. These elements, alongside with "Data Object" and "Sentry" elements inherited by the CMMN notation are considered as the key features of the ACM meta-model extension (Fig. 2). What is more, a "User" element was inserted to the meta-model replacing the "Role" element that existed before, as a specialization of the involved actors, with the characteristics and actions of a social process user (communication, notifications, publishing).

Moreover, "Data Object" element do not replace the Artifact element which exists in the ACM meta-model [23] but it is inserted into the model as a more general notion of data created during the case execution. For instance, in a daily registration of patient's data, there could be important information, or plain data about his/her case.

What was the most important change into the ACM meta-model [23] was the substitution of the Roadmap element with a more general notion, the one "Sentry", inherited from OMG's CMMN standard [21]. More analytically, the Sentry element was highlighted as it represents a combination of conditions and events that define the sequence of tasks to be implemented, alongside with the roles involved in the case execution and is divided into two parts (states). The first one, the "if part" represents a condition that needs to be satisfied in order to both lead to the second part, the "on part", that describes the event that itself will trigger an action, either a notify activity or a publish activity, and consider any case milestone as a reached one.

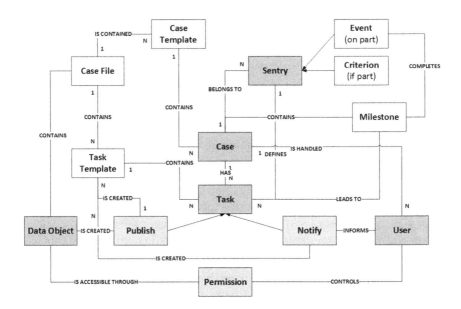

Fig. 2. Social-extended ACM meta-model

Moreover, the task and case templates were retained into this model as well. These elements represent the knowledge attained by the involved actors in any human-centric environment over the completion of similar cases. The data structure that contains all the data objects created for the case as well as the templates for tasks and the case itself, is named as Case File a notion that it was inherited from the CMMN standard and the main theory behind Adaptive Case Management.

However, despite the fact that the Task element was retained into the extended meta-model, it was also divided into two basic categories of human actions, the "notify" and "publish" actions, each one of them has a different result into the case execution. To be more specific, the "notify" action leads to the notification of the appropriate actor, which is to trigger the next user event, so the next appropriate action to be taken, as a result from a previously created task. On the other hand, the "publish" action is the one that is creating, updating or using artifacts and data for the case executed, data that are saved into the case file as data objects. The reason behind this extension of the Task element lies to the fact that these tasks are the most common human tasks to be executed in a social environment, as notification of involved actors and publishing of data can be characterized as a daily routine of social interaction.

Additionally, the "Permission" element was also inserted into the extended meta-model. Permissions were added to separate the access level that each case role has into the case files and useful data objects. It also describe the separation of users in categories within the collaborative ReWeee process, always having in mind the social environment in order to promote the automated execution of CMMN diagrams.

In the following (Table 1), we explain the way each meta-model element is used to describe the ReWeee social process as well as how each one of the Case Management extensions proposed, enhances the overall process. For this purpose, the below projected table was created, in order to match ReWeee process features with ACM meta-model elements.

Table 1. Meta-model elements applicability in the ReWeee process

Meta-model elements	Social and collaborative platform
Data objects	As far as data are concerned, what is stored in the platform are user platform data (uploaded products, region, rating), product data (category, state, quantity, type)
Publish tasks	Publish Tasks are performed during user registration, when user data are saved, or when a new product is offered in the platform and its details are being stored
Notify tasks	Notify Tasks are performed by the time a product state change takes place. For instance, when a user declares interest for a product, its owner gets notified
Users	Role categorization includes two different types of platform users and a platform administrator. These differentiation in roles is used for better user permission handling
Sentries	Sentries take place to the platform mainly when a decision is made. Moreover, sentries happen when user publishing and notification events are triggered
Events	Events are triggered when a user declares interest for a product or rates a product transaction. Then the appropriate users get notified for any product state change
Permissions	Permissions are granted to each user according to the category it belongs to. Different types of users have different types of permissions and different usage restrictions
Milestones	Milestones describe any activity that, when accomplished, causes a process state change. In the ReWeee process, a successful transaction could be defined as a milestone

4   Modeling ReWeee Social Process Using CMMN

In CMMN, ACM data objects and case artifacts are projected as CaseFile items, representing a piece of information of any nature, ranging from unstructured to structured, and from simple to complex [21]. On the other hand, tasks are handled as an atomic unit of work and are divided into different task types, namely, human tasks, process tasks or decision tasks. Sentries as a combination of an event and a condition (criterion) are not used as a standalone CMMN element. At first sentry criteria are used, categorized into two different types, entry and exit criteria of a sentry. Secondly, events are displayed and described into the CMMN standard as anything that can change the case state as far as either the case data or the sequence of case stages, case tasks or case milestones, is concerned, and are divided into two different types, as timer events and user events. Additionally, a milestone represents an achievable target, defined to enable evaluation of progress of the Case. No work is directly associated with a Milestone, but completion of set of tasks or the availability of key deliverables typically leads to achieving a Milestone [21].

4.1   A CMMN Model for the ReWeee Social Process

Having the main executable elements of the extended meta-model for ACM methodology linked with CMMN and described as Case Management entities, it is about time to make a modeling attempt of the ReWeee collaborative process, as the final act of transition from ACM to a social perspective or Case Management. For this purpose an open source CMMN modeling tool was used, Camunda Modeler modeling tool. Mainly, this tool was chosen because of its ease of use, its multi-platform availability and the fact that CMMN was directly available as a modeling language in this tool, without any need of parameterization.

ReWeee social process consists of the main human tasks that a ReWeee platform user can make during his/her interaction either with the platform itself or with other platform users. Having that taken as a premise for the CMMN model that was required to be designed, the below displayed Case Management model was created (Fig. 3) for the ReWeee social process, containing also for the appropriate task elements the notion of stereotype task, marked either as "publish" task or "notify task" as introduced in Fig. 2.

The first human-centric activity that a registered platform user can do is to view the created profile, alongside with whatever this activity includes (i.e. read messages, view notifications). Furthermore, from that point, users can take different human tasks such as searching for a product into the platform database, placing their own adverts of donating or exchanging equipment and directly communicate or interact with other platform users (place comment, submit rating).

If a user decides to place his own advert for an EE product, then this task has some characteristics except from being marked as a "publish" stereotype. Firstly, it has an entry criterion of an existing user profile, namely, no unregistered user can use the platform to place a product advert. Secondly, it has an exit criterion which creates the following case elements. It not only creates data objects containing information about the newly submitted product, but also the case milestone of the product publication is reached alongside with some notification events that happen based on some conditions. The first one is to have an exchange advert published, namely, a product is proposed for exchange with another product previously published in the platform. Thus, the owner of the previously published product is notified that there is interest about his/her advert. The second is a notification event triggered to users that have the newly published product on their wishing list and have them automatically notified that there is one available.

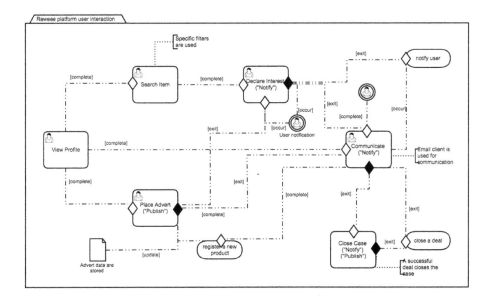

Fig. 3. CMMN model for ReWeee social process.

If the first one, the "product search" task is chosen, except from the entry criterion of an existing user profile, the user sees a list of available products referring to the filters applied, from which a product can be chosen for viewing and from this point he/she can declare interest upon that. That describes another human task marked as a "notify" stereotype that can be taken within the ReWeee collaborative process, the one of "declare interest" which also has some characteristics. Firstly, there are two separate entry criteria that are independent to each other. The first one, is the entry criterion of having searched a product in order to view it and declare interest for it, while the second one refers to the entry criterion of having placed an advert for product exchange which contains interest declaration of another product. Independently of these entry criteria, this task triggers a user event of notification to the product owner about which there is declared interest, after its exit criterion is satisfied. It, additionally, not only leads to the "communicate" human task marked as "notify" stereotype, but also after its execution, the process reaches the milestone of platform user notification.

Finally, if the communication task is chosen or triggered then an collaboration subprocess is initiated, alongside with continuous notification events for new incoming platform messages. More specifically, this task contains entry criteria of having interest declared for a product, having a registered user profile and have an advert place for a product. This human task's exit criterion is satisfied, then the "close case" task is triggered marked as both "publish" and "notify" stereotype that has as entry criterion a successful communication between platform users, and its exit criterion completes the milestone of closing the case, namely, the transaction for a product.

4.2 Discussion upon ReWeee Process CMMN Model

ReWeee process modeling using CMMN was a straight-forward task. There was no difficulty for the modeler in understanding the transformation of the abstract notions described into the extended ACM meta-model (Fig. 2) to related CMMN elements.

The use of task stereotypes promotes automation within the process, as it gets easier for a CMMN modeler to link these two stereotypes, namely, "publish" and "notify" labels, with specific following actions into the process life-cycle, or even with the triggering of specific events that can change the process state. Furthermore, the use of stereotypes promotes the notion of "publishing" information, which is considered as a vital activity in social environments as well as a basic feature of Adaptive Case Management theory.

Finally, the benefits of milestone and sentry elements included in the model of ReWeee process should be highlighted. The use of milestones and sentries provides a process designer with the ability to control the flow of the social process. Moreover, it gets easier to put restrictions to the process execution to make its implementation more efficient, while it helps process modeler to make the process representation more descriptive.

5   Conclusions and Future Work

The adaptation of CMMN language to model social processes was explored in the paper. The ReWeee collaboration process was used as a case study. Corresponding CMMN social process models should be executable ones and may be implemented within a social platform environment. As a first step, we discussed some minor extension of CMMN and the corresponding ACM meta-model to serve ReWeee process in an effort to establish executable social process models.

Regarding CMMN adaptation, we concluded that the social infusion of Adaptive Case Management leads to an enhancement of the collaboration between the involved case actors, a fact that facilitates automatic scheduling and design of a human-centric process. Furthermore, despite the fact that ACM may seem unstructured, with the appropriate event handling, a decent level of automation can be achieved. Finally, a future challenge for this work could be set so as to have an executable version of the above projected collaborative process model in order to have the presented extension of ACM, implemented.

Acknowledgement. This work is co-funded by the European Commission through the LIFE+ Funding program, LIFE14 ENV/GR/000858: LIFE REWEEE.

References

1. ISIS Papyrus. www.isis-papyrus.com
2. Ajay, K.: Managing Unpredictability using BPM for Adaptive Case Management (2013)
3. Alexopoulou, N., Nikolaidou, M., Anagnostopoulos, D., Martakos, D.: An event-driven modeling approach for dynamic human-intensive business processes. In: Rinderle-Ma, S., Sadiq, S., Leymann, F. (eds.) BPM 2009. LNBIP, vol. 43, pp. 393–404. Springer, Heidelberg (2010). https://doi.org/10.1007/978-3-642-12186-9_37
4. Alexopoulou, N., Nikolaidou, M., Stary, C.: Blending BPMS with social software for knowledge-intense work: research issues. In: Nurcan, S., Proper, H.A., Soffer, P., Krogstie, J., Schmidt, R., Halpin, T., Bider, I. (eds.) BPMDS/EMMSAD -2013. LNBIP, vol. 147, pp. 18–31. Springer, Heidelberg (2013). https://doi.org/10.1007/978-3-642-38484-4_3
5. Bider, I.: Towards process improvement for case management: an outline based on viable system model and an example of organizing scientific events. In: Reichert, M., Reijers, H.A. (eds.) BPM 2015. LNBIP, vol. 256, pp. 96–107. Springer, Cham (2016). https://doi.org/10.1007/978-3-319-42887-1_9
6. Bruno, G.: Tasks and assignments in case management models. Procedia Comput. Sci. **100**, 156–163 (2016). http://linkinghub.elsevier.com/retrieve/pii/S1877050916323031
7. Bruno, G., Dengler, F., Jennings, B., Khalaf, R., Nurcan, S., Prilla, M., Sarini, M., Schmidt, R., Silva, R.: Key challenges for enabling agile bpm with social software. J. Softw. Maint. Evol. **23**(4), 297–326 (2011). https://doi.org/10.1002/smr.523
8. Cognini, R., Hinkelmann, K., Martin, A.: A case modelling language for process variant management in case-based reasoning. In: Reichert, M., Reijers, H.A. (eds.) BPM 2015. LNBIP, vol. 256, pp. 30–42. Springer, Cham (2016). https://doi.org/10.1007/978-3-319-42887-1_3
9. Czepa, C., Tran, H., Zdun, U., Tran Thi Kim, T., Weiss, E., Ruhsam, C.: Towards structural consistency checking in adaptive case management. In: Reichert, M., Reijers, H.A. (eds.) BPM 2015. LNBIP, vol. 256, pp. 90–95. Springer, Cham (2016). https://doi.org/10.1007/978-3-319-42887-1_8
10. De Carvalho, R.M., Mili, H., Boubaker, A., Gonzalez-Huerta, J., Ringuette, S.: On the analysis of CMMN expressiveness: revisiting workflow patterns. In: Proceedings - IEEE International Enterprise Distributed Object Computing Workshop, EDOCW 2016-September, pp. 54–61 (2016)
11. Huber, S., Lederer, M., Bodendorf, F.: It-enabled collaborative case management - principles and tools. In: 2014 International Conference on Collaboration Technologies and Systems, CTS 2014, pp. 259–266 (2014)
12. Kurz, M., Schmidt, W., Fleischmann, A., Lederer, M.: Leveraging CMMN for ACM: examining the applicability of a new OMG standard for adaptive case management. In: Proceedings of the 7th International Conference on Subject-Oriented Business Process Management, p. 4 (2015)

13. Hauder, M., Kazman, R., Matthes, F.: Empowering end-users to collaboratively structure processes for knowledge work. In: Abramowicz, W. (ed.) BIS 2015. LNBIP, vol. 208, pp. 207–219. Springer, Cham (2015). https://doi.org/10.1007/978-3-319-19027-3_17

14. Marin, M., Hull, R., Vaculín, R.: Data centric BPM and the emerging case management standard: a short survey. In: La Rosa, M., Soffer, P. (eds.) BPM 2012. LNBIP, vol. 132, pp. 24–30. Springer, Heidelberg (2013). https://doi.org/10.1007/978-3-642-36285-9_4

15. Marin, M.A.: Introduction to the Case Management Model and Notation (CMMN) (2016)

16. Marin, M.A., Hauder, M., Matthes, F.: Case management: an evaluation of existing approaches for knowledge-intensive processes. In: Reichert, M., Reijers, H.A. (eds.) BPM 2015. LNBIP, vol. 256, pp. 5–16. Springer, Cham (2016). https://doi.org/10.1007/978-3-319-42887-1_1

17. Motahari-Nezhad, H.R., Spence, S., Bartolini, C., Graupner, S., Bess, C., Hickey, M., Joshi, P., Mirizzi, R., Ozonat, K., Rahmouni, M.: Casebook: a cloud-based system of engagement for case management. IEEE Internet Comput. **17**(5), 30–38 (2013)

18. Motahari-Nezhad, H.R., Bartolini, C., Graupner, S., Spence, S.: Adaptive case management in the social enterprise. In: Liu, C., Ludwig, H., Toumani, F., Yu, Q. (eds.) ICSOC 2012. LNCS, vol. 7636, pp. 550–557. Springer, Heidelberg (2012). https://doi.org/10.1007/978-3-642-34321-6_39

19. Motahari-Nezhad, H., Swenson, K.: Adaptive case management: overview and research challenges. In: 2013 IEEE 15th Conference on Business Informatics (CBI), July 2013, pp. 264–269 (2013)

20. Mundbrod, N., Kolb, J., Reichert, M.: Towards a system support of collaborative knowledge work. In: La Rosa, M., Soffer, P. (eds.) BPM 2012. LNBIP, vol. 132, pp. 31–42. Springer, Heidelberg (2013). https://doi.org/10.1007/978-3-642-36285-9_5. http://www.uni-ulm.de/dbis

21. Object Management Group: Case Management Model and Notation v1.1 (2016). http://www.omg.org/spec/CMMN/1.1/CMMN

22. Osuszek, L., Stanek, S.: Social networking influence on case management in enterprise business. Int. J. Acad. Res. Bus. Soc. Sci. **5**(2), 97–106 (2015). http://hrmars.com/index.php/journals/papers/IJARBSS/v5-i2/1465

23. Routis, I., Stratigaki, C., Nikolaidou, M.: Exploring ACM and S-BPM for modelling human-centric processes: an empirical comparison. In: Proceedings of the 8th International Conference on Subject-oriented Business Process Management, pp. 5:1–5:4 (2016). http://doi.acm.org/10.1145/2882879.2882893

24. Rychkova, I., Rychkova, I., Kirsch-pinheiro, M.: Automated Guidance for Case Management : Science or Fiction? March 2016 (2013)

25. Scheit, S., Ploom, T., O'Reilly, B., Glaser, A.: Automated event driven dynamic case management. In: Proceedings - IEEE International Enterprise Distributed Object Computing Workshop, EDOCW 2016-September, pp. 62–71 (2016)

26. Swenson, K.D.: Mastering the Unpredictable: How Adaptive Case Management Will Revolutionize the Way that Knowledge Workers Get Things Done. Meghan-Kiffer Press, Tampa (2010). http://www.worldcat.org/search?qt=worldcat_org_all&q=9780929652122

27. Swenson, K.D.: Demo: cognoscenti open source software for experimentation on adaptive case management approaches. Proceedings - IEEE International Enterprise Distributed Object Computing Workshop, EDOCW, pp. 402–405 (2014)
28. Tenschert, J., Lenz, R.: Towards speech-act-based adaptive case management. In: Proceedings - IEEE International Enterprise Distributed Object Computing Workshop, EDOCW 2016-September, pp. 25–32 (2016)

Speech Acts in Actual Processes: Evaluation of Interfaces and Triggers in ITIL

Johannes Tenschert[1]([✉]), Jana-Rebecca Rehse[2], Peter Fettke[2],
and Richard Lenz[1]

[1] Institute of Computer Science 6 (Data Management),
Friedrich-Alexander-Universität Erlangen-Nürnberg (FAU),
Erlangen, Germany
{Johannes.Tenschert,Richard.Lenz}@fau.de
[2] Institute of Information Systems (IWi),
German Research Center for Artificial Intelligence (DFKI),
Saarbrücken, Germany
{Jana-Rebecca.Rehse,Peter.Fettke}@iwi.dfki.de

Abstract. Today's organizations are socio-technical systems in which human workers increasingly perform knowledge work. Interactions between knowledge workers, clerks, and systems are essentially speech acts controlling the necessity and flow of activities in semi-structured and ad-hoc processes. IT-support for knowledge work does not necessarily require any predefined process model, and often none is available. To capture what is going on, a rising number of approaches for process modeling, analysis, and support classify interactions and derive process-related information. The frequency and diversity of speech acts has only been examined within delimited domains, but not in the larger setting of a reference model covering different types of work and domains, multiple takeholders, and interacting processes. Therefore, we have investigated interactions in the IT Infrastructure Library (ITIL). ITIL is a collection of predefined processes, functions, and roles that constitute best practices in the realm of IT service management (ITSM). For ITIL-based processes, we demonstrate the importance, prevalence, and diversity of interactions in triggers, and that further abstraction of interactions can improve the reusability of process patterns. Hence, at least in ITSM, applying speech act theory bears great potential for process improvement.

Keywords: Speech act theory · ITIL
Knowledge-intensive business process · Adaptive case management
Integration

1 Introduction

Today's organizations are socio-technical systems [1]. There, the human workers increasingly perform knowledge work rather than manual labor [2]. In collaboration, knowledge and transaction workers perform structured, semi-structured,

© Springer International Publishing AG 2018
E. Teniente and M. Weidlich (Eds.): BPM 2017 Workshops, LNBIP 308, pp. 348–360, 2018.
https://doi.org/10.1007/978-3-319-74030-0_26

and ad-hoc processes [3]. For these different requirements on flexibility, various techniques are applied, e. g. standardized processes, tailored information processing systems, adaptive case management systems (ACMS), and groupware [4]. Unsurprisingly, process-driven applications rely heavily on interaction with their environment [5]. Interactions between knowledge workers, clerks, and systems are essentially speech acts controlling the necessity and flow of activities and information in semi-structured and ad-hoc processes. We proposed to use speech-act-based classification for integration of systems [3]. With a communication-centric ACMS, we intend to improve collaboration between knowledge workers, clerks, and systems, exchange of information across processes, and overall performance and reliability of interfaces.

But *are* the speech acts in organizations with cooperating agents sufficiently homogeneous for such a support? We analyze the IT Infrastructure Library (ITIL) to assess this question. ITIL is a well-known suite for best practices in IT service management (ITSM) that consists of five core publications for 26 processes [6–10]. A wide range of organizations use ITIL as a blueprint to implement and manage their ITSM processes. This way, they are able to be audited and certified for ISO/IEC 20000. In addition, they may also save time and cost, and benefit from the experience and expertise dedicated to ITIL development. Reference models like ITIL are associated with a higher quality of the business process models and the related processes. They simplify internal communications by introducing a common terminology and considerably reduce the resources required for business process management [11].

The contribution of this paper is examining ITIL-based processes in regard to the importance, prevalence, and diversity of their interactions in triggers, and unveiling the strong coupling between actual processes indicated by their interfaces. The generic nature of many ITIL-based processes suggests that our findings can be extended onto other reference models and domains. This way, we both check the premises speech-act-based or generally communication-centric process support relies on and unveil potential for abstraction of elements in ITIL.

The following sections introduce our methods, speech act theory, and related work. In Sect. 5, we analyze triggers in ITIL in regard to prevalence and diversity of speech acts, and in Sect. 6, we analyze interfaces in ITIL to indicate that abstraction of interactions can improve reusability of process patterns. We discuss our approach in Sect. 7. Finally, we conclude and outline further work.

2   Methods

We intend to clarify the importance of speech acts in actual processes. Therefore, we first identified three research questions to answer:

RQ 1 Do speech acts have an important role in actual business processes?
RQ 2 Are the speech acts prevalent in actual processes diverse?
RQ 3 Could further abstraction of interactions improve processes in regard to reusability of process patterns?

One could (1) observe and model processes of some organizations, (2) use existing process models of some organizations, or (3) use a reference model explaining processes for a whole domain. Since a reference model not only contains processes valid for an entire domain, but also assures a certain degree of quality and distribution, we chose option (3). This way, we are able to abstract from the specificities of an individual organization, making our analysis more convincing and easily reproducible. ITIL is well-suited for our analysis due to its relevance for almost all organizations, its textual documentation and its scope, covering a wide range of processes, activities, and stakeholders within and between organizations. Every chapter for a process contains a section "Triggers, inputs, outputs and interfaces". We use these descriptions of touch points between processes.

RQ 1 clarifies whether speech acts are worth considering in actual processes at all. Since triggers of a process in ITIL are well documented, we classify triggers and distinguish classes that do or do not represent speech acts.

RQ 2 is based on RQ 1 being affirmed. In this case, all relevant speech acts in an actual process might be of the same type, e. g. some requests. Hence, RQ 2 checks whether just making transparent that there is an interaction might not be sufficient. To answer RQ 2, we evaluate the triggers that represent speech acts in regard to their potential pragmatic intentions.

RQ 3 unveils potential to apply speech act theory in process modeling. It might easily be answered with one positive effect. Answers could have a different focus, e. g. compliance [12], making contents of a process transparent [13], or integration [3]. Here, we focus on the latter. Without further abstraction of touch points, tight coupling of processes suggests tight coupling of the systems involved. We evaluate all interfaces between processes in ITIL.

3   Speech Act Theory

Speech act theory was introduced by Austin [14] and Searle [15]. Saying something is an action with a particular intention of the speaker. Not only utterances are speech acts, but rather all activities with the intention to send a message. Some types of interactions adhere to typical patterns, e. g. questions are usually answered. The speaker is well aware of the context and of his pragmatic intention, but the systems supporting him currently are not. A speech act consists of its illocutionary force F and its propositional content P. In Searle's $F(P)$ framework, a speech act can be the propositional content of some other speech act. Searle distinguishes five illocutionary points of illocutionary forces:

1. **Assertive**: Commit the speaker to something being the case, e. g. assert, inform, remind.
2. **Commissive**: Commit the speaker to some future course of action, e. g. commit, promise, accept.
3. **Declarative**: Change the reality according to the propositional content, e. g. approve, decline, judge.

4. **Directive**: Attempt to cause the hearer to take some particular action, e. g. request, ask, order.
5. **Expressive**: Express the attitude or emotions of the speaker, e. g. thank, congratulate, apologize.

The English language alone contains at least 4800 speech act verbs [16] which may have different meanings in different domains. Selections of illocutionary forces for a specific domain or purpose may be called a speech act library.

4   Related Work

Dietz [17,18] proposed the *Design & Engineering Methodology for Organizations* (DEMO) to analyze and model organizations and their processes. It considers speech acts and production acts. Making this dichotomy of actions explicit allows considering dependencies in coordination and production separately to unveil problems or opportunities. Van Nuffel et al. [19] saw a potential to use DEMO to verify completeness and consistency of BPMN models. Dietz and Van Nuffel et al. already assumed that speech acts are prevalent in and important parts of business processes. In this paper, we try to confirm these assumptions.

Decker and Barros [20] introduced extensions to BPMN for interaction modeling to describe the control flow of interactions. The approach emphasizes on interactions in a way that reduces redundancy and the risk of potentially incompatible behavior. Nearly all interactions in [20] contain speech act verbs. Even though the additional effort for annotations of illocutionary forces would be negligent, they are not made explicit. Nonetheless, Decker et al. suggested that interaction modeling allows faster creation and understanding by human modelers, and that interactions seem to be prevalent in some domains.

Auramäki et al. [21] already suggested a speech-act-based approach for office automation as early as 1988. In their model, offices are networks of commitments. Acts are classified into instrumental acts and speech acts. Discourse analysis is carried out in regard to coherency, completeness, and ambiguity or rather the quality of discourses. In this paper, we analyze interactions in ITIL in a higher granularity, i. e. up until the illocutionary force. This can help in extracting a speech act library that is applicable to ad-hoc processes of the same domain.

Alter [22] proposed system interaction patterns as a frame of reference and vocabulary. He identified 19 patterns and subdivided them into four categories, including speech act and coproduction patterns. The approach depends on interactions being prevalent in socio-technical systems, which for ITIL is examined in this paper. Compared to the level of abstraction in our analysis, Alter aggregates speech acts, and enriches them with information useful for support in practice.

KQML is a language to communicate attitudes about information, e. g. querying or offering [23]. It contains an extensible set of predefined performatives. Messages may contain other KQML queries, i. e. KQML supports the $F(P)$ framework. Kimbrough and Moore [24] introduced a speech-act-based formal language for business communication (FLBC) in the context of an army office. Paths of command and responsibility can easily be delineated and the illocutionary force is made explicit. KQML and FLBC show that (i) electronic communication can be semantically enriched with speech acts, (ii) Searle's $F(P)$ framework has practical use, and (iii) inference on speech acts is useful for integration.

Schoop [25] analyzed scenarios of communication problems in health care using both Habermas' Theory of Communicative Action as a taxonomy of communication breakdowns and speech act theory as a taxonomy of utterances. She presented a framework for cooperative documentation systems integrating both theories. Schoop showed that a small speech act library suffices to analyze and model most interactions in a hospital, i. e. she analyzed RQ 2 in health care. For ITIL, we outline how to extract a library of similar size, and evaluate the impact of interactions on processes as well as potentials in further abstraction.

Negoisst [26] is a speech-act-based negotiation support system for complex negotations of contracts. Schoop et al. provided a small speech act library to annotate messages and support the process flow. Negoisst shows that speech-act-based support can be implemented in a way that is accepted and properly applied by end users. We analyze speech acts in ITIL to unveil potential impact of support in a broader range of processes and in ITSM in particular.

Reference models in general have been a popular topic for BPM researchers for a long time [11]. What makes ITIL particularly interesting is that it is a reference model that was mainly developed by practitioners with the goal of meeting customer demands and does not contain innovative features, developed e. g. by researchers [27]. This practical orientation makes ITIL a de-facto industry standard [28]. Former analyses have shown that the amount of consolidated knowledge and experience and the promised methodological support for day to day governance practice is what makes ITIL attractive for organizations [29]. Its origins in practice and its widespread adoption suggest that speech acts identified in ITIL are in fact relevant to individual organizations as well.

5   Triggers in ITIL

ITIL describes interactions informally, but does not clearly highlight or classify them as speech acts. For all 26 ITIL processes, we examined the triggers subsection and, where necessary, descriptions of related artifacts. The analysis was performed in two steps: First, similar triggers were grouped and the groups were labeled as categories. In the second step, all immediately identifiable speech acts were examined to ascertain their illocutionary force(s). This two-step analysis is intended to safeguard both RQ 1 and RQ 2, by decoupling the observation of the importance and amount of speech acts in processes from their classification into illocutionary forces. Hence, regardless of the judgment of the individual classifications, the basic conclusions and answers to RQ 1 and RQ 2 should hold.

5.1   Categories of Triggers

First, we subdivided triggers into the categories immediately identifiable speech acts, artifact, observation, and periodic trigger[1]. Differences between artifact-based triggers become apparent after an iteration through all triggers. Even though decisions are artifacts, they are communicated to stakeholders and implicitly represent speech acts. Hence, artifacts were further subdivided into decisions, system events, and plain artifacts. We identified six categories of triggers:

1. **Immediately identifiable speech acts:** *Interactions* that are some sort of speech act, e. g. requesting, complaining, and triggers after review meetings[2].
2. **Decision made by other department:** A decision made by some other department has to be communicated to start a process. Even if it is only written in a document, commanding its execution definitely is a speech act.
3. **Artifact:** The creation of artifacts, e. g. utilization rates, a scheduled meeting, or a published customer satisfaction survey, may or may not be triggered by speech acts. For reasons of clarity, artifacts do not count as speech acts.
4. **System event:** Special case of artifact in which a computer system raises an exception or yields certain events and alerts. Even though some of those messages could be viewed as speech acts, for reasons of clarity they are not.
5. **Periodic trigger:** For example, monthly/quarterly reporting or budget planning cycles. Periodic triggers do not count as speech acts.
6. **Observation:** Some insight a member of an organizational unit gained about changes in business needs, opportunities, etc. This knowledge does not always have to be transferred and interactions are not always necessary to trigger a process. For reasons of clarity, observations do not count as speech acts.

Table 1. Categorization of triggers for each ITIL core publication

Trigger	Book					
	Service strategy	Service design	Service transition	Service operation	Continual Serv. improvement	Σ
Immediately identifiable speech acts	10	24	14	12	1	61
Decision	9	22	7	0	2	40
Artifact	5	6	3	3	1	18
System event	0	4	0	7	0	11
Periodic trigger	2	7	1	0	1	11
Observation	3	7	0	0	1	11
Σ	29	70	25	22	6	152

---

[1] These were identified by evaluating and grouping similar triggers in service strategy.
[2] They are raised through the interactions and decisions in a meeting. Even though in this step their exact categorization was not elaborated, the trigger *is* a speech act.

Table 2. Categorization of illocutionary forces for immediately identifiable speech acts

Illocutionary Force$_{\text{Illoc. Point(s)}}$	Service strategy	Service design	Service transition	Service operation	Continual service improvement	Σ
Accept$_{23}$		1				1
Authorize$_3$			4			4
Clarification$_1$		1				1
Complain$_{145}$	2	1		1	1	5
Compliment$_{45}$		1				1
Confirm$_2$		1				1
Counter-Offer$_{24}$		1				1
Inform$_1$	1		1			2
Inform$_{14}$	1		1			2
Initiate$_4$				1		1
Notify$_1$		5	2	5		12
Offer$_2$		1				1
Order$_4$			1			1
Question$_4$		1				1
Recognize$_{23}$		3				3
Recommend$_4$	1					1
Reject$_3$		1	1			2
Request$_4$	5	7	6	7		25
Review$_{14}$	1	12				13
Review$_{34}$		1				1
Suggest$_4$	1					1
Warn$_4$	2			4		6
Σ	14	37	16	18	1	86

For example, in business relationship management, the trigger "Customer complaint" is immediately identifiable as speech act, while the "Creation of a new or updated capacity plan" in knowledge management is a decision. Nearly all triggers of service portfolio management and all triggers of release and deployment management, change evaluation, request fulfillment, and access management were immediately identifiable speech acts. Only two out of 26 processes do not list any speech act as trigger: Service validation and testing (Service Transition) is triggered by scheduled activity on a release plan, and event management (Service Operation) is triggered by system events, e. g. exceptions and warnings.

Table 1 summarizes the categorization of all identified triggers. Each cell counts all triggers of the selected category and book. Purely quantitative, 66% of these triggers can be attributed immediately or implicitly to speech acts. The pragmatic intention of most triggers is obvious to modelers and stakeholders, but the systems processes are deployed in currently are not aware of this intention and therefore cannot make use of it. Due to size constraints, we did not include all classifications in this paper. However, we made them available online[3].

5.2   Immediately Identifiable Speech Acts

For each immediately identifiable speech act, one or more speech act verbs were chosen based on wording of the triggers and additional domain knowledge, e. g. contents of a negotiation. If the illocutionary point of the verb is not obvious, the pragmatic dimension is determined by consulting Ballmer and Brennenstuhl [16]. If the speech act verb can have more than one illocutionary point for a trigger, all possible illocutionary points are appended to the verb. They are read as OR, e. g. for x_{12}, the verb x can have assertive (1) and/or commissive (2) character.

From the 61 triggers that were immediately identifiable as speech acts, we extracted 86 illocutionary forces. Table 2 shows the distribution of different speech act verbs per book. Although requests, reviews and notifications present

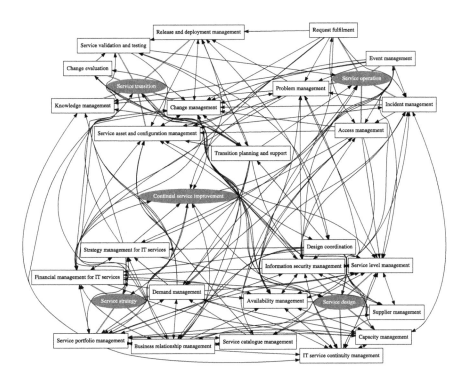

Fig. 1. Best practices: interfaces in ITIL

[3] Individual classifications and justifications are available as a report (PDF) and in XML. The XML files adhere to a DTD we provide for traceability and comprehensibility,and they were used to generate the report. https://www6.cs.fau.de/people/johannes-tenschert/bpms2-2017.

a large share of the identified speech act verbs, the entries show how many triggers we found, not how often triggers initiate a process. We extracted 20 different speech act verbs. Some represent more than one intention, e. g. *complain.* There, the speaker might have the intention to assert something being wrong, to trigger some action, or to express the feeling that he does not like the situation. We did not split those into smaller unambiguous speech act verbs. But for a speech act library, this has to be performed. On this abstraction level, the library extracted from ITIL triggers would consist of 22 to 29 illocutionary forces.

In summary, speech acts prevalent in actual processes are diverse (RQ 2). Similar to Sect. 5.1, we made the detailed classification of illocutionary forces in triggers available online (See footnote 3).

6   Interfaces in ITIL

RQ 3 is intended for unveiling potential to apply speech act theory in process modeling, in particular in regard to integration of processes. We examined all interfaces between processes to depict the coupling of processes and therefore the potential gains an abstraction of the intentions of interactions could provide in regard to reusability, integration, and transparency. For each process, all interfaces to other processes were extracted. Depending on the scope and focus of the process, an interface might be described either uni- or bidirectional.

As a result, Fig. 1 shows connections between all explicitly elaborated interfaces. Generally addressed interfaces were omitted, i. e. knowledge management and continual service improvement have interfaces addressed to all other processes, and only the explicitly addressed interfaces of these processes were added to the diagram. Each lifecycle stage is represented by an ellipse, and each process is drawn as a rectangle. Nodes are connected by edges $A \rightarrow B$ if there is an interface from process A to B, and $C \leftrightarrow D$ if there are two interfaces from process C to D and vice versa. The publications contain 216 explicitly elaborated descriptions of interfaces. Of these, 52 interfaces were described bidirectional, and 112 interfaces were elaborated only unidirectional. This huge amount of interfaces introduces a lot of complexity into a system aspiring to implement ITIL.

Figure 1 illustrates the tight coupling of processes. A focus on speech acts could lead to a much smaller set of different types of interactions users have to master. Schoop [25] demonstrated that 21 illocutionary forces sufficed to analyze communication in hospitals. Since a large amount of interfaces and a small library require that some interfaces share a category, this can improve reusability. Rules can be formulated on the illocutionary force instead of each message, which facilitates integration [3] and compliance [12]. If these techniques are applied, the pragmatic intention is made explicit in the process model or implementation. Hence, this approach can improve the transparancy and reusability (RQ 3).

7   Discussion

In this section, we discuss our chosen methods and scope, the results of Sects. 5 and 6, and implications in regard to semantic annotation, integration of systems in a socio-technical environment, and speech-act-based inference.

Methods and Scope. We formulated three research questions on the impact and diversity of speech acts in actual business processes, and the potential to use this information. One might challenge the chosen approach, the chosen reference model, and restricting the scope to triggers and interfaces. Examining a reference model that contains processes valid for an entire domain and assures a certain degree of quality and distribution, enables us to abstract from specificities of an individual organization. ITIL covers a wide range of processes, activities, and stakeholders within and between organizations. Restricting the scope to triggers and interfaces in ITIL should suffice to bear out our hypotheses: If speech acts are prevalent and important in triggers, they are prevalent and important in processes. If abstraction of interactions improves interfaces between processes in regard to reusability, then it improves processes in regard to reusability.

Speech Acts in Actual Processes. We identified six types of triggers of processes. Two types immediately or implicitly (decisions) represent speech acts, periodic triggers definitely represent no speech acts, and three types (artifact, system event, observation) could result in interactions, but are not required to. Out of 152 triggers in ITIL, 40% were immediately, and 26% were implicitly identifiable as speech acts. Hence, speech acts have an important role in actual business processes (RQ 1). Moreover, we show that interactions in ITIL are diverse (RQ 2). Our analysis indicates that a library on this abstraction level would consist of 22 to 29 illocutionary forces. The publications contain 216 explicit descriptions of interfaces, whereas 52 interfaces were described bidirectional, and 112 interfaces were elaborated only unidirectional. Hence, processes in ITIL are tightly coupled to other processes. A small speech act library offers an abstraction of interactions for more loose coupling and enables ad-hoc interactions triggering processes as well. Many illocutionary forces are identified as triggers in more than one core publication, i. e. at least for ITIL this library is fit for reusability of speech act triggers or rather process patterns (RQ 3).

Semantic Annotation. Francescomarino et al. [30] conducted an empirical study on the impact of semantic annotation in business process modeling. Semantically annotating processes enriches elements with annotations from an ontology defining their domain semantics. This enables additional tools and services that facilitate analysis, and support search and validation of processes. Semantic annotations increase the number of refinements, and improve the quality of the modeled processes without strongly affecting the time spent or rather modeling costs. Even if semantically enriched processes are not used in an environment that processes annotations, the modeler becomes aware of the speech acts involved and will use this knowledge to improve the quality of the model. Since we demonstrate that speech acts have impact on actual processes, we believe that semantically annotating interactions can improve process models.

Integration. Binding systems together requires some sort of mapping of corresponding artifacts. Since our results indicate a high importance and diversity of interactions at touch points of processes, performing this mapping considering speech acts seems natural. Speech-act-based annotation, mapping, or documentation allows to formulate business rules over interactions across systems. We already proposed how business rules over interactions could integrate these systems [3]. This can especially be beneficial to integrate ACMSs with structured processes or tailored information processing systems, since it allows to include ad-hoc documentation of activities that are not modeled at all.

Inference. Business rules can cover the pragmatic intention of the speaker, roles of speaker and hearer, and propositional content. At design time, annotations can be used to check compliance of a process in regard to legal and organization-specific regulations. If interaction patterns are made explicit, they can be analyzed in regard to intended outcomes of the process, goals of an organization, and best practices. Business rules can be agnostic to the system documenting the speech act. While automating parts of a process in the future, business rules might not need to change as long as the deep structure of a process remains [17]. Speech-act-based inference does not require new techniques for inference engines. We already outlined how it can be integrated into LTL and CTL rules, rete-based BRMSs, and complex event processing (CEP) [12].

8   Conclusion

For ITIL-based processes, we categorized triggers and revealed that 66% of those can immediately or implicitly be attributed to speech acts. From those interactions, we extracted 20 different speech act verbs. Our results demonstrate the importance, prevalence, and diversity of interactions in triggers, and that further abstraction of interactions can improve reusability of process patterns. We discussed our methods and scope in regard to alternatives and the selection of ITIL, and the prevalence and types of speech acts we found in ITIL processes. We conclude that semantically annotating processes with speech acts could improve the process models. Moreover, we discussed integration of systems based on semantic annotations of interactions, and speech-act-based inference.

Still, there are open questions. We assumed that other reference models and applied business processes show similar variety in regard to flexibility and interfaces. Schoop [25] analyzed interactions in hospitals and extracted a library of similar size. However, both domains cover a wide range of work types, so probably our results have a strong correlation with knowledge work being present. Therefore, processes covering different types of work should be analyzed to gain more confidence in generality. Moreover, we discussed how a focus on speech acts could facilitate integration of systems and a more holistic inference, but the approach has not yet been examined in practice.

Also, the connection between speech acts and reference models could be investigated in more depth. On the one hand, reference models are known to be rather generic, so it would be interesting comparing the analysis of ITIL with the analysis of an ITIL implementation. Analogies and differences could establish a certain relation between reference model and process model, which might help to further develop or adapt the reference model based on speech acts. On the other hand, analyzing speech acts, their roles assumed in a process, and dependencies between interactions in the reference model itself may also identify weaknesses and optimization potentials, for example if a planned request is never answered. In general, speech acts hold potential for the thorough analysis of reference models, which should be addressed by future research.

In summary, interactions and their pragmatic intention have significant impact in actual business processes. A focus on speech acts in modeling and inference can facilitate integrating the structured, semi-structured and ad-hoc processes today's socio-technical systems rely on.

References

1. Berg, M., Aarts, J., van der Lei, J., et al.: Ict in health care: sociotechnical approaches. Methods Archive **42**(4), 297–301 (2003)
2. Lund, S., Manyika, J., Ramaswamy, S.: Preparing for a new era of knowledge work. McKinsey Q. **4**, 103–110 (2012)
3. Tenschert, J., Lenz, R.: Towards speech-act-based adaptive case management. In: AdaptiveCM 2016–5th International Workshop on Adaptive Case Management and other Non-workflow Approaches to BPM (2016)
4. Swenson, K.D.: Innovative organizations act like systems, not machines. In: Fischer, L. (ed.) Empowering Knowledge Workers: New Ways to Leverage Case Management, pp. 31–42. Future Strategies Inc., New York (2014)
5. Stiehl, V.: Definition of process-driven applications. Process-Driven Applications with BPMN, pp. 13–41. Springer, Cham (2014). https://doi.org/10.1007/978-3-319-07218-0_2
6. Great Britain Cabinet Office: ITIL Service Strategy 2011. Best Management Practices. TSO (The Stationery Office), London (2011)
7. Great Britain Cabinet Office: ITIL Service Design 2011. Best Management Practices. TSO (The Stationery Office), London (2011)
8. Great Britain Cabinet Office: ITIL Service Transition 2011. Best Management Practices. TSO (The Stationery Office), London (2011)
9. Great Britain Cabinet Office: ITIL Service Operation 2011. Best Management Practices. TSO (The Stationery Office), London (2011)
10. Great Britain Cabinet Office: ITIL Continual Service Improvement 2011. Best Management Practices. TSO (The Stationery Office), London (2011)
11. Fettke, P., Loos, P.: Perspectives on reference modeling. In: Fettke, P., Loos, P. (eds.) Reference Modeling for Business Systems Analysis, pp. 1–20. Idea Group Publishing, Hershey (2007)
12. Tenschert, J., Michelson, G., Lenz, R.: Towards speech-act-based compliance. In: 2016 IEEE 18th Conference on Business Informatics (2016)
13. Tenschert, J., Lenz, R.: Supporting knowledge work by speech-act based templates for micro processes. In: Reichert, M., Reijers, H.A. (eds.) BPM 2015. LNBIP, vol. 256, pp. 78–89. Springer, Cham (2016). https://doi.org/10.1007/978-3-319-42887-1_7

14. Austin, J.L.: How to do Things with Words. Oxford University Press, Oxford (1975)
15. Searle, J.R.: Speech Acts: An Essay in the Philosophy of Language. Cambridge University Press, Cambridge (1969)
16. Ballmer, T.T., Brennenstuhl, W.: Speech Act Classification - A Study in the Lexical Analysis of English Speech Activity Verbs. Springer, Heidelberg (1981). https://doi.org/10.1007/978-3-642-67758-8
17. Dietz, J.L.G.: Enterprise Ontology: Theory and Methodology. Springer, Heidelberg (2006). https://doi.org/10.1007/3-540-33149-2
18. Dietz, J.L.G.: The deep structure of business processes. Commun. ACM **49**(5), 58–64 (2006)
19. Van Nuffel, D., Mulder, H., Van Kervel, S.: Enhancing the formal foundations of BPMN by enterprise ontology. In: Albani, A., Barjis, J., Dietz, J.L.G. (eds.) CIAO!/EOMAS -2009. LNBIP, vol. 34, pp. 115–129. Springer, Heidelberg (2009). https://doi.org/10.1007/978-3-642-01915-9_9
20. Decker, G., Barros, A.: Interaction modeling using BPMN. In: ter Hofstede, A., Benatallah, B., Paik, H.-Y. (eds.) BPM 2007. LNCS, vol. 4928, pp. 208–219. Springer, Heidelberg (2008). https://doi.org/10.1007/978-3-540-78238-4_22
21. Auramäki, E., Lehtinen, E., Lyytinen, K.: A speech-act-based office modeling approach. ACM Trans. Inf. Syst. **6**(2), 126–152 (1988)
22. Alter, S.: System interaction patterns. In: 2016 IEEE 18th Conference on Business Informatics (CBI), vol. 1, pp. 16–25, August 2016
23. Finin, T., Fritzson, R., McKay, D., McEntire, R.: KQML as an agent communication language. In: Proceedings of the Third International Conference on Information and Knowledge Management, CIKM 1994, pp. 456–463. ACM, New York (1994)
24. Kimbrough, S.O., Moore, S.A.: On automated message processing in electronic commerce and work support systems: speech act theory and expressive felicity. ACM Trans. Inf. Syst. **15**(4), 321–367 (1997)
25. Schoop, M.: Habermas and searle in hospital: a description language for cooperative documentation systems in healthcare. In: Proceedings of the Second International Workshop on Communication Modeling - The Language/Action Perspective (1997)
26. Schoop, M., Jertila, A., List, T.: Negoisst: a negotiation support system for electronic business-to-business negotiations in e-commerce. Data Knowl. Eng. **47**(3), 371–401 (2003)
27. Gerke, K., Tamm, G.: Continuous quality improvement of IT processes based on reference models and process mining. In: Proceedings of the Fifteenth Americas Conference on Information Systems, San Francisco, CA, USA (2009)
28. Rohloff, M.: A reference process model for IT service management. In: Benbasat, I., Montazemi, A.R. (eds.) 14th Americas Conference on Information Systems, AMCIS 2008, Toronto, Ontario, Canada (2008)
29. Looso, S., Goeken, M.: Application of best-practice reference models of IT governance. In: Alexander, P.M., Turpin, M., van Deventer, J.P. (eds.) 18th European Conference on Information Systems, Pretoria, South Africa, pp. 1375–1388 (2010)
30. Francescomarino, C.D., Rospocher, M., Ghidini, C., Valerio, A.: The role of semantic annotations in business process modelling. In: Proceedings of the 2014 IEEE 18th International Enterprise Distributed Object Computing Conference, EDOC 2014, pp. 181–189. IEEE Computer Society, Washington, DC (2014)

SLA-Based Management of Human-Based Services in Business Processes for Socio-Technical Systems

Mirela Riveni[1(✉)], Tien-Dung Nguyen[2], and Schahram Dustdar[1]

[1] Distributed Systems Group, TU Wien, 1040 Vienna, Austria
{m.riveni,dustdar}@infosys.tuwien.ac.at
[2] International University National University,
Ho Chi Minh city, Vietnam
ntdung@hcmiu.edu.vn

Abstract. Research and industry are focused on Collective Adaptive Systems (CAS) to keep up with the social changes in the way we work, conduct business and organize our societies. With advances in human computing, we can strengthen these systems with a crucial type of resources, namely *people*. However, while other resources in CAS are managed by Service Level Agreements (SLAs) in an automated way, human-based services are not. Considering the fact that not much work has been reported on SLAs in settings where human computation is an integral part of a process, this paper investigates SLAs for social computing, including non-functional parameters for human-based services, as well as privacy constraints. We investigate and evaluate SLA changes at runtime, which in turn influence elastic runtime social-collective adaptations, in processes designed to allow for elastic management of social collectives.

Keywords: Human-centric processes · Collective Adaptive Systems
Service Level Agreements · Human computation
Social Compute Units

1 Introduction

Traditional information systems are not good enough to keep up with the evolving societal needs, including the ever bigger complexities and changing dynamics in how work is organized, businesses are run and societies are governed. Thus, researchers and industry have began to focus on Collective Adaptive Systems (CAS), which are systems that include multiple resource-collectives (or sub-systems). They include heterogeneous type of resources and services that inter-operate and provide seamless service. Typical resources can be software services, Cloud services, different agents, sensors and Internet of Things (IoT) services. However, there is a crucial component of collectives that will make a key difference in CAS, namely people. With the advance of human computation in

© Springer International Publishing AG 2018
E. Teniente and M. Weidlich (Eds.): BPM 2017 Workshops, LNBIP 308, pp. 361–373, 2018.
https://doi.org/10.1007/978-3-319-74030-0_27

general and social computing in particular, today it is possible for human-based services/resources to be part of CAS. Examples of these systems are described in [1,3,19].

While other types of resources in CAS can, and are managed by automated Service Level Agreements (SLAs), human-based services are not. People providing their skills as services online are mainly managed by standard (human-language) contracts or no contracts at all. Nevertheless, worker performance should be monitored and managed in processes in an automated way, so that task management can be efficient, and both customer and worker needs can be fulfilled. Thus, the challenges that we tackle in this paper are modeling SLAs for human computation and process-based SLA adaptation mechanisms. To approach them, we utilize the term Social Compute Units (SCUs) introduced in [4]. SCUs represent human collectives, who act as resources and/or provide their skills as online services for processing tasks that can not be processed by software. We name individuals who work within these collectives, Individual Compute Units (ICUs). SCUs are formed mostly on customers' request and with customer-set constraints but they can also be formed voluntarily and in an ad-hoc way. They are invoked to reach a certain task-oriented goal, with a specific quality and in a specific context. Thus, they have their own execution lifecycle that we have previously discussed in [14].

Dustdar et al. in [5] have introduced elastic processes for cloud and human computation, and their fundamental principles including *resource elasticity*, *cost elasticity* and *quality elasticity*. In the particular context of elasticity for collectives such as SCUs, SLAs play a crucial role in managing their execution and controlling the quality of the offered services and thus the returned results. However, as far as we know, SLAs in human computation in general are not explicitly investigated in the context of processes with which collectives are managed elastically together with SLAs based on specific metrics designed specifically for collectives such as SCUs. In the paper at hand we investigate how SLAs fit in an SCU provisioning platform. In addition, we describe and present our experiment results of an SCU elastic management strategy based on SLAs, which is focused on cost optimization of SCUs while they are being elastically adapted at runtime. The contribution of this work is to show that the design and implementation of social-computing processes need to allow for elasticity regarding resources, customer requirements, and worker/ICU profile parameters, in order to be efficient. We justify this along with the justification for the need to design systems that provide the possibility of having elastic SLAs, the changes of which trigger elastic social-collective adaptations.

The remainder of this paper is organized as follows: In Sect. 2 we describe real-world motivation scenarios, and discuss an SCU provisioning-platform model including SLAs. In Sect. 3 we discuss parameters for modeling SLAs for human computation. In Sect. 4 we describe a proof-of-concept prototype demonstrating a process-based SCU adaptation, based on SLA changes. Section 5 presents related work, and we conclude the paper in Sect. 6.

2   Motivation and SCU Environment Setting

2.1   Illustrative Scenarios

Facility Management. A typical example of a Collective Adaptive System (CAS) is a smart city. Let us consider the case of facility management in a smart-city. To maintain multiple smart buildings and manage utilities effectively, such as power management, water management, temperature, and security management, the managing enterprise/organization utilizes applications, platforms and infrastructure, thus working with different software services, Cloud platforms and infrastructures, as well as different sensors and monitoring equipment (IoT). Data gathered from monitoring equipment needs to be analysed in real time so that actions can be taken in real time. The following human-based collective-types (SCUs) could included in the facility-management process: (a) *SCUs for data-analysis*, (b) *SCUs with domain experts*, and c) *SCUs with on-site technical-experts*.

Language Translation. Recently we have seen an increase in online platforms for translations e.g., documents, books, documentary(film) subtitles and available languages for software applications. Let us consider a publishing company that needs to translate a book within a short time in a specific language for which it does not have suitable translators or contractors. Today it is feasible to hire multiple translators online. Assuming that the book is submitted to be translated as a task to a Socio-technical CAS, it is possible that the platform first uses software translation-services as a first round of translating and then assigns appropriate people/ICUs to fix the software translation, and lastly it assigns an ICU or a collective of ICUs to do the final editing. Another possibility, is that the system just assigns the translation task to a set of ICUs (an SCU) from start and then in a second iteration assigns the translated material to a set of ICUs for reviewing and final editing. Two SCUs would be formed in this scenario: (a) a *language-translation SCU*, and (b) a *reviewing and final-editing SCU*.

The aforementioned scenarios validate the need of humans-in-the-loop within socially-enhanced CAS. The main concerns in these type of systems are the type of SCUs that are needed, with which capabilities, and more importantly, how can they be process-managed at runtime to keep up with clients' requests and requirements as well as worker needs and conditions. Meeting customer requirements as well as worker conditions in human computation platforms can be done more efficiently if there is a negotiation defining clear terms and conditions in the form of SLAs. To the best of our knowledge, there is no research in SLAs for human-computation specifically, which goes along with research on how to manage online work based on metrics. The current state of the art engages people in human-computation based on natural language contracts or no contracts at all. However, research in managing mechanisms, such as, collective formation, task delegations, proactive and elastic adaptation of human collectives online, need to address the challenge of SLAs because of the unpredictable nature of humans. To clarify, we identify some of the core characteristics of human-based resources that would impact SLA specification and management as follows:

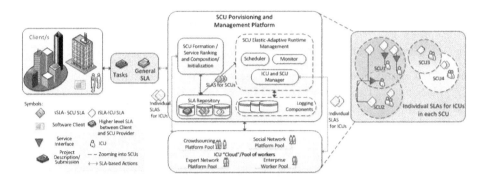

Fig. 1. A conceptual overview of an SLA-based ICU-SCU Provisioning Platform

– Human performance is a function of monetary or non-monetary *rewards* and
  penalties. Thus, human misbehavior can be prevented or managed by *incentives* respectively *sanctions*.
– People are *paid/rewarded* by the service provisioning platform as an intermediary and/or by the platform-customers; the platform in turn is paid by its
  customers. Thus, in human computation, contract negotiation is hierarchical,
  between a customer and the service-provisioning platform, and between the
  platform and people who provide their skills online.
– People need to be managed with their consent and guided by human-rights
  principles. Hence, *privacy* is a crucial element to be considered during system
  design, which is unfortunately often overlooked.

2.2   SCU Computational Environment Setting

In Fig. 1 we give an abstract overview of an SCU provisioning platform. The
figure shows that SLAs are established between a customer and the platform
for a specific project that requires a specific SCU, so there are SLAs for SCUs.
The platform can run a ranking algorithm to rank ICUs from an ICU Pool (a
proprietary one, or from another pool that it has access to) according to the
requirements within the SCU SLAs. In the next step, a negotiating component
negotiates individual SLAs with ICUs in the ranked list and forms an appropriate
SCU. All negotiations between all parties should be automated so that SLAs
can be automatically changed at runtime if customers or ICUs decide to change
specific SLA parameters. From the aspect of an SCU execution that is controlled
by SLAs, an SCU may elastically be adapted, such that: (a) *compliance with runtime changes in customer and/or worker requirements is achieved*, and (b) *an
SLA violation is prevented or managed*. A customer might want to change his/her
requirements at runtime, in this case the platform should enable automated
SLA changes at run-time. On the other hand, when adapting an SCU such that
an SLA violation is prevented, e.g., by substituting a misbehaving ICU with
another more appropriate one, from the customers requirements perspective the

Table 1. Notation and description of basic parameters for SCU SLAs

Properties	Description	Value
Consistency	Executed tasks as promised, e.g., via feedback messages	[0..1]
Socio-technical trust score [15]	Includes other metrics: social scores (subjective), and monitored performance metrics(objective), e.g.,: productivity, effort, success rate, nr. of successfully executed delegations, etc	[0..1]
Reputation	An average of the performance and social scores of ICUs across all SCUs in which it was invoked	[0..1]
Average queue time	An average of task waiting time in an ICUs queue; for tasks with specific skill requirements and comparable complexities (e.g., a page for translation)	[0..1]
Result timeliness	An average of on-time result-delivery score of ICUs	%
Availability	ICU availability (as time intervals.)	hour
Budget	The max price that the customer may pay	currency

same SLA terms and conditions for the substituted ICU must be valid. However, the new ICU may have its own requests. In these cases negotiations should be conducted at run-time as well. In consequence to these two cases, to best reflect human behavior we argue the need of *elastic SLAs*, because of the mere nature of people and their working dynamics. Thus a provisioning platform supporting elastic SCUs governed by SLAs, should also support elastic SLAs.

3   Modeling SLAs for SCUs

3.1   Human-Centric Properties and Metrics

Non-functional properties (NFPs) as key indicators of service performance are the parameters that are the most important in SLA negotiations. In Table 1 we list relevant metrics from our previous work [14] and some new ones that we have identified as crucial for SCUs, and which can be used as Service Level Objectives (SLOs) in SLAs. We denote an ICU as a member of an SCU with s_i. An ICU as a member of SCU, has the following set of properties, $ICU_{prop}^s = \{id, skill, metrics, iSLA\}$, while an SCU is defined as $SCU_{exec} = \{state, structure, metrics, sSLA\}$. For the mathematical definitions of *Socio-Tecnical Trust Score* and *Reputation* metrics we advise you to check our previous work presented in [15]. Currently, platforms in industry that involve social-computing, all have a common flaw, the lack of privacy. Apart form performance parameters for workers, the inclusion of privacy-related constraints for systems will provide systems that respect both workers and customers. While we leave the investigation of privacy-related parameters for future work, identified privacy-related conditions that can be included in SLAs are: *software to be used for task execution and communication, time allowed to keep the collected data (e.g. location), audit periods for third-parties to check if obligations are fulfilled, time period until an audit report is received, deadline for informing users after a security breach has happened, and privacy-breach penalties for the system.*

3.2   Penalties

We consider two cases regarding penalties: (1) when SLA parameters are changed at runtime, and (2) when there has been an SLA violation. We have previously defined a metric that we call Socio-technical Trust for ICUs (in [15]). This metric is comprised of two complex metrics, namely (1) *Performance trust*(PT) that we define and calculate with the help of other performance-measuring metrics, and (2) Membership Collaboration Trust Score (MCTS), which we defined as a weighted sum of satisfaction scores from each worker to another within the same collective, representing a *Social trust* metric in our model. The Socio-technical Trust score is calculated as: $STT(s_i) = w_{pt} * PT(s_i) + w_{mcts} * MCTS(s_i)$ - our equation from [15], where s_i denotes an ICU and $w(x)$ is the weight of a metric x. Now we extend this formula and define it as: $STT(s_i) = w_{pt}*PT(s_i)+w_{mcts}* MCTS(s_i)+w_{cps}*CPS(s_i)$, where CPS is the satisfaction score comprised of a satisfaction score from the customer and the platform (measuring consistency of behavior). We define CPS as: $CPS(s_i) = w_{css}*CSS(s_i)+w_{pss}*PSS(s_i)$, where CSS is Customer Satisfaction Score and PSS is Platform Satisfaction Score.

Let us take a concrete example, when an SLA change occurs at runtime, e.g., a customer requests a cost-change for a specific skill-type. In this case, ICUs already within the collective can be sent a notification message for the newly changed fee. In this example, those ICUs that are already within the SCU and do not accept the changes are excluded from the SCU and their Platform Satisfaction Score (PSS) is lowered by a preset value ∂ as a penalty (lines 10–11 in Algorithm 1). The tasks for the changed cost are executed by other appropriate ICUs that are newly included in the SCU from the available pool of ICUs. ICUs should always be informed that they will get lower scores for certain properties, which will influence their overall reputation score, so that this knowledge can also serve as an incentive for future behavior. The same penalties can be applied in SLA violations. In these cases, in addition to lowering performance values and satisfaction scores, there can be monetary penalties. As an extension to our cost model from previous work [14], the new cost of the SCU after adapting it because of a violation by ICUs will be: $Cost_{adapt}(scu_i) = Cost_{old}(scu_i) -$

$$\sum_{i=1}^{m}\sum_{x=1}^{j} c(s_i, t_x) + \sum_{i=1}^{m}\sum_{x=1}^{j} c(s_i^{nw}, t_x) - \sum_{i=1}^{m} c(s_i, vSLA_i), \text{ where } \sum_{i=1}^{m}\sum_{x=1}^{j} c(s_i^{nw}, t_x)$$

is the cost of the newly added ICUs, and $\sum_{i=1}^{m} c(s_i, vSLA_i)$ is the cost for SLA violation by ICUs; t_x denotes a task executed by an ICU s_i.

4 Programming Adaptations Based on Elasticity Principles

4.1 Algorithm: Runtime SLA-changes Invoking SCU Adaptation

In Algorithm 1 we describe an example of adapting an SCU at runtime when a specific SLA Parameter is changed by the customer at runtime, namely cost. The algorithm has a ranked pool of resources at its disposal, where ICUs are ranked by requirements pre-set from the customer. In addition, the algorithm supposes that an SCU is already formed from that ranked list, based on the type of skills needed for separate tasks. At runtime, in the event that a customer changes the budget for the SCU, the algorithm calculates new costs for each skill-type and sends a notification request-message to each ICU to get back an approval or rejection of the changes (lines 1–5 of Algorithm 1). In the case that not all ICUs approve the changes, new ICUs with the cost parameter fitting the request and the skill-type for which the cost was changed, are selected from the ranked list, and individual SLAs are set with them. They are then added to the SCU, and tasks which were previously assigned to ICUs in the SCU that rejected the changes are delegated to the new ICUs. Hence, the SCU is adapted. The Platform Satisfaction score, is lowered for those ICUs that have rejected the payment changes, and those ICUs are excluded from the collective (lines 7–13).

4.2 Prototype Implementation and Experiments

We ran a Java-based experiment to evaluate runtime adaptations of SCUs when requirements for parameters are changed at runtime. The goal of our experiments is to show that processes allowing for elastic adaptation of SCUs along with runtime SLA changes are more efficient, as workers can be added or excluded from execution as needed, and customers can get faster results with lower cost. We show this by showing that SCU performance is higher with each adaptation point (Fig. 2), as well as the time for total-task execution is lower with each elastic adaptation of social collectives (Fig. 3). Our proof-of-concept prototype consists of the following components: (1) *a model of ICUs, SCUs, ICU and SCU metrics, and Tasks*; (2) a component for ICU *ranking*, according to specific metric constraints from the customer; (3) *adaptation algorithms* that we feed to our process engine; (4) *a process execution engine*, with which we run the process described in Algorithm 1. More specifically, we designed and implemented ICUs with a single skill, for better overview of the results, and cost-per-task as pre-set properties. In addition, we designed performance metrics [15]. The following are only some of them: availability, total assigned tasks, delegated tasks, success-rate, productivity, reliability, social trust, performance trust, socio-technical trust. We designed tasks with skill and cost requirements, and SCUs with metrics specific to them. The source code for the implementation can be found at: https://mirusx@bitbucket.org/mirusx/ulyss.git.

Algorithm 1. Elastic Adaptation of SCUs based on SLA changes

Require: icu member of SCU, icu member of ICU Pool
1: **if** $scu.currentBudget == (scu.currentBudget - amount)$ **then**
2:     **for all** icu in SCU **do**
3:         calculateNewFees(taskList)
4:         sendFeeChangeNotification()
5:         $icuApprovalList \leftarrow getApproval(icu)$
6:     /* Adapt the SCU if not all ICUs approved the agreement changes */
7:     **for all** icu in icuApprovalList **do**
8:         **while** icuCurrent.Approval()== false **do**
9:             $icuNeededSkillsList \leftarrow icuCurrent.getSkill()$
10:            $icuCurrent.icuMetrics.PSS = getPSS(icuCurrent) - \partial$
11:            $updateAgreementReciprocity(icuCurrent)$
12:            $icuCurrent.updateMetrics()$
13:            $SCU \leftarrow removeICU(icuCurrent)$
14:     /*Add the next matching ICU from the ranked list that approves the SLA*/
15:     **for all** icuNeededSkillsList **do**
16:         **for all** icu in rankingList **do**
17:            **if** $icu.Approve() == true$ **then**
18:                $icuCurrent = icu$
19:                $SCU \leftarrow addICU(icuCurrent)$
20:            **break**

SCU Adaptation with SLA parameter changes at runtime. We implemented a variation of Algorithm 1 where the cost is the parameter to be changed, but instead of lowering the total budget of the SCU we lowered the fee for tasks of specific skill types. Thus, we simulated time points at which changes for specific skill-types are required as an input. Already assigned tasks for which the fee was lowered were delegated to other ICUs with matching skill and (lower) cost. In addition, tasks were also delegated at points when they were assigned but were not executed up until a pre-set time-threshold. In both cases tasks were delegated to ICUs already within the SCU or new ICUs were invoked form an ICU pool if no matches were found. Let us examine the results, which are shown in Fig. 2. Figure 2 (a) shows the socio-technical trust scores of four SCUs that we selected (as interesting cases to analyze). We selected to show the socio-technical trust score (STT), as it is a metric that encompasses all other important metrics, such as success rate, performance, productivity of ICUs as well as social-trust scores that ICUs assign to each other according to their collaboration satisfaction. We randomized the assigned social-trust scores, but biased in a way that we assigned higher scores to ICUs with higher number of executed tasks and lower scores to those that delegated their tasks. In Fig. 3(a) we can see that the trust scores are generally rising for all SCUs, but some low points exist. The cost change events for SCU1 and SCU2 happen at the last adaptation point, while the cost changes for SCU3 and SCU6 are at the sixth point of adaptation. Let us examine these cases more closely. SCU1 has a high STT after the end of the

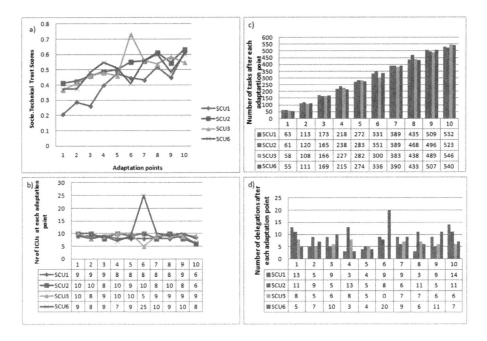

Fig. 2. Elastic SCU adaptation with SLA cost changes

last adaptation, and if we examine sub-figure (b) we will see that at that point it had a low nr of ICUs (6) but high delegations (sub-figure d). Thus, STT is high because the SCU performance is high, as a consequence of the fact that at the last adaptation point some ICUs are excluded form the SCU and more tasks are executed with lower number of people. Examining the case of SCU3 we see that at the sixth adaptation point, it has a spike on STT scores shown in sub-figure (a), while sub-figure (b) clearly shows that the number of ICUs at point 6 is five, and we do not have delegation of tasks at point 6 (shown in sub-figure (d)). In this case the cost changes were made at point 6, and all the new tasks to be assigned with those changes were accepted by ICUs already in the SCU. SCU6 on the other hand, has a low score of STT at point 6 in sub-figure (a). Sub-figure (b) shows that the number of its ICUs is considerably high, 25 ICUs at point 6, and the number of delegated tasks at point 6 is 20 (sub-figure d). We noticed that 18 new ICUs were added to adapt to the cost change requirements as we gave as an input a very low fee for two skill types. In addition, when ICUs are newly included in the SCU their social trust score is lower (we count invocations within an SCU at each adaptation point), which also ads to the lower value of the STT. In summary, this kind of flexibility at runtime is not possible without having a platform design that provides the possibility to change SLAs at run-time, but most importantly to include ICUs in negotiations when SLAs change, so that collective adaptations at run-time are more efficient and agreeable for all parties.

Comparison of elastic and fixed SCU adaptation. We ran another experiment comparing the variation of Algorithm 1 and a Base-Algorithm with which the SCU was adapted as tasks were scheduled and assigned according to skills, and they were delegated when they reached a time-threshold, as well as when a cost change requirement was given as an input at runtime. However, in the base-algorithm tasks are delegated only to ICUs that are already within the SCU, and no new ICUs were added to the SCU. Thus, in the variation of Algorithm 1 the adaptation is elastic, we exclude and add ICUs from the pool of ICUs, while in the base algorithm after the SCU is formed, ICUs are not changed. We compared the two algorithm in terms of time efficiency. Figure 3 shows the results. At each adaptation point a new bag of tasks is assigned to SCUs, sub figure (a) shows the number of tasks with each adaptation point. Sub-figure (b) shows the time units after each adaptation point, these time units are calculated as the time to execute the bag of tasks for (and after) each adaptation point, including the delegated tasks. Sub-figure (b) clearly shows that even if the task number assigned at each adaptation point is similar in both SCUs, the time to execute the tasks is not. In the fixed SCU the changes in cost for a skill resulted in delegating multiple tasks to a single ICU, thus the waiting time in queue and the total execution time of tasks has a higher value. On the other hand, in the elastic adaptation algorithm we scheduled each task for which the cost is changed to a new and different available ICU from outside the SCU and included them in the SCU at the needed adaptation points and excluded them when no longer needed. This clearly lowered the time to execute tasks,as shown in sub-figure (b).

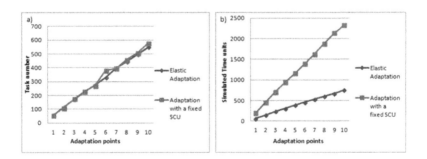

Fig. 3. Task-execution time-comparison in elastic and fixed adaptation algorithms

5   Related Work

The fundamental work that first introduced Social Compute Units is presented in [4], whereas [18] discusses a specific scenario detailing further the way of SCUs utilization and its benefits with a concrete real-life scenario. Proposals for provisioning human skills with the service oriented model are presented in [6].

There is very little work on SLAs in human computation environments. Initial work which concerns crowdsourcing environments in particular is presented in [8]. The authors give an example of an SLA that may be exchanged between customers and crowd-provider platforms and also present a few crowdsourcing specific SLOs concerning worker skills, quality of executed tasks and customer fees. Another work considering SLAs for human computation, is presented in [13] by Psaier et al. The authors discuss scheduling mechanisms for crowdsourcing environments, which will meet the requirements of contracts, between multiple entities/roles. Schall in [16] discusses protocols for human computation and supports our claim that platforms for provisioning social computation need to work with SLAs, for better task and resource management, just as in software services. A description of requirements and considerations for platforms that support QoS management with SLAs in human computation is introduced in [7]. Elastic management of properties such as cost and workload are presented in [17], where elasticity is used to define restrictions for performance metrics defined in SLA guarantee terms. On the other hand, because human behavior is highly unpredictable, the strict definition of time-related constraints is crucial for SCUs. The temporality aspect for SLAs is investigated in [12]. Müller et al. in [11] have presented a formalization model for creating SLAs with compensations such as rewards and penalties. The compensation functions that they propose can be used in Compensable Guarantees (a term the authors coin for Guarantee Terms with compensation functions). Interesting for our work is that this work concerns Cooperative Information Systems and the models would also be fit to be used for SLAs for human-based services. Authors of [10] also discuss a contract model and a negotiation process considering a social computing scenario where rewards and penalties are not only monetary. Key Performance Indicators mapped to SLOs as metrics for business processes, and specific metrics for services in an orchestration, aggregated in SLAs are discussed by Wetzstein et al. in [20]. Social BPMN is discussed in [2], as a mechanism that integrates social interactions in standard organizational business processes. An example of a process-based programming model for crowdsourcing is presented in [9].

6   Conclusion

We investigated process-based SLAs and human-based collective management, by providing key parameters to be included in SLAs for human computation. In addition, we provided an example mechanism for SCU adaptation based on SLA changes at runtime, and provided a prototype for process-based management of SCUs. Our experiments with a processed-based algorithm showed that social-computing mechanisms designed with elasticity in mind where SLA parameters can be changed at runtime and SCUs can be adapted based on those changes, are more efficient than traditional processed with fixed-resource management. In future work we plan to investigate privacy-enforcing parameters for SLAs, and the way that they can be included in processes for SCU management.

References

1. Andrikopoulos, V., Saez, S.G., Karastoyanova, D., Weiß, A.: Collaborative, dynamic & complex systems - modeling, provision & execution. In: CLOSER 2014 - Proceedings of the 4th International Conference on Cloud Computing and Services Science, Barcelona, Spain, 3–5 April 2014, pp. 276–286 (2014)
2. Brambilla, M., Fraternali, P.: Human computation for organizations: socializing business process management. In: Michelucci, P. (ed.) Handbook of Human Computation, pp. 255–264. Springer, New York (2013). https://doi.org/10.1007/978-1-4614-8806-4_21
3. Bucchiarone, A., Dulay, N., Lavygina, A., Marconi, A., Raik, H., Russo, A.: An approach for collective adaptation in socio-technical systems. In: Proceedings of the 2015 IEEE International Conference on Self-Adaptive and Self-Organizing Systems Workshops, SASOW 2015, pp. 43–48. IEEE Computer Society, Washington, DC (2015)
4. Dustdar, S., Bhattacharya, K.: The social compute unit. IEEE Internet Comput. **15**, 64–69 (2011)
5. Dustdar, S., Guo, Y., Satzger, B., Truong, H.L.: Principles of elastic processes. IEEE Internet Comput. **15**(5), 66–71 (2011)
6. Dustdar, S., Truong, H.L.: Virtualizing software and humans for elastic processes in multiple clouds- a service management perspective. IJNGC **3**(2), 109–126 (2012)
7. Kern, R., Zirpins, C., Agarwal, S.: Managing quality of human-based eServices. In: Feuerlicht, G., Lamersdorf, W. (eds.) ICSOC 2008. LNCS, vol. 5472, pp. 304–309. Springer, Heidelberg (2009). https://doi.org/10.1007/978-3-642-01247-1_31
8. Khazankin, R., Psaier, H., Schall, D., Dustdar, S.: QoS-based task scheduling in crowdsourcing environments. In: Kappel, G., Maamar, Z., Motahari-Nezhad, H.R. (eds.) ICSOC 2011. LNCS, vol. 7084, pp. 297–311. Springer, Heidelberg (2011). https://doi.org/10.1007/978-3-642-25535-9_20
9. Kucherbaev, P., Tranquillini, S., Daniel, F., Casati, F., Marchese, M., Brambilla, M., Fraternali, P.: Business processes for the crowd computer. In: La Rosa, M., Soffer, P. (eds.) BPM 2012. LNBIP, vol. 132, pp. 256–267. Springer, Heidelberg (2013). https://doi.org/10.1007/978-3-642-36285-9_31
10. Michalk, W., Haas, C.: Incentives in service level agreement establishment the case of economic and social aspects. In: 2011 First International Workshop on Requirements Engineering for Social Computing, pp. 30–33, August 2011
11. Müller, C., Gutiérrez, A.M., Martín-Díaz, O., Resinas, M., Fernández, P., Ruiz-Cortés, A.: Towards a formal specification of SLAs with compensations. In: Meersman, R., Panetto, H., Dillon, T., Missikoff, M., Liu, L., Pastor, O., Cuzzocrea, A., Sellis, T. (eds.) OTM 2014. LNCS, vol. 8841, pp. 295–312. Springer, Heidelberg (2014). https://doi.org/10.1007/978-3-662-45563-0_17
12. Müller, C., Martín-Díaz, O., Ruiz-Cortés, A., Resinas, M., Fernández, P.: Improving temporal-awareness of WS-agreement. In: Krämer, B.J., Lin, K.-J., Narasimhan, P. (eds.) ICSOC 2007. LNCS, vol. 4749, pp. 193–206. Springer, Heidelberg (2007). https://doi.org/10.1007/978-3-540-74974-5_16
13. Psaier, H., Skopik, F., Schall, D., Dustdar, S.: Resource and agreement management in dynamic crowdcomputing environments. In: 2011 15th IEEE International, Enterprise Distributed Object Computing Conference (EDOC), pp. 193–202, August 2011

14. Riveni, M., Truong, H.-L., Dustdar, S.: On the elasticity of social compute units. In: Jarke, M., Mylopoulos, J., Quix, C., Rolland, C., Manolopoulos, Y., Mouratidis, H., Horkoff, J. (eds.) CAiSE 2014. LNCS, vol. 8484, pp. 364–378. Springer, Cham (2014). https://doi.org/10.1007/978-3-319-07881-6_25

15. Riveni, M., Truong, H.L., Dustdar, S.: Trust-aware elastic social compute units. In: 2015 IEEE Trustcom/BigDataSE/ISPA, vol. 1, pp. 135–142. IEEE (2015)

16. Schall, D.: Service oriented protocols for human computation. In: Michelucci, P. (ed.) Handbook of Human Computation, pp. 551–559. Springer, New York (2013). https://doi.org/10.1007/978-1-4614-8806-4_42

17. Schulz, F.: Elasticity in service level agreements. In: 2013 IEEE International Conference on Systems, Man, and Cybernetics (SMC), pp. 4092–4097. IEEE (2013)

18. Sengupta, B., Jain, A., Bhattacharya, K., Truong, H.-L., Dustdar, S.: Who do you call? Problem resolution through social compute units. In: Liu, C., Ludwig, H., Toumani, F., Yu, Q. (eds.) ICSOC 2012. LNCS, vol. 7636, pp. 48–62. Springer, Heidelberg (2012). https://doi.org/10.1007/978-3-642-34321-6_4

19. Truong, H.-L., Dustdar, S.: Context-aware programming for hybrid and diversity-aware collective adaptive systems. In: Fournier, F., Mendling, J. (eds.) BPM 2014. LNBIP, vol. 202, pp. 145–157. Springer, Cham (2015). https://doi.org/10.1007/978-3-319-15895-2_13

20. Wetzstein, B., Karastoyanova, D., Leymann, F.: Towards management of slaaware business processes based on key performance indicators. In: 9thWorkshop on Business Process Modeling, Development and Support (BPMDS 2008) - Business Process Life-Cycle:Design, Deployment, Operation Evaluation (2008)

Using Smart Edge Devices to Integrate Consumers into Digitized Processes: The Case of Amazon Dash-Button

Michael Möhring[1(✉)], Barbara Keller[1], Rainer Schmidt[1], Lara Pietzsch[2], Leila Karich[2], Carolin Berhalter[2], and Karsten Kilian[2]

[1] Munich University of Applied Sciences, Munich, Germany
{michael.moehring,barbara.keller,rainer.schmidt}@hm.edu
[2] University of Applied Sciences Würzburg-Schweinfurt, Würzburg, Germany
karsten.kilian@fhws.de

Abstract. Integrating consumers into business processes has always created special challenges. However, smart edge devices improve the integration of consumers into processes significantly. Smart edge devices such as the Amazon Dash-Button allow to trigger processes with the simple press on a button from everywhere. In this empirical research we conducted a pre-study to explore the potential value provided by smart devices such as the Amazon Dash-Button to extend the reach of business processes. Thus, we present the design of the pre-study and first results.

Keywords: Smart edge devices · IoT · Retail processes · BPM · Amazon Dash

1 Introduction

Integrating consumers into business processes has always been a special challenge [45–47]. First, the inclination of consumers to use complicated interfaces is rather low. Any process interface that creates a high threshold to interact with the process is and will be refused by the consumers. Second, the consumers expect to interact with the process where they want. Thus, the devices have to placed where the consumers expect them to be. Third, the interaction should happen immediately or with a very low delay. Consumers do not want to wait on computers to boot etc. [42–44].

These three challenges could not be addressed properly by the devices so far. Personal computers have a complicated interface, they have a fixed place and booting them takes too much time. The same applies for mobile phones and tablets, although on a lower level. Most consumers have only one devices and thus these devices have to be moved to all possible places, where the consumer wants to interact with the process.

However, nowadays, a class of devices, called smart edge devices [1] is improving the integration of consumers into processes significantly. Smart edge devices [1] extend the reach of business processes. Smart Edge Devices such as Amazon Dash-Button [2] and Amazon Echo Dot [3] are internet-connected devices capable to interact with the user and integrate the user into digitized business processes. Up to now they are primarily used for sales support and simply queries such as: "Who will be the weather" today. However, the information provided by Smart Edge Devices is accessible via high-level

© Springer International Publishing AG 2018
E. Teniente and M. Weidlich (Eds.): BPM 2017 Workshops, LNBIP 308, pp. 374–383, 2018.
https://doi.org/10.1007/978-3-319-74030-0_28

APIs that are the foundation for business process integration [4]. In this way, Smart Edge Devices (SDEs) provide the technical means that users can start business processes, perform tasks, make decisions, and finish business processes. For instance, Alexa Skills embrace custom skill such as smart home skills and flash briefing skills [4]. Custom skills allow to response to requests for actions and information and can be integrated into shopping lists and to-do lists. Smart home skills allow the user to control cloud-enabled smart-home devices. Flash briefing skills provide content to the user. Furthermore, retailing business processes can be supported and triggered by smart edge devices such as an Amazon Dash-Button [2]. The process can be shortened and triggered at the place of need by the customer.

Although smart edge devices like Amazon Dash-Button [2] provide a number of technological solutions, their capabilities to extend the reach of business processes from a user's perspective have not been investigated in depth. Therefore, we conducted a pre-study on the potential value provided by smart devices such as the Amazon Dash-Button to extend the reach and customer orientation of business processes. We extend previous business process related research in the retailing sector, which is a very important sector for BPM cases (e.g., [36–40]).

The paper is structured as follows: Sect. 2 after the introduction a background of smart edge device is given, Sect. 3 the pre-study design of the potential value of an Amazon Dash-Button is defined, Sect. 4 Research methods and data collection are described, Sect. 5 results are shown and Sect. 6 a conclusion is given.

2   Background: Smart Edge Devices

Smart Edge Devices are part of the Internet of Things (IoT) [5]. They are devices that are either directly or indirectly connected with the internet, capable to interact with human users and able to integrate them within triggered processes. In general, IoT applications are important for Business Process Management [6–8]. The interaction with the user can be divided into different technological capabilities (Fig. 1). Depending on the used Smart Edge device, the categories can be organized quite different. While process interaction for the Amazon Dash-Button is clearly defined and immutable for the user, Amazon's Alexa is quite flexible in this pattern.

	Interaction with user	Feedback	Process Interaction	Start	Do Tasks	Decide	Finish
Amazon Dash	Push Button	Green light	Pre-Defined	x			
Amazon Alexa	Voice	Voice	Variable	x	x	x	x

Fig. 1. Technological capabilities of smart edge devices

The Amazon Dash-Button [2] reduces the effort of the purchase business process by reducing the amount of tasks done by the users. For instance, if the consumer uses the

Dash-Button for purchasing, he has only to click on it without specifying payment method, address and billing information etc. Therefore, the cancellation rate of the process can be also reduced. Furthermore, the complexity of the business process can be reduced from the viewpoint of the consumer.

Although, the potential value of the smart edge devices is obviously, no prior research about the potential value of such a device focussing on customer's viewpoint can be found. We conducted a broad literature review in leading databases like AISeL, IEEExplore, SpringerLink etc. following the guidelines recommended in literature [9].

However, it is quite important to get awareness about the perceived potential values of smart edge devices and related business processes with regards to the costumers and their behaviour. Otherwise, the technology will be developed without a real use case and could evoke unsatisfied business processes. With a view to discover first insights of the potential use of smart edge devices, we design in the following a short pre-study focussing on the potential benefits of the Amazon Dash-Button. We used Amazon Dash-Button [2], because currently it is more often used by customers than Amazon Alexa and there exists implemented and often used purchasing business processes. Additionally, its potential to improve business is enhanced. For instance, marketing activities and related business processes can be restructured and their effectiveness could be increased by sending promotions to the peer-group [10–13] and saving marketing expenses for common advertising can be saved. Besides, other business processes might achieve an advantage. We suppose particularly selling/buying processes to have a special benefit from smart edge device (Amazon Dash-Button). On the one hand, the processes are standardized and pass without intermediary. On the other hand, the buying process for the customer is quite comfortable [14, 15], because to buy the product just one push is needed. Therefore, the decision to buy goes fast and maybe without extensive evaluation of alternatives.

3   Pre-study Design: Potential Value of an Amazon Dash-Button

In our pre-study, we want to investigate which factors are influencing the potential value of the Amazon Dash Button and how high is the perceived potential value of such a smart edge device. This is a first step to better understand the use case of human trigger business processes via an Amazon Dash Button. The research model is defined in the following as well as depicted in Fig. 2.

Convenience is a very important factor for online shoppers [16, 25] compared to local retail shopping. If this basic aspect would be missing in an online shopping environment and related business processes, the effect would be quite negative. The more this factor is fulfilled, the better customers will respond. Therefore, convenience [25] is a typical satisfier for online. The Amazon Dash button allows the user easily to trigger the shopping process. Therefore, we suppose that the provided convenience has an influence on the potential value and hypothesize as follows 1:

H1: Convenience is influencing the potential value of a Dash-Button.

Dash-Buttons helps customer to buy products easily without going to the next shopping facility. Especially, in rural areas this aspect is quite important with regards to shop

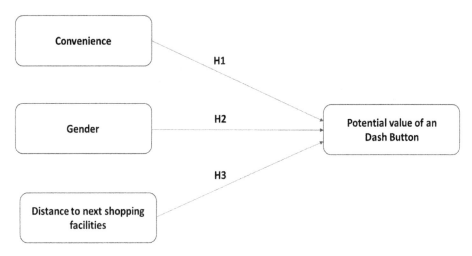

Fig. 2. Research model

online [17]. Therefore, the distance to the next supermarket could be influencing the potential use of an Amazon Dash-Button and related business processes. For instance, if the next supermarket is quite near and reachable without any effort, customers may not want to use the Amazon Dash-Button, because they do not perceive any benefit. Consequently, we created the following hypothesis:

H2: The distance to the next supermarket influences the potential value of a Dash-Button.

The product range available by the Dash- Button is wide. However, it is dominated by drugstore items [2]. This product category is intensified bought by female. Therefore, it might be the case, that the potential value depends on the user's gender. Also, gender aspects might be interesting for designing business processes and the adoption of such services [27–29]. Therefore, we construct a third hypothesis as follows:

H3: There are gender specific differences in the potential value of a Dash Button.

For the investigation of our research model, we used the research methods described in the following:

4   Research Methods and Data Collection

We collected our data from a quantitative web-based empirical study according to general research recommendations (e.g., [26, 30]).

At first, we pre-tested our study to ensure a high quality of the research and the data we want to gather. After, we conducted the study in the timeframe from October until December 2016 in Germany. The study was implemented as an online survey via the open source survey platform Limesurvey [31]. An excerpt about the items of the questions can be found in the appendix section of the paper. The questions were ranked on a four- point Likert scale [32]. Gender related questions are coded with "0" for man and "1" for female. Furthermore, the age of the participants and the household size were collected in whole numbers.

After cleaning our data (e.g., missing values) we got a finale sample size of n = 184. The sample consist of 65% female and 35% male participants with an average age about 28.1 years. The average household size of our sample is about approx. 2.58 persons.

With regards to our research model, we used a structural equation modelling (SEM) approach [18, 19] for the analysis of our empirical data. SEM links our research model (causal model) with empirical data. We are using the Smart PLS [20] approach based on partial least square regression and the calculation of significances via bootstrapping [18]. Furthermore, SEM is often applied in information systems research (e.g., [33–35, 49]). We used Smart PLS 3 as well as Microsoft Excel for the data preparation and analysis. All important quality criteria like Cronbach's alpha, R^2 etc. were extracted and evaluated like recommended in literature.

Overall, most of the participants (approx. 63%) perceived a good or high potential value of a Dash-Button. Therefore, consumer will use this smart edge device as a trigger for a purchasing business process.

The results are detailed described in the next section.

5   Results

The results of our SEM are shown in Fig. 3 and described in the following.

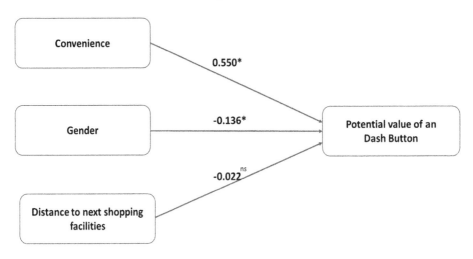

*: p<0.05

Fig. 3. Results of the SEM

Our model fits well in relation to the recommended quality metrics of the literature [21]. Only one factor is not significant (p > 0.05). The coefficient of determination (R^2) is also in a satisfying range (0.305 > 0.19) according to Chin [21]. All quality metrics are shown in Table 1.

Table 1. Quality metrics of the SEM

	Path coefficient	Significance (p-values)	AVE	Cronbach's alpha	Composite reliability
Convenience	0.550	0.000	0.65	0.734	0.848
Distance to next shopping facilities	−0.022	0.499	1 (1 item)	1 (1 item)	1 (1 item)
Gender	−0.136	0.037	1 (1 item)	1 (1 item)	1 (1 item)
Potential value of an Dash Button	–	–	0.69	0.773	0.867

Regarding the results of the SEM (Fig. 3 as well as Table 1), we can discuss our research model:

According to our results, H1 (*Convenience is influencing the potential value of a Dash-Button.*) can be confirmed, because of a positive path coefficient (+0.550). The path is also significant (p = 0.00). Therefore, convenience influences the potential value of a Dash-Button. Consequently, Dash Buttons can make the purchasing business process for customers much more convenient compared to traditional business process at an online store. Consumers can trigger the process at the place of need (e.g., in front of a washing machine). The new smarter business process must also integrate all the needed information (e.g., payment method or shipping address) as before. Therefore, traditional business processes must be re-designed.

Hypothesis 2 (*The distance to the next supermarket influences the potential value of a Dash-Button.*) can not be confirmed based on our empirical data, because of a missing significance (p > 0.05). Shoppers might not be influenced by the distance to the super-market. An explanation might be that people like to go to supermarkets because there they can get in touch with other people. Future research should take a deeper look into this interesting field of purchasing business processes for urban vs. village areas.

Furthermore, Hypothesis 3 (*There are gender specific differences in the potential value of a Dash-Button*) can be confirmed. A negative path coefficient (−0.136) indicates that male shoppers have a higher potential value of an Amazon Dash-Button, because man a coded with 0 and female coded with 1. The relation is also significant (p = 0.037). Therefore, there might be different use cases for purchasing processes as well as different variants of it. The aspect that male respond more favourable to this shopping process, because they are may more affected by gamification aspects (e.g., pushing a Dash-Button could be linked to gamification). Future research should discover these factor in relation to trigger aspects and process variants in detail.

In summary, the important details of the SEM are described below:

Regarding the foundations of smart edge devices and our results, we want to conclude as follows in the next section.

6   Conclusion

Smart edge devices, like Amazon Dash-Button or Amazon Alexa, generating new possibilities to integrate the consumer in digital processes. Research of this kind of human interaction with business process management is in an early stage. We show different insights in the use of smart edge devices from BPM viewpoint and implement a first pre-study of the potentials of the value of such devices.

Our research contributes to scientific literature in different ways. Our research defines foundations of smart edge devices in relation to business process management. Additionally, it reveals the high potentials and value of smart edge devices from the customer's point of view and shows simultaneously how these are influenced by customer's characteristics. Furthermore, practice can benefit from our research. We show how customers can be integrated via smart edge devices in the digitization process of business processes and how possible users see the value of such solutions. In this way enterprises can may save costs in terms of eliminating manually processes, increase the quality of the digitized processes, and reduce the cycle time due to a more automatically and standardized business process. Limitations of our work can be seen in the missing evaluation of all details of smart edge devices as well as the implementation of a short pre-study, which is very common in research (e.g., [26]).

Future research, should have a deeper look in detailed business process support by smart edge devices as well as broader empirical studies about the use and key success factors of it. If pictures are used by smart edge devices in the future, the possibilities of Image Mining [41] related to BPM should be observed. Furthermore, impacts on information system design [22] (e.g., CRM & cloud integration [48]), human-computer-interaction [23], enterprise architecture [24] and product returns behavior [50] are important to discover.

Appendix

See (Table 2):

Table 2. Excerpt of the items of our study

	Items*
Convenience	• "For me, the Dash Button facilitate the purchase of products." • "The use of the Dash Buttons decreases the effort I have to spend for shopping." • "My leisure time was increased noticeable since using the Dash Button."
Distance to next shopping facilities	• Distance to next shopping facilities
Gender	• Gender
Potential value	• "The functions of the Dash are quite interesting." • "The Dash Buttons offers a high volume of potential values." • "The use cases of the Dash Button are manifold."
Age	• Age
Household size	• Household size

*: translated from the German language.

References

1. The first international workshop on smart edge computing and networking 2017 - welcome and committeees. In: 2017 IEEE International Conference on Pervasive Computing and Communications Workshops (PerCom Workshops), pp. 1–2 (2017)
2. Amazon.com: Dash Button. https://www.amazon.com/b/?node=10667898011&sort=date-desc-rank&lo=digital-text. Accessed 4 Dec 2015
3. Warren, A.: Amazon Echo: The Ultimate Amazon Echo User Guide 2016 Become An Alexa and Echo Expert Now! CreateSpace Independent Publishing Platform, USA (2016)
4. Alexa Skills Kit - Build for Voice with Amazon. https://developer.amazon.com/alexa-skills-kit. Accessed 1 Jun 2017
5. Atzori, L., Iera, A., Morabito, G.: The Internet of Things: a survey. Comput. Netw. **54**(15), 2787–2805 (2010)
6. Bandyopadhyay, D., Sen, J.: Internet of Things: applications and challenges in technology and standardization. Wirel. Pers. Commun. **58**(1), 49–69 (2011)
7. Houy, C., Fettke, P., Loos, P., van der Aalst, W.M.P., Krogstie, J.: BPM-in-the-Large – towards a higher level of abstraction in business process management. In: Janssen, M., Lamersdorf, W., Pries-Heje, J., Rosemann, M. (eds.) EGES/GISP -2010. IFIP AICT, vol. 334, pp. 233–244. Springer, Heidelberg (2010). https://doi.org/10.1007/978-3-642-15346-4_19
8. Meyer, S., Sperner, K., Magerkurth, C., Pasquier, J.: Towards modeling real-world aware business processes. In: Proceedings of the Second International Workshop on Web of Things, p. 8 (2011)
9. Webster, J., Watson, R.T.: Analyzing the past to prepare for the future: writing A. MIS Q. **26**(2), 494–508 (2002)
10. Inman, J.J., McAlister, L., Hoyer, W.D.: Promotion signal: proxy for a price cut? J. Consum. Res. **17**(1), 74–81 (1990)
11. Laran, J., Tsiros, M.: An investigation of the effectiveness of uncertainty in marketing promotions involving free gifts. J. Mark. **77**(2), 112–123 (2013)

12. Dahlén, M., Lange, F., Smith, T.: Marketing Communications: A Brand Narrative Approach. John Wiley & Sons, London (2010)
13. Koski, N.: Impulse buying on the internet: encouraging and discouraging factors. Front. E-Bus. Res. **4**, 23–35 (2004)
14. Jarvenpaa, S.L., Todd, P.A.: Is there a future for retailing on the Internet. Electron. Mark. Consum. **1**(12), 139–154 (1997)
15. Burke, R.R.: Do you see what I see? The future of virtual shopping. J. Acad. Mark. Sci. **25**(4), 352–360 (1997)
16. Zhou, Y.C., Liu, X.P., Wang, X.N., Xue, L., Liang, X.X., Liang, S.: Business process centric Platform-as-a-Service model and technologies for cloud enabled industry solutions. In: 2010 IEEE 3rd International Conference on Cloud Computing (CLOUD), pp. 534–537 (2010)
17. Lennon, S.J., Kim, M., Johnson, K.K., Jolly, L.D., Damhorst, M.L., Jasper, C.R.: A longitudinal look at rural consumer adoption of online shopping. Psychol. Mark. **24**(4), 375–401 (2007)
18. Wong, K.K.-K.: Partial Least Squares Structural Equation Modeling (PLS-SEM) techniques using SmartPLS. Mark. Bull. **24**, 1–32 (2013)
19. Hooper, D., Coughlan, J., Mullen, M.R.: Structural equation modelling: guidelines for determining model fit. Electron. J. Bus. Res. Methods **6**, 53–60 (2008)
20. Ringle, C.M., Wende, S., Will, A.: SmartPLS 2.0 (beta), Hamburg, Germany (2005)
21. Chin, W.W.: The partial least squares approach to structural equation modeling. Mod. Methods Bus. Res. **295**(2), 295–336 (1998)
22. Gubbi, J., Buyya, R., Marusic, S., Palaniswami, M.: Internet of Things (IoT): a vision, architectural elements, and future directions. Futur. Gener. Comput. Syst. **29**(7), 1645–1660 (2013)
23. Kranz, M., Holleis, P., Schmidt, A.: Embedded interaction: interacting with the Internet of Things. IEEE Internet Comput. **14**(2), 46–53 (2010)
24. Zimmermann, A., Schmidt, R., Sandkuhl, K., Wissotzki, M., Jugel, D., Möhring, M.: Digital enterprise architecture - transformation for the Internet of Things. In: IEEE International Enterprise Distributed Object Computing Conference (EDOC 2015), Workshop Proceedings, Adelaide (2015)
25. Zhou, L., Dai, L., Zhang, D.: Online shopping acceptance model-a critical survey of consumer factors in online shopping. J. Electron. Commer. Res. **8**(1), 41 (2007)
26. Parasuraman, A., Grewal, D., Krishnan, R.: Marketing Research. Cengage Learning, Boston (2006)
27. Ilie, V., Van Slyke, C., Green, G., Lou, H.: Gender differences in perceptions and use of communication technologies: a diffusion of innovation approach. Inf. Resour. Manage J. **18**(3), 16–31 (2005)
28. Riquelme, H.E., Rios, R.E.: The moderating effect of gender in the adoption of mobile banking. Int. J. Bank Mark. **28**(5), 328–341 (2010)
29. Ahuja, M.K., Thatcher, J.B.: Moving beyond intentions and toward the theory of trying: effects of work environment and gender on post-adoption information technology use. MIS Q. **29**(3), 427–459 (2005)
30. Recker, J.: Scientific Research in Information Systems: A Beginner's Guide. Springer, Heidelberg (2012)
31. Schmitz, C.: LimeSurvey: an open source survey tool. LimeSurvey Project Hamburg, Germany (2012). http://www.limesurvey.org
32. Likert, R.: A technique for the measurement of attitudes. Arch. Psychol. **140**, 1–55 (1932)
33. Urbach, N., Ahlemann, F.: Structural equation modeling in information systems research using partial least squares. JITTA J. Inf. Technol. Theor. Appl. **11**(2), 5–40 (2010)

34. Wetzels, M., Odekerken-Schröder, G., Van Oppen, C.: Using PLS path modeling for assessing hierarchical construct models: Guidelines and empirical illustration. MIS Q. **33**, 177–195 (2009)
35. Münstermann, B., Eckhardt, A., Weitzel, T.: The performance impact of business process standardization: an empirical evaluation of the recruitment process. Bus. Process Manag. J. **16**(1), 29–56 (2010)
36. Becker, J., et al.: A framework for efficient information modeling - guidelines for retail enterprises. In: Proceedings of the Third Informs Conference on Information Systems and Technology, pp. 442–448 (1998)
37. Silvestro, R., Westley, C.: Challenging the paradigm of the process enterprise: a case-study analysis of BPR implementation. Omega **30**(3), 215–225 (2002)
38. Keller, B., Schmidt, R., Möhring, M., Härting, R.-C., Zimmermann, A.: Social-data driven sales processes in local clothing retail stores. In: Reichert, M., Reijers, H.A. (eds.) BPM 2015. LNBIP, vol. 256, pp. 305–315. Springer, Cham (2016). https://doi.org/10.1007/978-3-319-42887-1_25
39. Lefebvre, L.A., et al.: RFID as an enabler of B-to-B e-Commerce and its impact on business processes: a pilot study of a supply chain in the retail industry. In: Proceedings of the 39th Annual Hawaii International Conference on System Sciences, HICSS 2006. IEEE (2006)
40. Becker, J., Kugeler, M., Rosemann, M. (eds.): Process Management: A Guide for the Design of Business Processes. Springer, Heidelberg (2013)
41. Schmidt, R., Möhring, M., Zimmermann, A., Härting, R.-C., Keller, B.: Potentials of image mining for business process management. In: Czarnowski, I., Caballero, A.M., Howlett, R.J., Jain, L.C. (eds.) Intelligent Decision Technologies 2016, Part II. SIST, vol. 57, pp. 429–440. Springer, Cham (2016). https://doi.org/10.1007/978-3-319-39627-9_38
42. Houston, M.B., Bettencourt, L.A., Wenger, S.: The relationship between waiting in a service queue and evaluations of service quality: a field theory perspective. Psychol. Mark. **15**(8), 735–753 (1998)
43. Nah, F.F.-H.: A study on tolerable waiting time: how long are web users willing to wait? Behav. Inf. Technol. **23**(3), 153–163 (2004)
44. Hoxmeier, J.A., Dicesare, C.: System response time and user satisfaction: an experimental study of browser-based applications. In: AMCIS 2000 Proceedings (2000)
45. Laudon, K.C., Laudon, J.P.: Management Information Systems: Managing the Digital Firm, vol. 8. Prentice-Hall, Upper Saddle River (2006)
46. Chesbrough, H., Spohrer, J.: A research manifesto for services science. Commun. ACM **49**(7), 35–40 (2006)
47. Harrison-Broninski, K.: Human Interactions: The Heart and Soul of Business Process Management: how People Really Work and how They Can be Helpful to Work Better. Meghan-Kiffer Press, Tampa (2005)
48. Schmidt, R., Möhring, M., Keller, B.: Customer relationship management in a public cloud environment–key influencing factors for European enterprises. In: Proceedings of the 50th Hawaii International Conference on System Sciences (2017)
49. Schmidt, R., et al.: Benefits from using Bitcoin: empirical evidence from a European country. Int. J. Serv. Sci. Manag. Eng. Technol. (IJSSMET) **7**(4), 48–62 (2016)
50. Möhring, M., et al.: Präventives retourenmanagement im eCommerce. HMD Prax. der Wirtsch. **50**(5), 66–75 (2013)

Towards a Solution Space for BPM Issues Based on Debiasing Techniques

Maryam Razavian, Irene Vanderfeesten$^{(\boxtimes)}$, and Oktay Turetken

School of Industrial Engineering and Innovation Sciences,
Eindhoven University of Technology, Eindhoven, The Netherlands
{M.Razavian, I.T.P.Vanderfeesten, O.Turetken}@tue.nl

Abstract. In previous work, we discussed how cognitive biases may lead to issues in the design phases of the business process management lifecycle, such as the development of suboptimal process architectures and incomplete process models, the identification of irrelevant bottlenecks and weaknesses in a process, and the selection and implementation of confirmatory redesigns. This position paper makes a first step towards solving these issues through the use of debiasing techniques. Such techniques can be used to reduce or avoid the cognitive biases that potentially lead to BPM issues.

Keywords: Cognitive biases · Debiasing techniques · BPM issues
BPM lifecycle · BPM method

1 Introduction

During the continuous improvement of business processes many (re)design decisions have to be made by human decision makers. There are many approaches and best practices that aim at supporting good business process analysis and (re-)design, e.g. [4, 15]. However, these approaches and best practices often lack considerations of human factors in design, such as cognitive biases, proper design reasoning and communication, and reflection; all of which can affect the design decision making [5, 12, 17]. Consequently, outcomes of these activities may not be optimal. As argued in our previous work [13] and summarized in Table 1, cognitive biases may lead to a number of issues such as the design of incomplete process models and selection of confirmatory redesigns.

In this position paper, we explore the solution space for overcoming these BPM issues by mitigating the related cognitive biases. This is often called debiasing: the reduction of bias, particularly with respect to judgment and decision making (wikipedia). In the next section, we outline a number of debiasing techniques, inspired by related work in the Information Systems Design field. Next, we link these techniques to the earlier identified BPM issues through their related cognitive biases. This position paper is concluded by an outlook into further research on mitigating the cognitive biases that may lead to BPM issues.

© Springer International Publishing AG 2018
E. Teniente and M. Weidlich (Eds.): BPM 2017 Workshops, LNBIP 308, pp. 384–390, 2018.
https://doi.org/10.1007/978-3-319-74030-0_29

Table 1. BPM issues related to cognitive biases.

BPM issue	Influencing cognitive biases	Relevant BPM phases
Suboptimal process architecture	Completeness, overconfidence, anchoring, escalation	Process identification
Incomplete process model	Completeness, testimony	Process discovery
Irrelevant bottlenecks and weaknesses	Completeness, anchoring, confirmation	Process analysis
Confirmatory redesign	Anchoring, overconfidence, rule	Process redesign

2 Debiasing Techniques

Let us for now assume that the BPM designer is conscious of biases and tries to avoid, or what we call debias, them. This can be helped using some techniques and methods targeted at addressing specific biases. Although, to the best of our knowledge, there is no clear overview of debiasing techniques, we derived a list of promising debiasing practices from literature in the IS Design domain. Below, these techniques are briefly introduced together with the biases they can help to overcome. Section 3 discusses which of these techniques would be suitable for overcoming each BPM issue.

2.1 Assumption Analysis

Assumption analysis is about information gathering. It involves identifying the relevant context of a (business process) design and including factors that affect the design [12]. The validity and accuracy of a design decision is based on whether there are assumptions behind the decisions and if any hidden assumptions may inadvertently affect the design.

Anchoring bias happens when the first impression of relevant information for design that comes to mind anchors, and it may be difficult to adjust or change an assumption even when there is evidence to show that the initial assumption is not optimal. To debias anchoring, it is therefore prudent to carry out an assumption analysis, questioning possible tacit assumptions that may have been made, consciously or unconsciously. A suggestion is for the stakeholders to make explicit the assumptions of any key requirements and design [7].

2.2 Constraint Analysis

Like assumption analysis, constraint analysis is about identifying relevant context of the (business process) design. Context entails requirements, system environments, project environments and organizations which all exert some constraints on the way a solution may be designed. These constraints are often tacit and not explicitly discussed or documented. Van den Berg et al. [16] define design constraint as a limiting condition that a design concern imposes upon the outcomes. In searching for a solution in an infinitely large solution space, design constraints would help to reduce this space by limiting solutions to those that satisfy the constraints.

Completeness bias happens when constraints and their interdependencies are not identified adequately. For instance, constraints may conflict with each other, and when this happens some requirements may need to be compromised. Some constraints may influence the entire solution, e.g. certain time limits. To debias *completeness*, it is therefore useful for a designer to note the constraints of a requirement and a design. This would serve to detect conflicts in design. Additionally, at a decision point, BP designers may assess constraining requirements for tradeoff analysis.

2.3   Problem Structuring

Problem structuring is about formulating what the design problem is and what core problems need to be addressed. The articulation of design problems is a key component in the design task [3, 6]. Designers should also relate design issues to investigate how they influence each other. From a well-considered set of design problems, solution options can be devised. Although there are no exact recipes on how this should be done and the nature of the design problems and the knowledge of the designers dictate what is more useful and effective way of problem structuring in practice, some useful problem structuring techniques have been proposed in the IS research. These techniques are advocated as an effective way of dealing with biases. For instance, *Problem frames* are well-known technique to structure the problem and manage complexity and the interplay of design components.

Overconfidence bias happens when a designer overestimates his or her ability to identify and articulate the core design problems. An important aspect of confidence bias is the restriction of the search for new information about the design problems and the interdependencies among them (*completeness bias*). This is often followed by *escalation bias* when the designer commits to address the perceived core problems. As a debiasing strategy, IS problem structuring techniques can be used. Hall et al. [6] and Nuseibeh [10] suggest using *problem frames* to provide a means of analyzing and decomposing problems, and allowing a designer to iterate between the design problems and the solution options. Problem frames encourage designers to identify the core design problems and their relationships. As such, when BP designers formulate the problems, problem frames can alleviate the *overconfidence*, *completeness* and *escalation* biases.

2.4   Option Analysis

Option analysis is a technique for identifying solution options for design problems and potentially create new design problems. Option analysis assumes that problems and solutions can co-evolve. When design decisions are made or prior decisions are backtracked, new context and problems could surface. The explorations of problems and solutions are described by Maher et al. [9], Dorst and Cross [3], and Shaw [14]. There are many option analysis techniques proposed in IS research. Examples are a Questions, Options, and Criteria analysis (QOC) [8] and gIBIS [1].

Anchoring bias occurs when the first impression of a problem or solution dictates the thought process of the BPM designer. *Completeness bias* happens when BP designers do not consider solution options even when they exist. *Confirmation bias* leads to solutions that are accepted with no requirements or do not address any design problems. To alleviate these biases, option analysis can be used to evaluate the completeness and the appropriateness of a solution. Option analysis can create the habit of identifying various solution options for a certain requirement or problem in business process (re)design. In an empirical study, it was found that designers who are prompted to state solutions options create a better design [12].

2.5 Risk Analysis

Risk analysis is a technique to identify any risks or unknowns which might adversely affect the design. Risks can be of technical nature, such as stability of a software platform, or of non-technical nature, such as the ability of a team to successfully adopt and implement an information system [11]. When designing a BP, designers must be aware of the potential shortfalls of a design.

Testimony bias can happen when the designer is unable to recall some shortfalls or 'what-if scenarios'. For instance, BPM design might not be implementable because designers are unaware of the business domain, technology being used and the skill set of the team. *Completeness bias* occurs when important uncertainty and unknowns are not recognized during design and the design decisions. Risk analysis techniques such as the one proposed by Poort and van Vliet [11] help alleviating the abovementioned biases. More specifically they can enable recognizing the unknowns that could adversely affect your design; and if so, what their probability and impacts are.

3 Mitigating BPM Issues with Debiasing Techniques

With the above overview in mind, it is now discussed how the BPM issues potentially could be avoided using some of the debiasing techniques. Figure 1 depicts these relationships. Due to space limitations, only a selection of these links is elaborated below.

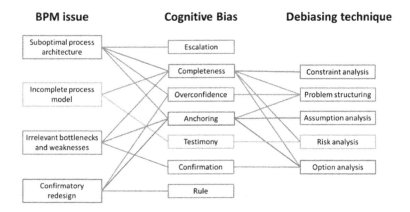

Fig. 1. Links between BPM issues and debiasing techniques through cognitive biases.

3.1 Suboptimal Process Architecture

The main challenge when creating a process architecture is to describe the processes landscape of a company, include all relevant processes and their relation, and to demarcate the beginning and end point of each process. When suffering from a completeness bias, a process architect may perceive a process architecture to be complete and stop searching for omissions or alternative designs. Using a *constraint analysis technique* may help to discover the boundaries of a process (landscape) and therefore help to design a better process architecture. For instance, the OTOPOP (One Time, One Place, One Person) principle that is often used in use case definition, may be applied when delineating activities and processes. It looks into the constraints of when, by whom, where, etc certain activities can be executed [2].

3.2 Incomplete Process Model

Incomplete process models miss alternative paths or business domain related exceptions, or lack sufficient level of detail in driving the subsequent efforts (for process automation or guiding process enactment). There are various reasons for arriving at an incomplete model. One of the most important ones occurs when information is elicited from a limited set of process participants and these participants have only partial knowledge of how the process runs (particularly about those parts that other process participants are performing). The *risk analysis* can be an effective debiasing technique to deal with incomplete process models, particularly with those that miss alternative paths and exceptions. Going over the process models using the scenario building technique and performing risk analysis to uncover events or conditions that may lead to alternative paths, would help in discovering such missing processing logic within the business context.

3.3 Irrelevant Bottlenecks and Weaknesses

The selection of which part(s) of a process to improve is done based on an analysis of bottlenecks and weaknesses, and their impact. Having a preliminary idea in mind may limit the open view of the analyst and may lead to the selection of a less important weakness or bottleneck, simply because other bottlenecks where not considered or identified. Making an *assumption analysis* a compulsory part of the process analysis phase will help to mitigate the risk of this anchoring bias.

3.4 Confirmatory Redesign

Business process (re)designers can make process redesign decisions based on partial information. Even if new information becomes available at a later stage, they may still be inclined to ignore that. They may prefer to stick with a redesign option that first came to mind without properly investigating all options. This is an effect of the anchoring bias. A solution to alleviate this is to explicitly conduct an *option analysis*, e.g. using the QOC-technique, as part of the redesign phase of the BPM lifecycle. Sufficient time and attention should be put to the systematic analysis of redesign options, their critical evaluation, and selection of the best one.

4 Conclusion and Outlook

In this paper, we explored solution directions for overcoming BPM design issues that may be related to cognitive biases in the business process designer. Several debiasing techniques to avoid these cognitive biases were studied, and through the cognitive biases they intend to mitigate, they were linked to the BPM issues. This exercise has given some insights into the solution space for BPM issues, but it is just a first stepping stone. The overview of applicable debiasing techniques is not complete, and these techniques should be further concretized with actual methods to be used, and integrated within the BPM lifecycle and existing BPM approaches. Therefore, we plan to first perform a thorough literature analysis on debiasing techniques, focusing on the biases with the designer as well as other stakeholders involved in a design effort. This analysis should cover various domains (e.g. IS design research, but also social sciences and behavioral economics) as design and decision related biases and their mitigation are studied in many domains. After a comprehensive inventory of debiasing techniques is made, we plan to evaluate which of those concepts and techniques are suitable to be applied in a BPM context, and plan to make concrete proposals to adopt these techniques in existing BPM methods.

References

1. Conklin, J., Begeman, M.: gIBIS: a hypertext tool for exploratory policy discussion. In: ACM Conference on Computer-Supported Cooperative Work, pp. 140–152 (1988)
2. Dijkman, R., Vanderfeesten, I., Reijers, H.A.: Business process architectures: overview, comparison and framework. Ent. Inf. Syst. **10**(2), 129–158 (2016)
3. Dorst, K., Cross, N.: Creativity in the design space: co-evolution of problem-solution. Des. Stud. **22**(5), 425–437 (2001)
4. Dumas, M., et al.: Fundamentals of Business Process Management. Springer, Heidelberg (2013)
5. Hadar, I., Soffer, T., Kenzi, K.: The role of domain knowledge in requirements elicitation via interviews: an exploratory study. Req. Eng. **19**(2), 143–159 (2014)
6. Hall, J., Jackson, M., Laney, R., et al.: Relating software requirements and architectures using problem frames. In: IEEE Joint International Conference on Requirements Engineering, pp. 137–144 (2002)
7. Lago, P., Van Vliet, H.: Explicit assumptions enrich architectural models. In: Proceedings of the 27th International Conference on Software Engineering. ACM (2005)
8. Maclean, A., Young, R., Bellotti, V., Moran, T.: Questions, options and criteria: elements of design space analysis. In: Moran, T., Carroll, J. (eds.) Design Rationale: Concepts, Techniques and Use, Ch. 3, pp. 53–106. Lawrence Erlbaum Associates (1996)
9. Maher, M.L., Poon, J., Boulanger, S.: Formalising design exploration as co-evolution: a combined gene approach. Technical report, University of Sydney (1996)
10. Nuseibeh, B.: Weaving together requirements and architecture. IEEE Comput. **34**(3), 115–119 (2001)
11. Poort, E.R., van Vliet, H.: Architecting as a risk- and cost management discipline. In: Proceedings of the 9th IEEE/IFIP Working Conference on Software Architecture, pp. 2–11 (2011)

12. Razavian, M., et al.: In two minds: how reflections influence software design thinking. J. Softw. Evol. Process **28**(6), 394–426 (2016)
13. Razavian, M., Turetken, O., Vanderfeesten, I.: When cognitive biases lead to business process management issues. In: Dumas, M., Fantinato, M. (eds.) BPM 2016 Workshops. LNBIP, vol. 281, pp. 147–156. Springer, Cham (2017). https://doi.org/10.1007/978-3-319-58457-7_11
14. Shaw, M.: The role of design spaces. IEEE Softw. **29**(1), 46–50 (2012)
15. Silver, S.: BPMN Method and Style. Cody-Cassidy Press (2011)
16. Van den Berg, M., Tang, A., Farenhorst, R.: A constraint-oriented approach to software architecture design. In: Proceedings of the QSIC 2009, pp. 396–405 (2009)
17. Van Vliet, H., Tang, A.: Decision making in software architecture. J. Syst. Softw. **117**, 638–644 (2016)

A Framework for Improving User Engagement in Social BPM

Vanisha Gokaldas[(✉)] and Mohammad Ehson Rangiha

Department of Computer Science, School of Mathematics,
Computer Science and Engineering, City, University of London, London, UK
{Vanisha.Gokaldas,Mohammad.Rangiha.2}@city.ac.uk

Abstract. Businesses are increasingly focused on becoming flexible and versatile, and are moving away from the traditional business process management approach. Social BPM supports this as it involves the interaction and collaboration of users, using social software to improve business process management. However, the success of Social BPM is dependent on the participation of users. Therefore, organisations need to ensure that users are motivated enough to participate in the use of social software for creating and maintaining process models. This paper presents a Social BPM User Engagement Framework that combines theories from psychology and computer science disciplines on user motivation and online participation in the workplace. The framework is designed in three levels – organizational, social software and task, to improve user participation in Social BPM.

Keywords: Social BPM · Online participation · Social software
User engagement · Motivation

1 Introduction

In order for businesses to remain profitable within global competition, it is important to ensure that organisations' business process management (BPM) are effective enough to provide flexible business operations [22]. BPM is a structured approach for improving business activities and operations [35]. However, traditional BPM suffers from the inability to be adapted to a less structured work environment [33]; processes are created using a top-down approach and workers are obliged to follow this. This creates a situation where processes are not being executed in the way they were originally planned to [26], as workers may not agree to or follow the processes. Further, due to the hierarchical structures of organisations, important knowledge that exists with workers is not shared or applied within its business processes [11]. Today, organisations are becoming increasingly flexible and versatile, making it difficult to maintain the structured and hierarchical approach offered by traditional BPM. According to Back et al. [5], organisations are moving from well managed and structured systems into network systems with less structured processes where decision making is shared, and activities are jointly managed.

Social BPM supports the notion of interaction, collaboration and flexibility; it allows for the integration of stakeholders within the BPM life cycle [12]. The purpose

© Springer International Publishing AG 2018
E. Teniente and M. Weidlich (Eds.): BPM 2017 Workshops, LNBIP 308, pp. 391–402, 2018.
https://doi.org/10.1007/978-3-319-74030-0_30

of Social BPM is based on the idea of adopting traditional BPM with the collaboration of participants to foster improved processes and utilization of knowledge management [2]. Therefore, rather than a limited group of experts creating a business process, Social BPM takes a bottom-up approach and involves a wide range of people through the use of social software. Social software can be described as "a set of web-based applications that allow users to interact and share data with each other in a free and unrestricted environment" [33]. According to Schmidt and Nurcan [26], social software enables both social interaction and social production. They describe social interaction as the interaction between individuals who are not hierarchically structured and do not necessarily know each other, and social production being the formation of artefacts such as content and context information. When applying social software within the context of BPM, Schmidt and Nurcan [26] suggests that it can be used for the designing, operating and improving business processes.

Social BPM therefore helps overcome the hierarchical nature of traditional BPM and brings about benefits to the organisation. The sharing of knowledge between expertise allows for problem solving to occur in a jointly manner, involving the views of all stakeholders to be considered. Fleischmann et al. [12] suggest that the organisation of work moves from central planning, to a more decentralised manner, were planning and execution is done in a self-organised manner by knowledge workers through social interaction and collaboration. Additionally, Social BPM allows for transparency of information and knowledge sharing between people. Consequently, the most important element required for the success of Social BPM is the participation of stakeholders [22]. Participation is a useful component for people to share their wisdom and exploit their tacit knowledge [2]. However, the availability of infrastructure within organisations, such as platforms and tools, does not necessarily mean users will adopt and utilise this to its full extent [22]. Adopting Social BPM goes beyond the use of a new software; it requires a shift in culture to the way employees work. To maintain engagement and contribution, organisations need to ensure that users are motivated enough to voluntarily participate and use social software for creating and maintaining process models [11, 23]. Pflanzl and Vossen [22] suggest that in order to ensure participation within Social BPM, research needs to be done to examine measures that allow re-enforcing of motivation to contributors.

Motivation is defined as "the willingness to exert high levels of effort toward organisational goals, conditioned by the effort's ability to satisfy some individual need" [25]. Human motivation can be categorised into intrinsic or extrinsic; intrinsic being where users are interested in the work itself and value it brings to them, and extrinsic being the desire to obtain outcomes such as monetary remuneration [3, 22]. With the advancement of technology, there is increased focus on online communities and participation. According to Galehbakhtiari and Hasangholi Pouryasouri [13] online participation can be seen as "a complex behaviour which rises in the heard of user's lived experience". There has been a lot of research into users and their satisfaction with using technology. For example, Davis [9] developed the Technology Acceptance Model (TAM) with the aim of predicting whether it will be used or not. He suggested that the users perceived usefulness (the degree to which using it enhances ones job) and perceived ease of use (the degree to which using the system would be free of effort) will determine their attitude and behavioural intention to use a technology. Therefore, this could be related to the social software being used in Social BPM, and whether employees perceive it as easy to use and useful; a possible measure for how participation could be enhanced.

This paper attempts to review literature from both a motivational workplace point of view, as well as an online participation point of view. The aim is to find relevant factors from psychology and computer science disciplines that suggest ways in which motivation can be increased in the workplace, both online and offline. These factors will then be aligned to the context of Social BPM by providing a solution through a preliminary framework that suggest ways in which the limitation of user participation can be reduced.

2 Theories on Motivation and Engagement

Employee motivation within organisations is an integral part to management in terms of performance, and has received a lot of attention from organisational research in terms of building theories for effective management [28]. There is an exhaustive amount of literature done to examine motivation and performance in the workplace. The theories stem back to the 1960s, whereby the focus was on identifying factors that determine how humans are motivated in the workplace. This period of time saw some of the most dominant theories being developed, some of which are shown on Table 1.

Table 1. Older theories on human motivation.

Author	Theory
Maslow [19]	**Hierarchy of needs.** Humans are motivated to satisfy a set of hierarchical needs, from physiological needs to self-actualisation, and if the needs are met at the workplace, it would result in job satisfaction
Herzberg [16]	**Two-factor theory.** There are certain factors in the workplace that can lead to job satisfaction; motivators such as more responsibility, interesting work, recognition, and hygiene factors such as salary, work conditions, job security. However the lack of hygiene factors these could result in dissatisfaction
Adams [1]	**Equity theory.** Employees are motivated by equality and fairness, in terms of salary, promotion, opportunities, and comparing the ratio of what is put into work (contribution) versus what is gained (benefits)
McGregor [20]	**Theory X-Y.** Two approaches to workplace motivation – Theory X suggests people who dislike work, prefers to be directed; and Theory Y – people have self-control and self-direction to achieve their needs and objectives

Since then, although there has not been many breakthrough developments in theory of motivation, there have been a number of extensions and refinements to existing theories, resulting in more sophisticated research methods and empirical findings [28]. In addition, the world of a workplace has evolved dramatically; organisations are now defined by increased diversity needs, and technology has also changed the mannerism and location of work [28]. These changes have had an effect on how managers within organisations attempt to retain staff and continue to keep them motivated, as well as how human behaviour has adapted in the workplace to reflect these new ways of working.

A systematic literature review was carried out in order to understand, from more recent academic research articles, what factors have been identified to increase motivation and online participation in the workplace. A variety of results were found, consisting of a number of factors which have proved to work within the psychology and computer science disciplines. The factors are grouped in the following four categories: Motivation, Individual Values, Community, and Technology.

Motivation (C1). Research has suggested a number of reasons why people tend to be motivated in the workplace. One of the most common factors found was the aspect of organisations providing performance feedback to employees [10, 15, 29, 32] such that it impacts their perceived empowerment and results in task motivation and performance. Similarly, having supervisor and organisational support [14] was also important to keep someone motivated.

In the same way, providing rewards in the workplace could result in motivation in employees [10, 24]. In particular, Yoon et al. [34] suggested that intangible rewards was positively related to intrinsic and extrinsic motivation towards employee creativity. Intangible rewards include verbal praise, acknowledgement, and social praise, such that employees feel appreciated for their efforts [34]. Surprisingly, it was found that tangible rewards, such as financial gains or promotions, is negatively related to extrinsic motivation for creativity.

Other factors that were found to increase motivation were having employees do tasks which were significant which also required a variety of skills [29]. Undoubtedly, individual differences play a role in motivation; for example, when analysing involvement in an online crowdsourcing campaign, Sultan [29] found that personal circumstances affects the level of involvement. This included barriers like location or disabilities. Additionally, having the time available to contribute also impacts the level of involvement [4].

Individual Values (C2). According to Bishop [7], individuals have desires which lead to their plans and action but these need to be compatible to their existing plans and goals, values and beliefs. Therefore people's motivation in the workplace may depend on their individual personalities, drives and motivations to engage in communities [4]. Individual beliefs can hold as a strong factor towards motivation. For example, Chung et al. [8] found that participants were likely to participate in discussion in two ways – firstly, as a corrective action to what they found against their belief, and secondly as a promotional action to share what they found in favour of their belief.

Some research papers suggested that altruism, the desire to help others or give back to the community, was a factor towards motivation [6, 21, 24]. This is also closely related with reciprocity – the willingness to reciprocate to what you have gained from others, an example of extrinsic motivation [17].

Further, to enrich one's self knowledge, it was found that employees are motivated to participate because it allows them to gain new perspectives [13, 21] and develop a reputation or status for themselves [7, 24] in a way that they are developing an identity for themselves [13]. Nov [21] suggests that factors that correlate with contribution levels were when individuals felt they needed to enhance their ego by exhibiting knowledge they knew to those who needed it. Relatedly, the sense of promoting their career was a motivator for participation [24].

Community (C3). The theory of social identity [30] suggests that part of an individual's self-concept is derived from their association with a social group. Therefore, a person who has greater similarities with other members in a group will have a greater sense of belonging to the group. Having a sense of belonging to a community was found to be a reoccurring factor in literature as a motive for participation [13, 18, 27], as well as being part of a group to build relationships [13]. It was also found that using narratives to keep people within the community informed of what is happening in the community, plays a significant role in participatory behaviour by increasing the sense of membership, and integration and fulfilment of needs for members. Despite this, the type of group and quality of responses from a group that someone participated in was also an important factor for continuous engagement [4].

Schroer and Hertel [27] also found that the perceived benefits of participating was a motive being volunteers on Wikepedia. On the other hand, people who do not participate in online communities believe they do not need to contribute because they are not helping if they do so [7]; in this case, they don't have the motivation to continue participation as there are no perceived benefits to them.

Technology (C4). According to Amichai-Hamburger et al. [4], technical reasons would impact the level of participation or contribution from users. For example, where there were no design flaws, users found it easier to use a system. This is consistent with the TAM, such that the perceived ease of use impacts an individual's behaviour to use a technology.

In addition, Amichai-Hamburger et al. [4] suggested that the psychological safety of participants was important, such that they are not ridiculed for posting their opinion or views, as well as privacy in terms of participants wanting to remain anonymous in an online community.

Other factors included accidental exposure and the use of cues. Research done by Valeriani and Vaccari [31] found that accidental exposure to news that individuals were not actively looking for contributed to people's online contribution. Cues, such as the number of postings, the velocity of comments appearing, the number of sources, has an impact on the way individuals form impressions on other users and has an influence on behavioural intentions to participate and comment [32].

3   Social BPM User Engagement Framework

Pflanzl and Vossen [22] mentioned that the first step towards accomplishing continuous participation in Social BPM is to reach a critical mass of users. Once organisations have selected the right individuals for initiation, some measures need to be in place to ensure continuous involvement.

Based on the findings from literature, a Social BPM User Engagement Framework has been proposed that provides a mechanism for organisations to increase their user participation and motivation for Social BPM. The framework groups together relevant factors that were found from research, in three different categories or levels where they are likely to occur. It starts with the wider level – the organisational level, to a more specific level which is the technology used for Social BPM, that is the social software, and finally the imminent task at hand that is being carried out for process improvement.

The formation of these three levels are discussed, along with the categories in which the factors were derived from as described above (C1, C2, C3 and C4).

Organisational level. Providing continuous feedback, support and rewards to employees for their efforts when performing Social BPM (C1). This would make employees feel valued for their work, allow them to understand the benefits of their work, and motivate them to continue participating in such activities. Rewards does not need to be monetary remuneration at the expense of the organisation, but as research also found, intangible rewards such social praise and acknowledgement is just as useful [34]. Additionally, using narratives such as virtual or verbal cascades within the organisation would create awareness and keep employees informed (C3), and consequently could lead to accidental exposure of information which results in participation (C4).

Social software level. This level is around the use of chosen social software technology. Ensuring the software used for Social BPM is free of flaws and is easy for participants to use and understand would makes it easier for users to use (C4). This meets the perceived ease of use as defined in the TAM. Also, using social software for Social BPM activities should create a community, such that employees have a sense of belonging (C3) and would not feel ridiculed for contributing their knowledge [4].

Task level. The final layer involves the imminent task that is being carried out using the role of Social BPM. As described in C1, the significance and importance of a task would be a motivational factor, which could subsequently result in employees understanding the perceived benefits, or lead to corrective action if they find something against their personal belief (C2). Additionally, if the task requires a variety of skills, this could give employees the opportunity to gain new perspectives on tasks and activities (C1).

The Social BPM User Engagement Framework (Fig. 1) is designed such that each level is responsible for a number of factors that will lead to the desired outcome of increasing Social BPM participation. The factors that are implemented may trigger one

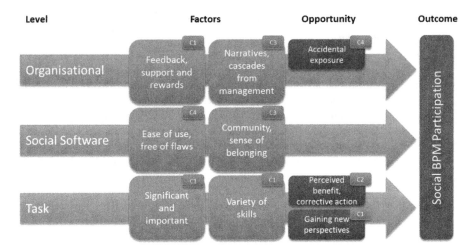

Fig. 1. Social BPM User Engagement Framework

or more opportunities of further benefits which strengthen the participation. For example, on the Organisational level, if narratives or cascades are implemented, it could result in accidental exposure being triggered as an opportunity factor, which could also enhance social BPM participation. Similarly, at the Task level, if the task includes a variety of skills, it may allow the user to gain new perspectives, a factor which can be elicited additionally and enhance participation.

It is expected that the more factors that are implemented across different levels, the greater the chance of increasing motivation and participation of Social BPM.

3.1   Application

The framework has been designed in an attempt to help organisations that suffer from the lack of user participation in Social BPM. In order for this to be successful, there are a number of responsibilities that need to take place within the organisation.

Although the framework has not been validated yet, it is anticipated that the implementation of it would require the following application measures to be in place, depending on the level(s) being executed. Without this, there is a likely risk that the execution of factors from the framework may not result in the desired outcome of increasing Social BPM participation. The procedures and responsibilities required for each level of the framework are outlined in Table 2 below.

Table 2. Procedures required for applying the Social BPM User Engagement Framework

Level	Procedure
Organisational	Managers should be responsible for monitoring the performance of their employees in order to provide support, feedback and rewards. Managers should have an active role in reviewing what employees are working on and how much contribution they provide, and consequently providing feedback and support for their work, and rewards in cases where it is deserved
	Managers should be responsible for circulating narratives and cascades to their team to keep them informed of company news such as processes or creation of processes such that it may engage the opinions of employees and encourage them to participate and contribute their knowledge
Social software	The social software chosen to be used in the organisation should be free of flaws and easy to use, with minimal training required. This may require some user testing to confirm employees perceived ease of use
	The social software should serve as a platform whereby employees are able to identify themselves to the group or community. Managers should play an active role in participating to create this environment
Task	The task being carried out should to add value to the organisation and/or be recognised as important by senior management. This will highlight the significance and consequence of employees participating
	The task should involve skills and input from a variety of people. Managers should play an active role to highlight and invite a diverse group of employees to participate

The above guidelines will now be described with a fictitious scenario of employees organising a corporate Christmas party. The process would begin with managers doing three tasks. Firstly, highlighting the importance of organising the event; for example to improve networking and the social and community spirit of the company. Secondly, managers should also highlight the variety of skills that employees will gain out of this – for example, project management skills for the organisation, sales skills for the negotiation of venues and marketing skills for the branding and promotion of the event to the rest of the organisation. Finally, managers should use narratives and cascades to highlight the event to other employees who may be interested in participating with the organisation of the event. This could be done through weekly stand-up sessions or weekly emails to make employees in their team aware of news and updates within the organisation.

At this point, the expectation is that a group of employees will be interested and motivated to perform the task of organising the Christmas party and using social software, will form a team of organisers. Although not a direct task in this process, however the organisation's social software should operate without any uneasiness from employees. Once the group of employees have formed a team, they should ensure they develop a sense of belonging to the community they are working within such that they can identify themselves as part of the group. While the team of organisers are planning the event, managers should have regular reviews of the progress that the employees are making, and provide support in terms of sponsorship, guidance and interest to their performances. They should also provide feedback along the way and if the event is a success, a reward could be given. The reward could be tangible in terms of monetary gains or intangible in the form of recognition. This would motivate the employees to take part in such an activity again.

A dynamic view of this is shown in Fig. 2, indicating the process and interaction that takes place. In a fictitious Company X, there are mainly two streams of resources within the process – managers and employees. Based on the scenario explained, the flow diagram indicates the tasks and responsibilities that managers within the organisation should aim to do in order to expect an increase in user engagement of Social BPM.

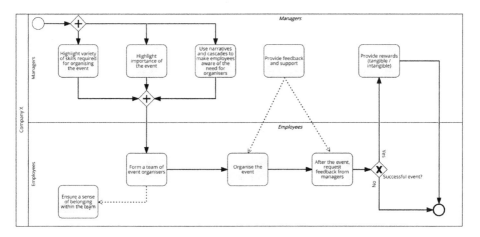

Fig. 2. Dynamic view of proposed framework being fictitiously implemented

As mentioned previously, the implementation of the framework could include one or more factors but it is expected that the more factors that are implemented, the greater the increase in participation. For example, using the scenario above, if only factors on the task level are implemented (highlighting the variety of skills required and highlighting the importance of the event), employees may not be motivated to do a similar activity in the future as they may not feel rewarded or supported for their efforts without any recognition or feedback to highlight their good work.

However, the framework does stand with some limitations. Firstly, since the framework has not been validated yet, it is difficult to determine whether the application of factors will truly lead to an increase in user participation for Social BPM. This may require longitudinal studies to understand participation levels over a period of time as a result of applying these factors. Secondly, the application of these procedures may result in additional time, effort and cost spent by the organisation. For organisations that do not already have some of these measures in place, it may require a change in culture to their current ways of working to adopt these procedures. For example, managers would be required to play an active role in determining what their employees are working on and assess how they are performing in order to provide feedback, support and rewards. Another example is that particularly for large organisations, it may be difficult to adopt factors on the social software level of the framework as they may already have software in place. Therefore implementing a new software to ensure it is free of flaws, for example, would incur a lot of time and cost. Additionally, it could be argued that some factors proposed in the framework could have varying effects depending on the size and nature of organisations. For example, the factor of narratives and cascades may work better in smaller organisations; whereas in larger organisations narratives and cascades may not be as effective since information could get lost within the several streams of managers and hierarchies that exist.

Finally, as the research used a specific search approach for finding factors that increase motivation and user engagement, it could not have captured all possible factors that exist in literature. There are other motivating factors that could have an impact on Social BPM participation. For example, if employees or managers have been set a target to achieve, or have deadlines to meet this would motivate them to complete their task for the reason complying with managerial authorities.

Despite these limitations, these measures are anticipated to be important and the framework remains open for validation.

4   Conclusion and Future Research

Social BPM has been identified as a possible way to overcome traditional BPM limitations. However, an important element in order for Social BPM to be successful is the engagement and participation of users [22]; a current drawback of Social BPM. Through a systematic literature review into theories within both the psychology and computer science disciplines, theories and solutions were found that suggest ways in which to increase motivation of employees in the workplace, and participation of employees in online communities. Factors found were categorised into four areas; *motivation, individual values, community and technology*. Based on this, a framework was designed with three levels – *organisational, social software and task*; each level has factors that lead to an increase in participation, some of which provide opportunities for other factors to be realised.

By proposing a Social BPM User Engagement Framework, the benefits of theories found from other disciplines are being utilised and aligned to Social BPM. Guidelines and procedures for implementing the framework are proposed. Additionally, the framework is explained through a fictitious scenario, along with a demonstration of a dynamic view on how the process would work if it was being implemented.

However the framework comes with limitations, some of which future research can address. As this research is still in progress, it has not yet been validated in the field. Therefore, the research will require validation with organisations that currently struggle or face the problem of user participation. Future research should particularly look into the change in number of users engaging in Social BPM when the factors are implemented in an organisation. Additionally, it could be looked at with each factor implemented to understand which has the strongest impact.

Overall, the framework proposed is based on factors that have been researched and suggested from other disciplines. Therefore, these factors are expected to have similar effects when applied to Social BPM to reduce the limitation of user participation.

References

1. Adams, J.S.: Inequality in social exchange. In: Advances in Experimental Social Psychology, vol. 2, pp. 267–299. Academic Press, New York (1965)
2. Alexopoulou, N., Nikolaidou, M., Stary, C.: Blending BPMS with social software for knowledge-intense work: research issues. In: Nurcan, S., Proper, H.A., Soffer, P., Krogstie, J., Schmidt, R., Halpin, T., Bider, I. (eds.) BPMDS 2013 and EMMSAD 2013. LNBIP, vol. 147, pp. 18–31. Springer, Heidelberg (2013). https://doi.org/10.1007/978-3-642-38484-4_3
3. Amabile, T.: Motivational synergy: toward new conceptualizations of intrinsic and extrinsic motivation in the workplace. Hum. Resour. Manag. Rev. **3**, 185–201 (1993)
4. Amichai-Hamburger, Y., Gazit, T., Bar-Ilan, J., Perez, O., Aharony, N., Bronstein, J., Sarah Dyne, T.: Psychological factors behind the lack of participation in online discussions. Comput. Hum. Behav. **55**, 268–277 (2016)
5. Back, A., von Krogh, G., Seufert, A., Enkel, E.: Putting Knowledge Networks into Action: Methodology, Development, Maintenance. Springer, Heidelberg (2005)
6. Baruch, A., May, A., Yu, D.: The motivations, enablers and barriers for voluntary participation in an online crowdsourcing platform. Comput. Hum. Behav. **64**, 923–931 (2016)
7. Bishop, J.: Increasing participation in online communities: a framework for human–computer interaction. Comput. Hum. Behav. **23**, 1881–1893 (2007)
8. Chung, M., Munno, G., Moritz, B.: Triggering participation: exploring the effects of third-person and hostile media perceptions on online participation. Comput. Hum. Behav. **53**, 452–461 (2015)
9. Davis, F.: Perceived usefulness, perceived ease of use, and user acceptance of information technology. MIS Q. **13**, 319 (1989)
10. Drake, A., Wong, J., Salter, S.: Empowerment, motivation, and performance: examining the impact of feedback and incentives on nonmanagement employees. Behav. Res. Account. **19**, 71–89 (2007)
11. Erol, S., Granitzer, M., Happ, S., Jantunen, S., Jennings, B., Johannesson, P., Koschmider, A., Nurcan, S., Rossi, D., Schmidt, R.: Combining BPM and social software: contradiction or chance? J. Softw. Maint. Evol. Res. Pract. **22**, 449–476 (2010)
12. Fleischmann, A., Schmidt, W., Stary, C.: Subject-oriented BPM = Socially Executable BPM. In: IEEE International Conference on Business Informatics, pp. 399–407 (2013)
13. Galehbakhtiari, S., Hasangholi Pouryasouri, T.: A hermeneutic phenomenological study of online community participation. Comput. Hum. Behav. **48**, 637–643 (2015)
14. Gillet, N., Gagné, M., Sauvagère, S., Fouquereau, E.: The role of supervisor autonomy support, organizational support, and autonomous and controlled motivation in predicting employees' satisfaction and turnover intentions. Eur. J. Work Organ. Psychol. **22**, 450–460 (2013)
15. Guo, Y., Liao, J., Liao, S., Zhang, Y.: The mediating role of intrinsic motivation on the relationship between developmental feedback and employee job performance. Soc. Behav. Personal. Int. J. **42**, 731–741 (2014)
16. Herzberg, F.: Work and the Nature of Man. World Publishing Company, Cleveland (1966)

17. Hung, S., Lai, H., Chang, W.: Knowledge-sharing motivations affecting R&D employees' acceptance of electronic knowledge repository. Behav. Inf. Technol. **30**, 213–230 (2011)
18. Lampe, C., Wash, R., Velasquez, A. Ozkaya, E.: Motivations to participate in online communities. In: Proceedings of the SIGCHI Conference on Human Factors in Computing Systems, pp. 1927–1936 (2010)
19. Maslow, A.H.: Motivation and Personality. Harper & Row, New York (1054)
20. McGregor, D.: The Human Side of Enterprise. McGraw-Hill, New York (1960)
21. Nov, O.: What motivates Wikipedians? Commun. ACM **50**, 60–64 (2007)
22. Pflanzl, N., Vossen, G.: Challenges of social business process management. In: Hawaii International Conference on System Science (HICSS), pp. 3868–3877. IEEE (2014)
23. Pflanzl, N., Vossen, G.: Human-oriented challenges of social BPM: an overview. In: 5th International Workshop on Enterprise Modelling and Information System Architectures (EMISA), vol. 222, pp. 163–176 (2013)
24. Preece, J., Shneiderman, B.: The reader-to-leader framework: motivating technology-mediated social participation. AIS Trans. Hum. Comput. Interact. **1**, 13–32 (2009)
25. Robbins, S.: Organisational Behaviour, 6th edn. Prentice-Hall, Englewood Cliffs (1993)
26. Nurcan, S., Schmidt, R.: Introduction to the First International Workshop on Business Process Management and Social Software (BPMS2 2008). In: Ardagna, D., Mecella, M., Yang, J. (eds.) BPM 2008 Workshops. LNBIP, vol. 17, pp. 647–648. Springer, Heidelberg (2009). https://doi.org/10.1007/978-3-642-00328-8_64
27. Schroer, J., Hertel, G.: Voluntary engagement in an open web-based encyclopedia: wikipedians and why they do it. Media Psychol. **12**, 96–120 (2009)
28. Steers, R., Mowday, R., Shapiro, D.: The future of work motivation theory. Acad. Manag. Rev. **29**, 379–387 (2004)
29. Sultan, S.: Examining the job characteristics: a matter of employees' work motivation and job satisfaction. J. Behav. Sci. **22**, 13–25 (2012)
30. Tajfel, H.: Social Identity and Intergroup Relations. Cambridge University Press, New York (1982)
31. Valeriani, A., Vaccari, C.: Accidental exposure to politics on social media as online participation equalizer in Germany, Italy, and the United Kingdom. New Media Soc. **18**, 1857–1874 (2015)
32. Velasquez, A.: Social media and online political discussion: the effect of cues and informational cascades on participation in online political communities. New Media Soc. **14**, 1286–1303 (2012)
33. Wohed, P., Henkel, M., Anderson, B., Johannesson, P.: Business Process Management with Social Software: An Integrated Technology for Work Organisations. Research Gate (2009). https://www.tuchemnitz.de/wirtschaft/wi2/wp/wpcontent/uploads/2009/12/VR09ProjectDescription2.pdf
34. Yoon, H., Sung, S., Choi, J., Lee, K., Kim, S.: Tangible and intangible rewards and employee creativity: the mediating role of situational extrinsic motivation. Creat. Res. J. **27**, 383–393 (2015)
35. Zairi, M.: Business process management: a boundaryless approach to modern competitiveness. Bus. Process Manag. J. **3**, 64–80 (1997)

A Systematic Literature Review of the Use of Social Media for Business Process Management

Jana Prodanova[1(✉)] and Amy Van Looy[2]

[1] Department of Economics and Business Administration, Faculty of Economics and Business, University of Burgos, C/Parralillos, s/n, 09001 Burgos, Spain
jprodanova@ubu.es

[2] Department of Business Informatics and Operations Management, Faculty of Economics and Business Administration, Ghent University, Tweekerkenstraat 2, 9000 Ghent, Belgium
Amy.VanLooy@UGent.be

Abstract. In today's expansion of new technologies, innovation is found necessary for organizations to be up to date with the latest management trends. Although organizations are increasingly using new technologies, opportunities still exist to achieve the nowadays essential omnichannel management strategy. More precisely, social media are opening a path for benefiting more from an organization's process orientation. However, social media strategies are still an under-investigated field, especially when it comes to the research of social media use for the management and improvement of business processes or the internal way of working in organizations. By classifying a variety of articles, this study explores the evolution of social media implementation within the BPM discipline. We also provide avenues for future research and strategic implications for practitioners to use social media more comprehensively.

Keywords: Business process management · Lifecycle management
Social BPM · Collaborative BPM · Social media · Business 2.0

1 Introduction

Nowadays, social media are one of the most important instruments to enhance the information flows and relationships between individuals and organizations. The main reasons for using social media in businesses are, among others, customer satisfaction, loyalty, engagement and sales increase [1]. Contributing to external and internal business objectives, a social media strategy affects the employees, internal communication, product/ service innovation, growth related to people capabilities, systems and organizational procedures, and the optimization and management of business processes or the internal way of working in organizations [2]. As such, social media management and business process management (BPM) are closely related disciplines.

While the current body of knowledge recognizes the uptake of social media in society [3, 4], a lack of research is perceived related to the impact of social media on business processes and BPM. To fill this gap, this study provides a Systematic Literature Review (SLR) of the social media use by organizations, and particularly their complementarity

E. Teniente and M. Weidlich (Eds.): BPM 2017 Workshops, LNBIP 308, pp. 403–414, 2018.
https://doi.org/10.1007/978-3-319-74030-0_31

with BPM. Our objective is to shed light on the state of the research on social media use in the BPM discipline and to present opportunities for future academic research and relevant recommendations for management practitioners.

Subsequently, Sect. 2 provides the theoretical background. The SLR research method is described in Sect. 3. Afterwards, the results are presented in Sect. 4 and discussed in Sect. 5, offering implications of scientific and applied relevance.

2 Theoretical Background

2.1 Theoretical Background of Social Media Use by Organizations

Since social media have a key position in the B2C communication, they are implemented in the traditional CRM (Customer Relationship Management) systems [6]. Additionally, it was confirmed that new technologies affect operational and management processes as well [6, 7]. Moreover, it is not required to use all social media tools, but rather the most suitable ones depending on corporate objectives and strategies [2, 5].

Various attempts explain the structure and purposes of social media. Most social media classifications are, however, customer-oriented [5, 8]. In addition to the more customer-oriented classifications, other classifications exist that consider implementation aspects of social media for organizations [9, 10]. One of the most complete classifications is the honeycomb model by Kietzmann et al. [10]. This framework not only focuses on the functional characteristics of social media tools, but also on the business implications. It is represented by a honeycomb with seven blocks. First, the (1) *"identity"* block refers to the extent to which users reveal personal information and subjective information (e.g. opinions), which is why companies have to control data privacy and security. The functional block related to (2) *"conversations"* is explained by the communication between social media users, so organizations can follow conversations on a certain topic. The (3) *"sharing"* block represents users who exchange content or are connected by a shared object (e.g. discount vouchers). This block calls for content management systems and building social graphs for business intelligence reasons. The (4) *"presence"* functional block refers to the extent to which users know about other users (e.g. their location or availability). The (5) *"relationships"* function denotes which users are related to each other and how (i.e. the structural and flow properties in a network), which does not require a formal relationship. The (6) *"reputation"* functional block offers users the possibility to identify their and others' reputation based on user-generated information, such as the number of followers or shares. Also sentiment analysis is implied in this block. Finally, the (7) *"groups"* block refers to the users' ability to form communities. This can be realized through membership rules and protocols.

Although studies acknowledge the usefulness of social media for business communication [11], customer communication [12] and managing business processes or the internal way of working [7], research on social media in organizations is still scarce.

2.2 Theoretical Background of BPM

Various BPM attempts explain the lifecycle through which each business process evolves. Although BPM lifecycles differ in the naming and number of phases, they closely relate to the established Plan-Do-Check-Act cycle [13]. This means that each business process should first be identified and modeled or designed ("PLAN") before it can be deployed or executed ("DO"), monitored and analyzed ("CHECK") in order to be improved, optimized or innovated ("ACT"). Only few BPM lifecycles include a "MANAGEMENT" phase around the PDCA. BPM lifecycles are increasingly criticized for being technology-oriented and neglecting the organizational success factors [14, 15]. Consequently, a more holistic view on BPM, which takes into account the organizational culture and structure, is called "Business Process Orientation" (BPO) [16, 17]. For instance, de Bruin and Rosemann [18] developed a maturity model including six capability areas (or critical success factors): (1) methods, (2) IT, (3) governance, (4) strategic alignment, (5) people, and (6) culture. Other holistic BPM scholars focused on one particular area, e.g. process-oriented values in the "culture" area [15]. A comprehensive overview of BPM capability areas is provided by Van Looy et al. [19], based on a theoretical validation in the literature and existing theories and an empirical validation based on 69 BPM/BPO maturity models. This BPM framework consists of six main capability areas with 17 sub capabilities: (A) process modeling, (B) process deployment, (C) process optimization, (D) process management, (E) a process-oriented culture, and (F) a process-oriented structure. The first three areas relate to the PDCA cycle, while the fourth considers the managerial aspects per business process. The final two areas cover organizational success factors, and transform BPM to BPO.

The capability areas presented above should be adopted based on their contingency with an organization's business context to reach an optimal level [14, 20]. Other researchers cover dynamic capabilities to achieve process changes [21]. The latter co-exist with the BPM capability areas, since process changes will be realized by changing one or more (operational) capability areas. Since our focus is rather on critical success factors for the state of BPM than on the change procedure itself, this study focuses on operational capability areas for BPM [19]. To our knowledge, no study digs deeper into the benefits of social media for BPM or an organization's process orientation with all its capability areas and subareas [24, 25].

3   Methodology

The systematic literature review (SLR) is "a form of secondary study that uses a well-defined methodology to identify, analyze and interpret all available evidence related to a specific research question in a way that is unbiased and (to a degree) repeatable" [22, p. 7]. It typically follows a protocol [23] (Table 1).

Table 1. The Systematic Literature Review protocol for this study.

Protocol elements	Translation to this study
Research objective and questions	What is the state of research on social media in the BPM discipline? (To be answered by SLR-RQ1 and SLR-RQ2)
Sources searched	Emerald, IEEE Explore, Science Direct, Scopus, Web of Science, AIS Electronic Library and ACM Digital Library
Search terms	Social media, business process*, (business) process management, (management) information systems
Search strategy	Peer-reviewed journals and conference papers; theoretical and empirical studies; no publication date limit, no sector limit, no topic limit; search terms contained in articles' title, abstract and keywords
Inclusion criteria	Business processes or BPM and social media implementation
Exclusion criteria	(a) Articles using "process*" with a different meaning than BPM (b) Articles without full access
Quality criteria	(a) Only peer-reviewed articles in the academic databases chosen (b) Following a validated and comprehensive BPM framework [19]

3.1 SLR Research Questions

We supplemented the social media classification by Kietzmann et al. [10] with a validated framework that expands business processes to BPM and BPO [19].

- **SLR-RQ1:** In which particular BPM matters (i.e. conventional areas and subareas of BPM) is social media use most frequently investigated?
- **SLR-RQ2:** Considering SLR-RQ1, what are the research avenues and business implications of social media use in BPM, as mentioned in the literature?

3.2 SLR Search and Selection Procedure

Figure 1 illustrates our procedure to select papers for our research objective.

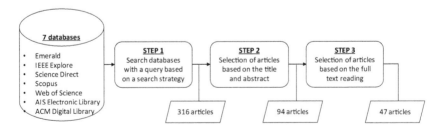

Fig. 1. The selection of articles considered for this research.

Step 1 was completed by browsing research papers in seven databases (Fig. 1). We considered all results until mid-2016, which is when the research started. To determine the link of the articles with our SLR-RQs, we decided to look for precise terms contained in the articles' title, abstract and keywords. Thus, our search was founded on combinations of: "social media" AND "business process*"; "social media" AND "business process management"; "social media" AND "process management"; "social media" AND "information systems"; and "social media" AND "management information systems". After Step 1, we came out with 316 papers.

In *Step 2*, we established inclusion and exclusion criteria that would help us determine which of the 316 articles were really relevant for our objective [22]. We decided to exclude all papers that used the terms "processing" or "to process" with a different meaning. For instance, based on their abstracts, we excluded every study that merely referred to a "diffusion process", "research process", "word processing", "sense making process" or "learning process". Every paper that simultaneously discussed business processes or BPM and social media would be included. Articles that we could not fully access were likewise discarded (i.e. six in total), resulting in 94 articles.

In *Step 3*, based on a full text reading, the previous criteria were applied to verify whether the articles were closely related to the social media implementation in BPM. A paper was only selected when it was related to at least one of the BPM categories that we considered with the applied framework [19]. Hence, we sampled 47 articles which can be found here: https://docs.google.com/document/d/1_N-8cyZSVLLrMUxS-Bujdd37y6d15BnZbX5wK07kB0Dw/edit?usp=sharing.

3.3 SLR Classification

We adopted a framework that offers a general overview of BPM, and which differentiates the narrow view on the traditional BPM lifecycle from the holistic view of business process orientation (BPO) by describing different layers of capability areas in clusters (i.e. BPM versus BPO), main areas and subareas [19]. Since the BPM framework and its capability areas were (theoretically and empirically) validated before, we classified the articles based on their understanding in previous research.

As part of the main capability area (A) *"Process modeling"*, we distinguished articles that explored: (1) *"Business process design"*: papers specifying the relationship between events, activities and decisions in a value chain, as well as the actors involved and the related chains. (2) *"Business process analysis"*: papers referring to the validation, simulation and verification of the designed business process models.

The main capability area (B) *"Process deployment"* included studies dedicated to: (3) *"Business process implementation and enactment"*: papers including the operational models, implemented procedures and software systems. (4) *"Business process measurement and control"*: papers referring to data collection and monitoring of running process instances for correcting deviations and providing status updates.

The main capability area (C) *"Process optimization"* embraced articles that examined: (5) *"Business process evaluation"*: papers that intent to quantify the performance of finished process instances and the operational environment. (6) *"Business process improvement"*: papers for making business processes conform to their process models and optimizing or innovating the models through redesign.

The main capability area (D) *"Process management"* contained studies that focused on: (7) *"Strategy and Key performance indicators":* papers aligning business processes to strategic objectives and customer needs. (8) *"External relationships and Service level agreements":* papers actively involving external parties, like partnering with suppliers and customers. (9) *"Roles and responsibilities":* papers discussing the process manager and his/her team responsible for the performance and improvements of a specific business process. (10) *"Skills and training":* papers elaborating on the acquisition of skills for the actors involved. (11) *"Daily management":* papers dealing with specific management domains to be executed by the process manager.

The main capability area (E) *"Process-oriented culture"* was represented by articles that aimed at organizational characteristics, instead of a specific business process: (12) *"Process-oriented values":* papers presenting values which facilitate the realization of the previous capability areas (e.g. customer focus, empowerment, innovation, multidisciplinary collaboration and trust). (13) *"Process-oriented attitudes and behaviors":* papers discussing attitudes and behaviors that facilitate BPM across business processes and so concretize the defined values, such as BPM awareness, knowledge sharing and acceptance of change. (14) *"Process-oriented appraisals and rewards":* papers related to the HR implications (e.g. combining team incentives with individual benchmarks related to process performance). (15) *"Top management commitment":* papers in which top managers also support BPM and create a process-related C-level leadership role with responsibilities.

The main capability area (F) *"Process-oriented structure"* likewise comprised articles focusing on the organization characteristics, yet with more structural interests: (16) *"Process-oriented organization chart":* papers determining changes in the organization structure to emphasize the cross-departmental business processes and the new roles. (17) *"Process-oriented bodies":* papers creating governance bodies across business processes, e.g. a BPM program management council and competence center.

4   Results

4.1   Results for SLR-RQ1

Figure 2 shows that most conventional BPM capability areas in the framework by Van Looy et al. [19] were covered by the sampled papers, except for the "Process-oriented structure" area. The attention of current studies was most frequently attained by the utility of social media in the "Process management" and "Process-oriented culture" areas of BPM (i.e. covered by respectively 37 and 30 sampled papers), followed by the areas related to the traditional BPM lifecycle (i.e. 28 papers for "Process optimization", 26 for "Process modeling" and 24 for "Process deployment").

Figure 3 refines Fig. 2 by illustrating differences among the BPM subareas. The "External relationships and Service Level Agreements" and the "Process-oriented attitudes and behaviors" were the most frequently captured practices in the reviewed literature (i.e. both with 17 papers). They are closely followed by the subareas of "Business process design" (i.e. 16 papers), "Business process implementation and enactment" (i.e.

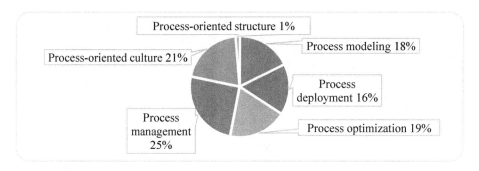

Fig. 2. The relative importance of the articles classified per BPM capability area (N = 47).

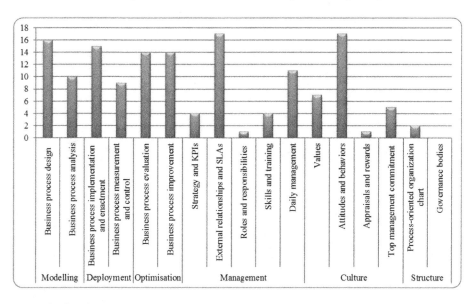

Fig. 3. The articles classified across BPM capability areas and subareas (N = 47).

15 papers), "Business process evaluation" (i.e. 14 papers) and "Business process improvement" (i.e. 14 papers) in the traditional BPM lifecycle.

4.2 Results for SLR-RQ2

A qualitative analysis of the sampled papers in Nvivo allowed us to obtain an overview of the most frequently used terms related to the social media implementation in a business context (Fig. 4). The word clouds were obtained by browsing the frequency of keywords specifically describing the relationship between BPM and social media per BPM capability area. The full texts of the 47 papers, including references, were analyzed for this purpose. Given the fact that slight differences in wordings were observed for the areas related to the BPM lifecycle, we summarized them as PDCA.

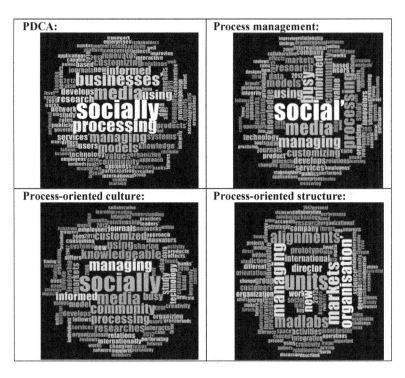

Fig. 4. The resulting word cloud per BPM capability area (N = 47).

In addition to the word clouds, we had a closer look at the word frequency queries to distinguish typical terms and gain more insight per BPM capability area (Table 2). The most recurrent terms are shown in Table 2; a list that could be further extended taking into account all terms that have been used to describe the objective.

Table 2. The word frequency query results per BPM capability area (N = 47).

BPM capability area	Frequently used terms typifying each area (word count)
PDCA	"Processing" (1781); "Models" (1197); "Customizing" (785); "Innovator" (668); "Network" (560); "Openness" (411); "Supports" (399); "Consuming" (392); "Platform" (331)
Process management	"Managing" (1487); "Busy" (1366); "Customizing" (1177); "Community" (1073); "Informing" (977)
Process-oriented culture	"Managing" (1422); "Community" (1271); "Knowledgeable" (1119); "Study" (816); "Sharing" (588); "Leadership" (507); "Values" (338); "Learning" (271)
Process-oriented structure	"Units" (238); "Markets" (187); "Managing" (180); "Organization" (167); "Alignments" (159); "New" (120); "Director" (85); "Integrative" (54)

5   Discussion

With respect to the conventional BPM areas, our study shows most recurrent attention for social media use in the "Process management" and "Process-oriented culture" areas. One explanation is that social media can improve communication, which is fundamental for the external relationships of an organization (e.g. involving contact with partners and customers) and for the process-oriented attitudes and behaviors (e.g. focusing on knowledge sharing and acceptance of change). Nonetheless, communication is likewise a necessity for process modeling, deployment and optimization. The latter BPM areas encompass activities and decisions regarding the procedures and systems implemented in a business process as well as the evaluation and optimization of model performance, requiring updated information and continuous communication.

Current academic research seems centered on the more general or management-related aspects (e.g. culture-related attitudes and behaviors, and the external relationships with customers, suppliers and other stakeholders), which are less specific to BPM. The more technical aspects of business processes in the BPM lifecycle (e.g. collaborative process modeling, execution and optimization with partners) are considered to a somewhat lesser extent, but also appear to be attractive for researchers so far. Thus, one of our main findings is the need to improve an organization's relationships with customers, suppliers, stakeholders and employees. This can more easily and effectively be done through social media by recognizing all implications presented in Kietzmann's honeycomb [10] (Sect. 2). Such implications could be used individually or combined, and their relevance is that social media can help organizations recognize and better understand their audience and engagement needs.

As a refinement to SLR-RQ1, SLR-RQ2 looked at any difference in focus when applying social media in a business process context. In the traditional process lifecycle, social media platforms and networks appear to be mainly advised to create openness for modeling, deploying and optimizing business processes. For instance, openness may refer to better serving the customer, and possibly includes the customization or tailoring of products and services. Furthermore, openness is also linked to opportunities for innovation. Sampled papers suggested that social media could be suitable to include customers in the planning process or for idea generation. Co-creation and crowdsourcing are likewise found to be useful strategies, counting on people's willingness to help for a project to be developed. Comments systems enable knowledge and information sharing by idea competition, innovation contests, ratings and reviews.

In the "Management" area, social media are ways to facilitate an informed community of process participants and to support customization. The sample alludes to the necessity of an interactive relationship with customers through social networks and social CRM. Forms to accomplish relationship management proficiency could be web characteristics: ease of networking, ease of participation, ease of collaboration.

The sampled articles dealing with a "Process-oriented culture" applied social media to better share knowledge or best practices across the organization. Studies suggested the use of communication and coordination tools, organization memory tools and project management tools, e.g. online fora, wikis and various intranet or communication software that enable multi-point conferencing, text chat, and file transfers.

Finally, the few papers on a "Process-oriented structure" mentioned the use of social media to align business units and markets, and to facilitate the coordination of process-related activities across all value chains throughout the organization by a central director. The idea is to create new units whose specialty would be social media management and the coordination of activities among internal processes and at different levels. As such, a better alignment between business units is expected.

With respect to the honeycomb [10], the functions and purposes within the capability areas of an individual business process (i.e. "PDCA" and "Process management" in Table 2) mainly relate to the blocks of customer "identity", "presence", "relationships" and "reputation". Such BPM areas include activities for creating, monitoring, controlling and managing the value chains and the related communication flows of especially external relationships. The more organizational areas of "Process-oriented culture" and "Process-oriented structure" are closely linked to the honeycomb blocks of (internal) "conversations", "sharing" and "groups". These BPM areas typically affect the entire organization by enabling internal communication and information sharing, and an enhanced contact between organizational groups and communities.

Thus, by comparing our results with the social media contributions proposed by the honeycomb [10], we anticipate how social media could be implemented in a certain BPM capability area to appreciate and enhance the social media functions (Table 3).

Table 3. Recommendations for using social media per BPM capability area.

BPM capability area	Examples of social media opportunities
1/Process modeling	• Collaborative process modeling with partners • Online requirements gathering with customers and stakeholders • Co-creation and crowdsourcing for Research & Development
2/Process deployment	• E-commerce by mobile or tablet applications • Streamlining information flows among employees and partners • Online monitoring of product/service feedback • Online complaint handling
3/Process optimization	• Creating value from online customer feedback and complaints • Collecting (improvement or innovation) ideas based on gamification techniques among employees (e.g. the number of bottom-up ideas, likes, badges, progress bars)
4/Process management	• Incorporating social media in business process strategies • Internal trainings on how social media and BPM can be combined • Collaborative process management and social media management • Social CRM and customer process management
5/Process-oriented culture	• Using wikis or the Intranet to share knowledge and best practices
6/Process-oriented structure	• Creating a competence center (or Center of Excellence) with expertise on synergies between BPM, social media and BI/big data • Formal coordination among business processes (e.g. by wikis)

Finally, we give recommendations to attain the implications derived from our study. First, we revealed a lack of research regarding social media in the area of "Process-oriented structure". This gap should empirically be investigated to verify whether managers do not recognize a necessity for social media in this area or whether another reason prevails. Moreover, managerial awareness about social media opportunities should be increased through informative workshops or seminars, since a number of executives do not truly understand their benefits and implications [10]. Furthermore, instead of focusing on one or a few BPM areas, the relationships inside and outside an organization can be improved by combining social media tools through different BPM areas. Such an omnichannel management strategy could provide seamless communication and control of business processes, anytime, from anyplace, with real-time visibility of systems, processes and people. More recommendations are given in Table 3.

6   Conclusion

This article focused on gaps in the literature and their practical implications, taking into account the limitations inherent to our SLR protocol. Although the current literature covers different sectors and topics to which BPM is related, many opportunities still seem to exist to further scrutinize the implications on BPM in more detail.

First, the literature on social media for BPM is increasing, but it is still relatively limited. Secondly, given the multidisciplinary approach that is inherent to our research subject, possible theories to underpin future BPM work might belong to disciplines such as communication science, psychology, management and business economics (e.g. marketing, innovation), and information systems. We particularly think of relationship management, innovation strategies, gamification theories, or social network techniques using big data. Thirdly, social media may provide more diverse opportunities (e.g. crowdsourcing and gamification) to organizations than the ones mostly covered in the investigated body of knowledge. Hence, future research could likewise consider social software as an alternative to the currently contemplated social media keyword. Next, we call for more research on omnichannel management strategies to let BPM evolve by social media in a more comprehensive manner, i.e. by examining how different business contexts profit from new technologies and digital innovation across the BPM capability areas to create synergies.

In the future, a more detailed analysis of the sampled papers could be enriching. An alternative starting point could be a smaller sample of articles from reputable journals and conferences only, based on more specific parameters to be reviewed. Another aim could be to look for BPM-related gaps in case studies to detect new ways for expanding BPM by social media implementation and, as such, improving an organization's business performance. Lastly, differentiating social media applications for BPM in large, medium-sized and smaller organizations could also be enriching.

References

1. Blanchard, O.: Social Media ROI. Pearson Education, Boston (2011)
2. Van Looy, A.: Social Media Management. Springer, Switzerland (2016)
3. Mangold, W.G., Faulds, D.J.: Social media: the new hybrid element of the promotion mix. Bus. Horiz. **52**(4), 357–365 (2009)

4. Kaplan, A.M., Haenlein, M.: Users of the world, unite! The challenges and opportunities of social media. Bus. Horiz. **53**(1), 59–68 (2010)
5. Safko, L., Brake, D.K.: The Social Media Bible. John Wiley & Sons Inc., Hoboken, New Jersey (2009)
6. Greenberg, P.: The impact of CRM 2.0 on customer insight. J. Bus. Ind. Mark. **25**(6), 410–419 (2010)
7. Choudhury, M.M., Harrigan, P.: CRM to social CRM. J. Strateg. Mark. **22**(2), 149–176 (2014)
8. Cook, N.: Enterprise 2.0. Gower Publishing, Ltd., Aldershot (2008)
9. Helms, R.W., Booij, E., Spruit, M.R.: Reaching out: Involving users in innovation tasks through social media. Reach. Out **5**, 15–2012 (2012)
10. Kietzmann, J.H., Hermkens, K., McCarthy, I.P., Silvestre, B.S.: Social media? Get serious! Bus. Horiz. **54**(3), 241–251 (2011)
11. Smits, M., Mogos, S.: The impact of social media on business performance. In: ECIS, p. 125 (2013)
12. Fuduric, M., Mandelli, A.: Communicating social media policies. J. Commun. Manag. **18**(2), 158–175 (2014)
13. Dumas, M., La Rosa, M., Mendling, J., Reijers, H.A.: Fundamentals of Business Process Management. Springer, Heidelberg (2013)
14. vom Brocke, J., Schmiedel, T., Recker, J., Trkman, P., Mertens, W., Viaene, S.: Ten principles of good business process management. Bus. Process Manag. J. **20**(4), 530–548 (2014)
15. vom Brocke, J., Sinnl, T.: Culture in business process management: a literature review. Bus. Process Manag. J. **17**(2), 357–377 (2011)
16. McCormack, K., Johnson, W.C.: Business Process Orientation. St. Lucie Press, Boca Raton, Florida (2001)
17. Van Looy, A., De Backer, M., Poels, G.: Defining business process maturity. Total Qual. Manag. Bus. Excell. **22**(11), 1119–1137 (2011)
18. de Bruin, T., Rosemann, M.: Using the Delphi technique to identify BPM capability areas. In: ACIS Proceedings, vol. 42, pp. 642–653 (2007)
19. Van Looy, A., Backer, M.D., Poels, G.: A conceptual framework and classification of capability areas for business process maturity. Enterp. Inf. Syst. **8**(2), 188–224 (2014)
20. vom Brocke, J., Zelt, S., Schmiedel, T.: On the role of context in business process management. Int. J. Inf. Manag. **36**(3), 486–495 (2016)
21. Pöppelbuß, J., Plattfaut, R., Ortbach, K., Niehaves, B.: A dynamic capability-based framework for business process management: theorizing and empirical application. In: HICCS Proceedings, pp. 4287–4296 (2012)
22. Kitchenham, B.: Guidelines for performing systematic literature reviews in software engineering. Technical report, Ver. 2.3 EBSE (2007)
23. Boell, S.K., Cecez-Kecmanovic, D.: On being 'systematic' in literature reviews in IS. J. Inf. Technol. **30**(2), 161–173 (2015)
24. Nascimento, A.M., da Silveira, D.S.: A systematic mapping study on using social media for business process improvement. Comput. Hum. Behav. **73**, 670–675 (2017). https://doi.org/10.1016/j.chb.2016.10.016. (in press)
25. Erol, S., Granitzer, M., Happ, S., Jantunen, S., Jennings, B., Johannesson, P., Koschmider, A., Nurcan, S., Rossi, D., Schmidt, R.: Combining BPM and social software: contradiction or chance? J. Softw. Maint. Evol. Res. Pract. **22**(6–7), 449–476 (2010)

1st International Workshop on Cognitive Business Process Management (CBPM 2017)

Introduction to the 1st International Workshop on Cognitive Business Process Management (CBPM'17)

Hamid R. Motahari Nezhad[1], Richard Hull[2], and Boualem Bentallah[3]

[1] IBM Almaden Research Center, San Jose, CA, USA
motahari@us.ibm.com
[2] IBM TJ Watson Research Center, Yorktown, CA, USA
hull@us.ibm.com
[3] University of New South Wales, Sydney, NSW, Australia
boualem@cse.unsw.edu.au

Abstract. Cognitive Computing is regarded as the next wave of computing that enables the evolution from the programmatic era into cognitive era. Advances in cognitive computing coupled with the availability of large amount of dark and unexploited data in the enterprise and in the public domain, related to the operation of businesses and their processes, has created an unprecedented opportunity for rethinking how business processes are supported in the enterprise, and in the automation of a large volume of business processes that have defined automation so far. This workshop consists of a series of papers that present and discuss the challenges, opportunities and progress in the broad research direction of cognitive business process management (CBPM).

Keywords: Knowledge-intensive process · Cognitive BPM

1 Introduction

Cognitive Computing is regarded as the next wave of computing that enables the evolution from the programmatic era into cognitive era. Programmatic era is characterized as a computing paradigm in which the computing logic is codified as deterministic programs in machine language by humans, and a paradigm in which humans proactively seeks information of interest in the personal and enterprise data spaces. Building on top of advances in artificial intelligence, machine learning and natural language processing, cognitive computing defines a paradigm where the computing logic is codified into probabilistic models that are learned by machines through ingesting big sea of structured and unstructured data, and where a proactive seek-approach on human side is replaced with proactive information push model by machines. Advances in cognitive computing coupled with the availability of large amount of dark and unexploited data in the enterprise and in the public domain, related to the operation of businesses and their processes, has created an unprecedented opportunity for rethinking how business processes are supported in the enterprise, and in the automation of a large volume of business processes that have defined automation so far.

This workshop provided a forum for researchers and professionals interested in understanding, envisioning and discussing the challenges and opportunities of moving from a programmatic era into a cognitive era on business processes, and on business process management as a discipline. The workshop attendees including academics, researchers and practitioners had an opportunity to share knowledge and exchange opinions and ideas around the requirements for and the need for investigating the transformation potential on a new research direction that is referred as "cognitive business process management" (CBPM).

The workshop's program [1] included six presentations, starting with a keynote talk, and 5 paper presentations each focused on different aspects of cognitive business process management as follows:

- Keynote talk entitled "AI Foundations for Cognitive BPM", by Prof. Giuseppe De Giacomo from Sapienza Universita' di Roma.
- The paper "Cognitive Computing: What's in for Business Process Management? An Exploration of Use Case Ideas" by Maximilian Roeglinger, Johannes Seyfried, Simon Stelzl, Michael zur Muehlen.
- The invite talk and paper entitled "Cognitive Business Process Management for Adaptive Cyber-Physical Processes" by Andrea Marrella and Massimo Mecella.
- The paper entitled "BPM for the masses: empowering participants of Cognitive Business Processes" by Aleksander Slominski and Vinod Muthusamy.
- The paper "Using Insights from Cognitive Neuroscience to Investigate the Effects of Event-Driven Process Chains on Process Model Comprehension" by Michael Zimoch, Tim Mohring, Rudiger Pryss, Thomas Probst, Winfried Schlee, and Manfred Reichert.
- The invited talk and paper entitled "Knowledge-intensive Processes: a research framework" by Flavia Maria Santoro and Fernanda Araujo Baião.

2 Workshop Co-organizers

Hamid R. Motahari Nezhad	IBM Almaden Research Center, USA
Richard Hull	IBM TJ Watson Research Center, USA
Boualem Benatallah	University of New South Wales, Australia

3 Program Committee

Han van der Aa	VU University Amsterdam
Seyed-Mehdi-Reza Beheshti	University of New South Wales, Australia
Moshe Barukh	University of New South Wales, Australia
Schahram Dustdar	TU Wien, Austria
Aditya Ghose	University of Wollongong, Australia
Farouk Toumani	Limos, Blaise Pascal University, Clermont-Ferrand, France

Massimo Mecella	SAPIENZA Università di Roma, Italy
Avigdor Gal	Technion, Israel
Henrik Leopold	VU University Amsterdam, the Netherlands
Fabio Casati	University of Trento, Italy
Anup Kalia	IBM TJ Watson, USA
Florian Daniel	Politecnico di Milano, Italy
Flavia Santoro	NP2Tec/UNIRIO, Brazil
Marlon Dumas	University of Tartu, Estonia
Hagen Voelzer	IBM Research, Zurich
Manfred Reichert	University of Ulm, Germany
Fernanda Araujo Baiao	UNIRIO, Brazil
Claudio Diciccio	Vienna University of Economics and Business, Austria

4   Reference

1. 1st International Workshop on Cognitive Business Process Management (CBPM) (2017). https://sites.google.com/site/cbpm2017/workshop-report

Cognitive Computing: What's in for Business Process Management? An Exploration of Use Case Ideas

Maximilian Roeglinger[1(✉)], Johannes Seyfried[1], Simon Stelzl[2],
and Michael zur Muehlen[3]

[1] FIM Research Center, University of Bayreuth, Bayreuth, Germany
{maximilian.roeglinger,johannes.seyfried}@fim-rc.de
[2] FIM Research Center, University of Augsburg, Augsburg, Germany
simon.stelzl@fim-rc.de
[3] Stevens Institute of Technology, Hoboken, NJ, USA
mzurmuehlen@stevens.edu

Abstract. Cognitive Computing promises to fundamentally transform corporate information processing and problem solving. Building on latest advances in cognitive, data, and computer science, Cognitive Computing aims to deliver autonomous reasoning and continuous learning under consideration of contextual insights and the natural interaction of humans and machines. Cognitive Computing is expected to offer significant application opportunities for business process management (BPM). While first studies have investigated the potential impact of Cognitive Computing on BPM, the intersection between both disciplines remains largely unexplored. In particular, little work has been done on identifying Cognitive BPM use cases. To address this gap, we develop an analysis framework that aims to assist researchers and practitioners in the development of Cognitive BPM use case ideas. This framework combines the most significant problem classes addressed by Cognitive Computing with central activities of the BPM lifecycle. We also used the framework as foundation of explorative workshops and report on the most interesting cognitive BPM use cases ideas we discovered.

Keywords: Cognitive Computing · Business process management
Cognitive BPM · Use cases

1 Introduction

In recent years, Cognitive Computing (CC) has received increasing interest from industry and academia, as it is seen as an emerging technology tied to a new era of computing [3]. Building on the latest advances of disciplines such as cognitive, data, and computer science, CC generates context-aware insights from structured and unstructured data by leveraging autonomous reasoning and continuous learning based on an ever-growing knowledge base [11]. CC is mimicking facets of the human brain, including the ability to analyze text, images, voice, and videos in context and the interaction with humans [4]. Domains well-suited for CC are characterized by high uncertainty and knowledge-intensive problems with many potential solutions [11]. As, in the

© Springer International Publishing AG 2018
E. Teniente and M. Weidlich (Eds.): BPM 2017 Workshops, LNBIP 308, pp. 419–428, 2018.
https://doi.org/10.1007/978-3-319-74030-0_32

area of BPM, topics such as flexibility, context awareness, or the automation of unstructured tasks receive ever more attention [17, 18], we believe that the determining features of CC have high transformational impact on BPM research and practice in the future [8].

Cognitive BPM has been introduced by Motahari-Nezhad and Akkiraju [15] as well as Hull and Motahari-Nezhad [8]. They claim that a new BPM lifecycle, based on the *plan-act-learn* paradigm, is necessary to realize the potential of CC in the context of BPM [8]. This new BPM lifecycle shall support processes ranging from highly standardized routine processes to less predictable ad-hoc processes. Cognitive BPM involves those facets of BPM where CC offers new opportunities, either by changing the way how data is processed, presented, or how processes are designed. That said, research in this area remains scarce except for the studies of Motahari-Nezhad et al. [8, 15, 16]. Hull and Motahari-Nezhad [8] call for a framework that helps operationalize their proposed high-level Cognitive BPM lifecycle, offering tangible insights into Cognitive BPM use cases. This is the starting point of our research. We analyze the following research question: *What are use cases of Cognitive Computing in the context of BPM?*

To answer this question, we propose an analysis framework that relates the most important problem classes addressed by CC to activities from the BPM lifecycle. Our framework builds on insights into existing definitions and constitutive characteristics of CC, which we developed through a literature review. The framework is designed to help researchers and practitioners in the identification and articulation of Cognitive BPM use cases. We illustrate the use of our framework by outlining a series of exemplary Cognitive BPM use cases. In line with the explorative nature of our study, these high-level use case ideas should be seen as a starting point for a community-wide discussion about how to exploit the technological opportunities of CC for BPM.

This paper is organized as follows: In Sect. 2, we lay the foundations of our analysis framework by developing a definition of CC and by summarizing ways of structuring the BPM discipline. In Sect. 3, we introduce our analysis framework. We also report on the exemplary Cognitive BPM use case ideas we identified. We conclude by summarizing our findings, by stating limitations, and by pointing to further research in Sect. 4.

2 Domain Background

2.1 Cognitive Computing

CC is an emerging field without a commonly accepted definition [8]. Attempts to define CC mainly occurred in industry [4, 6, 12]. Thus, there is a need for a common definition synthesizing technology- and domain-specific interpretations. In this section, we summarize CC definitions we found in the academic and practitioner-oriented literature. On this foundation, we derive constitutive characteristics and a working definition of CC. As CC has multiple origins, the term 'Cognitive Computing' has been defined and interpreted differently. Our literature review yielded 26 definitions. In two workshops, we discussed and selected the eight most comprehensive definitions from both academia and industry: academic publications [14, 19], books [7, 9], industry reports [11, 12], and interpretations from a consortium of researchers and practitioners dedicated to CC [4, 6]. While analyzing less technical definitions, we identified four topics that reoccurred frequently. These topics are: interaction, context awareness, reasoning, and learning. Based on this review and in line with extant literature, we define the constitutive characteristics of CC as follows [4, 10]: (1) *Interaction*: Natural communication between humans and machines as well as among humans, (2) *Context awareness*: Identification and extraction of contextual information from structured and unstructured data at large scale, (3) *Reasoning*: Generation, testing, and assessment of hypotheses based on context information and past learnings, and (4) *Learning*: Continuous expansion of the knowledge base by incorporating learnings of prior decisions and reasoning. Subsuming, we define CC as follows for the purposes of our study: *Cognitive computing is an umbrella term for new problem-solving models that strive for mimicking the cognitive capabilities of the human mind by autonomously reasoning and learning based on incomplete structured and unstructured contextual data, and through natural interactions with humans and machines.*

2.2 Business Process Management

As BPM is a vital dimension of our framework, we investigate which approaches have been proposed to structure BPM. The most common approaches are lifecycle models and capability frameworks [20]. We focus on these comprehensive structures, not on individual methods or tools to ensure a holistic picture of BPM. Lifecycle models structure BPM along the (management) activities that occur during the lifecycle of a business process [20]. Although there are many conceptualizations of the BPM lifecycle, the involved activities vary only slightly. Most BPM lifecycles cover the following activities: process design and modeling, process implementation and execution, process optimization and improvement [13]. In their recent work on Cognitive BPM, Hull and Motahari-Nezhad [8] propose a shift in the BPM lifecycle paradigm, anticipating the characteristics of CC in the context of BPM. Their Cognitive BPM lifecycle includes the activities 'plan', 'act', 'monitor', and 'analyze'. In this new BPM lifecycle, the differentiation between the activities of the traditional BPM lifecycle gets blurred. The iterative planning, continuous monitoring, integrated analysis, and refinement of processes are also supposed to blur the separation of process models and process instances.

3 Analysis Framework and Cognitive BPM Use Case Ideas

Below, we introduce our analysis framework for Cognitive BPM use cases ideas. We introduce our framework in Sect. 3.1 based on the literature review presented in Sect. 2. Having used our framework as foundation for explorative workshops, we also report on the most interesting use case ideas we discovered in the Sects. 3.2 to 3.5.

3.1 General Setting

Our analysis framework comprises two dimensions: a BPM and a CC dimension. When conceptualizing the BPM dimension, we had to choose between BPM capability frameworks and BPM lifecycle models. For this study, we selected lifecycle models as they are very tangible, reflecting how tasks within the lifecycle of a process can be supported by CC. BPM capability frameworks are more fine-grained and also include elements (e.g., people, or culture) that can only indirectly be enhanced by emerging technologies such as CC. In Sect. 2.2, we introduced the traditional BPM lifecycle and the Cognitive BPM lifecycle as proposed by Hull and Motahari-Nezhad [8]. As our framework aims to assist in discovering Cognitive BPM use case ideas, we adopted the traditional BPM framework for conceptualizing the BPM dimension. Reasons are that the traditional BPM lifecycle is very mature and captures the contemporary conceptualization of BPM from a lifecycle perspective. The Cognitive BPM lifecycle, in contrast, focuses more strongly on the target state after the traditional BPM lifecycle has been transformed. This makes the Cognitive BPM lifecycle less suitable for the purposes of our study. Following Macedo de Morais et al. [13], we cluster the activities included in the BPM lifecycle into *definition and modeling*, *implementation and execution*, *monitoring and controlling* as well as *optimization and implementation*.

When conceptualizing the CC dimension, we used the working definition and constitutive characteristics from Sect. 2.1. On this foundation, we derived the most important problem classes addressed by CC, i.e., *knowledge-intensive problems*, *human-computer interaction*, and *human collaboration*. Grounding this dimension on concrete CC functionalities and technologies would have been too fine-grained for developing Cognitive BPM use cases, as a previous version of our framework showed. Knowledge-intensive problems require extracting information, weighing its relevance and validity as well as generating and testing hypotheses. As noted by Aamodt [1], this includes sub-processes of inferring context, reasoning, and learning. These steps match three constitutive characteristics of CC. Building on the 'context awareness' and 'learning' characteristics, CC extracts knowledge and context from structured and unstructured data and continuously feeds its knowledge base with new insights. Human-computer interaction as a problem class includes several key elements such as the understanding of language, perception of intention, and domain knowledge [5]. The constitutive characteristic 'interaction' highly contributes to this problem class. Leveraging contextual information about humans to develop human-like empathy and communications skills, CC interacts with humans in a natural way, bringing advances in the field of human-computer interaction [4, 11]. The third problem class human collaboration is related to human-computer interaction. Perceiving and understanding humans, CC improves human collaboration by providing tools that can be adapted to the context of participants [2]. Interaction as the most contributing characteristic of these two problem classes is supported by the other characteristics, as understanding human language and intentions comprehensively requires context-based information and reasoning abilities. The 'learning' characteristic further improves the accuracy of interactions. Table 1 shows our analysis framework that puts the most important problem classes addressed by CC and the key activities of the BPM lifecycle into perspective. Below, we outline the initial Cognitive BPM use case ideas that we identified structured along the BPM dimension of our framework.

Table 1. Analysis framework of Cognitive Computing in the context of BPM

Activities of the traditional BPM lifecycle	Cognitive Computing problem classes		
	Solutions of knowledge-intensive problems (A)	Human-computer interaction (B)	Human collaboration (C)
Definition & modeling (1)			
Implementation & execution (2)			
Monitoring & controlling (3)			
Optimization & implementation (4)			

3.2   Use Case Ideas for 'Definition and Modeling'

Discover process models from unstructured data (A1). This use case idea refers to the automated discovery of process models from structured and unstructured, potentially non-process-related, data. The data processing features of CC could enhance process mining techniques to leverage unstructured data (e.g., emails, conversations, or documents). Thereby, CC uses contextual knowledge to generate hypotheses about new process models. *Example*: Suggestion of a new process model based on concepts extracted from regulatory documents.

Design and adaption of configurable process models considering organizational context (A1). CC could help derive context-specific models from configurable or reference process models. Based on the organizational structure, domain, available resources as well as other processes and dependencies, CC could automatically suggest configured process models by applying reasoning and learning techniques. *Example*: Adaptation of a company-wide invoice approval reference process for a department where invoices from certain partners require special approval. Based on its knowledge about the department's context, CC is aware of this requirement and adapts the reference process automatically.

Interactive process design support (B1). Building on information about the process (e.g., goal, purpose, stakeholder, resources), organizational context (e.g., industry, regulations, other processes, best practices), and information about the process modeler (e.g., experience, skills), CC could suggest process steps to be included, data elements to be used, role assignments to be made, and connections with other processes to be created. In a responsive manner, CC would react to the modeler's input. *Example*: Assistance in designing a customer support process. After a customer inquiry is categorized, CC may suggest modeling an XOR split to make a decision whether to automatically respond to this inquiry or to assign a user. Thereby, CC considers existing automated response systems within the organization.

Visualization of process models considering different stakeholders (C1). Different stakeholders and process model users may have different experience and skill levels. CC could incorporate the context and knowledge about users to evaluate the effectiveness of a process model's visualization. If the model does not seem clear to the user, CC could suggest a different visualization form. CC may support collaboration among humans by translating different user perspectives. *Example*: For a management meeting, a complex process model captured in BPMN is presented as a simple flowchart, including the most important process elements. Thereby, CC perceives information about the participants of the meeting and extracts important process elements according to participants' background knowledge and preferences.

Support in process design collaboration (C1). Cross-organizational process modeling involves linguistic barriers (e.g., different vocabularies and semantics) as well as coordination effort (e.g., time, distance). CC could support the translation among the involved process modelers by automatically designing a meta model that abstracts from the organization- or domain-specific context. CC could identify dependencies between departments or organizations at the process level. *Example*: In a joint venture, two organizations align their procurement processes. CC may support this by creating a meta model for mapping organization-specific names, abbreviations, systems, roles, and activities. Thereby, CC would leverage knowledge about both organizations and the domain-specific context.

3.3   Use Case Ideas for 'Implementation and Execution'

Dynamic resource allocation at runtime (A2). Dynamic resource allocation considers several criteria. It includes the allocation of individual tasks to humans or software services based on availability, capacity, workload, human's mental state (e.g., stress, concentration) and skills as well as context. Moreover, this mechanism may account for deviant behavior via dynamic re-planning or choosing alternative suitable process variants. *Example*: Resource allocation in a call center. Based on the language, a French-speaking caller with a complex problem is allocated to an experienced agent who can handle the problem due to his experience and mental state. Thereby, CC obtains contextual insights about the caller and his inquiry by analyzing the problem statement. CC may also redirect all inquiries that do not require human skills to an automated messaging system (e.g., chatbot) to handle peak loads.

Automatic execution or suggestions of next best task at runtime (A2). At runtime, CC could observe the execution of a process and predict the next possible tasks by analyzing structured and unstructured data. Based on these insights, CC may reason about each step in a process and propose the next best task. *Example*: In an automated process, a chatbot initially handles all customer inquiries. Based on an analysis of social media posts, CC detects negative feedback regarding a specific product of the company. To prevent damage to the company, CC suggests handling all inquiries regarding that product manually by customer service due the empathy of human agents.

Interactive task assignment assistant at runtime (B2). In addition to the use case idea above, CC could work as a personal assistant [16]. Accounting for their mental state, experience, and skill set, CC may guide human process participants through their worklist to effectively and efficiently meet process goals as well as performance targets. Considering that users interact with their cognitive assistants about their work schedule, this covers conversations about the scheduling, prioritizing, timing, skipping of tasks, or requesting additional auxiliary tasks. CC is responsive and learns the personal preferences over time. *Example*: User input: "What are tasks of higher priority today?". A cognitive assistant may prioritize tasks for investigating fraudulent payments leveraging knowledge about specific payment terms. As sensor data measures a rise of the user's stress level, complex fraud is automatically forwarded to a less busy user.

Support in decision-making at runtime (B2). Regarding decisions that require the analysis of large datasets and expert knowledge, CC could support decision-makers with contextual information and hypotheses about the decision at hand or relevant information. CC could also anticipate user input by adjusting context and iterative reasoning. In this case, CC heavily relies on its continuously expanding knowledge base, but also on perceiving the context of the decision process. *Example*: In the process of running a marketing campaign for a new product, CC may suggest different methods and propose interpretations of the campaign results by inferring the context and reasoning about structured and unstructured data (e.g., comments on social media).

Dynamic suggestions of collaboration at runtime (C2). Following up on the previous use case idea, CC could support decision-makers by automatically matching co-workers with complementary knowledge and experience to collaborate on a task. Moreover, CC could help match co-workers regarding their skill sets as well as personality. Thereby, CC extracts characteristics of workers from sensor data, written text, and past collaborations. *Example*: During a human resource process, applicants and interviewers are automatically assigned to each other based on same personality type and knowledge backgrounds in order to create a fair common ground.

3.4 Use Case Ideas for 'Monitoring and Controlling'

Automatic anomaly and deviant behavior detection at runtime (A3). This use case idea builds on process mining and predictive analytics. Reasoning about and learning from structured and unstructured data that is directly or indirectly produced during process execution (e.g., text, documents, sensor data, log data), CC could automatically detect and predict process anomalies and deviant behavior at runtime. Based on this ability, CC may consider actions of exception handling by automatically changing or stopping a process instance or notifying a process manager or other authorities for intervention. *Example*: In a customer service scenario, CC automatically checks the conformance of customer inquiries by identifying insufficient responses before being sent out. Therefore, CC matches topics of the inquiry and the response messages. As this is a deviant behavior in the process, CC warns a customer service worker accordingly.

Conversation-like process monitoring queries (B3). CC could support humans by providing insights into currently running processes and concurrent instances. In an interactive way, CC may process natural language queries and respond accordingly. Further, CC could reason about the conversation and respond in an intelligent way by interpreting requested data and suggesting further interesting insights. CC could also learn process-specific user preferences. *Example*: User input: "return all running processes that contain activities that need to be executed by someone with the role manager and that are exceeding the planned processing time". CC translates this natural language query and responds with the requested information. Additionally, it automatically informs the user about a specific process step that could harm the organization to a great extent if it is not investigated.

3.5   Use Case Ideas for 'Optimization and Improvement'

Proactive identification of process improvement opportunities (A4). CC could proactively help identify process improvement opportunities by analyzing process anomalies, deviant behavior, external information (e.g., best practices, novel designs), and insights from automated processes. Thereby, CC would rely on its ability of inferring the process context, continuously learning, and generating hypotheses. *Example*: In an organization, the first-level support for the order process is currently performed by a human process participant. CC perceives and automatically learns the steps of action of the first-level support at large scale. Thus, CC suggests the automation of this process as CC produces the same outcome at a shorter runtime.

Identification of need for training (B4). Accounting for the performance, skills, experience, and mental state of a process participant over time, CC could automatically identify the need of training. It may suggest and interactively guide process participants through individual training. Thereby, the progress and learning curve is dynamically monitored and the training is adjusted accordingly. *Example:* CC detects that a user's performance at investigating claims at an insurance company falls below the performance of his peers (e.g., same age, education, task assignments). Identifying the lack of knowledge about a specific type of claims, CC automatically suggests a training on the law underlying this type of claims.

Support of collaboration between process managers and participants (C4). CC could support process managers and participants in their collaboration to analyze and improve processes. As both parties might not have the same skills and background, CC could dynamically translate suggested process improvements at the process participant level (e.g., improvement of a distinct task) to the broader perspective of a process manager overseeing a portfolio of processes. *Example*: A process participant proposes to perform several tasks concurrently instead of sequentially. Thereby, CC automatically translates this idea into the process manager's perspective, checking the consequences of this idea regarding dependencies with other processes as well as compliance with regulations and company governance. It may also suggest a counter-proposal that is translated to the process participant's perspective again.

4   Conclusion, Limitations, and Further Research

In this study, we investigated the impact of CC on BPM. Our contribution is threefold: First, we derived constitutive characteristics of CC (i.e., interaction, context awareness, reasoning, and learning) based on extant literature and proposed a corresponding working definition. Second, we proposed an analysis framework that aims to assist researchers and practitioners in the systematic derivation of Cognitive BPM use case ideas. This framework builds on the BPM lifecycle and essential problem classes addressed by CC, i.e., knowledge-intensive problems, human-computer interaction, and human collaboration. Third, we reported on interesting high-level Cognitive BPM use case ideas using our framework as a foundation. We identified a large potential for CC in the BPM domain covering all activities of the BPM lifecycle and entailing a higher level of automation in BPM. This increasing level of automation enabled by CC also fosters the human centricity of BPM, as CC with its characteristics promotes user-aware assistance systems and a natural interaction between humans and BPM systems. In general, we expect it to play a central role for next-generation BPM systems.

Our study is beset with limitations that call for further research. First, we consider activities of the BPM lifecycle and CC problem classes in an aggregated view. To take a more detailed perspective on the impact of CC on BPM, further research is required that caters for a different and more fine-grained view on the BPM lifecycle. For a more detailed perspective on CC, further research could investigate more technical details of CC such as concrete CC functionalities or technologies. Second, our explorative approach of discovering use cases ideas comprises limitations. Our goal was to compile an initial set of Cognitive BPM use case ideas, motivating researchers and practitioners for further investigations. We do by no means claim that our list is exhaustive. To further develop and validate this initial compilation, we recommend conducting Delphi studies, focus groups, or expert interviews leveraging the knowledge of many BPM and CC experts. Third, in this study, we have not yet conducted a detailed investigation of the identified use case ideas, neither from a technical nor from a business case perspective. We call for further research in close collaboration with industry to probe into the feasibility of our and, of course, new Cognitive BPM use case ideas. In an ongoing research project, we are currently working on a reference architecture for Cognitive BPM and software prototypes of selected use case ideas.

Despite these limitations, we believe that our analysis framework and the exploration of initial use case ideas are first steps toward more grip on Cognitive BPM. With this study, we invite fellow researchers and practitioners to challenge and extend our ideas and help explore the technological opportunities of CC for BPM.

References

1. Aamodt, A.: A Knowledge-intensive, integrated approach to problem solving and sustained learning. Dissertation. University of Trondheim, pp. 27–85 (1991)
2. Aranda, G.N., Vizcaíno, A., Cechich, A., Piattini, M.: Applying strategies to recommend groupware tools according to cognitive characteristics of a team. In: Wang, Y., Zhang, D., Kinsner, W. (eds.) Advances in Cognitive Informatics and Cognitive Computing, pp. 105–119. Springer, Heidelberg (2010)
3. Brant, K.F., Austin, T.: Hype Cycle for Smart Machines 2016. https://www.gartner.com/doc/3380751/hype-cycle-smart-machines. Accessed 14 June 2017
4. Cognitive Computing Consortium: Cognitive Computing Defined. https://cognitivecomputingconsortium.com/resources/cognitive-computing-defined/-1467829079735-c0934399-599a. Accessed 14 June 2017
5. Dix, A.: Human-computer interaction. In: Liu, L., ÖZsu, M.T. (eds.) Encyclopedia of Database Systems, pp. 1327–1331. Springer, Boston (2009)
6. Feldman, S.: Defining Cognitive Computing. https://www.youtube.com/watch?v=MjrID_HmRY8. Accessed 14 June 2017

7. Gudivada, V.N.: Cognitive computing: concepts, architectures, systems, and applications. In: Gudivada, V.N., Raghavan, V.V., Govindaraju, V., Rao, C.R. (eds.) Handbook of Statistics, vol. 35, pp. 3–38. Elsevier, Amsterdam (2016)

8. Hull, R., Motahari Nezhad, H.R.: Rethinking BPM in a cognitive world: transforming how we learn and perform business processes. In: La Rosa, M., Loos, P., Pastor, O. (eds.) BPM 2016. LNCS, vol. 9850, pp. 3–19. Springer, Cham (2016). https://doi.org/10.1007/978-3-319-45348-4_1

9. Hurwitz, J., Kaufman, M., Bowles, A.: Cognitive computing and big data analytics. Wiley, Hoboken (2015)

10. IBM Research: Cognitive Computing. http://researcher.watson.ibm.com/researcher/view_group.php?id=7515. Accessed 14 June 2017

11. Kelly III, J.E.: Computing, cognition and the future of knowing. Whitepaper, IBM Research. http://www.research.ibm.com/software/IBMResearch/multimedia/Computing_Cognition_WhitePaper.pdf. Accessed 14 June 2017

12. KPMG: Embracing the cognitive Era. https://assets.kpmg.com/content/dam/kpmg/pdf/2016/03/embracing-the-cognitive-era.pdf. Accessed 14 June 2017

13. Macedo de Morais, R., Kazan, S., Inês Dallavalle de Pádua, S., Lucirton Costa, A.: An analysis of BPM lifecycles: from a literature review to a framework proposal. BPMJ 20(3), 412–432 (2014)

14. Modha, D.S., Ananthanarayanan, R., Esser, S.K., Ndirango, A., Sherbondy, A.J., Singh, M.P.: Cognitive computing. Commun. ACM 54, 62–71 (2011)

15. Motahari Nezhad, H.R., Akkiraju, R.: Towards cognitive BPM as the next generation BPM platform for analytics-driven business processes. In: Fournier, F., Mendling, J. (eds.) BPM 2014. LNBIP, vol. 202, pp. 158–164. Springer, Cham (2015). https://doi.org/10.1007/978-3-319-15895-2_14

16. Motahari-Nezhad, H.R., Gunaratna, K., Cappi, J.: eAssistant: Cognitive assistance for identification and auto-triage of actionable conversations. In: Proceedings of the 26th International Conference on WWW Companion, pp. 89–98. International WWW Conferences Steering Committee, Perth, Australia (2017)

17. Reichert, M., Weber, B.: Enabling Flexibility in Process-Aware Information Systems: Challenges, Methods, Technologies. Springer, Berlin Heidelberg (2012). https://doi.org/10.1007/978-3-642-30409-5

18. Swenson, K.: Mastering the Unpredictable: The Nature of Knowledge Work. Meghan-Kiffer Press, Tampa, FL (2010)

19. Taylor, J.G.: Cognitive computation. Cogn. Comput. 1, 4–16 (2009)

20. van der Aalst, W.M.P.: Business process management: a comprehensive survey. ISRN Softw. Eng. 2013, 1–37 (2013)

Cognitive Business Process Management for Adaptive Cyber-Physical Processes

Andrea Marrella$^{(\boxtimes)}$ and Massimo Mecella

Dipartimento di Ingegneria Informatica, Automatica e Gestionale,
Sapienza Università di Roma, via Ariosto 25, 00185 Rome, Italy
{marrella,mecella}@diag.uniroma1.it

Abstract. In the era of Big Data and Internet-of-Things (IoT), all real-world environments are gradually becoming cyber-physical (e.g., emergency management, healthcare, smart manufacturing, etc.), with the presence of connected devices and embedded ICT systems (e.g., smartphones, sensors, actuators) producing huge amounts of data and events that influence the enactment of the Cyber Physical Processes (CPPs) enacted in such environments. A Process Management System (PMS) employed for executing CPPs is required to automatically adapt its running processes to anomalous situations and exogenous events by minimising any human intervention at run-time. In this paper, we tackle this issue by introducing an approach and an adaptive Cognitive PMS that combines process execution monitoring, unanticipated exception detection and automated resolution strategies leveraging on well-established action-based formalisms in Artificial Intelligence, which allow to interpret the ever-changing knowledge of cyber-physical environments and to adapt CPPs by preserving their base structure.

Keywords: Cognitive business process management
Cyber-Physical Processes · Process adaptation and recovery
Situation calculus · IndiGolog · Automated planning

1 Introduction

In the last years, we have witnessed the emergence of new computing paradigms, such as Industry 4.0[1], Health 2.0 (e.g., cf. [1]) and mobile-based emergency management [2], in which the interplay of Internet-of-Things (IoT) devices, i.e., devices attached to the Internet, cloud computing, Software-as-a-Service (SaaS), and Business Process Management (BPM) create the so-called *cyber-physical environments* and give rise to the concept of *Cyber-Physical Systems* (CPSs).

---

[1] cf. H. Kagermann, W. Wahlster and J. Helbig: Recommendations for implementing the strategic initiative Industrie 4.0: Final report of the Industrie 4.0 Working Group, 2013, Frankfurt, http://www.acatech.de/fileadmin/user_upload/Baumstruktur_nach_Website/Acatech/root/de/Material_fuer_Sonderseiten/Industrie_4.0/Final_report__Industrie_4.0_accessible.pdf .

© Springer International Publishing AG 2018
E. Teniente and M. Weidlich (Eds.): BPM 2017 Workshops, LNBIP 308, pp. 429–439, 2018.
https://doi.org/10.1007/978-3-319-74030-0_33

The role of CPSs is to monitor the *physical processes* enacted in cyber-physical environments, create a virtual copy of the physical world and make decentralized decisions, by introducing automated and intelligent support of workers in their increasingly complex work [3].

A relevant aspect in these environments lies in the fundamental role played by the processes orchestrating the different actors (software, humans, robots, etc.) involved in the CPS. We refer to these processes as *cyber-physical processes* (CPPs), whose enactment is influenced by user decision making and coupled with contextual data and knowledge production coming from the cyber-physical environment. According to [4], *Cognitive Process Management Systems* (CPMSs) are the key technology for supporting CPPs. A PMS is said to be *cognitive* when it involves additional processing constructs that are at a semantic level higher than those of conventional PMSs. These constructs are called *cognitive BPM constructs* and include data-driven activities, goals, and plans [4]. Their usage can open opportunities for new levels of automation for CPPs, such as - for example - *the automated synthesis of adaptation strategies at run-time exploiting solely the process knowledge and its expected evolution.*

During the enactment of CPPs, variations or divergence from structured reference models are common due to exceptional circumstances arising (e.g., autonomous user decisions, exogenous events, or contextual changes), thus requiring the ability to properly *adapt* the process behavior. *Process adaptation* can be seen as the ability of a process to react to exceptional circumstances (that may or may not be foreseen) and to adapt/modify its structure accordingly. Exceptions can be either *anticipated* or *unanticipated*. An anticipated exception can be planned at design-time and incorporated into the process model, i.e., a (human) process designer can provide an *exception handler* that is invoked during run-time to cope with the exception. Conversely, *unanticipated exceptions* refer to situations, unplanned at design-time, that may emerge at run-time and can be detected only during the execution of a process instance, when a mismatch between the computerized version of the process and the corresponding real-world process occurs. To cope with those exceptions, a PMS is required to allow *ad-hoc process changes* for adapting running process instances in a context-dependent way.

The fact is that, in cyber-physical environments, the number of possible anticipated exceptions is often too large, and traditional manual implementation of exception handlers at design-time is not feasible for the process designer, who has to anticipate all potential problems and ways to overcome them in advance [5]. Furthermore, anticipated exceptions cover only partially relevant situations, as in such scenarios many unanticipated exceptional circumstances may arise during the process instance execution. Therefore, the process designer often lacks the needed knowledge to model all the possible exceptions at the outset, or this knowledge can become obsolete as process instances are executed and evolve, by making useless her/his initial effort.

To tackle this issue, in this paper we summarize the main ideas discussed in [6] and introduce our work on SmartPM[2], a CPMS able to *automatically adapt CPPs at run-time* when *unanticipated exceptions* occur, thus requiring no specification of recovery policies at design-time. The general idea builds on the dualism between an *expected reality* and a *physical reality*: process execution steps and exogenous events have an impact on the physical reality and any deviation from the expected reality results in a mismatch to be removed to allow process progression.

To that end, we have resorted to three popular *action-based formalisms* and *technologies* from the field of Knowledge Representation and Reasoning (KR&R): situation calculus [9], IndiGolog [10], and automated planning [11,12]. We used the situation calculus logical formalism to model the underlying domain in which processes are to be executed, including the description of available tasks, contextual properties, tasks' preconditions and effects, and the initial state. On top of such model, we used the IndiGolog high-level agent programming language for the specification of the structure and control flow of processes. Importantly, we customized IndiGolog to monitor the online execution of processes and detect potential mismatches between the model and the actual execution. If an exception invalidates the enactment of the processes being executed, an external state-of-the-art classical planner is invoked to synthesise a recovery procedure to adapt the faulty process instance.

The choice of adopting action-based formalisms from the KR&R field is motivated by their ability to provide the right *cognitive* level needed when dealing with dynamic situations in which data (values) play a relevant role in system enactment and automated reasoning over the system progress. In the field of BPM, many other formalisms (in particular Petri Nets-based and process algebras) have been successfully adopted for process management, but all of them are somehow based on synthesis techniques of the control-flow, when considering their automated reasoning capabilities. This implies the level of abstraction over dealing with data and dynamic situations is fairly "raw", when compared with KR&R methods in which automated reasoning over data values and situations is much more developed [9,13,14]. The choice of KR&R technologies allowed us to develop a principled, clean and simple-to-manage framework for process adaptation based on relevant data manipulated by the process, without compromising efficiency and effectiveness of the proposed solution.

The rest of the paper is organized as follow. Section 2 introduces conceptual architecture for CPMSs that manage CPPs. Such an architecture is then instantiated in the SmartPM approach and system outlined in Sect. 3. Finally Sect. 4 concludes the paper by discussing our approach in the larger context and presenting possible future evolutions.

---

[2] The reader interested to the very technical details may refer to [6–8].

2   A Conceptual Architecture for Managing CPPs

CPSs are having widespread applicability and proven impact in multiple areas, like aerospace, automotive, traffic management, healthcare, manufacturing, emergency management [15]. According to [16], any physical environment which contains computing-enabled devices can be considered as a cyber-physical environment.

The trend of managing CPPs, i.e., processes enacted in cyber-physical environments, has been fueled by two main factors. On the one hand, the recent development of powerful mobile computing devices providing wireless communication capabilities have become useful to support mobile workers to execute tasks in such dynamic settings. On the other hand, the increased availability of sensors disseminated in the world has lead to the possibility to monitor in detail the evolution of several real-world objects of interest. *The knowledge extracted from such objects allows to depict the contingencies and the context in which processes are carried out, by providing a fine-grained monitoring, mining, and decision support for them.*

We devise in the following a conceptual architecture to concretely build an adaptive CPMS in cyber-physical environments. The management of a CPP requires additional challenges to be considered if compared with a traditional "static" business process. On the one hand, there is the need of representing explicitly real-world objects and technical aspects like device capability constraints, sensors range, actors and robots mobility, etc. On the other hand, since cyber-physical environments are intrinsically "dynamic", a CPMS providing real-time monitoring and automated adaptation features during process execution is required.

To this end, the role of the data perspective becomes fundamental. Data, including information processed by process tasks as well as contextual information, are the main driver for triggering process adaptation, as focusing on the control flow perspective only - as traditional PMSs do - would be insufficient. In fact, in a cyber-physical environment, a CPP is genuinely knowledge and data centric: its control flow must be coupled with contextual data and knowledge production and process progression may be influenced by user decision making. This means that traditional imperative models have to be extended and complemented with the introduction of specific *cognitive constructs* such as *data-driven activities* and *declarative elements* (e.g., tasks preconditions and effects) which enable a precise description of data elements and their relations, so as to go beyond simple process variables, and allow establishing a link between the control flow and the data perspective.

Starting from the above considerations, coupled with the experience gained in the area and lessons learned from several projects involving CPSs, we have devised a conceptual architecture to build a CPMS for the management of CPPs, which supports the so-called *Plan-Act-Learn* cycle for cognitively-enabled processes [4]. As shown in Fig. 1, we identified 5 main architectural layers that we present in a bottom-up fashion.

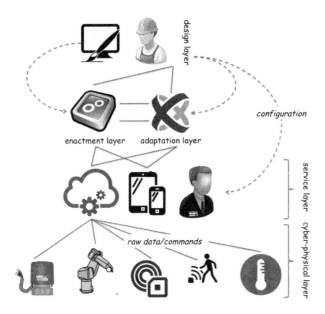

Fig. 1. A conceptual architecture for CPPs.

The ***cyber-physical layer*** consists mainly of two classes of physical compo-nents: *(i)* sensors (such as GPS receivers, RFID chips, 3D scanners, cameras, etc.) that collect data from the physical environment by monitoring real-world objects and *(ii)* actuators (robotic arms, 3D printers, electric pistons, etc.), whose effects affect the state of the physical environment. The cyber-physical layer is also in charge of providing a physical-to-digital interface, which is used to transform *raw* data collected by the sensors into machine-readable events, and to convert *high-level* commands sent by the upper layers into *raw* instructions readable by the actuators. The cyber-physical layer does not provide any intelligent mechanism neither to clean, analyse or correlate data, nor to compose high-level commands into more complex ones; such tasks are in charge of the uppers layer.

On top of the cyber-physical layer lies the ***service layer***, which contains the set of services offered by the real-world entities (software components, robots, agents, humans, etc.) to perform specific process tasks. In the service layer, available data can be aggregated and correlated, and high-level commands can be orchestrated to provide higher abstractions to the upper layers. For example, a smartphone equipped with an application allowing to sense the position and the posture of a user is at this layer, as it collects the raw GPS, accellerometer and motion sensor data and correlates them to provide discrete and meaningful information.

On top of the service layer, there are two further layers interacting with each other. The ***enactment layer*** is in charge of *(i)* enacting complex processes by deciding which tasks are enabled for execution, *(ii)* orchestrating the differ-ent available services to perform those tasks and *(iii)* providing an execution monitor to detect the anomalous situations that can possibly prevent the correct execution of process instances. The execution monitor is responsible for deciding if process adaptation is required. If this is the case, the **adaptation layer** will provide the required algorithms to *(i)* reason on the available process tasks and contextual data and to *(ii)* find a recovery procedure for adapting the process instance under consideration, i.e., to re-align the process to its expected behaviour. Once a recovery procedure has been synthesized, it is passed back to the enactment layer for being executed.

Finally, the **design layer** provides a GUI-based tool to define new process specifications. A process designer must be allowed not only to build the process control flow, but also to explicitly formalize the data reflecting the contextual knowledge of the cyber-physical environment under study. It is important to underline that data formalization must be performed without any knowledge of the internal working of the physical components that collect/affect data in the cyber-physical layer. To link tasks to contextual data, the GUI-based tool must go beyond the classical "task model" as known in the literature, by allowing the process designer to explicitly state what data may constrain a task execution or may be affected after a task completion. Finally, besides specifying the process, configuration files should also be produced to properly configure the enactment, the services and the sensors/actuators in the bottom layers.

3   The **SmartPM** Approach and System

SmartPM (Smart Process Management) is an approach and an adaptive CPMS implementing a set of techniques that enable to automatically adapt process instances at run-time in the presence of unanticipated exceptions, without requiring an explicit definition of handlers/policies to recover from tasks failures and exogenous events. SmartPM adopts a *layered service-based* approach to process management, i.e., *tasks are executed by services*, such as software applications, humans, robots, etc. Each task can be thus seen as a single step consuming input data and producing output data.

To monitor and deal with exceptions, the SmartPM approach leverages on [17]'s technique of adaptation from the field of agent-oriented programming, by specializing it to our CPP setting (see Fig. 2). We consider adaptation as *reducing the gap* between the *expected reality* **EXP**, the (idealized) model of reality used by the CPMS to reason, and the *physical reality* **PHY**, the real world with the actual conditions and outcomes. While **PHY** records what is *concretely* happening in the real environment during a process execution, **EXP** reflects what it is *expected* to happen in the environment. Process execution steps and exogenous events have an impact on **PHY** and any deviation from **EXP** results in a mismatch to be removed to allow process progression. At this point, a state-of-the-art automated planner is invoked to synthesise a recovery procedure that adapts the faulty process instance by removing the gap between the two realities.

To realize the above approach, the implementation of SmartPM covers the modeling, execution and monitoring stages of the CPP life-cycle. To that end, the architecture of SmartPM relies on five architectural layers.

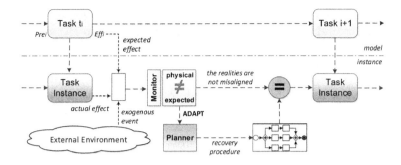

Fig. 2. An overview of the SmartPM approach.

The **design layer** provides a graphical editor developed in Java that assists the process designer in the definition of the process model at design-time. Process knowledge is represented as a *domain theory* that includes all the contextual information of the domain of concern, such as the people/services that may be involved in performing the process, the tasks, the data and so forth. Data are represented through some *atomic terms* that range over a set of *data objects*, which depict entities of interest (e.g., capabilities, services, etc.), while atomic terms can be used to express properties of domain objects (and relations over objects). *Tasks* are collected in a repository and are described in terms of *preconditions* - defined over atomic terms - and *effects*, which establish their expected outcomes. Finally, a process designer can specify which *exogenous events* may be caught at run-time and which atomic terms will be modified after their occurrence. Once a valid domain theory is ready, the process designer uses the graphical editor to define the process control flow through the standard BPMN notation among a set of tasks selected from the tasks repository.

The **enactment layer** is in charge of managing the process execution. First of all, the domain theory specification and the BPMN process are automatically translated into situation calculus [9] and IndiGolog [10] readable formats. Situation calculus is used for providing a declarative specification of the domain of interest (i.e., available tasks, contextual properties, tasks preconditions and effects, what is known about the initial state). Then, an *executable model* is obtained in the form of an IndiGolog program to be executed through an IndiGolog engine. To that end, we customized an existing IndiGolog engine[3] to *(i)* build a physical/expected reality by taking the initial context from the external environment; *(ii)* manage the process routing; *(iii)* collect exogenous events from the external environment; *(iv)* monitor contextual data to identify changes or events which may affect process execution. Once a task is ready for being executed, the IndiGolog engine assigns it to a proper process participant (that could be a software, a human actor, a robot, etc.) that provides all the required capabilities for task execution.

---

[3] http://sourceforge.net/projects/indigolog/.

The **service layer** acts as a middleware between process participants, the enactment layer and the cyber-physical layer. Specifically, in the service layer, process participants interact with the engine through a *Task Handler*, an interactive GUI-based software application realized for Android devices that supports the visualization/execution of assigned tasks by selecting an appropriate outcome. Possibly such an Android application can exploit sensors and actuators (e.g., an Arduino board connected through Bluetooth, as currently realized in our implementation), thus effectively offering services over the *cyber-physical layer*. Every step of the task life cycle - ranging from the assignment to the release of a task - requires an interaction between the IndiGolog engine and the task handlers. The communication between the IndiGolog engine and the task handlers is mediated by the *Communicator Manager* component (which is essentially a web server) and established using the Google Cloud Messaging service.

To enable the automated synthesis of a recovery procedure, the **adaptation layer** relies on the capabilities provided by a planner component (the LPG-td planner [18]), which assumes the availability of a classical planning problem, i.e., an initial state and a goal to be achieved, and of a planning domain definition that includes the actions to be composed to achieve the goal, the domain predicates and data types. Specifically, if process adaptation is required, we translate *(i)* the domain theory defined at design-time into a planning domain, *(ii)* the physical reality into the initial state of the planning problem and *(iii)* the expected reality into the goal state of the planning problem. The planning domain and problem are the input for the planner component. If the planner is able to synthesize a recovery procedure δ_a, the *Synchronization* component combines δ' (which is the remaining part of the faulty process instance δ still to be executed), with the recovery plan δ_a, builds an adapted process $\delta'' = (\delta_a; \delta')$ and converts it into an executable IndiGolog program so that it can be enacted by the IndiGolog engine. Otherwise, if no plan exists for the current planning problem, the control passes back to the process designer, who can try to manually adapt the process instance.

The **cyber-physical layer** is tightly coupled with the physical components available in the domain of interest. Since the IndiGolog engine can only work with defined discrete values, while data gathered from physical sensors have naturally continuous values, the system provides several web tools that allow process designers to associate some of the data objects defined in the domain theory with the continuous data values collected from the environment. For example, we developed several web tools to associate the data collected from sensors (GPS, temperature, noise level, etc.) to discrete values. We provided a concrete example of a location web tool that allows process designers to mark areas of interest from a real map and associate them to discrete locations. The mapping rules generated are then saved into the Communication Manager and retrieved at run-time to allow the matching of the continuous data values collected by the specific sensor into discrete data objects.

4   Concluding Remarks

We are at the beginning of a profound transformation of BPM due to advances in AI and Cognitive Computing [4]. Cognitive systems offer computational capabilities typically based on large amount of data, which provide cognition power that augment and scale human expertise. The aim of the emergent field of cognitive BPM is to offer the computational capability of a cognitive system to provide analytical support for processes over structured and unstructured information sources. The target is to provide proactivity and self-adaptation of the running processes against the evolving conditions of the application domains in which they are enacted.

In this direction, our paper summarizes the most interesting results reported in [6], which have been devoted to the realization of a general approach, a concrete framework and a CPMS implementation, called SmartPM, for automated adaptation of CPPs. Our purpose was to demonstrate that the combination of procedural and imperative models with cognitive BPM constructs such as data-driven activities and declarative elements, along with the exploitation of techniques from the field of AI such as situation calculus, IndiGolog and classical planning, can increase the ability of existing PMSs of supporting and adapting CPPs in case of unanticipated exceptions.

Existing approaches dealing with unanticipated exceptions typically rely on the involvement of process participants at run-time, so that authorized users are allowed to manually perform structural process model adaptation and ad-hoc changes at the instance level. However, CPPs demand a more flexible approach recognizing the fact that in real-world environments process models quickly become outdated and hence require closer interweaving of modeling and execution. To this end, the adaptation mechanism provided by SmartPM is based on execution monitoring for detecting failures and context changes at run-time, without requiring to predefine any specific adaptation policy or exception handler at design-time (as most of the current approaches do).

From a general perspective, our planning-based automated exception handling approach should be considered as complementary with respect to existing techniques, acting as a "bridge" between approaches dealing with anticipated exceptions and approaches dealing with unanticipated exceptions. When an exception is detected, the run-time engine may first check the availability of a predefined exception handler, and if no handler was defined it can rely on an automated synthesis of the recovery process. In the case that our planning-based approach fails in synthesizing a suitable handler (or an handler is generated but its execution does not solve the exception), other adaptation techniques need to be used. For example, if the running process provides a well-defined intended goal associated to its execution, we could resort to the van Beest's work [19] and do planning from first-principle to achieve such a goal. Conversely, if no intended goal is associated to the process, a human participant can be involved, leaving her/him the task of manually adapting the process instance. Future work will include an extension of our approach to "stress" the assumptions imposed by the usage of classical planning techniques for the synthesis of the recovery procedure, which frame the scope of applicability of the approach for addressing more expressive problems, including incomplete information, preferences and multiple task effects.

The current implementation of SmartPM is developed to be effectively used by process designers and practitioners[4]. Users define processes in the well-known BPMN language, enriched with semantic annotations for expressing properties of tasks, which allow our interpreter to derive the IndiGolog program representing the process. Interfaces with human actors (such as specific graphical user applications in Java) and software services (through Web service technologies) allow the core system to be effectively used for enacting processes. Although the need to explicitly model process execution context and annotate tasks with preconditions and effects may require some extra modeling effort at design-time (also considering that traditional process modeling efforts are often mainly directed to the sole control flow perspective), the overhead is compensated at run-time by the possibility of automating exception handling procedures. While, in general, such modeling effort may seem significant, in practice it is comparable to the effort needed to encode the adaptation logic using alternative methodologies like happens, for example, in rule-based approaches.

Acknowledgments. This work is partly supported by the projects Social Museum and Smart Tourism (CTN01_00034_23154), NEPTIS (PON03PE_00214_3), RoMA (SCN_00064), and by the Sapienza project "Data-aware Adaptation of Knowledge-intensive Processes in Cyber-Physical Domains through Action-based Languages".

References

1. Cossu, F., Marrella, A., Mecella, M., Russo, A., Bertazzoni, G., Suppa, M., Grasso, F.: Improving operational support in hospital wards through vocal interfaces and process-awareness. In: 25th International Symposium on Computer-Based Medical Systems (CBMS). IEEE (2012)
2. Humayoun, S.R., Catarci, T., de Leoni, M., Marrella, A., Mecella, M., Bortenschlager, M., Steinmann, R.: Designing mobile systems in highly dynamic scenarios: the WORKPAD methodology. Knowl. Technol. Policy **22**(1), 25–43 (2009)
3. Seiger, R., Keller, C., Niebling, F., Schlegel, T.: Modelling complex and flexible processes for smart cyber-physical environments. J. Comput. Sci. **10**, 137–148 (2014)
4. Hull, R., Motahari Nezhad, H.R.: Rethinking BPM in a cognitive world: transforming how we learn and perform business processes. In: La Rosa, M., Loos, P., Pastor, O. (eds.) BPM 2016. LNCS, vol. 9850, pp. 3–19. Springer, Cham (2016). https://doi.org/10.1007/978-3-319-45348-4_1
5. Reichert, M., Weber, B.: Enabling Flexibility in Process-Aware Information Systems - Challenges, Methods, Technologies. Springer, Berlin Heidelberg (2012). https://doi.org/10.1007/978-3-642-30409-5
6. Marrella, A., Mecella, M., Sardiña, S.: Supporting adaptiveness of cyber-physical processes through action-based formalisms. AI Communications (2017, to appear)

---

[4] See: http://www.dis.uniroma1.it/~smartpm.

7. Marrella, A., Mecella, M., Sardina, S.: SmartPM: an adaptive process management system through situation calculus, indigolog, and classical planning. In: Proceedings of the 14th International Conference on Principles of Knowledge Representation and Reasoning, KR 2014 (2014)

8. Marrella, A., Mecella, M., Sardiña, S.: Intelligent process adaptation in the SmartPM system. ACM TIST **8**(2), 1–25 (2017)

9. Reiter, R.: Knowledge in Action: Logical Foundations for Specifying and Implementing Dynamical Systems. MIT Press, Cambridge (2001)

10. De Giacomo, G., Lespérance, Y., Levesque, H., Sardina, S.: IndiGolog: a high-level programming language for embedded reasoning agents. In: El Fallah Seghrouchni, A., Dix, J., Dastani, M., Bordini, R. (eds.) Multi-Agent Programming: Languages, Tools and Applications. Tools and Applications, Springer, Boston (2009). https://doi.org/10.1007/978-0-387-89299-3_2

11. Nau, D., Ghallab, M., Traverso, P.: Automated Planning: Theory & Practice. Morgan Kaufmann Publishers Inc., San Francisco (2004)

12. Geffner, H., Bonet, B.: A Concise Introduction to Models and Methods for Automated Planning. Morgan & Claypool Publishers, San Rafael (2013)

13. Reichgelt, H.: Knowledge Representation: An AI perspective. Greenwood Publishing Group Inc., Westport (1991)

14. Brachman, R., Levesque, H.: Knowledge Representation and Reasoning. Elsevier, Amsterdam (2004)

15. Rajkumar, R.R., Lee, I., Sha, L., Stankovic, J.: Cyber-physical systems: the next computing revolution. In: Proceedings of the 47th Design Automation Conference, DAC 2010, pp. 731–736. IEEE (2010)

16. Lee, E.A.: Cyber physical systems: design challenges. In: Proceedings of the 11th IEEE International Symposium on Object and Component-Oriented Real-Time Distributed Computing, ISORC 2008, pp. 363–369. IEEE (2008)

17. De Giacomo, G., Reiter, R., Soutchanski, M.: Execution monitoring of high-level robot programs. In: Proceedings of the Sixth International Conference on Principles of Knowledge Representation and Reasoning, KR 1998, pp. 453–465 (1998)

18. Gerevini, A., Saetti, A., Serina, I., Toninelli, P.: LPG-TD: a fully automated planner for PDDL2.2 domains. In: Proceedings of the 14th International Conference on Automated Planning and Scheduling, ICAPS 2004 (2004)

19. van Beest, N.R., Kaldeli, E., Bulanov, P., Wortmann, J.C., Lazovik, A.: Automated runtime repair of business processes. Inf. Syst. **39**, 45–79 (2014)

BPM for the Masses: Empowering Participants of Cognitive Business Processes

Aleksander Slominski[(✉)] and Vinod Muthusamy

IBM T.J. Watson Research Center, Yorktown Heights, NY 10598, USA
{aslom,vmuthus}@us.ibm.com

Abstract. Authoring, developing, monitoring, and analyzing business processes has required both domain and IT expertise since Business Process Management tools and practices have focused on enterprise applications and not end users. There are trends, however, that can greatly lower the bar for users to author and analyze their own processes. One emerging trend is the attention on blockchains as a shared ledger for parties collaborating on a process. Transaction logs recorded in a standard schema and stored in the open significantly reduces the effort to monitor and apply advanced process analytics. A second trend is the rapid maturity of machine learning algorithms, in particular deep learning models, and their increasing use in enterprise applications. These cognitive technologies can be used to generate views and processes customized for an end user so they can modify them and incorporate best practices learned from other users' processes.

Keywords: BPM · Cognitive computing · Blockchain · Privacy
Machine learning

1 End Users and Business Process Management

Business Process Management (BPM) tools and techniques address the life cycle of developing, executing, and managing business processes [1]. By capturing process execution traces inside a BPM runtime, it was possible to get better visibility into process state and address problems faster. Moreover, gathered data could be used to improve processes and create better outcomes for process participants. However, the greater visibility into process execution was not available for participants that do not have direct access to the BPM runtime. Instead, they get what is made visible through UI interactions but rarely have any access to process state or historical data. When there are multiple participants in a business process from different organizations, each participant may have their own BPM system and have limited or no visibility into other systems.

One well-tested way to get better visibility into business operations is to record them in a shared ledger. Blockchain is a modern version of a traditional business ledger. Although it has roots in Bitcoin [9], the value of blockchain is not limited to electronic currency. For example, the Ethereum project allows building apps that run on "a custom built blockchain, an enormously powerful shared global infrastructure that can move value around and represent the ownership of property" [11], and the Hyperledger Fabric project is used to develop blockchain applications that use cross-industry blockchain

© Springer International Publishing AG 2018
E. Teniente and M. Weidlich (Eds.): BPM 2017 Workshops, LNBIP 308, pp. 440–445, 2018.
https://doi.org/10.1007/978-3-319-74030-0_34

technologies [10]. The value of blockchain in these systems is its role as a decentralized, immutable, and trusted database that can be shared among non-trusted parties across enterprises, and that contains a single source of truth about the historical record of business process transactions (Figs. 1 and 2).

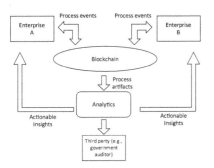 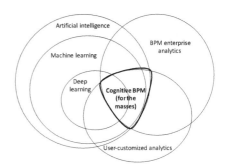

Fig. 1. Enterprise process events shared through a blockchain allow for end-to-end analytics that can be used by different parties such as business partners and government.

Fig. 2. Cognitive BPM lies in the intersection of artificial intelligence technologies, enterprise analytics tools, and user analytics capabilities.

A traditional challenge with constructing a business process is the specification of standard schemas that multiple parties can agree on, sometimes referred as the Process Reuse challenge [1]. We argue that the use of blockchain forces parties collaborating in a business process to develop, and more importantly, agree on the schema. This, in turn, offers end-to-end visibility into the process and the ability for new participants to devise ad-hoc auxiliary business processes without permission from a central coordinator. Our vision is that this could be as significant as the introduction of spreadsheets for PCs in creating a programming model that allows casual users to create processes and analyze data without becoming computer science or data science experts.

The emerging area of cognitive computing offers powerful technologies that may be useful in solving some BPM challenges. We define cognitive computing as algorithms and systems that augment human decision making. An important property is that "rather than being explicitly programmed, they learn and reason from their interactions with us and from their experiences with their environment." [8] Furthermore, cognitive computing is gaining adoption in enterprises where it can be described as applications that "introduce cognitive computing into the software enabling an organization's business processes. The goal … is to make business processes more efficient, accurate, relevant and reliable." [8] In this paper we define Cognitive BPM as systems that use cognitive techniques to help participants run business processes more efficiently.

To allow casual users to take full advantage of process analytics it is necessary to augment it with cognitive computing technologies. One possibility is for process participants to access process related data and use it exactly as in BPM systems, but with the help of cognitive systems it becomes feasible for casual process participants to create and run long-tail processes without IT expertise.

2   Use Case: Health Care

Consider a health care scenario where there are multiple parties involved in the care of a patient, including the hospital, physicians, nurses, lab technicians, insurance agency, government auditors, the patient, and their family members. A business process that manages a patient's hospital visit needs to coordinate activities among all these participants without violating privacy policies.

It is extremely unlikely that one business process will capture all the rules and requirements among these parties. What is more practical is for subsets of parties to devise processes that help them get their work done. For example, a hospital, insurance agency, and auditor may develop a process to manage insurance policy requests and payment invoices.

The power of using a shared blockchain to maintain process state using a public schema is that it allows new processes to be added without having to navigate a centralized governance process.

As data is added to the overall end-to-end process blockchain, multiple participants with access to the data can delegate it to pieces of code that run automatically. The code can be a smart contract that does something when conditions are met in the blockchain. For example, the trigger can be the event that a patient enters the hospital, as in Fig. 3. Alternatively, end users can directly consume updates to the blockchain by receiving events and having apps on their mobile device perform a custom action. Another option is for events to be passed to other services that a user has authorized to access some of their data in the blockchain.

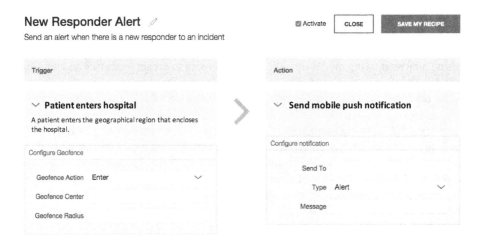

Fig. 3. This screenshot is an illustration of how a process end user can specify event triggers and generate relevant notifications that may send alerts to their mobile device.

Storing even a small amount of relevant data in a blockchain allows any participant to build end-to-end dashboards and analytics capabilities. For example, in this health care use case, the dashboard in Fig. 4 compares key metrics in two departments using data that could be stored in a blockchain.

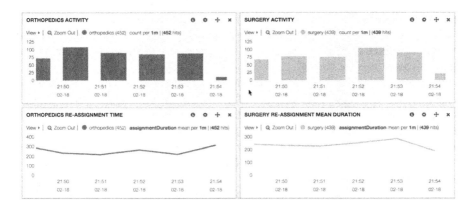

Fig. 4. This screenshot shows how blockchain data could be leveraged to build analytics solutions that cross traditional boundaries and silos. In this hypothetical example, data from two separate hospital divisions can be put into one analytics dashboard even though each division may run its own BPM system but they share state in blockchain.

We want to emphasize the importance of the data being in a shared blockchain that any authorized user can access. This supports innovative applications that make use of the data for discovering inefficiencies, enforcing smart contracts, auditing for compliance with regulations, or building reactive user experiences.

3   Privacy in Cognitive BPM

Privacy controls are a major challenge in many enterprise applications. Fortunately, it is possible to not only to store process state in a blockchain, but also related permissions and have the sensitive data encrypted and made accessible only to authorized parties. Encrypting data, however, makes it difficult to perform analytics, and hence decreases one of main advantages of sharing state. Consider a health case scenario that exemplifies some of the challenges: patients, health providers, insurance agencies, and government auditors all need different views of a patient's sensitive information.

There are well-known security technologies, including differential privacy [2], homomorphic encryption, role-based access control, and pseudo-anonymity, that can be used to address some of the important privacy issues. The same issues exist in cloud computing in general [3]. The use of these and other techniques to blockchain applications is still an open problem.

4   Machine Learning to Guide Process Development

Developing and refining a business process can be difficult and time consuming. It requires domain expertise to understand how to optimize the process logic, as well as technical proficiency to formally encode the process logic. We cannot assume that end users will be skilled in either of these dimensions, but there are machine learning techniques that can lower the barrier.

One technique takes advantage of the fact that we can observe in the blockchain the end-to-end behavior of processes. We can, therefore, extract individual process traces, find recurring patterns, and generate business rules or process models that capture this behavior. Existing process mining techniques can be applied here [4, 5]. There is still, however, an open problem on how to present these process templates at a level of abstraction that is both easy to understand for the user, and gives them the ability to customize it.

The above technique assumes that the process models themselves are not shared, but only their effects. In cases where users are willing to share their custom processes and rules, we can apply different techniques to find the ones that are performing well (according to some business metric) and suggest these to other users. Clustering and recommendation algorithms, among others, in the process analytics domain can be used here [6].

Consider an example of a coach helping a patient reach a fitness goal, such as recovering from a heart surgery. The coach develops a customized plan that defines concrete tasks and reminders related to their diet, physical therapy, medication and lab visits. This plan is essentially a set of business rules and processes that requires collaboration among the coach, patient, and other health providers. The participants may further refine the process as needed. While each patient will need a customized process, there are probably patterns that emerge on the kinds of plans that work well with certain classes of patients. Machine learning algorithms can help find these best practices and suggest them to the right users.

We believe that the visibility into the behavior of blockchain-based applications offers a rich, and unprecedented, amount of data that can be mined to help simplify and guide end users who want to develop a business process.

5   Conclusions

Using blockchain to orchestrate cognitive BPM processes not only make them easier for cognitive analytics but make them more accessible to process participants. By developing an ecosystem of apps and services that can get access to cognitive BPM state in blockchain, users will have a marketplace of solutions and new types of apps that are much better at understanding users and what they are doing by understanding how they participate in cognitive processes.

As these cognitive technologies and solutions are built they will make business process participants more engaged, they will understand better what is going on in business, and will be able to react with the information they need. These BPM systems would not be static, but instead end users could modify them and use analytics and cognitive techniques to help identify best practices learned from other users' processes.

We believe that these two technologies—blockchain for process visibility, and cognitive algorithms to guide the development and understanding of processes—together can make BPM more accessible to casual users.

References

1. Aalst, W.M.P.: Business process management: a comprehensive survey. ISRN Softw. Eng. **2013**, 37 pages (2013). Article ID 507984
2. Narayanan, A., Shmatikov, V.: Robust de-anonymization of large sparse datasets. In: Proceedings of the IEEE Symposium on Security and Privacy (2008)
3. Chen, D., Zhao, H.: Data security and privacy protection issues in cloud computing. In: International Conference on Computer Science and Electronics Engineering (2012)
4. Muthusamy, V., Slominski, A., Ishakian, V., Khalaf, R., Reason, J., Rozsnyai, S.: Lessons learned using a process mining approach to analyze events from distributed applications. In: Proceedings of the 10th ACM International Conference on Distributed and Event-based Systems (2016)
5. Evermann, J.-R., Fettke, P.: Predicting process behaviour using deep learning. Decision Support Systems (2017)
6. Ehrig, M., Koschmider, A., Oberweis, A.: Measuring similarity between semantic business process models. In: Proceedings of the Fourth Asia-Pacific Conference on Conceptual Modelling, vol. 67 (2007)
7. Kelly III, J.E.: Computing, Cognition and the Future of Knowing: How Humans and Machines Are Forging a New Age of Understanding. IBM (2015). https://www.research.ibm.com/software/IBMResearch/multimedia/Computing_Cognition_WhitePaper.pdf
8. Tarafdar, M., Beath, C., Ross, J.: Enterprise cognitive computing applications: opportunities and challenges. IT Prof. **19**(4), 21–27 (2017)
9. Nakamoto, S.: Bitcoin: A Peer-to-Peer Electronic Cash System. https://bitcoin.org/bitcoin.pdf
10. Hyperledger Fabric. https://www.hyperledger.org/projects/fabric
11. Ethereum Project. https://www.ethereum.org/

Using Insights from Cognitive Neuroscience to Investigate the Effects of Event-Driven Process Chains on Process Model Comprehension

Michael Zimoch[1]([✉]), Tim Mohring[1], Rüdiger Pryss[1], Thomas Probst[2], Winfried Schlee[2], and Manfred Reichert[1]

[1] Institute of Databases and Information Systems, Ulm University, Ulm, Germany
{michael.zimoch,tim.mohring,ruediger.pryss,manfred.reichert}@uni-ulm.de
[2] Department of Psychiatry and Psychotherapy, Regensburg University, Regensburg, Germany
thomas.probst@psychologie.uni-regensburg.de,
winfried.schlee@googlemail.com

Abstract. Business process models have been adopted by enterprises for more than a decade. Especially for domain experts, the comprehension of process models constitutes a challenging task that needs to be mastered when creating or reading these models. This paper presents the results we obtained from an eye tracking experiment on process model comprehension. In detail, individuals with either no or advanced expertise in process modeling were confronted with models expressed in terms of Event-driven Process Chains (EPCs), reflecting different levels of difficulty. The first results of this experiment confirm recent findings from one of our previous experiments on the reading and comprehension of process models. On one hand, independent from their level of expertise, all individuals face similar patterns, when being confronted with process models exceeding a certain level of difficulty. On the other, it appears that process models expressed in terms of EPCs are perceived differently compared to process models specified in the Business Process Model and Notation (BPMN). In the end, their generalization needs to be confirmed by additional empirical experiments. The presented experiment continues a series of experiments that aim to unravel the factors fostering the comprehension of business process models by using methods and theories stemming from the field of cognitive neuroscience and psychology.

Keywords: Business process model comprehension
Event-driven Process Chains · Eye tracking

1 Introduction

Many enterprise repositories comprise large collections of *process models*, which represent business processes serving to achieve specific goals with their corresponding actors, tasks, and decisions. Usually, process models vary in respect

© Springer International Publishing AG 2018
E. Teniente and M. Weidlich (Eds.): BPM 2017 Workshops, LNBIP 308, pp. 446–459, 2018.
https://doi.org/10.1007/978-3-319-74030-0_35

to their quality and level of granularity. As a consequence, these models face a wide range of challenges affecting model comprehensibility and error probability. However, the comprehensibility of process models is crucial for enterprises to enable an overall understanding of respective processes for all involved actors.

A process model may be represented either as a textual or a graphical documentation, whereas the latter provides specific advantages compared to textual descriptions [1]. Focusing on the graphical specification of processes, there exist various modeling languages like, for example, *EPCs* [2], *BPMN* [3], and *Flow Chart* [4]. Each of these languages is defined through its syntax and contains a set of graphical symbols for documenting processes. Many studies have shown that the use of graphical symbols foster process model comprehension [5].

Putting an emphasis on EPCs, this paper presents an experiment on process model comprehension using eye tracking. Usually, processes expressed in terms of EPCs consist of three different elements, i.e., functions, events, and logical connectors. In general, an EPC is a chain of alternating events and functions. Their specification, in turn, primarily describes the business logic of the process, thus having a positive impact on process model comprehension [2].

The presented experiment is part of a series of experiments using a conceptual framework that incorporates concepts from *cognitive neuroscience* and *psychology* for process model comprehension [6]. Regarding the influence of process modeling expertise on model comprehension, the results are similar compared to a priorly conducted experiment addressing BPMN process models. However, the results additionally revealed that process models expressed in terms of EPC are perceived differently than BPMN models. Due to the fact that the obtained results may be considered as preliminary, however, they provide promising insights with respect to the reading and comprehension of process models.

The paper is organized as follows: Sect. 2 describes the context of the experiment. The experimental setting and operation is introduced in Sect. 3. In Sect. 4, the obtained results are analyzed and hypotheses are tested for statistical significance. Finally, Sect. 5 discusses related work and Sect. 6 summarizes the paper.

2   Context of the Experiment

The complex biological and cognitive processes in the head of an individual, whether being conscious or subconscious, ultimately decide how the environment is perceived and, hence, influence the decisions on the further actions to be taken [7]. In the domain of process modeling, the application of *cognitive neuroscience* and *psychology* entails auspicious prospects [8–10]. Focusing on issues related to the comprehension of process models, we currently conduct a series of experiments, utilizing advantages of different concepts from cognitive neuroscience and psychology. Among others, we strive for a comparison between existing process modeling languages to yield the perceived pros and cons of respective languages. In order to achieve these objectives, we make use of a conceptual framework we developed [6]. Table 1 presents concepts, for which we already conducted experiments using the conceptual framework. Furthermore,

Table 1 illustrates the number of involved subjects and analyzed process models. Table 2, in turn, depicts the evaluated process modeling languages and relates them to the mentioned concepts, these languages have been evaluated with. In more detail, we observed and measured changes in the *electrodermal activity* as well as in the *heart rate* of subjects while confronting them with different modeling related issues (e.g., level of difficulty, used process modeling languages) with respect to the process models [11,12]. Furthermore, for maybe lowering the needed amount of mental effort and to may reduce the perceived difficulty, while reading and comprehending process models, we applied the *Cognitive Load Theory* and the *Construal Level Theory* in this context [13,14].

Table 1. Experiments using the conceptual framework

Concepts	No. Subjects	No. Models
Cognitive Load Theory	52	588
Construal Level Theory	136	262
Eye Tracking	41	492
Electrodermal Activity	8	128
Heart Rate	7	168

Table 2. Process modeling languages evaluated with the conceptual framework

Concepts	BPMN	EPK	Petri net	eGantt	Flow chart	UML AD
Cognitive Load Theory	●	●	●	●	●	●
Construal Level Theory	●					
Eye Tracking	●	●	●	●	●	
Electrodermal Activity	●	●	●		●	●
Heart Rate	●	●	●		●	●

3   Experimental Setting

The expertise in process modeling might be a factor influencing the comprehension of process models. This leads us to the following research question:

> **Research Question**
>
> *Does expertise in the domain of process modeling has a positive effect on reading and comprehending process models expressed in terms of EPCs?*

To investigate this research question, an experiment using eye tracking is conducted. Note that the experiment is conducted as a quasi-experiment since the subjects are assigned by judgment with respect to the level of expertise a subject has in process modeling [15]. Generally, eye tracking is a reliable measurement method to determine differences between subjects while comprehending a visual stimulus (e.g., picture) [16]. Furthermore, the use of eye tracking reveals insights about cognition as well as cognitive development of a subject and provides non-invasive indices of brain functions. By using this measurement method, we want to obtain insights into how subjects are comprehending process models, while, for example, monitoring attention shifts. In our context, the eye movements of the participating subjects were recorded, while three EPC process models had to be comprehended, along with a set of comprehension questions. In the experiment, the recorded types of eye movements were *fixations*, *saccades*, and *gaze paths* [17]. Fixations are very slow eye movements at a specific point in a stimulus, whereas saccades represent fast eye movements. To be more precise, a saccade constitutes a change of fixation of the eyes in a stimulus. The chronological order of fixations and saccades creates a gaze path. It is found in eye tracking experiments that experts comprehending a stimulus are more likely to have a smaller number of fixations, saccades, and consequently a shorter gaze path length compared to novices [18].

3.1   Hypothesis Formulation

The following six hypotheses were derived to investigate whether or not expertise in process modeling has a positive impact on process model comprehension. More precisely, we focus on the question whether intermediates (i.e., individuals with more expertise in process modeling) are more effective regarding the comprehension of process models expressed in terms of EPCs compared to novices:

$H_{0,1}$:	Intermediates need not less duration time for process model comprehension compared to novices.
$H_{1,1}$:	Intermediates need significantly less duration time for process model comprehension compared to novices.
$H_{0,2}$:	Intermediates do not achieve a better score for answering the questions compared to novices.
$H_{1,2}$:	Intermediates achieve a significantly better score for answering the questions compared to novices.
$H_{0,3}$:	Intermediates do not have a better response time for answering the questions compared to novices.
$H_{1,3}$:	Intermediates have a significantly better response time for answering the questions compared to novices.
$H_{0,4}$:	Intermediates do not have less fixations in process model comprehension compared to novices.
$H_{1,4}$:	Intermediates have significantly less fixations in process model comprehension compared to novices.
$H_{0,5}$:	Intermediates do not have less saccades in process model comprehension compared to novices.
$H_{1,5}$:	Intermediates have significantly less saccades in process model comprehension compared to novices.
$H_{0,6}$:	Intermediates do not have a shorter gaze path in process model comprehension compared to novices.
$H_{1,6}$:	Intermediates have a significantly shorter gaze path in process model comprehension compared to novices.

3.2   Experimental Setup

This section describes the subjects and objects of the experiment, together with the independent and dependent variables.

Subjects. There were no prerequisites for participating in the experiment. Therefore, subjects with diverse backgrounds (i.e., students, academics, and professionals) were invited to participate in the experiment. Subjects were informed that the experiment takes place in the context of process model comprehension and anonymity was guaranteed for all subjects. As done in other scientific fields, the categorization into groups (i.e., intermediates and novices), in turn, was accomplished by a median split, i.e., based on time spent on process modeling provided by self-reporting of the subjects.

Objects. In the experiment, three process models reflecting different levels of model difficulty (i.e., easy, medium, and hard) are presented to the subjects. In particular, the used process models were created in collaboration with several experts in the domain of process modeling. The models were expressed in terms of *EPCs*. The easy process model comprises only basic modeling elements (i.e., events and functions). With rising level of difficulty, new EPC elements were introduced and the total number of elements was increased. The eye movements were recorded throughout these comprehension tasks. After comprehending a process model, four *true-or-false* comprehension questions, referring solely to the scenario semantics, had to be answered by the subjects. The questions were used to evaluate whether or not the process models were correctly interpreted.[1] In addition, experts and novices in the domain of process modeling, who were not participating in the experiment, ranked the used process models with respect to their level of difficulty. Moreover, they were asked to compare the EPC models with the priorly used BPMN models to ascertain a comparability between these two modeling languages [6].

Independent variables. The experiment contains two independent variables; i.e., the ① *level of difficulty* and ② *expertise level* of subjects.

Dependent variables. Regarding the level of difficulty, the dependent variables include the ① *duration* needed for comprehending a process model, ② *achieved score* based on the comprehension questions, and ③ needed *response time* for answering the questions. In the context of eye tracking, the dependent variables include the ④ *number of fixations*, ⑤ *number of saccades*, and ⑥ *length of the gaze path*. Figure 1 summarizes the research model of the experiment.

In general, since the experiment consists of pure comprehension tasks, the results may be considered as preliminary, i.e., their generalization needs to be confirmed by additional experiments.

---

[1] Material downloadable from: https://www.dropbox.com/sh/th6wc0761ajlxcw/AABs_LXE8mh-ufzSp95lT66za?dl=0.

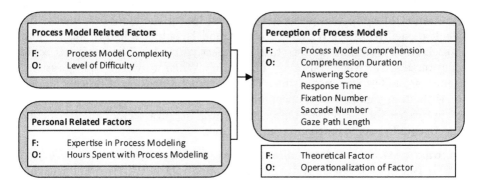

Fig. 1. Research model of the experiment

3.3 Experimental Design

The experimental setting is based on the guidelines set out by [19]. The experiment was conducted in a designated lab at Ulm University. Prior to the experiment, three pilot studies with 12 subjects were conducted to improve the experimental design and material as well as to eliminate potential ambiguities, e.g., optimization of the used process models. Figure 2 shows the procedure of the experiment: ① An introduction was given, ② subjects had to sign a consent form, and ③ demographic data (e.g., expertise in process modeling in general, familiarity with particular modeling languages) were collected. Subsequently, ④ the eye tracking appliance was calibrated and ⑤ subjects completed a tutorial in order to familiarize them with the functionality of the eye tracker and the procedure of the experiment. Therefore, an exemplary task based on the actual experiment was shown to the subjects. The experiment could be done either in English or German. After completing these mandatory steps, ⑥ subjects needed to comprehend three EPC process models. First, the process model reflecting an

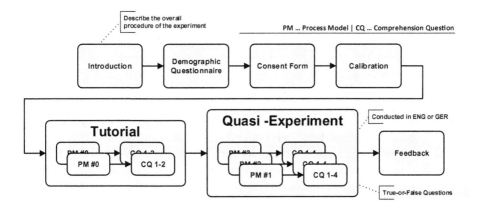

Fig. 2. Experimental design

easy level of difficulty was presented, followed by the medium and, finally, the difficult model. After studying each of the models, subjects had to answer four comprehension questions related to the respective model. The questions could be answered with 'true', 'false', or 'uncertain'. When answering the questions, the process models were not visible. The pure comprehension of process models (i.e., without any guidance) is uncommon, but we wanted to deliberately disclose the approaches for the pure comprehension on EPC process models. ⑦ Finally, subjects could provide textual or oral feedback.

Instrumentation and data collection procedure. For eye tracking, we used the SMI iView X Hi-Speed system at a sampling rate of 240 Hz.[2] The tracking appliance was placed in front of a monitor that provides the process models to subjects. Subjects used a keyboard with three predefined keys providing the options for answering the comprehension questions. Eye tracking data collected during the experiment were analyzed, visualized, and exported with SMI BeGaze software [20]. Demographic data and qualitative feedback were gathered with questionnaires.

3.4   Data Validation

Overall, data from 36 subjects (12 female participants) were collected, i.e., 20 students, 12 academics, and 4 professionals participated. Specifically, 19 of them were computer scientists; additionally, 4 psychologists, 4 economists, and 4 social workers participated. Finally, 5 subjects did not provide a statement. The median of hours spent for process modeling, which is used to divide the participants in intermediates and novices, was 20.5 h. This resulted in a number of 21 intermediates and 15 novices. Furthermore, eye tracking data from 4 subjects were excluded due to invalidity, i.e., eye movements were not recorded properly. Finally, the group of novices included 14 subjects and the one of intermediates consisted of 18 subjects.

4   Data Analysis and Interpretation

Table 3 presents mean and standard deviation (i.e., STD) for novices and intermediates. It shows the process model comprehension duration (in ms) as well as the achieved answering scores. Thereby, specific values to each answering option were assigned, i.e., 'true' = 1, 'false' = -1, and 'uncertain' = 0. Furthermore, response times for answering related questions (in ms), number of fixations as well as number of saccades and, finally, length of the gaze path (in px) are listed in Table 3 (i.e., theoretical factor and operationalization of factor).[3]

Generally, all values, except the answering scores, increase with rising level of difficulty, as expected by us. For the process model with an easy level of

---

[2] http://www.smivision.com/en/gaze-and-eye-tracking-systems/products/iview-x-hi-speed.html.

[3] Sample images downloadable from: www.dropbox.com/sh/th6wc0761ajlxcw/AABs_LXE8mh-ufzSp95lT66za?dl=0.

Table 3. Obtained experimental results

		Theoretical Factor	Operation. of Factor	Both Mean	STD	Novices Mean	STD	Intermediates Mean	STD
Difficulty	Easy	Comprehension	Duration	36398	23034	43481	29257	28304	7980
			Score	3.57	0.82	3.19	0.98	4	0
			Resp. Time	4933	1326	5254	1497	4575	1039
		Eye Tracking	Fixations	104	57	120	71	85	26
			Saccades	90	44	100	56	79	22
			Gaze Path	15149	12081	17927	15599	11974	4948
Difficulty	Medium	Comprehension	Duration	54360	18325	58228	21385	49940	13490
			Score	2.63	1.4	2.81	1.22	2.43	1.6
			Resp. Time	7985	2138	7931	1843	8045	2504
		Eye Tracking	Fixations	171	44	178	51	164	34
			Saccades	154	39	156	48	151	28
			Gaze Path	26197	7421	26666	8990	25661	5385
Difficulty	Hard	Comprehension	Duration	90355	30750	98771	36306	80737	20039
			Score	1.73	1.57	1.88	1.5	1.57	1.7
			Resp. Time	8358	2719	8528	3307	8163	1949
		Eye Tracking	Fixations	279	107	299	136	256	57
			Saccades	253	101	261	126	243	64
			Gaze Path	40511	17599	43566	21957	37020	10491

difficulty, the results reveal that intermediates are more effective in terms of process model comprehension compared to novices. The comprehension duration is shorter and, in average, intermediates made less mistakes in answering the questions. Furthermore, they needed less fixations and saccades, resulting in a shorter gaze path. Concerning the process model with the medium and the one with the highest difficulty, the results between novices and intermediates are approaching a similar level and only slight differences are observable. It appears to be that novices and intermediates perform equally regarding the comprehension of EPC process models. However, it is noteworthy that the results do not differ significantly considering the fact that few novices have had no experience in EPCs at all. Figures 3, 4, 5 and 6 show selected results of the experiment. Figure 3 indicates that the time needed for process model comprehension increases with rising level of difficulty. Figure 4 illustrates that the answering scores are decreasing with rising level of difficulty. Moreover, for the process models with the medium and highest level of difficulty, novices achieve a slightly better score. The response times for answering the questions increase with rising level of difficulty (cf. Fig. 5). Especially between the easy and medium process model, a difference is discernible. The fixation number for intermediates is always lower than the number for novices (cf. Fig. 6). Altogether, EPC process models seem to be fairly comprehensible without previous knowledge.

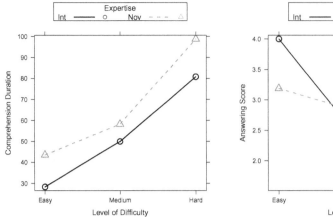

Fig. 3. Comprehension duration

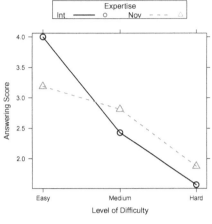

Fig. 4. Answering score

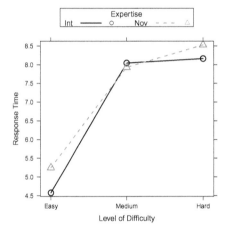

Fig. 5. Response time

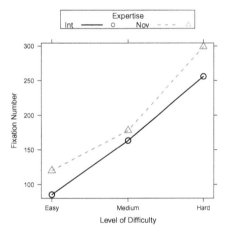

Fig. 6. Fixation number

4.1   Hypotheses Testing

The stated hypotheses (cf. Sect. 3.2) are tested for statistical significance using a *Student's t-test* (cf. Table 4). Particularly, we refer to the rule of thumb, according to the literature, that the use of the t-test is appropriate at a sample size of $n > 30$ [21]. In this context, a successful t-test (with $p < p_0$ at risk level $\alpha = 0,05$) will reject a null hypothesis [22]. Only for hypothesis $H_{1,2}$ for the easy process model, a significant result emerged. In addition, tendencies are discernible in $H_{1,1}$ and $H_{1,4}$. However, the results confirm our observations from the first experiment that process model comprehension not necessarily correlates with expertise in this domain [6]. This is also evident in the results obtained from the process models with a higher level of difficulty.

4.2   Threats to Validity

Any experiment bears risks and, hence, its levels of validity need to be checked and discussed. The level of difficulty reflected by the respective process models constitutes a threat to validity. The gap between the single difficulties might be not alike. Furthermore, the comprehension of process models without any concrete task or guidance is uncommon in practice. Moreover, process models must be memorized for answering the comprehension questions. Consequently, the risk arises that process models are wrongly memorized. Furthermore, process scenarios are perceived differently based on various factors (i.e., familiarity). Therefore, the considered scenarios constitute an additional risk. Finally, splitting and defining novices and intermediates, based on the hours spent on process modeling, is another threat to validity. Obviously, an individual with a high number of hours spent with process modeling can be considered as an expert, but it is questionable whether 20.5 h are sufficient to denote an intermediate. Moreover, the representativeness of the results is limited by the relatively small sample sizes. The sizes of the samples also limit the statistical power and there might be significant differences between novices and intermediates, which we could not show in this experiment, but which might become apparent in larger samples. Regarding the statistics, it has to be mentioned that multiple t-tests were performed and no correction for multiple testing was applied. We justify this by the fact that this experiment was an explorative one instead of an experiment aiming to replicate findings of a previous experiment on EPC.

4.3   Discussion and Comparison with Prior Results

Prior to this experiment, a similar experiment (i.e., same setting and operation) was conducted using process models expressed in terms of *BPMN 2.0* [6]. Furthermore, special care was taken that the used process models are comparable between the two modeling languages as well as their levels of difficulty.

Table 4. Hypotheses testing results

Theoretical factor	Operationalization of factor	Level of difficulty		
		Easy (p)	Medium (p)	Hard (p)
Comprehension	$H_{1,1}$ - Duration	0.062	0.210	0.100
	$H_{1,2}$ - Score	0.005*	0.473	0.610
	$H_{1,3}$ - Resp. time	0.162	0.890	0.712
Eye Tracking	$H_{1,4}$ - Fixations	0.084	0.365	0.262
	$H_{1,5}$ - Saccades	0.181	0.714	0.631
	$H_{1,6}$ - Gaze path	0.165	0.710	0.300

The main differences between both experiments were the various process scenarios captured in the used process models. However, prior experiments showed that a familiar or unfamiliar scenario does not have an influence on the comprehension of process models [23]. In general, two common effects could be observed in both experiments with respect to process model comprehension. First, the performance of subjects decreases with rising level of difficulty and, second, the performance of novices and intermediates approximates with each other as well with rising level of difficulty. It is noteworthy that the overall performance of subjects confronted with EPC process models indicates better results compared to BPMN models. Descriptively, comprehension duration and response times for answering related questions are almost the same, but the final answering scores are substantially higher for EPCs, despite expertise in process modeling and level of difficulty. It appears that subjects cope better with EPC process models than BPMN models. However, in the end, the stated observations need to be investigated by inferential statistics and by further research either through replication or similar experiments. Finally, based on a set of stated categories that can foster experiments on process model comprehension, and with the use of the conceptual framework, further experiments will be subject of future work [23].

5   Related Work

With a focus on process model comprehension, [24] evaluates different process modeling languages, whereas [25] presents factors that influence the comprehension of process models. The influence of process model complexity on related model comprehensibility is investigated in [26]. In [27], factors that affect the comprehension of process models are discussed. A state-of-the-art report on empirical research on process model comprehension can be found in [28].

Regarding cognitive aspects in process modeling, [29] discusses how the comprehension of conceptual models is influenced by a reduced cognitive load. [30] shows the difficulty of comprehending different relations between the elements in a process model. [31], in turn, discusses principles for the design of a cognitively effective visual modeling language, whereas [32] attempts to operationalize perceptual properties of modeling languages to improve their cognitive effectiveness. Moreover, [9] discusses individual preferences for process representations based on their cognitive style.

In line with eye tracking, [33] concludes that a higher mental effort can be measured by the change of pupil dilation for task-based process models. Furthermore, [34] presents results on how eye tracking leads to a more fine-grained understanding of process models. [35] studies the factors that influence process model comprehension using eye tracking.

Comparing process models expressed in terms of BPMN and EPC respectively, [36] describes an experiment in which subjects ranked the modeling languages according to their subjective comprehension difficulty. The results conclude that BPMN models are easier to comprehend. Finally, [37] investigates the differences between BPMN and EPC process models, yielding that no final message can be made regarding a positive perception.

6   Summary and Outlook

This paper investigates whether or not expertise in the domain of process modeling has a positive impact on process model comprehension. Particularly, an eye tracking experiment, using process models in terms of EPCs and related results were presented. The preliminary results indicate that expertise in process modeling not necessarily implies a better comprehension of process models. In detail, intermediates as well as novices are struggling similarly once process models exceed a certain level of difficulty. The obtained results are in line with results from a priorly conducted experiment using BPMN process models. However, it appears that EPC process models are easier to read and comprehend. Further, with broader and more detailed investigations, we attempting to confirm respective generalization. The presented experiment is part of a series of experiments, in which a conceptual framework is used that incorporates methods and theories from cognitive neuroscience and psychology [6]. Thereby, eye tracking constitutes only one measurement method from the pool of existing methods in this context. By using the conceptual framework, additional concepts from neuroscience and psychology (e.g., electrodermal activity, Cognitive Load Theory) will be used in future experiments with a particular focus on how to foster the reading and comprehension of process models.

References

1. Ottensooser, A., Fekete, A., Reijers, H.A., Mendling, J., Meicstas, C.: Making sense of business process descriptions: an experimental comparison of graphical and textual notations. J. Syst. Softw. **85**, 596–606 (2012)
2. van der Aalst, W.M.P.: Formalization and verification of event-driven process chains. Inf. Soft Tech. **41**(10), 639–650 (1999)
3. OMG: Business Process Management & Notation 2.0 (2017). www.bpmn.org. Accessed 27 Feb 2017
4. Schultheiss, L.A., Heiliger, E.: Techniques of flow-charting. In: Proceedings of 1963 Clinic on Library Applications of Data Processing, pp. 62–78 (1963)
5. Johansson, L.O., Wärja, M., Carlsson, S.: An evaluation of business process model techniques, using Moodys quality criterion for a good diagram. In: CEUR Workshop Proceedings, vol. 963 (2012)
6. Zimoch, M., Pryss, R., Probst, T., Schlee, W., Reichert, M.: Cognitive insights into business process model comprehension: preliminary results for experienced and inexperienced individuals. In: Reinhartz-Berger, I., Gulden, J., Nurcan, S., Guédria, W., Bera, P. (eds.) BPMDS/EMMSAD -2017. LNBIP, vol. 287, pp. 137–152. Springer, Cham (2017). https://doi.org/10.1007/978-3-319-59466-8_9
7. Schwarz, N.: Emotion, cognition, and decision making, pp. 433–440 (2000)
8. Zugal, S., Pinggera, J., Weber, B.: Assessing process models with cognitive psychology. In: EMISA, vol. 190, pp. 177–182 (2011)
9. Figl, K., Recker, J.: Exploring cognitive style and task-specific preferences for process representations. Requir. Eng. **21**(1), 63–85 (2014)
10. Recker, J., Reijers, H.A., van de Wouw, S.G.: Process model comprehension: the effects of cognitive abilities. Learn. Style Strategy **34**, 199–222 (2014)

11. Prokasy, W.: Electrodermal Activity in Psychological Research. Elsevier, Amsterdam (2012)
12. Camm, A.J., et al.: Heart rate variability: standards of measurement. Physiol. Interpretation Clin. Use **93**, 1043–1065 (1996)
13. Sweller, J., Ayres, P., Kalyuga, S.: Cognitive Load Theory. Springer, New York (2011)
14. Trope, Y., Liberman, N.: Construal-level theory of psychological distance. Psychol. Rev. **117**, 440–463 (2010)
15. Cook, T.D.: Quasi-Experimental Design. Wiley, Hoboken (2015)
16. Gegenfurtner, A., et al.: Expertise differences in the comprehension of visualizations: a meta-analysis of eye-tracking research in professional domains. Educ. Psychol. Rev. **23**(4), 523–552 (2011)
17. Salvucci, D.D., Goldberg, J.H.: Identifying fixations and saccades in eye-tracking protocols. In: Proceedings of 2000 Symposium on Eye Tracking Research & Application, pp. 71–78 (2000)
18. Raney, G.E., Campbell, S.J., Bovee, J.C.: Using eye movements to evaluate the cognitive processes involved in text comprehension. J. Vis. Exp. **10**(83), e50780 (2014)
19. Wohlin, C., Runeson, P., Höst, M., Ohlsson, M.C., Regnell, B., Wesslen, A.: Experimentation in Software Engineering - An Introduction. Kluwer, Norwell (2000)
20. SMI: iView X Hi-Speed (2016). http://www.smivision.com/en/gaze-and-eye-tracking-systems/products/iview-x-hi-speed.html. Accessed 27 Feb 2017
21. Hogg, R.V., Tanis, E.A.: Probability and Statistical Inference. Macmillan, New York (1977)
22. Sirkin, M.: Statistics for the Social Sciences, vol. 3. Sage, Thousand Oaks (2005)
23. Zimoch, M., Pryss, R., Schobel, J., Reichert, M.: Eye tracking experiments on process model comprehension: lessons learned. In: Reinhartz-Berger, I., Gulden, J., Nurcan, S., Guédria, W., Bera, P. (eds.) BPMDS/EMMSAD -2017. LNBIP, vol. 287, pp. 153–168. Springer, Cham (2017). https://doi.org/10.1007/978-3-319-59466-8_10
24. Kiepuszewski, B., ter Hofstede, A.H.M., Bussler, C.J.: On structured workflow modelling. In: Wangler, B., Bergman, L. (eds.) CAiSE 2000. LNCS, vol. 1789, pp. 431–445. Springer, Heidelberg (2000). https://doi.org/10.1007/3-540-45140-4_29
25. Melcher, J., Mendling, J., Reijers, H.A., Seese, D.: On measuring the understandability of process models. In: Rinderle-Ma, S., Sadiq, S., Leymann, F. (eds.) BPM 2009. LNBIP, vol. 43, pp. 465–476. Springer, Heidelberg (2010). https://doi.org/10.1007/978-3-642-12186-9_44
26. Mendling, J., Reijers, H.A., Cardoso, J.: What makes process models understandable? In: Alonso, G., Dadam, P., Rosemann, M. (eds.) BPM 2007. LNCS, vol. 4714, pp. 48–63. Springer, Heidelberg (2007). https://doi.org/10.1007/978-3-540-75183-0_4
27. Mendling, J., Strembeck, M., Recker, J.: Factors of process model comprehension-findings from a series of experiments. Decis. Support Syst. **53**(1), 195–206 (2012)
28. Figl, K.: Comprehension of procedural visual business process models-a literature review. Bus. Inf. Syst. Eng. **59**, 41–67 (2017)
29. Moody, D.L.: Cognitive load effects on end user understanding of conceptual models: an experimental analysis. In: Benczúr, A., Demetrovics, J., Gottlob, G. (eds.) ADBIS 2004. LNCS, vol. 3255, pp. 129–143. Springer, Heidelberg (2004). https://doi.org/10.1007/978-3-540-30204-9_9

30. Figl, K., Laue, R.: Cognitive complexity in business process modeling. In: Mouratidis, H., Rolland, C. (eds.) CAiSE 2011. LNCS, vol. 6741, pp. 452–466. Springer, Heidelberg (2011). https://doi.org/10.1007/978-3-642-21640-4_34
31. Moody, D.: The "Physics" of notations: toward a scientific basis for constructing visual notations in software engineering. Trans. Softw Eng. **35**(6), 756–779 (2009)
32. van der Linden, D., Zamansky, A., Hadar, I.: How cognitively effective is a visual notation? On the inherent difficulty of operationalizing the physics of notations. In: Schmidt, R., Guédria, W., Bider, I., Guerreiro, S. (eds.) BPMDS/EMMSAD -2016. LNBIP, vol. 248, pp. 448–462. Springer, Cham (2016). https://doi.org/10.1007/978-3-319-39429-9_28
33. Dobesova, Z., Malcik, M.: Workflow diagrams and pupil dilatation in eye tracking testing. In: Proceedings of 13th International Conference on Emerging eLearning Techniques & Applications, pp. 59–64 (2015)
34. Hogrebe, F., Gehrke, N., Nüttgens, M.: Eye tracking experiments in business process modeling: agenda setting and proof of concept. In: Proceedings of 4th International Workshop on Enterprise Modelling and Information Systems Architectures, pp. 183–188 (2011)
35. Petrusel, R., Mendling, J.: Eye-tracking the factors of process model comprehension tasks. In: Salinesi, C., Norrie, M.C., Pastor, Ó. (eds.) CAiSE 2013. LNCS, vol. 7908, pp. 224–239. Springer, Heidelberg (2013). https://doi.org/10.1007/978-3-642-38709-8_15
36. Gabryelczyk, R., Jurczuk, A.: The diagnosis of information potential of selected business process modelling notations. Inf. Syst. Manag. **4**, 26–38 (2015)
37. Recker, J.C., Dreiling, A.: The effects of content presentation format and user characteristics on novice developers understanding of process models. Commun. Assoc. Inf. Syst. **28**, 65–84 (2011)

Knowledge-intensive Process: A Research Framework

Flavia Maria Santoro[✉] and Fernanda Araujo Baião

Federal University of the State of Rio de Janeiro, Avenida Pasteur, 458,
Rio de Janeiro, RJ 22290-400, Brazil
{flavia.santoro,fernanda.baiao}@uniriotec.br

Abstract. Great value is now being credited to the so-called Knowledge-intensive Processes (KiP) benefiting from the advent and proliferation of social media, smart devices, real-time computing, and technologies for big data. Our research investigates the origin, formalization, and support for KiP towards what we call a Knowledge-intensive Process-Aware Information System (KiPAIS). We propose a research framework to address the following challenges, aligned with the pillars of the CBPM due to intrinsic relationships among them: (1) eliciting and discovering KiP; (2) representation and support to the implementation of KiP; (3) formal theory capable of explaining KiP; (4) measuring the performance of KiP.

Keywords: Knowledge-intensive Process · Cognitive BPM

1 Introduction

The focus of Business Process Management (BPM) research and development was for a long time on structured processes (represented by imperative models, such as BPMN), supported directly by Process-Aware Information Systems (e.g., Business Process Management Systems - BPMS). Nowadays, however, processes are in many cases supported by a variety of applications, which can also provide data in event logs but are not process-aware. Business processes are increasingly being conducted by organizations and customers networked on social media platforms and enabled by mobile devices. So, there is a need to integrate different kinds of data sources to obtain information on the performance and compliance of such processes, as well as take proactive or corrective measures to improve them. Therefore, great value is now being credited to poorly structured processes, or the so-called Knowledge-intensive Processes (KiP), benefiting from the advent and proliferation of social media, smart devices, real-time computing and technologies for big data.

Our research investigates the origin, formalization, support and management of KiP concerning the term defined in this paper as *Knowledge-intensive Process-Aware Information System (KiPAIS)*. Accordingly, it is necessary to analyze existing data volumes from a variety of sources (including stories, e-mail repositories, sensor monitoring data, blogs and social networks) to extract and generate knowledge that can contribute to a better understanding of the events carried out (together with the context in which they were executed) and consequent modeling of a KiP in different perspectives, besides

© Springer International Publishing AG 2018
E. Teniente and M. Weidlich (Eds.): BPM 2017 Workshops, LNBIP 308, pp. 460–468, 2018.
https://doi.org/10.1007/978-3-319-74030-0_36

providing technological support to these processes. We argue that the sharing of such knowledge may result in valuable benefits to the people and organizations involved with the processes.

In addition to addressing the volume and heterogeneity (not only in the syntactical and structural levels, but also - and more important - in the semantic level) of the data, new ways of accessing the data on the Web must be considered, since current tools are mainly focused on structured data management. These issues require new research efforts towards increasing semantic precision for defining and modeling a KIP, allowing access to unstructured and contextual data that may be used as data sources for discovering a KIP, as well as measuring its performance to improve the analysis and decision-making processes. In this whole context, Cognitive Computing (as emphasized by [9]) can bring benefits from several perspectives, and might further encourage the establishment of a new BPM paradigm.

In this scenario, we propose a research framework to address specifically the following challenges: (1) eliciting and discovering KiP, and consequently defining which information is relevant to the process; (2) representing and supporting the implementation of KiP, since traditional platforms do not meet the needs of flexibility; (3) specifying a formal theory capable of explaining KiP; (4) defining a system of appropriate indicators to measure the performance of KiP, in-line with its specific characteristics.

This paper relates our research framework with Cognitive BPM, summarizing the results already obtained in light of the framework proposed by Hull and Nezhad [9], and discusses open issues and future research perspectives.

2   Research Background

2.1   Knowledge-intensive Process

The management of Knowledge-intensive Process (KiP) is an emerging field within the Business Process Management area. According to DiCiccio *et al.* [5], Knowledge-intensity in business process is characterized by the presence of collaborative interactions among participants and flexibility to perform work, making the process less predictable than a structured routine. Moreover, KiPs are processes "whose conduction and execution are heavily dependent on knowledge workers performing several interconnected knowledge intensive decision-making tasks" [5].

A KiP is essentially goal-oriented, and typically collaborative, unpredictable, not repeatable, and strongly guided by events, constraints and rules. In addition, Marjanovic and Freeze [12] investigated the relevance of knowledge creation in a KiP and argued that the expansion and use of knowledge among organizations depend on formal and informal social processes through effective communication. Examples of KiP are customer support, new product/service design, marketing, data quality management, IT governance, strategic planning. In such scenarios, existing contextual data may pose a higher influence than regular normative power in guiding the flow of activities; moreover, social interactions among stakeholders also interfere in the flow of a process, allowing (sometimes even stimulating) variants to emerge.

2.2   Cognitive BPM

Hull and Nezhad [9] state that Cognitive Computing (CC) "*can accelerate the arrival of the next generation in BPM, by enabling the development of a fundamentally new family of process abstractions that will support much richer, more adaptive, more proactive, and more user-friendly styles of process coordination*". They highlight KiP as a scenario for Cognitive BPM and explained, for example, that the separation of the process model and its instances is too restricting for cognitively-rich KiP. The authors propose a framework for a Cognitive BPM (CBPM), which is founded both on traditional BPM and Case Management contexts, as well as the new Cognitive Process Abstractions, and likewise is composed by 4 pillars:

(i)   Cognitive Decision Support: CC will enable an increase in the quantity and breadth of human decisions based on an enormous volume of different types of data;
(ii) Cognitive Interaction: CC might improve interactions within processes by providing new channels and devices (including participation of cognitive agents);
(iii) Cognitive Process Learning: CC can benefit capturing and codifying process specifications, to support flexible automation;
(iv) Cognitive Process Enablement: Different types of business processes should be supported, in which the underlying process model is event-driven, and focused on ongoing goal formation, learning of relevant knowledge including constraints, planning and decision-making.

Furthermore, the authors also indicate that an appropriate process meta-model for CBPM will be based on a Plan-Act-Learn cycle. In this cycle, plans and decisions may lead to world side-effecting actions, and to learning activities, which in turn will feed into an ever-expanding knowledge base. This knowledge base could also be improved by events from the environment, and environmental reactions to process actions. And the cycle is closed once the knowledge base might lead to further decisions, goals, and plans. They argue that the high variability of Plan-Act-Learn-based process instances (which is an essential characteristic of KiP) demands new perspectives of how to support traditional BPM capabilities such as monitoring, auditing, and improvements through analytics based on history.

3   A Research Framework for KiP

We propose a research framework on KiP, which provides a basis for the lines of investigation in this domain. Although those lines could be developed independently, they all converge towards establishing the notion and components of Knowledge-intensive Process-Aware Information Systems. Those challenges are aligned with the pillars of the CBPM due to intrinsic relationships among them: (1) eliciting and discovering KiP are concerned to Cognitive Process Learning; (2) representing and supporting the implementation of KiP relate to Cognitive Process Enablement; (3) defining a formal theory to explain KiP is associated to Cognitive Process Abstractions and Cognitive Process

Interaction; and (4) defining a system of appropriate indicators to measure the perform-ance of KiP is linked to Cognitive Process Decision-Support. Our proposals and results are summarized according to these relationships.

3.1   Cognitive Process Abstractions

Since human knowledge and involvement are key to KiP execution [10], diverse elements beyond traditional workflow-oriented processes arise, such as beliefs, inten-tions, desires, feelings, decisions, collaboration, and contingency events. Given that the representation of knowledge-intensive aspects is far from trivial [6], the Knowledge-intensive Process Ontology (KiPO) [6] was proposed to identify all aspects involved within a KiP. KiPO is a well-founded task ontology [8] with definitions that enable a precise interpretation and a deeper exploration of all relevant concepts comprised within a KiP.

KiPs are complex and human-centered; thus, they generate value through the exchange of knowledge among participants, often involving decision-making tasks with different alternatives for the next step in the process flow. For this reason, the human factor is the main source of complexity, especially due to the difficulty of modeling its behavior. In this challenge, two key factors are explored: (i) the difficulty of under-standing the human factor, combining the advances of related research fields (such as Philosophy and Psychology) in a coherent theory to explain human behavior within a KiP, focusing on the concepts of Belief, Desire and Intention and their role in human action; (ii) a comprehensive semantic conceptualization, based on solid foundations provided by the Unified Fundamental Ontology (UFO) [8], providing the foundation to define a KiP with precise semantics, thus avoiding issues such as conceptual ambiguity and enabling its application in both modeling, discovery and execution support of a KiP. We described a formal specification of the Collaboration view of KiPO in [13].

In the most abstract level, we propose to address the problem of distinguishing instances and models by applying multi-level conceptual modeling [4] for representing elements with multiple classification levels, such as MLT (Multi-Level Theory) [4]. Moreover, we apply powertype patterns for representing KIP characterizations in KiPO [2]. In the visual representation level, we defined the Knowledge-intensive Process Notation (KIPN) [14], which addresses the representation of all relevant perspectives in KIPs, filling existing gaps in the literature with regard to integrating actors and roles into the definition of semi-structured processes, as stated by [5] as an important chal-lenge. KIPN provides adequate support for specifying collaboration and interactions among knowledge workers in the process enactment. Moreover, KIPN also concerns the understanding of the link between the evolution of data and the decision-making process during the execution of a KIP, as well as graphically presenting specific roles that workers interpret in the execution of activities.

3.2   Cognitive Process Learning

We investigate algorithms for knowledge discovery in structured logs, as well as in texts produced within collaborative tools. The KiP elements sought are aligned with KiPO:

collaboration, decision making, business rules, human aspects and objectives, basic flow of activities. Some results have already been reported in [3, 7, 21].

In [3], we concluded that some of the discovered decisions within a KiP are candidates to become business rules that might serve as strategic knowledge for the organization and support future decisions to be made. We used decision mining techniques to discover business rules within the flow of activities of a KiP associated with a log of textual messages exchanged by process participants. Previously, we investigated the application of NLP and Text Mining techniques on emails and histories told by participants, generating representations that partially explain a KiP [21].

3.3   Cognitive Interaction with Processes

Process participants perform activities and collaborate with each other, driven by their Beliefs, Desires and Intentions (BDI); therefore, the analysis of these elements is vital to the understanding, modeling and execution support of a KiP. In [18], we proposed a method based on Speech Act Theory [1] and Process Mining to discover the flow of speech acts related to BDI from event logs, and show how this relation fosters process performance analysis. When process participants interact through natural language, the three elements are present in communication, so we analyze human conversations, supported by the Speech Act Theory.

According to [20], an illocutionary act holds the pragmatics of an utterance and is characterized by a distinct illocutionary point. We argue that illocutionary points may be correlated to BDI, which opens a path to analyzing speech acts that may represent part of human knowledge and involvement in KiPs, as previously defined in KiPO [6]. KiPO comprises precise well-founded definitions of agents and interactions among them, and how the mental moments that are inherent to them (Beliefs, Desires, Intentions and Feelings) influence (or even drive) their decisions and the control-flow of the activities executed in each instance. The challenge addressed is the difficulty to analyze how human knowledge and involvement influence a KiP execution when this information is present only in unstructured natural language resources. The first results of this work may be learned in [18].

3.4   Cognitive Process Enablement

The computational support for the life cycle of a KiP is still an open issue [5], especially considering the Plan-Act-Learn cycle. However, most modern companies have systems that (at least partially) support the execution of KiPs. For example, in a health care setting, a patient's medical record may contain information about all events, decisions made, and people involved in the treatment over time. Because of inherent flexibility and unpredictability, instances of the same KiP may be different from each other, with no clear guideline standards for a single, complete model [9]. We argue that the set of KiP execution registers (KiP log) can be considered as a process model repository so that they can be properly maintained, analyzed and explored for long periods of time by various stakeholders [19]. Therefore, a KiP repository based on KiPO and physically stored on a NoSQL DBMS platform [11] has been implemented. This solution will be incorporated into the GCAdapt environment proposed in [15], which enables the execution of processes in a flexible way through dynamic adaptation, based on contextual information and a planning algorithm.

Context plays a fundamental role in this proposal. The flexible enactment of a KiP depends on its management, comprising modeling, capturing, analyzing, and continuously updating a context model for KiP. Thus, we developed a semi-automatic method to discover contextual elements associated to a KiP. The result is a decision tree that supports the choice of variables to be monitored, which determine the need for dynamic adaptation [17]. The evolution of this environment is also concerned.

3.5 Cognitive Decision Support for Processes

KiPs, as well as other types of business processes, need to be measured to continually improve performance. This is usually done by defining, calculating and evaluating Process Performance Indicators (PPI). Performance management has already been widely discussed in the context of structured business processes [16]. Existing solutions, however, are not directly applicable to KIPs because they are not able to measure their particular characteristics. Traditional structured business processes have a predefined behavior, including possible interactions between the different participants, but this is not the case in KIPs. Participants' behavior, their interactions and decisions are not known until the execution time. That is why, in addition to the kind of measures that are commonly used such as time, cost or quality, a new set of measures that explicitly refer to characteristics that play a significant role in the KiPs is needed, and therefore impact on their performance, such as collaboration between process participants, the explicit knowledge used or the constraints and rules that drive action and decision-making during the execution of the process.

4   Towards the Definition of KiPAIS

All the results presented in the previous sections compose the research framework on KiP and are the components of a generic architecture of a Knowledge-intensive Process-Aware Information System. A KiPAIS should allow modeling, running, and monitoring a KiP based on cognitive computing techniques. Figure 1 depicts the architecture of KiPAIS and highlights the support to the Plan-Act-Learn cycle [9]. The Work System Environment (WSE) embraces BPMS, Case-Based Management systems, but also integrates any collaborative system used by an organization. The process should be modeled with an adequate notation, such as KIPN. Within the WSE, contextual information about the running instances of a KiP is continuously captured through sensors, agents or services (Context Capturing Mechanisms). The Repositories of models and instances of KiP is stored in a NoSQL graph DBMS, using a KIPO-aware schema.

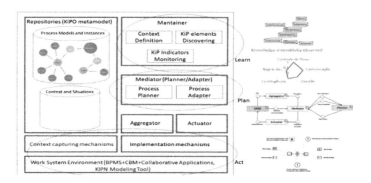

Fig. 1. KiPAIS architecture

The Mediator identifies the need for adaptation when it detects a situation that will prevent the process instance to achieve its goal. It uses intelligent behavior and decision-making support skills and is responsible for identifying possible adaptations during at runtime. When re-planning process instance, Mediator tries to fulfill goals and satisfy planning actions as its best achievement. It may find more than one possible adaptation, each of them satisfying goals in different degrees. The Actuator receives the decisions taken by the Mediator and triggers adaptations in the process instance through the Implementation Mechanisms. It involves sending commands to WSE to accomplish the necessary changes in the process instance. The Maintainer manages the context model to guarantee it will be always updated according to the current state of the KiP, and also implements the KiP PPI providing information for monitoring them, for example, in a dashboard. The right part of Fig. 1 shows some possible implementations for the Maintainer, Mediator and WSE.

5 Conclusions and Open Issues

This paper pointed research initiatives about KiP that ended up in the specification of KiPAIS, a new architecture to support the Plan-Act-Learn cycle of CBPM. Besides the results achieved so far, much work is still to be done. We list some items of an (open) agenda still based on the CBPM pillars:

- Abstractions: explore the possible associations among business rules and decision-making; establish formal relations between data elements in KiPO and domain ontologies; integrate KiPN with BPMN and commercial modeling tools;
- Learning: develop and test mining techniques to discover the diverse KiP elements; as well as perform case studies in real scenarios of big data settings;
- Interaction: investigate the BDI theory to define cognitive agents' behavior;
- Enablement: test planning algorithms to improve efficiency; implement a mechanism to capture contextual elements and complex situations, analyze multiple values and trends along time; apply mining techniques in a goal-oriented approach to continuously discover contextual elements that affect KiP;

- Decision-support: develop a method to support the definition of a system of PPI for KiP; relate the PPIs to elements of KiPO such as decision-making concepts.

References

1. Austin, J.L.: How to Do Things with Words. Oxford University Press, Oxford (1975)
2. Barboza, T., Baião, F.A., Santoro, F.M.: Applying Multi-level typing to Model Knowledge-intensive Processes. DSc and MSc Consortium on Ontologies, ONTOBRAS, Brasília (2017)
3. Campos, J.G., Richetti, P.H., Baião, F.A., Santoro, F.M.: Discovering business rules in knowledge-intensive processes through decision mining: an experimental study. In: Teniente, E., Weidlich, M. (eds.) BPM 2017 Workshops. LNBIP, vol. 308, pp. 554–565. Springer, Cham (2018)
4. Carvalho, V.A., Almeida, J.P.A., Fonseca, C.M., Guizzard, G.: Multi-level ontology-based conceptual modeling. Data Knowl. Eng. **109**, 3–4 (2017)
5. Di Ciccio, C., Marrella, A., Russo, A.: Knowledge-intensive processes: an overview of contemporary approaches. In: 1st International Workshop on Knowledge-Intensive Business Processes, pp. 33–47 (2012)
6. França, J.B.S., Netto, J.M., Carvalho, J.E.S., Santoro, F.M., Baião, F.A., Pimentel, M.: KIPO: the knowledge-intensive process ontology. Softw. Syst. Model. **14**(3), 1127–1157 (2015)
7. Gonçalves, J.C.A.R., Baião, F., Santoro, F.M., Revoredo, K.: Discovering intentions and desires within knowledge intensive processes. In: Reichert, M., Reijers, H.A. (eds.) BPM Workshops 2015. LNBIP, vol. 256, pp. 273–285. Springer, Cham (2016). https://doi.org/10.1007/978-3-319-42887-1_22
8. Guizzardi, G., Wagner, G.: A unified foundational ontology and some applications of it in business modeling. In: CAiSE 2004 Workshops, vol. 3, pp. 129–143 (2004)
9. Hull, R., Motahari Nezhad, H.R.: Rethinking BPM in a cognitive world: transforming how we learn and perform business processes. In: La Rosa, M., Loos, P., Pastor, O. (eds.) BPM 2016. LNCS, vol. 9850, pp. 3–19. Springer, Cham (2016). https://doi.org/10.1007/978-3-319-45348-4_1
10. Isik, O., Mertens, W., den Bergh, J.V.: Practices of knowledge intensive process management: quantitative insights. BPM J. **19**(3), 515–534 (2013)
11. Junghanns, M., Petermann, A., Neumann, M., Rahm, E.: Management and analysis of big graph data: current systems and open challenges. In: Zomaya, A.Y., Sakr, S. (eds.) Handbook of Big Data Technologies, pp. 457–505. Springer, Cham (2017). https://doi.org/10.1007/978-3-319-49340-4_14
12. Marjanovic, O., Freeze, R.: Knowledge intensive business processes: theoretical foundations and research challenges. In: 44th IEEE Hawaii International Conference System Sciences (HICSS), pp. 1–10 (2011)
13. Moura, E.V., Santoro, F.M., Baião, F.A.: XCuteKIP: support for knowledge intensive process activities. In: Baloian, N., Zorian, Y., Taslakian, P., Shoukouryan, S. (eds.) CRIWG 2015. LNCS, vol. 9334, pp. 164–180. Springer, Cham (2015). https://doi.org/10.1007/978-3-319-22747-4_13
14. Netto, J.M., França, J.B.S., Baião, F.A., Santoro, F.M.: A notation for knowledge-intensive processes. In: 2013 IEEE 17th International Conference on Computer Supported Cooperative Work in Design (CSCWD), Whistler, vol. 1, pp. 190–195 (2013)
15. Nunes, V.T., Santoro, F.M., Werner, C.M.L., Ralha, C.G.: Real-time process adaptation: a context-aware replanning approach. IEEE Trans. Syst. Man Cybern. Syst. **1**(99), 1–20 (2016)

16. Ortega, A.R., Resinas, M., Cabanilla, C., Ruiz-Cortés, A.: On the definition and design-time analysis of process performance indicators. Inf. Syst. **38**(4), 470–490 (2013). Special Section on BPM 2011 Conference (2013)
17. Ramos, E.C., Santoro, F.M., Baião, F.A.: A method for discovering the relevance of external context variables to business processes. In: International Conference on Knowledge Management and Information Sharing (KMIS), Paris (2011)
18. Richetti, P.H.P., Gonçalves, J.C.A.R., de Baião, F.A., Santoro, F.M.: Analysis of knowledge-intensive processes focused on the communication perspective. In: Carmona, J., Engels, G., Kumar, A. (eds.) BPM 2017. LNCS, vol. 10445, pp. 269–285. Springer, Cham (2017). https://doi.org/10.1007/978-3-319-65000-5_16
19. Rosa, M., Reijers, H., van der Aalst, W., Dijkman, R., Mendling, J., Dumas, M., Garcia-Bañuelos, L.: APROMORE: an advanced process model repository. Expert Syst. Appl. **38**(6), 7029–7040 (2011)
20. Searle, J.R.: A taxonomy of illocutionary acts. Linguistic Agency University of Trier (1976)
21. Soares, D., Santoro, F., Baião, F.: Discovering collaborative knowledge-intensive processes through e-mail mining. J. Netw. Comput. Appl. **36**, 1451–1465 (2013)

1st International Workshop on Cross-cutting Aspects of Business Process Modeling (CCABPM 2017)

Introduction to the Workshop on Cross-cutting Aspects of Business Process Management

Jörg Becker[1], Stefanie Betz[2], Cristina Cabanillas[3], Claudia Cappelli[4],
Leticia Duboc[5], and Nadine Ogonek[1]

[1] WWU Münster, Münster, Germany
{joerg.becker,nadine.ogonek}@ercis.uni-muenster.de
[2] Karlsruhe Institute of Technology, Karlsruhe, Germany
stefanie.betz@kit.edu
[3] Vienna University of Economics and Business (WU), Vienna, Austria
cristina.cabanillas@wu.ac.at
[4] UNIRIO, Rio de Janeiro, Brazil
claudia.cappelli@uniriotec.br
[5] State University of Rio de Janeiro, Rio de Janeiro, Brazil
leticia@ime.uerj.br

Abstract. The Workshop on Cross-cutting Aspects of Business Process Management explored problems related to the areas of resource management, governance and sustainability in business process management (BPM). Specifically, it was a joint workshop gathering together the 2nd Workshop on Resource Management in Business Processes (ReMa), the International Workshop in BPM Application to the e-Government domain (BPMGOV) and the 2nd International Workshop on Sustainability-Aware BPM (SABPM). A total of 9 papers were submitted, of which four full papers and one short paper were accepted and presented in a half-day setting that aimed at motivating discussions on topics that could be of interest for the three workshops involved.

Keywords: BPM · Resource management · Team management
Human resources, Sustainability

1 Aims and Scope

The Workshop on Cross-cutting Aspects of Business Process Management, which was held in conjunction with the BPM'17 conference in Barcelona (Spain), addressed several areas of business process management (BPM) related to the three workshops it comprised, namely, the 2nd Workshop on Resource Management in Business Processes (ReMa)[1], the International Workshop in BPM Application to the e-Government domain (BPMGOV)[2] and the 2nd International Workshop on Sustainability-Aware BPM (SABPM)[3]. The joint workshop was structured in two sessions and lasted half a

---

[1] https://ai.wu.ac.at/rema2017/.

[2] https://www.wi.uni-muenster.de/department/is/bpmgov2017.

[3] http://sustainabilitydesign.org/second-international-workshop-on-sustainability-aware-business-process-management/.

day. Four full papers and one short papers were presented, and time was left for open discussions with the audience and a working session on topics related to the workshop.

The *ReMa workshop* was concerned with how human resources are involved and can be managed in processes with intensive resource needs, including resource assignement at desing time, resource allocation at run time, and resource analysis at any phase of the BPM lifecycle. One paper was accepted for presentation at the joint workshop. The paper, entitled "Towards a Taxonomy of Human Resource Allocation Criteria" and written by Michael Arias, Jorge Munoz-Gama and Marcos Sepúlveda, addresses resource allocation and proposes a taxonomy of criteria for selecting the most appropriate resource for executing a work item at run time. A broader systematic literature review on resource allocation in business processes is behind this work.

e-Government comprises the use of information and communication technologies (ICT) to broaden and deepen governance by enabling citizens to connect with government entities. Wisely applied, it can lead to new methods of obtaining services and produce public policies and services. In this scenario BPM can be a strong tool in terms of coproduction-based approaches, making citizens part of the conception, design, steering, and management of public policies and services. BPM has not only become a powerful means but a necessity to streamline governmental processes, making them more efficient, effective, customer-friendly and prepared for the challenges that are to come. This set the scene for the *BPMGOV workshop*, where two papers that show the use and benefit of BPM in the public sector were presented. The first paper, entitled "E-Government Services: Comparing Real and Expected User Behavior", depicts a novel approach for the analysis of e-government services with the help of process mining techniques. Those can be applied to real-life event logs of e-government systems and assist in finding discrepancies between real and expected user behavior. The second paper with the title "Establishing transparent interorganizational relationships through shared goals for anti-corruption in Brazil", explores the challenge of managing interorganizational relationships, especially those related to public transparency and describes an approach to support the interorganizational relationships management based on shared goals.

Sustainability is, simply put, the capacity to endure and it can be differentiated in the five dimensions: environmental, individual, societal, technical, and economic. Sustainability, the capacity to endure within these five dimensions, is recognized as one of the most important political and economic challenges faced by the world community in the 21st century. Currently, business starts to recognize the sustainability trend followed by a shift in customer values. Business Process Management offers the possibility to integrate sustainability into day-to-day operations. This has led to a growing interest in sustainability within the BPM research and practice. The *SABPM* workshop provided a forum for researchers and practitioners interested in sustainability and BPM to discuss, analyze and exchange ideas about incorporating sustainability into BPM. Two papers were selected to base the discussion in the joint workshop. The first one, authored by Lübbecke and Loos, presents guidelines for ecology-aware process design (Towards Guidelines of Modeling for ecology-aware Process Design).

The guidelines are a set of design principles, which allows process models to be assessed and optimized against ecological sustainability. The second one was a short paper authored by Couckuyt, Looy and Backer (Sustainability Performance Measurement: A Possible Classification of Models and Indicators). It presents a preliminary framework to classify sustainability models and indicators based on their focus (organizational, single activities) or their operationalization (operationalized, non-operationalized).

The last session of the Workshop on Cross-cutting Aspects of Business Process Management was a working session on sustainability analysis. There, with the help of an example process and an analysis template, the 17 workshop participants analyzed, identified and presented several possibilities to enhance individual, social, environmental, technical and economical sustainability of the process. We would like to thank all the participants for the lively discussion, their valuable input and their interesting ideas.

We also sincerely thank the Program Committee Members of the ReMa, BPMGOV and SABPM 2017 workshops for their time and support throughout the reviewing processes.

2   Workshop Chairs

Jörg Becker	WWU Münster, Germany
Stefanie Betz	Karlsruhe Institute of Technology, Germany
Cristina Cabanillas	WU Vienna, Austria
Claudia Cappelli	UNIRIO, Brazil
Leticia Duboc	State University of Rio de Janeiro
Andréa Magalhães	Fluminense Federal University, Brazil
Nanjagud C. Narendra	Ericsson Research Bangalore, India
Alex Norta	Tallinn University of Technology, Estonia
Nadine Ogonek	WWU Münster, Germany
Manuel Resinas	University of Seville, Spain
Oktay Turetken	Eindhoven University of Technology, Netherland

3   Program Committee Members

3.1 ReMa 2017

Ahmed Awad	Cairo University, Egypt
Fabio Casati	University of Trento, Italy
Florian Daniel	Politecnico di Milano, Italy
Adela del Río Ortega	University of Seville, Spain
Claudio Di Ciccio	Vienna University of Economics and Business, Austria
Schahram Dustdar	Vienna University of Technology, Austria

Stefanie Rinderle-Ma	University of Vienna, Austria
Anderson Santana De Oliveira	SAP Labs, France
Daniel Schall	Siemens Corporate Technology, Austria
Stefan Schulte	Vienna University of Technology, Austria
Stefan Schönig	University of Bayreuth, Germany
Marcos Sepúlveda	Pontifical Catholic University of Chile, Chile
Mark Strembeck	Vienna University of Economics and Business, Austria
Luis Jesús Ramón Stroppi	National Technological University of Santa Fe, Argentina

3.2 BPMGOV 2017

Naheb Azab	American University in Cairo, Egypt
Michael Räckers	University of Muenster, Germany
Robert Krimmer	Tallinn University of Technology, Estonia
Adegboyega Ojo	National University of Ireland, Ireland
Crisiano Maciel	Universidade Federal do Mato Grosso, Brazil
Flavia Bernadini	Universidade Federal Fluminense, Brazil
Hendrik Scholta	University of Muenster, Germany
Gustavo Almeida	Fundação Getúlio Vargas, Brazil
Marcello la Rosa	Queensland University of Technology, Australia
Soon Ae Chun	City University of New York, USA
Vanessa Nunes	Universidade de Brasilia, Brazil
Ana Cristina Bicharra García	Universidade Federal Fluminense, Brazil

3.3 SABPM 2017

Andreas Oberweis	KIT Karlsruhe, Germany
Agnes Koschmider	KIT Karlsruhe, Germany
Carina Alves	UFPE, Brazil
Claudia Maria Lima Werner	COPPE-UFRJ, Brazil
Colin C. Venters	University of Huddersfield, UK
Fernanda Baião	UNIRIO, Brazil
Guillermo Rodriguez-Navas	Mälardalen University, Sweden
Juliana Jansen	IBM Research, Brazil
Leonardo Dutra	EY, Brazil
Leonardo Murta	UFF, Brazil
Mario Lima	TerraCap, Brazil
Martina Kolpondinos-Huber	University of Zürich, Switzerland

Michael Fellmann	University of Rostock, Germany
Norbert Seyff	University of Applied Sciences and Arts Northwestern, Switzerland
Pratyush Bharati	University of Boston, USE
Priscila Engiel	PUC-Rio, Brazil
Sedef Akinli Kocak	Ryerson University, Canada
Thais Vojvodic	Coca-Cola Institute, Brazil
Victor Almeida	UFF and Petrobrás, Brazil

Towards a Taxonomy of Human Resource Allocation Criteria

Michael Arias[(✉)][iD], Jorge Munoz-Gama[iD], and Marcos Sepúlveda[iD]

Computer Science Department, School of Engineering, Pontificia Universidad
Católica de Chile, Santiago, Chile
{m.arias,jmun}@uc.cl, marcos@ing.puc.cl

Abstract. Allocating the most appropriate resource to execute the
activities of a business process is a key aspect within the organizational
perspective. An optimal selection of the resources that are in charge of
executing the activities may contribute to improve the efficiency and the
performance of the business processes. Despite the existence of resource
metamodels that seek to provide a better representation of resources,
a detailed classification of the allocation criteria that have been used
to evaluate resources is missing. In this paper, we provide an initial proposal for a resource allocation criteria taxonomy. This taxonomy is based
on an extensive literature review that yielded 2,370 articles regarding
the existing resource allocation approaches within the business process
management discipline, from which 95 articles were considered for the
analysis. The proposed taxonomy points out the most frequently used
criteria for assessing the resources from January 2005 to July 2016.

Keywords: Human resource allocation · Resource management
Allocation criteria · Business processes management

1 Introduction

Business process management (BPM) is a discipline that combines distinct
approaches that can be used for the design, execution, control, measurement
and optimization of business processes [2]. According to [2], there are four business process perspectives: (a) control-flow perspective; (b) organizational perspective; (c) case perspective; and (d) time perspective. Traditionally, research
efforts have been focused on the control-flow perspective [23]. Recent research
has evidenced the need to provide better support to the organizational perspective [6,18,19], also known as resource perspective [10]. This need is motivated
due to the focus that this perspective has on the analysis of resources that participate in the execution of process activities (whether they are human or not
human resources [17]), and how this analysis could help to improve the process
efficiency [5]. Typically, the management of resources in BPM could be separated
into two task: resource assignment and resource allocation [9]. On the one hand,
resource assignment has to do with the definition at design-time of the conditions that resources must fulfill in order to become candidates to work on the

© Springer International Publishing AG 2018
E. Teniente and M. Weidlich (Eds.): BPM 2017 Workshops, LNBIP 308, pp. 475–483, 2018.
https://doi.org/10.1007/978-3-319-74030-0_37

process activities. On the other hand, resource allocation refers to the designation of the actual process activities executors at run-time. Specifically, the task of human resource allocation (we focus on human resources, hereinafter referred to as 'resources') represents a key aspect within the organizational perspective, seen as an important challenge from the BPM discipline [21,23]. The Process-Aware Information Systems (PAISs) [1] provide several information systems that support the execution of business processes. One particular type of information systems is Business Process Management Systems (BPMSs). BPMSs focus on coordinating and automating business processes in such way that the work is executed at the right time and the allocated resources are available and authorized to perform the work [10]. For instance, Bizagi (bizagi.com) provides an organizational metamodel including properties such as: role, organizational position, and expertise criteria. Moreover, Bonita BPM (bonitasoft.com) presents an organizational metamodel considering properties such as: role, organizational position, and authorization criteria. One salient feature of Bonita BPM is the use of memberships and organizational groups to handle resource allocation. Distinct articles have focused on supporting the organizational perspective through metamodels that perform the modeling and visualization of requirements related to the resources. Within these proposed metamodels, there has been an important interest in the relationship between resources and their competencies (e.g., expertise), and the organizational structure (e.g., role or organizational position) [8,16,19]. Although the proposed metamodels have sought to represent resources, they have not considered a broader set of criteria for assessing resources and determining their suitability to participate in the execution of process activities. Despite the focus on process management and the adequate selection of resources to be allocated, the currently provided support by BPMSs to the organizational perspective has room for improvement [11,19], as a way to advance PAISs towards the concept of Process- and Resource-Aware Information System(PRAIS) [7].

In order to contribute to improve this shortfall, our work is a first step towards a taxonomy of resource allocation criteria. We conducted a Systematic Literature Review (SLR) [13] of the research area of resource allocation within BPM. Further details about the systematic review process performed can be found in [4]. From 2,370 articles, we systematically analyzed a set of 95 articles that pertain to the period between January 2005 and July 2016. This work may serve as a reference map of resource-related criteria that are commonly assessed in existing allocation approaches, a classification that has not been reported so far. This proposed classification may help those in charge of the process-oriented systems to identify what other resource-related information is relevant to capture, a frequent question from the point of view of the BPMSs [10]. This paper is organized as follows: Sect. 2 presents the resource metamodels found in the reviewed literature. In Sect. 3 we identify and classify distinct types of resource allocation criteria based on the 95 articles. Finally, Sect. 4 outlines the conclusions and future work.

2 Resource Metamodels in Human Resource Allocation

Diverse approaches have been presented to face the challenge of improving the resource allocation task. These approaches have proposed allocation methods using techniques and algorithms belonging to different fields, such as machine learning [12], dynamic programming [14], or computational optimization [22]. Within these allocation methods, different metamodels (see Table 1) have been proposed with the aim of providing a better representation of resource-related information, identifying criteria and other properties that are considered when allocating resources to activities.

Table 1. Identified resource metamodels

Name	Description	Criteria Used
Human Resource MetaModel (HRMM) [16]	Allows the association of roles and resources. Provides a competence metamodel for the modeling of resources, considering their competences, skills and knowledge	Role, Authorizations, Organizational position, and Expertise
Resource perspective extension to the BPMN 2.0 metamodel [19]	Supports the resource requirements modeling and visualization. It includes three main aspects: structure, authorization, and work distribution, focuses on the distribution of work corresponding to atomic activities among resources	Role, Authorizations, Organizational position, Experience, and Expertise
Organizational metamodel [17]	Is an organizational metamodel used to define a set of workflow resource patterns	Role, Organizational position, Experience and Expertise
Resource Perspective Implementation Metamodel (RPIMet) [20]	Enables the representation of entities provided by WfMSs to implement the resource perspective aspects. Is based on the generic elements: Resource, ResourceParameter and ResourceRole defined by BPMN	Role, Authorizations, Organizational position, Experience, and Expertise
Organisational metamodel [18]	Metalmodel used to express organisational information, which is able to cover the workflow patterns	Identity, Roles or Groups, and Relation
Metamodel for resource modeling [15]	Represents a hybrid meta model, which is based on a previous analysis of organizational metamodels within workflow management systems	Role, Organizational position, Organizational unit, Privilege, and Expertise
UML organizational model [3]	Is a UML class diagram that includes it corresponding XML rendition, which can be used for the specification of workflow resources	Roles, Organization structure, Availability, Location, and Expertise

After reviewing these proposed metamodels, we found that they have prioritized the inclusion of criteria such as: organizational structure, roles, authorization aspects, experience, and expertise level as well as resource constraints. However, in the literature, we found that there are other criteria being used by the allocation methods to assess resources. These criteria have not been mapped to date yet, and they need to be reported in order to suggest information that should be recorded in the BPMSs to those in charge of the process-oriented systems.

3   Types of Resource Allocation Criteria

In this paper, we conducted a SLR following the guidelines proposed by Kitchenham [13] in order to identify, evaluate, and classify the resource allocation criteria followed to allocate resources. The guidelines include four main steps. First, the definition of the research question. In our case, we created the following research question: What resource-related criteria have been used to perform the resource allocation? The second step refers to conduct the search. This step involves the definition of the keywords to perform the search. The set of selected keywords was: *resource patterns*, *resource allocation*, *resource assignment*, *staff assignment*, *task allocation*, *task assignment*, *process mining*, and *business process management*. Third, we proceed with the screening of papers. We reviewed the title, abstract and keywords of the selected papers, and evaluated them according with our predetermined inclusion or exclusion criteria. Fourth, the data extraction step focused on answering the aforementioned research question. Initially, we evaluated 2,370 articles. We excluded any duplicate papers identified. Thus, a set of 1,950 papers was obtained. Then, a total of 95 articles met our selection criteria, which were used for further analysis in the data extraction step. For details on the SLR protocol, we refer the reader to [13]. Our aim is to propose a classification that may serve as a reference to improve the capture of information that is currently carried out through the BPMSs. Our classification gather criteria associated with resource properties, which have been proposed by methods of resource allocation throughout the analysis period. It should be noted that we only considered resource-related information. Attributes related with task information, time information, and process information are not part of the scope of this work, but will be studied in greater detail in order to extend the proposed taxonomy. The proposed classification is presented in Fig. 1.

We considered the following allocation criteria:

Amount: Number of resources required.
Experience: Resource experience executing process activities (e.g., years).
Expertise: The expertise category includes the following criteria:

- *Cognitive attributes:* Cognitive characteristics a resource might possess, such as sentience, volition and causability.
- *Expertise:* Resources capabilities, competences, skills, and knowledge.
- *Functional attributes:* Resource behavior characteristics (e.g., adaptability).
- *Non-functional attributes:* Other attributes that may influence the performance of the resources (e.g., environmental factors and technical aids).
- *Work variety:* Analyses similar and dissimilar tasks done by a resource in a day.

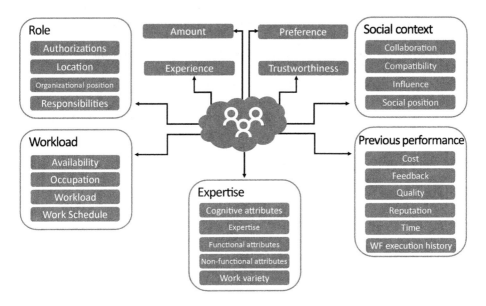

Fig. 1. Proposed taxonomy

Preference: Resource preference for executing certain types of activities.

Previous performance: The previous performance category includes the following criteria:

- *Cost:* Evaluates cost attributes such as resource total cost.
- *Feedback:* resources give their feedbacks in order to accept or refuse the work done by other resources.)
- *Quality:* Evaluates the satisfaction level of the executed process activities (e.g., customer satisfaction).
- *Reputation:* Evaluates resource social standing within a resource network based on previous performance.
- *Time:* Evaluates time attributes such as execution time.
- *WF execution history:* Audit trails provided by workflow management systems.

Role: The role category includes the following criteria:

- *Authorizations:* Constraints regarding to a specific person or role to allocate, and authorization privileges.
- *Location:* Resources has attributes to describe its location and the structure of activities that it can perform in a workflow.
- *Organizational position:* Constraints regarding to a specific organizational position.
- *Responsibilities:* Set of responsibilities on a resource to perform specific activities.

Social context: The social context category includes the following criteria:

- *Collaboration:* Measures resource collaboration and cooperation.
- *Compatibility:* Measures resource compatibility.
- *Influence:* Degree of the influence that on resource has on some other resources.
- *Social position:* Resources form various social communities and take different social positions while participating in business processes.

Trustworthiness: Notion of trust degree that a resource may have to execute activities.

Workload: The workload category includes the following criteria:

- *Availability:* An existing resource is available, busy or not available.
- *Occupancy:* Consider the actual idle level of a resource. consider how a resource is occupied executing activities.
- *Workload:* The capacity of resources to perform specific activities is constrained.
- *Work Schedule:* Refers to different types of work schedules (e.g., shift plan, part time or full time).

We have classified the selected articles according to these proposed categories. Figure 2 shows the distribution of the articles according to each criterion. It should be noted that more than one criterion might have been used in a single article. We can see that eight criteria (30% of total) are the most relevant ones considering their occurrence frequency. We found that *Authorizations, Availability,* and *Expertise* are the most frequently used criteria, which are consistent with the criteria priorities proposed by the existing metamodels (shown in Sect. 2). However, we found that 18 criteria (70% of total) correspond to criteria that, despite their occurrence frequency are not very high (5 times or less), represent a key insight in regard to the importance of evaluating other criteria when selecting resources.

Regarding these categories, we note that Role (68 times) and Workload (66 times) group criteria are the most often used by resource allocation approaches

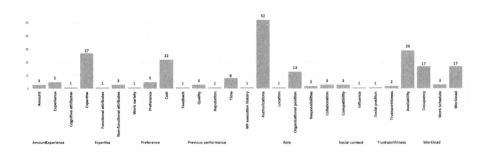

Fig. 2. Amount of articles per resource allocation criteria

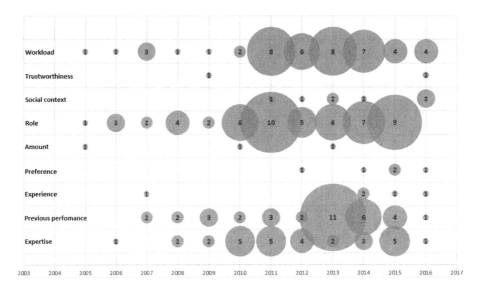

Fig. 3. Breakdown of the resource allocation categories per year

(59%, 134 times). Meanwhile, the remaining 7 categories represent 41% (92 times), where we can highlight *Previous performance* (36 times), *Expertise* (33 times), and *Social context* (8 times) as categories that are becoming prominent within approaches to allocate resources. Figure 3 provides a breakdown of the resource allocation categories per year. We can see that *Workload* and *Role* categories have been present in studies throughout the period of analysis. It is possible to highlight that in the period 2010 to 2016, there is a high concentration of allocation methods that used criteria regarding *Workload*, *Role*, and *Expertise* categories. This concentration confirms the preference to use criteria associated with the organizational position, availability and workload, and the resource suitability for allocation considering their expertise level. In addition, it is relevant to see how other criteria such as *Previous performance*, *Social context*, and *Preference* begin to be more popular criteria in the last 5 years of the period analysis. Specifically, we can highlight *Cost* as an emerging criterion (22 times). Some human resources are more expensive than others, and cost being used as an important property to assess within the allocation approaches, and as a major decision criteria in large companies.

We can conclude that resource allocation approaches have mainly considered criteria such as *Authorization* or *Expertise* because these criteria are commonly found in meta models and through BPMSs. However, there is a trend towards multi-criteria approaches to allocate resources, where using distinct criteria than those frequently used to assess resources is an increasingly common practice in order to select the most suitable resources for executing process activities. Evaluate *Previous performance* and consider *social context* attributes are two categories that being more popular in recent years. From our point of view, *Expertise*,

Role and *Workload* are criteria that will continue to be used. Nonetheless, there has been a need to combine these criteria with other criteria in order to optimize the resource allocation task, due to the evolution about how resources are being evaluated in organizations. Identify other criteria associated with time, process and task information, and propose allocation mechanisms that allow prioritizing and recommending resources are challenges that require further research.

Due to space constraints, further details about followed review process and the entire classification can be found in [4].

4   Conclusions and Future Work

In this paper, we have identified and classified the main criteria used in resource allocation approaches in order to improve this task within organizations. We focused on considering those criteria that are related to the properties of human resources. We compiled a SLR of a set of 95 articles that proposed resource allocation approaches. We intend that the proposed classification can help those in charge of the process-oriented systems to discover common information used to evaluate resources. In addition, this classification may suggest the capture and integration of new resource-related information as part of BPMS systems, which may serve to improve the support currently given to the organizational perspective. As future work, we plan to extend the proposed classification, including other evaluation criteria, evaluate the effectiveness of the criteria on resource allocation, as well as formalize the allocation criteria identified in a taxonomy of resource allocation criteria.

Acknowledgments. This project was partially funded by the Ph.D. Scholarship Program of CONICYT Chile (Doctorado Nacional/2014-63140181) and by Universidad de Costa Rica.

References

1. van der Aalst, W.M.P.: Process-aware information systems: lessons to be learned from process mining. ToPNoC **2**, 1–26 (2009)
2. van der Aalst, W.M.P.: Process Mining - Data Science in Action. Springer, Heidelberg (2016). https://doi.org/10.1007/978-3-662-49851-4
3. van der Aalst, W.M.P., Kumar, A., Verbeek, H.M.W.E.: Organizational modeling in UML and XML in the context of workflow systems. In: Proceedings of the 2003 ACM Symposium on Applied Computing (SAC), Melbourne, FL, USA, 9–12 March 2003, pp. 603–608 (2003)
4. Arias, M., Munoz-Gama, J., Sepúlveda, M.: Introducing a taxonomy of human resource allocation criteria. Technical report, Pontificia Universidad Católica, Santiago, Chile, June 2017. http://processmininguc.com/publications/
5. Arias, M., Rojas, E., Lee, J., Munoz-Gama, J., Sepúlveda, M.: ResRec: a multi-criteria tool for resource recommendation. In: BPM Demos 2016, pp. 17–22 (2016)
6. Cabanillas, C.: Enhancing the management of resource-aware business processes. AI Commun. **29**(1), 237–238 (2016)

7. Cabanillas, C.: Process-and resource-aware information systems. In: EDOC 2016, Vienna, Austria, 5–9 September 2016, pp. 1–10 (2016)
8. Cabanillas, C., Resinas, M., Ruiz-Cortés, A.: RAL: a high-level user-oriented resource assignment language for business processes. In: Daniel, F., Barkaoui, K., Dustdar, S. (eds.) BPM 2011. LNBIP, vol. 99, pp. 50–61. Springer, Heidelberg (2012). https://doi.org/10.1007/978-3-642-28108-2_5
9. Cabanillas, C., Resinas, M., del-Río-Ortega, A., Cortés, A.R.: Specification and automated design-time analysis of the business process human resource perspective. Inf. Syst. **52**, 55–82 (2015)
10. Dumas, M., Rosa, M.L., Mendling, J., Reijers, H.A.: Fundamentals of Business Process Management. Springer, Heidelberg (2013). https://doi.org/10.1007/978-3-642-33143-5
11. Havur, G., Cabanillas, C., Mendling, J., Polleres, A.: Resource allocation with dependencies in business process management systems. In: La Rosa, M., Loos, P., Pastor, O. (eds.) BPM 2016. LNBIP, vol. 260, pp. 3–19. Springer, Cham (2016). https://doi.org/10.1007/978-3-319-45468-9_1
12. Kim, A., Obregon, J., Jung, J.-Y.: Constructing decision trees from process logs for performer recommendation. In: Lohmann, N., Song, M., Wohed, P. (eds.) BPM 2013. LNBIP, vol. 171, pp. 224–236. Springer, Cham (2014). https://doi.org/10.1007/978-3-319-06257-0_18
13. Kitchenham, B.: Procedures for performing systematic reviews. Keele, UK, Keele Univ. **33**(2004), 1–26 (2004)
14. Koschmider, A., Yingbo, L., Schuster, T.: Role assignment in business process models. In: Daniel, F., Barkaoui, K., Dustdar, S. (eds.) BPM 2011. LNBIP, vol. 99, pp. 37–49. Springer, Heidelberg (2012). https://doi.org/10.1007/978-3-642-28108-2_4
15. zur Muehlen, M.: Organizational management in workflow applications - issues and perspectives. Inf. Technol. Manag. **5**(3–4), 271–291 (2004)
16. Oberweis, A.: A meta-model based approach to the description of resources and skills. In: Sustainable IT Collaboration Around the Globe, AMCIS 2010, Lima, Peru, 12–15 August 2010, p. 383 (2010)
17. Russell, N., van der Aalst, W.M.P., ter Hofstede, A.H.M., Edmond, D.: Workflow resource patterns: identification, representation and tool support. In: Pastor, O., Falcão e Cunha, J. (eds.) CAiSE 2005. LNCS, vol. 3520, pp. 216–232. Springer, Heidelberg (2005). https://doi.org/10.1007/11431855_16
18. Schönig, S., Cabanillas, C., Jablonski, S., Mendling, J.: A framework for efficiently mining the organisational perspective of business processes. DSS **89**, 87–97 (2016)
19. Stroppi, L.J.R., Chiotti, O., Villarreal, P.D.: A BPMN 2.0 extension to define the resource perspective of business process models. In: XIV CIbSE, pp. 25–38 (2011)
20. Stroppi, L.J.R., Chiotti, O., Villarreal, P.D.: Defining the resource perspective in the development of processes-aware information systems. Inf. Softw. Technol. **59**, 86–108 (2015)
21. Yaghoubi, M., Zahedi, M.: Resource allocation using task similarity distance in business process management systems. In: ICSPIS 2016, IEEE . pp. 1–5 (2016)
22. Zhao, W., Yang, L., Liu, H., Wu, R.: The optimization of resource allocation based on process mining. In: Huang, D.-S., Han, K. (eds.) ICIC 2015. LNCS (LNAI), vol. 9227, pp. 341–353. Springer, Cham (2015). https://doi.org/10.1007/978-3-319-22053-6_38
23. Zhao, W., Zhao, X.: Process mining from the organizational perspective. In: Wen, Z., Li, T. (eds.) Foundations of Intelligent Systems. AISC, vol. 277, pp. 701–708. Springer, Heidelberg (2014). https://doi.org/10.1007/978-3-642-54924-3_66

E-Government Services: Comparing Real and Expected User Behavior

A. A. Kalenkova[1]([✉]) [iD], A. A. Ageev[1], I. A. Lomazova[1] [iD],
and W. M. P. van der Aalst[2] [iD]

[1] National Research University Higher School of Economics, Moscow, Russia
aaageev@edu.hse.ru, {akalenkova,ilomazova}@hse.ru
[2] Eindhoven University of Technology, Eindhoven, The Netherlands
w.m.p.v.d.aalst@tue.nl

Abstract. E-government web services are becoming increasingly popular among citizens of various countries. Usually, to receive a service, the user has to perform a sequence of steps. This sequence of steps forms a service rendering process. Using process mining techniques this process can be discovered from the information system's event logs. A discovered process model of a real user behavior can assist in the analysis of service usability. Thus, for popular and well-designed services this process model will coincide with a reference process model of the expected user behavior. While for other services the observed real behavior and the modeled expected behavior can differ significantly. The main aim of this work is to suggest an approach for the comparison of process models and evaluate its applicability when applied to real-life e-government services.

Keywords: E-government services · BPM mining
Increasing citizen acceptance · Business process quality
Process discovery · Comparing process models
BPMN (Business Process Model and Notation)

1 Introduction

Today, e-government web services play an important role in everyday life of people all over the world. Well-designed information systems with intuitive interfaces help users to receive services, dramatically reducing the rendering time. Even if the information system is mature enough, users may face different problems when receiving a service. For instance, the user may have difficulties when filling specific web forms. Analysis of the user behavior can assist in finding redundant process steps, which take extra time or lead to the cancellation of the process, thus, giving additional information for the process improvement.

In comparison to the existing practices of e-government service analysis [1–3], in this work we will take the detailed history of services' executions as a starting point and will try to explicitly reveal deviations from the expected user behavior.

Usually information systems record their behavior in the form of event logs. These event logs can be further analyzed using *process mining* techniques [4].

© Springer International Publishing AG 2018
E. Teniente and M. Weidlich (Eds.): BPM 2017 Workshops, LNBIP 308, pp. 484–496, 2018.
https://doi.org/10.1007/978-3-319-74030-0_38

Process mining is a relatively new field in data science offering algorithms and tools for the *discovery* and *conformance checking* of processes using the event data.

Discovered process models reveal users' steps and give an overall view of the real user behavior. Over the last 15 years many algorithms for the discovery of process models from event logs have been developed [5–11]. Most of them produce Petri nets, causal nets, or process trees. In this work we consider BPMN (Business Process Model and Notation) [12] – one the most popular process modeling languages among analysts and developers (including e-government domain), as a basis language for the process models representation. For this purpose we apply conversion techniques for the transformation of process models, discovered from event logs, to BPMN [13,14].

In contrast to the existing conformance checking approaches aimed at finding discrepancies between a model and a log (M2L) [15,16], or between two event logs (L2L) using footprints [4], in this paper we propose to compare process models (M2M) reflecting the real and expected user behavior. Once the process model representing the real user behavior was discovered, it can be verified against a reference model created by a services analyst. Since process models are considered as graphs, the *minimal graph edit distance* [17] (minimal number of steps needed to transform one graph to the other) can be used as a desired metric for the comparison of process models. Indeed, this metric can be easily used by experts to explicitly show models' discrepancies, highlighting matched and differentiating model elements (tasks, routing constructions, or connections).

In order to make finding minimal graph distance technique applicable for the analysis of process models constructed from real-life event logs, avoiding exponential growth of possible matching solutions, several improvements were suggested and implemented within a web-tool called BPMNDiffViz [18]. The proposed comparison approach was verified on event logs of a real-life e-government system and demonstrated its applicability to the analysis e-government services.

The paper is organized as follows. Section 2 introduces core BPMN elements used to model processes within this work. Section 3 gives a brief overview of process mining and in particular process discovery techniques. A detailed description of a method for the BPMN models comparison is presented in Sect. 4. Section 5 contains results of application of the proposed technique to real-life event logs generated by a e-government system. And finally, Sect. 6 concludes the paper.

2   Core BPMN Modeling Constructs

In this section we introduce semantics of core BPMN constructs (events, tasks, gateways, and sequence flows) used to model processes within this work.

Although BPMN offers various modeling primitives, not all of them are frequently used in practice [19]. In this work, we stick to the most popular ones, which define the control-flow perspective of a process.

Tasks (Fig. 1c) model atomic process steps, *gateways* (Fig. 1d and e) are routing constructions, *start* (Fig. 1a) and *end events* (Fig. 1b) specify the beginning and termination of the process respectively. *Sequence flows* (Fig. 1f) connect nodes (tasks, events, and gateways) irrespective of their type.

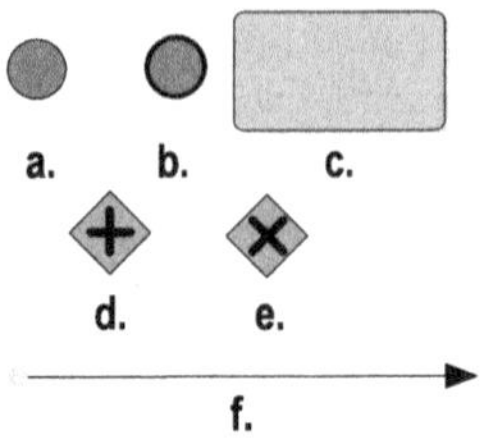

In the remainder we assume that core BPMN models are directed graphs with one start and multiple end events, such that the start event and end events do not have incoming and outgoing sequence flows respectively, and each node of the graph is on a path from the start to an end event.

Fig. 1. Core BPMN elements

Operational semantics of core BPMN models are based on model states (markings). In each state of a model, its *sequence flows* may carry *tokens*. In an *initial marking* each outgoing sequence flow of the start event contains a token, while other sequence flows do not.

Each node (except start event) of the BPMN model may be *enabled* and may *fire*. A task is *enabled* if and only if at least one of its incoming sequence flows contains a token. Suppose that a task is enabled, then it may fire, consuming a token from the incoming sequence flow and producing a token for each of the outgoing sequence flows.

Similarly, an exclusive gateway (Fig. 1e) is enabled if and only if there is an incoming sequence flow, which contains at least one token. When firing, an exclusive gateway produces a token to only one of the outgoing sequence flows. According to the specification the *cross marker* can be omitted. Gateways without any markers used later are also exclusive.

A parallel gateway (Fig. 1d) is enabled if and only if each incoming sequence flow contains at least one token. An enabled parallel gateway may fire, consuming a token from each incoming sequence flow and producing a token to each outgoing sequence flow.

An end event in a current marking is enabled if and only if one of the incoming sequence flows carries a token. When the end event fires, it consumes a token from an incoming sequence flow and does not produce any tokens, i.e., an end event consumes tokens as they arrive. It is assumed that the *final marking* is reached, when no tokens are left.

A toy example of a service rendering process is shown in Fig. 2. First, the user registers or signs in, then he or she fills the form and selects a payment

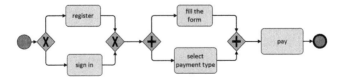

Fig. 2. An example of a service rendering process

type (these operations can be executed in any order, and thus, are represented as parallel using the appropriate BPMN gateways), and finally pays.

The core BPMN modeling elements described in this section will be used throughout the paper. Note that since tasks act as exclusive merge and parallel splitting gateways, according to their semantics, these types of gateways connected with tasks via outgoing and incoming sequence flows respectively can be omitted.

3 Process Discovery

The proposed approach for the analysis of service rendering processes uses the process discovery techniques. In this section, we will sketch the main principles and challenges of the process discovery.

Process mining offers various algorithms [5–11] discovering process models from event logs. The main discovery challenge is to synthesize a model, which represents the event log in a best possible way.

Let us illustrate a process discovery procedure by an example. Consider an event log containing a small portion of the history of a service rendering process executions (Table 1).

Table 1. An event log of a service rendering process

Case ID	Activity name	Timestamp	Price	Client IP
1	Register	2017-04-22 10:45:12:123	10	192.168.1.39
1	Fill the form	2017-04-22 10:51:32:367	10	192.168.1.39
2	Sign in	2017-04-22 10:55:34:739	5	192.168.1.35
1	Select payment type	2017-04-22 10:56:33:524	10	192.168.1.39
3	Sign in	2017-04-22 10:56:45:673	10	192.168.1.11
1	Pay	2017-04-22 10:57:23:854	10	192.168.1.39
...	...	...	...	...

To receive a service, people register or sign in using a special web site, where they fill web forms, choose a type of payment and pay for the service. Timestamps usually reveal the order of events. In this example an identifier of each process execution corresponds to a particular IP address of the user. Beside that, additional information, such as price (or personal data), can be also presented in the event log.

To learn the control flow, an event log can be represented as a multiset of runs (or traces), which are the sequences of activity names, corresponding to case identifiers:

$$L = [\langle register, fill\ the\ form, select\ payment\ type, pay \rangle^{10}, \langle register, select\ payment\ type,$$
$$fill\ the\ form, pay \rangle^{7}, \langle sign\ in, fill\ the\ form, select\ payment\ type, pay \rangle^{9}, \langle sign\ in,$$
$$select\ payment\ type, fill\ the\ form, pay \rangle^{1}].$$

A BPMN model discovered from this event log, using the inductive mining algorithm [11], is presented in Fig. 2. This model perfectly *fits* the given event log, i.e., each trace can be replayed by the model, starting from the initial marking, and reaching a final marking with no tokens left. In this case we say that the model has a *perfect fitness*.

4   A Technique for Process Models Comparison

Our graph comparison algorithm aims to find discrepancies between reference and discovered process models. The problem of graph comparison is known to be NP-hard [20] in time complexity. However, we do not consider arbitrary graphs, but deal with the process models discovered from event logs. These models are labeled directed graphs with certain types of nodes supposed to have much in common: contain the same activities or even identical sub-parts. Under such assumptions additional heuristic rules used for processes comparison can be applied.

4.1   Preliminaries

From now on we will deal with *business process graphs* – a certain abstraction of core BPMN models. Let us define them formally. A *business process graph* is a tuple $G = (V, E, t, l)$, where V is a set of vertices (or nodes), $E \subseteq (V \times V)$, is a set of directed edges, $t : (V \cup E) \rightarrow T$, where $T = \{start\ event, end\ event, task, exclusive\ gateway, parallel\ gateway, sequence\ flow\}$, is a function, which defines element types, $l : (V \cup E) \rightarrow L$, is a labeling function, where L is set labels.

Now let us consider two business process graphs $G_1 = (V_1, E_1, t_1, l_1)$ and $G_2 = (V_2, E_2, t_2, l_2)$. The *edit relation* between G_1 and G_2 is defined as $R \subseteq (V_1 \cup E_1 \cup \{\epsilon, \tau\}) \times (V_2 \cup E_2 \cup \{\epsilon, \tau\})$, such that for each element $i_1 \in V_1 \cup E_1$ ($i_2 \in V_2 \cup E_2$) exists one and only one element $i_2 \in V_2 \cup E_2 \cup \{\epsilon, \tau\}$ ($i_1 \in V_1 \cup E_1 \cup \{\epsilon, \tau\}$), such that $(i_1, i_2) \in R$. If for some element $i \in V_1 \cup E_1$ ($i \in V_2 \cup E_2$) holds that $(i, \epsilon) \in R$ ($(\epsilon, i) \in R$), we say that this element is *deleted* (*inserted*), if $(i, \tau) \in R$ ($(\tau, i) \in R$), we say that there is no *corresponding* element for i. Moreover, for any two edges $e_1 = (v_1, v_1') \in E_1$ and $e_2 = (v_2, v_2') \in E_2$, holds that $(e_1, e_2) \in R$ if and only if $(v_1, v_2) \in R$ and $(v_1', v_2') \in R$ and each element $r \in R$ is defined as follows:

$$r = \begin{cases} (i_1, i_2), & i_1 \in (V_1 \cup E_1), i_2 \in (V_2 \cup E_2), t_1(i_1) = t_2(i_2), \\ (i_1, \epsilon), & i_1 \in V_1 \cup E_1, \\ (\epsilon, i_2), & i_2 \in V_2 \cup E_2, \\ (i_1, \tau), & i_1 \in V_1 \cup E_1, \\ (\tau, i_2), & i_2 \in V_2 \cup E_2. \end{cases}$$

For each $r = (i_1, i_2) \in R$ we define its cost as:

$$cost(r) = \begin{cases} lev(l_1(i_1), l_2(i_2)) * c_{lev}, & i_1 \neq \epsilon, i_2 \neq \epsilon, \\ c_{delete}^t, & i_2 = \epsilon, t_1(i_1) = t, \\ c_{insert}^t, & i_1 = \epsilon, t_2(i_2) = t, \\ 0, & i_1 = \tau \lor i_2 = \tau. \end{cases}$$

By *lev* we denote a function, which calculates Levenshtein distance [21] between two strings, c_{lev} is a special coefficient. This function will be used to match tasks with similar (not necessarily coinciding) labels.

By c_{delete}^t and c_{insert}^t we denote costs of the deletion and insertion operations respectively for elements with the type $t \in T$.

The *graph edit distance* (or *cost*) for *edit relation* R is a sum of costs of each pair belonging to this relation: $cost(R) = \sum_{r \in R} cost(r)$.

4.2   Calculating Minimal Graph Edit Distance

As an underlying technique for calculating minimal graph edit distance we will take A* algorithm [22]. Let us consider two business process graphs $G_1 = (V_1, E_1, t_1, l_1)$ and $G_2 = (V_2, E_2, t_2, l_2)$ and present an algorithm, for their comparison.

Algorithm 1. Business process graph comparison

Data: $G_1 = (V_1, E_1, t_1, l_1)$ and $G_2 = (V_2, E_2, t_2, l_2)$
Result: R_{min} – an edit relation with minimal cost
\\R_{init} – *the initial edit relation;*
$R_{init} \leftarrow \{\}$;
for $i_1 \in V_1 \cup E_1$ **do**
  | $R_{init} \leftarrow R_{init} \cup \{(i_1, \tau)\}$;
end
for $i_2 \in V_2 \cup E_2$ **do**
  | $R_{init} \leftarrow R_{init} \cup \{(\tau, i_2)\}$;
end
\\Q - *is a queue of element relations ordered by their costs;*
$Q \leftarrow \langle R_{init} \rangle$;
while *true* **do**
  | *take an edit relation with a minimal cost;*
  | $R_{min} \leftarrow takeMinRelation(Q)$;
  | **if** *(R_{min} contains pairs with τ)* **then**
  | | $i \leftarrow takeElementWithoutPair(R_{min})$;
  | | *expand all kinds of combinations for element i in relation R_{min};*
  | | $Q.remove(R_{min})$;
  | | $Q.add(expand(R_{min}, i))$;
  | **else**
  | | **return** R_{min};
  | **end**
end

At each step the algorithm takes an edit relation with a minimal cost from the queue and constructs novel edit relations by adding all possible correspondences for a graph element, which has no corresponding element yet. After that, novel edit relations are added to the queue of intermediate results. The algorithm terminates when there is an edit relation in the queue, which contains no pairs with τ (for all elements corresponding elements are identified) and has a minimal cost. This edit relation will be considered as a result of comparison.

4.3   Heuristic Function and Correctness of the Algorithm

In order to reduce the computational time of calculating graph edit distance, the total cost can be extended by adding a heuristic function, which does not overestimate it: $Cost(R) = \sum_{r \in R} cost(r) + H(R)$. Where heuristic function $H(R)$ is defined as:

$$H(R) = \sum_{t \in T} \begin{cases} |I_1^t| - |I_2^t| \cdot c_{delete}^t, & |I_1^t| \geq |I_2^t|, \\ |I_2^t| - |I_1^t| \cdot c_{insert}^t, & |I_2^t| > |I_1^t|. \end{cases}$$

$I_1^t \subseteq V_1 \cup E_1$ and $I_2^t \subseteq V_2 \cup E_2$ denote sets of process models elements of type t, which were not mapped yet (correspond to τ within the edit relation R).

The edit relation R between two business process graphs $G_1 = (V_1, E_1, t_1, l_1)$ and $G_2 = (V_2, E_2, t_2, l_2)$ calculated using Algorithm 1 and the heuristic function H is an edit relation with the minimal cost. Indeed, since Algorithm 1 constructs a one-to-one edit relation, the minimal distance between G_1 and G_2 will be not less than the insertion (deletion) coefficient multiplied by the difference between number of elements in G_1 and G_2, if the number of elements in G_1 is less (equal or greater) than in G_2. Thus, heuristic function $H(R)$ does not overestimate the cost of a transformation from G_1 to G_2, and, according to [22], Algorithm 1 will find the minimal graph edit distance.

4.4   Algorithm Improvements

Although the proposed algorithm is exponential in number of nodes, it can show good results in practice, when dealing with specific types of graphs. Considering business process models discovered from the event logs and those that were constructed manually (or discovered from another part of the event log) one may conclude that usually they are weakly connected graphs with many coinciding tasks, and have few differences. These observations may help to adjust the business process graph comparison algorithm.

The first idea for the algorithm *improvement* is to stick to the best tasks' matching and use it as a starting point for other matchings. Indeed, usually tasks represent specific activities from the event log and thus can be easily compared to each other without exponential growth of possible solutions. If the best possible matching between tasks can be found, then there is no need to recalculate tasks' matchings later, thus, the overall time can be dramatically reduced. This could be achieved if costs for the task deletion and insertion are significantly higher than other deletion and insertion costs, assuming that c_{lev} coefficient is also high.

Another idea for *improving* the algorithm is to first consider nodes, which are connected with already matched nodes, taking into account sequence flow addition or deletion costs. More formally, consider two business process graphs $G_1 = (V_1, E_1, t_1, l_1)$ and $G_2 = (V_2, E_2, t_2, l_2)$, and impose additional constraints on the nodes being expanded in the course of the algorithm. Suppose an edit relation with a minimal cost R_{min} was constructed. Then select a node $v_1 \in V_1$, such that $(v_1, \tau) \in R_{min}$, and $\exists v_1' \in V_1$, such that $(v_1, v_1') \in E_1$ or $(v_1', v_1) \in E_1$, and $(v_1', i_2) \in R_{min}$, where $i_2 \in V_2 \cup \{\epsilon\}$, and expand node v_1 along with incident edges. In other words, the next node to be expanded is a node, for which the correspondence is not defined yet, but it is connected by an edge with a node, which has a corresponding element from $V_2 \cup \{\epsilon\}$. If there is no such a node, we select an arbitrary node from N_1. Edge $(v_1, v_1') \in E_1$ or $(v_1', v_1) \in E_1$ should be expanded in the next step.

The proposed *improvements* impose restrictions on the costs and the order of element processing, preserving the correctness of Algorithm 1.

5 Experimental Results

Five event logs were analyzed to test the proposed comparison approach using a web-based tool [18]. These logs reflect the real-life behavior of citizens using an e-government portal. All the logs were created using Yandex.Metrica – a free tool for evaluating site traffic. Every record of each log contains the following features: IP-address (network part), age, gender, city, URL-address, date, and time.

5.1 Preprocessing

Every user was identified with the following attributes: IP-address, age, gender, city. The URL-address of a web-page visited by the user was considered as an *activity name*, the date and time of the web-page visit are considered as a *timestamp*. Thus, a sequence of web-page visits forms a *trace*. If there was a period of more than one hour between two actions of one user, then the next action was considered as a new user's visit and as a start of a new trace.

In order to model the behavior of users based on the aforementioned assumptions, the data was prepared and cleaned. After that, URL addresses were replaced with new names in accordance with their meaning. A new feature "ID of user's visit" was added to every row and after that Case ID was determined with the set of the following attributes: IP-address (network part), age, gender, ID of user's visit (Table 2).

The first event log is a log of a car registration service (CR), collected during March 2017. After data preprocessing the log contained 29063 events. There were

Table 2. Fragment of log after preprocessing

IP-address	ID of Visit	Age	City	Gender	URL-address	Date and time
109.110.70.xxx	1	25–34	Male	Moscow	PAGE 1	2017-03-07 09:41:30
109.110.70.xxx	1	25–34	Male	Moscow	PAGE 2	2017-03-07 09:42:10
...	...	...	...	...	...	
109.110.70.xxx	2	25–34	Male	Moscow	PAGE 3	2017-03-07 10:42:11
...	...	...	...	...	...	

228 unique actions, 1361 unique traces and 1898 cases. Since the goal was to model the most general users' behavior, only TOP-6 the most important actions were considered. Other actions were removed from all the traces. Finally, traces that appeared less than 10 times were filtered out and as a result, 1192 events, 5 activities, 29 unique traces, and 263 cases were left.

The second event log is a log of users applying for a foreign passport (FP) during February 2017. Since the application steps are the same as in the first case, the second log was preprocessed in the same way. Similarly, only TOP-6 of the most important actions were considered. Traces that appeared less than 38 times were filtered out and, thus, there were 14772 events, 5 activities, 31 unique traces, and 6947 cases.

The remaining three logs are the logs of the following services: appointment with a doctor (AD), a fine payment service (FPS), and a tax payment (TP). These logs were preprocessed exactly in the same way.

5.2   Modeling and Analyzing

In this section BPMN models discovered from the event logs (CR, FP, AD, FPS, TP) are compared to a reference model describing the expected user behavior. Figure 3 presents a result of comparison of a BPMN model discovered from the CR event log and a reference model. Deleted and added sequence flows are highlighted in red and green respectively.

As the comparison reveals, the expected behavior differs from the actual mainstream behavior, recorded during one month. According to the reference model, users fill the application form in two ways: either via the main page by creating a new form, or via the personal account by opening the draft of the form created earlier. The real user behavior seems to be more chaotic. Some users start to fill a new form, then refresh the web-page and continue filling it. In addition to this, users visit their personal pages at the end of process when the form is filled.

Figure 4 presents the result of comparing the BPMN model discovered from the FP event log and the reference model.

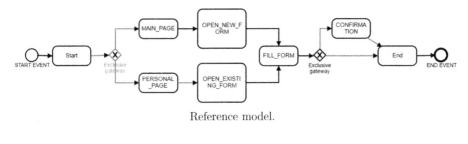

Reference model.

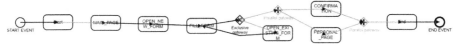

Inductive miner. Noise threshold = 0.6.

Fig. 3. Comparing the reference BPMN model with a model discovered from the event log CR (Color figure online)

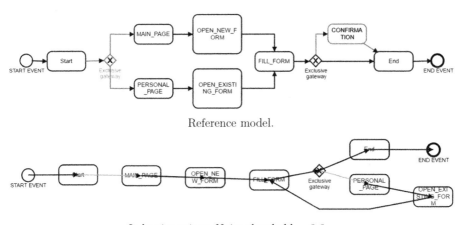

Reference model.

Inductive miner. Noise threshold = 0.6.

Fig. 4. Comparing the BPMN model describing the normative/expected behavior with the model discovered from the event log FP

It turned out that action CONFIRMATION is not frequent, so most of the users fill the application form and drop it. Some users make two or more attempts to fill the form, and then finally leave.

Each row of Table 3 presents one particular comparison involving a reference BPMN model and a BPMN model discovered using the technique listed in the same row. The first column presents an event log identifier, the second column specifies a discovery algorithm. Beside that, each row lists the distance between the two BPMN models and the number of steps performed in order to find it.

Table 3. Comparison of a reference BPMN model with models discovered using different algorithms

Event log	Discovery algorithm	Distance	Number of steps
CR	Inductive miner (threshold 0.2) [11]	28	66
CR	Inductive miner (threshold 0.6) [11]	18	47
CR	Heuristic miner [7]	22	56
CR	Alpha miner [6]	69	104
FP	Inductive miner (threshold 0.2)	24	40
AD	Inductive miner (threshold 0.2)	27	51
FPS	Inductive miner (threshold 0.2)	57	455
TP	Inductive miner (threshold 0.2)	55	494

For all the comparisons insertion and deletion costs were set to 1 (except tasks, for them insertion and deletion costs were identified as 10).

The algorithm improvements applied allowed to obtain valuable results in a small number of steps. In comparison to that, direct application of the original algorithm will lead to the explosion of intermediate solutions, and even for compact models, no result can be obtained in a reasonable amount of time.

As it follows from the table the processes of making an appointment with a doctor and applying for a car registration or passport are closer to the reference model, while the processes of payment differ significantly from the expected behavior. This can be explained by the fact that these services imply visiting payment pages that are missed in the reference model.

6   Conclusion

In this paper we proposed a novel approach for the analysis of e-government services. Our experimental results show that the process mining techniques can be applied to real-life event logs of e-government systems and assist in finding discrepancies between real and expected user behavior. Although algorithms and tools for finding minimal graph edit distance existed before, in this work we applied them within a process mining context and demonstrated their applicability for the comparison of business process models constructed from real-life event logs of an e-government system. The proposed approach takes into account specific features of analyzed process models and helps to reduce the computational time. Nevertheless, since the computational complexity of graph edit distance is exponential in number of nodes, in the future we plan to adopt existing suboptimal comparison algorithms in order to analyze even larger process models, constructed from real-life e-government event logs. We also plan to analyze correlations between behavioral (M2L and L2L) discrepancies, calculated using existing conformance checking techniques, and structural differences of corresponding process models (M2M), measured by the technique proposed in this paper.

Acknowledgments. This work was supported by the Basic Research Program at the National Research University Higher School of Economics and funded by RFBR and Moscow city Government according to the Research project No 15-37-70008 "mol_a_mos".

References

1. Gebba, T., Zakaria, M.: E-government in Egypt: An analysis of practices and challenges. Int. J. Bus. Res. Dev. **4**(2), 11–25 (2012)
2. Rorissa, A., Demissie, D.: An analysis of African e-government service websites. Gov. Inf. Quart. **27**(2), 161–169 (2010)
3. Ha, H., Coghill, K.: E-government in Singapore - a SWOT and PEST analysis. Asia Pac. Soc. Sci. Rev. **6**(2), 103–130 (2008)
4. van der Aalst, W.M.P.: Process Mining: Data Science in Action, 2nd edn. Springer, Heidelberg (2016). https://doi.org/10.1007/978-3-662-49851-4
5. Bergenthum, R., Desel, J., Lorenz, R., Mauser, S.: Process mining based on regions of languages. In: Alonso, G., Dadam, P., Rosemann, M. (eds.) BPM 2007. LNCS, vol. 4714, pp. 375–383. Springer, Heidelberg (2007). https://doi.org/10.1007/978-3-540-75183-0_27
6. van der Aalst, W.M.P., Weijters, A., Maruster, L.: Workflow mining: discovering process models from event logs. IEEE Trans. Knowl. Data Eng. **16**(9), 1128–1142 (2004)
7. Weijters, A., Ribeiro, J.: Flexible heuristics miner (FHM). In: IEEE Symposium on Computational Intelligence and Data Mining (CIDM 2011), Paris, France, IEEE. pp. 310–317, April 2011
8. van der Aalst, W.M.P., Rubin, V., Verbeek, H., van Dongen, B., Kindler, E., Günther, C.: Process mining: a two-step approach to balance between underfitting and overfitting. Softw. Syst. Model. **9**(1), 87–111 (2010)
9. Carmona, J., Cortadella, J.: Process mining meets abstract interpretation. In: Balcázar, J.L., Bonchi, F., Gionis, A., Sebag, M. (eds.) ECML PKDD 2010. LNCS (LNAI), vol. 6321, pp. 184–199. Springer, Heidelberg (2010). https://doi.org/10.1007/978-3-642-15880-3_18
10. Günther, C.W., van der Aalst, W.M.P.: Fuzzy mining – adaptive process simplification based on multi-perspective metrics. In: Alonso, G., Dadam, P., Rosemann, M. (eds.) BPM 2007. LNCS, vol. 4714, pp. 328–343. Springer, Heidelberg (2007). https://doi.org/10.1007/978-3-540-75183-0_24
11. Leemans, S.J.J., Fahland, D., van der Aalst, W.M.P.: Discovering block-structured process models from incomplete event logs. In: Ciardo, G., Kindler, E. (eds.) PETRI NETS 2014. LNCS, vol. 8489, pp. 91–110. Springer, Cham (2014). https://doi.org/10.1007/978-3-319-07734-5_6
12. OMG: Business Process Model and Notation (BPMN). Object Management Group, formal/2013-12-09 (2013)
13. Kalenkova, A., de Leoni, M., van der Aalst, W.M.P.: Discovering, analyzing and enhancing BPMN models using ProM. In: Proceedings of the BPM Demo Sessions 2014 Co-located with the 12th International Conference on Business Process Management, p. 36 (2014)
14. Kalenkova, A., van der Aalst, W.M.P., Lomazova, I., Rubin, V.: Process mining using BPMN: relating event logs and process models. Softw. Syst. Model. **16**(4), 1019–1048 (2017)

15. Adriansyah, A., van Dongen, B., van der Aalst, W.M.P.: Conformance checking using cost-based fitness analysis. In: Proceedings of the 2011 IEEE 15th International Enterprise Distributed Object Computing Conference, EDOC 2011, Washington, DC, USA, pp. 55–64. IEEE Computer Society (2011)
16. Adriansyah, A., van Dongen, B.F., van der Aalst, W.M.P.: Towards robust conformance checking. In: zur Muehlen, M., Su, J. (eds.) BPM 2010. LNBIP, vol. 66, pp. 122–133. Springer, Heidelberg (2011). https://doi.org/10.1007/978-3-642-20511-8_11
17. Sanfeliu, A., Fu, K.: A distance measure between attributed relational graphs for pattern recognition. IEEE Trans. Syst. Man Cybern. SMC **13**(3), 353–362 (1983)
18. Ivanov, S., Kalenkova, A., van der Aalst, W.M.P.: BPMNDiffViz: a tool for BPMN models comparison. In: Proceedings of the BPM Demo Session 2015 Co-located with the 13th International Conference on Business Process Management. pp. 35–39 (2015)
19. Muehlen, M., Recker, J.: How much language is enough? Theoretical and practical use of the business process modeling notation. In: Bellahsène, Z., Léonard, M. (eds.) CAiSE 2008. LNCS, vol. 5074, pp. 465–479. Springer, Heidelberg (2008). https://doi.org/10.1007/978-3-540-69534-9_35
20. Zeng, Z., Tung, A., Wang, J., Feng, J., Zhou, L.: Comparing stars: on approximating graph edit distance. Proc. VLDB Endow. **2**(1), 25–36 (2009)
21. Levenshtein, V.: Binary codes capable of correcting deletions, insertions and reversals. Sov. Phys. Dokl. **10**, 707 (1966)
22. Hart, P., Nilsson, N., Raphael, B.: A formal basis for the heuristic determination of minimum cost paths. IEEE Trans. Syst. Sci. Cybern. SSC **4**(2), 100–107 (1968)

Establishing Transparent Interorganizational Relationships Through Shared Goals for Anti-corruption in Brazil

Bruna Diirr[(✉)] and Claudia Cappelli

Programa de Pós-Graduação em Informática, Universidade Federal do Estado do Rio de Janeiro (PPGI/UNIRIO), Rio de Janeiro, Brazil
{bruna.diirr,claudia.cappelli}@uniriotec.br

Abstract. Market changes and the need to remain competitive have lead organizations to establish partnerships that allow them to share resources with each other for better handling an identified opportunity. Such associations are also established at the government level, where the limited resources are shared at the service of society's common good. However, besides having a mutual or compatible goal, it is common that partner organizations have distinct characteristics, which may lead to several challenges to be faced. The present research explores the interorganizational relationship management. For this, this paper outlines an approach based on shared goals, making the involved organizations more transparent, integrated, prepared to interoperate their processes and information and able to develop skills to act and achieve shared goals even with existing differences. An anti-corruption partnership in Brazil illustrates the approach application.

Keywords: Interorganizational relationships · Process integration
Shared goals

1 Introduction

Transformations in the economy, globalization, technological innovations, fast dissemination of information, the need for reducing time and cost of product development cycle, the urge for providing more transparency in existing processes and information, have led to organizations' adaptation to remain competitive. A new organizations approach to face this dynamic and unpredictable environment has been crossing their own borders and establishing partnerships with other organizations, which may be their rivals or operate in businesses than their own [1–5].

When organizations need to collaborate to achieve a common goal, they can establish different types of relationships, such as alliances, networks, joint ventures, outsourcing etc. Choosing the most appropriate type of interorganizational relationship considers the goals to be achieved and common interests; confidence arising from the transparency between those involved; cooperation scope; relationship duration; communication structure; decisions autonomy; relationship formalization etc. [4, 6–9]. Regardless the adopted interorganizational relationship type, they all aim to combine

© Springer International Publishing AG 2018
E. Teniente and M. Weidlich (Eds.): BPM 2017 Workshops, LNBIP 308, pp. 497–509, 2018.
https://doi.org/10.1007/978-3-319-74030-0_39

resources, knowledge or power in benefit of organizations, and share the results achieved from partnership [5, 10]. These relationships allow organizations to share human and financial resources, systems, work processes and information, enabling them having access to a wider range of tools at a more favorable cost than they would on their own. In addition, they can organize information that is related to the same business process, but as each part of this process is performed by a different organization, it is distributed in isolated databases. Moreover, these organizations can respond to market challenges more quickly by complementing existing skills [4, 10].

Such investment in interorganizational relationships is not restricted to organizations that aim to remain competitive in an increasingly challenging environment. It is something that also occurs in the governmental scope, where the limited resources and isolated processes and information are shared at the service of society's common good to solve complex public problems [11, 12]. In several Brazilian anticorruption actions [13–18], it is possible to notice the articulation of different governmental agencies. As they are responsible for some investigation aspect, whether in terms of responsibility or access to information, these agencies must collaborate and make decisions to identify and punish those involved in corruption acts.

However, besides having a mutual or compatible goal, it is common that these organizations have distinct characteristics, which can act both positively and negatively to the partnership. Positive aspects of the interorganizational relationship were already mentioned. Nevertheless, this new interorganizational group stimulates the interaction of several organizations, thus increasing the probability of facing misunderstandings and influencing the partnership performance alignment [19, 20]. Partner organizations should develop skills to work in this new dynamic, identifying, integrating and managing all shared elements to ensure that these elements favor the execution of activities supporting the group strategy integration and shared goals achievement [5, 19–22]. If it is not possible to establish a compromise between all existing dynamics, the interorganizational relationship may fail [9, 21].

Present research identifies a challenge in the interorganizational relationship management. This paper explores the challenges related to public transparency, outlining the approach for managing interorganizational relationships based on shared goals. By understanding the opportunity that has generated the organizations' joint work, it is possible to detail what is required to achieve the shared goal. After analyzing each participating organization, and consequently determining the existing and missing inputs in this partnership, responsibilities should be assigned. We claim that this approach makes the involved organizations more transparent, integrated, prepared to interoperate their processes and information and able to develop skills to act and achieve shared goals even with existing differences. The approach is applied to an anti-corruption partnership and is critiqued to guide further work.

The paper is organized as follows: Sect. 2 characterizes the interorganizational relationships dynamics when handling anti-corruption actions, besides discussing related papers. Section 3 describes the proposed approach for managing interorganizational relationships based on shared goals, discussing the importance of transparency on it. Section 4 illustrates the approach application to an anti-corruption partnership and discusses the obtained results. Finally, Sect. 5 concludes the paper.

2 Interorganizational Relationship Dynamics for Anti-corruption

According to the Transparency International assessment [23], corruption remains a "plague" around the world, making clear that Society no longer tolerates these actions and demands that this problem is tackled. Brazil does not show up well in this scenario, obtaining 40 points (from 0, highly corrupted to 100, very clean), thus occupying the 79th place of 176 countries assessed. This organization further states that the best performers countries share key characteristics such as access to information and citizen participation. Access to information allows creating an open democratic society with engaged citizens, providing them with tools to understand and use information and stimulating critical thinking about the information and services provided [24].

For this reason, transparency, or lack thereof, has been at the top of the Brazilian public agenda in many aspects and its importance has been highlighted by the increasing demand for e-Gov provision [25]. The Brazilian Strategy of Digital Governance (EGD) [26] aims to guide and integrate the Federal Government digital governance initiatives. It contributes to generating benefits for the Society by expanding access to government information, improving digital services and increasing social participation. One of its principles is openness and transparency. It states that, accordingly to the Access to Information Law [27] and Transparency Law [28], data and information are public assets that should be available to Society to promote transparency and publicity to the public resources use in programs and services, thus generating social and economic benefits. These laws have encouraged active transparency, which has resulted in greater availability of information in public organizations' institutional websites. This information availability also allows Society to identify corruption cases, such as the most recent case involving misappropriation of scholarships at UFPR [29].

Anti-corruption actions often demand the articulation of different governmental agencies. As they are responsible for some investigation aspect, these agencies need to collaborate so those involved in corruption acts are identified and punished. Therefore, these agencies can share existing human and financial resources, systems, equipment, work processes, skills and information to solve complex public problems, thus having access to a wider range of tools at a more favorable cost than they would on their own. In addition, they can organize information that is often related, but as each information is created and/or controlled by a different agency, it is distributed in isolated databases. Moreover, they are able to respond to the identified challenges more quickly and with more quality by complementing existing skills [4, 10–12].

However, these agencies should be prepared to face the challenge of working in this new dynamic, or this ecosystem regarding involved software [30, 31]. By acting alone, each agency only manages its own personnel, processes, information, systems, equipment and financial resources to achieve the desired goals. Establishing an interorganizational relationship stimulates the interaction of agencies with different characteristics, cultures and values, thus influencing the different agencies performance alignment [5, 10, 20]. In this new interorganizational group, agencies' members, who operated according to a set of values and basic assumptions, need to cooperate, thus evidencing existing differences [5, 10, 11, 20]. It increases the probability of facing misunderstandings and conflicts since agencies have different forms of planning, decision-making, resource allocation, process and information definition [19]. If it is not possible to establish a compromise between the variety of existing dynamics, with agencies facing relationship problems and conflicts, the interorganizational relationship may fail, even leading to an enormous loss to Society [9, 21].

Thus, for an effective collaborative work, ensuring that the existing elements favor the execution of activities supporting anti-corruption, it is necessary to minimize beliefs and work styles differences. Governmental agencies need to identify, integrate and manage all shared elements, developing skills that enable them to cope even with existing differences so that joint work is effective, shared goals are achieved and the relationship can be considered successful [19–22]. However, making these agencies integrated and with skills to cope besides their differences is not a trivial task. Appropriate methods, techniques and tools are required.

Authors argue that organizations' members are key elements for interorganizational alignment, acting as channels through which all organizational dynamics are shared with the partner organization [20–22]. In a mixed and multicultural context, these organization's members must cooperate, evidencing differences. Thus, it is necessary to encourage them to be committed to shared goals, be aware of differences and learn to behave in a new organizational dynamic to ensure the joint work effectiveness.

Enterprise Architecture is another aspect to be considered [1, 3, 19, 32–35]. The way organizations establish their ways of planning; decision-making; existing knowledge and skills identification, security and application; resource allocation; process, systems and information designing etc., differs from each other. With the need to work together, participating organizations should explain how to model and organize the information flow, facilitate the knowledge exchange and maintain intellectual property rights. Within Enterprise Architecture, a key factor for an interorganizational relationship is the Process Architecture [36, 37]. Organizations develop and maintain a great collection of business process models to represent the complex system that they are, becoming necessary to capture and analyze these processes and their interdependencies with each other. It becomes a greater challenge when handling cross-organizational processes. Thus, understanding the existing architecture allows understanding how each organization can best contribute to the partnership, establish agreements to use the available resources and better integrate them.

Business and shared goals modeling also become more complex in this context, with definitions and changes impacting the entire partnership [22]. Methodologies for understanding an organization should be improved and extended to deal with interorganizational relationship particularities [9, 19, 38–40].

Besides that, it is also necessary to think over systems adopted by participant organizations (Interorganizational Information Systems – IOIS) so that they support the necessary activities for the partnership, are not impacted by the existing differences and are aligned with business strategy [41–44].

Although describing proposals that could help organizations involved in partnerships, the cited studies are only interested in some aspect of interorganizational relationships management, also neglecting the transparency aspect. Partner organizations should develop skills to work in this new dynamic, identifying, integrating and managing all shared elements to ensure that these elements favor the execution of activities supporting the group strategy integration and shared goals achievement. Mechanisms that makes these organizations more transparent, integrated, prepared to interoperate their processes and information and able to act and achieve shared goals even with existing differences are still required.

3 Managing Interorganizational Relationships Based on Shared Goals

The main goal of this research is to explore the interorganizational relationship management. It is proposed that the integration of organizations involved in interorganizational relationships be oriented to the shared goals (Fig. 1). The need for joint work arises from the identification of an opportunity of interest to both organizations, with the goal being the common element of these organizations. From this goal, the other aspects can be designed. This orientation is possible since the goals created for a certain aspect at organizational processes modeling time can be implemented in execution time, guaranteeing this alignment [45].

Fig. 1. Approach for managing interorganizational relationships based on shared goals

The first step is to understand the opportunity that has generated the organizations' joint work. It is necessary to determine the shared goal to be achieved, from which the other aspects will be designed. It is important to highlight that this shared goal could be challenging and complex, being necessary to detail it into different subgoals and explicit the relationship between them. From this, it is possible to detail how the partnership will perform to achieve the shared goal (Fig. 2). It involves identifying processes, roles, information, systems, skills, considering the rules that define them.

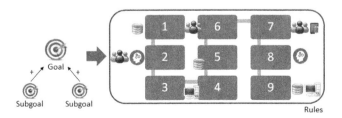

Fig. 2. Required inputs for shared goal achievement

With this detailing, it is possible to map what each participating organization can supply to achieve the shared goal (Fig. 3). It involves (a) the culture characterization, (b) the existing skills identification and (c) the identification of processes, information, systems, technology and financial resources of each organization.

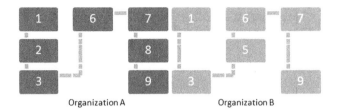

Organization A   Organization B

Fig. 3. Organizations' available inputs

By understanding the participating organizations, it is possible to identify the available inputs in the partnership (Fig. 4). This is possible from the intersection between each participating organizations' mapping (Fig. 3) and the detailing of how the partnership should act to achieve the shared goal (Fig. 2).

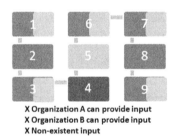

X Organization A can provide input
X Organization B can provide input
X Non-existent input

Fig. 4. Partnership' available inputs

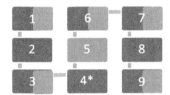

X Organization A can provide input
X Organization B can provide input

* Changed, so Organization A can provide input

Fig. 5. Partnership process

After this step, the partnership processes should be defined (Fig. 5). Organizations already have procedures to handle some required circumstances. In case the partnership does not have the necessary inputs and according to the related aspects, it is necessary: (a) process: finding an organization to integrate the partnership, searching for external training, developing (or even replacing) the required process; (b) personnel: hiring staff or training available staff that are not involved in partnership; (c) system: acquiring or developing system; (d) information: generating information or searching for information externally; (e) technology: acquiring or renting equipment.

Finally, it is necessary to assign the responsibility of each organization involved in partnership (Fig. 6). It requires identifying what is under the responsibility of each organization and what is under shared responsibility. If more than one partner can provide the same input, a negotiation should take place to determine which of them will be responsible for performing the activity. It should consider the service capacity, the involvement in other activities and the activity strategic importance. For shared activities, cultures, processes and information must be integrated.

X Organization A responsability
X Organization B responsability
X Shared responsability

Fig. 6. Responsibilities assignment

It is argued that such an approach allows organizations involved to become more integrated, prepared to interoperate their processes and information and able to act and achieve shared goals. When applied at the governmental level, it could also provide organizational transparency, adding social values related to characteristics such as auditability, accessibility, comprehension, clarity, reliability, etc. [25, 46].

4 Proposal Application: Anti-corruption Partnerships in Brazil

In Brazil, the Federal Police conducts investigations of criminal offenses against the political and social order or in detriment to Federal Government interests, as well as infractions with interstate or international repercussions. According to the investigation direction, the Federal Police reaches agreements with other governmental agencies to conduct the investigation by sharing resources and exchange private data. Regarding corruption and public money misappropriation, the Internal Revenue Service (IRS) should be involved. This agency is responsible for federal taxes administration, tax evasion combat and illegal acts related to international trade (smuggling, piracy, trafficking etc.).

Given this context, the first step for proposed approach application is to understand the opportunity that has generated the agencies collaboration. Thus, the shared goal is to combat corruption, identifying, investigating and punishing people and/or organizations who have misappropriated public money (Fig. 7).

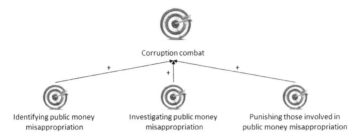

Fig. 7. Understanding opportunity for agencies collaboration

From this shared goal, it is possible to detail the necessary inputs for the partnership operation. During corruption combat, Federal Police agents (intelligence and operational) and IRS agents should perform a set of activities (irregularity identification, irregularity verification, operation planning, operation execution and sending evidence to Justice). These actions require some information, captured from Federal Police database, income tax return, banking operations, authorized wiretapping, court orders etc.; use some systems from Federal Police and IRS; need equipment as vehicles, guns etc.; and must follow the rules defined by Brazilian laws (Fig. 8).

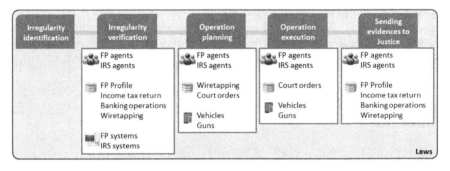

Fig. 8. Required inputs for corruption combat

With this detailing, it is possible to map what both Federal Police and IRS can supply to the partnership, thus identifying how it will operate to combat corruption (Fig. 9). For instance, Federal Police (blue) can provide agents, information retrieved from its database and authorized wiretapping, existing systems, vehicles and guns. IRS (yellow) can provide agents, information from income tax return, existing systems and vehicles. However, those agencies cannot provide some required inputs (red). For instance, the banking operations must be retrieved from Financial agencies and court orders must be determined from Justice agents. Therefore, it is necessary to gather this information externally.

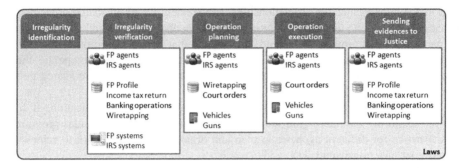

Fig. 9. Partnership available inputs for corruption combat (Color figure online)

Finally, it is necessary to assign responsibilities for involved agencies (Fig. 10). Considering the activities nature, the IRS will be responsible for irregularity verification since this agency administrates federal taxes. The Federal Police will be responsible for the operation execution since it has the power to comply court orders in its attributions and sending information to Justice. Both agencies will be responsible for irregularity identification, either through internal investigation or Society's denunciations, besides operation planning.

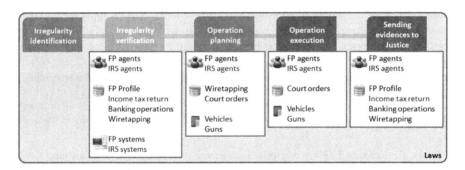

Fig. 10. Responsibilities assignment for corruption combat

5   Conclusion

The need to remain competitive in an increasingly dynamic and challenging environment lead organizations to establish partnerships for combining resources, knowledge and power in benefit of participating organizations. Such need also increases at the governmental level, with different agencies collaborating for Society's interests. However, it is a challenge to make these organizations being integrated and working in a new dynamic, thus ensuring the joint work are effective and successful.

This paper has explored the challenge of managing interorganizational relationships, especially those related to public transparency. It has described an approach to support the interorganizational relationships management based on shared goals, from which it is possible to determine what is required to achieve this shared goal and what the partnership can provide to this end. It is argued that proposed approach makes the involved organizations more transparent, integrated, prepared to interoperate their processes and information and able to develop skills to act and achieve shared goals even with existing differences.

The anti-corruption partnership has illustrated the approach application. It has allowed exploring the interorganizational relationship management, besides providing an alternative for different agencies to plan and organize their joint work to achieve a shared goal in a transparent way.

However, it is important to highlight some research challenges. The first challenge concerns the goal detailing. It cannot be so clear to organizations describe the goal shared by the partnership, besides subgoals arising from it. It is also possible to handle organizations that know how to handle required circumstances, but their internal processes may not be described in a structured way. Besides that, organizations may face difficulties in providing non-existent inputs for the partnership because they cannot find a way to do it. Finally, negotiation for responsibilities assignment can be arduous. Solutions to overcome these challenges still need to be developed.

As future work, the specification of a prototype to support approach application is under development, taking into account the public administration peculiarities. In addition, we will work on mechanisms to evaluate the partnership performance and the transparency degree achieved to both involved organizations and Society. Besides that, we aim to evaluate the proposed approach in a real scenario, besides applying it in other domains than the government. The obtained results will provide inputs for a better understanding of approach application impact and help its improvement and evolution.

Acknowledgments. The authors wish to thank CAPES by grants supporting the research.

References

1. Drews, P., Schirmer, I.: From enterprise architecture to business ecosystem architecture. In: International Enterprise Distributed Object Computing Conference, Ulm, pp. 13–22 (2014)
2. Guanciale, R., Gurov, D.: Privacy preserving business process fusion. In: Fournier, F., Mendling, J. (eds.) BPM 2014. LNBIP, vol. 202, pp. 96–101. Springer, Cham (2015). https://doi.org/10.1007/978-3-319-15895-2_9
3. Mueller, T., Schuldt, D., Sewald, B., Morisse, M., Petrikina, J.: Towards inter-organizational enterprise architecture management. In: Americas Conference on Information Systems, Chicago (2013)
4. Ranaei, H., Zareei, A., Alikhani, F.: Inter-organizational relationship management: a theoretical model. Int. Bull. Bus. Adm. **9**, 20–30 (2010)
5. Van Fenema, P., Keers, B., Zijm, H.: Interorganizational shared services: creating value across organizational boundaries. In: Shared Services as a New Organizational Form, vol. 13, pp. 175–217 (2014)

6. Del-Río-Ortega, A., Gutiérrez, A.M., Durán, A., Resinas, M., Ruiz-Cortés, A.: Modelling service level agreements for business process outsourcing services. In: Zdravkovic, J., Kirikova, M., Johannesson, P. (eds.) CAiSE 2015. LNCS, vol. 9097, pp. 485–500. Springer, Cham (2015). https://doi.org/10.1007/978-3-319-19069-3_30

7. Del-Río-Ortega, A., Resinas, M., Durán, A., Ruiz-Cortés, A., Toroa, M.: Visual PPINOT: a graphical notation for process performance indicators. Decis. Support Syst. 1–25 (2015). https://doi.org/10.1007/s12599-017-0483-3

8. Khalfallah, M., Figay, N., Ghodous, P., Da Silva, C.F.: Cross-organizational business processes modeling using design-by-contract approach. In: van Sinderen, M., Oude Luttighuis, P., Folmer, E., Bosems, S. (eds.) IWEI 2013. LNBIP, vol. 144, pp. 77–90. Springer, Heidelberg (2013). https://doi.org/10.1007/978-3-642-36796-0_8

9. Sebu, M., Ciocârlie, H.: Merging business processes for a common workflow in an organizational collaborative scenario. In: International Conference on System Theory, Control and Computing, Cheile Gradistei, pp. 134–139 (2015)

10. Harrison, J.: Strategic Management of Resources and Relationships. Wiley, New York (2003)

11. Luna-Reyes, L., Picazo-Vela, S., Luna, D., Gil-Garcia, R.: Creating public value through digital government: lessons on inter-organizational collaboration and information technologies. In: Hawaii International Conference on System Sciences, pp. 2840–2849 (2016)

12. Markus, M.L., Bui, Q.: Governing public sector interorganizational network infrastructures: the importance of formal and legal arrangements. In: Hawaii International Conference on System Sciences (2011)

13. BBC: Brazil's Odebrecht corruption scandal. http://www.bbc.com/news/business-39194395

14. The Guardian: Brazil: explosive recordings implicate President Michel Temer in bribery. https://www.theguardian.com/world/2017/may/18/brazil-explosive-recordings-implicate-president-michel-temer-in-bribery

15. The Guardian: Brazil's former richest man sought by police in vast corruption inquiry. https://www.theguardian.com/world/2017/jan/26/brazil-corruption-investigation-eike-batista-bribes

16. The New York Times: Sérgio Cabral, Ex-Governor of Rio de Janeiro, Arrested on Corruption Charges. https://www.nytimes.com/2016/11/18/world/americas/sergio-cabral-rio-governor-corruption.html

17. The Wall Street Journal: Brazil's Former House Speaker Eduardo Cunha Arrested in Corruption Investigation. https://www.wsj.com/articles/brazils-former-house-speaker-eduardo-cunha-arrested-in-corruption-investigation-1476895613

18. Time: Brazil Prosecutor Says Massive Corruption Probe Could Double in Size. http://time.com/4651414/brazil-corruption-probe-car-wash/

19. Legner, C., Wende, K.: The challenges of inter-organizational business process design – a research agenda. In: European Conference on Information Systems (2007)

20. Zhu, Z., Huang, H.: The cultural integration in the process of cross-border mergers and acquisitions. Int. Manag. Rev. 3(2), 40–44 (2007)

21. Hofstede, G., Hofstede, G.J.: Culture and Organizations: Software of the Mind. McGraw-Hill, New York (2005)

22. Bocanegra, J., Pena, J., Ruiz-Cortes, A.: Interorganizational business modeling: an approach for traceability of goals, organizational models and business processes. IEEE Lat. Am. Trans. 9(1), 847–854 (2011)

23. Transparency International: Corruption perception index. https://www.transparency.org/news/feature/corruption_perceptions_index_2016

24. Fung, A., Graham, M., Weil, D.: Full Disclosure, the Perils and Promise of Transparency. Cambridge University Press, Cambridge (2007)

25. Denis, J.A., Nunes, V., Ralha, C., Cappelli, C.: E-gov transparency implementation using multi-agent system: a brazilian study-case in lawsuit distribution process. In: Hawaii International Conference on System Sciences, pp. 2772–2781 (2017)
26. Planejamento: Portaria no 68 – Estratégia de Governança Digital da Administração Pública Federal. http://www.planejamento.gov.br/EGD/arquivos/portaria-68-07-03-2016.pdf
27. Brasil: Lei no 12.527 – Lei de acesso à informação, http://www.planalto.gov.br/ccivil_03/_ato2011-2014/2011/lei/l12527.htm
28. Brasil: Lei complementar no 131 – Disponibilização em tempo real de informações. https://www.planalto.gov.br/ccivil_03/Leis/LCP/Lcp131.htm
29. Gazeta do Povo: Estudante detectou sozinha desvio milionário de bolsas que a UFPR não viu. http://www.gazetadopovo.com.br/vida-e-cidadania/estudante-detectou-sozinha-desvio-milionario-de-bolsas-que-a-ufpr-nao-viu-52c7c52x896li4rb2qkrjeona
30. Kutvonen, L.: Enhancing the maturity of open service ecosystems and inter-enterprise collaborations. In: van Sinderen, M., Oude Luttighuis, P., Folmer, E., Bosems, S. (eds.) IWEI 2013. LNBIP, vol. 144, pp. 6–21. Springer, Heidelberg (2013). https://doi.org/10.1007/978-3-642-36796-0_3
31. Santos, R., Cappelli, C., Maciel, C., Leite, J.C.: Transparência em Ecossistemas de Software. In: Workshop de Desenvolvimento Distribuído de Software, Ecossistemas de Software e Sistemas-de-Sistemas, Maringá, pp. 75–79 (2016)
32. Choi, T., Kröschel, I.: Challenges of governing inter-organizational relationships: insights from a case study. In: European Conference on Information Systems, Münster, pp. 1–16 (2015)
33. Köpke, J., Eder, J., Künstner, M.: Projections of abstract interorganizational business processes. In: Decker, H., Lhotská, L., Link, S., Spies, M., Wagner, Roland R. (eds.) DEXA 2014. LNCS, vol. 8645, pp. 472–479. Springer, Cham (2014). https://doi.org/10.1007/978-3-319-10085-2_43
34. Montarnal, A., Wang, T., Truptil, S., Bénaben, F., Lauras, M., Lamothe, J.: A social platform for knowledge gathering and exploitation, towards the deduction of inter-enterprise collaborations. Procedia Comput. Sci. **60**, 438–447 (2015)
35. Dang, D., Pekkola, S.: Systematic literature review on enterprise architecture in the public sector. Electron. J. e-Gov. **15**(2), 132–154 (2017)
36. Eid-Sabbagh, R.-H., Dijkman, R., Weske, M.: Business process architecture: use and correctness. In: Barros, A., Gal, A., Kindler, E. (eds.) BPM 2012. LNCS, vol. 7481, pp. 65–81. Springer, Heidelberg (2012). https://doi.org/10.1007/978-3-642-32885-5_5
37. Eid-Sabbagh, R.-H., Hewelt, M., Weske, M.: Business process architectures with multiplicities: transformation and correctness. In: Daniel, F., Wang, J., Weber, B. (eds.) BPM 2013. LNCS, vol. 8094, pp. 227–234. Springer, Heidelberg (2013). https://doi.org/10.1007/978-3-642-40176-3_19
38. Bouchbout, K., Alimazighi, Z.: Inter-organizational business processes modelling framework. In: Conference on Advances in Databases and Information Systems, Vienna (2011)
39. Lawall, A., Schaller, T., Reichelt, D.: Restricted relations between organizations for cross-organizational processes. In: Conference on Business Informatics, Lisbon (2014)
40. Lin, D., Ishida, T.: Coordination of local process views in interorganizational business process. IEICE Trans. Inf. Syst. **E97-D**(5), 1119–1126 (2014)
41. Hsu, C., Lin, Y.-T., Wang, T.: A legitimacy challenge of a cross-cultural interorganizational information system. Eur. J. Inf. Syst. **24**, 278–294 (2015)
42. Kauremaa, J., Tanskanen, K.: Designing interorganizational information systems for supply chain integration: a framework. Int. J. Logist. Manag. **27**(1), 71–94 (2016)

43. Sun, K., Lai, W.C.: ISAM-based inter-organization information systems alignment process. In: International Conference on Computer Science and Service System, Nanjing, pp. 1358–1361 (2011)
44. Sun, K., Yu, K.: Research on project management for inter-organizational information systems. In: International Conference on E-Business and E-Government, Shanghai, pp. 1–4 (2011)
45. José, H.S.S., Gonçalves, F.E., Cappelli, C., Santoro, F.M.: Providing semantics to implement aspects in BPM. In: Dumas, M., Fantinato, M. (eds.) BPM 2016. LNBIP, vol. 281, pp. 264–276. Springer, Cham (2017). https://doi.org/10.1007/978-3-319-58457-7_20
46. Cappelli, C., Leite, J.C.: Software transparency. Bus. Inf. Syst. Eng. 2(3), 127–139 (2010)

Towards Guidelines of Modeling for Ecology-Aware Process Design

Patrick Lübbecke[(✉)], Peter Fettke, and Peter Loos

German Research Center for Artificial Intelligence, Saarland University, Saarbrücken, Germany
{patrick.luebbecke,peter.fettke,peter.loos}@iwi.dfki.de

Abstract. During the last couple of years, the ecological impact of manufacturing and logistical processes gained importance. Many global players, including Walmart, Kellogg's and L'Oréal, oblige their suppliers to eliminate greenhouse gas emissions in their operations. Contrary to this development, traditional process models are mostly compiled using entities, roles, or resources that are relevant for a mere economical description of a business process. To account for the rising importance of ecological matters, with this work in progress paper we provide ecology-oriented Guidelines of Modeling (EGoM). These design principles allow to compile process models, that can be assessed and optimized in terms of their ecological footprint. The principles are tested through the application of a technique for compliance checking, where we implement a search algorithm for ecological weakness patterns in process models to the BPM software ARIS. The results indicate that the Guidelines foster the application of Green BPM methods that help achieving ecology-friendly process design.

Keywords: Process modeling · Modeling principles · Green BPM

1 Introduction

During the last couple of years, the ecological impact of manufacturing and logistical processes gained importance. Many global players, including Walmart, Kellogg's, or L'Oréal, oblige their suppliers to eliminate a share of greenhouse gas (GHG) or carbon-dioxide (CO_2) emission in the years to come [1]. To comply with these ambitious demands, suppliers must expand their environmental policies from a sheer intra-organizational and manufacturing-centric point-of-view to a holistic approach that includes internal support processes (e.g. administrative processes) and processes in a supply chain (e.g. logistic).

The field of Green IS is devoted to fostering environmentally responsible action by means of Information Systems [2]. In Green IS, many concepts were developed over the last decade, that addressed the analysis and optimization of traditional corporate tasks with respect to ecological goals, for example in manufacturing [3] or administration [4]. Green BPM is a specialization of Green IS, in which established methods from the area of Business Process Management are applied to meet sustainability related goals [5].

© Springer International Publishing AG 2018
E. Teniente and M. Weidlich (Eds.): BPM 2017 Workshops, LNBIP 308, pp. 510–519, 2018.
https://doi.org/10.1007/978-3-319-74030-0_40

The plethora of methods in Green IS and Green BPM come with limitations. Many methods aim at the optimization during the actual execution of tasks, e.g. by measuring exhaust emissions rather than incorporating ecological aspects at early stages of process design. The latter, however, offers more potential for eco-friendly alignment of processes because ex post optimization of processes might be restricted by explicit design choices made during process design which lead to a lock-in to certain technologies or behavior. And most importantly, many of the existing methods in Green BPM are not applied in a context that was adapted to the domain of ecology, but under the same premises that economic-oriented BPM is applied to instead.

With this paper, we present our ongoing work on Guidelines of Modeling for ecological process design (EGoM) that are based on the Guidelines of Modeling proposed by [6, 7]. Therefore, we use the term EGoM for the remainder of this paper. The EGoMs are a set of design principles that identify relevant information or construction principles that should be present in process models (e.g. BPMN, EPC, UML activity diagram). If adhered to the principles, automated process optimization methods such as *compliance checking* or *simulation* can be applied to process models to identify violations of "good ecological practice".

The remainder of the paper is structured as follows: in Sect. 2, we discuss the influence of ecology on process design. In Sect. 3, we introduce to the GoMs by Becker et al. and discuss important concepts of the GoMs. After that, we construct our EGoM framework in Sect. 4 and emphasize the contribution of the design principles to ecological sustainability. To demonstrate the contribution of the framework to ecological process improvement, we redesign a process according to the EGoMs and apply a compliance checking technique on the redesigned process that can automatically identify unecological parts of the process. After that, we conclude the paper with a detailed discussion and conclusion of our work in Sect. 6.

2 Ecological Process Design

When speaking of sustainability, the understanding of the term is very vague. Through its baffling ambiguity, the use of the term has become very inflationary. The term was originally coined in the early 18th century by Carl Hans von Carlowitz, who was a chief mining official at the time. In his book, *Sylvicultura oeconomica*, von Carlowitz advocated prudent management of forest resources, that were necessary to fuel the surrounding ore-smelts. Von Carlowitz pled for a systematic reforestation that would allow the forest to recover and urged to plant new trees according to the amount of wood that was logged [8]. The idea of preserving resources for future generations has since gained even more importance and was later adopted by the Brundtland commission [9] and has become a major objective in *society, environment,* and *economy* [10].

Today, as society and economy are moving closer, these three dimensions have started influencing each other. The rising environmental consciousness in society has urged firms to question the resource consumption and pollution emission of their processes. Many firms have started to add environmental sustainability to the dimensions they consider when designing new processes, i.e., *quality, cost, time, and flexibility* [11].

Sustainability can be introduced to processes during the *engineering* (design of new processes from scratch) or in the *reengineering* phase, when existing processes are analyzed and improved.

The basic idea of *process engineering* is to translate the needs of the customers into products or services that convert raw materials or other resources into value-added components or workflows. New processes that are designed from scratch offer the highest potential for sustainability, as no lock-in to certain technology choices already took place. On the other hand, only few concepts and methods support sustainability at this early stage of process design. A prominent method is reference modeling, which provides best practice for process design that can be implemented "as-is" or be adapted to the specific application context.

Business Process reengineering (BPR), on the other hand, aims at the redesign of existing processes to optimize end-to-end workflows. The term was first introduced in the IT domain, hence BPR is supported by a considerable amount of methods and software tools. In the research field of Green BPM, many methods were developed that support environmental objectives at every stage of the BPM lifecycle [5]. *Monitoring and controlling*, for example, is addressed by [12] who propose a methodology to estimate the carbon foot-print of business processes. For the *optimization and improvement*, traditional methods of BPM like simulation [13] or patterns [4, 14] were adapted to the ecological domain. Opitz et al. [15] discuss the contribution of modeling languages for environmental process improvement.

The major downside is that, while the methods were adapted from traditional BPM, they are applied in the context of sustainability without adjusting the underlying information models to the new context. Most process models are aligned with economic objectives and lack the kind of information that is needed to analyze and optimize the environmental impact of a process with these methods. To exploit the potential of Green BPM methods, we have to make sure that the right kind of information at the right level of granularity is available in process models [2]. With this paper, we provide process modeling conventions that align process models with Green BPM methods, thus supporting environmental objectives.

3   The Guidelines of Modeling

One of the earliest work on modeling conventions were the Guidelines of Business Process Modeling by [6]. Borrowing from the German *Generally Accepted Accounting Principles (GAAP)*, Becker et al. proposed six guidelines that determine model quality: *correctness, relevance, economic efficiency, clarity, comparability*, and *systematic design*.

Correctness means, that the model follows semantical and syntactic rules that are provided by the meta model of the modeling language. In order to be semantically correct, the structure and the behavior of the model have to be consistent with the real world [7].

To accomplish *relevance,* a model must include all the elements that are required to properly describe the context of the model. From a viewpoint of critical realism, [16] plead for the use of the terms *construction adequacy* and *language adequacy* instead of *correctness* and *relevance*. The authors argue that a definite representation of the real world is not possible, therefore a definite judgement about whether a model is "correct" is not feasible.

Language adequacy means that the modeler provides his/her model using the right modeling language. For process modeling, an UML class diagram would not be adequate because it does not allow to represent all the necessary information.

Economic Efficiency refers to the policy, that only elements are modelled that are necessary for the objective of the model.

Clarity refers to the understandability of the model by the addressee. This can be very subjective which is why many authors proposed auxiliary criteria that support clarity in models.

Comparability means, that all models of the same modeling project comply with the same set of rules to be able to compare the models. This can relate to the use of language, grammar, or granularity [16].

Systematic Design relates to the coupling of information models of different views, for example, data models with process models [7] by identifying corresponding elements in both model types.

4   Construction of the EGoM Framework

In this section, the implications of the GoM for modeling eco-friendly processes will be analyzed and condensed to the EGoM framework. This will help to find an answer to an important issue in Green BPM: what information at what level of granularity is necessary to optimize a given process model [2]. To the best of our knowledge, there are no GoMs in the field of ecology available yet.

Starting point for ecological process improvement is the *strategy* of the firm that determines the sustainability strategy (ecology-centric, economy-centric, or something in between). The strategy leads to ecological objectives, e.g. reduction of GHG emissions or substitution of problematic substances [17]. The strategy is implemented through *processes* that comprise of single *activities* in which *tools* are used to process *resources*. These four elements are domain-specific. For example, the tools that are used in manufacturing processes (mill, lathe) can be different from tools used in administration (photocopier) or logistics (forklift). Independently from the domain, information about these four elements must be provided in models. The *data* layer captures necessary information of the mentioned elements that go beyond actual activities and involved tools, which are normally provided in process models. Therefore, this paper will focus on the process layer (Fig. 1).

As was mentioned before, models comprise the elements that are necessary for the objective of the model, which is traditionally to ensure *quality, cost, time,* or *flexibility*. To assess the environmental impact of a process, more specific information is necessary to be provided in process models. Part of syntactic and semantic *correctness,* or *construction adequacy* [16], is a consensus about the naming convention and the information objects that are used in the model. To assess the environmental impact of

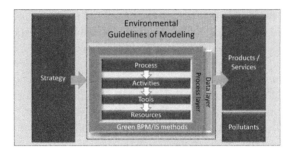

Fig. 1. The EGoM framework

activities, these two facets are of particular importance. A more precise naming helps to find substitutes for existing activities that are eco-friendlier. The label of the administrative activity "create invoice", for example, does not put any emphasis on what tools were used to create the invoice and what resources were consumed during the activity. A more precise label "print invoice (laser)" allows to understand what tools were involved, what resources were consumed (paper, electrical energy for the thermal process), and what the result of the process is. The same applies for the information objects. Traditional objects of BPMN or EPC do not sufficiently allow to model the ecological impact of an activity. Hence, we postulate the three EGoMs G1, G2a, and G2b (Table 1).

Table 1. The ecological guidelines of modeling

Correctness/Const. adequacy	*G 1: Labels should include precise information about whether something is created, modified or depleted. This applies to physical and non-physical objects*
	G 2a: Every tool that is used for an activity (ICT, machine) shall be represented by its own information object to allow assessing the environmental impact
	G 2b: For every tool, the necessary type of resources should be modelled accordingly
Relevance/Lang. adequacy	*G 3: For every resource, the extent of its consumption per activity should be modelled to allow comparison with alternative techniques*
Economic efficiency	*G 4: Evaluate, what information in process models is necessary for the possible application of traditional BPM methods*
Clarity	*G 5a: Prior to the modeling project, an assessment of tools used in the domain and their possible pollutants should be available*
	G 5b: For the modeling of pollutants and resource consumption, a consistent unit system should be used throughout all processes
Comparability	*G 6: The comparability of as-is with to-be models should be secured by providing a semantic representation of the activities in that domain*
Systematic design	*G 7: For the data layer, a domain specific data model identifying relevant elements and their relationships should be used*

The original guideline of *relevance*, or *language adequacy* refers to the selection of a relevant object system [7], which means to identify objects that are relevant for the discourse. For an object to be *relevant* means, that the absence of a specific object would come along with a loss of meaning for the model user. We believe that, from an ecological standpoint, it is essential for a model to provide information not only about the type of resource consumption, but about the extent as well [5]. With this information provided, activities can be benchmarked with tasks that serve the same purpose but are more ecological (G3).

The guideline of economic efficiency addresses the level of detail that models should have in order to provide a means of improvement. The rule of thumb is to make sure the extent of the modeling process does not exceed the benefit that arises through the modeling project. This guideline can be perceived as a "meta guideline" that defines the boundaries for all the other guidelines. The relationship between ecological improvement and economic efficiency is far from trivial. Some ecological improvements go along with economic benefits. Some eco-improvements, on the other hand, can only be implemented with additional costs. Economic efficiency vastly depends on the strategy and what the implementing firm is willing to invest to pursue ecological objectives. As we will demonstrate in the following section, the ecological improvement of processes with BPM methods can be fostered, if the proposed guidelines are considered. For now, we formulate the guideline G4.

The understanding of the model by the addressee by means of hierarchical decomposition, layout design, and filtering of information is subject to the principle of *clarity* [16]. Clarity in the sense of understandability must be differentiated from the semantical correctness of G1. Clarity means that the modeler draws from a consistent system of objects in all related processes. Such an object system must be provided by domain experts. In the administrative domain, for example, a photocopier, a laser printer, and an ink-jet printer can be used to print documents, or in more general terms "to create an information output". To be consistent in the model system would require using the proper label (i.e. "print document") along with the correct object type (output type), and most importantly the correct tool (photocopier, laser, or ink-jet printer). Regarding the tools, an assessment about relevant tools of a certain domain or for the firm that is modeled must be performed prior to the process design (G5).

The guideline of *comparability* was proposed to allow the comparison of as-is with to-be models both, semantically and syntactically. In the ecological domain, the relevant aspects that should be compared are not primarily analogies in process constructs. Instead, constructs of the as-is process should be compared to constructs of to-be processes, e.g., reference models or process patterns in terms of their semantic meaning. For example, if a function for the mailing of an invoice to the customer should be improved, the model system must support the filtering of similar process constructs and a hypothetical automatic recommender system should only suggest (a) *send invoice by mail*, (b) *by e-mail*, (c) *enclose with the goods* as possible alternatives (G6). To accomplish that, a semantic description of the activities should be provided for a specific domain. This, however, is subject to further research.

Systematic design relates to the consistent modeling of two or more layers, e.g. data- and process model. Important for the ecological improvement of processes is the design of the process model on the one hand, and the mapping of related resource and emission data to activities and tools on the other hand. To be consistent, a meta model for the data layer must be provided that represents relevant tools, emissions, and most importantly their relationship among each other (G7).

5   Application of the Framework

In this section, the contribution of the EGoM framework for ecological process improvement will be demonstrated. During earlier research we identified 26 ecological weakness patterns for administrative processes that should be avoided in process models [4]. Weakness patterns represent formalized knowledge about adverse process constructs (activities or tools) that can be performed in a more ecological way (i.e. with fewer resource consumption). The patterns have been synthesized from knowledge of domain experts. By using techniques of compliance checking, these adverse process constructs can be found automatically, and the process can then be revised. Unfortunately, every modeler has his/her own preferences in terms of naming conventions or granularity. Therefore, most real-world process models do not comply with the EGoM principles and therefore cannot be processed automatically due to missing or malformed information. For example, if no information is provided in terms of what tool was used during a certain activity, an automated recommendation of more sustainable ways to do it is not possible.

To demonstrate our reasoning, we have used as-is process models that were collected from German municipal administrations and we transformed the models based on the EGoM principles. Among others, we have added missing information and followed a consistent naming scheme. We then applied process mining techniques such as pattern matching and recognition or Natural Language Processing (NLP) to identify the constructs in these models that are considered to be a weakness. Figure 2 shows the initial process on the left side, and a version that was adjusted based on the EGoM principles on the right.

As can be seen in the initial process model on the left side, most elements are denoted in a very generic way and important information is missing. There is no information present in terms of what "generate permission" means, which tools are involved in this activity, and what kind of information is created by this activity. The same problem applies to "send permission to applicant". This activity does not provide any information, if the permission is sent by mail, e-mail, or fax. Only an expert who is familiar with the domain can understand the activities and come up with ecological optimization. Without this kind of information, an automated compliance checking against weakness patterns would not be possible.

With the required information added to the revised process, we could implement an automated compliance check in the BPM software ARIS and actually find instances of the 26 weakness patterns in the revised process models. Figure 3 exemplarily shows two identified weaknesses along with suggestions for improvement that were defined for each weakness pattern (Lübbecke u. a. 2016). The results indicate that a recommendation of more ecological process patterns can be successfully implemented if the process is annotated with information according to the EGoMs.

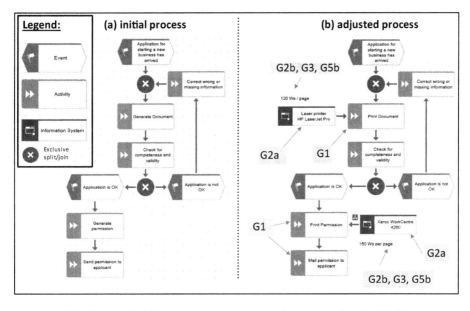

Fig. 2. An administrative process prior to and after applying the EGoM

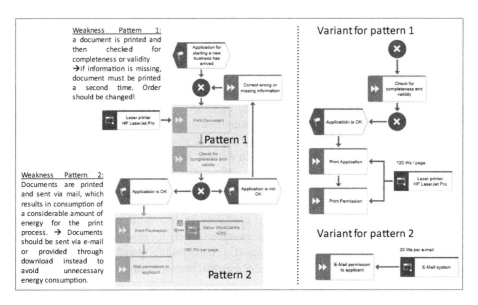

Fig. 3. Identification of weakness patterns in the process model

6   Discussion and Conclusion

It belongs to human nature that every modeler has his/her own preferences and style when it comes to modeling. The results are a varying level of detail, diverging name schemes, and even a lack of important information in process models. Unfortunately, this prevents from the application of many Green BPM methods that draw from information in process models. Process models are the springboard for process improvement. They allow to assess weaknesses and to replace adverse constructs with more favorable solutions. In this paper, we presented our ongoing work on creating Ecological Guidelines of Modeling. The intention of the guidelines is to provide a set of rules on how processes should be modeled to be able to analyze and optimize the models with Green BPM methods and make them more ecological.

To apply our design principles, we have used processes that were collected from public administration and revised the processes according to the EGoM principles. We then applied *compliance checking* on the models, which is a prominent technique from the field of Green BPM that requires process models to provide a certain set of information. With compliance checking and 26 ecological weakness patterns that were synthesized in prior research, we could find instances of the weaknesses in the revised models and propose improvement measures for each weakness. The results of the experiment indicate that following the EGoM principles can indeed contribute to models that allow the application of automated Green BPM methods and therefore foster ecological improvement.

Our approach, however, comes with some limitations. First and foremost, we have evaluated the approach in a synthetic scenario where we transformed existing process models according to the design principles. To increase the validity of the evaluation, the EGoMs should be applied in a case study where an unbiased process designer creates new process models referring to the principles. The execution of compliance check or any other Green BPM method on the process models will lead to more meaningful results and the number of identified weaknesses could serve as a metric.

A second limitation comes with the nature of the applied research method. The guidelines were extracted deductively through an argumentative analysis. Some assumptions made during the process of deduction could be wrong. To eliminate this second limitation, the guidelines should be exposed to experts in the field of (Green) BPM, modeling professionals, and domain experts (e.g. in manufacturing, logistics, administration) to discuss the results and to define some of the guidelines more specifically. An evaluation of that kind shall be the next steps of our work.

Future research will focus on the integration of the ecological guidelines to BPM software tools to support the process of modeling at design-time. For example, if a process modeler misses to use proper information objects for involved tools (guideline G2a) or uses inconsistent naming schemes (G1, e.g. "print", "printed", "printing"), an instant notification could pop up to avoid these mistakes instantly.

Acknowledgement. The research described in this paper was supported by a grant from the German Ministry of Education and Research (BMBF), project name GreenFlow, support code 01IS12050.

References

1. Pyper, J.: Business: 64 companies follow Wal-Mart's effort to reduce suppliers' emissions (2013). http://www.eenews.net/climatewire/stories/1059981502
2. Watson, R.T., Boudreau, M.-C., Chen, A.J.: Information systems and environmentally sustainable development: energy informatics and new directions for the IS community. Manag. Inf. Syst. Q. **34**, 23–38 (2010)
3. Lee, H., Ryu, K., Son, Y.J., Cho, Y.: Capturing green information and mapping with MES functions for increasing manufacturing sustainability. Int. J. Precis. Eng. Manuf. **15**, 1709–1716 (2014). https://doi.org/10.1007/s12541-014-0523-6
4. Lübbecke, P., Fettke, P., Loos, P.: Sustainability patterns for the improvement of IT-related business processes with regard to ecological goals. In: Dumas, M., Fantinato, M. (eds.) BPM 2016. LNBIP, vol. 281, pp. 428–439. Springer, Cham (2017). https://doi.org/10.1007/978-3-319-58457-7_31
5. Houy, C., Reiter, M., Fettke, P., Loos, P., Hoesch-Klohe, K., Ghose, A.: Advancing business process technology for humanity: opportunities and challenges of green BPM for sustainable business activities. In: vom Brocke, J., Seidel, S., Recker, J. (eds.) Green Business Process Management, pp. 75–92. Springer, Heidelberg (2017). https://doi.org/10.1007/978-3-642-27488-6_5
6. Becker, J., Rosemann, M., Schütte, R.: Grundsätze ordnungsmäßiger Modellierung. Wirtschaftsinformatik **37**, 435–445 (1995)
7. Becker, J., Rosemann, M., von Uthmann, C.: Guidelines of business process modeling. In: van der Aalst, W., Desel, J., Oberweis, A. (eds.) Business Process Management. LNCS, vol. 1806, pp. 30–49. Springer, Heidelberg (2000). https://doi.org/10.1007/3-540-45594-9_3
8. von Carlowitz, H.C.: Sylvicultura Oeconomica (1713)
9. World Commission on Environment and Development: Report of the WCoED: our common future. Med. Confl. Surviv. **4**, 300 (1987)
10. Adams, W.M.: The Future of Sustainability (2006). https://doi.org/10.1007/1-4020-4908-0
11. Seidel, S., Recker, J., vom Brocke, J.: Green business process management. In: vom Brocke, J., Seidel, S., Recker, J.C. (eds.) Green Business Process Managementt, pp. 3–13. Springer, Heidelberg (2012). https://doi.org/10.1007/978-3-642-27488-6_1
12. Recker, J.C., Rosemann, M., Hjalmarsson, A., Lind, M.: Modeling and analyzing the carbon footprint of business processes. In: vom Brocke, J., Seidel, S., Recker, J.C. (eds.) Green Business Process Management, pp. 93–109. Springer, Heidelberg (2012). https://doi.org/10.1007/978-3-642-27488-6_6
13. Wesumperuma, A., Ginige, J.A., Ginige, A., Hol, A.: A framework for multi-dimensional business process optimization for GHG emission mitigation. In: Australasian Conference on Information Systems, pp. 1–10 (2011)
14. Nowak, A., Leymann, F., Schleicher, D., et al.: Green business process patterns. In: Proceedings of the 18th Conference on Pattern Languages of Programs - PLoP 2011, pp. 1–10 (2011)
15. Opitz, N., Erek, K., Langkau, T., et al.: Kick-starting green business process management–suitable modeling languages and key processes for green performance measurement. In: Americas Conference on Information Systems, pp. 1–10 (2012)
16. Schuette, R., Rotthowe, T.: The guidelines of modeling – an approach to enhance the quality in information models. In: Ling, T.-W., Ram, S., Li Lee, M. (eds.) ER 1998. LNCS, vol. 1507, pp. 240–254. Springer, Heidelberg (1998). https://doi.org/10.1007/978-3-540-49524-6_20
17. Reiter, M., Fettke, P., Loos, P.: Towards a reference model for ecological IT service management. In: International Conference on Information Systems, Milan, pp. 1–20 (2013)

Sustainability Performance Measurement

A Preliminary Classification Framework of Models and Indicators

Dries Couckuyt[1](✉) ⓘ, Amy Van Looy[1] ⓘ, and Manu De Backer[1,2] ⓘ

[1] Department of Business Informatics and Operations Management, Faculty of Economics and Business Administration, Ghent University, Tweekerkenstraat 2, 9000 Ghent, Belgium
{Dries.Couckuyt,Amy.VanLooy,Manu.DeBacker}@ugent.be
[2] Department of Management Information Systems, Faculty of Economics and Business Administration, University of Antwerp, Prinsstraat 13, 2000 Antwerp, Belgium

Abstract. In this position paper we focus on the diversity of sustainability measurements. Based on existing research on performance measurement, we propose a preliminary classification framework summarizing sustainability models and indicators. By describing illustrative examples, we claim that several models and indicators can be distinguished with their own peculiarities. Having such a framework is interesting for both academia and business to structure the range of models and indicators and to ultimately select the appropriate sustainability measurement approach. The proposed framework should be validated by further research.

Keywords: Sustainability · Green BPM · Performance measurement

1 Introduction

Sustainability gained momentum for practice and research in recent years. Organizations introduce sustainable technologies and focus on optimizing their operations in a sustainable way in order to meet customer demands and jurisdictive requirements [13]. For this purpose, business process management (BPM), which refers to a body of methods, techniques and tools to discover, analyze, redesign, execute and monitor business processes [1], is a key starting point. The objectives of conventional BPM methods and techniques, however, typically refer to cost, quality, time, and flexibility improvements - the so-called 'devil's quadrangle' [16]. Since sustainability also gained importance in the field of BPM, [18] call for an extended version that also includes sustainability as a fifth process performance dimension.

This position paper aims at providing a way to classify performance measurements from the perspective of sustainability. We will propose a preliminary classification framework to better understand the variations in sustainability performance measurement. This framework can provide a foundation for future sustainability performance measurement research striving to advance a sustainable perspective in BPM research and practice.

© Springer International Publishing AG 2018
E. Teniente and M. Weidlich (Eds.): BPM 2017 Workshops, LNBIP 308, pp. 520–524, 2018.
https://doi.org/10.1007/978-3-319-74030-0_41

2   Sustainability in BPM

The Triple Bottom Line defined three interdependent dimensions of sustainability (i.e. economic, social and environmental) on which organizations should focus in order to succeed in the long run [2]. It could be argued that the economic dimension of sustainability is already included in conventional BPM as it optimizes processes along time, cost, quality, and flexibility. Concerning the other two dimensions, the existing approaches in the area of sustainability and BPM are primarily focusing on the environmental scope, and often overlooking the social aspect [11, 19]. Previous literature reviews mention several contributions to the field of Green BPM which concerns the ecological impact of business processes [5, 15, 19]. We are aware that a full integration of sustainability in the BPM discipline also requires research and practice on the social dimension. However, because sustainability is interpreted in environmental terms by a majority of authors in the BPM discipline, the present article continues with the same comprehension and thus discusses measurements of sustainability in light of ecology.

3   Sustainability Performance Measurement

As organizations aim to achieve outstanding results, performance measurement is of crucial importance. By reviewing extant literature, [20] identified two groups of papers on this topic. The first group focused on performance measurement models; the second on performance indicators. For the first group, they distinguished models focusing on the entire organization from models focusing on a single business process. For each type of performance measurement model, indicators should be defined. A performance indicator describes how it is measured and how it can be compared against a target value. For instance, an ecological performance indicator can be measured in CO_2 per process instance with a target value of zero, meaning that the aim is to eliminate carbon completely. For this second group, a further distinction was made between indicators found with operationalization (i.e., concretization by means of a formula) and those without operationalization. Based on these findings, we now propose a framework to classify several sustainability performance measurement models and indicators (Table 1). The remainder of this section describes each category with an illustrative example.

Table 1. Preliminary classification framework of sustainability models and indicators.

Sustainability		Performance measurement model	
		Organizational	Business process
Performance indicators	Operationalized	e.g. GRI, CSDI	e.g. gCO$_2$eq/kWh, Ws
	Non-operationalized	e.g. SBSC, ISO 14001:2015	e.g. 'Green KPIs', 'sustainability'

First, organizational performance measurement models typically intend to provide a holistic view of an organization's performance. In case indicators are concretized by means of a formula, the model includes operationalized performance indicators. For instance, The Global Reporting Initiative (GRI) [4] offers a voluntary reporting framework that contains goals and operationalized indicators with respect to environmental but also economic and social sustainability, i.e. as stated in the Triple Bottom Line [2]. The dimensions are not correlated so the ecological indicators can act as environmental sustainability guidelines for organizations. This is also the case for the Composite Sustainable Development Index (I_{CSD}) [10] which is composed of the economic sub-index ($I_{S,1}$), the environmental sub-index ($I_{S,2}$) and the social sub-index ($I_{S,3}$). These sub-indices are in turn composed of respectively normalized economic, environmental and social indicators extracted from other frameworks (including GRI). For both GRI and CSDI holds that by means of extensive standardization (i.e. operationalization of indicators), companies are offered a tool which allows them to benchmark with other businesses. However, this standardization also impedes the integration of company-specific sustainability endeavors because a general reporting frame is set. Moreover, other dimensions that could be important for the company or stakeholders are neglected if they are not included in the framework.

Secondly, more flexibility is offered by organizational performance measurement models without operationalized indicators. For instance, the Balanced Score Card (BSC) provides four perspectives (i.e. financial, customer, internal processes, learning and growth) for which objectives and performance indicators ensure alignment between strategies and operations [8]. The BSC is an open system which means that all stakeholder interests can be included if they are vital for the success of a strategy. Therefore, it is possible to develop a variant, the Sustainability BSC (SBSC), that integrates strategically relevant environmental goals [12]. Such goals can be integrated into the four existing performance perspectives, or a new key perspective can be added. These SBSCs and associated strategy maps constitute an open framework that comprises sustainability-oriented indicators [6]. Similarly, ISO 14001:2015 [7] provides a framework for environmental management in organizations. It provides guidelines that can be applied to the environmental aspects of activities, products or services in line with organizational objectives. As for SBSC, also these indicators should be defined and operationalized at the individual company level, which means that they are less appropriate for benchmarking.

Thirdly, in addition to organizational models, performance measurement can focus on a single business process. This approach is generally less holistic and provides concrete measurements. For instance, [3] introduced a framework that models the relationship between resources and activities to inform the business process with its carbon emission impact. Besides emissions, literature on the reduction of energy consumption in business processes also presents operationalized performance indicators [17]. These sort of indicators are operationalized with figures and formulas provided by environmental authorities (e.g. grams of CO_2 equivalent per kilowatt hour of generation, [gCO_2eq/kWh]) [3], or rely on accepted measurement units (e.g. Watt seconds [Ws]) [17].

The fourth category covers performance indicators without operationalization for single business processes. For instance, [14] presentes a list of 'Green KPIs' at the business process level (e.g. Power Usage Effectiveness, Data Center Infrastructure Efficiency, Data Center Energy Productivity) without any operationalization of the measurements. Similarly, [18] states that business processes should be optimized in light of 'sustainability'. However, to the best of our knowledge, examples of this category are rare in the literature, indicating that scholars mainly propose concrete performance indicators at the business process level. This category has added value from the descriptive point of view and with the intention for operationalization in later work.

4   Discussion and Conclusion

Based on existing research on performance measurement, we propose a framework summarizing: (1) sustainability models at the level of an entire organization or a single business process, and (2) indicators, operationalized or non-operationalized by means of a formula. By describing illustrative examples, we show that several models and indicators can be distinguished with their own peculiarities. For instance, a sustainable organizational performance measurement model with operationalized indicators (e.g. GRI) allows to compare between companies in a specific business sector but does not offer the flexibility to integrate company-specific sustainability endeavors (e.g. SBSC).

Since this is a position paper, no structured approach was used to identify all possible sustainable performance models and indicators. Therefore, we recommend to retrieve all available literature on sustainability performance measurement by means of a systematic literature review [9]. Subsequently, the preliminary classification framework can be validated by mapping the identified performance measurement models and indicators. This exercise should also give answer to the question if the proposed categories are sufficient to guide academics and practitioners. Moreover, it will also clarify if double classifications, e.g. a model for both the organizational and process level, exist.

Instead, we now merely focused on the diversity of sustainability measurements to foster further research in the area. A sustainable transition of organizations and underlying business processes will require multiple measurement methods and techniques. Our proposed framework is interesting for both academia and business to structure the range of models and indicators and to ultimately select the appropriate sustainability measurement approach.

References

1. Dumas, M., La Rosa, M., Mendling, J., Reijners, H.A.: Fundamentals of Business Process Management. Springer, Heidelberg (2013). https://doi.org/10.1007/978-3-642-33143-5
2. Elkington, J.: Cannibals with Forks: the Triple Bottom Line of Sustainable Development, 1st edn. Capstone Publishing, Oxford (1997)
3. Ghose, A., Hoesch-Klohe, K., Hinsche, L., Le, L.-S.: Green business process management: a research agenda. Australas. J. Inf. Syst. 16(2), 103–117 (2009)
4. Global Reporting Initiative. https://www.globalreporting.org/standards/gri-standards-download-center. Accessed 30 May 2017
5. Gohar, S.R., Indulska, M.: Business Process Management: Saving the Planet? In: Australasian Conference on Information Systems, Adelaide, pp. 1–14 (2015)
6. Hansen, E.G., Schaltegger, S.: The sustainability balanced scorecard: a systematic review of architectures. J. Bus. Ethics 133(2), 193–221 (2016)

7. International Organization for Standardization (ISO): ISO 14001:2015 Environmental Management Systems – Requirements with Guidance for Use. https://www.iso.org/obp/ui/#iso:std:iso:14001:ed-3:v1:en
8. Kaplan, R.S., Norton, D.P.: The Balanced Scorecard: Translating Strategy into Action, 1st edn. Harvard Business School Press, Boston (1996)
9. Kitchenham, B.: Procedures for performing systematic reviews. Keele University, Keele, Technical report, pp. 1–28 (2004)
10. Krajnc, D., Glavič, P.: A model for integrated assessment of sustainable development. Res. Conserv. Recycl. **43**, 189–208 (2005)
11. Magdaleno, A.M., Duboc, L., Betz, S.: How to Incorporate Sustainability into Business Process Management Lifecycle? In: Dumas, M., Fantinato, M. (eds.) BPM 2016. LNBIP, vol. 281, pp. 440–443. Springer, Cham (2017). https://doi.org/10.1007/978-3-319-58457-7_32
12. Möller, A., Schaltegger, S.: The sustainability balanced scorecard as a framework for eco-efficiency analysis. J. Ind. Ecol. **9**(4), 73–83 (2005)
13. Nowak, A., Leymann, F., Schumm, D.: The differences and commonalities between green and conventional business process management. In: International Conference on Dependable, Autonomic and Secure Computing Proceedings, pp. 569–576. IEEE, Sydney (2011)
14. Opitz, N., Erek, K., Langkau, T., Kolbe, L., Zarnekow, R.: Kick-starting green business process management – suitable modeling languages and key processes for green performance measurement. In: Americas Conference on Information Systems Proceedings, Seattle, pp. 1–10 (2012)
15. Opitz, N., Krüp, H., Kolbe, L.M.: Green business process management - a definition and research framework. In: Hawaii International Conference on System Sciences Proceedings, pp. 3808–3817. IEEE, Waikoloa (2014)
16. Reijers, H.A., Mansar, S.L.: Best practices in business process redesign: an overview and qualitative evaluation of successful redesign heuristics. Omega **33**(4), 283–306 (2005)
17. Reiter, M., Fettke, P., Loos, P.: Towards green business process management: concept and implementation of an artifact to reduce the energy consumption of business processes. In: Hawaii International Conference on System Sciences Proceedings, pp. 885–894. IEEE, Waikoloa (2014)
18. Seidel, S., vom Brocke, J., Recker, J.: Call for action: investigating the role of business process management in green IS. In: All Sprouts Content, pp. 1–6. AISeL (2011)
19. Stolze, C., Semmler, G., Thomas, O.: Sustainability in business process management research - a literature review. In: Americas Conference on Information Systems Proceedings, pp. 1–10. AISeL, Seattle (2012)
20. Van Looy, A., Shafagatova, A.: Business process performance measurement: a structured literature review of indicators, measures and metrics. SpringerPlus **5**(1), 1–24 (2016)

5th International Workshop on Declarative/Decision/Hybrid Mining and Modeling for Business Processes (DeHMiMoP 2017)

5th International Workshop on Declarative/Decision/Hybrid Mining and Modeling for Business Processes (DeHMiMoP'17)

Jan Vanthienen[1], Claudio Di Ciccio[2], Hajo A. Reijers[3], Tijs Slaats[4],
Dennis Schunselaar[3], and Søren Debois[5]

[1] Department of Management Informatics, KU Leuven, Naamsestraat 69, 3000
Leuven, Belgium
jan.vanthienen@kuleuven.be
[2] Institute for Information Business, WU Vienna, Welthandelsplatz 1, 1020
Vienna, Austria
claudio.di.ciccio@wu.ac.at
[3] Department of Computer Science, VU University Amsterdam, De Boelelaan
1081, 1081 Amsterdam, The Netherlands
{h.a.reijers,d.m.m.schunselaar}@vu.nl
[4] Department of Computer Science, University of Copenhagen, Emil Holms
Kanal 6, 2300 Copenhagen, Denmark
slaats@di.ku.dk
[5] Department of Computer Science, IT University of Copenhagen, Rued
Langgaards Vej 7, 2300 Copenhagen, Denmark
debois@itu.dk

Abstract. Rules, decisions, and workflows are intertwined components depicting the overall process. So far imperative workflow modelling languages have played the major role for the description and analysis of business processes. Despite their undoubted efficacy in representing sequential executions, they hide circumstantial information leading to the enactment of activities, and obscure the rationale behind the verification of requirements, dependencies, and goals. This workshop aimed at providing a platform for the discussion and introduction of new ideas related to the development of a holistic approach that encompasses all those aspects. The objective was to extend the reach of the business process management audience towards the decisions and rules community and increase the integration between different imperative, declarative and hybrid modelling perspectives. Out of the high-quality submitted manuscripts, three papers were accepted for publication, with an acceptance rate of 50%. They contributed to foster a fruitful discussion among the participants about the respective impact and the interplay of decision perspective and the process perspective.

Keywords: Decision models · Business process models
Decision mining · Declarative processes · Hybrid processes

1   Aims and Scope

Business Process Management (BPM) and its life cycle activities – design, modelling, execution, monitoring and optimisation of business processes – have become a crucial part of business management. Most processes and business process models incorporate rules and decisions of some kind that describe the premises and possible outcomes of a specific situation. In particular, Knowledge-intensive Processes (KiPs) such as checking creditworthiness in a financial process, claim acceptance in an insurance process, eligibility decisions in social security, etc., highly rely on such rules and decisions to guide the workflows of all process stakeholders. The high variability of the situations in which those processes are enacted makes them very flexible by nature, and calls for the explicit description of all the involved rules and decisions to properly depict their behaviour.

While traditional imperative notations such as BPMN excel at describing happy paths, they turn out to be rather inadequate for modelling rules and decisions. Imperative notations indeed tend to describe possible behaviour as alternative, restricted flows. As a consequence, encompassing all possible variations makes imperative models very cluttered and thus proves to be impractical in highly flexible scenarios. Against this background, a new declarative modelling paradigm has been proposed that aims to directly capture the business rules or constraints underlying the process. Academic interest in the approach has grown in recent years, leading to the development of several declarative notations, such as Declare, DCR Graphs, DMN, CMMN, GSM and eCRG.

However, declarative notations have struggled with industrial adoption. A common hypothesis is that the declarative paradigm requires modellers to think in radically new ways, which makes them hesitant to abandon the imperative approaches that they are used to. Preliminary research has indeed shown that users are by far more perceptive towards the idea of combining their imperative work practices with the new declarative approach. Grounded in these observations, a hybrid paradigm has been proposed, which aims to combine the strengths of both the imperative and declarative approaches.

The presented papers emphasised the role of decisions both in the modelling of processes, and in the automated discovery of KiPs' business rules. The separation of concerns and subsequent integration of decision and process modelling has been the main challenge tackled by the works of Hasić et al. and of Bazhenova et al. In particular, Hasić and his colleagues discuss in the paper entitled "Challenges in Refactoring Processes to Include Decision Modelling" insights and challenges for an integrated, holistic modelling approach of processes and decisions that does not confine decisions to single gateways or activities, but rather considers their spanning over multiple process model fragments. They further elaborate on possible inconsistencies arising when integrating the two perspectives. In their work entitled "Data-Centric Extraction of DMN Decision Models from BPMN Process Models", Bazhenova and her colleagues investigated patterns of BPMN modelling process-data-driven decisions, and subsequently formally mapped them onto DMN decision requirements diagram fragments. Thereupon, they validate the proposed mapping through its application on an example taken from the healthcare domain. Finally, Campos and his colleagues present in their manuscript "Discovering Business Rules in Knowledge-intensive Processes Through Decision Mining: an Experimental Study" the results of the application of a decision mining technique on data-sets registered during the executions of a KiP. The event log consists both of structured and non-structured data. The mined rules pertain to the resolution of ICT incidents related to client's assets, such as e-mail server outages and network connection problems.

2   Workshop Co-organisers

Jan Vanthienen	KU Leuven, Belgium
Claudio Di Ciccio	WU Vienna, Austria
Hajo A. Reijers	VU University Amsterdam, Netherlands
Tijs Slaats	University of Copenhagen, Denmark
Dennis Schunselaar	VU University Amsterdam, Netherlands
Søren Debois	IT University of Copenhagen, Denmark

3   Program Committee

Bart Baesens	KU Leuven, Belgium
Fernanda Baião	Universidade Federal do Estado do Rio de Janeiro, Brazil
Andrea Burattin	University of Innsbruck, Austria
Massimiliano de Leoni	Eindhoven University of Technology, Netherlands
Riccardo De Masellis	Fondazione Bruno Kessler, Italy
Johannes De Smedt	KU Leuven, Belgium
Chiara Di Francescomarino	Fondazione Bruno Kessler, Italy
Thomas Hildebrandt	IT University of Copenhagen, Denmark
Amin Jalali	Stockholm University, Sweden
Fabrizio M. Maggi	University of Tartu, Estonia
Andrea Marrella	Sapienza University of Rome, Italy
Marco Montali	Free University of Bozen-Bolzano, Italy
Jorge Munoz-Gama	Pontificia Universidad Católica de Chile, Chile
Stefan Schönig	University of Bayreuth, Germany
Seppe K. L. M. vanden Broucke	KU Leuven, Belgium
Barbara Weber	Technical University of Denmark, Denmark
Richard Weber	Universidad de Chile, Chile
Qiang Wei	Tsinghua University, China
Mathias Weske	Hasso Plattner Institute, University of Potsdam, Germany

Challenges in Refactoring Processes to Include Decision Modelling

Faruk Hasić[1(✉)], Lesly Devadder[1], Maxim Dochez[1], Jonas Hanot[1],
Johannes De Smedt[2], and Jan Vanthienen[1]

[1] Leuven Institute for Research on Information Systems (LIRIS),
KU Leuven, Leuven, Belgium
{faruk.hasic,jan.vanthienen}@kuleuven.be

[2] Management Science and Business Economics Group, University of Edinburgh
Business School, Edinburgh, Scotland
johannes.desmedt@ed.ac.uk

Abstract. Until lately decisions were regularly modelled as a part of
the process model, negatively affecting the maintainability, comprehensibility and flexibility of processes as well as decisions. The recent establishment of the Decision Model and Notation (DMN) standard provides
an opportunity for shifting in favour of a separation of concerns between
the decision and process model. However, this challenge of separation
of concerns and subsequently consistent integration has received limited
attention. This work discusses difficulties and challenges in separating
the concerns and then integrating the two models. The most challenging scenario is when integrating decision models which entail the process
holistically, rather than merely focusing the decisions on local decision
points. In this work, we shed a light on the importance of the separation
of concerns and identify inconsistencies that might arise when separating
and integrating processes and decisions.

Keywords: Decision modelling · DMN · Process modelling · BPMN
Integrated modelling · Separation of concerns

1 Introduction

An increased interest in separating the decision and process concerns in modeling and mining is present in scientific literature, as illustrated by the vast body
of recent literature on Decision Model and Notation [1–4]. DMN consists of two
levels. Firstly, the decision requirement level in the form of a Decision Requirement Diagram (DRD) is used to portray the requirements of decisions and the
dependencies between the different constructs in the decision model. Secondly,
the decision logic level is used to specify the underlying decision logic. The standard also provides an expression language S-FEEL (Simple Friendly Enough
Expression Language), as well as boxed expressions and decision tables for the
notation of the decision logic. Representing decision logic in decision tables is a
core concept in DMN. Decision tables have extensively been adapted in previous

© Springer International Publishing AG 2018
E. Teniente and M. Weidlich (Eds.): BPM 2017 Workshops, LNBIP 308, pp. 529–541, 2018.
https://doi.org/10.1007/978-3-319-74030-0_42

works, as shown in [5]. The DRD depicts decisions and subdecisions, business knowledge models, input data, and knowledge sources. The decision logic is usually represented in the form of decision tables. A link can be made between a decision task in the process model and the actual decision model. An example of a DRD is given in Fig. 5. DMN is a declarative decision language. Hence, DMN provides no decision resolution mechanism, this is left to the invoking context. The same holds for the processing and storage of outputs and intermediate results. This is a burden of the invoking entity (e.g. a business process).

Literature has focused little on integrating process and decision models, while most research in the field revolves around automated discovery of decision models. The DMN standard was developed with the apparent aim to be used in conjunction with BPMN [6–8]. Since the establishment of DMN as a standard, the general consensus is to model decisions and decision logic outside business processes. The Business Process Management field is shifting in favour of this *separation of concerns* paradigm [9] by exteriorising the decisions and the decision logic from the process flow. Numerous tool developers have constructed new or adapted their existing tools to support the DMN standard: IBM, Signavio, Camunda, Decision Management Solutions, and FICO among others.

This paper is structured as follows. In Sect. 2, a framing of integrated modelling is constituted and a literature review of decision modelling is provided. Section 3 outlines challenges of integration by providing possible integration scenarios as well as a list of possible inconsistency concerns which should be remedied in future work. In Sect. 4 the issue of decisions encompassing an entire process in modelling tools is discussed. Finally, Sect. 5 concludes and discusses future work.

2   Decision Modelling and Separation of Concerns

In this section we present a motivation and related work for the separation of process and decision models, and their subsequent integration. Additionally, we provide a formalisation for some key DMN constructs.

2.1   Related Work

Decisions were often imitated using process control flows, which can result in cascading gateways and complex processes. In the trend towards separation of concerns and the subsequent integration of the separated models, the hidden decisions must be identified in the process and externalised in a decision model. After identifying and separating these decisions the resulting model must be integrated consistently with the process model. In more complex processes several decisions might influence both the flow and the result. Representing these decisions and invoking them correctly in the process is of paramount importance for a proper understanding of the process. However, these more complex situations have received little attention in literature.

Works revolving around process-data consistency have already been proposed. Extensions regarding data-awareness in process modelling have been researched as well. In [10] an ontology-based knowledge-intensive approach is suggested, while [11] proposes an enhancement of declarative process models [12] with DMN logic. Furthermore, countless works concerning data-aware coloured Petri Nets are available as well, offering a formally sound approach to data and process integration, such as in [13,14]. However, merely focusing on data fragments is not sufficient to holistically incorporate decision-awareness in processes, which DMN aims to achieve.

The decision modelling approaches present in process management literature often breach the separation of concerns between control and data flow, resulting in cascading gateways and spaghetti-like processes, hence negatively influencing maintenance and reusability [4,15–18]. They do this by hard coding and fixing the decisions in processes [7]. Consequently, splits and joins in processes are misused to represent typical decision artifacts such as decision tables. Recently, more attention was given to the separation of processes and decision logic, as such an approach is supported by the DMN standard [1,6] that can be used in conjunction with BPMN [19]. Decoupling decisions and processes to stimulate flexibility, maintenance, and reusability, yet integrating decision and process models is therefore of paramount importance [20].

The Separation of Concerns (SoC) and Service-Oriented Architecture (SOA) paradigms offer firm motivation for keeping multi-perspective modelling tasks, such as control flows and decision making, isolated and founded on a basis which can be used to ensure consistency. The integrated modelling and externalisation was already considered in terms of business rules [21,22]. With DMN, externalisation of decisions has become a possibility, since decisions can be encapsulated in separate decision models and linked to the invoking context.

2.2 Basic Definitions

Here, we provide a formalisation for key DMN concepts to avoid ambiguity in the discussion and to enhance the understanding of the problem statement around consistent integration. We adhere to the definitions of *decisions* and *decision requirement diagrams* from [20] and extend them to present a more profound formalisation for DMN. A DMN model can formally be represented as follows.

Definition 1. *A decision requirement diagram DRD is a tuple (D_{dm}, ID, IR) consisting of a finite non-empty set of decision nodes D_{dm}, a finite non-empty set of input data nodes ID, and a finite non-empty set of directed edges IR representing the information requirements such that $IR \subseteq (D_{dm} \cup ID) \times D_{dm}$, and ($D_{dm} \cup ID$, IR) is a directed acyclic graph (DAG).*

The DMN standard allows a DRD to be an incomplete or partial representation of the decision requirements in a decision model. The set of all DRDs in the decision model constitutes the exhaustive set of requirements.

Definition 2. *The decision requirements level R_{DM} of a decision model DM is the set of all decisions requirement diagrams in the model.*

The information contained in this set can be combined into a single DRD representing the decision requirements level holistically. The DMN standard refers to such a DRD as a decision requirement graph (DRG). We expand the notion of a DRG, in such a way that a DRG is a DRD which is self-contained, i.e. for every decision in the diagram all its requirements are also represented in the diagram.

Definition 3. *A decision requirement diagram $DRD \in R_{DM}$ is a decision requirement graph DRG if and only if for every decision in the diagram all its modeled requirements, present in at least one diagram in R_{DM}, are also represented in the diagram.*

According to the DMN specification a decision is the logic used to determine an output from a given input. In BPMN a decision is an activity, i.e. the act of using the decision logic, represented by the business rule task. Another common meaning is that a decision is the actual result, which we call the output of a decision, or simply the decision result. We define a decision using its essential elements.

Definition 4. *A decision $d \in D_{dm}$ is a tuple (I_d, O_d, L), where $I \subseteq ID$ is a set of input symbols, O a set of output symbols and L the decision logic defining the relation between symbols in I_d and symbols in O_d.*

3   Process-Decision Continuum and Incompatibilities

In this section we elaborate upon possible integration scenarios along the process-decision continuum and consequently deduce integration incompatibilities that might reveal themselves in the relevant scenarios.

3.1   Process-Decision Continuum

This subsection theoretically explains the possible integration scenarios that can occur between processes and decisions. A first theoretical description of possible process-decision integration scenarios is outlined by [20]. Here, we expand on those scenarios to identify challenges and difficulties for integration and to confer integration opportunities.

A first case is the **process only occurrence without decisions**. This scenario is illustrated in Fig. 1. This case is rather straightforward as it is only relevant for very simple processes concerned with completing a simple task. Examples are simple subprocesses of an overarching, bigger and more complicated process, the process of sending an email or the process of querying and updating a database. In this case decision management is not required and DMN models cannot be used to complement the process model. Hence, no concerns regarding process-decision integration are present in this case.

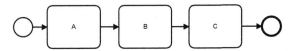

Fig. 1. Process only occurrence without decisions

Secondly, a **process only occurrence with embedded decisions** is a possibility as well. A theoretical example is given in Fig. 2. In process models decisions are often represented using complex control flows. By doing so, identifying which decisions are being made and how they are made becomes a cumbersome task. Consequently, the flexibility and maintainability of process models is negatively influenced by the hard coded decision logic. Besides, hard coding decisions in XOR-gates separates related decisions and assumes a process-first approach. The result is a non-integrated process-decision model and a process model that resembles a decision tree rather than the actual business process. Difficulties regarding the interpretation of both processes and decisions arise, as decisions and control flow are intertwined in the same model. Therefore, applying changes to the decision structure or the decision logic becomes a burdensome endeavour, as these decisions are fixed in the process flow. Moreover, evaluating the outcomes and enactment of decisions becomes challenging as well [7]. Additionally, hard coding decisions in a control flow contrasts the declarative nature decisions should have. Hence, this scenario of encoding the decisions within the process model should be avoided, as it breaches the separation of concerns.

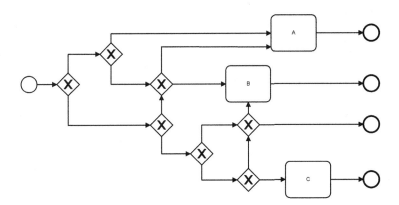

Fig. 2. Process only occurrence with embedded decisions

In the third scenario, the **process-decision occurrence with decisions as a local concern**, the modeller opts for simple separation of control flow and data in the form of local decisions whose outcomes influence the control flow through XOR-splits. These models treat decisions as a local concern. However, they do encapsulate part of the decision logic separately in a decision model.

The decisions in the process are situated before an XOR-split and the outcome of the decision influences the control flow of the process after the XOR-split. However, a decision only influences the control flow around the XOR-split it pertains to and it has no influence on the rest of the process, as the decisions do not span across the entire process. Compared to the *process only occurrence with embedded decisions* scenario, an increase in flexibility and maintainability of the process is achieved, as the control flow is relieved from incorporating the decision inputs. This partially simplifies and untangles the decision structure and control flow of the process. Integration of the process and decision models is achieved by providing the correct process control flow for each decision outcome at the decision point. However, the process is still burdened with incorporating the decision outcomes in the control flow through the use of XOR decision points. Additionally, these models do not support an overarching and holistic decision model that spans over multiple decision points and activities, or even over the entire process. Figure 3 provides an example of decisions as a local concern in processes.

These are the typical models retrieved in automated discovery techniques of decision point mining. Automatic retrieval of such models was done in [23] or more advanced in [24]. These works start off by building process models that can replay the control flow behaviour and next annotate the decision points, XOR-splits, with a decision activity. Additionally, [25] provides a method for extracting the underlying decisions and simplifying the corresponding process models. A more holistic and process encompassing automatic decision discovery technique is provided in [26].

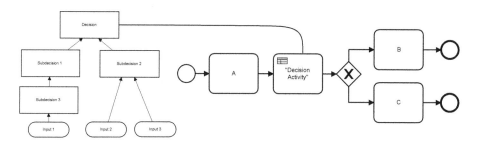

Fig. 3. Process-decision occurrence with decisions as a local concern

A more interesting scenario occurs when, instead of viewing decisions as a local concern, interrelated decisions extend over multiple process constructs, influencing the process in various ways, not only in terms of control flow at gateways as in the previous scenario. These decisions arching over the process in it's entirety will shape the flow and the outcome of the process, as well as the process modelling itself. The current scenario formulates long-distance dependencies between activities, data, control flow, and decisions in the process model. This scenario of **process-decision occurrence with decisions as a global concern** is illustrated in Fig. 4. Sound data management will be paramount to link the data provided by process activities and the decision requirements in the decision model. This data management will be key to model integration, as according to the DMN standard, decisions should be invoked through an interface by providing the necessary data to the decision model interface. To alleviate this concern, data objects and the results of subdecisions will be stored during the process, which can increase the complexity of the process model. Furthermore, special attention must be extended to the ordering of decision activities based on their requirements in the decision model. By maintaining a correct order of subdecisions, redundant re-enactment of certain subdecisions is avoided. Likewise, no added data management is required as it is handled internally by the process, reducing the process complexity.

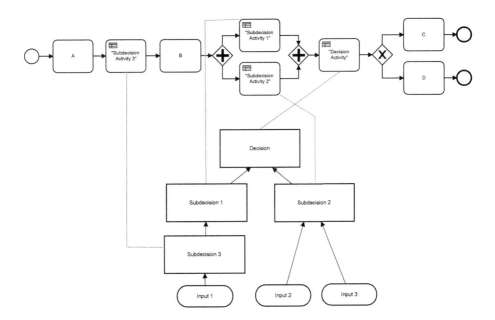

Fig. 4. Process-decision occurrence with decisions as a global concern

A final possibility is a scenario where only a decision model is present without a process model, or the **decision only occurrence without the process** scenario. This scenario is the complement of the first scenario. Hence, there is no requirement for integration since there is no concrete process. If needed, an associated process model, based on the decision model, can be generated from this decision model. The generated process will simply outline the execution order of the decisions made, rather than constitute a real business process.

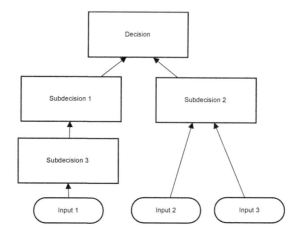

Fig. 5. Decision only occurrence without the process

3.2   Universe of Discourse

Note that the *process only occurrence without decisions* and *decision only occur-rence without the process* scenarios are not relevant for consistent model inte-gration, as only one model is present representing solely a process or a deci-sion respectively. The *process only occurrence with embedded decisions* is a sce-nario which also contains one model. However, the model represents not only the process, but also incorporates the underlying decisions in the control flow of the process. Hence, this modelling approach clearly breaches the separation of concerns paradigm as it convolutes decisions with process constructs, hence obstructing a clear view and interpretation of both decisions and the process itself. Ergo, modelling decisions as intricate control flows using process gate-ways should be avoided in order to enhance maintainability, comprehensibility, scalability, and flexibility of both the decisions and the process. The two remain-ing scenarios *process-decision occurrence with decisions as a local concern* and *process-decision occurrence with decisions as a global concern* are more interest-ing for the endeavour of consistent process-decision integration, as both scenarios contain two models, a process and a decision model, that need to be integrated with each other. Possible difficulties and incompatibilities between the two mod-els might arise. These incompatibilities must be remedied to achieve a consistent process-decision integration. We discuss some of the incompatibilities in the fol-lowing subsection.

3.3   Potential Incompatibilities

In this subsection, possible incompatibilities that might arise between the process model and the decision model are discussed, as the goal is to identify potential Incompatibilities and subsequently to alter the process to restore consistency [18].

An inappropriate way to model parts of the decision logic is to embed business decisions or logic in gateways. We call this the *Decision Logic Incompatibility*. In cases where a business process contains parts of the business logic of the decision model, the business process is incapable of accommodating to changes in the underlying decision model. When changes occur, the business process itself needs to be adapted. This occurs when the separation of concerns is not adapted strictly and thus the business logic is not separated and encapsulated in an independent decision model. Hence, this inconsistency does not ensure a more agile whole and does not allow evolution of both models disjointedly.

A second issue might be the *Decision Outcome Incompatibility*. In this case, not all outcomes from the decisions are included in the process model. Decisions can (re)direct the flow of the process. In an integrated process-decision model, all outcomes of the decision should be represented in the control flow if that decision redirects the process. Modelling all possible decision outcomes in the process is vital for a correct conclusion of the process. A first attempt to remedy this incompatibility is provided by [27].

Inconsistencies arise when subdecisions are not modelled in the process, despite the fact that the process uses the outcome of said subdecisions. Therefore, certain parts of the flow could be disturbed and render the process model inconsistent. Hence, a process model that is consistent with the decision model should ensure that all the subdecisions that contain an intermediate result which is relevant for the process execution are explicitly invoked as well. We call this concern the *Intermediate Result Incompatibility*.

Opposite to the *Intermediate Result Incompatibility*, more subdecisions than necessary can be included in the process. This inconsistency occurs when subdecisions which do not contain relevant intermediate results for the process are modelled within the process itself and we call this the *Subdecision Inclusion Incompatibility*. In this case the process becomes unclear and overly complex. Along with that, by modelling every subdecision in the process, the decision enactment or execution steps become fixed. This contradicts the declarative nature of decision modelling and reduces the flexibility provided by the decision model.

Depending on the outcome of certain subdecisions the control flow of the process may be diverted to include additional activities, to generate exceptions or even to lead to process termination. Excluding these subdecisions that have an influence on the control flow of the process leads to process-decision inconsistency. This inconsistency we call the *Subdecision Exclusion Incompatibility* and it is closely related to the *Intermediate Result Incompatibility*: while the *Intermediate Result Incompatibility* focuses on the exclusion of generated data by certain subdecisions, this inconsistency focuses on the change of control flow as a result of a subdecision.

Another inconsistency occurs when the order of the decision activities in the process model is contradictory to the hierarchy of the decisions in the decision model. Consequently, the process cannot function correctly, as decisions are forced to enact without the prerequisite enactment of the necessary subdecisions. We call this the *Decision Hierarchy Incompatibility*. This order of decisions and subdecisions, and consequently decision activities and subdecision activities, introduces a partial order on the decision activities in the process.

Decision activities also require prerequisites in order to function correctly. These prerequisites can be the outcome of certain subdecisions, as illustrated in the *Intermediate Result Incompatibility*, but also take the form of for instance user-generated input data. The inconsistency in this case occurs when the required input data is not available in a process when a certain decision task needs to be executed. This is called the *Input Data Incompatibility*.

4   Facing Holistic Decision Support in Tools

Multiple vendors have incorporated DMN modelling in their software packages, among others IBM, Signavio, Camunda, FICO and Trisotech. However, they all approach integrated process and decision modelling from a rather straightforward point of view. The tools allow to link a top-level decision from a decision model to one activity in the process. This activity invokes the decision and depending on the decision outcome, an XOR-split will guide the flow after this activity. Hence, all tools confine the decision model to a single activity in one particular location in the process. This is an incomplete and insufficient approach towards decision modelling known as decision points within the process. Indeed, decisions can span over multiple activities and over different segments of the process, especially in knowledge-intensive processes [28, 29]. Tools should allow to distinguish between DRDs and the DRG and to link a single decision model to multiple decision activities in the process, while also linking the information requirements in the decision model with data object management in the process model. That way, a holistic approach to decision modelling can be achieved. Additionally, tools should be able to recognise whether two linked models are consistently integrated and to suggest why integration is not consistent or sufficient, and suggest steps to remedy any inconsistencies. Finally, tools should strictly adhere to the principle of **Separation of Concerns** and observe the decision model not as an application itself, but as an external service, according to the **Service-Oriented Architecture** paradigm. Thus a more rigorous approach to decision modelling is advised, rather than considering decision models as mere add-ons for process modelling.

5   Conclusion and Future Work

This work discusses insights and challenges for an integrated modelling approach of processes and decisions. Though most works approach the problem from a simplistic point of view by merely dealing with decision points and allocating decisions to a single gateway in the process, we approach decisions in a holistic manner as they can span over multiple constructs of a process model, not only a single gateway or activity. We discussed integration scenarios and elaborated upon inconsistencies that might arise during the endeavour of model integration.

Consistent integration should rely on a profound data management of intermediate results of subdecisions and on correctly matching process data necessary for decision enactment to the information requirements in the decision model.

In future work, we will further investigate how data management needs to be organised in order to reach consistency in integration. Besides, the modelling complexity of integrated models is an interesting question for research as well. Additionally, integrated modelling in cooperative information systems and in distributed processes is of particular interest for Internet of Things (IoT) applications [30]. Furthermore, the modelling complexity of DMN will be addressed [3]. Finally, we will attempt to incorporate case models in the endeavour of consistent integration of processes, decisions, and cases.

References

1. OMG: Decision Model and Notation 1.1 (2016)
2. Bazhenova, E., Weske, M.: Deriving decision models from process models by enhanced decision mining. In: Reichert, M., Reijers, H.A. (eds.) BPM 2015. LNBIP, vol. 256, pp. 444–457. Springer, Cham (2016). https://doi.org/10.1007/978-3-319-42887-1_36
3. Hasić, F., De Smedt, J., Vanthienen, J.: Towards assessing the theoretical complexity of the decision model and notation (DMN). In: International Working Conference on Business Process Modeling, Development and Support (BPMDS), pp. 64–71. CEUR (2017)
4. Hasić, F., De Smedt, J., Vanthienen, J.: A service-oriented architecture design of decision-aware information systems: decision as a service. In: Panetto, H., et al. (eds.) OTM 2017. LNCS. Springer, Cham (2017). https://doi.org/10.1007/978-3-319-69462-7_23
5. Vanthienen, J.: What business rules and tables can do for regulations. Bus. Rules J. 8(7) (2007)
6. Taylor, J., Fish, A., Vanthienen, J., Vincent, P.: Emerging standards in decision modeling. In: Intelligent BPM Systems: Impact and Opportunity. BPM and Workflow Handbook Series, iBPMS Expo, pp. 133–146 (2013)
7. Vanthienen, J., Caron, F., De Smedt, J.: Business rules, decisions and processes: five reflections upon living apart together. In: Proceedings SIGBPS Workshop on Business Processes and Services (BPS 2013), pp. 76–81 (2013)
8. Biard, T., Le Mauff, A., Bigand, M., Bourey, J.P.: Separation of decision modeling from business process modeling using new decision model and notation (DMN) for automating operational decision-making. In: Camarinha-Matos, L.M., Bénaben, F., Picard, W. (eds.) PRO-VE 2015. IAICT, vol. 463, pp. 489–496. Springer, Cham (2015). https://doi.org/10.1007/978-3-319-24141-8_45
9. Gordijn, J., Akkermans, H., van Vliet, H.: Business modelling is not process modelling. In: Liddle, S.W., Mayr, H.C., Thalheim, B. (eds.) ER 2000. LNCS, vol. 1921, pp. 40–51. Springer, Heidelberg (2000). https://doi.org/10.1007/3-540-45394-6_5
10. Rao, L., Mansingh, G., Osei-Bryson, K.M.: Building ontology based knowledge maps to assist business process re-engineering. Decis. Support Syst. 52(3), 577–589 (2012)

11. Mertens, S., Gailly, F., Poels, G.: Enhancing declarative process models with DMN decision logic. In: Gaaloul, K., Schmidt, R., Nurcan, S., Guerreiro, S., Ma, Q. (eds.) CAISE 2015. LNBIP, vol. 214, pp. 151–165. Springer, Cham (2015). https://doi.org/10.1007/978-3-319-19237-6_10

12. Goedertier, S., Vanthienen, J., Caron, F.: Declarative business process modelling: principles and modelling languages. Enterp. Inf. Syst. **9**(2), 161–185 (2015)

13. Cabanillas, C., Resinas, M., Ruiz-Cortés, A., Awad, A.: Automatic generation of a data-centered view of business processes. In: Mouratidis, H., Rolland, C. (eds.) CAiSE 2011. LNCS, vol. 6741, pp. 352–366. Springer, Heidelberg (2011). https://doi.org/10.1007/978-3-642-21640-4_27

14. Serral, E., De Smedt, J., Snoeck, M., Vanthienen, J.: Context-adaptive petri nets: supporting adaptation for the execution context. Expert Syst. Appl. **42**(23), 9307–9317 (2015)

15. Weber, B., Reichert, M., Mendling, J., Reijers, H.A.: Refactoring large process model repositories. Comput. Ind. **62**(5), 467–486 (2011)

16. van der Aa, H., Leopold, H., Batoulis, K., Weske, M., Reijers, H.A.: Integrated process and decision modeling for data-driven processes. In: Reichert, M., Reijers, H.A. (eds.) BPM 2015. LNBIP, vol. 256, pp. 405–417. Springer, Cham (2016). https://doi.org/10.1007/978-3-319-42887-1_33

17. Hu, J., Aghakhani, G., Hasić, F., Serral, E.: An evaluation framework for design-time context-adaptation of process modelling languages. In: Poels, G., Gailly, F., Serral, A.E., Snoeck, M. (eds.) PoEM 2017. LNBIP, pp. 112–125. Springer, Cham (2017). https://doi.org/10.1007/978-3-319-70241-4_8

18. Hasić, F., De Smedt, J., Vanthienen, J.: An Illustration of Five Principles for Integrated Process and Decision Modelling (5PDM). Technical report, KU Leuven (2017)

19. OMG: Business process model and notation (BPMN) 2.0 (2011)

20. Janssens, L., Bazhenova, E., Smedt, J.D., Vanthienen, J., Denecker, M.: Consistent integration of decision (DMN) and process (BPMN) models. In: CAiSE Forum, CEUR Workshop Proceedings, Vol. 1612, pp. 121–128 (2016). CEUR-WS.org

21. Goedertier, S., Vanthienen, J.: Compliant and flexible business processes with business rules. In: Proceedings of the CAISE Workshop on Business Process Modelling, Development, and Support BPMDS (2006)

22. Wei, W., Indulska, M., Sadiq, S.: Guidelines for business rule modeling decisions. J. Comput. Inf. Syst. 1–11 (2017)

23. Rozinat, A., van der Aalst, W.M.P.: Decision mining in ProM. In: Dustdar, S., Fiadeiro, J.L., Sheth, A.P. (eds.) BPM 2006. LNCS, vol. 4102, pp. 420–425. Springer, Heidelberg (2006). https://doi.org/10.1007/11841760_33

24. de Leoni, M., van der Aalst, W.M.P.: Data-aware process mining: discovering decisions in processes using alignments. In: Proceedings of the 28th Annual ACM Symposium on Applied Computing, pp. 1454–1461. ACM (2013)

25. Bazhenova, E., Buelow, S., Weske, M.: Discovering decision models from event logs. In: Abramowicz, W., Alt, R., Franczyk, B. (eds.) BIS 2016. LNBIP, vol. 255, pp. 237–251. Springer, Cham (2016). https://doi.org/10.1007/978-3-319-39426-8_19

26. De Smedt, J., Hasić, F., vanden Broucke, S.K.L.M., Vanthienen, J.: Towards a holistic discovery of decisions in process-aware information systems. In: Carmona, J., Engels, G., Kumar, A. (eds.) BPM 2017. LNCS, vol. 10445, pp. 183–199. Springer, Cham (2017). https://doi.org/10.1007/978-3-319-65000-5_11

27. Batoulis, K., Weske, M.: Soundness of decision-aware business processes. In: Carmona, J., Engels, G., Kumar, A. (eds.) BPM 2017. LNBIP, vol. 297, pp. 106–124. Springer, Cham (2017). https://doi.org/10.1007/978-3-319-65015-9_7

28. Di Ciccio, C., Marrella, A., Russo, A.: Knowledge-intensive processes: characteristics, requirements and analysis of contemporary approaches. J, Data Semant. **4**(1), 29–57 (2015)
29. Moura, E.V., Santoro, F.M., Baião, F.A.: XCuteKIP: support for knowledge intensive process activities. In: Baloian, N., Zorian, Y., Taslakian, P., Shoukouryan, S. (eds.) CRIWG 2015. LNCS, vol. 9334, pp. 164–180. Springer, Cham (2015). https://doi.org/10.1007/978-3-319-22747-4_13
30. Horita, F.E., de Albuquerque, J.P., Marchezini, V., Mendiondo, E.M.: Bridging the gap between decision-making and emerging big data sources: an application of a model-based framework to disaster management in brazil. Decis. Support Syst. **97**, 12–22 (2017)

Data-Centric Extraction of DMN Decision Models from BPMN Process Models

Ekaterina Bazhenova[1(✉)], Francesca Zerbato[1,2], and Mathias Weske[1]

[1] Hasso Plattner Institute, University of Potsdam, Potsdam, Germany
{ekaterina.bazhenova,mathias.weske}@hpi.de
[2] University of Verona, Verona, Italy
francesca.zerbato@univr.it

Abstract. Operational decisions in business processes can be modeled by using the Decision Model and Notation (DMN). The complementary use of DMN for decision modeling and of the Business Process Model and Notation (BPMN) for process design realizes the separation of concerns principle. For supporting separation of concerns during the design phase, it is crucial to understand which aspects of decision-making enclosed in a process model should be captured by a dedicated decision model. Whereas existing work focuses on the extraction of decision models from process control flow, the connection of process-related data and decision models is still unexplored. In this paper, we investigate how process-related data used for making decisions can be represented in process models and we distinguish a set of BPMN patterns capturing such information. Then, we provide a formal mapping of the identified BPMN patterns to corresponding DMN models and apply our approach to a real-world healthcare process.

Keywords: Business process models · Process-related data
Decision models

1 Introduction

Decision-making plays a crucial role in business processes, as decision activities are included in process models, and process participants often act as decision-makers. The Decision Model and Notation (DMN) [12] was developed by the Object Management Group for modeling decisions and can be used to complement the Business Process Model and Notation (BPMN) [11] for decision design. Thereby, the "separation of concerns" [8,14,16] is supported, as decision logic is stored in independent decision models, separated from process models.

However, the extraction of decision models from process-related information is left to stakeholders, who need to choose which are the crucial aspects of decision-making and to which extent they should be modeled. In the context of easing the extraction of decisions from process models, existing work explores the discovery of DMN decision models from the process control flow [1]. Yet,

© Springer International Publishing AG 2018
E. Teniente and M. Weidlich (Eds.): BPM 2017 Workshops, LNBIP 308, pp. 542–555, 2018.
https://doi.org/10.1007/978-3-319-74030-0_43

the correspondence between process-related data used for decision-making and decision models remains unexplored.

This paper addresses this issue, by providing a mapping between process-related data used for decision-making, and decision models. Such mapping is tailored to consider different kinds of process-related data used for making decisions, spanning from structured operational data to informally defined domain knowledge. To this end, we firstly distinguish a set of data-centric decision patterns that are commonly found in BPMN process models. Then, we map the identified data-centric decision patterns to dedicated DMN decision models. The mapping, which constitutes the core contribution of this paper, is then applied to a real-world example taken from the healthcare domain, to illustrate the validity and applicability of our approach.

The remainder of this paper is structured as follows. In Sect. 2 we provide an example which motivates the conducted research. In Sect. 3 we consider how process-related data used for making decisions is represented in process models and introduce a set of BPMN data-centric decision patterns. Section 4 discusses how data is represented in DMN decision models. Section 5 provides the mapping of the distinguished patterns to decision models. Further, we apply the presented mapping to a real-world example in Sect. 6. Related work is then discussed in Sect. 7, followed by the conclusion in Sect. 8.

2   Motivating Example

In order to provide a better understanding of data-centric process decisions, let us consider the BPMN process model depicted in Fig. 1, which presents the initial steps for managing patients with suspected diabetes, at a high level.

As shown in Fig. 1, the process is triggered by the start message event *Patient Request*, which contains the patient's personal data and the request reason. The request is further evaluated by a *Nurse*, who is the process resource responsible for conducting the *Evaluate Request* decision activity. The nurse evaluates

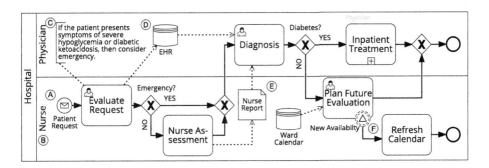

Fig. 1. Diagnostic process for patients with suspected diabetes conducted in a healthcare institution. Several data kinds are used by process activities for decision-making: Ⓐ start message events; Ⓑ resources; Ⓒ text annotations; Ⓓ data stores; Ⓔ data objects; and Ⓕ boundary non-interrupting events.

the degree of emergency, according to guidelines extracted from clinical documents and specified in the text annotation attached to the decision activity. If the patient does not require urgent treatment, the nurse proceeds with *Nurse Assessment*, which consists in examining the patient and writing the *Nurse Report*, represented as a data object. This report is later used by the *Physician* for conducting the *Diagnosis* decision activity, in conjunction with the patient data retrieved from the *Electronic Health Record (EHR)*, shown as a data store. The outcome of the diagnostic decision determines if the patient requires *Inpatient Treatment* or if the he/she is out of danger and the nurse can *Plan Future Evaluation*. In this latter case, the nurse decides when the evaluation should be scheduled, based on the patient conditions and on the physician's availability, recorded in the *Ward Calendar*. If a *New Availability* notification arrives during the scheduling of the appointment, as depicted by the corresponding boundary non-interrupting even, the time slot is rescheduled.

The process model from Fig. 1 considers data carried by BPMN data objects, text annotations, data stores, events, and resources. For supporting the separation of concerns, the process decision logic need to be externalized to a dedicated decision model. However, it is not immediately clear which process data carry decisional value, and how the various kinds of process-related data should be differentiated when externalized to a decision model. In order to help stakeholders to overcome this problem, we distinguish a set of data-centric patterns representing process decisions and show how to map such patterns to decision models, based on examples of BPMN and DMN.

3 Representation of Decision-Related Data in Process Models

In this section, we introduce the definitions of process model and decision activities in Sect. 3.1 and distinguish a set of BPMN data-centric decision patterns in Sect. 3.2.

3.1 Process and Decision Activity Models

For our work, we rely on the following notions of process model and decision activity.

Definition 1 (Process Model). A process model $m = (N, DN, C, TA, F, T, R, \alpha_k, \alpha_t, \beta, \rho)$ consists of a finite non-empty set of control flow nodes N, a finite set of data nodes DN, a finite non-empty set C of control flow edges, a finite set of text annotations TA, a finite set of data associations F, a finite set of undirected associations T, and a finite set of resources R. The set $N = A \cup G \cup E$ of control flow nodes consists of the disjoint sets A of activities, G of gateways and E of events. $E_B \subseteq E$ is the set of boundary events. Control flow $C \subseteq N \times N$ defines a partial ordering between the elements of N. $DN = DO \cup DS$ is the set of data nodes, consisting of the disjoint sets DO of data objects and DS of data stores.

$F \subseteq (DN \times A) \cup (A \times DN)$ is the set of data associations that connect data nodes with activities. $T \subseteq (TA \times A) \cup (A \times TA)$ is the set of symmetric associations that connect activities with text annotations. α_k, and α_t are functions that associate a type to the elements of A. $\alpha_k : A \rightarrow \{task, subprocess\}$ distinguishes activities into tasks and sub-processes. $\alpha_t : A \rightarrow \{abstract, manual, user, business\ rule, service, script\}$ associates to each activity a specific type. $\beta : A \rightarrow 2^{E_B}$ is a function that associates to each activity $a \in A$ a set of boundary events. Finally, $\rho : A \rightarrow R$ is a function that assigns to each activity the resource responsible for its execution. ◇

Process models can contain activities that involve decision-making. Generally speaking, a decision activity is any process task or sub-process that represents a reasoned choice among several alternatives, based on decision logic [16]. In our work, we assume that process or decision analysts identify which process tasks are modeled as decision activities. The DMN standard recommends that decision activities are executed either manually as in *user tasks*, or in a business-rule-driven manner, as in *business rule tasks* [12]. We consider both types of decision activity, as reflected in the definition below.

Definition 2 (Decision Activity). Let $DA \subseteq A$ be the set of decision activities for a process model. If $da \in A$ is a *decision activity*, then $\alpha_t(da) \in \{user, business\ rule\}$. ◇

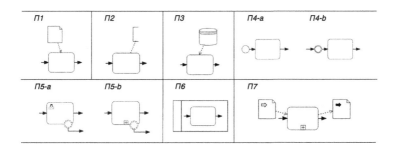

Fig. 2. BPMN data-centric decision patterns showing decision activities identified by experts (cf. Definition 2). The possible plurality of the elements connected to a single decision activity (data artifacts, annotations or boundary events) is omitted for better readability.

3.2   BPMN Data-Centric Decision Patterns

Below, we introduce a set of data-centric decision patterns $\Pi 1 - \Pi 7$, which we derived by systematically analyzing the BPMN standard and a set of real-world process models from previous research [1,5]. We use BPMN as it is the most consolidated standard for business process modeling [8,13] and propose a formalization of the patterns.

The overview of the patterns is presented in Fig. 2. Each pattern corresponds to a process fragment, that is, a subgraph of a process model, which represents a data-centric decision that can be externalized into a separate decision model. For better readability, we always show one process elements of a kind, thus omitting the representation of multiple elements connected to a single decision activity. For example, pattern $\Pi 1$ shows only one data object connected with a decision activity, although the formalization of $\Pi 1$ provides that multiple data objects can be attached to a single decision activity.

$\Pi 1$ *Data objects used by a decision activity.* A decision activity $da \in DA$ uses the set of data objects $DO' \subseteq DO$ iff for each $do \in DO'$, $(do, da) \in F$. $\Pi 1$ is a process fragment that consists of decision activity da, a set of data objects $DO' \subseteq DO$, and a set of data associations $F_{DO'} = \{(do, da) \mid do \in DO'\} \subseteq F$.

$\Pi 2$ *Text annotations used by a decision activity.* A decision activity $da \in DA$ uses the set of text annotations $TA' \subseteq TA$ iff for each $ta \in TA'$, $(ta, da) \in T$. $\Pi 2$ is a process fragment that consists of decision activity da, a set of text annotations $TA' \subseteq TA$, and a set of undirected associations $T' = \{(ta, da) \mid ta \in TA'\}$.

$\Pi 3$ *Data stores used by a decision activity.* A decision activity $da \in DA$ uses the set of data stores $DS' \subseteq DS$ iff for each $ds \in DS'$, $(ds, da) \in F$. $\Pi 3$ is a process fragment that consists of decision activity da, a set of data stores $DS' \subseteq DS$, and a set of data associations $F_{DS'} = \{(ds, da) \mid ds \in DS'\} \subseteq F$.

$\Pi 4$ *Event data used by a subsequent decision activity.* A decision activity $da \in DA$ uses the data from a previously occurred event $e \in E$ iff $(e, da) \in C$. According to the kind of triggered event, we distinguish two variants of this pattern:

$\Pi 4 - a$ is a process fragment that consists of decision activity da, start event e, and control flow $(e, da) \in C$.

$\Pi 4 - b$ is a process fragment that consists of decision activity da, intermediate event e, and control flow $(e, da) \in C$.

$\Pi 5$ *Boundary event data used by a decision activity.* A decision activity da may be influenced by the occurrence of the set of boundary events $E'_B \subseteq E_B$ during its execution iff for each $e_b \in E'_B$, $(e_b, da) \in \beta'$, $\beta' : A \rightarrow 2^{E'_B}$, and e_b occurs while da is being executed. This holds only for human decision-making, or for subprocesses, which usually take a certain amount of time to be executed, during which an event can occur and provide data that influence decision outcomes. Indeed, a standalone business-rule decision activity is assumed to invoke a decision model which is executed instantly [12]. Accordingly, we distinguish two pattern variants:

$\Pi 5 - a$ is a process fragment that consists of decision activity da, the boundary event e_b such that $e_b \in \beta(da)$, where $\alpha_k(da) = \{task\}$ and $\alpha_t = \{user\}$.

$\Pi 5 - b$ is a process fragment that consists of decision activity da, boundary event e_b such that $e_b \in \beta(da)$, where $\alpha_k(da) = \{subprocess\}$.

$\Pi 6$ *Decision activity associated to a specific resource.* A decision activity da is based on data related to the associated process resource $r \in R$ iff $(da, r) \in \rho$.

$\Pi\,6$ is a process fragment that consists of decision activity da and associated resource r.

$\boldsymbol{\Pi}\,\boldsymbol{7}$ *Decision subprocess.* A decision activity da is a decision subprocess iff $\alpha_k(a)$ $\in \{subprocess\}$, data object $do \in DO$ is an input for da and $(do, da) \in F$, data object $du \in DO$ is an output for da and $(da, du) \in F$, and a set of data attributes of du is a subset of a set of data attributes of do. Decision subprocesses have a hidden internal logic, which is used to determine the subset of decision outputs from a set of *alternatives* given in input to the subprocess[1]. $\Pi\,7$ is a process fragment that consists of the decision subprocess da, the data input do, the data association (do, da), the data output du and the data association (da, du).

It is worth noticing that not all kinds of data specified within process models need to be externalized in dedicated decision models. For instance, let us consider event-based gateways. In this case, the event occurrence drives instantaneous decisions, which are managed by process engines. Such kind of *process-engine decisions* should not be included in decision models, as they are rather based on process logic and routing rules, which are out of the control of a (human) decision maker. On the contrary, information used for decision-making and included in decision models may not be explicitly represented in process models. For instance, domain knowledge, Key Performance Indicators (KPIs), or process execution logs often drive decision making, but they are represented as meta-information rather than being included in process models.

4   Representation of Data in DMN Decision Models

In order to externalize the decision logic from process-related data, it is necessary to link decision-related information to the corresponding elements of a decision model. For providing such linkage, in this section, we present an overview of the elements that compose DMN decision models, focusing on discussion of their data aspects.

DMN decision models consist of two logical layers – the decision requirements level, and the decision logic level. The decision requirements level specifies decisions and their interconnections. The decision logic level describes how exactly decisions are made, e.g., by defining a decision table, a textual description, a program code, etc. [12]. In this paper, we abstain from a detailed discussion of the decision logic layer, as our goal is to detect which process-related data can be included in DMN decision models in principle, and such information has to be present at the decision requirements level.

As reflected in Fig. 3, the decision requirements level is described in DMN by a Decision Requirements Diagram (DRD), which consists of DRD elements (column I) and their interdependencies (column II). As it can be seen from column I of Fig. 3, the following DRD elements can be discerned:

---

[1] This pattern is proposed and explained in more detail in our previous work [3].

– A *Decision (d)* is an act of determining an output from a number of inputs, using decision logic. Therefore, *Decisions* are acts on process-related data.
– A *Business knowledge (b)* denotes a function encapsulating business knowledge, e.g., a decision table. Hence, *Business knowledge* elements represent functional requirements for acting on process-related data.
– A piece of *Input data (i)* denotes the information used as an input by one or more *Decisions*. Thereby, *Input data* directly represent process-related data.
– A *Knowledge source (k)* denotes an authority for a *Business knowledge* or *Decision*, which can be represented as domain experts responsible for maintaining decisions or source documents from which decisions are derived. Thus, *Knowledge source* elements represent non-functional requirements for acting on process-related data.
– A *Text annotation (t)* denotes an explanatory comment linked to a DRD element. Thus, *Text annotations* express non-functional requirements for acting on process-related data.

Further, as shown in column *II* of Fig. 3, the dependencies between DRD elements express three kinds of requirements:

– An *Information requirement (ir)* denotes either *Input data* being used as input to a *Decision* (rule *i–d*) or *Decision* output being used as input to a *Decision* (rule *d–d*). In DMN, there is no distinction between input data that is always required and input data that is only sometimes required. However, information requirements may be optional when the decision is made for a specific transaction. Thereby, *Information requirements* express functional requirements for acting on process-related data.
– A *Knowledge requirement (kr)* denotes the invocation of *Business knowledge* by the decision logic either of a *Decision* (rule *b–d*), or of another *Business knowledge* element (rule *b–b*). In this way, *Knowledge requirements* also represent functional requirements for acting on process-related data.

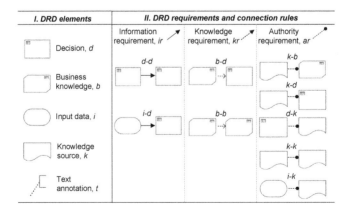

Fig. 3. Decision Requirements Diagram (DRD) elements and their connections.

- An *Authority requirement (ar)* denotes the dependence of a DRD element on another DRD element that acts as a source of guidance or knowledge (rules *k–b, k–d, d–k, k–k,* and *i–d*). This may be used to record the fact that a set of business rules must be consistent with a published document (e.g., a piece of legislation or a statement of business policy), or that a specific person or organizational group is responsible for defining some decision logic. To conclude, *Authority requirements* represent non-functional requirements for acting on process-related data.

Other types of dependencies than those described above are not allowed according to the DMN standard. All the rules governing the permissible ways of connecting elements with requirements in a DRD are summarized in three sub-columns of column *II* from Fig. 3. Our formalization of DMN decision requirement diagrams is presented below.

Definition 3 (Decision Requirement Diagram). A *Decision Requirement Diagram (DRD)* is a tuple (D, B, I, K, IR, KR, AR) consisting of:

- a finite non-empty set of *Decision* nodes D;
- a finite set of *Business knowledge* nodes B;
- a finite non-empty set of *Input data* nodes I;
- a finite set of *Knowledge source* nodes K;
- a finite non-empty set of directed edges IR representing *Information requirements* such that $IR \subseteq (I \cup D) \times D$;
- a finite set of directed edges KR representing *Knowledge requirements* such that $KR \subseteq B \times (D \cup B)$;
- a finite set of directed edges AR representing *Authority requirements* such that $AR \subseteq (D \cup I \cup K) \times (D \cup B \cup K)$.

  Herewith, $(D \cup B \cup I \cup K, IR \cup KR \cup AR)$ is a directed acyclic graph (DAG). ◇

To sum up, in this section we formalized the elements of the DMN decision requirements diagram. In the next section, we introduce the mapping of the extracted BPMN data-centric decision patterns to the DRD elements of DMN decision models.

5   Mapping BPMN Data-Centric Decision Patterns to DRD Models

In this section, we introduce the formal mapping between the set of BPMN decision patterns introduced in Sect. 3.2 and DRD models.

Below, we introduce a set of DRD fragments $\Delta = \{\Delta 1, \ldots, \Delta 7\}$ which corresponds to the set of extracted BPMN data-centric decision process patterns $\Pi = \{\Pi 1, \ldots, \Pi 7\}$. Further, we provide a *correspondence relation* $\Gamma = \{\Gamma 1, \ldots, \Gamma 7\}$, such that $\Gamma \subset \Pi \times \Delta$. All the DRD fragments are subgraphs of a DRD (c.f. Definition 3), such that $d \in D, I' \subseteq I, K' \subseteq K, IR' \subseteq IR, AR' \subseteq AR$. The

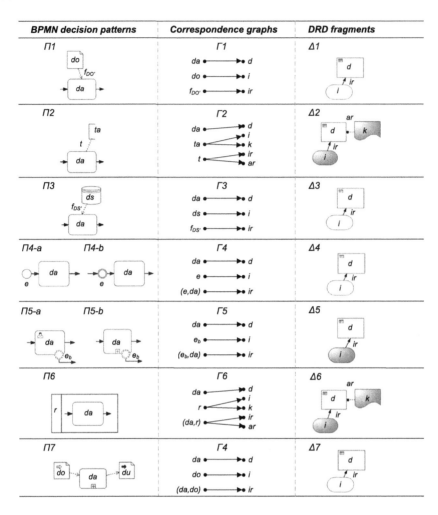

Fig. 4. Mapping of the introduced BPMN patterns to DRD fragments. Shading of the DRD shapes means that the elements are optional for modeling and execution.

correspondence relation Γ is visualized with the help of *correspondence graphs* in Fig. 4, and discussed below. In Fig. 4, the shading of the shapes denotes that the corresponding DRD elements are optional for both decision representation and execution. Again, for readability reasons, we do not show the possible plurality of elements of the same kind, connected to a decision activity. In detail, the correspondence relation Γ maps decision activity da of each BPMN decision pattern to decision d of the corresponding DRD pattern. Bearing this in mind, below we discuss only the additional correspondences for each mapping.

$\Gamma 1$. A mapping $\Gamma 1$ is a correspondence relation between the BPMN pattern $\Pi 1 = (da, DO', F_{DO'})$ and the DRD fragment $\Delta 1 = (d, I', IR')$. Thereby, each data object $do \in DO'$ corresponds to input data $i \in I'$ as they both represent operational data used by the decision. Each corresponding data association $f_{do'} \in F_{DO'}$ corresponds to information requirement $ir \in IR'$.

$\boldsymbol{\Gamma 2}$. A mapping $\Gamma 2$ is a correspondence relation between the BPMN pattern $\Pi 2 = (da, TA', T')$ and the DRD fragment $\Delta 1 = (d, I', K', IR', AR')$. Text annotation $ta \in TA'$ corresponds to input data $i \in I'$ if it represents operational data needed for making the decision. In this case, undirected association $t \in T'$ should be mapped to information requirement $ir \in IR'$. Alternatively, text annotation $ta \in TA'$ may be mapped to knowledge source $k \in K'$ if it represents a non-functional requirement for decision-making. In this latter case, undirected association $t \in T'$ should be mapped to authority requirement $ar \in AR'$. However, as text annotations do not always represent data used for making decisions, both input data and knowledge source in the DRD fragment are represented as optional for representation and execution.

$\boldsymbol{\Gamma 3}$. A mapping $\Gamma 3$ is a correspondence relation between the BPMN pattern $\Pi 3 = (da, DS', F_{DS'})$ and the DRD fragment $\Delta 3 = (d, I', IR')$. Each data store $ds \in DS'$ corresponds to input data $i \in I'$ as it represents operational data used by the decision. Each data association $f_{DS'} \in F_{DS'}$ corresponds to information requirement $ir \in IR'$.

$\boldsymbol{\Gamma 4}$. A mapping $\Gamma 4$ is a correspondence relation between the BPMN pattern $\Pi 4 = (da, e, (e, da))$ and the DRD fragment $\Delta 4 = (d, i, ir)$. Both $\Pi 4 - a$ and $\Pi 4 - b$ are implied, as they have the same formal structure. Event e carries process data influencing the decision, and it corresponds to input data $i \in I$. The corresponding control flow edge $(e, da) \in C$ is mapped to information requirement $ir \in IR$.

$\boldsymbol{\Gamma 5}$. A mapping $\Gamma 5$ is a correspondence relation between the BPMN pattern $\Pi 5 = (da, E'_B, \beta')$ and the DRD fragment $\Delta 5 = (d, I', IR')$. Each boundary event $e_b \in E'_B$ carries data influencing the decision, and it corresponds to input data $i \in I'$. As well, the corresponding relation $(e_b, da) \in \beta'$ is mapped to information requirement $ir \in IR'$. As the boundary event in both cases might not occur at all, the corresponding input data element in the DRD fragment is shown as optional.

$\boldsymbol{\Gamma 6}$. A mapping $\Gamma 6$ is a correspondence relation between the BPMN pattern $\Pi 6 = (da, r, (da, r))$ and the DRD fragment $\Delta 6 = (d, i, k, ir, ar)$. Resource r can be mapped to input data $i \in I$ if it represents role information used for decision-making. In this case, $(da, r) \in \rho$ should be mapped to information requirement ir. Alternatively, resource r can be mapped to knowledge source $k \in K$ if it represents a non-functional requirement for making the decision, and then $(da, r) \in \rho$ should be mapped to authority requirement $ar \in AR$.

$\boldsymbol{\Gamma 7}$. A mapping $\Gamma 7$ is a correspondence relation between the pattern $\Pi 7 = (da, do, (do, da), du, (da, du))$ and the DRD fragment $\Delta 7 = (d, i, ir)$. The decision subprocess is mapped to the decision activity, data object do corresponds to input data $i \in I$ as it represent operational data needed for making the decision. Data association (do, da) is mapped correspondingly to the information requirement $ir \in IR$. The output data object du and the data association (da, du) is not mapped to the DRD as it does not participate in decision making.

It can be added that *Decisions* in the DRD fragments presented above can additionally reference *Business knowledge* nodes reflecting decision logic. This is recommended if the decision logic is reused by multiple *Decisions*. In this case, corresponding *Knowledge requirements* should be provided. Also, text commentaries can be added in the output DRD fragments to additionally specify the DRD elements.

Construction of the complete DRD. When extracting a DRD from a given BPMN process model, the analyst can consequently apply the introduced mappings. This should be followed by a compilation of the derived DRD fragments into a comprehensive decision requirement diagram and by the reduction of repeated elements. When a decision activity produces data objects or text annotations that are reused by another decision activity, an information requirement should be added between these two decisions.

6   Application of the Mapping to a Real-World Example

In order to validate the mapping introduced in Sect. 5, now we apply it to the real-world example introduced in Sect. 2.

1. Extraction of DRD fragments. Firstly, we analyzed the BPMN process model from Fig. 1 and detected that there are three decision activities which reflect the process of choosing an output from some alternatives, namely *Evaluate Request*, *Nurse Assessment*, and *Plan Future Evaluation*. According to the presented classification of the BPMN data-centric decision patterns from Sect. 3, an aggregate of the process fragments corresponding to these decision activities was detected (see Fig. 5): (I) The *Evaluate Request* decision activity and the start message event *Patient Request*; (II) The *Evaluate Request* decision activity and the text annotation connected to it; (III) The *Evaluate Request* decision activity and the *Nurse* resource; (IV) The *Diagnosis* decision activity and the *Nurse Report* data object; (V) The *Diagnosis* decision activity and the *EHR* data store; (VI) The *Diagnosis* decision activity and the process resource *Physician*; (VII) The *Plan Future Evaluation* decision activity and the *Ward Calendar* data object; (VIII) The *Plan Future Evaluation* decision activity and the *New Availability* notification; and (IX) The *Plan Future Evaluation* decision activity and the *Physician* resource.

Further, we applied the mapping from Sect. 5 on the detected fragments and provided the corresponding DRD fragments, as shown in Fig. 5. The cases including optional mappings were treated as follows. In case II, the text annotation is mapped to the knowledge source and not to input data, as it is a non-functional requirement for executing decision activity *Evaluate Request* decision activity. All the resources were mapped to knowledge sources (cases III, VI, and IX), as they are constant for the exemplified process model, whereas input data rather reflects changeable operational data. Accordingly, the *New Availability* boundary event (case VIII) is mapped to the input data.

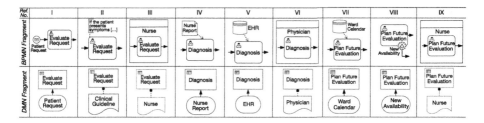

Fig. 5. Detected data-centric decision patterns for the BPMN process model from Fig. 1 and corresponding DRD fragments.

2. Construction of Complete DRDs. The discovered fragments were consequently combined together and compiled into two decision requirements diagrams, presented in Fig. 6. As the output of the *Evaluate Request* decision activity is written in the *EHR* data store which is read by the *Diagnosis* decision activity, an information requirement between the corresponding two decisions was added, as shown in Fig. 6a. Since the *Plan Future Evaluation* decision activity is connected with the other decision activities only through the process control flow, it is designed as an independent DRD in Fig. 6b.

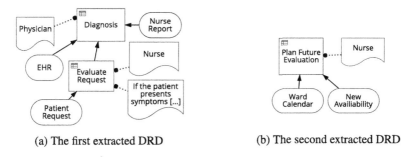

(a) The first extracted DRD   (b) The second extracted DRD

Fig. 6. Extracted DMN model corresponding to the BPMN process model from Fig. 1

The extracted DMN model in Fig. 6 serves as an explanatory decision model for the BPMN process model from Fig. 1, as it incorporates explicitly the decision-related process data. Herewith, the extracted decision model can be executed complementary to the process model, and thus, the principle of separation of concerns [16] is supported. Thereby, the original BPMN process model should contain the undirected association links of decision activities to the corresponding elements of the extracted DMN decision model, which can be done at the implementation level.

7   Related Work

Most approaches dealing with the extraction of decision logic from processes are based on traditional activity flows and they do not consider data or business artifacts present in the models. Accordingly, extraction of decision logic from the process control flow is considered in [1]. Mining of decision models relying on complex decision logic discovered from process event logs was done in [2,6,10]. An overview summarizing approaches dedicated to mining decision models from events logs is presented in [7]. In [15], the authors propose a methodology for derivation of DMN models complementary to BPMN process models from product data models (PDM), which support the design of complex data dependencies (in contrast to BPMN). However, PDM do not distinguish between different types of data artifacts, as addressed by us.

In this paper, we consider to construct decision models from data that is *explicitly* represented in process models. However, additional information can be used by decision activities, such as reference data and domain knowledge used for non-operational decision-making [5]. Another example of meta-information are KPIs, used to make decisions about the process, according to specific objectives and milestones [4,9,17]. The approach presented in this paper is purely model-based and, thus, it does not require additional information, which is planned for future work.

8   Conclusion

The paper presents a pattern-based approach to externalize process-related data used for making decisions into dedicated decision models. In detail, we chose BPMN for modeling processes and DMN for modeling decisions, as their complementary usage realizes the separation of concerns principle. In order to achieve this, we firstly distinguished a set of BPMN patterns capturing process data used for making decisions. Secondly, we introduced a formal framework for expressing DMN decision requirements diagrams. This allowed us to formalize the mapping between the detected set of BPMN patterns and corresponding fragments of the DMN decision requirements diagrams. The mapping considers the different nature of data artifacts specified in process models. We validated the mapping by applying it to an example taken from the healthcare domain. The limitation of the introduced mapping is that its correct application largely relies on the process expert, e.g., for identifying which activities are involved in decision-making. The automation of this step will be considered in future, for instance by analyzing the text of activity labels and annotations. In future work, we plan to extend our mapping with additional information about decision-making in processes and we aim to consider a wider range of process-related data, such as domain knowledge and process KPIs. As well, we plan to apply the mapping on a larger set of real-world process models.

References

1. Batoulis, K., Meyer, A., Bazhenova, E., Decker, G., Weske, M.: Extracting decision logic from process models. In: Zdravkovic, J., Kirikova, M., Johannesson, P. (eds.) CAiSE 2015. LNCS, vol. 9097, pp. 349–366. Springer, Cham (2015). https://doi.org/10.1007/978-3-319-19069-3_22

2. Bazhenova, E., Buelow, S., Weske, M.: Discovering decision models from event logs. In: Abramowicz, W., Alt, R., Franczyk, B. (eds.) BIS 2016. LNBIP, vol. 255, pp. 237–251. Springer, Cham (2016). https://doi.org/10.1007/978-3-319-39426-8_19

3. Bazhenova, E., Weske, M.: A data-centric approach for business process improvement based on decision theory. In: Bider, I., Gaaloul, K., Krogstie, J., Nurcan, S., Proper, H.A., Schmidt, R., Soffer, P. (eds.) BPMDS/EMMSAD -2014. LNBIP, vol. 175, pp. 242–256. Springer, Heidelberg (2014). https://doi.org/10.1007/978-3-662-43745-2_17

4. Bazhenova, E., Weske, M.: Deriving decision models from process models by enhanced decision mining. In: Reichert, M., Reijers, H.A. (eds.) BPM 2015. LNBIP, vol. 256, pp. 444–457. Springer, Cham (2016). https://doi.org/10.1007/978-3-319-42887-1_36

5. Combi, C., Oliboni, B., Zardini, A., Zerbato, F.: Seamless design of decision-intensive care pathways. In: IEEE International Conference on Healthcare Informatics, pp. 35–45. IEEE (2016)

6. de Leoni, M., Dumas, M., García-Bañuelos, L.: Discovering branching conditions from business process execution logs. In: Cortellessa, V., Varró, D. (eds.) FASE 2013. LNCS, vol. 7793, pp. 114–129. Springer, Heidelberg (2013). https://doi.org/10.1007/978-3-642-37057-1_9

7. De Smedt, J., vanden Broucke, S.K.L.M., Obregon, J., Kim, A., Jung, J.-Y., Vanthienen, J.: Decision mining in a broader context: an overview of the current landscape and future directions. In: Dumas, M., Fantinato, M. (eds.) BPM 2016. LNBIP, vol. 281, pp. 197–207. Springer, Cham (2017). https://doi.org/10.1007/978-3-319-58457-7_15

8. Debevoise, T., Taylor, J.: The micro-guide to process modeling and decision in BPMN /DMN. CreateSpace Independent Publishing Platform (2014)

9. Dumas, M., Rosa, M., Mendling, J., Reijers, H.A.: Fundamentals of Business Process Management. Springer, Heidelberg (2013). https://doi.org/10.1007/978-3-642-33143-5

10. Mannhardt, F., de Leoni, M., Reijers, H.A., van der Aalst, W.M.P.: Decision mining revisited - discovering overlapping rules. In: Nurcan, S., Soffer, P., Bajec, M., Eder, J. (eds.) CAiSE 2016. LNCS, vol. 9694, pp. 377–392. Springer, Cham (2016). https://doi.org/10.1007/978-3-319-39696-5_23

11. OMG: Business Process Model and Notation (BPMN), v. 2.0.2 (2013)

12. OMG: Decision Model And Notation (DMN), v. 1.1 (2016)

13. Silver,B.: BPMN Method and Style. Cody-Cassidy Press, 2nd edn. (2011)

14. Taylor, J., Purchase, J.: Real-World Decision Modeling with DMN. Meghan-Kiffer Press, Tampa (2016)

15. van der Aa, H., Leopold, H., Batoulis, K., Weske, M., Reijers, H.A.: Integrated process and decision modeling for data-driven processes. In: Reichert, M., Reijers, H.A. (eds.) BPM 2015. LNBIP, vol. 256, pp. 405–417. Springer, Cham (2016). https://doi.org/10.1007/978-3-319-42887-1_33

16. Von Halle, B., Goldberg, L.: The Decision Model. Taylor and Francis Group, Abingdon (2010)

17. Wetzstein, B., Zengin, A., Kazhamiakin, R., Marconi, A., Pistore, M., Karastoyanova, D., Leymann, F.: Preventing KPI violations in business processes based on decision tree learning and proactive runtime adaptation. J. Syst. Integr. **3**(1), 3–18 (2012)

Discovering Business Rules in Knowledge-Intensive Processes Through Decision Mining: An Experimental Study

Júlio Campos[✉], Pedro Richetti, Fernanda Araújo Baião, and Flávia Maria Santoro

Federal University of the State of Rio de Janeiro (UNIRIO), Rio de Janeiro, Brazil
{julio.campos,pedro.richetti,fernanda.baiao,
flavia.santoro}@uniriotec.br

Abstract. Decision mining allows discovering rules that constraint the paths that the instances of a business process may follow during its execution. In Knowledge-intensive Processes (KiP), the discovery of such rules is a great challenge because they lack structure. In this context, this experimental study applies a decision mining technique in an event log of a real company that provides ICT infrastructure services. The log comprises structured data (ticket events) and non-structured data (messages exchanged among team members). The goal was to discover tacit decisions that could be potentially declared as business rules for the company. In addition to mining the decision points, we validated the discovered rule with the company w.r.t. their meaning.

Keywords: Decision mining · Business rules · Knowledge-intensive Processes

1 Introduction

Process mining techniques aim to discover business process models from events recorded in data logs. Most algorithms used for this purpose generate models that show the flow of activities, but do not identify or detail how decisions are made along it. Nonetheless, recent advances in techniques produce models that are more adjusted to the event log, since they contemplate decisions that regulate the activities flow [2]. Decision mining allows discovering decision points to explain how different paths are taken during a process execution [12]. However, the discovery of decisions is not trivial, especially for the so-called Knowledge-intensive Processes (KiP), which are weakly structured and are not driven by pre-established rules. KiPs are mostly carried out based on knowledge and experience of actors involved in its execution [11].

Despite these characteristics, some of the discovered decisions within a KiP are candidates to become business rules that might serve as strategic knowledge for the organization and support future decisions to be made. The literature shows that few works address the relationship between the logic of decisions made and the process. So, the problem investigated here is whether a decision mining technique allows to discover business rules within the flow of activities of a KiP. The main goal of this paper is to

© Springer International Publishing AG 2018
E. Teniente and M. Weidlich (Eds.): BPM 2017 Workshops, LNBIP 308, pp. 556–567, 2018.
https://doi.org/10.1007/978-3-319-74030-0_44

discuss the results from an experimental study made with a log of a com-pany that provides IT services to several clients.

2   Background Knowledge

2.1   Knowledge-Intensive Processes

According to [11], business processes can be classified into three types: structured, semi-structured and unstructured. Structured processes are completely predefined, i.e., there are fixed rules that cannot be changed to perform each activity. Semi-structured processes contain structured and unstructured parts; not all the activities have pre-defined rules regarding the next steps in the flow. Unstructured processes are completely unpredictable and pose no pre-defined order for the execution of the activities, being commonly called knowledge-intensive processes (KiP). KiPs are not suitable for auto-mation and are conferred to a great degree of freedom to achieve their goals. For [11], value is created within KIP by fulfilling the participants' knowledge requirements. Thus, the decisions that are made during the accomplishment of the tasks are directly influenced by the knowledge of who performs them.

Di Ciccio et al. [4] defined KiP as processes "whose conduct and execution are heavily dependent on knowledge workers performing various interconnected knowledge intensive decision-making tasks." The authors also elicited key typical characteristics of KiPs: knowledge-driven; collaboration-oriented; unpredictable; emergent; goal-oriented; event-driven; constraint-and rule-driven; and non-repeatable. Examples of KiPs include customer support, design of new products/services, marketing, IT governance or strategic planning. Besides, they concluded that the way organizations deal with this kind of processes has changed over time, e.g. the customer support processes in several organizations have evolved from highly structured to knowledge-intensive, and personalized, flexible individual cases.

França [5] proposed an ontology named KiPO (Knowledge-Intensive Process Ontology) aimed at comprising the key concepts and relationships that are relevant for understanding, describing and managing a knowledge-intensive process. KiPO provides a common, domain-independent understanding of KiPs and, as such, it may be used as a metamodel for structuring KiP concepts. KiPO is composed of 5 sub-ontologies, which reflect the main KiP perspectives. The Business Process Ontology (BPO) comprises the traditional elements of business process modeling (such as activ-ities, event flows, input/output data objects). The Collaboration Ontology (CO) de-picts concepts to explain how knowledge artifacts are exchanged among process participants, and how the collaboration takes place. The Decision Ontology (DO) aims at describing the rationale of the decisions made by the process agents (i.e., the "why" and "how" decisions were made by the people involved in the process) thus allowing the tracking of what motivated a decision and the outcomes from it. The Business Rules Ontology (BRO) provides the means to describe some parts of the KiP from a declarative perspective, since describing the rules that govern a KiP is especially useful for describing the parts of the process which are very flexible and not subject to predefined event flows. Finally, the Knowledge Intensive Process Core Ontology (KiPCO) comprises the core concepts of a KiP (mainly

Agents, Knowledge-intensive activities and contextual elements involved in their execution). KiPO argues that in a KiP the flow of activities (especially decision-making) is deeply influenced by tacit elements from its stakeholders. In this paper, we explore the decision making associated to business rules perspectives of KiPO.

2.2 Decision Mining

Business processes are established and structured upon business rules. A business rule "is a statement that defines or restricts some aspect of an organization's business" [6]. Process mining discovers how business processes are structured through two techniques: process discovery and conformance checking. The first one builds a process model that reflects the behavior observed in event logs. The second one tries to detect deviations in the existing model [12]. According to [12], in the early days of process mining, most algorithms supported only the control flow perspective. Slight attention has been given to values of data attributes which can affect the routing of an instance during a process execution. Decision mining research was advanced in this context. The term was first used by researchers who described so-called decision points in models. Decision points are "parts of the model in which the process is divided into alternative branches" [12]. The researchers created a decision tree algorithm, provided in ProM[1] framework, which retrieves test results at a split point to analyze how choices are made in a business process model.

De Leoni and Van der Aalst [2] stated that the technique proposed by [12] was not able to discover the conditions associated with split exclusive or and loops. Another limitation was that the event log required to be in full compliance with the flow control modeled, i.e., the order in which the activities are executed would never be different from the order of the idealized model. The authors proposed a new approach in which an alignment between an event log and a process model is performed first, and then a decision tree algorithm is applied. This solution was implemented in ProM framework through the Multi-Perspective Process Explorer (MPE) plugin [8]. The "Discover Data Perspective" mode of MPE allows discovering guards associated with a transition (process activity). A guard can be any Boolean expression that uses logical operators such as conjunction ($\wedge$), disjunction ($\vee$) and negation ($\neg$). The user selects one among five decision tree algorithms to discover the guards as well as the attributes to be considered. The user has also to configure two parameters: minimum of elements associated with the leaves of the decision tree (min cases) and the mini-mum adjustment of the flow of control for each instance to be considered. The min instances parameter is important because it influences whether the guards are over-adjusted, or poor-adjusted [8]. Recently, Mannhardt et al. [9] developed a new technique that allows discovering overlapping decision rules, since the algorithm proposed by [12] was able to mine only mutually exclusive splits (XOR). The solution was added to the Multi-Perspective Process Explorer (MPE) plugin.

De Smedt et al. [3] argue that most of the literature on decision mining focuses on increasingly refined techniques on the retrieval of decision information in business

---

[1] http://www.promtools.org/

process models. However, for the authors, only a few works are dynamically capable of discovering the stages of the decision-making process. The authors created a framework with four perspectives to evaluate decision mining techniques. They concluded that few studies cover this perspective that explains how the decisions discovered are related to the own decision-making process.

2.3   Decision Model and Notation (DMN)

Due to the emergence of decision mining, several organizations started to address a need for a standardized notation to represent decisions in business process models. In 2015, The Object Management Group (OMG) released the first version of the DMN (Decision Model and Notation). DMN goal is to ensure that a decision model is inter-changeable between entities through an XML representation [10]. Through the DMN, "decisions can be modeled so that a decision-making in an organization can be easily represented in diagrams, accurately by business analysts and (optionally) automated".

According to the [10], decision-making is addressed by two different perspectives by existing model standards: First in BPMN process models, which represent tasks in which decisions occur. The second perspective refers to a decision logic, i.e. a specific logic to make particular decisions, for example, in business rules, decision tables or in executable analytical models. For several authors, a decision making has an internal structure that is not conveniently captured by either of the two perspectives cited. Thus, the purpose of the DMN is to provide a third perspective, the Decision Requirements Diagram (DRD), which forms the bridge between business models and decision logic models. OMG (2015) suggests three possible uses of DMN in order to understand and define how decisions are made in a company or organization: (i) modeling human decision making, (ii) modeling requirements for automated decision making, (iii) implementing automated decision making.

3   Experimental Study

We accomplished an experimental study comprising using an event log of an ICT company. The company has around a hundred contracts with different firms to pro-vide ICT infrastructure services. One of their business processes is the resolution of ICT incidents related to client's assets, such as e-mail server outages and network connection problems. An incident is an unexpected, unplanned episode, which if not solved correctly can cause loss, damage or even some kind of accident. The activities to address the incidents involve the application of technical skills, troubleshooting abilities, collaboration, and information exchange among technicians and between the team of technicians and the client. There is no strictly structured process to be followed, since most of the problems are situational and several ad-hoc decisions may be taken. These points characterize knowledge-intensive aspects in such a way that traditional control flow oriented business process would be not adequate to manage the scenario. The goal of this experimental study was to evaluate decision mining techniques to discover decision points in a KiP. Due to the characteristics of KiP, we show the relevance of also considering textual content in the analysis. Thus, we divided the study into 2 steps.

3.1 Experimental Setting: The Log

The log contains records of 6.337 instances of the process and 246.283 events, distributed by 32 activities. We filtered a sample of the log with all tickets opened in the 2nd semester of 2015. This sample included structured data about the tickets logged by the process-aware CRM system (explored in the first step), together with all e-mail messages exchanged between employees and customers for discussions about the problem to be solved (explored in the second step).

3.2 First Step: Discovering Rules Within the Structured Log

Method. The first step was shaped to mine the log attempting to find decision points. Although we are dealing with a KiP, since our focus in this paper is on decisions, we did not choose techniques commonly used to discover unstructured models (such as declarative ones). The MPE plugin was chosen to support this task because the technique fits into the third perspective of the framework proposed by [3]. The content of the conversations was not included in this step because it consists unstructured data and, as so, not eligible for analysis with the technique. Before performing the process mining, we filtered the event log. Filtering is an important preprocessing procedure because it allows for the discovery of error-free process models. It is also useful for selecting a subset of data of greater relevance or interest for an analysis. In our case, the purpose was to select tickets that generated e-mail exchanges between employees and customers, as the examples in Fig. 1. Thus, all instances involving null fields in the column "article_id" were filtered out. After filtering, the number of instances remained in 6.337. The number of events dropped to 63.424 and activities to 13.

	A	B	C	D	E	F	G	H	I
1	ticket_id	eventName	article_id	priority_id	ticketState	serviceType	solution_time	SLAMissed	eventDateTime
2	160431	NewTicket		3	new	Redes::Diagnósticos	720	N	2015-06-01 07:20:09
3	160431	ServiceUpdate		3	new	Redes::Diagnósticos	720	N	2015-06-01 07:20:09
4	160431	SLAUpdate		3	new	Redes::Diagnósticos	720	N	2015-06-01 07:20:09
5	160431	CustomerUpdate		3	new	Redes::Diagnósticos	720	N	2015-06-01 07:20:09
6	160431	EmailCustomer	605130	3	new	Redes::Diagnósticos	720	N	2015-06-01 07:20:09
7	160431	SendAutoReply	605131	3	new	Redes::Diagnósticos	720	N	2015-06-01 07:20:10
8	160431	SendAgentNotification		3	new	Redes::Diagnósticos	720	N	2015-06-01 07:20:10
9	160431	SendAgentNotification		3	new	Redes::Diagnósticos	720	N	2015-06-01 07:20:11
10	160431	SendAgentNotification		3	new	Redes::Diagnósticos	720	N	2015-06-01 07:20:11
11	160431	Lock		3	new	Redes::Diagnósticos	720	N	2015-06-01 09:05:13
12	160431	Misc		3	new	Redes::Diagnósticos	720	N	2015-06-01 09:05:13
13	160431	OwnerUpdate		3	new	Redes::Diagnósticos	720	N	2015-06-01 09:05:13
14	160431	TypeUpdate		3	new	Redes::Diagnósticos	720	N	2015-06-01 09:08:10
15	160431	ServiceUpdate		3	new	Redes::Diagnósticos	720	N	2015-06-01 09:08:10
16	160431	SLAUpdate		3	new	Redes::Diagnósticos	720	N	2015-06-01 09:08:10
17	160431	AddNote	605208	3	new	Redes::Diagnósticos	720	N	2015-06-01 09:08:10
18	160431	TicketDynamicFieldUpdate		3	new	Redes::Diagnósticos	720	N	2015-06-01 09:08:10
19	160431	TicketDynamicFieldUpdate		3	new	Redes::Diagnósticos	720	N	2015-06-01 09:08:10
20	160431	SendAnswer	605210	3	new	Redes::Diagnósticos	720	N	2015-06-01 09:08:32

Fig. 1. Sample of the selected events by the filter

Among the most executed activities in the process, "TimeAccount" standed out as the top most frequent, and "AddNote" the second. The "TimeAccount" activity is a manual record of the time an employee spent to interact with a customer. The "Add-Note" activity is an internal note used to exchange information between employees. this activity reinforces classifying this process as knowledge-intensive, since it represented exchanges of knowledge among employees to guide decision-making about problems and incidents that occur during the provision of company services.

After filtering, the process mined from the event log was represented as a Petri net, using the "Petri Net Mine with Visual Inductive Mining" plugin. In this plugin, the only adjustment was setting the value of the "Noise threshold" parameter from 0.20 to 0.0, in order to guarantee a perfect adjustment of the log. The Petri net model and filtered event log served as inputs for a "Multi-Perspective Process Explorer" plug-in. After the plugin was executed, it generated the base model in Fig. 2.

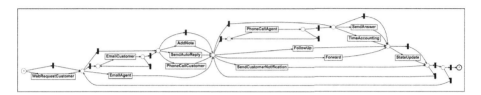

Fig. 2. Model generated by MPE

When selecting the "Discover Data Perspective" mode, the plugin performed a computational alignment between the event log and the Petri net model. To perform the alignment, the simple configuration mode, which uses standard parameters, has been chosen. A new base model was generated with an average adjustment rate of 100%, meaning no violations, missed or lost events. All 63.424 event log events have been correctly aligned to the Petri net model.

Results. Three scenarios were carried out to discover the rules on the Petri net model. In all of them, the value of the "min instances" parameter was modified, keeping the value of the "min fitness" parameter equal to 1. In the first scenario, the lowest possible value was selected for the "min instances" parameter (0.001), which allowed the discovery of very large and complex rules related to some activities. In the second scenario, the highest possible value was selected for the "min instances" parameter (0.5), and no rules were found in the model. In the third scenario, the value of the "min instances" parameter was changed between 0.001 and 0.5. In this scenario, guards were found in the "sink 6" position for the "AddNote" activity. AddNote is considered a knowledge intensive activity because it is when people interact (through messages) to discuss the problems. The results are shown in Table 1.

In Table 1, the accuracy of the rules found is measured by the parameter "Guard F-Measure". The higher the value of this parameter, the greater the accuracy of the uncovered guard. All rules discovered presented high accuracy. The difference among them lies in the degree of complexity. The algorithm could find simple rules for the "min instances" values below 0.2, but for the value 0.3 a large and complex rule related to the status of the ticket was found, as shown in Fig. 3.

Table 1. Rules related to the "AddNote" activity

AddNote	Decision tree (default false)		
Min instances	0.11	0.2	0.3
Min fitness	1.0	1.0	1.0
Guard F-measure	85.7%	81.7%	85.8%
Guard	article_id > 605709.0	article_id > 622246.0	(((((((ticketState == "Agendamento" ‖ ticketState == "closed successful") ‖ ticketState == "closed with workaround") ‖ ticketState == "merged") ‖ ticketState == "new") ‖ ticketState == "open") ‖ ticketState == "pending auto close +") ‖ ticketState == "pending auto close-") ‖ ticketState == "pending reminder")
Correct events	12570	12570	12570
Wrong events (data)	0	0	0
Missing events	0	0	0

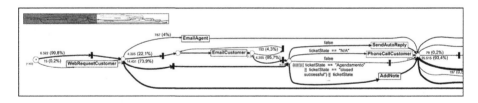

Fig. 3. Rule discovered by decision tree algorithm related to "AddNote" at "sink 6" (Color figure online)

In Fig. 3, we notice that some activities are highlighted in color. This is an indication that the discovery of rules for such activities is more difficult according to the value selected for the "min instances" parameter. Figure 4 shows the same model without the rules. "AddNote" is a transition of the "sink 6" location, as well as the "SendAutoRe-play", "PhoneCallCustomer" transitions, and also an invisible transition (black rectangle). According to the model, 18.736 events have passed from the local "sink 6". Of these, 12.570 (67.1%) performed the "AddNote" activity. In the model, thicker arcs indicate the main flow followed by most instances of the process.

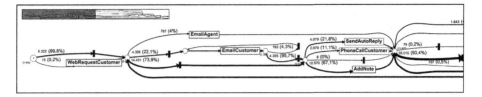

Fig. 4. Control flow with the percentage of events that executed each activity (Color figure online)

It is also possible to evaluate the accuracy of the base model, generated after the computational alignment. Table 2 shows a comparison of the accuracy of the base model with "sink 6", from which the rules were found. The accuracy of the base mod-el is low (37.6%). The local precision of "sink 6" is somewhat lower (27.7%). Although the base model has 100% fitness, the accuracy is low, i.e., it allows a behavior not observed in the event log.

Table 2. Comparison between precision of the model and "sink 6" place

Sink 6		General	
Local place precision	27.7%	Avg precision	37.6%
# Observed locally	20.747	# Observed	243.736
# Possible locally	74.944	# Possible	647.571
Global place precision	27.7%	Avg fitness	100%

The rules discovered in all three scenarios evaluated with the structured log were then validated with the company staff. Two managers responsible for keeping up with the tasks executed by the technical team were interviewed. The rules were presented to them and they were asked to analyze them and answer about the meaning and appropriateness. The goal was to understand if the rules could be considered correct and as well if they are really applied within this process. Both agreed with the rules, telling that they make sense for them, but in fact, they are not surprising. We concluded that the method applied may possibly discover correct rules; however, this is not enough in this case to provide insights to the company staff. So, we proceeded to explore the unstructured information available in the log.

3.3 Second Step: Discovering Rules Within the Unstructured Log

Method. The fourth scenario of the experiment explored the event log through text mining techniques. The goal was to seek decision rules derived from the knowledge of the employees that guided the decision making during the execution of the activities of the process within the records of conversations exchanged between the employees and the clients. These conversations were extracted from the original event log and copied to a new text file. Using the free software R, the text file was loaded and scanned with the grep command through a regular expression to filter tickets in which the words "incident", "rule", "procedure" and the radical "amos" and "soluc" (in Portuguese) were mentioned in order to find records of incidents that were solved.

Results. We found 421 results related to the word incident. Among the results obtained, four related to the solution of a problem were selected for analysis (Table 3).

Table 3. Report of an incident discussion

Ticket	Article	Activity	Message
165027	623276	*EmailAgent*	"We arrived at the place where there was no internet access. We did an analysis of the environment to detect the source of the problem. We identified that the up-link cable did not allow connection to the internet. We used another preexisting connection in the store, changed the up-link and solved the incident."
218683	690455	*SendAnswer*	"(…) We inform you that your request regarding 'Printer has stopped working' was completed by the FOT team who took great pleasure in helping you. We performed environment analysis, and we detected divergence of configurations. The stations were pointing to an address that differed from the address set on the printer. We made the correction in the printer, entering the address to which the stations pointed. This procedure solved the problem reported by our client, who received us on the spot, validated the conclusion of our call with success."
234964	745175	*TimeAc-counting*	"(…) According to the phone contact, a reboot procedure was performed on the server and it did not load the system correctly. After this episode, the server was shut down and reconnected without the physically connected off-board network cards. Since it was not successful, we are migrating the call to head-on service, which will be arranged by scheduling with the service center. We are aware of the criticality of the incident and are placing the call with a high urgency level."
166513	629237	*SendAnswer*	"(…) We have already corrected the Firewall rule that was identified by the support of yesterday. Rules loading tests were all performed successfully. The call will be ended."

The first incident, for example, records reports the decision made by company technician to solve a problem of internet access interruption. The need to analyze the environment indicates that the activity is knowledge-intensive, because it demands tacit knowledge of the employees. The analysis may have been based on a business rule defined by the company or derived from the employees' tacit knowledge. Business rule inferred by the conversation: "The first alternative to be tested in a case of internet interruption must be the pre-existent connection". The source of the problem was detected and a decision was made based on a workaround, that is, use another pre-existing connection in the client store, which solved the incident. Figure 5 shows a decision model of the treatment of that incident using the DMN notation.

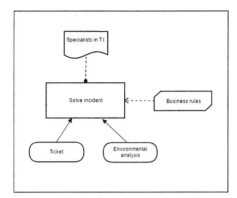

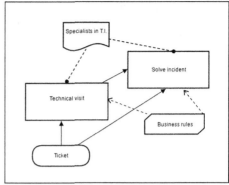

Fig. 5. Decision Requirement Diagram of the 1st and 2nd incident (Table 3).

Fig. 6. Decision Requirement Diagram of the 3rd incident (Table 3).

In Fig. 5, the input data from the Decision Requirement Diagram (DRD) is the conversation about the ticket and the analysis of the environment. This data provides the information needed to guide the decision, a Business Knowledge Model, which may relate to a business rule predefined by the company or represent a situation based on the employee tacit knowledge. The decision is made by a Knowledge Source, which is an authority defined to make the decisions. In the case under analysis, the company employees detain technical and tacit knowledge to solve complex problems.

The record of the second incident describes a situation similar to the first one. A staff performed an on-site environment analysis to solve a printer operating problem. The result of the analysis allowed them to identify the source of the problem to apply an appropriate correction procedure. The problem solution was validated by the customer who contacted the team. For this incident, the same DRD of Fig. 5 applies.

The third incident log describes an episode where a remote procedure was performed to try to solve a troubleshoot on a server. After two unsuccessful attempts, the employee requests a head-on service with a high level of urgency. In Fig. 6, this incident is modeled in a DRD. We notice that the decision about how to solve an incident depends on another decision, i.e. assessing the need for a technical visit. Ac-cording to the report, the decision to request face-to-face attendance was made due to the failure of the procedure performed, in addition to the high degree of criticality. The procedure is a clear evidence that the employee followed a business rule established by the company to try to solve the problem. A rule that can be inferred from the decision made by the employee: "A technical visit must be requested for highly critical incidents". The last incident log is a response to a client to inform about fixing a problem detected in a Firewall rule. The problem was solved after testing. This scenario describes a situation that is very frequent in the company: the solution of problems depends on the adjustment of rules of service configuration, which demands a high level of tacit knowledge of the employees.

Once more, we conducted a validation with the company staff. The same two managers were invited to analyze those rules based on the same criteria: meaning and appropriateness. This time their perception about the results were very positive. They recognized the rules as tacit knowledge of the team; therefore, they agreed that it would be very relevant to make them explicit, and also the possibility to disseminate to the other technicians, and finally institutionalize them.

4   Discussion

The results obtained in the experiment point to limitations of business rules discovery in knowledge-intensive processes. The decision mining technique discovered few rules due to the low number of event attributes in the company log. In addition, the decision rules discovered by the decision tree algorithm only informed the necessary conditions for the execution of certain activities. The decision mining algorithm used could discover three distinct rules for the "AddNote" activity (two simple and one complex), with good accuracy. However, the model from which the decision rules were discovered presented low precision, which significantly affected the quality of the discovered model. In addition, little understanding, or knowledge relevant to decision making was obtained from the rules. To find out how decisions were made during the execution of a knowledge-intensive activity, we analyzed the conversations exchanged between the employees and clients of the company. In the last scenario, it was clear how decision-making is contextualized and can vary from instance to instance. In this scenario, the discovery of general rules that guides decision making and that could be established as business rules becomes a major research challenge.

Some rules could be identified from the incident records. These rules were based on procedures and guided the technicians' decision-making process. The DRDs of the incidents illustrate how to explicit that the decision to solve an incident depends on another decision: to evaluate the necessity of a technical visit to the place. According to the records, the technical visit is requested when the level of criticality of the incident is high. In this situation, an environmental analysis is performed and it requires a more intensive level of employee knowledge to identify the cause of the problem. It seems to be aligned with the conclusions of [7], for whom in a Kip scenario, a decision model is required, and it is more relevant than the BPMN model.

A few limitations of the experiment are observed. In the first step, we chose one technique to discover rules, but we did not compare with other approaches, as for example [1]. One clear threat to validity was the validation, which was made qualitatively through interviews with only two participants of the process. Although they are the most experienced participants in this process, collect the perception of other members of the staff could improve the conclusions.

5   Conclusions

Mining decisions in knowledge-intensive processes is not an easy task, as their activities are poorly structured. Moreover, each instance of a knowledge-intensive process is executed in a different way, which further complicates the automatic extraction of decision rules associated with the execution of its activities. In the experiments performed with the support of a decision mining technique, few rules were found for the activities of the process. The model in which the rules were discovered showed low precision, which significantly affected their quality. The scenario with text mining showed the existence of decision-making records cannot yet be easily incorporated by current process mining techniques in order to enrich the mining decision models. Future work is enriching log events with complementary Natural Language Processing and text mining techniques, besides applying other approaches such as [1] to compare the results.

References

1. Bazhenova, E., Buelow, S., Weske, M.: Discovering decision models from event logs. In: Abramowicz, W., Alt, R., Franczyk, B. (eds.) BIS 2016. LNBIP, vol. 255, pp. 237–251. Springer, Cham (2016). https://doi.org/10.1007/978-3-319-39426-8_19
2. De Leoni, M., van der Aalst W.M.P.: Data-aware process mining: discovering decisions in processes using alignments. In: Proceedings of the 28th Annual ACM Symposium on Applied Computing, pp. 1454–1461. ACM (2013)
3. De Smedt, J., vanden Broucke, S.K.L.M., Obregon, J., Kim, A., Jung, J.-Y., Vanthienen, J.: Decision mining in a broader context: an overview of the current landscape and future directions. In: Dumas, M., Fantinato, M. (eds.) BPM 2016. LNBIP, vol. 281, pp. 197–207. Springer, Cham (2017). https://doi.org/10.1007/978-3-319-58457-7_15
4. Di Ciccio, C., Marrella, A., Russo, A.: Knowledge-intensive processes: characteristics, requirements and analysis of contemporary approaches. J. Data Semant. **4**(1), 29–57 (2015)
5. França, J.: Uma ontologia para definição de processos intensivos em conhecimento. Tese de Doutorado. M. Sc. dissertation, Departamento de Informática Aplicada (DIA), Universidade Federal do Estado do Rio de Janeiro (UNIRIO), Rio de Janeiro (2012)
6. Hay, D., et al.: Defining business rules-what are they really. Final report (2000)
7. Janssens, L., Bazhenova, E., De Smedt, J., Vanthienen, J., Denecker, M.: Consistent integration of decision (DMN) and process (BPMN) models. In: Proceedings of the CAiSE 2016 Forum at the 28th International Conference on Advanced Information Systems Engineering, Ljubljana, Slovenia (2016)
8. Manhardt, F., De Leoni, M., Reijers, H.A.: The multi-perspective process explorer. In: BPM (Demos), pp. 130–134 (2015)
9. Mannhardt, F., de Leoni, M., Reijers, H.A., van der Aalst, W.M.P.: Decision mining revisited - discovering overlapping rules. In: Nurcan, S., Soffer, P., Bajec, M., Eder, J. (eds.) CAiSE 2016. LNCS, vol. 9694, pp. 377–392. Springer, Cham (2016). https://doi.org/10.1007/978-3-319-39696-5_23
10. Object Management Group (OMG): Decision Model And Notation (DMN) version 1.1 (2016). Accessed 01 Jun 2016
11. Richter-Von Hagen, C., Ratz, D., Povalej, R.: Towards self-organizing knowledge intensive processes. J. Univ. Knowl. Manag. **2**, 148–169 (2005)
12. Rozinat, A., van der Aalst, W.M.P.: Decision mining in ProM. In: Dustdar, S., Fiadeiro, J.L., Sheth, A.P. (eds.) BPM 2006. LNCS, vol. 4102, pp. 420–425. Springer, Heidelberg (2006). https://doi.org/10.1007/11841760_33

1st International Workshop on Quality Data for Process Analytics (QD-PA 2017)

Introduction to the First International Workshop on Quality Data for Process Analytics (QD-PA 2017)

Moe Thandar Wynn[1], Marco Comuzzi[2], Minseok Song[3], and Lijie Wen[4]

[1] Queensland University of Technology, Brisbane, Australia
m.wynn@qut.edu.au
[2] Ulsan National Institute of Science and Technology, Ulsan, Republic of Korea
mcomuzzi@unist.ac.kr
[3] Pohang University of Science and Technology, Pohang, Republic of Korea
mssong@postech.ac.kr
[4] Tsinghua University, Beijing, China
wenlj@tsinghua.edu.cn

Abstract. The objective of the first edition of the Quality Data for Process Analytics (QD-PA 2017) workshop was to provide a forum to exchange findings and ideas on technologies and practices to achieve high-quality data for process analysis. The program of the workshop included three accepted research papers (50% acceptance rate); each paper was refereed based on its full content by two or three program committee members. They contributed to foster a fruitful discussion among the participants about the data quality challenges encountered in process analysis and possible approaches to address these challenges. The workshop took place on September 11, 2017, in Barcelona, Spain.

Keywords: Process management · Process mining · Data quality

1 QD-PA 2017 Aims and Background

The First International Workshop on Quality Data for Process Analytics (QD-PA 2017) aims to provide a high quality forum for researchers and practitioners to exchange research findings and ideas on technologies and practices to achieve high-quality data for process analysis. Despite the fact that organizations collect a plethora of process data, poor data quality still poses significant hurdles to successfully translating process data into actionable business insights. Data-driven process analysis techniques (e.g., process mining) rely on relevant, reliable and accurate historical records on business operations to be available as an event log. In the current approach to event log preparation for process analytics, process-related source data is massaged into event log format using manual, or, at best one-off, domain-specific, analysis-specific, semi-automated support (e.g., Extract-Transform-Load). This has proven to be time-consuming and error-prone, relying heavily on the experience of the analyst to recognize and address data quality issues. The advent of "big data" makes such manual, bespoke data pre-processing completely impractical. Moreover, there is a need to better understand the effects of data manipulation (e.g., anonymization, filtering, abstraction, etc.) on the reliability of analysis results. This lack of systematic and guided approaches to evaluate the health of process data and to assure high-quality data in data-driven process analysis lifecycle warrants further research.

2   QD-PA 2017 Main Topics

The main topics relevant to the DQ-PA workshop include, but are not limited to:

- Identification of data quality issues in Business Process Management (BPM)
- Categorization of data quality issues encountered in process analysis
- Criteria for quality process data that align with specific process analysis techniques
- Formalizations of quality metrics for data attributes recorded in an event log
- Algorithms to detect analysis-agnostic data quality problems in an event log
- Algorithms to detect analysis-specific data quality problems (alignment of analysis and data)
- Remediation approaches for event log quality assurance
- Methodology for anonymization of process-related data
- Algorithms to trace data quality issues to analysis insights
- Tools to detect the health of process data
- Approaches to keep track of data manipulations performed on process data
- Experience reports from tools that support data pre-processing
- Case studies in process analytics and process mining with data quality challenges

3   QD-PA 2017 Workshop Organizers

Moe Thandar Wynn	Queensland University of Technology, Australia
Marco Comuzzi	Ulsan National Institute of Science and Technology, Republic of Korea
Minseok Song	Pohang University of Science and Technology, Republic of Korea
Lijie Wen	Tsinghua University, China

4   QD-PA 2017 Program Committee

Wil van der Aalst	Eindhoven University of Technology, The Netherlands
Robert Andrews	Queensland University of Technology, Australia
Hyerim Bae	Pusan University, Republic of Korea
Marco Comuzzi	Ulsan National Institute of Science and Technology, Republic of Korea
Cinzia Cappiello	Politecnico di Milano, Italy
Jochen De Weerdt	Katholieke Universiteit Leuven, Belgium

Jidong Ge   Software Institute, Nanjing University, China
Haibin Liu   Yanshan University, China
Faming Lu   Shandong University of Science and Technology, China
Anne Rozinat   Fluxicon, The Netherlands
Minseok Song   Pohang University of Science and Technology, Republic
                             of Korea
Wei Song   Nanjing University of Science and Technology, China
Suriadi   Queensland University of Technology, Australia
Arthur ter Hofstede   Queensland University of Technology, Australia
Seppe vanden Broucke   Katholieke Universiteit Leuven, Belgium
Yiping Wen   School of Information Science and Engineering, Central
                             South University, China
Lijie Wen   Tsinghua University, China
Moe Wynn   Queensland University of Technology, Australia

Redo Log Process Mining in Real Life: Data Challenges & Opportunities

E. González López de Murillas[1]([✉]), G. E. Hoogendoorn[1],
and Hajo A. Reijers[1,2]

[1] Department of Mathematics and Computer Science,
Eindhoven University of Technology, Eindhoven, The Netherlands
{e.gonzalez,h.a.reijers}@tue.nl, g.e.hoogendoorn@student.tue.nl
[2] Department of Computer Science, Vrije Universiteit Amsterdam,
Amsterdam, The Netherlands

Abstract. Data extraction and preparation are the most time-consuming phases of any process mining project. Due to the variability on the sources of event data, it remains a highly manual process in most of the cases. Moreover, it is very difficult to obtain reliable event data in enterprise systems that are not process-aware. Some techniques, like redo log process mining, try to solve these issues by automating the process as much as possible, and enabling event extraction in systems that are not process aware. This paper presents the challenges faced by redo log, and traditional process mining, comparing both approaches at theoretical and practical levels. Finally, we demonstrate that the data obtained with redo log process mining in a real-life environment is, at least, as valid as the one extracted by the traditional approach.

Keywords: Process mining · Databases · Redo logs · Event logs
Data quality

1 Introduction

Data extraction and preparation are among the first steps to take in any business intelligence or data analysis project. In many cases, up to 80% of the time and effort, and 50% of the cost is spent during the data extraction and preparation phases [1]. This is due to the fact that the original sources of data come in great variety, differing in structure depending on the nature of the application or process under study. The standardization of this phase represents a challenge, given that a lot of domain knowledge is usually required in order to carry it out. It is because of this that most of the work is done by hand, in an ad-hoc fashion, requiring a lot of iterations in order to obtain the proper data in the right form.

In process mining the situation is not much different. Studies have been carried out, focusing on SAP [2–4], or ERPs in general [5]. Also, efforts have been made to achieve a certain degree of generalization with the tool XESame [6], which assists in the task of defining mappings between database fields on the one

ⓒ Springer International Publishing AG 2018
E. Teniente and M. Weidlich (Eds.): BPM 2017 Workshops, LNBIP 308, pp. 573–587, 2018.
https://doi.org/10.1007/978-3-319-74030-0_45

side, and events, traces and logs on the other. However, these solutions, which we refer to as part of the classical or traditional approach, are tightly coupled to the specific IT system or data schema they were designed to analyze. Moreover, they do not support the extraction of event data from systems that are non-process aware and do not explicitly record historical information.

For this reason, other techniques exist that try to leverage on the existence of alternative sources of data. A very promising approach is redo log process mining [7]. Most modern relational database management systems (RDBMSs) implement different mechanisms to ensure consistency and fault tolerance. One of these mechanisms is redo log recording, which consists of a set of files in which database operations are recorded before being applied to the actual data. This allows to rollback the state of the database to previous points in time, undoing the last operations recorded in the redo log files. Redo log process mining exploits the information stored in database redo log files in order to obtain event data. This event data can be analyzed to understand the behavior of processes interacting with the database. One of the benefits of this approach is its independence of the specific application or process in execution, being able to extract behavioral information from both process and non-process aware systems. Also, the event extraction is carried out automatically, without the need for domain knowledge to know how to build events from database tables, as is the case in the traditional approach. However, the prerequisites of this approach are that (a) the redo log system needs to be explicitly configured and enabled in order to record the events, and that (b) special database privileges are required to be able to read the content of the redo log files from the RDBMS.

With respect to the traditional and redo log process mining approaches, we face two main questions. (1) Is redo log process mining feasible in a real-life environment? (2) Are the results of both approaches comparable in terms of data quality? Based on our intuition and experience with sample datasets, we propose the following hypothesis: *the data obtained by the redo log process mining approach is at least as rich as the data obtained by traditional methods.* The goal of this paper is to answer these questions and find support for this hypothesis by comparing the results of both process mining approaches on a real-life dataset. The content of this paper is based on the work developed in [8] as part of one of the author's Master project.

The remainder of this paper is organized as follows. First, Sect. 2 provides some background on the event data extraction techniques about to be compared. After that, a theoretical comparison is presented in Sect. 3. Then, Sect. 4 proposes the practical comparison, introducing the business case, explaining the execution of the data extraction, and showing the results. Section 5 compares the results of the application of both approaches, discussing their validity and equivalence. Finally, Sect. 6 presents the conclusion of this paper.

2   Background

We want to compare two approaches for event data extraction: *traditional*, and *redo log process mining*. These two approaches differ with respect to the source of data, as well as the procedure they follow to extract it. This section provides some background on the particularities of both approaches, explaining the process to follow for their application, while focusing on the data extraction and processing stages.

2.1 Traditional Process Mining

In traditional process mining, event logs are constructed from the plain files or the database tables of the IT system under study. The main event attributes (activity name, case id, timestamp, etc.) are identified by hand, while making use of domain knowledge, and extracted in order to build an event log. This is a rather laborious task, as described in the procedure in [9], but very common during the first stages of a process mining project.

In some scenarios, data is obtained directly from the original IT systems that drive the process being analyzed. On other occasions, data has already been preprocessed and gathered in data warehouses or similar systems, alleviating somewhat the data extraction issue. In these cases, the work of extracting and processing the data cannot be avoided altogether; it must be performed in a previous phase. The complexity of the task is tackled before the analysis is done, but the decisions made during the data warehouse design can dramatically affect the kind of analysis that can be performed on the resulting data. In order to apply process mining techniques, it is required to have access to event data that includes, at least, timestamps, activity names, and case identifiers. However, not all the data models of data warehouses guarantee that these aspects of data are being preserved. In order to ensure that enough information is being collected, process-aware meta models like the one proposed in [10] can be adopted.

Regardless of the location of the data, it is necessary to obtain a valid *event log* in order to do the process mining analysis. Different methodologies exist in the literature that describe the steps to take in a process mining project. For our purpose, we decided to focus on the process mining methodology PM^2 [11], a recent methodology that covers all the stages in the life cycle of a process mining project, which has been verified in a real-life environment. PM^2 divides the project in six stages: planning, extraction, data processing, mining & analysis, evaluation, and process improvement & support. Given that we are interested in obtaining an event log, we must focus on the first three stages: planning, extraction, and data processing. Each of these stages has sub-steps as described in Table 1. In traditional process mining, these three stages are carried out manually by the analyst or the process mining team. Usually, these stages require substantial domain knowledge to define the business questions, select the right database tables, determine the case notion, and include interesting event and case attributes, among other tasks. This domain knowledge is often obtained through interviews with the process owners and users.

The data is usually retrieved from database tables, executing SQL queries to build the events and, finally, extract the event logs. However, the quality of the event data that can be obtained is constrained by the existence of historical information, timestamps, status changes, modifications, additional attributes, etc.

Table 1. First three steps of the PM^2 process mining methodology.

Stage 1: Planning	Stage 2: Extraction	Stage 3: Data Processing
Selecting business processes	Determining scope	Creating views
Identifying research questions	Extracting event data	Aggregating events
Composing project team	Transferring process knowledge	Enriching logs
		Filtering logs

As has been noted before, the structure of the data model strongly determines the usefulness of the resulting event logs. Other event data retrieving techniques, such as redo log process mining, try to mitigate these issues exploiting the existence of historical data automatically recorded by the database systems. Section 3 presents some of the challenges to face with the traditional process mining approach, and compares them to the ones faced by redo log process mining.

2.2 Redo Log Process Mining

Redo log process mining is a more automatic technique than the traditional approach. It requires less domain knowledge, and is independent of the system under study. It tries to exploit the execution information stored in database redo logs in order to extract event data. The database redo log system is a functionality of the database management systems that, in order to ensure consistency and fault tolerance, records all the data modification actions executed on the database before they are actually applied. Generally, a set of files are configured to store the redo logs. The RDBMS stores the actions in the redo log files and, when a file is full, it passes to the next file. When all the redo log files are full (according to a specific maximum size), the first file of the set is overwritten. This means that only a recent window of events could be retrieved from the redo logs when assuming a default setting. However, database systems usually allow to archive the completed redo log files in a separate location for subsequent analysis. This is a crucial aspect to take into account in order to collect enough data to perform a meaningful analysis. In general, any modification action on the database is recorded in the redo logs. This means that we cannot only observe insert, update, and delete operations performed on every piece of data, but also modifications on the data schema, transactions, rollback and commit operations, etc.

The main advantage of this technique is that it allows to analyze systems that are not process aware and do not explicitly record any execution information. Also, deleted data, not present in the database anymore, can be recovered from the redo logs. This has a great value from the forensic audit point of view. On the other hand, the technique presents challenges in terms of data availability, permissions and performance. The following section explores some of these difficulties and compares them to the ones faced by traditional process mining.

3   Theoretical Comparison

In the previous sections we have described the fundamentals of both the traditional and redo log process mining approaches. In this section we point out the main differences from a theoretical point of view, clarifying the challenges to face in order to apply either technique in a process mining project.

Table 2. Requirements for traditional and redo log process mining approaches.

	Aspect	Traditional PM	Redo Log PM
Data elements	Timestamps	Required	Guaranteed
	Case Notion	Required	Required
	Activity names	Required	Guaranteed
Technical aspects	Event recording	Application dependent	Automatic
	Completeness of data	Desirable	Desirable
	DB read access	Required	Required
	Special privileges	Not Required	Required
	Snapshot of DB	Desirable	Required

Table 2 shows the requirements of both approaches with respect to the availability of data elements, and some technical aspects to take into account. For each data element, the approaches present different levels of exposure. Something is *required* when it must be explicitly recorded and available in the database tables. If it is *guaranteed*, this means that it is assured to be available, regardless of the data schema or the application under study. With respect to technical aspects, something is *required* when it must be available at the extraction time. If it is *automatic*, this means that it is guaranteed to be available. *Desirable* means that it will positively affect the data quality, but is not critical for the technique to work. An aspect is *application-dependent* when it depends on the application under study to be available; therefore, some uncertainty exists. Finally, an aspect is *not required* if it is not necessary for the technique to work and, in fact, will not affect the quality of data. We will discuss these elements in more detail soon.

3.1   Data Elements

Looking at the top part of Table 2, we can identify that several data elements are needed to extract event logs at all, and we can see how differently these approaches obtain them. The presence of **timestamps**, a **case notion**, and **activity names** are required by the traditional approach. This means that these elements must be recorded by the application and be available in the database tables at the moment of extraction. This represents the first and most important challenge to face in a process mining project. Without these three elements, we cannot construct events and, therefore, no event log. In case that these elements are not explicitly available, we cannot apply the traditional approach, and must find different ways to obtain events. Redo log process mining has a partial solution for this situation. Thanks to the automatic recording of redo logs by the RDBMS, we can automatically obtain database events, which contain timestamps, activity names, and implicitly within the data, one or several case notions.

3.2   Technical Aspects

With respect to the technical aspects, the first challenge to face is the actual **event recording**. As mentioned before, in traditional process mining we depend on the application to actively record the events and store them in the database tables. Without this, we cannot build the event logs. However, redo log process mining relies on the automatic recording of events in the database redo logs. The fact that this is an automatic system means that event recording in redo logs is application-independent. Yet, it needs to be enabled. Many RDBMSs have this functionality, but it is not properly configured or even enabled by default. Therefore, despite being automatic, it is useless if it is not activated. The events in redo log process mining will be available as long as the recording is enabled, properly configured, and the redo logs are archived instead of being overwritten in a rotary manner.

Due to different reasons, the **completeness of data** available to be extracted cannot be guaranteed in either of the approaches. Missing events would lead to incomplete traces that could affect the quality of the resulting analysis. With respect to the traditional approach, incomplete data can be caused by clean-up activities performed in the database, removing batches of historical information for space saving purposes for example. Also, recording failures could cause completeness issues in the data. Based on our experience, this problem is present even more often when dealing with redo logs. As pointed out previously, redo log recording needs to be enabled and properly configured in order to work well for our purpose. The redo logs will only start to be recorded from the moment they are enabled. Any event that happened before that moment will be unknown to us. Also, the redo log archiving must be configured so the redo logs do not get overwritten or discarded. If that is not the case, gaps in the data could appear. This would be the reason for incomplete or missing traces that will affect the quality of the resulting event log.

Normally, when extracting event data, **read access to the database** is required in order to execute queries and read the content of tables. This requirement is independent of the technique used for the data extraction. Nonetheless, because of how critical the original files are, the redo log approach needs **special privileges** in order to load and read the content of redo log files. These privileges are not easy to obtain when dealing with production systems in a real-life environment. In our experience, it is safer to perform the data extraction from a cloned instance of the database system. This can be desirable for traditional process mining as well, but here it would not be critical since the extraction method is less computationally intensive and intrusive than redo log process mining.

Additionally, the extraction of events from redo logs has another relative drawback: it requires a **snapshot of the database**. This is due to the fact that the events recorded correspond to insertions, modifications or deletions of rows, and only the affected fields are reflected in the events. Therefore, unless we possess the complete set of redo log files since the system creation (which is extremely rare), it is not possible to reconstruct the content of the additional fields exclusively from the redo logs. To solve this issue, a snapshot of the database content is required, such that the values of the missing fields can be queried.

To summarize, the main challenges to face when extracting event data from a database system are determined by (a) the presence of the event data in the database, (b) the correct configuration of the event recording systems, and (c) the access and connectivity to the data systems with sufficient privileges to obtain the necessary information.

Until now, we have explained the particularities of two data extraction approaches, together with the challenges they face, at a theoretical level and in a very general way. The next section presents a practical comparison performed with data from a real-life system, using both data extraction approaches, to see how these issues work out in real life.

4 Practical Comparison

In the previous sections, the advantages and challenges of extracting events from redo logs with respect to database tables have been presented. However, these claims have no value without a proper validation. The aim of performing a case study with both traditional and redo log process mining approaches in this section is twofold. First, to show that applying redo log process mining in a real-life scenario is possible. Second, to demonstrate that, in situations that satisfy certain minimum requirements, the results of redo log process mining are of at least as much quality as the ones obtained from the traditional approach.

4.1 Business Case

In order to carry out this case study in a fair manner, it was important to select a system that fulfilled the minimum requirements of both process mining techniques. That is, a system that explicitly records events in the database tables, and that allows to enable redo log recording at the RDBMS level. The software system selected for this study is the OTRS[1] ticketing system. OTRS is a web-based open source process aware information system, commercialized by the OTRS Group, used for customer service, help desk, and IT service management.

---

[1] OTRS: https://www.otrs.com/.

It offers ticket creation and management, automation, time management, and reporting among other functionalities.

The specific instance of OTRS to be analyzed is a production installation within a well known ICT company set in The Netherlands. The company has been using this instance of OTRS for at least for two years now, since the end of 2014, with the purpose of managing the incidents of the IT systems of their clients. In fact, only a subset of the whole plethora of functionalities that OTRS offers are being actively used within the company. In the daily use of the OTRS system, customers send messages reporting issues. This triggers the creation of tickets in the system, that will be followed up by IT specialists. After some interaction between customers and specialists, trying to determine the root cause of the issue, the ticket status will evolve until it gets, hopefully, solved. The goal of the system is to help the company with their customer support in order to maintain a high level of service availability and quality.

There are several reasons to choose this specific instance of the OTRS ticketing system. First, the fact that it is a PAIS makes it very attractive to apply the traditional process mining approach. In addition to that, it runs on an Oracle RDBMS, with the possibility to enable redo log recording, which is a basic requirement to apply redo log process mining. Also, the system was being used in production, with real-life customers. And finally, the company owning the instance was interested in applying process mining to asses the quality of their service. This means that they were willing to cooperate and provide access to the required data and domain knowledge to carry out this case study. The next section describes the execution of the study and how both process mining approaches were applied on the OTRS data.

Table 3. Steps in the execution of the traditional and redo log process mining approaches to obtain an event log.

Traditional PM	Redo Log PM
1. Query the database (SQL Developer)	1. Connection to DB (PADAS)
2. View of events and cases (SQL Developer)	2. Extraction of Data Model (PADAS)
3. Export log to disk (SQL Developer)	3. Extract events for each table (PADAS)
4. Add trace attributes to log (RapidProM)	4. Build log (PADAS)
5. Load log for analysis (ProM)	5. Export log to XES format (PADAS)
	6. Load log for analysis (ProM)

4.2 Execution

To obtain an event log from the system under study, it is necessary to follow a specific set of steps, depending on the approach used to extract the event data. However, in both cases, first we must define the scope of the analysis. The company is interested in answering business questions related to the incident solving process. In particular, these related questions are about the service-level agreements (SLAs) they have with their customers. When looking at the data model[2] of the OTRS system, we observe that the table *TICKET* plays a central role in the general schema. This table contains the main attributes of a ticket in OTRS. Also, the table *TICKET_HISTORY* holds the historical information related to each ticket. This means that the changes in the tickets are stored in the form of events in that table. Additionally, messages and extra data linked to each ticket is stored in the table *ARTICLE*. In conclusion, we consider the table *TICKET* as the case table, and *TICKET_HISTORY* and *ARTICLE* as event tables.

With the scope being defined, it is possible to proceed with the data extraction to build an event log. Starting with traditional process mining, we executed the steps in the left column of Table 3. The details regarding the execution of these steps are outside the scope of this paper. However, extensive information about the full study can be found in [8]. The result has been an event log of which the characteristics can be observed in Table 4, under the column *Traditional PM*.

The data extraction process for the redo log process mining approach differs from the traditional mainly in the source of data, which are the redo log files instead of the database tables. This means that special tools need to be used, in this case the Process Aware Data Suite[3] (PADAS). This tool allows to connect to an Oracle database, and is able to extract the data model, and the events contained in the redo log files for any table of the schema. Also, once the events have been extracted, the tool supports the log creation step, grouping events in traces according to the desired case notion. More details on the log building creation are available in [7]. The steps followed in the data extraction and log building phase for the redo log approach are listed under the right column of Table 3. The log exported from the PADAS tool presents the characteristics observable in Table 4, under the column *Redo Log PM*.

4.3   Results

To discuss the results, we will take a look at the aspects of the event logs obtained by traditional and redo log process mining, to evaluate their main differences. Analyzing Table 4, it is clear that there is a big difference on the covered period of time, as well as to the size of the event logs obtained by the two data extraction approaches. The redo log data is not as extensive as the one obtained by the traditional method. This is due to the fact that the redo log recording on the Oracle database hosting the OTRS data schema was enabled at the beginning of the project, around March 2016, and continued until July of the same year. However, the traditional approach was able to extract all the events in the *TICKET_HISTORY* table, which was never deleted or purged since the OTRS system was setup at the end of 2014. That is the main reason for the big difference in data quality between both approaches.

---

[2] http://ftp.otrs.org/pub/otrs/doc/database-schema/otrs-3.3-database.png.

[3] PADAS: https://www.win.tue.nl/~egonzale/projects/padas/.

Table 4. Metrics of the resulting logs for both approaches on all the available data.

Metric	Traditional PM	Redo Log PM
Time window captured (days)	629	124
Magnitude (# of cases)	17822	3019
Support (# of events)	281906	31877
Number of distinct event classes	31	23
Granularity of timestamps	seconds	seconds

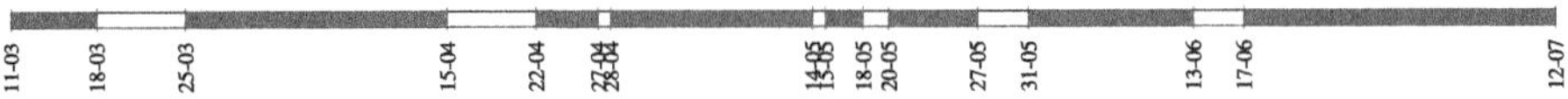

Fig. 1. Missing archived logs over time in 2016. Shaded areas indicate the availability of archived logs, and white areas indicate the gaps.

Table 5. Metrics of the resulting logs for the period from June 17th to July 12th.

Metric	Traditional PM	Redo Log PM
Time window captured (days)	26	26
Magnitude (# of cases)	907	907
Support (# of events)	6342	7866
Number of distinct event classes	22	22
Granularity of timestamps	seconds	seconds

Additionally, after observing the resulting event log from the redo log process mining approach, one more data quality issue was identified. Big time gaps were spotted in the extracted data, as shown in Fig. 1. However, this problem did not exist in the data obtained by the traditional approach, which was complete. Further investigation of the root cause showed that the reason for this was a misconfiguration of the cloned server used in the study. In this server, a daily script would archive the already filled redo log files to a storage location. However, in some cases, a race condition occurred with another script in charge of cleaning up storage for space saving purposes. This caused the loss of redo log files for full days, and consequently incomplete cases and data quality issues. The issue was fixed as soon as it was detected and, fortunately, data continued to be recorded, this time without interruption. In order to ensure a fair comparison of the process mining approaches, the following strategy was adopted: from the time line of redo log data observable in Fig. 1, the largest uninterrupted period was selected to be compared between both logs. This period is from June 17th to July 12th. The resulting event logs were then compared, and the metrics are presented in Table 5. The following section provides a discussion on the equivalence of these two event logs, looking at them from the structural and behavioral point of view.

5 Discussion

It has been previously stated that the goal of this work is to find support for the hypothesis that says that *the data obtained by the redo log process mining approach is at least as rich as the data obtained by traditional methods*. Section 3 shows the intuition behind this hypothesis from the theoretical point of view. Then, Sect. 4 takes a practical perspective on the evaluation, applying both process mining approaches in a real-life environment. The aim of this section is to analyze the results of the practical comparison, in order to support the aforementioned hypothesis, and explain the possible differences between the event logs obtained by both process mining approaches.

5.1 Event Labels Comparison

Table 5 shows that, when focusing on a period of time during which data is available for both approaches, the event logs coincide in the number of cases. Also, the number of events extracted by the redo log approach is higher than the amount obtained by the traditional one. However, this does not guarantee that the former is a superset of the latter. To find evidence of it, we have to look at the event labels in both logs. Table 6 shows a list of event labels ordered by frequency for both event logs. At first sight, the event labels seem disjoint. However, further analysis shows that the two most frequent event labels in the redo log process mining event log, namely *NewEventNoMsg*, and *NewEventWithMsg*, correspond to the redo log events obtained from the *TICKET_HISTORY* table. This table is the source of events for the traditional process mining approach. In fact, the sum of the frequencies of these two event labels, 5032 and 1310 respectively, is equal to 6342 events, the total number of events in the event log obtained with the traditional approach. The reason for which the 22 event types of one log are grouped in only two in the other one is that, as to the latter, the event classifier is automatically provided by the approach. This classifier takes into account the table in which the event occurred, and which fields were affected. However, in the traditional approach, the event classifier takes into account the value of the *ticket_state_id* field, which maps integer values to the event labels on the left side of Table 6. Therefore, using this event classifier in the events *NewEventNoMsg* and *NewEventWithMsg* of the redo log process mining event log, would result in the same set of event labels, with the same frequencies. To be precise, the events from the redo log approach with the label *NewEventNoMsg* correspond to a subset of the events obtained through the traditional method with the following event labels: *Misc, OwnerUpdate, StateUpdate, CustomerUpdate, NewTicket, SendAgentNotification, Lock, Unlock, Merged, TicketLinkAdd, Move, PriorityUpdate, TypeUpdate, SetPendingTime*. With respect to the events with the label *NewEventWithMsg*, they correspond to a subset of the events with the labels: *OwnerUpdate, StateUpdate, AddNote, FollowUp, EmailCustomer, SendAutoReply, SendAnswer, SendCustomerNotification, TimeAccounting, SendAutoFollowUp*. Therefore we see that there is not a 1:n mapping between the event classes obtained by both approaches. On the

Table 6. Event labels and frequencies with the default classifiers for the two event logs.

Traditional PM			Redo Log PM		
Activity label	Freq	Rel Freq	Activity label	Freq	Rel Freq
Misc	1182	18.64%	**NewEventNoMsg**	**5032**	63.97%
OwnerUpdate	976	15.39%	**NewEventWithMsg**	**1310**	16.65%
StateUpdate	919	14.49%	MessagePhoneOrNote	800	10.17%
CustomerUpdate	883	13.92%	MessageTicketMerged	117	1.49%
NewTicket	882	13.91%	AutoReplyTicketReceived	74	0.94%
SendAgentNotif	398	6.28%	NewArticleA	70	0.89%
AddNote	168	2.65%	NewArticleB	63	0.80%
Lock	165	2.60%	NewArticleC	51	0.65%
Unlock	164	2.59%	UpdateMsg-TicketId-Time-User	49	0.62%
Merged	102	1.61%	UpdateEvent-TicketId-Time-User	49	0.62%
TicketLinkAdd	102	1.61%	NewEmail_Note-Customer_Agent	40	0.51%
FollowUp	94	1.48%	NewArticleD	38	0.48%
EmailCustomer	82	1.29%	UpdateEvent-TicketId-Time	34	0.43%
SendAutoReply	72	1.14%	UpdateMessage-TicketId-Time	31	0.39%
SendAnswer	63	0.99%	UpdateMessage-TicketId-User	29	0.37%
Move	42	0.66%	UpdateEvent-TicketId-User	29	0.37%
PriorityUpdate	26	0.41%	NewMessage-CustomerOrAgent	19	0.24%
TypeUpdate	11	0.17%	NewEmailExternal	10	0.13%
SendCustomerNotif	4	0.06%	UpdateMessage-TicketId	9	0.11%
TimeAccounting	3	0.05%	UpdateEvent-TicketId	9	0.11%
SetPendingTime	2	0.03%	EmailFromCustomerWithoutCC	2	0.03%
SendAutoFollowUp	2	0.03%	EmailFromCustomerWithCC	1	0.01%
Total	**6342**	100.00%	Total	7866	100.00%

contrary, it is a n:m relation, with cases like the activity *OwnerUpdate* from the log of the traditional approach that groups events that can either correspond to the activity *NewEventWithMsg* or the activity *NewEventNoMsg* of the log of the redo log approach.

It is important to note that the fact that in Table 5 the number of distinct event classes is the same for both logs (22) is just a coincidence. Actually, the real number of event classes in the redo log process mining event log using an appropriate event classifier should be 42, since two of the event classes of this log correspond to the 22 obtained with the traditional method.

5.2   Control Flow Comparison

The equivalence of both event logs has been analyzed from the event labels point of view. However, without mining the traces, we cannot guarantee that the two event logs represent equivalent behavior. To check this aspect, we mined the event logs using the same event classifier in both cases. As discussed previously, the event log obtained from the redo logs contains a superset of the events in the one extracted by the traditional approach. In order to compare the behavior of both logs, we must focus on the same subset of activities. Therefore, the event log obtained from the redo log was filtered, to only include events corresponding to the labels *NewEventNoMsg* and *NewEventWithMsg*. Then, the same classifier as in the traditional approach was used, so both event logs would have the same set of event classes. After this preparatory step, we mined both logs using Inductive Miner Infrequent. The resulting process models can be observed in Fig. 2.

From observing both models we see that they mostly represent the same control flow. However, some differences can be spotted immediately. First, the activities *EmailCustomer* and *SendAutoReply* occur in parallel in Fig. 2a, while they are in a sequence in Fig. 2b. Second, the activity *SendAnswer* is part of a

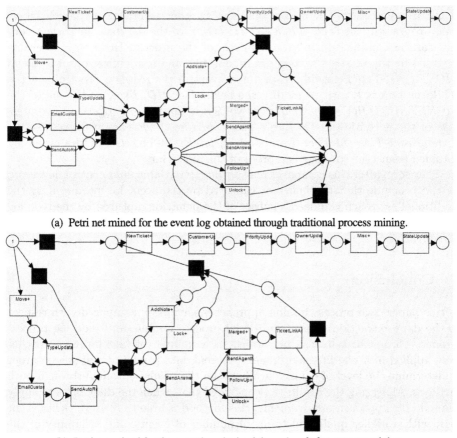

(a) Petri net mined for the event log obtained through traditional process mining.

(b) Petri net mined for the event log obtained through redo log process mining.

Fig. 2. Process models mined with Inductive Miner, Infrequent (noise threshold = 0.2).

choice in Fig. 2a, while it happens before the choice in Fig. 2b. Third, activities *NewTicket* and *CustomerUpdate* always happen in 6th and 5th position from the end of the trace in Fig. 2b, while in Fig. 2a they can only be executed in mutual exclusion with the bottom part of the process.

These differences, though graphically subtle, can mean a big difference in behavior. Fortunately, there is an explanation for them. There are two main reasons for this disagreement in control flow between both event logs. (1) The event timestamps obtained by both approaches are set by different mechanisms. In the traditional approach, the timestamps of each event correspond to the ones written by the OTRS system in the *timestamp* field of the *TICKET_HISTORY* table. In the redo log approach, the timestamps correspond to the ones recorded by the Oracle RDBMS when processing the SQL statements sent by the OTRS system. Therefore, a difference in order between the events of the traditional and the redo log approach could occur given that the timestamps in the former corresponds to the behavior enforced by OTRS, while the timestamps in the latter correspond to the actual execution of the associated statements in the database. (2) The events obtained by the traditional approach correspond to rows in the table *TICKET_HISTORY* of the database, and their content can be modified during the life-cycle of the process. However, the events recorded by the redo log system are immutable, and a modification of a row in *TICKET_HISTORY* would create a new event in the redo log files. In fact, the OTRS system is known to modify the fields *TicketID*, *User*, and *Time* of the *TICKET_HISTORY* rows whenever two tickets are merged together. The presence of the activities *UpdateEvent-TicketId-Time*, *UpdateEvent-TicketId-User*, *UpdateEvent-TicketId-Time-User*, and *UpdateEvent-TicketId* in the event log obtained from the redo logs is a proof of this behavior.

Therefore, after this comparison at both activity label, and control flow level, we can conclude that the behavior captured by the event log produced by the traditional approach is indeed a subset of the behavior captured by the redo log approach, and the latter can be easily filtered in order to achieve a high degree of equivalence.

6   Conclusion

In this paper, two process mining approaches have been compared with respect to the data extraction phase: traditional process mining, and redo log process mining. The evaluation was performed in a unique setting: both approaches were applied in a real-life environment, on real data from real systems, in order to determine the level of equivalence between the results obtained through both methods. Analyzing the results, we concluded that, when the difficulties to apply the redo log approach are overcome, this method is able to retrieve richer event logs, with a higher quality in terms of number of events, and reliability of the captured behavior. Additionally, it has been shown that traditional approaches are vulnerable to event manipulation, which can alter the results of the analysis, while the redo log approach ensures the immutability of the events, being therefore more robust to data manipulation and fraud. In addition to these benefits, redo log process mining, unlike the traditional approach, can be applied to non-process aware systems, in which events are not explicitly recorded at the application level, but they still use a RDBMS as a data storage. However, this comes at a price. The need for special privileges to configure and enable redo log recording makes it not easy to set up in all environments, while the traditional approach only requires read access to the relevant database tables. All things considered, redo log process mining must be considered as a viable alternative to traditional process mining. As future work, new sources of event data will be explored, in order to tackle the limitations of the redo log approach, and improve the quality of the extracted event logs with respect to traditional methods.

References

1. Watson, H.J., Wixom, B.H.: The current state of business intelligence. Computer **40**(9), 96–99 (2007). https://doi.org/10.1109/MC.2007.331
2. Ingvaldsen, J.E., Gulla, J.A.: Preprocessing support for large scale process mining of SAP transactions. In: ter Hofstede, A., Benatallah, B., Paik, H.-Y. (eds.) BPM 2007. LNCS, vol. 4928, pp. 30–41. Springer, Heidelberg (2008). https://doi.org/10.1007/978-3-540-78238-4_5
3. Roest, A.: A practitioner's guide for process mining on ERP systems: the case of SAP order to cash. Master's thesis, Technische Universiteit Eindhoven, The Netherlands (2012)
4. Segers, I.: Investigating the application of process mining for auditing purposes. Master's thesis, Technische Universiteit Eindhoven, The Netherlands (2007)
5. Yano, K., Nomura, Y., Kanai, T.: A practical approach to automated business process discovery. In: 2013 17th IEEE International Enterprise Distributed Object Computing Conference Workshops (EDOCW), pp. 53–62, September 2013
6. Verbeek, H.M.W., Buijs, J.C.A.M., van Dongen, B.F., van der Aalst, W.M.P.: XES, XESame, and ProM 6. In: Soffer, P., Proper, E. (eds.) CAiSE Forum 2010. LNBIP, vol. 72, pp. 60–75. Springer, Heidelberg (2011). https://doi.org/10.1007/978-3-642-17722-4_5
7. de Murillas, E.G.L., van der Aalst, W.M.P., Reijers, H.A.: Process mining on databases: unearthing historical data from redo logs. In: Motahari-Nezhad, H.R., Recker, J., Weidlich, M. (eds.) BPM 2015. LNCS, vol. 9253, pp. 367–385. Springer, Cham (2015). https://doi.org/10.1007/978-3-319-23063-4_25
8. Hoogendoorn, G.E.: A comparative study for process mining approaches in a real-life environment. Master's thesis, Eindhoven University of Technology (2017)
9. Jans, M.J.: From relational database to valuable event logs for process mining purposes: a procedure. Technical report, Hasselt University (2017)
10. de Murillas, E.G.L., Reijers, H.A., van der Aalst, W.M.P.: Connecting databases with process mining: a meta model and toolset. In: Schmidt, R., Guédria, W., Bider, I., Guerreiro, S. (eds.) BPMDS/EMMSAD -2016. LNBIP, vol. 248, pp. 231–249. Springer, Cham (2016). https://doi.org/10.1007/978-3-319-39429-9_15
11. van Eck, M.L., Lu, X., Leemans, S.J.J., van der Aalst, W.M.P.: PM^2: a process mining project methodology. In: Zdravkovic, J., Kirikova, M., Johannesson, P. (eds.) CAiSE 2015. LNCS, vol. 9097, pp. 297–313. Springer, Cham (2015). https://doi.org/10.1007/978-3-319-19069-3_19

From Relational Database to Event Log: Decisions with Quality Impact

Mieke Jans[1]([✉]) and Pnina Soffer[2]

[1] Hasselt University, Hasselt, Belgium
mieke.jans@uhasselt.be
[2] University of Haifa, Haifa, Israel
spnina@is.haifa.ac.il

Abstract. This paper addresses the topic of 'Remediation approaches for event log quality assurance'. The assumption of having readily minable event logs is often not fulfilled. This paper addresses, from an end-user's perspective, the quality issues that arise when an event log needs to be built from a relational database. The decisions that are taken when building the event log, have an impact on the quality of the event log. Namely, these decisions impact the suitability of an event log for the planned analyses. The goal of this paper is to provide an overview of the decisions that impact the quality of the event log, along with a realistic running example. Based on this overview of decisions, a procedure is presented. This procedure provides guidance to build the event log in a conscious manner, taking into account all the decisions and their impact on quality. This work relates to other studies on how to build an event log from relational databases, but puts more emphasis on how the technical decisions have a direct impact on the analyses of the practitioner that will use the event log afterwards.

1 Introduction

The field of process mining has received increasing attention in both academic research studies and in industry. In order to apply this data-driven approach in business process management, a minable event log is required. Guidance on how to build such an event log on the other hand, is highly under-investigated.

Where previous work focused on the technical challenges of creating an event log, this paper addresses the decisions that need to be supported and/or taken by the end user. Although some decisions might be guided by technical limitations, the consequences they ultimately carry for the log quality should not be considered as less relevant. Quality in this work is seen from an applicability standpoint. In case the event log is built in such a way that the log structure allows for the intended analyses, the quality of the log can be classified as 'high'.

This paper showcases the different decisions that need to be taken during the creation of an event log, along with their possible analysis consequences. The showcase is based on a traditional procure-to-pay process in an ERP-system to explain possible difficulties. Next, a procedure to build an event log in a conscious way, is presented. The aim of the procedure is to increase the level of support

© Springer International Publishing AG 2018
E. Teniente and M. Weidlich (Eds.): BPM 2017 Workshops, LNBIP 308, pp. 588–599, 2018.
https://doi.org/10.1007/978-3-319-74030-0_46

the final event log can offer for the end user, hereby providing the highest level of quality possible.

This study aims to contribute both to practice and to the research community. Practitioners are provided with real-life situations and clearly explained consequences. The research community gains more insight in which degrees of freedom remain absolutely necessary for the industry to build meaningful event logs.

The remainder of the paper is structured as follows: Sect. 2 provides a possible categorization of decisions taken when building an event log; Sect. 3 introduces the running example that will be used throughout the remainder of the paper; Sect. 4 discusses the different decisions explicitly, along with their consequences; Sect. 5 presents a procedure to build the event log and take step-by-step decisions; Sect. 6 discusses related work, and Sect. 7 concludes the paper.

2   Event Log Building Decisions - Classification

Building an event log holds two major phases: deciding on the architecture of the log at first, and feeding the architectural frame with data next. This work focuses on the first phase and its related decisions. Taking the XES-standard as the format an event log should adhere to, there are three levels of main elements: a complete log, a trace, and an event. These levels can be used for categorizing the decisions while building the log architecture. A first category of decisions is related to the process as a whole. Questions concerning which process should be selected and the exact scope of the process would make part of this category. The second category of decisions would entail all decisions that are related to the selection of the process instance. The third category holds the decisions that are related to the event level, like which activities to include. The focus of the paper is on taking the second and third category of decisions.

3   Running Example

In order to provide a good understanding of the discussed decisions, a running example is provided. The example relates to a realistic, traditional procurement process, as depicted in Fig. 1. This is a classic procurement process, like these are often supported by ERP systems. The presented BPMN model depicts a simplified version of the designed model the way it is generally understood by both managers and employees, involved in purchasing. Starting a process mining analysis, this level of abstraction of a designed process model is often the basis, along with some understanding of the supporting database. From a business perspective, mainly two types of high-level goals can be set: gaining assurance on compliance (f.e. when an auditor is requesting the project), and gaining insights into efficiency (e.g., in terms of performance indicators, often when the process owner is requesting the project). These goals are not mutually exclusive, but in reality the project sponsor is mostly skewed towards one of them.

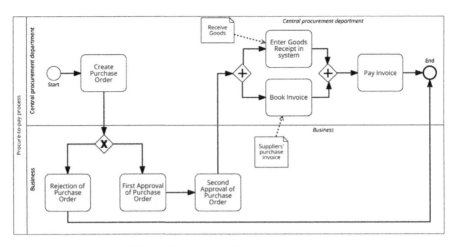

Fig. 1. Procurement process example

4   Event Log Building Decisions and Their Consequences

A decision that greatly impacts the final event log structure, is the selection of the process instance. In case the organisation already uses process models, it is an option to use the assigned process instance in their models as starting point. If there are no models with instances to start from, a process instance needs to be selected, based on the data structure. In our running example, a purchase could, at a first glance, be thought of as a process instance. In business processes how-ever, activities are often related to documents, and process instances naturally reflect one of those documents. As a result, a 'purchase' is too vague as a process instance. In our example, possible candidates could be a Purchase Order (PO), a goods receipt document, or an invoice. Activities in the event log should then relate to the selected document. These activities are typically extracted from a change log. Change logs capture changes to documents that could be translated into understandable events. However, a log does not always capture changes to a full document; changes might only affect a part of a document. In this light, there are two dimensions related to the selection of a process instance: (1) which document to select, and (2) on which level of granularity. These dimensions and their impact are discussed in the following subsections.

4.1   Document Selection of Process Instance

To select a document, two aspects have to be considered: the goal of the process analysis, and the cardinality of the relations between the candidate documents. Both aspects will be discussed in the two next subsections.

Document Selection and Goal of Process Analysis. As explained before, we envision two types of business goals of the analysis: efficiency (in terms of business objectives) and compliance. From an efficiency point of view, fall-out is often a key performance indicator. Fall-out refers to cases that started the process, but are not terminated. By selecting an instance document that is created in the beginning of the process, fall-out will be possible to identify.

Cases that do not start with the first step as specified by the model, are on the other hand often of interest for compliance-driven projects. These cases however will not be part of the event log in case the process instance is a start document. These process executions will only be identified if a document near to the end of the process is used as the process instance. Therefore, selecting such an 'end document' as the process instance is better suited in combination with a goal of compliance. The term 'end document' refers to the document of interest of the compliance officer, and often is more *towards* the end of a process. In a financial auditing setting for example -a traditional compliance context-, this is the invoice.

Selecting a possible start document to include fall-out identification or selecting an end document for compliance purposes, implies a trade-off between identifying fall-out and non-compliance. A possible solution that counters this trade-off, is to use a dynamic process instance. A dynamic process instance is a process instance that, depending on the individual case, may have a different underlying document as process instance. This dynamic process instance uses the first document that is present for an individual case as the process instance. The disadvantage of this approach, however, is a more difficult interpretation of the analysis results in the next phase. A case can refer in 1 case to a certain document, and in another case to another type of document. This approach is related to the artefact-centric approach, like described in [7].

Running example. *In case of an efficiency-inspired project, the PO would suit best as process instance. That way, POs that still did not result in an invoice, can be identified. In case of a compliance-driven project, the invoice might be a better candidate. Particularly since the invoice is the direct link to financial reporting, mostly the subject that needs to be provided assurance on in compliance-driven projects. Selecting the invoice as process instance allows for identifying invoices that do not stem from a PO. In case a dynamic process instance is used, a 'purchase', the case, will be represented by a PO in case a PO is present, or by an invoice, when no PO is present. Given that a goods receipt document can only exist after creating a PO, only the PO and the invoice will be valid representations of the dynamic instance.*

Document Selection and Cardinality Between Candidate Documents.
In case the candidate documents show cardinalities that are different than one-to-one, the process analyst should be aware of the consequences of choosing one document over the other document. Having many-to-many relationships can be disentangled to one-to-many and many-to-one relationships. Both the many documents-to-one instance relationship and the many instances-to-one document relationship hold their own consequence.

In case of a relationship where multiple other documents of the same type can be related to a single process instance, the event log will link a repetition of transactions on these other documents to a single instance. The resulting traces will contain repetitions or self-loops of these activities, representing the many-to-one relationship correctly from a technical point of view. In case of a relationship where multiple process instances can be related to a single document, however, the event log will artificially duplicate the transaction(s) on that document (under the assumption that these activities are included in the event log). This would result in a misrepresentation of reality. In case of a many-to-many relationship between two documents, one has to decide for which document an artificial duplication of the related transactions is least harmful or more in line with the goal of analysis. Specifically, if the analysis also relates to the load of resources, duplication of events might present a biased picture. Notice that, although the decision relates to the selection of a *document*, the consequences are on the level of the related *transactions* that take place on the involved documents.

Running example. *In a purchase process one PO can result in multiple invoices and vice versa. So there is a many-to-many relationship between PO and invoice. Different activities can be linked to these documents: 'Create PO' is related to the PO, while 'Book invoice' relates to the invoice. Making abstraction of other activities, selecting a unique PO document as process instance, would result in the following:*

- *In case a PO indeed resulted in multiple invoices, traces like <'Create PO', ... , 'Book Invoice', 'Book Invoice', 'Book Invoice', ...> will be created. The self-loop on the 'Book invoice' activity reveals the one-to-many relationship between the selected process instance document (PO) and the invoice. Notice that in case multiple events would be related to the invoice, for example also 'Pay invoice', these will all result in multiple events in a single trace. As such, the event log will uncover the repetition of the same activity(-ies) 'Book Invoice' and/or 'Pay Invoice' on a single PO, adequately representing reality at the level of the process instance.*
- *If an invoice combines purchases that were instantiated by multiple POs, the 'Book Invoice' activities will be mentioned in the traces of each of these PO-id's. Namely, since the PO is the process instance, followed throughout the process, the event of booking the invoice is artificially multiplied for all POs where this invoice is related to. This poses a discrepancy between reality -the invoice is only booked once- and the number of times an activity is included in the event log. This effect does not have to pose a great problem, it only should be taken into consideration when analyzing results. It is mostly important for avoiding an interpretation of the summary statistics on activity frequency.*

If, in this example, the invoice would have been selected as process instance, similar consequences would hold, only starting from a different point of view. A self-loop on the activity Create PO would show up, uncovering the creation of multiple POs, preceding the booking of a single invoice. Further, the activity Create PO would be artificially multiplied, because of the many-to-one relationship between PO and invoice.

To take many-to-many relationships and its consequences into account in the event log architecture, it is important to turn to the key questions and the goal of the project. What exactly do you wish to get an answer to? For example, a financial auditor is primarily interested in whether invoices were preceded by an approved PO. Therefore, selecting the invoice as the process instance, avoiding a possible artificial multiplication of financial documents, is highly recommended.

4.2   Granularity Level: Parent Versus Child

In case the selected document is represented by a parent-child relationship in the database, one has to decide which level of the document should serve as a process instance. Will the document as a whole be followed throughout the process, or will it be divided in sub-parts to follow separately throughout the process? This decision on process instance selection is highly related to the selection of activities to be included in the event log. Activities, at least in a business process, are recorded transactions on documents. Some transactions affect a full document (for example the approval of a document). Other transactions only affect part of a document. Think for example on entering the goods receipt document of some items in the system. This activity would relate to the item-level of a purchase. It stores, for example, information that the ordered 20 chairs have arrived (a line on a certain PO), but not that a PO has arrived. So ultimately, both at the process instance and at the activities selection, the event log builder needs to decide on the level of granularity. These decisions are intertwined with each other. In the following paragraph, we take the point of view from the process instance selection. However, bear in mind that process instance and activity selection influence each other and can be taken in different following order.

Three scenarios on granularity level and their consequences should be clear, in order to take a conscious decision when building a qualitative event log. These scenarios are discussed in the following paragraphs.

The Mix of a Process Instance at Parent Level with Activities on Child Level. When the process instance is set at the parent-level, activities that refer to transactions on that parent document do not pose a problem. However, the activities that relate to sub-parts of the selected document, the children, require more attention. It can be beneficial to define an aggregation function of activities that relate to transactions at the child level. In case no aggregation function is specified, and all distinct activities on the child-level are captured in the event log, the one-to-many relationship between parent and child will be copied in the process trace. A repetition or self-loop will be present on the more granular activities, while there is no information on whether the repetition of this activity refers to one activity that was repeated for different children, or to multiple executions of the same activity on a single child.

Running example. In case the selected process instance is a PO document (parent level) and a transaction 'Goods Receipt' is related to the level of PO items (child level), the following can be part of a trace in the event log: < 'Goods Receipt', 'Goods Receipt', 'Goods Receipt'>. This string can have an ambiguous *interpretation. At one end of the spectrum, this can suggest that this PO has three line items of which the goods have been received. This would imply an efficient process execution. At the other end of the spectrum, this can reflect three partial deliveries of one PO line item, implying inefficiency. All combinations in between can also be reflected by this string. The event log builder needs to decide on whether or not this ambiguity is a problem for the planned analyses afterwards.*

In essence, this is similar to the one-to-many relation between candidate documents for process instances, as described before. As a result, an important decision is whether all these finer granularity level activities should be part of the event log, or would an aggregation suffice too. For example, one could decide to only take the first recorded transaction that relates to the child-level of the process instance (or another related document).

The Mix of a Process Instance at Child Level with Activities on Parent Level. The opposite situation also has its own characteristics. In case the child level of a document is selected as the process instance, higher level activities have to be copied to all underlying children, resulting in artificial multiplication. Think for example about the multiplication of a single approval of a PO to all the items that are captured in this PO. This is again the same principle as with the one-to-many table relationships.

Running example. Returning to the previous example, selecting a PO line as process instance, instead of the PO as a whole, would take away the ambiguity of before. In case a subtrace < 'Goods Receipt', 'Goods Receipt', 'Goods Receipt'> would show up, this is for sure a repetition of 'Goods Receipt' on a single item. On the other hand, the activities that relate to the parent level, like 'Create PO' will be present in each trace of the related children. This is mostly important to take into account when aggregating information in descriptive statistics on the activity 'Create PO', since doubles are present in the log.

Exceptional Situation: Child Level Activity Triggers Parent Level Activity. One particular situation requires special attention: when an activity at a finer level of granularity triggers an activity of a higher level. Such a situation results in a dependency between activities, that are stored at different levels of granularity. If this construct is present, any resulting event log would require specific interpretation and assumptions to be analyzed. Nevertheless, if a log is still desired, easier interpretation would be enabled by selecting the process instance at the parent level. In that case, both the triggering activity and the reaction will at least be tied together in a single trace, where a process instance at the child level would result in meaningless traces. The disadvantages of the mix, as described above, remain also present. The situation is demonstrated in the running example below.

Running example. In our example process, the approvals of the PO are activities on the parent level: a PO document is approved as a whole. Such an approval is demanded upon creation of a PO, but might also be triggered when someone adds a line to the PO. This last transaction is related to the child level of the PO. Hence, we have a child level activity, 'Add PO line', that triggers a parent level activity, 'Approve PO'. Assume a PO is originally created with four lines. After approval, the goods receipts of all separate lines take place. In parallel, a fifth line is added. This activity triggers a new approval of the PO. While the invoices are received for the first 4 lines, the goods receipt of the 5th line takes place in parallel.

If the event log would follow a PO at the parent level and include all activities at a lower level i.e. using no aggregation function, the trace of this specific PO would look like follows: <Create PO, Approve PO, Goods Receipt, Goods Receipt, Add line, Goods Receipt, Goods Receipt, Approve PO, Receive Invoice, Receive Invoice, Receive Invoice, Goods Receipt, Receive Invoice, Receive Invoice>. If one would analyse this trace, questions like 'Why is there a goods receipt before the PO is approved?' and 'Why are invoices received before goods are received?' could raise. In reality, the interventions of the activity 'Add line' and 'Goods Receipt', relating to line 5, are in parallel with the flow of the first 4 lines. By following a process instance at the parent level, one cannot uniquely link these parent level activities to the child activity, but at least they all appear on one trace. When using such an event log in the analysis phase, assumptions on the correct functioning of the triggering mechanism need to be made. Selecting the process instance at the finer level of granularity on the other hand, splits the activities at the child level to different traces and no connection can be made anymore. The trace of the first item line in this example would be <Create PO, Approve PO, Goods Receipt, Approve PO, Receive Invoice>. The presence of the second 'Approve PO' is, in this trace, isolated from the trigger (present in the trace that relates to fifth item line). Therefore, a process instance at the parent level might be experienced as the lesser of two evils in this situation.

4.3   Activity Selection

Activities in the log reflect changes that were made to process documents. Such changes are typically captured by a timestamp attached to the document. One possibility, suggested by [1], is to include every timestamped activity in the log. This, however, might result in a mixture of entries at different abstraction levels, some of which introduce noise and hamper the analysis. A decision needs to be made on which activities to include. This can be guided by two concerns: (a) select only relevant events, considering the goal of analysis, and (b) maintain an appropriate granularity level.

As a last step related to activity selection, event attributes can be considered as potential variants of activities. For example, if a certain attribute only has a limited number n of possible values, it might be beneficial to create n variants of that activity.

Running example. *In the context of an efficiency goal, the end user might be interested in the activity 'Goods Receipt', an activity at the level of a PO line. However, a compliance-driven project might only be interested in whether the goods are received or not (not when, by whom...). This type of information can* *be stored as an attribute of the PO, which takes away the need to create an activity at a lower abstraction level of the PO. This will certainly be of interest when an invoice is taken as process instance, since multiple POs and consequently more goods receipt documents will be entered in the system for one invoice. Including all these transactions as activities would create irrelevant information in the light of the business goal.*

With regards to including event attributes as activities, the information on invoice document types could for example be included in the 'Book Invoice' activity. Instead of storing the document type as attribute, it could be interesting to create different 'Book xxx Invoice' activities in case the number of types is manageable. That way, the dimension of this characteristic is immediately included in the control-flow component. For the end user, this will be clearly visible in the process discovery step, where activities, along with their weights, will be visible for the different document types.

5   Procedure

In order to take the aforementioned decisions in a structured and efficient manner, conscious of all related consequences, guidance is needed. To this end, we suggest the use of a procedure. The procedure, proposed in this section, is a sequence of steps that could be followed during the architecture-step of creating an event log for process mining, starting from a relational database. The procedure aims at building an event log of a quality level as high as possible. The following procedure is suggested:

1. **Set primary business goal of project and identify key cornerstones**
   The primary business goal is the goal that is considered as a 'must have' by the project sponsor. This is often either compliance or efficiency. Both goals can be pursued, but the goal that has most leverage must be identified in this phase. The key cornerstones are the key activities in the process, according to the different project stakeholders. At this stage, these cornerstones are not yet verified as possible activities in the final event log, hence the differentiating terminology to avoid confusion.
2. **Gather information on the underlying database**
   This step requires the identification of the tables that capture information of the key cornerstones (of step 1). Starting from these tables, the database structure of these tables, along with other technically related tables, should be communicated with and understood by all stakeholders.
3. **Select process instance**
   Starting from the agreed upon goal, the key cornerstones, and the underlying database structure, this step requires a decision on which process instance to select: which document (see Sect. 4.1), and on which level of granularity (see Sect. 4.2). All consequences, presented in the previous section, should be taken into account. During this discussion, inevitably the list of key cornerstones and their possible inclusion as activity will already be discussed.

4. **Select activities**

   After selecting the process instance, the activities can be selected. As stated in Sect. 4.3, this selection can be guided by two concerns: select only relevant events, considering the goal of analysis, and maintain an appropriate granularity level.

5. **List attributes**

   The last step entails listing all case and event attributes. This selection is again influenced by the business goal. Also considering attributes to be included as activity, is part of this step (see Sect. 4.3).

6   Related Work

The challenges of extracting event logs from relational databases are briefly discussed in [2], mostly from a technical point of view. A broader view of transforming relational data into an event log is provided in [3]. The approach includes three main operations: scope, bind, and classify. Classification allows the creation of multiple event logs, to be mined separately or in a comparative manner. However, the business considerations taken in these steps are not elaborated. Hence, the guidance we propose in the current paper can be considered as complementary to these three operations, as it specifically deals with the business objectives and considerations of the related decisions.

Our work also relates to Calvanese et al., who present an ontology-driven approach to extract event log information from legacy relational databases, supporting domain knowledge [4]. Also Murillas et al. address related challenges in [8,9], and Lu et al. focus on the generation of an artefact-centric approach [7]. All these papers endorse the challenge of extracting relevant event log data out of information systems with a relational structure.

Logs of non-process-aware information systems are addressed by [13], attempting to correlate events into process instances based on similarity of their attributes. Yet, the question of how these instances should be selected is not addressed. [14] address merging logs produced by disintegrated systems that cover parts of the same process. Their implicit decision is to use the 'main' process, namely, the initiating entity, as a lead of the process instance. In case of multiple instances of other involved entities, they are all embedded within the higher-level entity which marks a process instance.

Instance selection is of particular interest for artifact-centric processes [6,7,10]. Artifact-centric processes are centered around artifacts or business entities, which encompass data and life-cycle models. Artifacts may enact and trigger other artifacts in a hierarchical manner, often including multiple instances. [11,12] address the possibility of many-to-many relationships between artifacts, and relate each event to a relevant artifact. As a result, mining artifact-related logs can produce a life-cycle model of these artifacts. While dealing with similar business decisions, these works build on logs produced from specific artifact-centric processes, and are not suitable for ordinary relational databases.

Several process mining algorithms consider the existence of entities at different abstraction levels in a process. Building on primary and foreign keys that have been extracted from database tables to event logs, [5], and [15] support discovery of BPMN models with sub-processes and multiple-instances. These are performed assuming an appropriate log already exists, while our approach supports the creation of such log.

7   Conclusion

This paper addresses the issues that arise when an event log needs to be constructed from data that is stored in a relational database. During the event log building, numerous decisions need to be taken in a context of both the business goal and the technical database structure. While no right or wrong decisions exist, different options have different consequences for the type of questions that can be answered afterwards, using the event log. An event log that allows better for running the planned analyses, is seen as a higher quality event log. In this paper, we demonstrate the different decisions by means of a realistic running example, based on a traditional ERP database structure. After the demonstration of the different types of decisions, along with their consequences on the analysis phase afterwards, a procedure to take these decisions is presented. This procedure provides guidance to both practitioners (to build the event log) and researchers (to better understand the necessary degrees of freedom when creating event logs).

The contribution of this paper is two-fold. Firstly, a clear recognition for the practitioner's role in the phase of creating an event log is given. The practitioner is faced with the consequences of decisions that are, at least currently, taken primarily on a technical basis. The presented procedure aims to involve the practitioner more actively in this important phase. Secondly, the paper contributes to the literature by making the different decisions during the event log building phase, along with its practical consequences, explicit. They can support researchers in providing technical solutions in an industry-compatible way, providing configuration freedom where it is useful. This way, the communication between scientists and practitioners improves, leading to a faster and better adoption and maturity cycle of the process mining field as a whole.

As with all research studies, some limitations and suggestions for future research hold. While the decisions are categorized in a similar way as the XES-format, some other type of decisions might exist in certain circumstances, that were not addressed in this overview. Also, although the procedure is based on objective technical settings on the one hand side, and on experience of the authors on the other hand side, it still needs to be evaluated in the field. Future research could evaluate whether the use of this procedure yields indeed a better understanding of the implications of taken decisions.

References

1. van der Aa, H., Leopold, H., Reijers, H.A.: Detecting inconsistencies between process models and textual descriptions. In: Motahari-Nezhad, H.R., Recker, J., Weidlich, M. (eds.) BPM 2015. LNCS, vol. 9253, pp. 90–105. Springer, Cham (2015). https://doi.org/10.1007/978-3-319-23063-4_6

2. van der Aalst, W.M.P.: Process Mining: Discovery, Conformance and Enhancement of Business Processes. Springer, Heidelberg (2011). https://doi.org/10.1007/978-3-642-19345-3
3. van der Aalst, W.M.P.: Extracting event data from databases to unleash process mining. In: vom Brocke, J., Schmiedel, T. (eds.) BPM - Driving Innovation in a Digital World. MP, pp. 105–128. Springer, Cham (2015). https://doi.org/10.1007/978-3-319-14430-6_8
4. Calvanese, D., Montali, M., Syamsiyah, A., van der Aalst, W.M.P.: Ontology-driven extraction of event logs from relational databases. In: Reichert, M., Reijers, H.A. (eds.) BPM 2015. LNBIP, vol. 256, pp. 140–153. Springer, Cham (2016). https://doi.org/10.1007/978-3-319-42887-1_12
5. Conforti, R., Dumas, M., García-Bañuelos, L., La Rosa, M.: Beyond tasks and gateways: discovering BPMN models with subprocesses, boundary events and activity markers. In: Sadiq, S., Soffer, P., Völzer, H. (eds.) BPM 2014. LNCS, vol. 8659, pp. 101–117. Springer, Cham (2014). https://doi.org/10.1007/978-3-319-10172-9_7
6. Hull, R., Damaggio, E., Masellis, R.D., Fournier, F., Gupta, M., Fenno Terry Heath, I., Hobson, S., Linehan, M., Maradugu, S., Nigam, A., Sukaviriya, P.N., Vaculin, R.: Business artifacts with guard-stage-milestone lifecycles: managing artifact interactions with conditions and events, pp. 51–62. ACM (2011). 2002270
7. Lu, X., Nagelkerke, M.Q.L., van de Wiel, D., Fahland, D.: Discovering interacting artifacts from ERP systems (extended version). BPM Reports 1508 (2015)
8. González López de Murillas, E., Reijers, H.A., van der Aalst, W.M.P.: Connecting databases with process mining: a meta model and toolset. In: Schmidt, R., Guédria, W., Bider, I., Guerreiro, S. (eds.) BPMDS/EMMSAD -2016. LNBIP, vol. 248, pp. 231–249. Springer, Cham (2016). https://doi.org/10.1007/978-3-319-39429-9_15
9. de Murillas, E.G.L., van der Aalst, W.M.P., Reijers, H.A.: Process mining on databases: unearthing historical data from redo logs. In: Motahari-Nezhad, H.R., Recker, J., Weidlich, M. (eds.) BPM 2015. LNCS, vol. 9253, pp. 367–385. Springer, Cham (2015). https://doi.org/10.1007/978-3-319-23063-4_25
10. Nigam, A., Caswell, N.: Business artifacts: an approach to operational specification. IBM Syst. J. **42**(3), 428–445 (2003)
11. Nooijen, E.H.J., van Dongen, B.F., Fahland, D.: Automatic discovery of datacentric and artifact-centric processes. In: La Rosa, M., Soffer, P. (eds.) BPM 2012. LNBIP, vol. 132, pp. 316–327. Springer, Heidelberg (2013). https://doi.org/10.1007/978-3-642-36285-9_36
12. Popova, V., Fahland, D., Dumas, M.: Artifact lifecycle discovery. arXiv:1303.2554 (2013)
13. Prez-Castillo, R., Weber, B., Garca-Rodrguez de Guzmn, I., Piattini, M., Pinggera, J.: Assessing event correlation in non-process-aware information systems. Softw. Syst. Model. **13**(3), 1117–1139 (2014)
14. Raichelson, L., Soffer, P.: Merging event logs with many to many relationships. In: Fournier, F., Mendling, J. (eds.) BPM 2014. LNBIP, vol. 202, pp. 330–341. Springer, Cham (2015). https://doi.org/10.1007/978-3-319-15895-2_28
15. Weber, I., Farshchi, M., Mendling, J., Schneider, J.G.: Mining processes with multiinstantiation, pp. 1231–1237. ACM (2015). 2699493

Design-Time Analysis of Data Inaccuracy Awareness at Runtime

Yotam Evron[✉], Pnina Soffer, and Anna Zamansky

University of Haifa, Mount Carmel, 3498838 Haifa, Israel
{yevron,spnina,annazam}@is.haifa.ac.il

Abstract. Analyzing potential data inaccuracy is an important aspect of business process design that has been mostly overlooked so far. To this end, process models should express the relevant information to support such analysis. In this paper we propose a formal framework for design-time analysis of potential data inaccuracy situations. In particular, we define a property of Data Inaccuracy Awareness which indicates the ability to know at runtime whether data values are accurate representations of real values. We propose an algorithm for analyzing this property at design time based on a process model. A preliminary evaluation of the applicability and scalability of the algorithm using a benchmark collection of process models is reported.

Keywords: Data inaccuracy · Business process management
Business process modeling

1 Introduction

Business processes and process aware information systems are designed with the assumption that the data used by the process is an accurate reflection of reality. Based on this assumption, humans do not need to physically sense the current state for deciding what activity to perform at a given moment and how to perform it. However, this assumption does not always hold, and situations of data inaccuracy might occur and bear substantial risks to the process and to business goals. Inaccurate data may lead process participants to make wrong decisions which will hamper the goal reachability of the process or cause other negative consequences. Since the data used by a process has such an impact on business goals, it is essential to track data errors as early as possible to avoid negative consequences.

Until now, data inaccuracy has mainly been addressed in the area of business process management as a possible exception at runtime, resolved through exception handling mechanisms [19]. We postulate that some of the negative consequences of potential data inaccuracy can be reduced by analyzing and anticipating data inaccuracy already at design time. In particular, in this paper we aim to support a design-time identification

The first and second authors were supported by the Israel Science Foundation under grant agreement no. 856/13.

E. Teniente and M. Weidlich (Eds.): BPM 2017 Workshops, LNBIP 308, pp. 600–612, 2018.
https://doi.org/10.1007/978-3-319-74030-0_47

of process parts where the existence of data errors might remain unrecognized. Such analysis can form a basis for facilitating an early detection of data inaccuracy at runtime to reduce its implications. To the best of our knowledge, no systematic design time approach has been proposed so far for this purpose.

We start by proposing a formal framework which enables considering potential data inaccuracy situations at design time. The main idea of our approach is a "bird-view" that extends a process model, acknowledging that part of the modelled process takes place in an (unmodelled) environment, the state of which is reflected (possibly incorrectly) in an information system (IS) used by the process. A main question is if and when can one know at runtime whether the data used by the process is an accurate reflection of real values. If this is analyzed at design time, measures can be taken so decisions and actions that are performed at runtime are better informed. The analysis approach we propose, based on our formal process view, is targeting this question.

The paper is structured as follows. Section 2 presents some preliminaries and a motivating example. Based on these, the formal framework is presented in Sect. 3. In Sect. 4 we provide the analysis approach and algorithm. In Sect. 5, a preliminary evaluation of the algorithm is reported. Section 6 discusses related work. Finally, Sect. 7 draws conclusion and discusses future work.

2   Preliminaries

As a basic representation of process models, we use Workflow nets with data (WFD-nets) [22]. A WFD-net is a Workflow net extended with conceptual read/write/delete data operations assigned to its transitions, as well as transition guards. These are logical expressions defined over data items, which guard transition execution.

Definition 1 (WFD-net) [22]. Let D be a finite set of data items. A workflow net with data $N = \langle P, T, F, \vartheta_D, rd, wt, del, grd \rangle$ for D consists of: A WF-net $\langle P, T, F \rangle$, A set of guard expressions ϑ_D, A reading data labelling function $rd:T \rightarrow 2^D$, A writing data labelling function $wt:T \rightarrow 2^D$, A deleting data labelling function $del:T \rightarrow 2^D$, and A guard function $grd:T \rightarrow \vartheta_D$ assigning guards to transitions.

To explicitly address data inaccuracy, we complement WFD-nets with concepts based on the Generic Process Model (GPM) [23, 25]. In the GPM view, a process takes place in a domain which is typically captured by a set of state variables, whose values at a given moment reflect the domain state at that moment. A process is then viewed as a sequence of state transitions, which are governed by a transformation law. Business processes typically involve information systems, where data items correspond to state variables and reflect their values.

Definition 2 (Domain, Sub-domain, corresponding data items). A domain D is represented by a set of state variables $SV = \{x_1, x_2, \dots, x_n\}$. A sub-domain is part of the domain described by a subset of D. For $\{x_1, x_2, \dots, x_n\}$, a reflection in the information system is a set $DI = \{d_1, d_2, \dots, d_n\}$ of corresponding data items, where d_i reflects x_i, and $<x_i, d_i>$ are termed a *corresponding couple*.

The common assumption is that data items correctly reflect state variable values. Lifting this assumption leads us to explore the phenomenon of data inaccuracy.

Definition 3 (Data inaccuracy). We say that a domain is accurately reflected by an information system at a given moment t if for every i the values of the corresponding couple $<x_i, d_i>$ match. We refer to any violation of this condition in a data item d_i as data inaccuracy with respect to d_i (at t).

Running Example: Consider a harbor which handles cargo containers arriving on ships, some for local customers and some need to be reloaded to other ships. A container reloading process starts when a ship is about to arrive, and a message is received from the shipping company, instructing to reload a container (given a container ID and its location on board) from ship A to ship B, given its estimated date of arrival at the harbor. The harbor operational unit determines where to store the container meanwhile. When ship A arrives the container is offloaded and stored; when ship B arrives the container is loaded to it. Now assume the container ID has been wrongly recorded in the harbor. When ship A arrives, it might not have a container whose ID matches the ID data, in which case the process is stuck. To avoid such situations, container ID is selected from a list of existing containers on board. However, if the ID selection was faulty, the wrong container is reloaded to ship B. In this case, the mistake will not be recognized until ship B arrives or until the container gets to its destination. This demonstrates that different data inaccuracy situations may bear different consequences, and the point at which the inaccuracy is discovered matters.

3   A Formal Framework for Data Inaccuracy

In GPM a process is modelled by one or more sequences of transitions which form a path from an initial state to a goal state. Each of these transitions relies on a subset of state variables as a decision base, so we have subsequences of decision bases. When all decision bases in one sequence depend only on the IS data and in another only on external state variables (in the real world), they form independent sequences. In general, there are infinitely many possible choices to split the domain into sub-domains, depending on the purpose and meaning one wishes to emphasize. Our approach focuses on a decomposition into two sub-domains: (i) Process (modelled) sub-domain, where an IS-managed process takes place, and decisions are made relying only on the data items in the IS (i.e., its transformations are IS-dependent, assuming the state of the world based on its reflection in IS data), and (ii) External (non-modelled, or environment) sub-domain which holds and acts upon real state variables.

Returning to our motivating example, let us assume the harbor received and recorded a message with the ID of the container to be reloaded and its destination ship. From now on, it can plan the temporary storage on its own, in its process sub-domain based on the IS data, without a need for any further information in order to proceed. Figure 1(a) provides a WF-net (without referring to data items for the sake of simplicity) of the harbor's process as viewed and managed internally (in the process sub-domain). In contrast, Fig. 1(b) shows a "bird-view" of the process, decomposed into two sub-domains (process and environment sub-domains) as explained above. Moreover, the figure illustrates an unmodelled part of the process where the container can be handles without interacting with the harbor. At this part of the process, the sub-domains employ independent (and concurrent) threads.

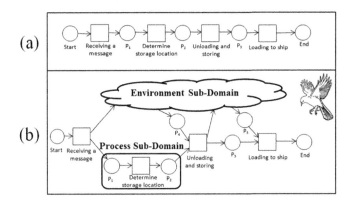

Fig. 1.   (a) The organization view of the party arrangement process modelled as a WF-net. (b) A "bird-view" of the process, including two independent sub-domains

While Fig. 1(a) shows a sequential process, Fig. 1(b) shows that the two sub-domains synchronize at Unloading. As opposed to the standard notion of synchronization (where several threads in the model converge), here synchronization applies to an unmodelled thread (sub-domain), outside the process control. When an external, uncontrolled sub-domain synchronizes with a sub-domain controlled by the process, a main concern is to have matching values for data items and corresponding state variables. We term such synchronization an 'external synchronization point'.

Definition 4 (External Synchronization Point). Let $<x_i, d_i>$ be a corresponding couple. An external synchronization point with respect to a data item d_i is a transition where two or more independent sub-domains synchronize, one of which includes x_i (in other words, it is a transition that relies on x_i to be enacted).

Herein, for the sake of brevity, we refer to external synchronization points as synchronization points. What makes synchronization points pivotal in our approach is that only at these points are situations of data inaccuracy certain to be detected at runtime. Returning to our example, when receiving the message, the IS data item of Container ID is updated. The process sub-domain and the external sub-domain then part ways, converging again at the point of unloading, which is an external synchronization point. In case of an incorrectly recorded container ID, unloading is the first point in the process where the error is certain to be discovered. Indeed, as long as the external and the process sub-domains function independently, there is no way of knowing whether the data item correctly reflects the state variable value.

Based on the (internal) process view only, for each corresponding couple $\langle x_i, d_i \rangle$ at a given state in the process at runtime, it may be known or not known which of $x_i = d_i$ or $x_i \neq d_i$ holds, leading to the following notion:

Definition 5 (Data Inaccuracy Awareness). A sub-domain state is data inaccuracy aware (DIA) with respect to a data item d_i iff it is known at this state whether $x_i = d_i$ where $\langle x_i, d_i \rangle$ form a corresponding couple.

Clearly, a state s following an external synchronization point with respect to d_i is DIA with respect to d_i. However, if no synchronization occurs, the state may not be DIA, which is an indicator for potential problems related to data inaccuracy. In what follows we propose an approach and an algorithm for the detection of such cases.

We now turn to an operational view of the process using a formal representation, in this case a WFD-net. In this model the synchronization points are not explicitly revealed. We thus add a marking of these points to the WFD-net formalism, allowing a label of the form $\Delta(d)$ at synchronization points for each data item d.

Definition 6. (S-WFD net) Let $D = \{d_1, \ldots, d_m\}$ be a finite set of data items. A synchronizing workflow net with data (S-WFD-net) $N = \langle P, T, F, \Delta_D, \vartheta_D, rd, wt, del, grd, sp \rangle$ for D consists of

- A WFD-net $\langle P, T, F, \vartheta_D, rd, wt, del, grd \rangle$,
- A set Δ_D of synchronization points of the form $\Delta(d)$ where $d \in D$,
- A synchronizing data labelling function $sp: T \rightarrow 2^{\Delta_D}$ assigning synchronization points to transitions.

With the formalism introduced above, we can now enrich the Petri-net based model of Fig. 1(a) as presented in Fig. 2. For example, *Unloading and storing* is an (external) synchronization point with respect to the data item *location on board*, since at this point the actual location of the container on board is sensed, thus it is marked with Δ (*location on board*). Note that *ID* has no synchronization point at this process.

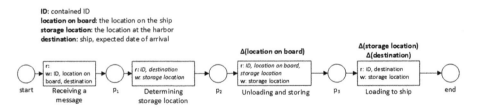

Fig. 2. The motivating example modelled using S-WFD net

Using S-WFD nets enables analyzing the process with respect to the DIA property and potential data inaccuracy, as presented next.

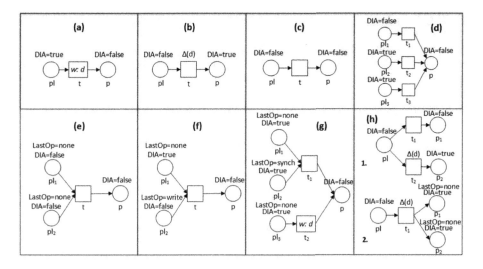

Fig. 3. Illustration of the cases in Table 1.

4   DIA Analysis

To operationalize the notion of DIA, we extend the definition of Data Inaccuracy Awareness (Definition 5) to WFD-nets.

Definition 7 (Data Inaccuracy Awareness in S-WFD net). Let N be a S-WFD net. A place p is DIA with respect to d_i iff when N is enacted the sub-domain state represented by p is DIA.

Definition 7 formally ties the concept of DIA (related to runtime) to a specific element (place) in a S-WFD net. Following this, it is possible to classify places in a S-WFD net as DIA or non-DIA for each data item already at process design. The importance of such analysis is in enabling the prediction of potential data inaccuracy situations. In particular, if the S-WFD net has a Read operation or a Guard expression for a data item following a non-DIA place, data inaccuracy might lead to erroneous actions that may have further consequences.

In what follows we propose an algorithm which analyzes a S-WFD net and establishes the DIA property for each place for a given S-WFD net and a data item d_i. We start by presenting the main premises which complement the definition of DIA. These premises are derived from Definition 5 (DIA) and our view of the real world.

Premise 1: The state variables represented as data items in the IS keep stable values that do not change by themselves during the process, unless explicitly written to[1]

---

[1] Premise 1 restricts our approach to state variables whose volatility is not high. For example, a state variable depicting the location of a moving car does not subscribe to this premise, while a state variable depicting the address of a customer does. This premise entails that the designer has to identify for which data items this analysis approach would be appropriate.

Premise 2: When a change is made to the value of a data item through a Write operation, errors are possible, so there is no certainty that the new value matches the value of the corresponding state variable. Hence, a Write operation induces a non-DIA state.

Premise 3: We will know whether the value of the data item matches the corresponding state variable or not if and only if we reach a synchronization point at runtime (although the real value is not necessarily known). Hence, a synchronization operation induces a DIA state, and lack of it induces a non-DIA state (see case 1 in Table 1 below).

Table 1. $DIA_d(p)$ cases calculation

1	if $|\bullet p| = 0$, then $DIA_d(p) = false$
	p is a source place, and thus is non-DIA by Premise 3 above.
2	if $(|\bullet p| = 1$ and $\forall\, t \in \bullet p : |\bullet t| = 1)$, then
	$$DIA_d(p) = propagate(t, DIA_d(pl))$$
	p has a single ingoing transition t which has a single ingoing place pl, as part of a sequence (see Fig. 3(a/b/c), or as part of a split (see Fig. 3 (h)). To explain the above calculation, consider, e.g. Fig. 3 (a), where the place pl is DIA, and transition t has a *write* operation. By Premise 2, p is non-DIA (regardless of pl). In Fig. 3 (b) while pl is non-DIA, due to the *synch* operation on t, p is DIA by Premise 3. Fig. 3 (c) shows a case when t has no operation, thus the DIA-value of pl is propagated to p. In Fig. 3 (h)-1, p_1 is non-DIA due to propagation from pl through t_1. But in the case of p_2, things are different due to the *synch* operation on t_2. In Fig. 3 (h)-2, p_1 and p_2 are DIA since we propagate using t_1 and the value *false* from pl. In Fig. 3 (h), the examples use the same mechanism as in Fig. 3(a/b/c).
3	if $(|\bullet p| > 1$ and $\forall\, t \in \bullet p : |\bullet t| = 1)$, then
	$$DIA_d(p) = \min\{propagate(t, DIA_d(pl)) : t \in \bullet p, pl \in \bullet t \}$$
	p has more than one incoming transition (see Fig. 3 (d)) and each one of them has only one place pl_i; the minimal (with respect to Boolean values) DIA-value is propagated.
4	if $(\forall t : t \in \bullet p, |\bullet t| > 1)$
	(i.e, all transitions leading to p have more than one input place, namely, they join parallel paths.)

	a	if $(\forall pl \in \bullet t, t \in \bullet p : (MaxLastOp(pl, Block(pl)) = none))$, then
		$$DIA_d(p) = \min\{propagate(t, DIA_d(pl)) : t \in \bullet p, pl \in \bullet t\}$$
		None of the blocks on parallel paths have operations $(MaxLastOp(pl, Block(pl)) = none)$. We take here the minimum (with respect to Boolean value) of the DIA-values of the places on these paths. Such example is shown on Fig. 3 (e): both pl_1 and pl_2 are non-DIA, and there is no operation on transition t. Accordingly, p is non-DIA.
	b	Else, (at least one of the previous places has a *synch/write* operation in a block, e.g. Fig. 3 (f))
		$$DIA_d(p) = false,$$ if one of the following cases holds:
		- at least one operation on transition preceding p is *write*
		- at least one last operation in block preceding p is *write* (i.e., the operation is in the set $\{MaxLastOp(pl, Block(pl)) : t \in \bullet p, pl \in \bullet t\}$), and the operation on transition corresponding to that block is not *synch*.
		$$DIA_d(p) = true,$$ otherwise.
		Consider, e.g., Fig. 3 (f), where pl_1 is DIA with no operation on that path, and pl_2 is non-DIA with a last operation of *write*. At least one operation in a block preceding p is *write* and the transition t which corresponds to that block has no *synch* operation, thus p is non-DIA.
5	Else,
	a combination of previous cases, see, e.g. Fig. 3 (g), which is a combination of case 2 and case 4b. In this case we compute the DIA of p for each path separately, and then take the minimal obtained value.

We also limit the proposed analysis to processes whose S-WFD net representation N has the following properties: (1) The process represented by N is sound [1]. (2) The S-WFD net N is well-structured[2] [1]. (3) For a given data item, each transition of N can have either Write or Δ but not both. (4) For two or more parallel sequences of N and a given data item d, only one[3] can include a Write operation of d.

Notation: Following [1], we use $\bullet t$ and $t\bullet$ to denote the set of input and output places for a transition t respectively. Similarly, we use $\bullet p$ and $p\bullet$ for a place p.

The input to our algorithm is a S-WFD net $N = \langle P, T, F, \Delta_D, \vartheta_D, rd, wt, del, grd, sp \rangle$ for D and a data item $d \in D$ We adopted the notion of block-structured workflow net [16] which is a workflow net that can be divided into segments having single entry and single exit points. In a well-handled WF-net [10], a pair of two nodes (x, y) are called matching operator nodes iff: (1) x is an AND-split and y is an AND-join or x is an XOR-split or a place, and y an XOR-join or a place. (2) There is a pair of elementary paths (threads) C_1 and C_2 leading from x to y such that: $\alpha(C_1) \cap \alpha(C_2) \Rightarrow \{x, y\} C_1 \neq C_2$. We say that $b = (x, y)$ is a block if x (entry point) and y (exit point) are matching operator nodes.

We denote by EP_b the set of elementary paths of a block b, where b is a block (x, y). Let p be a place on $ep \in EP_b$, then $Block(p) = b$. Let $LastOpOnPath(ep, b)$ denote the last data operation occurring in ep in block b. If no data operation occurs in ep, then $LastOpOnPath(ep, b) = none$.

We define a total order on the possible operations on transitions $\{write, synch, none\}$: $none < synch < write$. Using this order, we can determine the maximal last operation among all the paths in a block. We denote the maximum function induced by this order by max*, while reserving max for the usual order on Boolean truth values. For a block $b = (x, y)$ where x and y are transitions and a place $p \in y\bullet$, let $MaxLastOp(p, b) = max^*\{LastOpOnPath(ep, b)| ep \in EP_b\}$.

Finally, we define a function *Propagate* for assigning a truth value $v \in \{true, false\}$ of the DIA property when passing through transition $t \in T$, such that *Propagate(t, v) = false* if t has a write operation, *true* if t has a synchronization operation and v otherwise.

The DIA analysis algorithm walks through the S-WFD net in a Breadth-First Search (BFS)-like manner, setting $DIA_d(p)$ for each place p as specified in Table 1, which is divided to cases capturing different situations of paths preceding p.

---

[2] Our purpose is to reason about potential data inaccuracy situations at design-time while acknowledging runtime situations. Thus, exploring concurrency cases where data operations can execute simultaneously or at any order is vital. For that reason, a clear assignment of data operations to blocks in the net is essential.

[3] This is because parallel threads imply independence of the relevant sub-domains irrespectively of the order in which transition takes place. It is hence not possible for parallel threads to update the same data item [9, 23].

5   Preliminary Evaluation

We implemented the algorithm described above on top of a S-WFD net modeling tool and conducted a preliminary experimental evaluation, aiming to assess the ability of the algorithm to address non-trivial processes of different sizes and varying amounts of data operations. We also assessed its scalability with respect to certain characteristics of the process (structure, data operations and synchronization points). As a basis, we used a benchmark collection of 10 BPMN process models, which we converted to Petri-nets characterized in Table 2, using a ProM plugin [15]. The models were created for the Process Discovery Contest@BPM 2016. We chose this collection since the models were specifically designed to exhibit a variety of structural features.

Table 2. The investigated models (Num of elements: places + transition; Num of loops/elements in loop in total; Transitions max degree: max t● (concurrency); Places max degree: max p●)

Model	Num of elements	Num of loops	Num of elements in a loop	Transitions – max degree	Places – max degree
1	55	1	4	4	3
2	66	0	0	2	3
3	87	0	0	4	3
4	106	0	0	3	3
5	80	2	29	3	2
6	48	0	0	2	3
7	81	0	0	4	3
8	69	0	0	3	4
9	101	0	0	2	2
10	75	0	0	3	3

While the base collection used included non-trivial models of varied structural features, they did not bear any business meaning nor did they entail any data operations. We hence artificially added data operations to the models in a controlled manner. We randomly added 1–10 Write operations and 1–10 Synchronization points (all combinations of these) to every model (each considering one data item). Each of the combinations of Write/Synch operations was repeated 10 times. In each of these repetitions the specific transitions where these operations were added were selected randomly. For all these configurations, we measured the execution time of the algorithm. Figure 4 shows two typical examples of results obtained for all configurations.

These typical examples show no clear correlation between the numbers of Write operations/Synchronization points and the execution time (whose variation range is very small – within 4 ms). We also checked whether the execution time is affected by various features of the model, such as the number of elements, number and magnitude of loops, and maximal degrees of transitions and of places. To this end we sorted all the models according to the relevant feature. We observed no clear trend for most of the features, except for the number of elements, whose increase resulted in longer execution

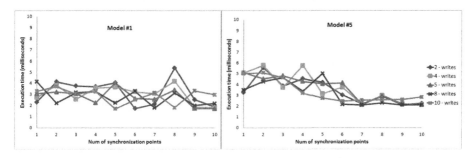

Fig. 4. Average execution times (in milliseconds) for Models #1 and #5 for different numbers of Synchronization points (x-axis) and a sample of numbers of Write operations (series)

times. This is not surprising because the algorithm checks every element in the model. We conclude that the proposed algorithm is applicable for DIA analysis in S-WFD nets. It is scalable for non-trivial processes and processes which have various amounts of data operations and elements (transitions and places).

6 Related Work

Data quality has been extensively addressed outside the context of business processes [2, 3, 26]. For instance, Wang and Strong [27] define the notion Data Quality as "fitness for use", i.e., the ability of the data to meet users' requirements. Moreover, they explained that in order to improve data quality, we have to understand what the data means to those who consume it. Orr [17] highlighted the importance of data quality, claiming that it can be improved by changing data usage. Some studies discuss the quality of data simply as the ability to meet requirements [12, 27], while others stressed the importance of referring to the data elements during database design in order to ensure data quality at runtime.

In the context of business processes, limited work concerning data quality has been done. Sadiq et al. [20] propose a method for supporting process designers in the selection of actions, considering data quality requirements when designing business processes. Rodriguez et al. [7, 18] introduced a BPMN-based data quality-aware process model. The approach is not formally anchored and does not support a systematic or automated analysis of potential data quality issues in a business process. Gharib and Giorgini [13] introduced a goal-oriented approach to model and analyze information quality (IQ) requirements in business processes from a socio-technical perspective. Automated soundness verification of the resulting models is possible in the proposed approach, but it does not allow an in-depth analysis of possible manifestations of data problems. Some studies [4, 5, 14] attempt to address data quality concepts at process design, aiming to help predicting how changes in the business processes would influence the data quality and support process designers in the selection of actions that should be adopted in the (re)design of business processes, with respect to data quality requirements. Their methods do not involve automated actions and require high human involvement. Cappiello et al. [6, 8] propose a data quality improvement strategy of inserting Data

Quality blocks [21], namely, designated monitoring points, to the process. Decision support and guidance is needed for designing the blocks and positioning them along a process.

Analyzing potential data inaccuracy at design time has been suggested by Soffer [24] who also provided formal definitions of the problem and an explanation of its underlying mechanism. Moreover, [24] discussed potential consequences of data inaccuracy in business processes and demonstrated possible scenarios of such. Following Soffer's ideas by extending WFD-nets with synchronization points [11], this paper introduces the notion of DIA. Furthermore, we propose an automated analysis of DIA in process models and show its applicability and scalability.

7   Conclusion and Future Work

Data inaccuracy may be manifested in business processes at runtime, when its consequences might be severe. In this paper we propose a design time analysis approach that can serve for anticipating manifestation of data inaccuracy at runtime. The approach builds on a formal model of data inaccuracy in business processes.

The contribution of this paper is twofold. First, we provide a formal framework defining the concept of DIA and supporting its representation and analysis. Since data values are not available at design-time, DIA is instrumental for analysis of potential risks due to data-inaccuracy. In particular, it can be expected that any use of data in a non-DIA state may lead to erroneous actions in data inaccuracy situations. Using S-WFD nets at design time, these can be observed as Read operations or Guard expressions following places whose DIA is False. Second, we propose an automated method for DIA analysis in S-WFD nets. We provide a preliminary evaluation of the applicability and scalability of our tool using a benchmark collection of process models.

The main limitation of the model is that it builds on a low-volatility assumption regarding the state variables addressed by the process. This assumption limits the applicability of the approach to processes that take place in environments where relevant variables are not subject to continuous and spontaneous value changes, but are rather under control. Even with this limiting assumption, a variety of processes in daily and common domains (e.g., logistics) can appropriately be addressed. Future research will address more unstable environments. The initial evaluation reported in this paper is merely a technical evaluation of the ability to perform DIA analysis. Future research will use real process data to evaluate the uses of DIA analysis, particularly its ability to predict business consequences of data inaccuracy and its associated risks.

References

1. Aalst, W.M.: The application of Petri nets to workflow management. J. Circ. Syst. Comput. 8(01), 21–66 (1998)
2. Aebi, D., Perrochon, L.: Towards improving data quality. In: CiSMOD, pp. 273–281 (1993)
3. Agmon, N., Ahituv, N.: Assessing data reliability in an information system. J. Manag. Inf. Syst. 4(2), 34–44 (1987)

4. Bagchi, S., Bai, X., Kalagnanam, J.: Data quality management using business process modeling. In: IEEE International Conference on Services Computing, SCC 2006 (2006)
5. Bringel, H., Caetano, A., Tribolet, J.M.: Business process modeling towards data quality: an organizational engineering approach. In: ICEIS, vol. 3, pp. 565–568 (2004)
6. Cappiello, C., Pernici, B.: Quality-aware design of repairable processes. In: ICIQ (2008)
7. Cappiello, C., Caro, A., Rodriguez, A., Caballero, I.: An approach to design business processes addressing data quality issues. In: ECIS, p. 216 (1987)
8. Cappiello, C., Pernici, B., Villani, L.: Strategies for data quality monitoring in business processes. In: Benatallah, B., Bestavros, A., Catania, B., Haller, A., Manolopoulos, Y., Vakali, A., Zhang, Y. (eds.) WISE 2014. LNCS, vol. 9051, pp. 226–238. Springer, Cham (2015). https://doi.org/10.1007/978-3-319-20370-6_18
9. Davidson, S.B., Garcia-Molina, H., Skeen, D.: Consistency in a partitioned network: a survey. ACM Comput. Surv. (CSUR) **17**(3), 341–370 (1985)
10. Eckleder, A., Freytag, T., Mendling, J., Reijers, H.A.: Realtime detection and coloring of matching operator nodes in workflow nets. In: Algorithms and Tools for Petri Nets, pp. 56–61 (2009)
11. Evron, Y., Soffer, P., Zamansky, A.: Incorporating data inaccuracy considerations in process models. In: Reinhartz-Berger, I., Gulden, J., Nurcan, S., Guédria, W., Bera, P. (eds.) BPMDS/EMMSAD-2017. LNBIP, vol. 287, pp. 305–318. Springer, Cham (2017). https://doi.org/10.1007/978-3-319-59466-8_19
12. Falge, C., Otto, B., Österle, H.: Data quality requirements of collaborative business processes. In: 2012 45th Hawaii International Conference on System Science (HICSS). IEEE (2012)
13. Gharib, M., Giorgini, P.: Detecting Conflicts in Information Quality Requirements: the May 6, 2010 Flash Crash (2014)
14. Heravizadeh, M., Mendling, J., Rosemann, M.: Dimensions of business processes quality (QoBP). In: Ardagna, D., Mecella, M., Yang, J. (eds.) BPM 2008. LNBIP, vol. 17, pp. 80–91. Springer, Heidelberg (2009). https://doi.org/10.1007/978-3-642-00328-8_8
15. Kalenkova, A., De Leoni, M., van der Aalst, W.M.: Discovering, analyzing and enhancing BPMN models using ProM. In: BPM (Demos), p. 36 (2014)
16. Leemans, S.J.J., Fahland, D., van der Aalst, W.M.P.: Discovering block-structured process models from event logs - a constructive approach. In: Colom, J.M., Desel, J. (eds.) PETRI NETS 2013. LNCS, vol. 7927, pp. 311–329. Springer, Heidelberg (2013). https://doi.org/10.1007/978-3-642-38697-8_17
17. Orr, K.: Data quality and systems theory. Commun. ACM **41**(2), 66–71 (1998)
18. Rodríguez, A., Caro, A., Cappiello, C., Caballero, I.: A BPMN extension for including data quality requirements in business process modeling. In: Mendling, J., Weidlich, M. (eds.) BPMN 2012. LNBIP, vol. 125, pp. 116–125. Springer, Heidelberg (2012). https://doi.org/10.1007/978-3-642-33155-8_10
19. Russell, N., van der Aalst, W., ter Hofstede, A.: Workflow exception patterns. In: Dubois, E., Pohl, K. (eds.) CAiSE 2006. LNCS, vol. 4001, pp. 288–302. Springer, Heidelberg (2006). https://doi.org/10.1007/11767138_20
20. Sadiq, S., Orlowska, M., Sadiq, W., Foulger, C.: Data flow and validation in workflow modeling. In: Proceedings of the 15th Australasian Database Conference, ADC 2004, vol. 27, pp. 207–214 (2004)
21. Shankaranarayanan, G., Wang, R.Y., Ziad, M.: IP-MAP: representing the manufacture of an information product. In: IQ (2000)
22. Sidorova, N., Stahl, C., Trčka, N.: Workflow soundness revisited: checking correctness in the presence of data while staying conceptual. In: Pernici, B. (ed.) CAiSE 2010. LNCS, vol. 6051, pp. 530–544. Springer, Heidelberg (2010). https://doi.org/10.1007/978-3-642-13094-6_40

23. Soffer, P., Kaner, M., Wand, Y.: Assigning ontological meaning to workflow nets. J. Database Manag. **21**(3), 1–35 (2010)

24. Soffer, P.: Mirror, mirror on the wall, can I count on you at all? Exploring data inaccuracy in business processes. In: Bider, I., Halpin, T., Krogstie, J., Nurcan, S., Proper, E., Schmidt, R., Ukor, R. (eds.) BPMDS/EMMSAD-2010. LNBIP, vol. 50, pp. 14–25. Springer, Heidelberg (2010). https://doi.org/10.1007/978-3-642-13051-9_2

25. Soffer, P., Wand, Y.: Goal-driven analysis of process model validity. In: Persson, A., Stirna, J. (eds.) CAiSE 2004. LNCS, vol. 3084, pp. 521–535. Springer, Heidelberg (2004). https://doi.org/10.1007/978-3-540-25975-6_37

26. Wand, Y., Wang, R.Y.: Anchoring data quality dimensions in ontological foundations. Commun. ACM **39**(11), 86–95 (1996)

27. Wang, R.Y., Strong, D.M.: Beyond accuracy: what data quality means to data consumers. J. Manag. Inf. Syst. **12**(4), 5–33 (1996)

3rd International Workshop on Interrelations Between Requirements Engineering and Business Process Management (REBPM 2017)

Introduction to the 3rd International Workshop on Interrelations Between Requirements Engineering and Business Process Management (REBPM'17)

Rüdiger Weißbach[1], Banu Aysolmaz[2], and Robert Heinrich[3]

[1] HAW Hamburg University of Applied Sciences, Hamburg, Germany
ruediger.weissbach@haw-hamburg.de
[2] School of Business and Economics, Maastricht University, Maastricht, The Netherlands
b.aysolmaz@maastrichtuniversity.nl
[3] Karlsruhe Institute of Technology (KIT), Karlsruhe, Germany
robert.heinrich@kit.edu

Abstract. Many challenges in business process management (BPM) are attributed to the first phase of the BPM lifecycle; process discovery. Process discovery is found to be an essential part of requirements engineering (RE) in software development. Typically, process models are developed as the deliverables of this phase. In recent years, agile principles are being used to overcome challenges in software development projects. As BPM and RE activities are so tightly-knit, agile principles may also be used to address challenges in process modeling projects. The purpose of this workshop was to examine the interrelations between requirements engineering and business process management with an agility perspective and investigate how agile principles may change the BPM phases. Out of the eight high-quality submissions we received, three of them were accepted as regular papers and three as short papers. The presenters and participants created a fruitful discussion environment on examining agility in the context of BPM.

Keywords: Business process management
Requirements engineering · Agile

1 Aims and Scope

While requirements engineering (RE) is concerned with eliciting and managing requirements related to a particular (software) system, Business Process Management (BPM) deals with modeling and managing organizational processes and business objectives. Information technology is an enabler of business change, and business processes are often a starting point of RE. Namely, process discovery, the first phase of the BPM lifecycle, is an essential part of RE. Thus, both domains are strongly interrelated while methods and processes differ. The results from our previous showed the strong interrelations between the BPM and the software engineering domains, and the potential benefits by integrating concepts and approaches from both.

Agile development principles are used for a long time in software engineering to manage uncertainties in the development life cycle. In BPM field, agile principles are just being embraced. The use of terms agility and flexibility have started to be commonly used. Agile principles can be referred to in different ways in the context of BPM. The organizations can benefit from agile principles to implement steps of the BPM life cycle, which we refer to as core part of BPM. Additionally, development of a business process management system (BPMS) may benefit from agile approaches, which we refer to as software part of BPM. Moreover, agile principles may help organizations to cope with flexible processes, which are processes that can change and adapt to different situations.

The presented papers emphasized the role of agile practices in the intersection of BPM and RE. The first group of papers focused on managing flexibility in processes. Blaukopf and Mendling, with their work entitled *"An Organizational Routines Perspective on Process Requirements"*, evaluated BPMN and CMMN for their capability to express routines characteristics based on the organizational routines theory. Their comparison highlights the need to capture process flexibility requirements in process models, specifically evolutionary aspects, with a different view borrowed from organization science. Kirchner and Laue presented the idea of business process variability patterns that deal with typical variable scenarios in processes with their work entitled: *"Variability Patterns for Analyzing Flexible Processes"*. The aim of these patterns is to speak about variable parts of a process (model) using a vocabulary that is close to domain experts' understanding. The patterns lead to typical questions and suggestions for optimization which can be helpful for a requirements engineer. Examples from the medical domain show how the patterns can be used in requirements and process analysis.

The second group of papers provided various ideas and approaches to implement RE practices in BPM in an agile way. Turban and Schmitz-Lenders suggested a method to elicit requirements more effectively and early by using process models, with their work entitled *"A Pattern-Based Question Checklist for Deriving Requirements from BPMN Models"*. Their work relies on the examination of patterns in process models that potentially signifies the need for detailed analysis of automation requirements. Their running example depicts, for a set of patterns, how the predefined questions guide the analysts to reveal requirements in a complete way. Cewe et al. discussed an approach for employing continuous testing and iterative development as part of Test Driven Development for Robotic Process Automation (RPA), with their work titled: *"Minimal Effort Requirements Engineering for Robotic Process Automation with Test Driven Development and Screen Recording"*. This approach shows that agile practices can be employed in the software development part of BPM. RPA allows flexibility in processes within the constraints of the process model by algorithmic changes. Lastly in this group, Pastrana et al. suggested a technique to elicit requirements in SCRUM with business process models rather than textually defined user stories, with their work titled: *"Optimization of the Inception Desk Technique for Eliciting Requirements in SCRUM through Business Process Models"*. The authors experiences from four projects show the improved communication and requirements management by using process model-based user story definition and planning in SCRUM.

As the last topic, Schönreiter presented findings from a survey study on the challenges of process harmonization for post-merger and acquisition situations with the paper named: *"Successful Post Merger Process Harmonization in the Triangle of Methodologies, Capabilities and Acceptance"*. The results indicate that while a certain methodology is not correlated with process harmonization success, a number of beneficial capabilities can be identified. Moreover, employee acceptance can be improved by implementing specific structure, information, and team composition policies.

The discussions during the workshop revealed that there are a lot of opportunities but also a need for approaches to integrate agile principles in BPM life cycle. To show some possibilities for agile development of business processes and for the development of flexible processes for changing requirements, a discussion paper was prepared in collaboration with the organizers and participants of the workshop entitled: *"A reflection on the Interrelations between Business Process Management and Requirements Engineering with an Agility Perspective"*.

2 Workshop Co-organizers

Rüdiger Weißbach   HAW Hamburg University of Applied Sciences, Germany
Banu Aysolmaz   Maastricht University, the Netherlands
Robert Heinrich   Karlsruhe Institute of Technology, Germany

3 Program Committee

Maya Daneva   University of Twente, the Netherlands
Nico Herzberg   SAP, Dresden, Germany
Jennifer Horkoff   Chalmers University of Technology, Gothenburg, Sweden
Matthias Kunze   Zalando SE, Berlin, Germany
Ralf Laue   University of Applied Sciences Zwickau, Germany
Sinisa Neskovic   University of Belgrade, Serbia
Rainer Schmidt   Munich University of Applied Sciences, Germany
Hannes Schlieter   TU Dresden, Germany
Inge van de Weerd   Vrije Universiteit Amsterdam, the Netherlands

An Organizational Routines Perspective on Process Requirements

Sabrina Blaukopf and Jan Mendling[✉]

University of Economics, Welthandelsplatz 1, 1020 Vienna, Austria
{sblaukop,jan.mendling}@wu.ac.at

Abstract. Process modeling notations are meant to capture the requirements of a business process in a correct and complete way. Up until now,
their assessment has been driven more by general modeling principles
than by specific characteristics of business processes. In this paper, we
turn to the theory of organizational routines in order to discuss mutual
strengths and weaknesses of classical workflow management and recent
adaptive case management as exemplified by their respective notations
BPMN and CMMN.

Keywords: Organizational routines · Process requirements · CMMN
BPMN

1 Introduction

There is a continuous debate about how the execution of business processes
can be best supported by process-aware information systems. In essence, this
debate is centered around the question in how far it is feasible to fully describe
certain business processes in a way that all execution options are anticipated or
whether such an ambition is doomed to fail. Classical workflow systems tend to
be more on the side of determinism [1,2] while authors like Hirschheim [3] or
Melão and Pidd [4] emphasize the risk of underestimating non-determinism in
socio-technical systems.

While this debate may seem theoretical at a first glance, it has important
consequences, which can be directly observed in practice. First, there are various
scenarios in which it was observed that the deterministic workflow model does
not work well. Such scenarios have been often referred to as being knowledge-
intense. These scenarios have stimulated research into making process execution
more flexible [5]. Second, there are also industry initiatives that aim to establish
a competing paradigm next to deterministic workflow execution. These are often
referred to as Adaptive Case Management (ACM) [6]. Modeling standards such
as the Case Management Model and Notation (CMMN) are meant to support
ACM in a better way than Business Process Management Notation (BPMN)
does [7].

In this paper, we follow Recker's call [8] to strengthen the theoretical core of
business process management, by discussing the example of a business process

© Springer International Publishing AG 2018
E. Teniente and M. Weidlich (Eds.): BPM 2017 Workshops, LNBIP 308, pp. 617–622, 2018.
https://doi.org/10.1007/978-3-319-74030-0_48

and how it can be described using BPMN and CMMN. We then employ the theory of organizational routines to compare these languages in how far they are able to express routines characteristics. Beverungen [9] identified the interplay between organizational routines [10] and business process management, but the consequences for modeling processes for the "messy everyday world of business" [9] were not discussed. We find that certain requirements are difficult to capture in BPMN and more easily in CMMN.

The rest of the paper is structured as follows: Sect. 2 describes how a example scenario can be modeled with BPMN and CMMN. Section 3 summarizes the essence of research on organizational routines and discusses how requirements can be described using each of these languages. Section 4 concludes the paper.

2   Background

In this section, we introduce the essential ideas and concepts of Adaptive Case Management as incorporated in CMMN. First, we present a typical insurance claims handling scenario and how it can be specified using BPMN and implemented using classical workflow concepts. Second, we use CMMN to illustrate how the process can be more flexibly specified and supported by ACM concepts.

2.1   Typical Workflow Scenario in BPMN

Figure 1 shows an insurance claims process. It starts by the claim being received and continues with data entry if the policy is valid. Then follows an investigation before the claim is approved if covered. The process is modeled using the BPMN notation elements event, activity, and gateway. At a first glance, this claims process looks straight forward, but it hides assumptions that might make it look too deterministic. For example, the data that was provided by customer might not be complete. It could miss the policy number that is required for a decision so that the clerk cannot proceed until this has been entered in the system. Such contingencies can be modeled in BPMN as additional choices. Case handling, however, introduces dedicated elements as we explain in the following.

Fig. 1. Typical claims process adapted from Cimino and Vaglini [11] (shortened)

2.2   Scenario Specified Using CMMN

Case management provides transparency, data integration and collaboration around the notion of the case. All involved participants for this claim will collect data and communicate in the case context. We have depicted the "Claims process" as a case in Fig. 2 using CMMN. The basic concepts are planned and mandatory activities, like "Identify responsibility". Milestones like "Responsibility identified" ensure that the worker does not miss these required stages. Several elements can have entry or exit criteria called Sentries, visualized using a diamond shape. Optional tasks like "Review Documents" can be triggered at any point in time, as long as its entry criterion is met.

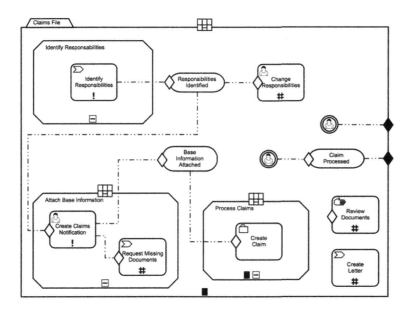

Fig. 2. Typical claims file adapted from OMG CMMN specification [12]

3 Characteristics of Routines in BPMN and CMMN

In this section, we discuss how CMMN and BPMN address characteristics of organizational routines. We structure our discussion by the help of the characteristics of routines as established by Becker [13]. Table 1 provides an overview.

Reoccurring Patterns: Reoccurring patterns describe the essence of organizational routines. They include both patterns of behavior (reoccurring activity patterns) and patterns of cognitive regularities (rules). The key characteristic is the interaction between workers. Routines can be subject to change when conditions change. BPMN allows to define a complete set of tasks with all possible dependencies in processes from the start to the end. Process paths can be optional when modeled with gateways. CMMN uses stages to define a set of tasks and represent a smaller set of tasks that can be optional to the case depending on the data-driven context. In this way, CMMN permits changes that emerge over time.

Table 1. Routines characteristics in BPMN and CMMN

Routines	BPMN	CMMN
Reoccurring Patterns	Processes	Stage
Multiple Actors	Participants, Swim lanes	Case file, Case Roles
Effortfull accomplishments	Not available	Milestones
Routines as processes	Not available	Not available
Context-dependence	Gateways	Sentries, Discretionary Tasks
Path dependence	Not available	Discretionary task
Triggers	Events	Event listeners

Multiple Actors: Routines or just certain parts of it are usually carried out by several individuals, who sometimes reside in different locations. Through interactions, reoccurring patterns are linked with each other throughout the organization. Depending on the degree of specialization, the overlap of knowledge between the actors can be limited. In BPMN participants are defined for a set of tasks. CMMN offers human-executed tasks that require a role definition. Originally, there are no mechanism to represent communication between the participants except for the explicit handing off of tasks. However, CMMN enables transparency through the case context, which is accessible to all case roles. In this way, actors can explicitly coordinate not only via completing tasks, but also by sharing data.

Mindlessness versus Effortful Accomplishment: Both aspects of executing routines have been discussed in literature, whereas empirical studies mostly observed the latter. Becker [13] proposes that most probably both behaviors (rule following vs. creative) will be existent, maybe even for one actor within the same routine. In BPMN, the complete operation of the process has to be studied and incorporated in the process flow at design time. The participants are not able to influence the flow of the process. In CMMN, tasks are mostly designed around milestones which allows for control and flexibility. In this way, both aspects are acknowledged.

Routines as Processes: When routines are observed in a process context, they can be analyzed for their degree of change. Measuring processual characteristics of routines like speed of decay, speed of execution, time lags, frequency of repetition, and fashion of shifting from one routine, are indicators for organizational and economic change. Even when not initially represented in the BPM notation, key process indicators play a part in the execution of BPMN processes in the monitoring phase. CMMN process flows are not easy to compare as they allow for different paths to complete the case file. Indications of change are not directly incorporate in neither BPMN nor CMMN, but could be more naturally observed in the latter.

Embeddedness and Context Dependence: Routines are embedded in their environment, from which they receive input and support. This includes artifacts as inputs as well as factors like time and space shaping the process. Therefore, it is expected that routines will change when their surroundings change. BPMN defines data and event rules for gateways while the overall flow of the process is expected to be anticipated. CMMN includes triggering tasks and stages at any point in time of the execution of the case through sentries. This concept is broader and has more affinity to the notion of context.

Path Dependence: Routines evolve dependently on where they started out from. If there is feedback about the outcomes, they may adapt incrementally building on this experience, but always in relation to their previous state. This characteristic cannot be directly related to a specific part of BPMN. Nevertheless, BPM does address change in the process management life cycle. CMMN defines a plan that is expected to be adapted during execution through discretionary tasks.

Triggers: In empirical studies actor-related and external triggers were identified as possible start conditions of routines. Actor-related triggers are often connected to the aspiration levels of the executing person that tries to achieve a certain outcome. External triggers could be other routines or specific events that start a certain procedure. BPMN only defines a limited amount of actor-related events. Events are mostly used for process automation. In CMMN, besides the process automation events, the actor is able to react to his own triggers by altering the process to his needs. This is a much broader coverage of the notion of trigger.

4 Conclusion

In this paper, we investigated the potential of BPMN and CMMN to capture process requirements. We turned to the characteristics of organizational routines in order to analyze this potential. We found that CMMN provides concepts to address characteristics that have been identified in the research field of organizational routines better. This finding is important for research on process requirements for the following reasons. First, our analysis highlights the potential to establish a theoretical foundation of business process modeling based on the requirements described in research on organizational routines. Second, it points to processes with specific characteristics that appear to face challenges when being modeled with BPMN. In particular, evolutionary aspects of processes are difficult to capture with BPMN. Third, the literature on organizational routines might provide a new perspective to inform the discussion of declarative and imperative process models and their maintenance [14].

References

1. Hollingsworth, D.: Workflow management coalition the workflow reference model. Workflow Management Coalition (1993)
2. Georgakopoulos, D., Hornick, M., Sheth, A.: An overview of workflow management: from process modeling to workflow automation infrastructure. Distrib. Parallel Databases **3**(2), 119–153 (1995)
3. Hirschheim, R.A.: Office Automation: A Social and Organizational Perspective. Wiley, Hoboken (1986)
4. Melão, N., Pidd, M.: A conceptual framework for understanding business processes and business process modelling. Inf. Syst. J. **10**(2), 105–129 (2000)
5. Reichert, M., Weber, B.: Enabling Flexibility in Process-Aware Information Systems. Springer, Heidelberg (2012). https://doi.org/10.1007/978-3-642-30409-5
6. Swenson, K.: Case management: contrasting production vs. adaptive, September 2012
7. Breitenmoser, R., Keller, T.: Case management model and notation-a showcase. Eur. Sci. J. **11**(25), 332–347 (2015)
8. Recker, J.: Suggestions for the next wave of BPM research: strengthening the theoretical core and exploring the protective belt. JITTA: J. Inf. Technol. Theory Appl. **15**(2), 5 (2014)
9. Beverungen, D.: Exploring the interplay of the design and emergence of business processes as organizational routines. Bus. Inf. Syst. Eng. **6**(4), 191–202 (2014)
10. Pentland, B.T., Feldman, M.S.: Designing routines: on the folly of designing artifacts, while hoping for patterns of action. Inf. Organ. **18**(4), 235–250 (2008)
11. Cimino, M., Vaglini, G.: An interval-valued approach to business process simulation based on genetic algorithms and the BPMN. Information **5**(2), 319–356 (2014)
12. Object Management Group Specification: Case Management Modeling Notation (CMMN), Beta 1. Object Management Group pct/07-08-04 (2016)
13. Becker, M.C.: Organizational routines: a review of the literature. Ind. Corp. Change **13**(4), 643–678 (2004)
14. Fahland, D., Mendling, J., Reijers, H.A., Weber, B., Weidlich, M., Zugal, S.: Declarative versus imperative process modeling languages: the issue of maintainability. In: Rinderle-Ma, S., Sadiq, S., Leymann, F. (eds.) BPM 2009. LNBIP, vol. 43, pp. 477–488. Springer, Heidelberg (2010). https://doi.org/10.1007/978-3-642-12186-9_45

Variability Patterns for Analyzing Flexible Processes

Kathrin Kirchner[1(✉)] and Ralf Laue[2]

[1] Berlin School of Economics and Law, Alt-Friedrichsfelde 60, 10315 Berlin, Germany
kathrin.kirchner@hwr-berlin.de

[2] Department of Computer Science, University of Applied Sciences of Zwickau,
Dr.-Friedrichs-Ring 2a, 08056 Zwickau, Germany
Ralf.Laue@fh-zwickau.de

Abstract. In practice, flexible processes often do not follow a fixed sequence of steps. Instead, process participants can rely on their knowledge and can decide for additional steps, change the execution order or skip a task. If a requirements engineer discusses such a process together with the stakeholders, variability has to be taken into account. The challenge is to translate the stakeholders' explanations into a process model by asking the right questions. In order to cope with this problem, we propose the use of patterns. Patterns are a well-known way in systems engineering to describe solutions for frequently occurring problems. In this paper, we present the idea of business process variability patterns and give examples how they can be used in requirements and process analysis.

Keywords: Flexible process · Process modeling · Process analysis
Pattern

1 Introduction

In real-world-scenarios, processes can be highly flexible. Process participants as domain experts decide on the execution of activities as well as their order during process execution. Therefore, it is not possible to fully prescribe such a process beforehand as it can be done for, e.g., a standardized purchasing process. A typical scenario for variability is a medical treatment process. Physicians decide dynamically which treatment is needed depending on the health status of the patient. A certain order of steps is known in advance, e.g., that some preliminary examinations have to be carried out before an operation. The sequence of these examinations might change due to several criteria. In order to document such processes, to optimize and to support them by information technology, the following question arises: How we can support the dialog between requirements engineer and domain experts as effectively as possible?

We propose the use of patterns that deal with typical variable scenarios. The aim of our patterns is to speak about variable parts of a process (model) using a vocabulary that is close to domain experts' understanding instead of speaking in terms such as "gateways" or "timer events" - which would be the

© Springer International Publishing AG 2018
E. Teniente and M. Weidlich (Eds.): BPM 2017 Workshops, LNBIP 308, pp. 623–629, 2018.
https://doi.org/10.1007/978-3-319-74030-0_49

vocabulary of a modeling language (in this case BPMN). The patterns lead to typical questions and suggestions for optimization which can be helpful for a requirements engineer. While the abstraction level of our patterns is higher than it is the case for the workflow patterns [1], it is still lower than it is the case for patterns that address the actual meaning of an activity [2].

In this paper, we draw on experiences from the research project PIGE (Process Intelligence in Healthcare) at the University Hospital Jena (Germany) [3]. We analyzed 47 process models, namely in the area of liver transplantation, living liver donor transplantation, liver cell cancer and hepatocellular carcinoma that were modeled together with medical experts. The examples in Sect. 2.2 are taken from this project.

2 Business Process Variability Patterns

2.1 Selected Related Work

In business process modeling, the most popular work on patterns are workflow patterns [4]. These patterns have been widely used by practitioners, vendors and academics. They are helpful for selecting a workflow system that corresponds to the needs of an organization and for evaluating the expressiveness of modeling languages. Workflow patterns provide an answer to the question which elements a workflow engine or modeling language should support in order to be useful in a given context. However, they do not deal with the topic of process variability.

To address process variability, the patterns need to have a somewhat higher abstraction level than the traditional workflow patterns [4]. For example, we wish to say that certain activities are executed in parallel instead of referring to the pattern PARALLEL SPLIT in combination with another pattern SYNCHRONIZATION as it would be the case in the terminology of workflow patterns. Also, the patterns should be named in a language that can be used intuitively by domain experts. For example instead of speaking about the INTERLEAVED ROUTING pattern, we use the term IN ANY ORDER.

Process variability can be expressed by means of patterns for differences/ changes between business process models. Such patterns have been discussed for various purposes and using different terminologies (an example is [1]). However, such change patterns have been developed for other purposes than the patterns in our paper. They describe variability on a rather technical level while the aim of our patterns is to use typical business vocabulary only. For example, our pattern PERFORM IN ANY ORDER corresponds to the change pattern MOVE PROCESS FRAGMENT in [1]. While our pattern describes a business situation, the change pattern deals with abstract operations on a model. The exception handling patterns in [5] are closer to our work (and were used as an important source for identifying our patterns). However, the focus of [5] is restricted to exception handling which is just one aspect of variability.

2.2   Variability Patterns

We collected typical variable situations in a process that are described by the stakeholders verbally. Based on a literature review and our modeling experience, we found three main classes for variability in business processes: First, the number of executions of an activity can be unknown until execution time. Activities may be optional, i.e. they can be skipped or it can be necessary to repeat an activity until the desired outcome has been achieved. Second, it can turn out that additional activities have to be added to the process. Third, the order between activities might not be known before the process is actually executed.

Using patterns allows a requirements engineer to communicate with stakeholders in a language that is very close to their domain vocabulary. In addition, each pattern leads to typical questions and optimization options. For illustrating the principle, we selected three patterns related to a situation where a series of tests must be passed. Such situations are common in medical processes, but also in other domains (e.g., approval procedures). Depending on the context, the most appropriate pattern has to be found. For its selection, a requirements engineer can already make use of a list of typical questions:

- Which activities lead to the situation where the object has to be tested?
- In which order the tests are usually performed? Why? Which resources are used? Are there dependencies between the tests that restrict the order?
- Which tests can be performed in parallel?
- What are the non-functional requirements for the tests? How important are speed, cost and accuracy?
- Is it always necessary to perform all tests? Are there cases where only the "most important" tests are performed (e.g., to save time)?
- What should be done in case of a failure of a test? (Patterns for possible consequences are discussed in [6].)
- Have any preparatory steps (such as collecting or organizing data) to be done before the actual tests can be carried out? Can data collected in advance become invalid before actually being used in a test?
- What are the percentage of failure and the probability distributions of cost and execution time for each individual test? How we can get access to logs of previously executed tests to answer these questions?
- Should we monitor the parameters described in the previous question at the runtime of future process instances? How should this information be used?

We can associate questions with certain patterns, so that one of the patterns described later can be selected depending on the questions' answers. Till now, we found a number of ten flexibility patterns, but due to space limitations, we present only three patterns here in more detail.

Pattern "Pass All Tests (Ordered)" *Problem:* Some object has to undergo a series of tests. The order in which the tests should be executed is known beforehand. The process can proceed only if all tests have been passed successfully.

As soon as the first test fails, another path in the process flow is taken, and no further tests are necessary.

Example: A prospective liver donor needs to undergo a blood test to investigate the compatibility to the patient's blood. In case the blood is not compatible, the donor is not suitable, and no further tests are necessary. Otherwise, an investigation by a psychologist is planned. Here, too, the donor can be found to be not suitable if there are any contraindications.

Questions: What are the reasons for defining the order of the individual tests? Which tests depend on the output of other tests?

Considerations for Optimization: If no dependencies between the tests exist, the issue of finding the optimal order between the tests is discussed in detail for the next pattern "Pass All Tests (Any Order)". Once the order is fixed, we can ask whether it corresponds to the spatial and organizational layout of the organization in order to avoid long ways for transport and unnecessary handovers.

If preparatory steps (such as collecting data) are executed before the first test starts, we can ask whether indeed all those steps are necessary *before* the sequence of tests. As it can be possible that the preparation for test 2 becomes irrelevant if already test 1 fails, it can be a good decision to prepare for test 2 only after test 1 succeeded.

Pattern "Pass All Tests (Any Order)" *Problem:* Some object has to undergo a series of tests. The order in which the tests should be executed has to be decided at run time. The process can proceed only if all tests have been passed successfully.

Example: In order to decide whether a person is a suitable liver donor for a near relative, a number of investigations and tests are necessary according to an evaluation check list. A transplant operation can only be planned if the whole evaluation procedure is completed successfully.

Questions: The tests can be performed in any order – but is there a typical/preferred order anyway? Why?

Considerations for Optimization: If for each test, the cost for executing the test is known as well as the probability that the test succeeds (i.e. delivers a "positive" result), the question for the optimal ordering of the tests is the well-known filter ordering problem (in the area of business process management also known as optimal ordering of knock-out processes [7]). The tests should be executed in increasing order of the ration between cost of a test and the probability that the test reveals a problem [8]. "Cost" can be replaced by another indicator which should be minimized (e.g. the radiation exposure from medical imaging). If the probabilities that a test succeeds are not known beforehand, they can be learned at runtime from previously executed process instances. If the tests are executed by different roles, it can also be helpful to take into account organizational handovers and transport ways between the performers of the tests. Both should be minimized.

Pattern "Pass All Tests (Parallel)" *Problem:* Some object has to undergo a series of tests. These tests can be done in parallel (to be executed by different actors).

Example: While the prospective liver donor is undergoing medical investigations, medical consultations (meetings among physicians from different disciplines) can take place to discuss the case and discuss possible contraindications.

Questions: What advantages do we expect from performing the tests in parallel?

Are there really no dependencies between the tests such that one test would need (or at least profit from) the results of another test?

How do we ensure that all tests have access to the same up-to-date data of the object under test?

If one of the tests fails, is it desirable to prevent (or even to stop) the execution of the remaining tests?

Considerations for Optimization: In general, this pattern should be chosen if execution time is more important than costs for potentially unnecessary tests. The disadvantage for this pattern is that it can happen that some tests are executed unnecessarily, because another test already reveals an error. Therefore, the requirements engineer should discuss whether/how the failure of a test should be communicated such that the roles performing the other tests do not have to do them anymore (see the discussion of the CANCEL ACTIVITY pattern in [4]).

2.3   Combining Patterns

When patterns are combined, more questions can emerge. We will illustrate this for the combination of any of the PASS ALL TESTS patterns with the pattern REPEAT ACTIVITY UNTIL SUCCESS: Some activity A must be executed repeatedly until several tests which follow its execution confirm that A has been executed correctly. A single failed test shows that A has to be repeated. Afterward, a new round of tests is necessary until all of them succeed (proving that A was carried out successfully). There are two possible ways to deal with this situation: Either all tests have to be executed again or only the tests that did not already succeed before. To decide between those options, we have to ask whether the rework can introduce new problems concerning features that have already passed a test.

In addition, there might be a time limit for executing A and the corresponding tests. This leads to a third pattern MUST SUCCEED IN A GIVEN TIME. The requirements engineer would have to ask what happens in the case of a timeout (a typical option would be that a person in a higher position in the organization becomes responsible for achieving the result of A). In case of time-consuming tests, it can also be an option only to perform a subset of the tests in order to exclude the most important risks only.

3   Pattern Application

For the process of a living liver donor in the PIGE project, medical experts were involved in the modeling and discussion of the as-is-process. Some parts of the process are flexible, e.g., the evaluation of possible donors mentioned before. Several examinations have to be carried out before a decision can be made whether the examined person can donate his/her liver. During the modeling of the process, the physicians showed a paper checklist containing all examinations that the possible donor has to undergo. Only if the person's results are positive in every examination, she/he is a suitable candidate for a donation. This calls for our pattern PASS ALL TESTS (IN ANY ORDER). According to this pattern, the modeling expert should ask whether a preferred sequence exists that is typically used. This way, additional conditions can be defined.

In the detailed discussion of the evaluation process, it turned out that after the blood compatibility and psychological testing, a number of further investigations and tests are necessary according to the check list. Their sequence is decided by the medical personal. If a mayor contraindication is found, the donor is not suitable, and no further tests are necessary.

Additionally, data was collected from the clinical information systems and analyzed alongside the evaluation procedure. It turned out that especially one examination often leads to a dropout of the possible donor. With this knowledge, the process could be optimized following the optimization suggestions from the pattern (see Sect. 2.2). The examination procedure was changed so that the examination with the high dropout rate is done at the beginning of the evaluation process. This way, the evaluation procedure for non-suitable donors is shorter, and hospital stay for further examinations can be avoided [9].

In the future, we plan to extend our set of variability patterns to cover more variable situations in business processes in other application areas. Additionally, we will work on the representation of our patterns in different modeling languages.

References

1. Weber, B., Reichert, M., Rinderle-Ma, S.: Change patterns and change support features - enhancing flexibility in process-aware information systems. Data Knowl. Eng. **66**(3), 438–466 (2008)
2. Thom, L.H., Reichert, M., Iochpe, C.: Activity patterns in process-aware information systems: basic concepts and empirical evidence. Int. J. Bus. Process Integr. Manage. **4**(2), 93–110 (2009)
3. Kirchner, K., Malessa, C., Scheuerlein, H., Settmacher, U.: Experience from collaborative modeling of clinical pathways. In: Modellierung im Gesundheitswesen: Tagungsband des Workshops im Rahmen der Modellierung 2014, pp. 13–24 (2014)
4. van der Aalst, W.M.P., ter Hofstede, A.H.M., Kiepuszewski, B., Barros, A.: Workflow patterns. Distrib. Parallel Datab. **14**(3), 5–51 (2003)
5. Staudt Lerner, B., Christov, S., Osterweil, L., Bendraou, R., Kannengiesser, U., Wise, A.: Exception handling patterns for process modeling. IEEE Trans. Softw. Eng. **36**(2), 162–183 (2010)

6. Namiri, K., Stojanovic, N.: Pattern-based design and validation of business process compliance. In: Meersman, R., Tari, Z. (eds.) OTM 2007. LNCS, vol. 4803, pp. 59–76. Springer, Heidelberg (2007). https://doi.org/10.1007/978-3-540-76848-7_6

7. van der Aalst, W.M.P.: Re-engineering knock-out processes. Decis. Support Syst. **30**(4), 451–468 (2001)

8. Garey, M.: Optimal task sequencing with precedence constraints. Discrete Math **37**(4), 37–56 (1973)

9. Kirchner, K., Scheuerlein, H., Malessa, C., Krumnow, S., Herzberg, N., Krohn, K., Specht, M., Settmacher, U.: Was ein klinischer Pfad im Krankenhaus bringt. Evaluation klinischer Pfade am Uniklinikum Jena am Beispiel des PIGE-Projekts, Chirurgische Allgemeine Zeitung, **15**(7+8), pp. 475–478 (2014)

A Pattern-Based Question Checklist for Deriving Requirements from BPMN Models

Bernhard M. Turban[1](✉) ⓘ and Johannes Schmitz-Lenders[2] ⓘ

[1] DCSM WI, University of Applied Sciences RheinMain, 65195 Wiesbaden, Germany
bernhard.turban@hs-rm.de
[2] parcs IT-Consulting GmbH, 57072 Siegen, Germany
jsl@parcs.de

Abstract. Concerning the realization of planned *Business Processes* through IT-Systems (IT Alignment), *Requirement Elicitation* is of central importance. One decisive factor for success is the experience knowledge of the *Requirement Engineers*, especially concerning the know how to ask the right questions for eliciting unknown requirements as early as possible in a project. This article presents an approach how this experience knowledge can be captured in a *Pattern-based Question Checklist* providing a structured set of questions to derive requirements from *Business Processes* modeled in BPMN. This captured knowledge can thus be provided to less experienced *Requirement Engineers* supporting them in their *Requirement Elicitation* activities.

Keywords: Business Process · BPMN · Question Checklist · Pattern ·
Requirement Elicitation · Requirement Engineering · IT Business Alignment

1 Introduction

In most cases when new *Business Processes* (BPs) are designed or existing BPs are redesigned, IT-Systems need to be adapted (as part of the *IT Business Alignment* at the operational level). Concerning the implementation of the tasks modeled within a BP, it is important to discover all requirements, as far as possible, for all involved systems. This activity usually turns out to be very complex and very dependent on the experience of the involved *requirement engineers*.

Often, BPs are modeled in the modeling language BPMN [10]. This article shows how the characteristics of BPMN collaboration diagrams and the pattern concept can be used to capture the experience of *requirement engineers* into a *Pattern-based Question Checklist* (*PQC*). The *PQC* enables to pass this experience knowledge to less experienced *requirement engineers* in order to improve their effectiveness.

For better illustration of the context, Sect. 2 discusses the general connection between the proposed approach and *Business Process Modeling* (*BPM*). Section 3 then discusses the basic ideas behind the proposed approach. Before the *PQC* is introduced in Sect. 5, Sect. 4 presents an exemplary BPMN diagram to better illustrate the *PQC*. Finally, Sect. 6 summarizes the results.

© Springer International Publishing AG 2018
E. Teniente and M. Weidlich (Eds.): BPM 2017 Workshops, LNBIP 308, pp. 630–641, 2018.
https://doi.org/10.1007/978-3-319-74030-0_50

2   Business Process Modeling and *PQC*

Business Process Modeling (BPM) can serve different purposes:

- As a means for designing new, or redesigning existing BPs
- For documenting BPs and instructing employees
- For automating BPs via *workflow engines*

This article mainly addresses the first purpose. In this scenario, after having designed the BPs from business perspective, the IT related tasks are mapped to specific IT-systems for implementation. Often, multiple IT-systems – existing, or new systems to be developed in-house or procured – are affected. This assignment can be considered part of *operative IT Business Alignment* [7] or *Enterprise Architecture* [6].

In many situations where BPs are newly designed or existing ones are redesigned, BPs are still not completely automated, but rather implemented as manual tasks supported by computers, or implemented without using a workflow engine. Therefore, it is important to derive the requirements for affected IT-systems from the available BP designs, in order to ensure that the corresponding functionalities for enabling the BPs are implemented accordingly. For an effective and efficient implementation, it is very important that a smooth transition between *BP models* and *IT-system requirements* is feasible.

For BP-centered requirement analysis, [3, 12] point out that BP descriptions alone are insufficient and detailed requirements must be worked out. They also describe which elements are typically used (e.g., Business Objects/Classes, Use Cases, Business Rules, Dialogs), but rather leave open how BPs can/should be used to elicit requirements based on and consistent with BPs.

In contrast to approaches for automated implementation via *workflow engines*, which are addressed by extensive portions of research endeavors (see e.g. [3, 9]), only very limited hints are provided by literature to address the above described scenario. Thus, it rather remains a question of experience of the involved *requirement engineers* how good or bad a seamless transition to requirements for the affected IT-Systems can be achieved.

3   Principles of the Approach

Concerning an approach to the above described transfer problem, it needs to be emphasized that a simple and general solution description is impossible, as this process is of a diffuse, situation and experience dependent nature.

An alternative approach to address such complex, experienced dependent problems in computer science is the *Pattern*[1] idea inspired by Christopher Alexander [1]. Besides *Software Design Patterns* (cf. e.g. [4]) a variety of different types of *Patterns* have been established. These also include e.g. *Requirement* [15] and *Analysis Patterns* [2]

---

[1]   [14; Ch. I.6.2.4] provides a comprehensive description of the basic ideas behind the *pattern* approach and the diverse types of *patterns*, its effects, disadvantages and the connections to other design theories in computer science.

providing support for *requirement elicitation*. *Patterns* try to describe recurring problems within complex interrelations and try to offer an adaptive solution for those problems, being also open for adaption to other similar problems. To achieve this, *Patterns* are documented in a structured way using *Pattern Templates*, and several *Patterns* are collected in *Pattern Catalogues*.

As an essential advantage, *Patterns* encode experience and expert knowledge [8; p. 139], [14; p. 101], thus making it easier to pass this knowledge over to less experienced persons, but also allowing experts to exchange ideas more easily [11], [14; p. 101]. E.g., Prechelt and Unger [11] show that inexperienced software developers could significantly improve the quality of their software designs by using patterns. This same effect is used by the approach proposed here.

Concerning *requirement elicitation*, discovering as much relevant requirements as possible (resp. missing as less requirements as possible) as early as possible in a project is a success-critical factor. This can be achieved by a search as systematic as possible. For this purpose, literature recommends to use checklists to elicit the information as systematically as possible. E.g., Rupp et al. propose using checklists for identifying stakeholders [13; p. 79] or nonfunctional requirements [13; p. 270ff] or in general for validating requirements [13; p. 329].

Section 5 shows a possibility how experience knowledge of requirement analysts can be documented and used to help less experienced requirement analysts. For this, a *Pattern-based Question Checklist* (*PQC*) has been developed. The *PQC* analyzes and describes different modelling situations within a *BPMN collaboration diagram* in a *pattern*-like way and provides a checklist with questions for *requirement elicitation* to each described modelling situation. Similar pattern based approaches are used by Maiden et al. [16] for deriving requirements automatically from "Strategic Dependency" and "Strategic Rationale" models, and by Laue et al. [17] for automated detection of quality problems in business process models.

4 Running Example

Figure 1 shows a modeled example of several BPs for elicitation, rating and implementation processes concerning requirements in an IT-department.

The example itself has no direct connection to this article, but is rather used as an understandable example showing all modeled elements and situations that are discussed by the *PQC* in Sect. 5. The *PQC* will refer to the corresponding modeling situation in Fig. 1 to provide a concrete example for a discussed situation. To ease this referencing, an id is assigned to each element in Fig. 1. To also indicate the element types of the modeled objects, the ids' letters also indicate the corresponding element type[2] – for a more detailed description of the elements and semantics of BPMN it is recommended to consult [10] or e.g. [3].

---

[2] SE = Start Event, T = Task/Activity, IE = Intermediate Event, ICE = Intermediate Cancel Event, IEE = Intermediate Error Event, EE = End Event, DS = Data Store, P = Pool, L = Lane.

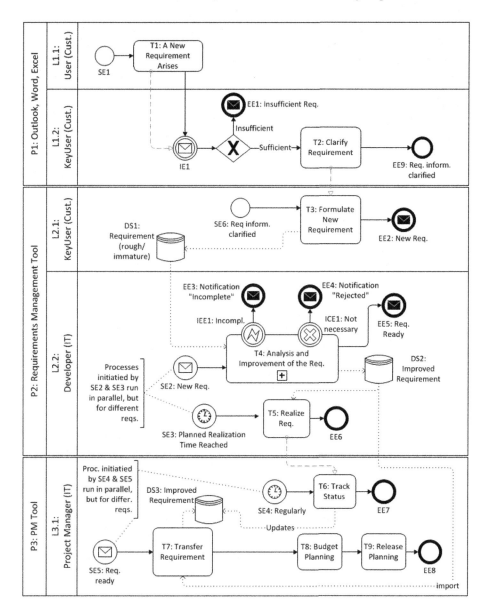

Fig. 1. Example of a BPMN diagram for requirements management BPs in an IT-Department

Besides the modelling of *Tasks*/activities and the *sequence flow* (solid arrows in Fig. 1), BPMN also offers the following characteristics helping to establish the proposed *PQC*:

- In BPMN, *Pools* can be used and *Lanes* can be nested in *Pools*. As shown in Fig. 1, this can be used to determine the IT-system and user role involved by a *Task*. E.g., task T1 is performed by a normal user using Outlook, Word or Excel. It is also possible to see that during the progress of some BPs, changes of the IT-system[3] and/ or the user role happen. These stipulations are especially important concerning requirement analysis, because they provide important information concerning *system boundaries* and the *user roles* to be supported by a system. *System boundaries* are discussed in Rule R5 (Sect. 5.5), and *user role changes* are explicitly discussed in Rule R6 (Sect. 5.6).

- Through *Events*, special events or milestones within a BP are modeled. BPs start with a *Start Event* and end in an *End Event*. Within a BP, *Intermediate Events* can be modeled. Furthermore, event symbols are semantically differentiated into symbols *raising* and symbols *receiving* an event. This leads to implicit dependencies between BPs, as a BP can *raise* an *Event* which can be *received* by some other BP which is thus triggered to proceed (see e.g. Fig. 1: EE2 triggers SE2 and thus activates this BP). This characteristic is covered later by Rule R3 (Sect. 5.3).
- *Intermediate Cancel Events* (see ICE1) representing BP-abortions by the user and *Intermediate Error Events* (see IEE1) representing error handling behavior and strategies within BPs are another important aspect modeled in *BPMN collaboration diagrams*. This topic is covered by Rule R4 (Sect. 5.4).
- Besides *sequence flow*, BPMN also allows to model *message/data flow* via dotted arrows (*Input* and *Output Data Associations* to *Tasks*). In connection to this, *important data artefacts* can be modeled by *Data Objects* or *Data Stores* (see Ids DS…). These modeling options in combination with Rule R2 (Sect. 5.2) help to discover important requirements concerning data artefact related operations.

5   The Pattern-Based Question Checklist (PQC)

This chapter presents the *Pattern-based Question Checklist (PQC)*. The idea to embed the questions into a description structure has been taken over from the *Pattern* concept (cf. *Pattern Catalogue* and *Pattern Template*). Thus, the whole description structure (equivalent to a–easy to remember– rule forming an embracing general topic for the contained *Pattern Catalogue*) is partitioned into chapters (the following subchapters), where each chapter is partitioned into the following substructure (equivalent to a *Pattern Template*):

- The caption consists of a general question. To ease referencing a rule at other parts of the document, each rule has a rule id (R…) at the beginning.
- In the first section beneath the caption, information about the rule's rationale and the connection to the questions following this section is provided.
- Afterwards, the questions are listed as a checklist. Each question starts with a question-id (Rx.Q…) making referencing to a specific question easy. The actual question

---

[3] The example assumes that certain decisions concerning which IT-system is used for which task have already been made before and the BPMN diagram has been modeled accordingly. In the BPMN-standard, *Pools* are originally dedicated for expressing organizations. However, in the authors' experience *Pools* can and are also used in practice to model the break-down of BPs to involved systems.

follows after the question-id. If the question is not self-explanatory enough, a further explanation –separated from the question by a hyphen– is provided.
- After the question section, a short explanatory section provides further practical explanations with reference to Fig. 1, in case uncertainties might still exist.

5.1   R1: Each IT-Related *Task* May Need at Least One *User Interface-Element*

R1 describes the fact that almost every IT-related *Task* (see T1-T9 in Fig. 1) is at least once somehow related to a user interaction. For this, usually at least one *User Interface-Element*[4] (UI-E) must be present in the corresponding IT-System[5].

Therefore, for each *Task* the following must be clarified:

- R1.Q1: Which information is necessary for the task? – It must be clarified[6] which information is essential and which additional information is possibly useful to perform the task.
- R1.Q2: How shall the needed information be presented and –potentially– edited? – For each information, it must be clarified with which UI-Es the corresponding information shall be presented and the –potential– user input of data be handled.
- R1.Q3: How many UI-Es are involved[7]? – Are there no, one or several UI-Es? The following questions must be answered for each identified UI-E.
- R1.Q4: Is a special (possibly new) UI-E necessary?
- R1.Q5: Is it possible to reuse and extend a UI-E already present (e.g. a GUI form)? – This question is a counter-question to R1.Q4.

Here, the following subquestions mainly addressing usability are recommended:

- R1.Q5.1: Do the data entities handled in the UI-E actually fit together?
- R1.Q5.2: Do the different tasks/activities handled by the UI-E fit together well?
- R1.Q5.3: Is the already present UI-E not overloaded by the new aspects?
  - R1.Q6: How is the user guided to the UI-E? – Are further search, filter, … operations necessary?
  - R1.Q7: Is the navigation concerning the other tasks/activities consistent? – Usually a task/activity has one or several pre- and post-activities. It must be ensured that a consistent transition/navigation between the activities and the UI-E's associated to them is happening.

These questions must be asked and answered for all *Tasks* (see T1-T9 in Fig. 1), because these questions lead to essential information for each task.

---

[4] E.g., a GUI form (GUI = Graphical User Interface), or a part of a GUI form, or a GUI control, or something else (e.g. a steering gesture, voice command, …).
[5] See also [3; p. 156f] where modeling the flow of GUI forms and BPMN is described.
[6] As much as needed for the level of detail wanted for the intended purpose of the requirements specification; not every requirement specification needs to address the user interface, this might be done later in the process; this also applies to all other questions in the PQC.
[7] This and the following questions address the requirements for the logical structure/the logical design of UI-E's. The concrete UI design depends –in addition to the chosen level of detail– on concrete targeted technologies, e.g. web browser(s), desktop, mobile devices.

5.2   R2: Each *Data Artefact* May Represent a Separate, Self-contained Artefact, or Some Other Important Aspect of Another Artefact

R2 describes the fact that *data artefacts* (*Data Objects* or *Data Stores*, see DS1-DS3 in Fig. 1) modeled in a BP most often indicate a self-contained data unit (artefact), or some other important data-related aspect/operation concerning another artefact.

Therefore, for each *data artefact,* the following must be clarified:

- R2.Q1: Is the modeled *data artefact* a separate, self-contained artefact, or is it part of some other artefact?

For *data artefacts*, incoming arrows (e.g. T3 → DS1 in Fig. 1) indicate that write access is needed for the involved task/activity, and outgoing arrows (e.g. DS1 → T4 in Fig. 1) indicate read access is needed. Therefore, for each *incoming arrow,* the following must be clarified:

- R2.Q2[8]: How is the input resp. creation of the data performed?
  - R2.Q2.1: Which information needs to be gathered (cf. R1.Q1)?
  - R2.Q2.2: According to which rules must the data be created, if any (cf. R1.Q1)?
  - Perform questions R1.Q2 – R1.Q7 for each identified type of information.

For each *outgoing arrow*, the following must be clarified:

- R2.Q3: Is a search-dialog necessary, or how else is the user getting to the indicated data?
- R2.Q4: How is the information/data processed?
  - R2.Q4.1: Which information needs to be presented (cf. R1.Q1)?
  - R2.Q4.2: Are some parts of the information also changed, perhaps? – I.e., are there not yet discovered write accesses?
  - Perform questions R1.Q2 – R1.Q7 for each identified type of information.

These questions must be asked and answered for all *data artefacts* (*Data Objects* or *Data Stores* (see DS1-DS3 in Fig. 1)). However, these questions are especially important for *Data Stores* because they rather represent persistent data units.

5.3   R3: *Start* and *End Events* Must Be Analyzed in Detail

Start Events start a process, and processes end in *End Events*. In BPMN, however, it is possible that an *End Event* triggers the *Start Event* of another process. Thus, implicit dependencies exist possibly disguising more complex connections.

Therefore, concerning *Start Events*, the following must be clarified:

- R3.Q1: How is a *Start Event* triggered? – Is a *Start Event* automatically triggered by a computer system or rather manually by a person?

---

[8] Here and concerning question block R2.Q4, overlaps to the questions of R1 do exist. In requirement analysis, it has proven successful to ask questions from different perspectives in order to elicit possibly still undiscovered aspects and requirements (cf. e.g. [13; p. 101f]).

- R3.Q2: Which kinds of information must be at hand and which other preconditions need to be fulfilled?

Concerning *End Events* the following must be clarified:

- R3.Q3: Which information will be created (if any) and to whom propagated?
- R3.Q4: Are there any *Start Events* triggered? – If yes, the following subquestions need to be clarified:
  - R3.Q4.1: How is this triggering done in detail? – Automatically via a system, or manually, or … (also cf. R3.Q1-Q3)?
  - R3.Q4.2: Are there any *system* or *user role boundaries* traversed? – *System boundaries* are especially critical (cf. R5 (Sect. 5.5)), but also *user role boundaries* should be carefully considered (cf. R6 (Sect. 5.6)).
  - R3.Q4.3: Does this seem to be consistent, or are there still contradictions and ambiguities? – Contradictions and ambiguities indicate that there might be hidden assumptions or hidden requirements.

An analysis using this rule and questions shows in Fig. 1 that *End Event* EE2 triggers the *Start Event* SE2 and *End Event* EE5 again triggers *Start Event* SE5. As the first case represents a traversal of a *user role boundary* and the second case a traversal *system boundary* (incl. a subsequent data transmission (see R5 (Sect. 5.5))), both situations must be identified and analyzed with special care.

Concerning this rule *Intermediate Events* in the normal *sequence flow*[9] are an open issue. In principle, this rule and the questions can also be applied for *Intermediate Events*. BPMN provides ten possible event type variants with "throw-catch" semantics. It remains a topic of further research whether these kinds of *Intermediate Events* imply further rules and questions beyond the here discussed ones.

5.4 R4: *Error* and *Cancel Events* Must Be Analyzed in Detail

In BPMN it is also possible to model *Error* and *Cancel Events*. *Error Events* are used to indicate a *business process* really has failed as a whole or in part [3; p. 162], and *Cancel Events* are used to model cancellation of a *business process* by a user.

Therefore the following aspects have to be clarified for *Error* and *Cancel Events*:

- R4.Q1: Is an *Error* or *Cancel Event* actually reasonable in the given context? Can an *Error* or *Cancel Event* be dealt with in this location in a reasonable manner, or would a delegation to another location be more reasonable?
- R4.Q2: Are there any hidden transitions to other systems, e.g. if an *Error Event* triggers a *Start* or *Intermediate Event* at some other location?
- R4.Q3: How is data consistency ensured? Previously generated and persisted data, depicted e.g. by data objects (see R2, Sect. 5.2), have to be brought into a clean state. The following aspects have to be clarified in this context:
  - R4.Q3.1: Can the generated data simply be deleted?

---

[9] Concerning *Intermediate Events* at the border of tasks consult the next chapter discussing *Intermediate Error* and *Cancel Events*.

   – R4.Q3.2: Does the generated data have to be set to an "invalid" status?
   – R4.Q3.3: Does related data in other systems have to be treated accordingly?

- R4.Q4: Is it necessary to store the *Error* or *Cancel Event* itself as a data artefact?

In Fig. 1, arrow IEE1 → EE3 depicts an *Error Event* and arrow ICE1 → EE4 depicts a *Cancel Event* that can be analyzed using the question checklist suggested in this article. Both arrows indicate the sending of messages (hidden system switch). In addition, it needs to be clarified if the generated data are deleted or set to "invalid".

5.5   R5: Arrows Across *System Boundaries* are Particularly Critical

Arrows across *system boundaries* are particularly critical because they indicate media discontinuity[10]. The definition of the *system boundaries* and of the surrounding context is an essential part of requirements analysis (see e.g. [13; p. 86ff.]). The fact that implementation projects are often divided into subprojects at the *system boundaries* increases the risk that important issues are neglected.

Sequence flows between *Pools* are forbidden in BPMN. Thus, the following question arises concerning *sequence flows*[11]:

- R5.Q1: Are there any *sequence flows* crossing system boundaries? - If yes, these *sequence flows* must be resolved to an end-start-event-relationship and a *message flow* between the *Tasks* must be modeled and the questions in R3 and R5.Q2-Q3 must be asked.

Message/Data flow arrows indicate that messages/data has to cross a *system boundary*. In this context, the following aspects must be clarified:

- R5.Q2: How exactly does this happen? This raises the following sub-questions:
   – R5.Q2.1: What is the trigger for the message/data transfer? – When is the trigger activated, and by whom or by what?
   – R5.Q2.2: Which interface is used to transfer the message/data? – Interfaces greatly influence complexity, time and effort/costs of a project.
   – R5.Q2.3: What is the data format (e.g. XML, …)?
   – R5.Q2.4: In case of a data transfer, are the internal data models of the involved systems compatible with each other?

- R5.Q3: In case of a data transfer, is the data transfer really necessary? In many cases, a data transfer between systems causes data redundancy. If such redundancy is indispensable, it raises following sub-questions:

---

[10] It is assumed here that the technical *system boundaries* have already been determined. In general, areas of responsibility (pools) can also be used to split a planned larger system into logical subsystems. During implementation, each area can result in a technical *system boundary*, or not in case the logical subsystems are merged into a common technical system.
[11] In Fig. 1, instead of the end-start-event-relationship EE9 → SE6 and the *message flow* T2 → T3, developers might come to the idea to model a direct *sequence flow* T2 → T3, because L1.2 and L2.1 denote the same user role. This would violate the BPMN standard, however. These situations are addressed by R5.Q1 and must be resolved as shown in Fig. 1.

- R5.Q3.1: Should both systems store exactly the same information, or would it be sufficient to only keep part of the information or surrogate objects of the artifacts? – The latter reduces redundancy and usually should be preferred.
- R5.Q3.2: What happens if the data is changed at a later date? – Do those changes have to be performed in both systems? Is it at all possible to make changes to the data in both systems? Is there a data master (in which system)?
- R5.Q3.3: How will data synchronization be done? – Question R5.Q2.1 also has to be taken into account in this context.

In Fig. 1, the *data flow* DS2 → T7 → DS3 is a critical *system boundary* crossing which requires carefully answering questions R5.Q2 and R5.Q3.

5.6   R6: Arrows Across *User Boundaries* are also Very Important

This paragraph addresses crossing of areas of responsibility between different user roles. These transitions are not as critical as crossing the boundaries between different IT-systems, but still they should not be underestimated. The main problem is that in this situation there usually is a discontinuity in the UI navigation, as the existence of different user roles usually means that there are different users.

This is why the following aspects need to be clarified for *sequence flow* arrows:

- Ask questions R5.Q1 – R5.Q2 for the cross-boundary transition
- R6.Q1: Are there any problematic discontinuities? – All potentially problematic discontinuities should be identified and analyzed.
- R6.Q2: Which measures can be applied to achieve a smooth transition? – I.e. how can discontinuities be minimized?

The transitions T1 → IE1, DS1 → T4 (also, see implicitly EE2 → SE2), T5 → T6 in Fig. 1 are typical situations to be analyzed using this rule.

6   Conclusion

This article has shown how a *Pattern-based Question Checklist (PQC)* can be used to capture *experience knowledge* of *requirements engineers* in transferring *Business Processes* modeled in BPMN into *IT-system requirements*, in an accessible format.

Since modeled *Business Processes* tend to be fairly abstract the transfer of such processes into IT-systems can be done in various ways – especially regarding surrounding conditions – and it is difficult to establish general applicable rules and formulae for deriving requirements. The present approach therefore focuses on an exemplary list of questions for different modeling situations of a BPMN model which is designed to help to disclose preferably all relevant requirements at an early stage. As Maiden et al. [16; p. 286] indicate, stakeholders more easily can discover problems in existing requirements than discover missing requirements. Therefore, the checklist concentrates on discovering missing requirements, as well as extensibility if new situations/good questions are encountered.

This *PQC* allows to go systematically through all *BPs* modeled with BPMN and to derive as many requirements as possible from the questions related to each of the modeled situations.

Furthermore, a *PQC* also can be useful in the following contexts:

- For detecting inaccuracies and inconsistencies within BPMN models (see R.5.Q1), also taking into account existing or to be introduced modeling conventions,
- to challenge/scrutinize/double-check decisions taken with regard to *system bounda-ries*, e.g. to reduce data redundancies, or to create additional possibilities for the realization of requirements in an IT-environment through the segmentation into logical areas of responsibility,
- for detecting missing requirements during the review process,
- to identify all aspects relevant for a cost estimate in an early stage of a project,
- for detecting potential gaps in the documentation after the implementation of *BPs*,
- or to help detecting essential aspects or requirements in case the implementation is automated via a *workflow engine*.

Besides capturing and providing the knowledge in an accessible format, a *Pattern-based Question Checklist* also has the advantage of encouraging a holistic view, beyond *system boundaries*.

The question checklist in its current scope does not claim to be complete and exhaus-tive. Rather it is expected that further questions will be added to the list over time. It is as yet unclear if rules including questions can be established for *Branching Gateways*, other *Gateways* and *Intermediate Events*. Furthermore, it still has to be discussed if the present approach may also be used effectively for BPMN charts other than *BPMN collaboration diagrams*.

Finally, the *requirements engineer* also has to find a way to document the require-ments in a suitable structure. While *process models* tend to be aligned with the needs of the respective *business processes*, the (IT-) requirements will have to be structured according to the IT-systems implementing these requirements. Often, *Use Cases* are used for this purpose, which in turn raises the question to what extent *Business Processes* and *Use Cases* [5] can be linked with each other (cf. also [3; p. 158f.], [13; p. 170f.]), and if a similar *Pattern*-based approach can be applied for this transformation.

References

1. Alexander, C., Ishikawa, S., Silverstein, M.: A Pattern Language: Towns, Buildings. Construction. Oxford University Press, New York (1977)
2. Fowler, M.: Analysis Patterns: Reusable Object Models. Addison-Wesley, Reading (1997)
3. Freund, J., Rücker, B.: Praxishandbuch BPMN. Hanser Verlag, München (2017)
4. Gamma, E., Helm, R., Johnson, R., Vlissides, J.: Design Patterns: Elements of Reusable Object-Oriented Software. Addison-Wesley, Reading (1995)
5. Jacobson, I., Christerson, M., Jonsson, P., Övergaard, G.: Object Oriented Software Engineering: A Use Case Driven Approach. Addison-Wesley, Wokingham (1992)
6. Op't Land, M., Proper, E., Waage, M., Cloo, J., Steghuis, C.: Enterprise Architecture: Creating Value by Informed Governance. Springer, Heidelberg (2008). https://doi.org/ 10.1007/978-3-540-85232-2

7. Heinrich, R.: Aligning Business Processes and Information Systems: New Approaches to Continuous Quality Engineering. Springer Vieweg, Wiesbaden (2014). https://doi.org/10.1007/978-3-658-06518-8
8. Moro, M.: Modellbasierte Qualitätsbewertung von Softwaresystemen. Books on Demand GmbH (2004)
9. Ouyang, C., van der Aalst, W., Dumas, M., ter Hofstede, A.: From business process models to process-oriented software. In: ACM Transactions on Software Engineering and Methodology (TOSEM), vol. 19, no. 1, pp. 1–37. ACM, New York (2009)
10. Object Management Group (OMG): Business Process Model and Notation (BPMN) V. 2.0 (2011). (http://www.omg.org/spec/BPMN/2.0. Accessed 02 May 2017
11. Prechelt, L., Unger, B.: Methodik und Ergebnisse einer Experimentreihe über Entwurfsmuster. In: Informatik Spektrum, vol. 14, no. 3 (1999)
12. Rohfleisch, F.: Geschäftsprozessorientierte Anforderungsanalyse – Business Analyse mit ARIS und UML. MV-Verlag, Münster (2011). Edition Octopus
13. Rupp, C.: Sophist Group: Requirements-Engineering und Management, 6th edn. Hanser Verlag, München (2014)
14. Turban, B.: Tool-Based Requirement Traceability between Requirement and Design Artifacts. Vieweg + Teubner Verlag, Wiesbaden (2013)
15. Withall, S.: Software Requirement Patterns. Microsoft Press, WA (2007)
16. Maiden, N., Manning, S., Jones, S., Greenwood, J.: Generating requirements from systems models using patterns: a case study. Requirements Eng. **10**(4), 276–288 (2005)
17. Laue, R., Koop, W., Gruhn, V.: Indicators for open issues in business process models. In: Requirements Engineering: Foundation for Software Quality (REFSQ), Gothenburg, Sweden (2016)

Minimal Effort Requirements Engineering for Robotic Process Automation with Test Driven Development and Screen Recording

Christoph Cewe[1](✉), Daniel Koch[2], and Robert Mertens[3]

[1] E.ON IT UK Limited, Nottingham, UK
christoph.cewe1@eon.com
[2] Valeroo GmbH, Munich, Germany
daniel.koch@valeroo.com
[3] HSW University of Applied Sciences, Hamelin, Germany
mertens@hsw-hameln.de

Abstract. Robotic Process Automation (RPA) can be regarded as a special kind of Business Process Management (BPM) that relies on GUI automation adaptors instead of regular interfaces for intersystem communication. Another difference between RPA and standard BPM is that RPA processes do not need to be defined from scratch as RPA is basically transforming an operator's implicit process knowledge into a workflow definition to be executed by the robot's workflow engine. In this context, the basic idea of Test Driven Development (TDD) can be used to jump start requirements engineering and process definition by leveraging the operator's interaction with the workflow targeted for automation. This paper presents a conceptual approach for integrating TDD with RPA development: In a first step, the manual process is enriched with probes that record input and output values for each execution of the workflow. Users then manually perform the process using screen recording and the probes to define annotated test cases stored in a backlog. TDD works by selecting an arbitrary recording from the backlog and using it for automating the equivalence class of test cases to which it belongs. All test cases from this equivalence class are then removed from the backlog. These two steps are iteratively repeated until all test cases are covered by the robot process definition.

1 Introduction

Robotic Process Automation (RPA) offers a highly promising approach for automating business processes with the aim to realize significant cost savings while simultaneously increasing the efficiency and quality of the process performance. In addition, RPA can be developed in a much shorter time than comparable IT solutions based on Business Process Management Systems (BPMS). Nevertheless, successful implementation of RPA solutions is not a matter of course. According to Ernst & Young 30 to 50 percent of initial RPA projects fail [1]. Related to this, the authors underline that one of the most common issues in failed RPA projects is the applying of traditional development methodologies: "Quite often companies try to apply an over-engineered software delivery

© Springer International Publishing AG 2018
E. Teniente and M. Weidlich (Eds.): BPM 2017 Workshops, LNBIP 308, pp. 642–648, 2018.
https://doi.org/10.1007/978-3-319-74030-0_51

method to RPA, with no-value documentation and gates, leading to extended delivery times – often months where weeks should be the norm" [1]. Most approaches for implementing RPA software are based on a classical waterfall development approach. They often begin with identifying all the requirements of the process to be automated associated with a comprehensive process documentation. According to Ernst & Young, these approaches are often over-engineered, extend the delivery times and include comprehensive process documentation with no value. "Companies should look to challenge and simplify existing methods and use an agile delivery approach to deliver at pace." [1]. Especially the process specification and documentation of traditional delivery methodologies offer considerable potential for optimization in terms of saving time and costs as well as increasing delivery efficiency. Therefore, the objective of this paper is to introduce an agile delivery methodology for improving the efficiency of RPA development. For this purpose, the idea of Test Driven Development (TDD) and documenting a process with screen recording can be used to deliver RPA incrementally based on an agile approach. To demonstrate the agile approach, the paper gives a brief overview of general process automation and a short description and distinction of RPA and BPMS (Sect. 2). Subsequently the idea and benefits of TDD will be described (Sect. 3). After this, there is a short introduction to traditional RPA delivery methodologies, the advantages of an agile RPA delivery methodology and the explanation of the agile RPA delivery methodology based on TDD and screen recording is given (Sect. 4).

2   Distinction between Robotic Process Automation and Business Process Management Systems

For process automation, it is necessary to connect the modelled business processes with the applications needed for performing the business process. This can be done by connecting to the application business logic (business logic integration) or by interacting with the user interface (presentation integration) [2] *(see Fig. 1)*. The typical approach for automating processes via business logic integration are BPMS. In such systems, a process is defined as a rule-based workflow which will be executed in a process engine [3]. When an executed process requires information or services of an application, the BPMS will communicate with the application through an application programming interface (API) [3, 4]. The development of APIs can be associated with comprehensive development effort. This is mostly time- and cost-consuming. Another approach for automating processes uses presentation integration. This approach, in contrast, does not connect the process with the application through APIs, but through interacting with the user interface [2]. This approach is used in RPA, which describes the usage of software to mimic the interaction of a human user. It is therefore often called "software robots" or "digital labour" [5]. This technology is becoming increasingly interesting for enterprises due to its short implementation time, the relatively low costs (in comparison to comparable IT solutions) and the high return on investment as a result [6]. But RPA is not an entirely new kind of technology. In principle, RPA is a further development of screen scraping, which is much more mature than classical screen scraping and consequently suitable for enterprise use [5]. In RPA, the process is defined as a rule-based workflow, which includes Business Objects, where elements of the user interface (like buttons or text fields) of the application are defined. A workflow is then executed in a process engine and performs the process interactions according the defined rules. In contrast to BPM, cost savings are achieved by reducing user interaction, not by process optimization.

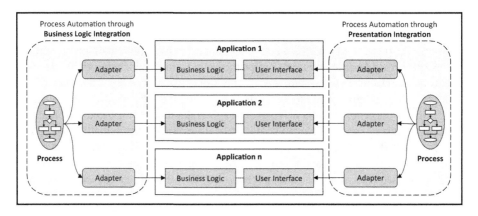

Fig. 1. Business logic integration vs. presentation integration. Adapted from [2]

Both approaches for process automation have different goals (see *Table 1*). In BPMS the definition of process logic and the integration of existing (parts of) programs through hard coded APIs result in a new application within a service oriented architecture. Therefore, BPMS allow to bring functionalities of disparate applications (e.g. from different back-end applications) together to develop reusable components [7]. For this purpose, advanced programming skills are usually necessary. In RPA, these programming skills are mostly not needed, because all interactions with the applications are done through the user interface [8]. The approaches are not contrary but complementary. In fact, Forrester Research recommends combining both for achieving the strategic corporate objectives in a long-term point of view [8].

Table 1. How RPA differs from BPMS at a glance. Adapted from [8].

Aspect	RPA	BPMS
Integration method	Presentation integration	Business logic integration
Business goal	Cost reduction by automating existing business processes	Drive efficiency through re-engineering existing processes
Technical goal	Use existing business applications	Create new business applications

3   Test Driven Development

Software testing is defined as "the process of executing a program with the intention of finding errors" [9]. It has been present since the beginning of software development and started as a manual task, documented on paper and performed manually by specially trained testers at the end of the development of an application [10]. Nowadays, software testing has become an important aspect in the software development life cycle and is therefore tightly integrated in the process of creating an application, usually early applied in a project and supported by software for automation [11]. The attempt to perform software testing as early as possible has led to different approaches towards both, software development itself and supporting tasks like project management. In fact, writing tests before implementing the actual logic is a fundamental principle with Extreme Programming, an agile software development methodology, developed by Kent Beck while working as a project leader in the late 1990s [12]. While Beck referred to this practice as "Test-First Programming", it is mostly known today as "Test Driven Development" (TDD) [13]. TDD is an iterative practice, consisting of writing a number of tests that fail initially, implementing the code unit under test and re-executing the tests to ensure they pass with the new implementation [14]. If tests still fail, developers go back to the implementation step, returning to the verification afterwards again. Once the tests pass successfully, the process is repeated for another part of the system.

4   Agile Development of Robotic Process Automation with Screen Recording and Test Driven Development

Every click a human user performs in the process to be automated must be defined for RPA development [2]. This is the reason why processes are often highly discussed and documented on a click-through level with marked screenshots and comprehensive text based descriptions (which is often called "Process Definition Document"). Usually, the actual development of RPA processes in projects using traditional delivery methodologies only starts when the process is completely documented and qualitatively reviewed. Also, robots will only be used when the process is developed completely according to the Process Definition Document. Consequently, robots which are fit for purpose will not be deployed when special cases documented in the Process Definition Document are still pending. Hence, an agile development approach offers a lot of potential for optimizations. Especially the requirements engineering part can be reduced by documenting the process with videos recorded using screen recording instead of comprehensive, text based and non-value process documentations. A process consists of inputs, a series of related activities and outputs. This information is sufficient for developing RPA processes. The inputs and outputs of the process to be automated can be represented by defined value pairs of the process while the series of related activities can be described with a video of the process performed by a human user. Combined with the idea of TDD, this can be used to greatly improve the way of developing RPA processes. The agile development approach described in the following is illustrated in *Fig.* 2. For developing an RPA process, the input and output value pairs must be defined and the video of the process execution must be recorded. As a first step, the manual process performed by human labor will be recorded via screen recording. The resulting video will be enriched with input and output values representing the desired results of the process. Furthermore, additional information of the process can be added for allowing later prioritizing. The information whether the process is a recurrent standard process or an exceptional case is an example of such additional information. The video, input and output values and further information are stored as a video backlog item in the video backlog list (see *Fig.* 2). RPA developers pick an item from the video backlog and create a test case based on the given input and output values. They then check if the test case is already fulfilled. In case it is not, developers will develop the RPA process based on the video of the video backlog item until the test case is fulfilled. Once a video backlog item is fully completed, the developers continue with the next video backlog item and automate further process scenarios. This process is iteratively repeated until all test cases are covered by the definition of the input and output values of the video backlog item. A single iteration step might solve more than one test case if it falls into the same equivalence class as others. Also, all test cases can be used as regression tests for continuous quality verification. Each automated process scenario documented in a video backlog item represents a single deliverable increment. Thus, these increments can be deployed as a completed usable increment if desired.

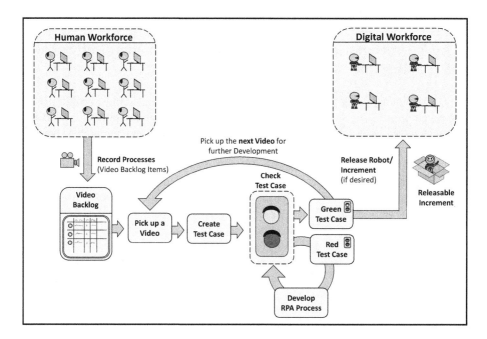

Fig. 2. RPA development approach with screen recording and TDD.

5   Summary

Using traditional delivery methodologies for automating business processes with RPA is often inefficient. The reason for this is that classical waterfall project management approaches are often over-engineered for RPA development, especially in terms of non-value process documentations. Because RPA performs the process using the user interface like a human user does, the process does not need to be defined from scratch for transforming user interactions into an RPA workflow definition. In this context, the basic idea of screen recording and TDD can be used to develop RPA processes in an agile and efficient way. For this, users will record activities of the process execution during their daily work, enrich the video with input and output values and store this in a video backlog. RPA developers pick backlog items, create a test cases based on the input and output values and check if the test case is already fulfilled. If the test case is not already fulfilled, RPA developers will automate the RPA process until the test case is covered by the process definition based on the video and the in- and output values. When the test case is fulfilled, developers move on to the next video. This process is iteratively repeated until all videos or test cases are covered by the process definition. The development methodology is currently at concept stage, implementing and evaluating it in a corporate setting is planned future work.

References

1. Ernst & Young: Get ready for robots: Why planning makes the difference between success and disappointment (2016). http://www.ey.com/Publication/vwLUAssets/Get_ready_for_robots/%24FILE/ey-get-ready-for-robots.pdf. Accessed 15 May 2017
2. Chappell, D.: Introducing Blue Prism: Automating business processes with presentation integration (2010). http://www.davidchappell.com/writing/white_papers/Introducing_Blue_Prism_v1.0-Chappell.pdf. Accessed 15 May 2017
3. Van der Aalst, W.M.P.: Business process management: a comprehensive survey. ISRN J. **2013**, 1–37 (2013)
4. Weske, M.: Business Process Management - Concepts, Languages, Architectures. Springer, Heidelberg (2012). https://doi.org/10.1007/978-3-642-28616-2
5. Willcocks, L.P., Lacity, M.C.: Service Automation: Robots and the Future of Work. Steve Brookes Publishing, Warwickshire (2016)
6. McKinsey and Company: The next acronym you need to know about: RPA (robotic process automation) (2016). http://www.mckinsey.com/business-functions/digital-mckinsey/our-insights/the-next-acronym-you-need-to-know-about-rpa. Accessed 15 May 2017
7. Forrester Research: The Role of IT in Business-Driven Process Automation (2011). https://www.blueprism.com/wpapers/forrester-report-role-business-driven-process-automation. Accessed 15 May 2017
8. Forrester Research: Building a Center of Expertise to Support Robotic Automation: Preparing for the Life Cycle of Business Change (2014). http://neoops.com/wp-content/uploads/2014/03/Forrester-RA-COE.pdf. Accessed 15 May 2017
9. Myers, G.J., Badgett, T., Sandler, C.: The Art of Software Testing. Wiley Association, New Jersey (2012)
10. Dustin, E., Rashka, J., Paul, J.: Automated Software Testing: Introduction, Management and Performance. Addison-Wesley, Boston (1999)
11. Lewis, W.E.: Software Testing and Continuous Quality Improvement, pp. 291–292. Auerbach Publications, Boca Raton (2009)
12. Beck, K.: Embracing change with extreme programming. IEEE Comput. **32**(10), 70–77 (1999)

13. Beck, K., Andres, C.: Extreme Programming Explained: Embrace Change. Addison-Wesley, Boston (2005)
14. Bhat, T., Nagappan, N., Maximilien, E., Williams, L.: Realizing quality improvement through test driven development: results and experiences of four industrial teams. Empir. Softw. Eng. J. **13**, 289–302 (2008)

Optimization of the Inception Deck Technique for Eliciting Requirements in SCRUM Through Business Process Models

Manuel Pastrana[1]([✉]), Hugo Ordóñez[1], Armando Ordonez[2], Lucinéia Heloisa Thom[3], and Luis Merchan[1]

[1] Research Laboratory in Development of Software Engineering, Universidad San Buenaventura, Cali, Colombia
{investigadorlidis03,haordonez,lerchan}@usbcali.edu.co
[2] Intelligent Management Systems, University Foundation of Popayán, Popayán, Colombia
jaordonez@unicauca.edu.co
[3] Institute of Informatics, Federal University of Rio Grande do Sul (UFRGS), Porto Alegre, Brazil
lucineia@inf.ufrgs.br

Abstract. Some techniques such as Inception Deck are used in software elicitation phase of the agile methodologies to unify the vision of all the stakeholders. This vision is stored in artifacts such as user stories. However, these artifacts are written in natural language and may, therefore, be ambiguous. Regarding SCRUM, the primary artifact is the product backlog, which in turn contains user stories. This paper describes how software development process with SCRUM can be improved by replacing the user stories with business process models to solve the ambiguity issue.

Keywords: Scrum · Inception · Business process model
Business process model and notation

1 Introduction

Agile methodologies have been in constant evolution to improve the software development [1]. This evolution has allowed identifying some success factors in the development of software such as the inclusion of the opinions of all the stakeholders, the constant communication, the collaborative work and the dynamics of retrospective [1]. One of these improvements is the inception deck technique, which was mentioned for the first time by Rasmusson [2] and detailed by Comba [3].

Inception Deck seeks to create a unified vision of the project from the consensus of the opinions of all involved: client, development team, sponsors and project managers. This unification is done in a meeting that can take from 4 h to 2 days. In this meeting it is established what is required to be done, the risks and strategies to mitigate these risks and the recommendations for the software architecture. The result of this exercise is a document called product backlog that includes the user stories.

© Springer International Publishing AG 2018
E. Teniente and M. Weidlich (Eds.): BPM 2017 Workshops, LNBIP 308, pp. 649–655, 2018.
https://doi.org/10.1007/978-3-319-74030-0_52

User stories are a very useful artifact, given its simplicity [4]. However, these stories are written in natural language, and their interpretation depends on who reads or writes the document. The above can lead to loss of information since some important details can be omitted while writing this artifact [5]. In this context, a strategy is required to reduce the ambiguity of these documents (user stories or use cases) [6].

This paper describes the use of inception deck technique as a requirement eliciting mechanism in SCRUM combined with business process models (BPs) instead of user stories.

The rest of the article is organized as follows: Sect. 2 presents an example of the requirements elicitation using inception deck. Section 3 describes the proposed strategy. Section 4 describes the estimation and planning of the sampling project. Section 5 evaluates the results obtained. Finally, Sect. 6 presents the conclusions and the future work.

2   Example of Requirements Elicitation Using the Inception Deck Technique in an Example Project

The implementation of the Inception Deck technique in the example project was based on the 10 dynamics described in [3]. The example project is called ENGINE. The objective of the application is to be an intermediary between software companies and potential clients. Next, the 10 steps are mentioned, but only relevant results will be shown.

Initial dynamic. Identify who is in the auditorium and define the rules

- *First dynamic: Why are we here?*
- *Second dynamic: Elevator Pitch*
- *Third dynamic: Design the product box*
- *Fourth dynamic: Generate the "no" list*
- *Fifth dynamic: Knowing the Neighborhood*
- *Sixth dynamic: What prevents us from sleeping at night?.* This dynamic focuses on identifying potential project risks, the strategies to solve them and the responsible PO (Product Owner), ST (Scrum Team) or SM (Scrum Master). (see Table 1).

Table 1. Project risks, Risk (R), Responsible (Rs), Action (A).

R	Rs	A
Commercial strategy still unclear	PO	Define commercial strategy in detail
The project must be finished in 5 months. Project Costs	ST	Planning with precise dates and follow through daily report
The team is not experienced with some components of the bootstrap graphics	ST	Perform concept tests during sprint 0

- *Seventh dynamic: Show the solution:* The wireframing option was selected for this project. The objective is to sketch the system screens, the flow of interactions, the allowed actions, and what aspects are relevant for the proper functioning of the system. Figure 1 shows on of the prototyping results.

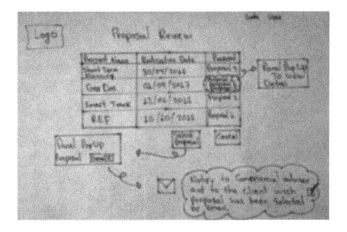

Fig. 1. Offers review form (Engine project)

- *Eighth dynamic: What should I give?*
- *Ninth dynamic: How much will it cost?*
- *Tenth dynamic: Summary*

3   Proposed Strategy

Requirements elicitation is performed through questions. The answers are recorded and in artifacts. These artifacts are then interpreted in different ways depending on who reads the document as these artifacts are written in natural language.

Some approaches have combined traditional elicitation with business process models (BP) to improve the requirements elicitation. These works [7, 8] have evidenced that a software analysis process can be improved when the final result of the elicitation is represented using BP, since the representation and the interpretation of the information are unique, thus eliminating the ambiguity of the natural language. Nevertheless, these works do not consider techniques like inception deck in SCRUM.

The purpose of this paper is to show the advantages of using inception deck in conjunction with BP within a project developed using SCRUM [9].

To apply the present approach, it is necessary to define the option to be used during the inception Deck (7th Dynamic "show the solution") [3]. Once the option has been defined, the relationship between this representation and BPMN is established. In the case of SCRUM, each component of user stories must be associated to BPMN. A user story (defined by [4]) may have the following structure: I as [Role] + I require [a functionality] + For [a reason] + Acceptance criteria. The relationship between the components of the user stories and the BP is detailed in Table 2.

Table 2. Comparison between user stories and BP

User stories		BPMN	
I as [role]	Who executes the actions in the system [4]	User	Determined by the lines of division of activities performed by a role within a process (Lanes)
I need [functionality]	The required functionality [4]	User Task Service Task Receive Task Send Task Scritp Task Manual Task Business rule Task	They are functionalities that must be done, taking into account the semantics that given by the type of element
For [reason]	Reason and importance for the business [4]	Does not apply	*For [reason]*
Acceptance criteria	The required elements of the system [4]		Business rules

For explanatory purposes, the approach is described in the example project (Engine). The Engine platform connects customers and suppliers of software solutions. The three primary processes of the platform are (I) publish a customer need to receive offers from suppliers. (II) Reviewing the customer's need by the commercial adviser who matches the suppliers. (III) Send a proposal to the supplier.

For this project, the wireframing technique was used instead of story mapping.

In Wireframing a basic prototype that shows the screens, roles, and functionalities are built. After the inception deck meeting, the BP is designed by the team (Fig. 2 shows

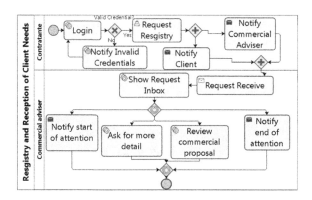

Fig. 2. BP Registry and Reception of customer requests (Engine project)

one of the processes generated) and is exposed to the client to start the next stages of development.

4   Estimation and Planning

Once the BPs are generated, the decomposition of the BP task into development task is done in the sprint planning. BP allows visualizing the functionalities to be performed, the interacting roles, the dependencies and the business rules. For explanatory purposes, the decomposition of the activities for the login task exposed on Fig. 2 is shown in Table 3. The same decomposition was performed for each of the other tasks. The latter does not affect the scrum poker estimation since the teams estimate the development task. This technique presents an improvement in the representation of the information for the estimation.

Table 3. Decomposition of development activities for the login task

Task		Development task
Login	1.	The screen must include a field to Enter the User Name
		1.1. The screen must include a field to Enter the Password
	2.	The password field information must be visible only with the character ($\bullet$)
	3.	The screen must have a button to access the system, once the user presses this button the system must perform the following validations:
		3.1. The User field must be filled in (Cannot be empty). In case the system is not fulfilled the following message must be generated (connection was not successful. Please enter the user)
		3.2. The Password field must be filled in (Cannot be empty). In case the system is not fulfilled, the following message must be generated (connection was not successful. Please enter the password)
		3.3. The User and Password must correspond to a record in the DB. Otherwise, the system must generate the following message (connection was not successful, User or Password invalid)

5   Results

Table 4 shows the results of the application of the approach in 4 projects. The table describes the number of requirements elicited, modified and added and no identified per project. From the Table 4, it can be clearly seen that the number of refined requirements for 4 projects is zero and the only project that had refined requirements was GesDos (with 10 refined requirements). This situation is because GesDos was not a development project from scratch like the other project, but a migration project from a desktop to a Web environment. In the desktop version of GesDos, some outdated processes were detected using inception deck + BP

Table 4. Requirements elicited, modified and added and no identified per project

Requirements	Smart Track	GesDos	Teachers ranking	REF	Engine
Elicited	25	130	41	13	20
Refined	0	10	0	0	0
Added	1	20	0	2	0
No identified	0	0	0	0	0
Total	26	160	41	15	20

These improvements in the processes are reflected in 20 new requirements to optimize the operation of the company and 10 requirements outdated in the current desktop application (these outdated processes modified to align them with the new processes of the company). It is also evident that the number of "no identified requirements" is 0, which demonstrates the effectiveness of the technique to collect the information.

6   Conclusions

The inception deck technique allows generating a unified vision of the project. For its part, the BP are an ideal complement to keep the information collected by Inception deck, since they offer an artifact with a unique interpretation. The latter increases the possibility identifying requirements and equally the number of requirements modified or added after the elicitation is decreased. The present approach allows to quickly identifying if the project processes are outdated or some improvements are required. On the other hand, the planning and estimation in scrum, which is carried out with the scrum poker method, does not require any drastic change. The only difference is that user stories are no longer considered but development activities are included. Thus scrum poker does not need to be replaced by another technique. Future work includes the analysis of the impact of using BP in the estimation using the poker scrum technique.

References

1. Hastie, S., Wojewoda, S.: Standish Group 2015 Chaos Report - Q&A with Jennifer Lynch (2015). http://www.infoq.com/articles/standish-chaos-2015
2. Rasmusson, J.: Agile project initiation techniques - the inception deck & boot camp. In: Proceedings - AGILE Conference, 2006, vol. 2006, pp. 337–341 (2006)
3. Enrique Comba Riepenhausen: Inception starting a new project. Madrid, España (2014)
4. Cohn, M.: User Stories Applied: For Agile Software Development. Addison Wesley Signature Series, vol. 1, no. 0 (2004)
5. Wautelet, Y., Heng, S., Kiv, S., Kolp, M.: User-story driven development of multi-agent systems: a process fragment for agile methods. Comput. Lang. Syst. Struct. (2017)
6. Rola, P., Kuchta, D., Kopczyk, D.: Conceptual model of working space for Agile (Scrum) project team. J. Syst. Softw. **118**, 49–63 (2016)
7. Decreus, K., El Kharbili, M., Poels, G., Pulvermueller, E.: Bridging requirements engineering and business process management. In: Gi-Edition. Proceedings-Lecture Notes in Informatics, p. 215 (2009)

8. Decreus, K., Poels, G., El Kharbili, M., Pulvermueller, E.: Policy-enabled goal-oriented requirements engineering for semantic business process management. Int. J. Intell. Syst. **25**(8), 784–812 (2010)
9. Heikkilä, V.T., Paasivaara, M., Rautiainen, K., Lassenius, C., Toivola, T., Järvinen, J.: Operational release planning in large-scale Scrum with multiple stakeholders – a longitudinal case study at F-Secure Corporation. Inf. Softw. Technol. **57**, 116–140 (2015)

Successful Post Merger Process Harmonization in the Triangle of Methodologies, Capabilities and Acceptance

Irene M. Schönreiter[✉]

Technical University Dresden, Dresden, Germany
Irene@schoenreiter.de

Abstract. Due to an increasing number of merger and acquisitions (M&As) and the high failure rate post merger integration (PMI) is still of growing interest. An effective process harmonization (PH) is essential during the PMI. The objective of this article is to identify crucial requirements for a successful post merger PH in the triangle of methodologies, capabilities and acceptance. The study assessed methodologies and capabilities during the various PH phases and the attainment of employee acceptance. 12 hypotheses affecting the three fields were developed and tested. A number of central results can be provided: there is no statistically significant evidence for a methodology to be applied, whereas a number of beneficial capabilities during the PH phases were identified. Conducing issues for employee acceptance especially in the areas structure, information policy and team composition are presented.

Keywords: Process harmonization · Post merger integration · Phase model

1   Introduction

Companies can widen their competitive advantage and position by combining the strengths of two companies to extend growth and business agility, so numerous mergers and acquisitions (M&A) are taking place [1]. Business Process Management (BPM) plays a vital role in organizational changes [2, 3], hence process harmonization (PH) is a key element in post merger integration (PMI). Especially when symbiosis or absorption is the company's chosen integration approach, two worlds of business processes and management systems have to be harmonized efficiently in a rapidly changing process activities environment. However value creation after M&A is rather scarce [4] and companies often fail to achieve performance goals [1]. After an M&A common processes are of prime importance as knowledge sharing occurs via a set of structures, people, processes and systems [5].

This study investigates potential and distinguished methodologies for a successful post merger PH, however the result demonstrates clearly the demand for developing a new one. Besides a methodology the capabilities for the application of a methodology and the enforcement of PH in the PMI are required [6, 7]. To set up a PMI project without the control of qualified persons and proper capabilities may lead to failures and

© Springer International Publishing AG 2018
E. Teniente and M. Weidlich (Eds.): BPM 2017 Workshops, LNBIP 308, pp. 656–668, 2018.
https://doi.org/10.1007/978-3-319-74030-0_53

unsuccessful PH. As the PMI is a change situation, it has to be handled with sensitivity regarding the employee acceptance and the organizational responsiveness since a change always affects both company level and the individual level of employees [8]. To minimize any adverse effects of changes made, the involvement and consultation of affected employees to trigger a willingness to change and achieve acceptance is essential [9]. So acceptance is identified as the third field in the PH triangle.

Figure 1 illustrates the dependencies of a methodology, the desired capabilities and employee acceptance. A methodology needs suitable capabilities to be applied, but in the end each PH will collapse if the acceptance for the changes is not assured. So a successful process harmonization needs all three elements, two elements might appear incident without the third one. The triangle is regarded predominantly of a managerial focus on the strength of intended success of the process harmonization project. Success means the unification of the double or multi available processes of the merged entities into a common management system that is accepted and active. This paper aims to provide comprehensive results for any domain with a documented process management system that facilitate decision making concerning PH in a merged company.

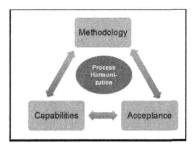

Fig. 1. Process harmonization triangle

This leads to following research question (RQ): *Which methodologies, capabilities and conditions for acceptance are required for a successful post merger PH?*

The paper is structured as follows. After this introduction the theoretical background is explained followed by the development of hypotheses for the triangle. Section 4 describes the research method and Sect. 5 presents the results followed by a discussion. A summary, limitations and ideas for future research close this article.

2   Theoretical Background

Business process management covers a wide range of activities within a company [10] and provides concepts, methods, techniques and tools for the whole process lifecycle [11]. Since BPM is closely linked to quality management and management systems [12], PH considers both single processes and the management system as a whole. PH is defined as "*process of designing and implementing business process standardization across different regions or units so as to facilitate achievement of the targeted business benefits arising out of standardization whilst ensuring a harmonious acceptance of the new processes by different stakeholders"* [13] without trying to unify processes when they are different.

PMI describes the integration planning and implementation after the signing of an M&A in order to realize the desired appreciation successfully [14]. PH is activated most by the integration approaches absorption (acquired company is absorbed by acquirer) and symbiosis (evolution from existing), whereas in preservation and holding approach PH scarcely occurs. Aside from post merger IT integration there is no identified methodology to harmonize processes in the post merger situation [15], although a rigorous methodology is needed. However a PH phase model defining the milestones of PH has been developed [16]. The PH phases exist of analysis phase, conception phase, realization phase and verification phase (see Fig. 2). In respect to different responsibilities and execution levels in a company, the model is divided into management system (MS) level and process level. In the analysis phase a maturity assessment on MS level and process analysis on process level is recommended, followed by strategic alignment and process optimization in the conception phase. The realization phase describes the implementation of the concept. Finally the PH attainment is evaluated in the verification phase. The PH phase model sets the basis of the hypothesis tests in the fields methodologies and capabilities.

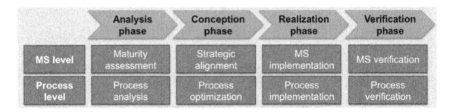

Fig. 2. Process harmonization phase model (MS – management system level)

3   Hypothesis Development

In this section the hypotheses for a successful PH are developed and sub classified in the fields methodology, capabilities and acceptance. The corresponding dependent and independent variables are shortly presented and concretized in Sect. 4.

Methodology:
As for PH no methodology exist [15], a set of methodologies adapted for each PH phase is assumed to provide a good solution. A questionnaire was set up and the respondents were asked for the applied methodology in their PH experience according to the phases of the PH phase model. The methodologies inquired by default were a mix of common and established tools and techniques and the result of previous executed expert interviews. Balanced Scorecard, Six Sigma and Benchmarking are quite popular, so the methodologies could be chosen as options in the questionnaire. Further to investigate the "better" management system maturity models were added: EFQM and other self-assessment models. For evaluating intermediate results in a fast and simple way and

work out future strategies, the SWOT analysis was adjoined. Process changes and optimizations are generally proceeded evolutionary or revolutionary. The evolutionary change corresponds to the process of continuous improvement in which the corporate structure and strategy is maintained [17]. For the revolutionary approach the Business Process Reengineering (BPR) was identified. Process audits are an instrument for verifying the PH success [16]. To involve customers in customer related processes, Service Blueprint was included as well. Quality Function Deployment (QFD) was identified due to its strong process focus and orientation on requirements. It is assumed the application of a special methodology in a particular PH phase leads to a significant better result. So the first hypothesis states:

H1: The results in the particular phases are better if a certain methodology in the phase has been applied.

Capabilities:
A perfectly executed methodology not sufficiently managed cannot lead to success. So capabilities and the resources for managing the PH project are needed. In the following the hypotheses for required capabilities are developed. The hypotheses H2–H7 follow the PH phase model.

Analysis phase: It is imperative to observe the maturity of business processes before any re-designing or optimizations to a higher level are started [18]. A maturity model enables the identification of weak points and unqualified developments [3]. So on management system level a maturity assessment is constituted for an overall analysis, whereas on process level a systematic process analysis is advised [16]. The hypothesis indicates:

H2: Using process analysis and maturity level identification leads to a better result in the analysis phase.

The results of the analysis phase give valuable input for the conception phase. Distinguished strengths and weaknesses are considered in the PH concept.

Conception phase: During the conception phase all planning takes place, responsibilities and actions are defined. The process design is an important process enabler [18]. The unification of processes does not mean that all processes must be unified and equal - local and individual solutions are allowed [7]. Companies with more harmonized processes have more standardized processes, synchronistic organizations that have less complex processes have more harmonized processes [19]. Companies should regard the complexity of their processes as a key factor and separate routine and non-routine processes to avoid useless overstandardization [20]. Especially in service companies selective flexibility is needed and heterogeneity in processes are a signal for high quality - in contrary to traditional manufacturing companies where process follow the one best way [21]. Very complex business processes can not be standardized independent from the invested effort [10]. Allowing for complex processes or knowledge intensive processes, individual solutions are of convenience. Thus the hypothesis states:

H3: Organizations that allow a mixture of standardized and individual processes achieve better results in the conception phase.

Contemporary processes demand supportive IT tools. The technical design of a business process and the information technology used for the execution of a process impact PH [19]. IT intensity in business processes contributes to process performance in terms of time, quality and efficiency and simplifies process standardization [22]. Jurisch et al. (2013) confirm in their survey the positive effect of IT capabilities on the final process performance and process change [23]. With IT influence processes are modernized e.g. with implemented workflows and automated procedures. This leads to following hypothesis:

H4: The modernization of processes lead to better results in the conception phase.

Realization phase: The chronological introduction of the adapted processes and management system has to be decided. The longer taking step-by-step implementation with lower risks and reduced complexity or a pilot-operated launch in one department after another with additional administrative expenses through parallel worlds opposes the Big-bang introduction on "Day X" with high speed, but the risk of a high failure rate [24]. Although the degree of integration and the speed of integration are not related [25], all three opportunities are accompanied with advantages and disadvantages. It is assumed that the success of PH correlate to the chosen introduction approach.

H5: The chronological introduction correlates with the results in the realization phase.

Besides the chronological introduction a clear structure and strong project management in a systematic way contributes to a successful PMI [26]. So H6 states:

H6: A clear defined implementation plan leads to better results in the realization phase.

Verification phase: In the verification phase the success of the PH is evaluated. Saarti et al. (2011) describe the merger of Finnish universities and their libraries using the ISO 9001 quality management system. The ISO 9001 requirements were used as a tool for the transition, amongst others because of the set schedules and required documentation [27]. The successful implementation of a common management system according to any common norm contains the most important and usual requirements and an affiliated certification evidences PH success. A certification body as an independent organization assures objective assessment:

H7: A common certification leads to better results in the verification phase.

Acceptance:

Previous chapters regarded PH on a factual way. But subjective issues for convincing the employees and achieve stakeholder support is the element deciding about the success or fail of the PH. The employee acceptance and their performance as a (new) team is of vital significance. Employees often invested years and hard work into their status, know-how, privileges and habits as well as their roles and task definitions that are with the M&A changes in danger of being lost combined with issues like anxiety, blockades, increased illness times, job security uncertainty, fluctuation or (passive) resistance [28, 29]. As a result valuable time investments and efforts can be lost. Power struggle and selfish behavior leave no room for strategic consideration [30]. PH is affected negatively when power struggles and conflicts of interests arising of the own position versus the best for company are primary motivations of key personnel. This leads to following hypotheses:

H8: Power struggles of single employees lead to a worse employee acceptance.

H9: Conflicts of interests lead to a worse employee acceptance.

To affect in opposition to this appearances preventive measures have to be taken. Communication is the key PMI success factor. Comprehensive communication and information for employees giving them the opportunity to express their concerns and ask questions is an essential prerequisite for a successful change [30]. Involving the entire staff combined with a honest communication is of vital importance [7]. On the one hand regular information is required for employee involvement, on the other hand the quality of the information is crucial. This leads to following hypotheses:

H10: Regular information about the PH progress increase employee acceptance.

H11: A systematic information policy leads to a higher employee acceptance.

Maire and Collerette (2011) identified the social factor as one of the M&A best practices. Bringing people together and pick out organizational, social and cultural differences as central topics helps to integrate the different company cultures [26]. Although knowledge sharing is difficult in merged companies due to a newly formed context, a personal interchange for the tacit dimension of knowledge sharing is a worthy strategy [28]. In conclusion the PH project teams should be composed of employees of both merged company parts.

H12: Composing the project team with employees of both merged companies leads to a higher employee acceptance.

4 Research Methodology

A questionnaire existing of 23 questions was developed and open to be filled in a period of three months in the tool SurveyMonkey. The questionnaire design was based on a previous qualitative analysis of a series of expert interviews. The experts were interviewed in depth about their experience in PH involved. With a valid sample representativeness and generalizability of the theses is achieved, therefore the hypotheses obtained can be validated. Derived from these results the questionnaire was composed and pre-tested several times with various managers involved in a post merger PH. The responses were received from Germany, Austria and Switzerland. In respect to an indeterminate basic population the sample size cannot be specified. There were five sections in the questionnaire:

- Section 1 (Questions 1–6): background of the company
- Section 2 (Questions 7–10): integration approach
- Section 3 (Questions 11–17): independent variables -> information about procedure of PH (methodologies in the single phases with a multi-item scale, time schedule, statements to be assessed in a Likert scale 1–5)
- Section 4 (Questions 18–21): dependent variables -> results (statements to be assessed in a Likert scale 1–5), measuring the individual success of PH (success is derived of a 4 or 5 assessment by the respondent)
- Section 5 (Questions 22–23): demographic response, background information

A neutral answer option in the Sections 3 and 4 "do not know" is offered to the participant to avoid distorting the results of the survey by selecting an arbitrary random option. In total, 91 responses were received. The respondents have following positions: Quality/process management (17%), executive board and management (52%), external consultant (17%), employees (14%). The branch of the respondents is divided in service companies (35%), production (24%), Transport, communication, energy (8%), whole sale (7%), others (26%).

Due to the high relevance of absorption and symbiosis integration approach for PH only respective responses were considered. In total 61 responses are integrated in the analysis, all other cases concerning preservation and holding had been excluded.

All data analysis was conducted with the IBM SPSS Statistics program. To check representativeness of the samples, 25% of the responses were compared with the result of all responses with no significant differences between the groups. For testing the hypothesis the regression analysis was conducted to investigate the coherence between variables. For the assessment of the predictive power of the hypotheses each result is presented with the R^2-value and F-value, as well as the regression coefficient b and beta value. The p-value describes the significance.

5   Results

Regression analysis for Methodology

H1 assumes the application of a certain methodology correlates with the PH results in a particular phase. The questionnaire respondents were requested to assign in each phase one or more methodologies used in their PH project. Table 1 shows the result of the regression analysis allocated to the single phases.

Table 1. Regression analysis methodologies (*partly excluded variables)

Phase	Analysis phase -> maturity assessment MS level R^2= .601; F=1.076			Analysis phase -> process analysis on process level R^2= .852; F=1.912			Conception phase -> process optimization R^2= .656; F= .955			Implementation phase -> process harmonization R^2= .564; F= .646		
Dependent variable	Results analysis phase			Results analysis phase			Results conception phase			Results implementation phase		
Predictor variable	**B**	**beta**	**Sig. (p)**	**B**	**beta**	**Sig. (p)**	**B**	**beta**	**Sig. (p)**	**B**	**beta**	**Sig. (p)**
(constant):			*.485*			*.323*			*.559*			*.749*
Benchmarking	1.813	.666	.748	-.500	-.212	.624	-.727	-.237	.661	-.189	-.077	.881
SWOT *				-.821	-.302	.619	1727	.563	.401	-.892	-.362	.628
EFQM *				-.179	-.042	.912						
Self-assessment	-2.000	-.735	.280	-.071	-.022	.958	-2.455	-.877	.186	-.378	-.097	.881
Process audit	.813	.256	.713	-2464	-.992	.114	.455	.148	.766	.297	.121	.885
Six Sigma *	1.625	.378	.615									
BPR	-1.187	-.374	.393	-1.643	-.382	.458	-3.455	-.713	.234	.703	.313	.734
Continuous Improvement *				.571	.210	.737	-2000	-.652	.199	-1486	-.661	.513
Balanced Scorecard	-.125	-.046	.906	.607	.223	.586	.727	.237	.511	-.189	-.084	.881
Service Blueprint *												
QFD	.875	.322	.426	-.393	-.124	.720	2636	.736	.206	2784	.968	.146

Surprisingly the results show that none of the methodologies is statistically significant. The beta values seem valid as some of the methodologies are not suitable for each phase. But regarding the significance ($p > 0.05$) a conclusion to a general valid prediction is impossible. No methodology in any phase is positively associated with PH, so all of these established methodologies are rejected. Each variable has a strong R^2, which means these results are based on a sufficient amount of cases. As R^2 have to be adjusted by the researcher [31], it is interpreted that all R^2 values are strong and confirm the conclusion: H1 is explicitly rejected.

Regression analysis and hypotheses testing for capabilities

Table 2 shows the results for capabilities structured due to the hypotheses H2–H7.

Table 2. Regression analysis of capabilities hypotheses

Hypotheses	Dependent variable	Predictor variable	R square	F	B	beta	Sig.
H2	Results analysis phase	Process analysis, maturity level	.186	8.901	.453	.431	.005
H3	Results conception phase	Standardized and individual processes	.153	6.483	.362	.391	.015
H4		Modernization of processes	.179	8.075	.575	.423	.007
H5	Results realization phase	Chronological introduction	.047	1.904	−.193	−.216	.175
H6		Clear defined implementation plan	.042	1.689	.209	.204	.201
H7	Results verification phase	Common certification	.115	5.068	.281	.339	.030

Analysis phase: H2 states that using process analysis and the identification of the maturity level leads to better results in the analysis phase. The data show strong support for H2. Process analysis and maturity level assessment statistically significant predict PH satisfaction (F = 8.901, p < 0.01), the standardized coefficient beta is .431, the fit of the model R2 is .186. It can be concluded that H2 is empirically fully supported as predicted.

Conception phase: H3 votes for better PH results if specific processes are consciously not standardized and individual solutions are allowed. The results statistically significant predict PH satisfaction (F = 6.483, p < 0.05), the standardized coefficient beta is .391, the fit of the model R^2 is .153. It can be concluded that H3 is empirically fully supported as predicted. This result confirms previous studies as already referenced in Sect. 3.

H4 states that PH results are evaluated superior when processes are modernized with more automatism, workflows etc. and are closer to applied IT tools. The data affirm the prediction (F = 8.075, p < 0.01), the standardized coefficient beta is .575, the fit of the model R^2 is .179. H4 is empirically fully confirmed.

Realization phase: H5 suggests a direct link between PH results in the realization phase and the chronological introduction (step-by-step, big-bang, pilot introduction), whereas the results negate this (F = 1.904, p > 0.05). The standardized coefficient beta is −.193, the fit of the model R^2 is .047. The standardized beta value revealed by the result is negative indicating that the chronological introduction does not allow any conclusions the realization phase success, though this association is not significant at p > 0.05. So H5 is not supported by the regression result, in contrary to prediction H5 is rejected.

A clear structured implementation plan with defined actions leads to better PH results is predicted in H6. The results demonstrate that the implementation plan does not influence the realization phase significantly (F = 1.689, p > 0.05), the standardized coefficient beta is .204, the fit of the model R^2 is .042. The prediction H6 is empirically not facilitated.

Verification phase: H7 proposes that a common certification of the entire management system leads to better PH results in the verification phase. The results confirm this relationship (F = 5.068, p < 0.05), the standardized coefficient beta is .339, the fit of the model R^2 is .115. It can be concluded that H7 is empirically fully endorsed as predicted.

Regression analysis and hypotheses testing for acceptance

Table 3 presents the results for acceptance hypotheses H8 to H12.

Table 3. Regression analysis of acceptance hypotheses

Hypotheses	Dependent variable	Predictor variable	R square	F	B	beta	Sig.
H8	Employee acceptance	Power struggles	.191	8.969	−.899	−.437	.005
H9		Conflicts of interests	.156	7.002	−.881	−.394	.012
H10		Regular information	.384	24.269	.537	.619	.000
H11		Systematic information policy	.212	10.515	.455	.461	.002
H12		Project team composition	.048	1.970	.216	.219	.168

H8 and H9 predict a low employee acceptance if PH has been influenced by power struggles of single employees or conflicts of interests affecting the own position. Both hypotheses H8 (F = 8.969, p < 0.01) and H9 (F = 7.002, p < 0.05) are statistically significant confirmed. The standardized coefficient beta in H8 is −.899 and in H9 −.394, the fit of the model R2 in H8 is .191, respectively .156 in H9. The standardized beta values evidence with the negative results that power plays and conflicts of interests lead to an incident harmonization process and thus to the risk of employee disacceptance.

H10 and H11 state that employee acceptance increases if they are regularly informed within a systematic information policy. The results show that a regular and systematic information policy significantly predict employee acceptance (H10: F = 24.269, p < 0.001; H11: F = 10.515, p < 0.01), the standardized coefficient beta is .619 in H10 and .461 in H11, the fit of the model R^2 is .384 in H10, respectively .212 in H11. Both hypotheses are empirically fully confirmed as predicted.

H12 predicts higher employee acceptance if project teams are composed of both companies. The results demonstrate that the composition of company mixed teams is not statistically significant to employee acceptance (F = 1.970, p > 0.05), the standardized coefficient beta is .219, the fit of the model R^2 is .048. H12 is rejected.

6   Discussion

A series of hypotheses have been developed and analyzed to reach knowledge about suitable methodologies, capabilities and factors to achieve employee acceptance. The results are among the first to systematically and empirically show the impact on PH capabilities and employee acceptance.

The first hypothesis expected a better result in a particular PH phase by applying a certain methodology. A clear statement could not be demonstrated. This confirms the results of a literature review surveying the interface of quality/process management, change management and M&A for finding an adequate methodology, an interface scarcely covered in the literature [32]. The prevenient qualitative study conducting expert interviews affirms the result as well. The experts had mentioned isolated methodologies, but especially in the analysis phase most experts described the comparability of the double existing processes and management systems as self-evident with the use of "horse-sense". The results of the quantitative analysis did not even demonstrate any tendency. One of the reasons is potentially caused in the fact of inadequate knowledge of methodologies. Another reason might be determined in the quality management nature of some methodologies that are supposed to be implemented for process optimization or process redesign. Obviously these same methodologies reach not the intended performance in another context or were simply not deployed.

The analysis of the capabilities hypotheses identified a mixture of expected and surprising results. To the effect that a beneficial harmonization allows a mixture of individualization and standardization confirms previous studies [10, 19–21].

A clearly negative impact on PH is determined in power struggles and conflicts of interests. This raises the question of the sequence of the PMI in the entire organization aside from PH. Right after an M&A most positions are duplicated and a part of them is redundant when combining departments. An early and clear solution of redundant positions with definition of a structure avoids negative effects.

As a managerial implication, managers can use the results as starting point for PH projects. Although a certain methodology cannot be recommended, the results indicate premises of a resource-based view on team composition, planning and communication strategies.

The scientific value of the study lies in widening the interface of the fields BPM, M&A and change management and filling a gap in the existing literature. Further promising surveys can be executed constitutively.

7   Conclusion and Limitations

The aim of this study was to find answers in decision-making and the definition of requirements which aspects are supportive to a successful post merger PH. The study assessed methodologies and capabilities during the single PH phases and the attainment of employee acceptance. 12 hypotheses referring to the triangle were developed and evaluated.

The study makes a number of valuable contributions. First, there is no statistically significant evidence for any established methodology to be applied in a certain PH phase. Second, general capabilities during the PH phases were identified principally in the fields of process analysis, appliance of a maturity model, flexibility in PH, modernization of processes and verification with a common certification. Third, for acceptance some issues have been investigated: before starting any PH a defined structure accounts for PH success besides a regular and systematic information policy. Merged organizations might profit enormous of the identified success factors in the PH triangle based on the experience of numerous PH participants in the survey. The current poor PMI success rate could improve.

While the findings of this study are expected to have important practical and theoretical implications, some limitations must be acknowledged. The first limitation refers to the focus of the analysis on the main methodologies and capabilities detected in PH related literature. Additional contextual factors like company size, number of actors involved and level of automation depend on company and branch are not part of this paper. Further research should analyze the effect of additional aspects on PH success. Second, the R^2-values in testing the capabilities and acceptance hypothesis are predominantly low compared with the good results of the R^2-value in testing the methodology hypotheses. For better R^2-values a greater sample size might lead to more explicit values. Third, reactivity results like acquiescence and social desirability response set might adulterate the responses, but were reduced through anonymous participation.

For future research this study implies following suggestion: A precise methodology tailored exactly on PH in the PMI phase is needed with the focus on the input of the detected capabilities and contributing factors for employee acceptance. Further deepened empirical research and explorative survey with the usage of case studies conveying additional contextual factors might be of value.

References

1. Lohrke, F.T., Frownfelter-Lohrke, C., Ketchen, D.J.: The role of information technology systems in the performance of mergers and acquisitions. Bus. Horiz. **59**, 7–12 (2016)
2. Rohloff, M.: Advances in business process management implementation based on a maturity assessment and best practice exchange. Inf. Syst. E-bus. Manag. **9**, 383–403 (2011)
3. Jochem, R., Geers, D., Heinze, P.: Maturity measurement of knowledge-intensive business processes. TQM J. **23**, 377–387 (2011)
4. Toppenberg, G., Henningsson, S.: Taking stock and looking forward: a scientometric analysis of IS/IT integration challenges in mergers. In: ECIS 2014 Proceedings, pp. 0–16 (2014)
5. Garud, R., Kumaraswamy, A.: Vicious and virtuous circles in the management of knowledge: the case of Infosys technologies. MIS Q. **29**, 9–34 (2005)
6. Ruess, M., Voelpel, S.C.: The PMI scorecard: a tool for successfully balancing the post-merger integration process. Organ. Dyn. **41**, 78–84 (2012)
7. Osarenkhoe, A., Hyder, A.: Marriage for better or for worse? Towards an analytical framework to manage post-merger integration process. Bus. Process Manag. J. **21**, 857–887 (2015)
8. Almaraz, J.: Quality management and the process of change. J. Organ. Chang. Manag. **7**, 6–14 (1994)

9. Vedenik, G., Leber, M.: Change management with the aid of a generic model for restructuring business processes. Int. J. Simul. Model. **14**, 584–595 (2015)
10. Schäfermeyer, M., Rosenkranz, C., Holten, R.: The impact of business process complexity on business process standardization: An empirical study. Bus. Inf. Syst. Eng. **4**, 261–270 (2012)
11. Dumas, M., La Rosa, M., Mendling, J., Reijers, H.: Fundamentals of Business Process Management. Springer, Heidelberg (2013)
12. Møller, C., Maack, C.J., Tan, R.D.: What is business process management: a two stage literature review of an emerging field. In: Xu, L.D., Tjoa, A.M., Chaudhry, S.S. (eds.) Research and Practical Issues of Enterprise Information Systems II. ITIFIP, vol. 254, pp. 19–31. Springer, Boston (2007). https://doi.org/10.1007/978-0-387-75902-9_3
13. Fernandez, J., Bhat, J.: Addressing the complexities of global process harmonization. In: Handbook of Research on Complex Dynamic Process Management: Techniques for Adaptability in Turbulent Environments: Techniques for Adaptability in Turbulent Environments, pp. 368–385. IGI Global (2010)
14. Müller-Stewens, G.: Mergers & Acquisitions Analysen, Trends und Best Practices. Schäffer-Poeschel, Stuttgart (2010)
15. Schönreiter, I.: Bedarfe zur Prozessharmonisierung in fusionierten Dienstleistungs unternehmen im Zeitalter Quality 4.0. In: Winzer, P. (ed.) Herausforderungen der Digitalisierung, pp. 35–49. Shaker, Aachen (2016)
16. Schönreiter, I.: Process harmonization phase model in post merger integration. In: Proceedings of the 6th International Symposium on Data-driven Process Discovery and Analysis (SIMPDA 2016), pp. 3–22, Graz, Austria (2016)
17. Jones, G.R.: Organizational Theory, Design, and Change. Pearson, Boston (2012)
18. Hammer, M.: The Process Audit. Harvard Business Review, pp. 1–15, April 2007
19. Romero, H.L., Dijkman, R.M., Grefena, P.W.P.J., van Weele, A.J., de Jong, A.: Measures of process harmonization. Inf. Softw. Technol. **63**, 31–43 (2015)
20. Schäfermeyer, M., Rosenkranz, C.: "To standardize or not to standardize?" - Understanding the effect of business process complexity on business process standardization. In: ECIS 2011 Proceedings (2011)
21. Wimble, M., Pentland, B., Hillison, D., Tripp, J.: Want pudding? An analytic model of the benefits and constraints of process standardization in services. In: ICIS 2010 Proceedings, Paper 170 (2010)
22. Beimborn, D., Joachim, N., Gleisner, F., Hackethal, A.: The role of process standardization in achieving IT business value. In: Proceedings of 42nd Annual Hawaii International Conference System Science, HICSS, pp. 1–10 (2009)
23. Jurisch, M., Palka, W.: Which capabilities matter for successful business process change? Bus. Proc. Manage. J. **20**, 47–67 (2014)
24. Hansmann, H., Laske, M., Redmer, L.: Einführung der Prozesse - Prozess-Roll-out. In: Becker, J., Kugeler, M., Rosemann, M. (eds.) Prozessmanagement: ein Leitfaden zur prozessorientierten Organisationsgestaltung, pp. 269–298. Springer, Berlin (2005)
25. Bauer, F., Matzler, K.: Antecedents of M&A success: the role of strategic complementarity, cultural fit, and degree and speed of integration. Strateg. Manag. J. **35**, 269–291 (2014)
26. Maire, S., Collerette, P.: International post-merger integration:l from an integration project in the private banking sector. Int. J. Proj. Manag. **29**, 279–294 (2011)
27. Saarti, J., Juntunen, A.: The benefits of a quality management system: The case of the merger of two universities and their libraries. Libr. Manag. **32**, 183–190 (2011)
28. Yoo, Y., Lyytinen, K., Heo, D.: Closing the gap: towards a process model of post-merger knowledge sharing. Inf. Syst. J. **17**, 321–347 (2007)

29. Binner, H.F.: Pragmatisches Changemanagement - Umsetzung über das MITO-Modell. Heider, Bergisch Gladbach (2010)
30. Doppler, K., Lauterburg, C.: Change Management den Unternehmenswandel gestalten. Campus-Verl, Frankfurt am Main (2008)
31. Cohen, J., Cohen, J.: Applied Multiple Regression/Correlation Analysis for the Behavioral Sciences. L. Erlbaum Associates, New York (2003)
32. Schönreiter, I.: Methodologies for process harmonization in the post merger integration phase - a literature review (2017, accepted)

A Reflection on the Interrelations Between Business Process Management and Requirements Engineering with an Agility Perspective

Banu Aysolmaz[1], Mehmet Gürsul[2], Kathrin Kirchner[3], Ralf Laue[4], Robert Mertens[5], Felix Reher[6], Irene M. Schönreiter[7], Bernhard M. Turban[8], and Rüdiger Weißbach[9(✉)]

[1] School of Business and Economics, Maastricht University, Maastricht, The Netherlands
b.aysolmaz@maastrichtuniversity.nl
[2] STM Defence Technologies Engineering and Trade Inc., Ankara, Turkey
mehmetgursul@gmail.com
[3] Berlin School of Economics and Law, Berlin, Germany
kathrin.kirchner@hwr-berlin.de
[4] University of Applied Sciences Zwickau, Zwickau, Germany
ralf.laue@fh-zwickau.de
[5] HSW University of Applied Sciences, Hameln, Germany
mertens@hsw-hameln.de
[6] University of the West of Scotland, Paisley, UK
felix.reher@uws.ac.uk
[7] Technical University Dresden, Dresden, Germany
irene@schoenreiter.de
[8] University of Applied Sciences RheinMain, Wiesbaden, Germany
bernhard.turban@hs-rm.de
[9] Hamburg University of Applied Sciences (HAW), Hamburg, Germany
ruediger.weissbach@haw-hamburg.de

Abstract. The paper points out some aspects of the interrelations between business process management, agility, flexibility, and requirements engineering. It shows some possibilities for agile development of business processes and for the development of flexible processes for changing requirements.

Keywords: Business process management · Requirements engineering · Agile
Flexible processes · Business process life cycle

1 Agility and Flexible Processes

Agile principles [1] (AP) are common in Software Engineering (SE) and became more important in Business Process Management (BPM) in the last years [2]. The terms of agility as well as flexibility are widespread and mutual in BPM literature [3–5]. The definitions of flexibility have a great similarity to those of agility. Unfortunately, agility

---

Authors are listed in alphabetical order to reflect their equal contributions.

© Springer International Publishing AG 2018
E. Teniente and M. Weidlich (Eds.): BPM 2017 Workshops, LNBIP 308, pp. 669–680, 2018.
https://doi.org/10.1007/978-3-319-74030-0_54

is often misused as a synonym for flexibility [3]. Agility can have varied interpretations in the field of BPM. On the one hand, it can refer to organizational agility, which can be an outcome of BPM deployment in an organization. It can also refer to the usage of agile approaches to BPM deployment efforts. Additionally, "agile BPM" may refer to business process management systems (BPMS) that incorporate agile development methodology in process automation. We proclaim the use of different terms for the clear differentiation of the actually intended contents. In our understanding, "agility" denotes agile methods according to AP, while "flexibility" refers to the adaptability, responsiveness, and context dependency of the processes.

In this paper, we present various aspects of the interrelations between BPM and requirements engineering (RE) with an agile perspective. In the rest of this chapter, we discuss on flexibility and organizational agility, the concept of flexible business processes (BPs), and how to apply agile methods in BPM. In Sect. 2, we look into the relationship between SE, RE, and BPM. Section 3 focuses on agile development processes and BPM. In Sect. 4, we discuss on some organizational circumstances that require agility and flexibility in processes. Lastly in Sect. 5, we conclude the paper.

1.1 Flexibility and Organizational Agility

A large body of knowledge around the definitions of agility and flexibility exist in literature. Eardley et al. suggest that flexibility is the ability to change direction rapidly or deviate from a predetermined course of action [6]. Doz et al. define strategic agility as "the capacity to continuously adjust and adapt strategic direction in a core business to create value for a company" [7]. Organizational agility is referred to as "the capacity of an organization to efficiently and effectively redeploy/redirect its resources to value creating and value protecting (and capturing) higher-yield activities as internal and external circumstances warrant" [8].

Singh et al. indicate that an organization's stimuli-response actions considered agile can be explained using a bi-fold idea of "magnitude of variety" change (named flexibility) and "rate of generating variety" change (named responsiveness) [9]. The "magnitude of variety change" refers to the architectural aspect of change and indicates the extent to which an organization can change and the amount of changes made in its products or practices. The other aspect, namely the "rate of variety change", refers to the impermanence of change and charts the relationship of the change in variety with time. Thus, organizational agility consists of the amount of change the firm makes to its products or processes in response to environmental stimuli as well as the rapidity with which such changes are made. This flexibility and responsiveness create "a meta-capability" that deploys a dynamic balance between sensing opportunities, enacting complementarities, and capturing value over time [10].

Organizational agility is necessitated by changes in an organization's environment such as technological shifts, talent pool skills shifts, resource limitations, emerging consumer markets, and changes in consumer expectations. An organization's agile capabilities govern how it integrates, builds, and reconfigures its resources both internal and external in response to these changing environments. From the organizational perspective, the flexibility of processes is different from the organization's ability to change processes. When processes are flexible, the organization can be more stable, at least until the number of process changes had changed the organization itself.

1.2 Flexible Business Processes

BPM needs to support, enact, and integrate organizational agility in order to dynamically manage BPs [11]. The aim of BPM is to improve the efficiency and effectiveness of the processes through process redesign and incorporate automation where feasible. BPM deployment typically involves implementing process-based, long-range business applications as part of a BPMS. Contexts change over time; thus, processes require continuous adaptation to the given context. This, in some cases, leads to obsolescence of the re-designed process in the implementation stage due to the rapidly changing requirements in a dynamic environment.

The key benefits of BPM can be summarized as "efficiency, effectiveness and agility" [12]. In the current dynamic and globalized environment, organizations need to be able to assimilate and counter changes on a real-time basis. The control on processes and the platform for rapid workflow modifications provided by BPM provides this agility. The feedback on operations management occurs in real time, with information secured and made available instantaneously, keeping productivity on track and efficient. In this way, company operations can react with greater agility, enacting operations change more easily and often. BPM empowers organizations to react better to times of quick change. The re-design of the workflow is then the instrument through which the association can react to this change. Therefore, BPMSs have to go beyond their classic features and incorporate, beside others, a contextual process management. Such systems have to support that process participants decide on the execution of activities as well as their order during process execution. Thus, it is not possible to fully prescribe such a process beforehand as it can be done for, e.g., a standardized purchasing process.

Some approaches have been developed to support flexible processes on design time as well as on runtime. During *runtime*, the ADEPT system offers the functionality of making dynamic changes during the execution time of a pathway. Running process instances can be migrated to new process model versions [13]. Based on ADEPT, MinAdept provides techniques for mining flexible processes [14]. Till now, the ADEPT concept does not include a concept for monitoring flexible processes.

For modeling variability at *design time*, several approaches exist: Declarative process modeling is an activity-centered approach in which constraints are used to prevent certain behavior [15]. During run-time, only allowed activities are shown in the work list of the user, and he decides about next activity to be executed. For specifying variants in procedural process models at design time, Hallerbach et al. identified two solutions in traditional tools [16]. The *multi-model approach* requires a separate model for each variant. The *single-model approach* makes use of one big model which covers all possible variants. Decisions that could be made at design-time appear as conditional branching that takes place at run-time. Both approaches have obvious disadvantages. The requirement is to have models that can be configured at design time. Torres et al. discusses two solution approaches [17]. *Behavioral approaches* model a superset of all variants and derive a particular variant by hiding or blocking elements. *Structural approaches* start from a "base" process model and derive variants by applying a set of change operations. Configurable Event-Driven Process Chains [18] is an example of behavioral approaches, whereas Provop [19] and vBPMN [20] follow the structural approach.

In the case handling approach, activities can be executed based on data dependencies [21]. For example, if an activity is still running but already produced data necessary to execute the next step, the following activity can start. In the same vein, the concept of Proclets allows the division of a process into several process parts. These snippets can be executed one after the other or interactively [22]. *Case management,* covered by CMMN, requires modeling that can express the flexibility of a knowledge worker during run-time while selecting and executing tasks for a specific case [23]. Tasks are modeled and can be specified as either mandatory or discretionary during design-time and serve as recommendations during run-time.

1.3   Application of Agile Methods in BPM

Traditionally, the BPM deployment lifecycle emphasizes detailed up-front planning of process analysis, process design and modeling, process implementation, monitoring, and improvement activities. In other words, traditional BPM adopts a waterfall approach. Von Rosing et al. have proposed the use of the agile method through the various stages of the BPM lifecycle [24]. In an agile approach, analysis, planning, and architecture design are the beginning phases (corresponding to BPM's design and modeling phase). On the other hand, the build, test and deployment phases have a circular approach in multiple short iterations instead of the linear execution, monitoring and improvement phases of traditional BPM.

Agile BPM targets an initial high-level blueprint detailing the estimates for project releases, resources, risks, and cost and benefits. From this, a BPM deployment roadmap is outlined as to when and which requirements can be met as the project advances through small releases. In the next phase, instead of a detailed up-front process model, a high-level design for the "to-be" processes is developed at the start of the project that sets the foundation for the agile BPM project choices and options. This high-level design guides the detailed design in each iteration as specified by the stakeholders, within the budgeted time and cost parameters.

Agile BPM in the deployment phase is based on cooperative and information-aligned sharing of accountability and governance. Traditional BPM project governance uses a gated approach to release and monitor the fixed up-front project funding and outcomes. Since agile BPM breaks up the effort into short releases, the funding is also made available on the successful release of each iteration.

2   Relationship Between SE, RE, and BPM

2.1   The Software Part of BPM

Development of a BPMS is an essential step in the BPM lifecycle. A BPMS is typically developed in implementation stage, which helps businesses to automate and manage

BPs and roles [25]. A BPMS is one of the most recommended investments for process improvement [26]. Since the BPM life cycle includes the development of a software system as an essential step, SE approaches are naturally utilized as part of the BPM activities. The development of a BPMS may differ from the traditional software development life cycle due to the use of process automation tools driven by process models rather than code [27]. Still, many SE practices are implemented during the development of a BPMS [28]. For example, process analysis is seen as an essential RE activity [29]. Table 1 presents the common SE activities in the first column [30], and provides a summary of similarities and differences of these activities for SE and the software part of BPM (i.e. development of a BPMS).

Table 1. SE activities and their relation to software part of BPM

Activities	Software Engineering	Software part of BPM
Requirement	BPs can be used for starting point of software requirements elicitation [31]	Requirements can be elicited traditionally or an agile approach can be used [32]
Design	Developing UML diagrams	Developing detailed process models and enriching process models with execution-related properties [33]
Implementation	Various implementation methodologies can be used	Configuring process automation tools [27]
Testing	Manual or automated tests	Manual or automated tests, flow analysis, and simulation [33]
Optimization	Refactoring the code	Process improvement through redesign of processes [34]
Integration	Integrating/communicating with other systems via services	Integrating/communicating with other systems via services [35]

In addition to the activities, roles in SE and BPM are also comparable. Despite the view that BPM roles are mostly related to business [36], many roles perform similar functions, such as domain experts, analysts, and developers [31]. It is essential that technical people are also involved in the development of BPMSs, and process models provide a good communication environment with them and non-technical ones [27].

2.2 The "Core" BPM

The activities at the intersection of BPM with SE do not constitute the main part of BPM. Actually, BPM is a discipline for which the focus is more on the humans and processes rather than technology. Technology, or the BPMS to be developed, is only a facilitator to improve processes. Process improvement can be achieved in many other ways such as implementation of various redesign heuristics [34]. BPM as a holistic approach covers six core factors; strategic alignment, governance, methods, information systems, people, and culture [37]. When we use the term "core BPM", we refer to all these factors other than the software development part of BPM in this paper. These encompasses all the activities in BPM lifecycle including process analysis, design, implementation, monitoring, and improvement [38].

3   Agile Development Processes and BPM

3.1   Agile Values and Terminology

In recent years, a number of agile methods have emerged, all based on the core values stated in the Manifesto for Agile Software Development [1]. The Manifesto expresses the core values in the form of "Important Part" and "Less Important Part". It should be noted, however, that the Manifesto does not deem the less important part to be unimportant. The following Table 2 shows the applicability of the values to BPM.

Table 2.   Agile values and their applicability to BPM

Value #	Important part	Less important part	Software part of BPM	Core BPM
1	Individuals and interactions	Processes and tools	+~	~
2	Working software	Comprehensive documentation	~	~
3	Customer collaboration	Contract negotiation	++	+
4	Responding to change	Following a plan	+	??

The two rightmost columns in this table reflect our evaluation about the applicability of the agile values to both the software part of BPM as well as the core BPM. Value #1 cannot easily be adapted, since individuals and interactions are controlled by management who usually want to have formalized overviews of implementation progress. This problem can be assumed to be even harder to solve when core BPM facets, such as employee workflows, are considered as more parties are involved. This is, however, more a question of company culture than BPM itself, since software developed in a BPM context is intertwined with BPs that might be regulated by laws and that have to be transparent to management. Value #2 could be the value that is most difficult to apply for both the software development as well as the core part of BPM. BPM is closely connected to business concerns [28]. Hence, Value #3 should not pose any serious problems in neither part, as customer collaboration plays an essential role in BPM anyway. It could, however, be more problematic when third party systems or workforce personnel are involved. Responding to change (Value #4) can easily be followed in the software part of BPM, provided that is not hindered by problems resulting from the realization of Value #2. If documentation has to be updated and complex change processes involving many stakeholders have to be run before implementation changes, fast responses to change can become difficult. Realizing Value #4 in the core BPM part might prove even more difficult as the workforce is affected by change as well. Possible strategies could be to decompose work steps into minimalistic actions that could easily be recombined or to qualify an agile workforce with employees who can handle constant change. Both strategies might bring their own hurdles and would have to be explored in different organizational contexts.

3.2   Development of an Agile Method for BPM

Agile methods can be seen as a collection/language of process patterns [39 (p.98), 40]. In this article, we use the term development process pattern (DPP) for what is referred to as process pattern in SE theory. Each practice of an agile method can be seen as an individual DPP, and all practices of an agile method can be rather seen as a DPP catalogue (or even pattern language) describing the patterns with their interdependencies/tradeoffs.

Figure 1 shows a schematic illustration of agile DPPs and their interrelations/tradeoffs. The figure is separated into four dimensions:

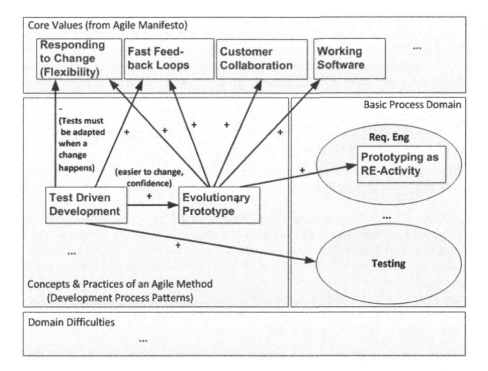

Fig. 1.   Illustration on the interactions of agile development process patterns such as Test Driven Development or Evolutionary Prototype with other aspects.

At the top, the Core Values of agile methods expressed by the Agile Manifesto are listed. In the left-center area, the DPP-dimension shows typical agile methods to be analyzed for their tradeoffs. At the right area, the dimension of Basic Processes enlists the basic processes of any software development method (requirements engineering, design, implementation and testing as the basic process set every agile method must at least address in some way. At the bottom, the Domain Difficulties are indicated as a further dimension. As we will point out later, this dimension is an important aspect that must be considered in order to ensure an agile method can be applied in a certain domain because the DPPs used must also address these domain-specific difficulties.

In the DPP-dimension, the two DPPs "Test Driven Development" (TDD) and "Evolutionary Prototype" (EP) are shown as examples. The tradeoffs of each DPP are modeled by arrows, where arrows annotated with '+' indicate a positive effect, whereas arrows annotated with '−' indicate negative effects. In this way, e.g. TDD has a positive effect on EP because changes can be made with less risk of undesired function side effects, as these would be discovered with automated testing, giving developers more trust in their code. These aspects are a fundamental basis to ensure that the EP DPP can work at all. It can also be seen that EP has positive effects to a lot of Core Values. The impact of TDD on the "Responding to Change" Core Value is two sided: TDD makes changes more work intensive as all test cases affected by the change have to be adapted but it also makes changes safer. TDD also has positive effects on other Core Values and EP and TDD also have positive effects on Basic Processes.

Figure 1 gives an impression on how two DPPs interact with each other and the other dimensions discussed here. It could be argued that, if a working agile method with its DPPs is completely analyzed in this way, any aspect in any of the shown dimension should somehow be covered by positive effects, thus forming a synergistic network that is described above by the metaphor of interlocked cogwheels.

Concerning the implementation of BPs through IT-Systems (IT Business Alignment), a simple adaption of existing agile methods and techniques will not be successful because of the following factors:

- BPs are often essentially complex, so they need an essential amount of upfront planning (e.g. by modelling them).
- BPs are often highly interconnected with other BPs and different users, so it is very difficult to acquire fast feedback that is essential for agility.
- BPs are often implemented into complex systems where workflow control is taken over by a workflow engine and the workflow triggers a number of other systems.
- Many BPs involve workforce personnel. People can hardly be involved in automated test procedures nor does it make sense to change people-based workflows too often and without proper change management.
- Legal constraints, business needs and organizational aspects do in some cases require higher degrees of documentation and planning.

These aspects could be seen as *domain difficulties* that should also be taken into account when considering agile methods. This follows the intuition that agile development is based on a set of assumptions that do not hold true in all kinds of projects [41]. Recent approaches have, however, successfully faced problems such as distributed development with organizational adaptations [42] or complex large scale projects with Large Scale Scrum [43]. So it might be possible to customize an agile method for BPM. There are some promising agile methods like TDD or EP that may at least in part be adaptable to implementation of BPs. However, this does not mean that this is a working agile method. To really establish an agile method, several of these principles must work together in a synergistic way addressing all important aspects to consider (Core Values,

Basic Processes and Domain Difficulties). In this way, it might be possible to identify a set of development process patterns forming a promising agile method for the BPM-domain. To achieve this, Fig. 1, might indicate a kind of conceptual skeleton for evaluating the fit of agile methods to BPM.

4   Organizational Circumstances

4.1   Process Life Cycles and Disruptions

While the idea of process modeling and process management is to develop a stable environment for tasks, especially for transactional tasks, every process model has its life cycle. We can differentiate between planned and unplanned, but unavoidable process life cycles. These "unavoidable" life cycles result from their development process: Processes will be initiated, they have a phase of stability, and typically also a declining phase. In some cases, they have a "sudden death" caused by an unexpected change in the frame condition. The initial phase is the phase of establishing a process in an organization. Typically there are some uncertainties in this phase, for example an incomplete or wrong documentation or missing trainings. This phase is also a learning phase for individuals as well as for the whole organization. In the following phase of stability, the process can be operated successfully.

Typically over the long run major changes will influence the stability of the process. Some of these changes will only have a restricted influence, for example the substitution of one machine by another. In this case, the relevance and quality of the process description will decline. Other influences will be game changers and disrupt the traditional processes by replacing them.

In many organizations, "planned life cycles" as iterative development of processes are implemented. Those iterations refer to the Deming (PDCA) cycle [44] and reflect limited stability of the environment and internal learning processes. Changes in the Deming cycle are thought as an optimization of the existing, less as a reaction on (or the creation of) dramatic disruptive changes (caused for example by constitutional and legal changes or technologically driven changes).

4.2   Flexible Processes in the Post-merger Integration

After mergers and acquisitions (M&A), most BPs are duplicated. During post-merger integration (PMI), an organization needs to decide how to proceed with doubled or redundant processes respectively and how to unify them in a common process map. An efficient process harmonization (PH) in a neutral and structured way with achieving the employee commitment promises a beneficial PMI. PH in the PMI context combines a common global management system across different regions or units with the allowances of defined variations at the process level [45].

Although PH has the highest relevance in the integration approaches absorption (acquired company is absorbed by acquirer) and symbiosis (evolution from existing), particular processes might be unified in preservation (acquired company retains independent) and holding (integration not intended) as well. Regardless of the various integration approaches, the need for action after an M&A is out of question. The question however is the intensity of PH.

While full benefit of process management is only apparent when the strategy, design, implementation and controlling of the processes are viewed holistically and coordinated with each other [46, 47], over-standardization with too strong regulations would lead to a loss of necessary flexibility and competitive advantages. The right balance between standardization and individualization must be evolved. Process flexibility in context of PH means a combination of standardized processes with necessary variants - in other words: process flexibility respects specific and necessary process characteristics.

For such scenarios the "gate principle" is proposed as the best option. The "gate" variant is focused on a common (intermediate) output of a process necessary for the next process (step) (=gate), e.g. particular KPIs, reports, defined status etc. Process execution is secondary as long as the defined output is generated. The focus on a unified output allows a company keeping flexibility within a certain process combined with the advantage of an overall harmonized process map. In the long-term, the processes can be redesigned with the gate used as a requirement for agile process development. This stepwise unification gives flexibility in a tense situation, while it can be seen as an analogy of growing together to a unique organization.

5   Conclusion

In this discussion paper, we looked into the concepts of flexible processes in organizational environments, approaches for agile development of processes and management of flexible processes in an agile way, relationship between the areas of SE, RE, and BPM, and opportunities to incorporate agile principles in BPM. With the arising importance of agility in BPM field, this paper may inspire the researchers to initiate new works in these areas.

Different trends are currently influencing the requirements for BPM. On the one side, flexibility and agility are important to develop modern business models and they need adequate methods for planning and implementing as shown in this paper. On the other side, stability, transparency and documentation are important for those business processes that have strong legal constraints (indemnification). The parallelism of both trends and their combination will be a challenge for further research.

References

1. Beck, K., Beedle, M., Van Bennekum, A. et al.: Manifesto for agile software development (2001)
2. Weißbach, R., Kirchner, K., Reher, F., Heinrich, R.: Challenges in business processes modeling – is agile BPM a solution? In: Dumas, M., Fantinato, M. (eds.) BPM 2016. LNBIP, vol. 281, pp. 157–167. Springer, Cham (2017). https://doi.org/10.1007/978-3-319-58457-7_12
3. Conboy, K.: Agility from first principles: reconstructing the concept of agility in information systems development (2009)
4. Sethi, A.K., Sethi, S.P.: Flexibility in manufacturing: a survey. Int. J. Flex. Manuf. Syst. 2, 289–328 (1990)

5. Chen, Y., Wang, Y., Nevo, S., et al.: IT capability and organizational performance: the roles of business process agility and environmental factors. Eur. J. Inf. Syst. **23**, 326–342 (2014)

6. Eardley, A., Avison, D., Powell, P.: Strategic information systems: an analysis of development techniques which seek to incorporate strategic flexibility. J. Organ. Comput. **7**, 57–77 (1997)

7. Doz, Y.L., Kosonen, M.: Fast Strategy: How Strategic Agility Will Help You Stay Ahead of the Game. Pearson Education, Harlow (2008)

8. Teece, D., Peteraf, M., Leih, S.: Dynamic capabilities and organizational agility. Calif. Manage. Rev. **58**, 13–35 (2016)

9. Singh, J., Sharma, G., Hill, J., Schnackenberg, A.: Organizational agility: what it is, what it is not, and why it matters. In: Academy of Management Proceedings, p. 11813 (2013)

10. Fourné, S.P.L., Jansen, J.J.P., Mom, T.J.M.: Strategic agility in MNEs. Calif. Manage. Rev. **56**, 13–38 (2014)

11. Triaa, W., Gzara, L., Verjus, H.: Organizational agility key factors for dynamic business process management. In: 2016 IEEE 18th Conference on Business Informatics (CBI), pp. 64–73. IEEE (2016)

12. Rudden, J.: Making the case for BPM-a benefits checklist. In: BPTrends, pp. 1–4 (2007)

13. Dadam, P., Reichert, M.: The ADEPT project: a decade of research and development for robust and flexible process support. Comput. Sci. Res. Dev. **23**, 81–97 (2009)

14. Li, C., Reichert, M., Wombacher, A.: The Minadept clustering approach for discovering reference process models out of process variants. Int. J. Coop. Inf. Syst. **19**, 159–203 (2010)

15. Pesic, M., van der Aalst, W.M.P.: A declarative approach for flexible business processes management. In: Eder, J., Dustdar, S. (eds.) BPM 2006. LNCS, vol. 4103, pp. 169–180. Springer, Heidelberg (2006). https://doi.org/10.1007/11837862_18

16. Hallerbach, A., Bauer, T., Reichert, M.: Issues in modeling process variants with provop. In: Ardagna, D., Mecella, M., Yang, J. (eds.) BPM 2008. LNBIP, vol. 17, pp. 56–67. Springer, Heidelberg (2009). https://doi.org/10.1007/978-3-642-00328-8_6

17. Torres, V., Zugal, S., Weber, B., Reichert, M., Ayora, C., Pelechano, V.: A qualitative comparison of approaches supporting business process variability. In: La Rosa, M., Soffer, P. (eds.) BPM 2012. LNBIP, vol. 132, pp. 560–572. Springer, Heidelberg (2013). https://doi.org/10.1007/978-3-642-36285-9_57

18. Rosemann, M., van der Aalst, W.M.P.: A configurable reference modelling language. Inf. Syst. **32**, 1–23 (2007)

19. Hallerbach, A., Bauer, T., Reichert, M.: Capturing variability in business process models: the provop approach. J. Softw. Maint. Evol. Res. Pract. **22**, 519–546 (2010)

20. Döhring, M., Zimmermann, B., Karg, L.: Flexible workflows at design- and runtime using BPMN2 adaptation patterns. In: Abramowicz, W. (ed.) BIS 2011. LNBIP, vol. 87, pp. 25–36. Springer, Heidelberg (2011). https://doi.org/10.1007/978-3-642-21863-7_3

21. van der Aalst, W.M.P., Weske, M., Grünbauer, D.: Case handling: a new paradigm for business process support. Data Knowl. Eng. **53**, 129–162 (2005)

22. van der Aalst, W.M.P., Barthelmess, P., Ellis, C.A., Wainer, J.: Proclets: a framework for lightweight interacting workflow processes. Int. J. Coop. Inf. Syst. **10**, 443–482 (2001)

23. OMG: Case Management Model and Notation (CMMN) Version 1.0 (2014)

24. Von Rosing, M., Von Scheel, J., Gill, A.Q.: Applying agile principles to BPM. In: The Complete Business Process Handbook. Body Knowledge from Process Model to BPM (2014)

25. Pourshahid, A., Amyot, D., Peyton, L., et al.: Business process management with the user requirements notation. Electron. Commer. Res. **9**, 269–316 (2009)

26. McCoy, D.W., Cantara, M.: Hype Cycle for Business Process Management (2010)

27. Dumas, M., van der Aalst, W.M.P., ter Hofstede, A.H.M.: Process-Aware Information Systems: Bridging People and Software Through Process Technology. Wiley, New Jersey (2005)
28. van der Aalst, W.M.P.: Business process management: a comprehensive survey. ISRN Softw. Eng. **2013**, 1–37 (2013)
29. Aysolmaz, B., Leopold, H., Reijers, H.A., Demirörs, O.: A semi-automated approach for generating natural language requirements documents based on business process models. Inf. Softw. Technol. **93**, 14–29 (2017)
30. IEEE Computer Society: Guide to the Software Engineering Body of Knowledge (SWEBOK Guide V3.0). IEEE Computer Society (2014)
31. Heinrich, R., Kirchner, K., Weißbach, R.: Report on the 1st International Workshop on the Interrelations between Requirements Engineering & Business Process Management (REBPM), RE 2014. IEEE (2014)
32. Aukema, C.: Managing requirements In: Business Process Management Suite Projects. University of Twente (2011)
33. Dumas, M., La Rosa, M., Mendling, J., Reijers, H.A.: Fundamentals of Business Process Management. Springer, Heidelberg (2013). https://doi.org/10.1007/978-3-642-33143-5
34. Reijers, H.A., Mansar, S.L.: Best practices in business process redesign: an overview and qualitative evaluation of successful redesign heuristics. Omega **33**, 283–306 (2005)
35. van der Aalst, W.M.P.: Process-aware information systems: lessons to be learned from process mining. In: Jensen, K., van der Aalst, Wil M.P. (eds.) Transactions on Petri Nets and Other Models of Concurrency II. LNCS, vol. 5460, pp. 1–26. Springer, Heidelberg (2009). https://doi.org/10.1007/978-3-642-00899-3_1
36. Swenson, K.: BPM is Not Software Engineering. https://social-biz.org/2008/11/25/bpm-is-not-software-engineering. Accessed 4 Oct 2017
37. Vom Brocke, J., Rosemann, M.: Handbook on Business Process Management 1. Springer, Heidelberg (2010). https://doi.org/10.1007/978-3-642-00416-2
38. Weske, M.: Business Process Management: Concepts, Languages, Architectures, 2nd edn. Springer, Heidelberg (2012). https://doi.org/10.1007/978-3-642-28616-2
39. Turban, B.: Tool-Based Requirement Traceability between Requirement and Design Artifacts. Springer, Wiesbaden (2013). https://doi.org/10.1007/978-3-8348-2474-5
40. Bozheva, T., Gallo, M.E.: Defining Agile Patterns. In: Dutoit, A.H., McCall, R., Mistrík, I., Paech, B. (eds.) Rationale Management in Software Engineering, pp. 373–390. Springer, Heidelberg (2006). https://doi.org/10.1007/978-3-540-30998-7_18
41. Turk, D., France, R.B., Rumpe, B.: Limitations of Agile Software Processes. CoRR abs/1409.6 (2014)
42. Ramesh, B., Cao, L., Mohan, K., Xu, P.: Can distributed software development be agile? Commun. ACM **49**, 41–46 (2006)
43. Larman, C., Vodde, B.: Large-Scale Scrum: More With Less. Addison-Wesley, Boston
44. Moen, R., Norman, C.: Evolution of the PDCA Cycle (2010). http://www.westga.edu/~dturner/PDCA.pdf
45. Fernandez, J., Bhat, J.: Addressing the complexities of global process harmonization. In: Handbook of Research on Complex Dynamic Process Management: Techniques for Adaptability in Turbulent Environments, pp. 368–385. IGI Global, Hershey (2010)
46. Stähler, D.: Standardisierung als Erfolgsvoraussetzung im Geschäftsprozessmanagement. Zeitschrift Führung + Organ. ZfO 75 (2006)
47. Becker, J., Kahn, D.: The process in focus. In: Becker, J., Kugeler, M., Rosemann, M. (eds.) Process Management. SE - 1, pp. 1–12. Springer, Heidelberg (2003). https://doi.org/10.1007/978-3-540-24798-2_1

1st Workshop on Security and Privacy-Enhanced Business Process Management (SPBP 2017)

Preface to the Workshop on Security and Privacy-Enhanced Business Process Management (SPBP'17)

Raimundas Matulevičius[1], Nicolas Mayer[2], and Ingo Weber[3]

[1] Institute of Computer Science, University of Tartu, Tartu, Estonia
`raimundas.matulevicius@ut.ee`
[2] Luxembourg Institute of Science and Technology, 5 Avenue des Hauts-Fourneaux, 4362 Esch-sur-Alzette, Luxembourg
`nicolas.mayer@list.lu`
[3] Data61, CSIRO and UNSW, Sydney, Australia
`ingo.weber@data61.csiro.au`

Despite the growing demand for business processes that comply with security and privacy policies, security and privacy incidents caused by erroneous workflow specifications are regrettably common. This is, in part, because business process management, security and privacy are seldom addressed together, thereby hindering the development of trustworthy, privacy and security-compliant business processes. The central theme of the workshop is the interplay between business process management (BPM), security, and privacy management.

The SPBP'17 program included one keynote and two papers. The keynote speech is given by Prof. Jan Mendling from the Vienna University of Economics and Business (Austria). The paper by Alshammari and Simpson presents a model for personal data management and privacy preservation. Using the proposed model, one is able to determine critical business process activities that potentially could harm the used personal data. In the next paper by Olifer *et al.* authors propose an alignment of the security requirements cost measurement model to the business process notations, such as BPMN and EPC. Authors report on the observations regarding the business model properties and applicability of the measurement model.

We wish to thank all those who contribute into making SPBP a success: the authors who submitted papers, the members of the Program Committee who carefully reviewed and discussed the submissions, and the speakers who presented their work at the workshop. We also express our gratitude to the BPM 2017 Workshop Chairs for their support in preparing the workshop.

1 Program Committee Chairs

Raimundas Matulevičius	University of Tartu, Estonia
Nicolas Mayer	Luxembourg Institute of Science and Technology, Luxembourg
Ingo Weber	Data61, CSIRO and UNSW, Sydney, Australia

2 Program Committee

Nicola Atzei	University of Cagliari, Italy
Massimo Bartoletti	University of Cagliari, Italy
Tiziana Cimoli	University of Cagliari, Italy
Christophe Feltus	LIST, Luxembourg
Virginia Franqueira	University of Derby, UK
Luciano García-Bañuelos	University of Tartu, Estonia
Ralph Holz	University of Sydney, Australia
Christos Kalloniatis	University of the Aegean, Greece
Jan Mendling	Vienna University of Economics and Business, Austria
Haralambos Mouratidis	University of Brighton, UK
Moussa Ouedraogo	LIST, Luxembourg
Günther Pernul	Universität Regensburg, Germany
Mark Staples	Data61, CSIRO and UNSW, Sydney, Australia
Sherry Xu	Data61, CSIRO and UNSW, Sydney, Australia

Personal Data Management: An Abstract Personal Data Lifecycle Model

Majed Alshammari[✉] and Andrew Simpson

Department of Computer Science, University of Oxford, Wolfson Building,
Parks Road, Oxford OX1 3QD, UK
{majed.alshammari,andrew.simpson}@cs.ox.ac.uk

Abstract. It is well understood that processing personal data without
effective data management models may lead to privacy violations. Such
concerns have motivated the development of privacy-aware practices and
systems, as well as legal frameworks and standards. However, there is a
disconnect between policy-makers and software engineers with respect to
the meaning of privacy. In addition, it is challenging: to establish that a
system underlying business processes complies with its privacy requirements; to provide technical assurances; and to meet data subjects' expectations. We propose an abstract personal data lifecycle (APDL) model to
support the management and traceability of personal data. The APDL
model represents data-processing activities in a way that is amenable to
analysis. As well as facilitating the identification of potentially harmful
data-processing activities, it has the potential to demonstrate compliance
with legal frameworks and standards.

Keywords: Data privacy · Data lifecycle model
Privacy-aware data processing · Compliance demonstration

1 Introduction

Privacy is typically articulated at a high level of abstraction. Thus, its concrete
manifestations are ambiguous to those concerned with data protection and to
those responsible for developing and maintaining systems [1,2]. Further, incorporating privacy requirements into the early stages of the development process
requires an appropriate interpretation of legal, social and political concerns [3].
These challenges lead to a disconnect between policy-makers and software engineers with regards to conceptualisations of privacy, its related concepts, and the
ways in which systems can be developed to comply with legal frameworks and
standards and to meet data subjects' expectations [4]. As such, there is a need for
generalised techniques that support the effective translation of abstract privacy
principles, models and mechanisms into implementable requirements [1,4].

The dominant approach to embedding privacy into the early stages of the
development process is Privacy by Design (PbD) [5]. The principles of PbD are
given at a high level of abstraction, which leads to challenges with regards to

© Springer International Publishing AG 2018
E. Teniente and M. Weidlich (Eds.): BPM 2017 Workshops, LNBIP 308, pp. 685–697, 2018.
https://doi.org/10.1007/978-3-319-74030-0_55

translation into engineering activities [3]. Data minimisation has been proposed as a necessary and foundational step to engineer systems in line with the principles of PbD [3] — but ensuring data minimisation is itself a challenge.

To achieve the aim of PbD, detailed privacy impact and risk assessments need to be conducted with the aim of identifying and addressing potential privacy risks [6]. A Privacy Impact Assessment (PIA) provides non-technical guidelines for stakeholders on identifying high level privacy requirements; however, it does not provide guidelines on translating these into technical system requirements [2]. In order for a PIA to be holistic and effective in supporting such translation, it needs to be complemented by an appropriate privacy risk management model; it also needs to be complemented by a sufficiently robust model that serves as the basis for the identification, analysis and assessment of potential privacy risks in a proactive, comprehensive and concrete manner. The representation of such a model tends to be relatively straightforward, capturing possible states and possible changes in these states brought about by operations [7].

Often, legal frameworks and standards are given at a high level of abstraction without relying on rigorous models that explicitly specify privacy-related concepts [8]: types and sensitivity of personal data; the purposes for, and the manner in which, this data is processed; involved actors; and assigned roles and responsibilities. An abstract data model can play a crucial role in providing a privacy-aware data lifecycle model in the context of data protection. Further, such a model can be a stepping stone for translating privacy requirements into system requirements by defining a foundation for contextual analysis.

The Abstract Data Lifecycle Model (ADLM) [9] was developed to serve as a generic data lifecycle model for data-centric domains, and can be used as a means to classify, compare and relate other data lifecycle models, as well as to provide the basis for new data lifecycle models [9]. It is the starting point for our contribution. We present an Abstract Personal Data Lifecycle (APDL) model that represents the personal data lifecycle in terms of lifecycle stages, along with associated activities and involved actors. The APDL model can be used to complement a PIA for describing the planned, actual and potential processing of personal data, which, in turn, helps facilitate the management and traceability of the flow of personal data, as well as the identification of data-processing activities that may lead to privacy violations or harms in a comprehensive and concrete manner. Furthermore, it has the potential to help demonstrate privacy compliance with legal frameworks and standards. Finally, it has the potential to underpin a conceptual framework for privacy engineering with the aim of helping stakeholders reason about design decisions.

2   Foundations

In the context of data-centric domains, data undergoes a variety of actions — including creation, use, publication and destruction — by several actors for various purposes. These actions in combination constitute a data lifecycle. It is understandable that each domain is concerned with a specific type of data and each data lifecycle model has its own specific focus. Most importantly, they all consider the same item of interest — data, which is a "living thing" that moves though various stages during its lifecycle and is at the heart of these systems [9]. In the context of data protection, personal data often moves through various stages that are governed by laws, regulations or standard principles. Accordingly, personal data should be at the heart of methods, techniques and tools that systematically and proactively identify and address privacy risks at the early stages of the design process.

The Abstract Data Lifecycle Model (ADLM) [9] was derived from specific instances of data lifecycle models to ensure broad coverage and wide applicability [9]. For each domain, a list of models was analysed in terms of their lifecycle phases, features, roles, actor features and metadata features. By analysing, comparing and contrasting these models, the ADLM was derived as an abstract data lifecycle model for data-centric domains. It establishes five areas of classification: lifecycle phases, features and roles, actor features, and metadata features. The ADLM provides a means to classify, compare and relate other data lifecycle models, and provides the basis to develop new lifecycle models [9].

The ADLM considers some aspects pertaining to data-centric domains that are of central importance, such as metadata. It provides a set of features to describe the primary data in relation to its sources, contents and domains, with the aim of supporting data production, retrieval and consumption. Metadata features require additional features for characterising the data lifecycle and involved actors, such as those features explained in [9]. As such, we illustrate only the parts of the ADLM that are relevant to representing personal data processing activities in a way that is amenable for analysis: lifecycle phases and roles. The ADLM consists of the following lifecycle phases: ontology development, planning, creation, archiving, refinement, publication, access, external use, feedback and termination [9]. Further, the ADLM considers the following roles: ontology designers, data creators, metadata creators, administrators and end users.

3   The APDL Model

The APDL is an abstract model that represents personal data processing in terms of states (data items), operations (processing activities), and roles (actors). It identifies a set of stages through which personal data moves during its lifetime and indicates the order and depth in which associated activities can occur. We will use and adapt features of the ADLM as points of reference for analysis.

3.1   Lifecycle Stages

As there are obligations and limitations on the stages of the personal data lifecycle and associated activities, our analysis of the ADLM has to consider such concerns. Some stages will be combined — generalised — according to their characteristics and associated activities, while others will be defined — specialised — to limit associated activities to particular privacy principles. Those stages not relevant to personal data will be discarded.

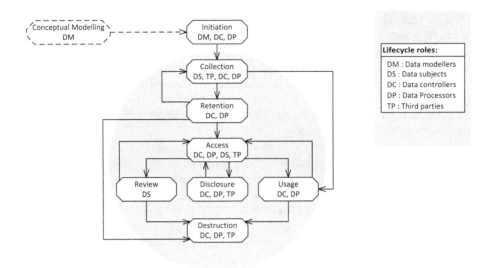

Fig. 1. The Abstract Personal Data Lifecycle (APDL) model.

It is essential to adopt a set of universal privacy principles that can be applied in a variety of contexts in various jurisdictions. As an example, the Fair Information Practice Principles (FIPPs) were developed as core principles of the Code of Fair Information Practice [10]. In 2006, at the 28th International Data Protection and Privacy Commissioners Conference, the Global Privacy Standard (GPS) [11] was accepted as a unified set of principles that reflects appropriate variants of the FIPPs. The GPS principles harmonise various sets of the FIPPs into universal privacy principles. We adopt the GPS principles to impose constraints on the stages of the lifecycle and associated activities.

Figure 1 illustrates the main stages of the APDL model, their logical dependencies on other stages, and relevant lifecycle roles. Table 1 summarises the personal data lifecycle stages, associated activities and dependencies in terms of inputs and outputs. We describe only the Collection stage in detail.

Conceptual Modelling constitutes a preliminary stage: it involves activities to develop a conceptual model that describes the problem and its solution in terms of the domain vocabulary. It represents key and relevant concepts, associated meanings, properties, relationships and constraints that restrict the semantics of the concepts and their conceptual relationships. Such a model can be used to facilitate communication with multiple stakeholders. It is represented by a dotted line to distinguish that it is not a core part of the personal data lifecycle.

Next, the *Initiation* stage involves activities to define a 'complete processing plan' that specifies the purpose for, and the manner in which, personal data is collected and processed in relation to the context. This is the basis for establishing a privacy notice to be communicated to data subjects.

Table 1. The stages, associated activities and dependency of the APDL model.

#	Stage	Activities	Input	Output
0	Conceptual Modelling	Specification Conceptualisation Representation	Domain knowledge	A conceptual model
1	Initiation	Specification	A conceptual model Privacy policies and procedures	A processing plan A privacy notice
2	Collection	Specification Collection or Acquisition	A processing plan A set of data values	A set of collected data values
3	Retention	Specification Storage Archiving Backup	A processing plan A set of collected data values	A set of stored data values
4	Access	Specification Retrieval	A processing plan A set of data items	A set of retrieved data values
5	Review	Review Rectification	A processing plan A set of retrieved data values	A set of rectified data value
6	Usage	Specification Manipulation Presentation Use	A processing plan A set of retrieved data values	A set of manipulated data values
7	Disclosure	Preparation Dissemination or Transmission or Make available	A processing plan A set of retrieved data values	A set of disclosed data values
8	Destruction	Erasure or Disposal or Destruction or Anonymisation	A processing plan Retention policies Destruction policies A set of data values	A set of destroyed data values

The Collection stage involves activities pertaining to recording, capturing and collecting personal data values, whether these values are directly collected from data subjects or have been acquired from external sources. The most important aspects in this stage are the set of personal data values, associated sources (i.e. whether they are primary or secondary sources), and the methods of collection (i.e. whether they actively or passively collect data values). In terms of dependencies, the input is a data-processing plan from the Initiation stage, and the output is a set of recorded, captured or collected personal data values to be used as inputs to the Retention stage. The relevant GPS principles are: Openness, Purposes, Consent and Collection Limitation. According to these principles, we identify four essential assessment criteria: personal data items need to be adequate, relevant and not excessive in relation to the specified purpose; the specified personal data has been collected by lawful and fair methods; the privacy notice has been communicated to data subjects at or before the time of collection; and the consent of data subjects has been obtained. If satisfied, data values can be recorded or collected; otherwise, corrective actions can be carried out. Depending on the unsatisfied criterion or criteria, the process will continue, either by specifying the minimum amount required of data items, specifying lawful and fair collection methods, communicating the privacy notice to data subjects at time of collection, or by obtaining their explicit or implicit consent.

The *Retention* stage follows the Collection stage, and involves organising, structuring and storing personal data values for a specific period of time in repositories or digital storage media. The *Access* stage follows the Retention stage, and involves specifying and retrieving personal data.

The *Review* stage follows the Access stage, and involves activities for implementing the access right and rectifying personal data values by data subjects to ensure that their data is accurate, complete and up-to-date. The *Usage* stage follows the Access stage, and involves activities for manipulating and using personal data values in conformance with the specified purpose.

The *Disclosure* stage follows the Access stage, and involves activities for disseminating, making available or transmitting the previously accessed and retrieved personal data for external use by third parties. The *Destruction* stage follows the Retention, Collection, Review and Usage stages, and involves activities for erasing, destroying, redacting or disposing of personal data in accordance with relevant retention and destruction policies.

3.2   Lifecycle Roles

A lifecycle role is a set of logically-related activities that are expected to be conducted together and assigned to different actors according to their capabilities. They define the ways in which actors participate in the activities of the lifecycle stages. We analyse the lifecycle roles of the ADLM and specialise these roles for the purposes of the APDL model. Each actor may be assigned to one or many roles and each role will typically be associated with one or more lifecycle stages.

1. *Data modellers* are involved in the Conceptual Modelling stage and establish the context in which personal data is processed. In addition, they are involved in the Initiation stage and are responsible for developing logical and physical models for the application domain.
2. *Data subjects* are involved in the Collection stage with the capability of providing their personal data. Such actors can actively participate in the collection of the personal data values. Data subjects may be involved in the Access and Review stages with the capability of accessing and rectifying their personal data. Actors with this ability can access, review, update or correct their personal data to ensure that the retained personal data is accurate.
3. *Data controllers* are actors who specify the purpose for, and the manner in which, personal data is to be collected and processed. They are involved in the

Initiation, Collection, Retention, Access, Usage, Disclosure and Destruction stages with administrative capabilities. Such actors are responsible for handling personal data items without changing its format or meaning. If the data controller is a data processor, administrators are responsible for archiving, making backup copies, disclosing and destroying personal data items. The administrative capabilities may also include other activities, such as those related to compliance monitoring and audit trails. Data controllers are also involved in the Access and Usage stages with different levels of user capabilities. Such actors manipulate and use the retained personal data items according to the purpose for which this data is collected. They perform data-processing activities, including classification, analysis, manipulation, combination or other actions as per the processing plan.

4. *Data processors* are actors who process the collected personal data on behalf of the data controller. They are involved in the Retention, Access and Usage stages and process personal data items without changing their format or meaning. Such actors are responsible for archiving, making backup copies and destroying personal data items according to the data controller instructions. The role of data processors may also include other responsibilities, such as those related to operations and performance monitoring.

5. *Third parties* are actors other than data subjects, data controllers or data processors. They may be involved in the Collection stage with data-providing capabilities, i.e. they may be secondary sources other than data subjects. Such actors actively participate in the collection of the personal data values. In addition, third parties may be involved in the Disclosure stage of the lifecycle with data-receiving capabilities. Such actors receive and use the disclosed personal data items only for the purposes specified in the processing plan and with the consent or knowledge of data subjects.

4 Case Study

4.1 Overview

Our concern is the ePetition system, the aim of which is to implement the European Citizens' Initiative (ECI)[1]. The ECI is used to support a formal request to an authority for submitting a proposal for a legal act. It enables EU citizens to invite the European Commission to propose a legal act on issues where it has competence to legislate. The main purpose of the ePetition system is to verify and certify the number of valid signatures that support a certain initiative. In order for signatories to support a specific initiative, they need to provide 'identifying' personal data, which is typically retained in databases. In compliance with applicable regulations, data controllers are required to apply appropriate security measures to protect the collected personal data, and ensure that it is used only for the specified purposes and retained only as long as necessary.

---

[1] http://ec.europa.eu/citizens-initiative/public/welcome.

The first step involves establishing a citizens' committee of at least seven EU citizens. All of the committee's members need to be permanent residents or citizens of the EU Member States and old enough to vote in elections to the European Parliament. This committee acts in its capacity as the official organiser of the initiative and is responsible for preparing and managing the initiative. Second, the organisers need to prepare an initiative and register it with the European Commission. The organisers also need to find a hosting provider when signatures are intended to be collected electronically by an online collection system — either using an instance of the open source software that is provided by the European Commission and hosting it at its site, or by developing their own collection system and using a hosting service provider. For both, organisers need to obtain a certificate from the competent national authority to verify its compliance with minimum technical requirements[2]. Then, the certificate should be posted in the online collection system. Next, individuals, who act as signatories, are able to submit their personal data and their statements of support. To give their support for the initiative, signatories need to provide the specified personal data. It is important to ensure that duplicate signatures by the same individual are avoided. Having reached the required number of signatures, organisers should send this personal data to relevant competent national authorities for verification and certification. Having received all certificates from competent national authorities, organisers should submit the initiative by sending these certificates to the European Commission.

In accordance with the EU Data Protection Directive[3] and the Regulation (EU) No. 211/2011 on the Citizens' Initiative[4], organisers and competent national authorities act as data controllers. In particular, organisers are required to notify the Data Protection Authority in the EU Member State where the personal data will be processed. They are also required to apply appropriate measures to protect personal data in compliance with the Directive and relevant regulations. This includes ensuring that personal data must be "adequate, relevant and not excessive" in relation to the purpose of supporting the initiative and verifying the statements of support. Accordingly, the organisers and the competent national authorities must ensure that collected personal data is not used for purposes other than those specified for supporting the initiative and verifying the statement of support respectively. In addition, the data controllers must destroy all statements of support and any copies one month after submitting the initiative to the Commission or issuing the certificate respectively.

---

[2] http://eur-lex.europa.eu/LexUriServ/LexUriServ.do?uri=OJ:L:2011:301:0003:0009:EN:PDF.

[3] http://eur-lex.europa.eu/legal-content/EN/TXT/PDF/?uri=CELEX:31995L0046&from=EN.

[4] http://eur-lex.europa.eu/legal-content/EN/TXT/PDF/?uri=CELEX:02011R0211-20131008&from=EN.

4.2   Lifecycle Stages

Initiation. In accordance with the aforementioned EU Data Protection Directive and the Regulation (EU) No. 211/2011 on the Citizens' Initiative, organisers are required to notify the Data Protection Authority in the EU Member State where the personal data will be processed before the collection of statements of support. This requires a complete processing plan that may serve as the basis of developing a privacy notice. The processing plan needs to outline: the elements of personal data to be collected, along with its sources; the purposes for, and the manner in which, this data is processed; the methods of collection, retention, retrieval, disclosure and destruction; the choices available to data subjects and the consent to be obtained; the involved actors and their assigned roles and responsibilities; relevant regulations and standards; and the domain-specific constraints. The specification of the required data is driven by the specification of purposes for which personal data is to be processed. In this case, the main purpose of collecting and processing signatories' personal data is to verify and certify the valid number of the submitted statements of support. In addition, logical and physical models need to be developed.

Collection. Once an initiative's registration has been confirmed, the relevant Data Protection Authority has been notified and the online collection system has been certified, the organisers may use an online collection system to collect the specified personal data from at least one million EU citizens who act as signatories. The specified data is collected within a specific time limit (no longer than 12 months from the date of registration). Importantly, the collected personal data values must not exist in the lifecycle before the collection to prevent duplicate statements of support. In order for organisers to collect adequate, relevant and not excessive personal data, the collection system must generate statements of support in an appropriate form.

Retention. The statements of support that have been submitted by signatories are required to be persistently stored in a primary storage media for operational purposes. One might also assume the existence of copies of the original personal data for operational recovery purposes. Once the collection period is finished and the personal data is sent for verification and certification, competent national authorities have three months to certify the number of valid statements of support. During this period, the retained data is no longer needed for regular use by the organisers and can be archived as historical data for compliance purposes. Having submitted the received certificates, organisers have one month to destroy the retained personal data and any copies thereof or 18 months from the date of the registration of the initiative, whichever is the earlier. Signatories' personal data or any copies thereof may be retained beyond the specified retention time for the purpose of legal or administrative proceedings relating to an initiative. This requires retaining statements of support and any copies thereof for one week after the date of conclusion.

Access. During the collection period, organisers need to monitor the collection of statements of support that have been submitted. Once the collection of the statements of support have been de-activated at the end of the collection period, organisers need to export signatories' personal data from statements of support and display the current signatures distribution, which are classified according to the Member State of signatories or the date of submission. These activities require specifying and retrieving the retained statements of support. In particular, signatories' personal data needs to be made accessible for use by involved actors, in this case, internal users who acting as organisers.

Review. Data subjects cannot access their personal data once they have submitted their statements of support. Thus, the ePetition system that implements the ECI does not provide signatories with full control over their personal data.

Usage. Signatories' personal data is collected and processed for verifying and certifying the number of valid statements of support. In this case, organisers manipulate, classify and use this data to fulfil the specified purpose. These include monitoring, deleting, exporting, preparing and sending statements of support to relevant competent authorities. The actual use of signatories' personal data is accomplished by relevant competent authorities as they conduct the verification process and produce certificates for valid statements of support.

Disclosure. Statements of support are used only for verification and certification; they cannot be disclosed to any other parties.

Destruction. Removing statements of support is the final stage. Signatories' personal data are required by law to be destroyed after a specific time limit. Statements of support need to be completely and permanently erased, and digital storage media needs to be destroyed. Original, archived or backup copies of the retained statements of support need to be disposed in accordance with relevant retention and destruction policies.

4.3   Lifecycle Roles

The *data modeller* role may be assigned to capable actors who are able to define appropriate conceptual, logical and physical data models for the context of participatory democracy and, in particular, for the ePetition system.

Citizens or permanent residents of the EU Member States act in their capacities as *data subjects* who are able to provide their personal data. They actively participate in the collection of personal data with the aim of supporting an initiative. However, data subjects are not able to access and review their personal data once they have submitted statements of support. Data subjects are mainly involved in the Collection stage. Organisers and competent national authorities act in their capacities as *data controllers*. Organisers are responsible for specifying the purpose for the required personal data, and the manner in which it is to be collected and processed. They are responsible for collecting, monitoring, preparing and sending personal data to competent national authorities. The competent national authorities are responsible for verifying and certifying the number of valid statements of support for an ECI.

Data controllers may act in their capacity as data controllers and processors at the same time if they are capable of operating the online collection system. Second, the European Commission may act in its capacity as a hosting service provider by providing the OCS. The third case is a third party that acts in its capacity as a hosting service provider. In all cases, data processors are responsible for handling personal data without changing its format or meaning. They are responsible for archiving, making backup copies and destroying this data according to the data controllers' instructions.

5   Conclusions

The integration of privacy into the early stages of the design process of business processes and their underlying systems is increasingly important — PIAs and PbD are now mandated by, for example, the EU General Data Protection Regulation (GDPR)[5]. Crucially, a PIA needs to be complemented by a sufficiently robust model that represents data-processing activities in a way that is amenable to risk analysis and compliance checking. To this end, we have introduced the APDL. Each stage is an abstraction of a set of logically related data-processing activities. This classification is based on the GPS principles, the nature and order of processing activities, and the role type of involved actors and their possible responsibilities. This gives the APDL model the potential to be applied in various domains, including dynamic and interconnected scenarios where data is collected from different sources with different formats. In addition, it supports the applicability of the model when there is more than one domain, as well as when data is collected and processed collaboratively by multiple stakeholders by determining who is responsible for which lifecycle stage and their level of authority with respect to the decisions and activities performed.

The APDL model distinguishes between the types of operations that can be performed on personal data. For each operation, it outlines various distinct activities in relation to the GPS principles with the aim of governing the behaviour of these operations. The separation is important for several reasons: it helps support the manageability and traceability of the flow of personal data during its lifecycle; it is necessary for ensuring and demonstrating compliance with legal frameworks and standards; it reflects the extent to which the flow of personal data is appropriate in terms of involved actors and their assigned roles and responsibilities; and it facilitates the identification of data-processing activities that may lead to privacy violations or harms.

We limit our model to those terms that are necessary to define the fundamental concepts of the personal data lifecycle. These might be further refined

---

[5] http://www.eugdpr.org/.

and extended by developing a conceptual model that precisely represents all relevant concepts, associated meanings, properties and relationships. For example, the lifecycle may be characterised by properties that help support its application in various domains, such as the type of the lifecycle, the openness of the processed data, and the centrality of the underlying system. Furthermore, the APDL model has been informally represented, which, in turn, affects the possibility of integrating such a model into an appropriate engineering process to elicit and model system requirements and to provide technical assurance.

We will next define a conceptual model that describes the problem and its solution in terms of the domain vocabulary as a prerequisite to any data lifecycle in the context of data protection. We intend to define a profile that allows the APDL model to be represented in the Unified Modeling Language (UML)[6] to illustrate how to address the complexity of practice, provide technical assurance and facilitate reasoning about compliance. Such a UML profile for the APDL model has the potential to complement the contributions of [3,12,13], by providing foundations for analysing functional requirements and assessing potential privacy risks. We also plan to use additional case studies with the aim of further validating the applicability of the model.

References

1. Shapiro, S.S.: Privacy by design: moving from art to practice. Commun. ACM **53**(6), 27–29 (2010)
2. Kost, M., Freytag, J.C., Kargl, F., Kung, A.: Privacy verification using ontologies. In: Proceedings of the Sixth International Conference on Availability, Reliability and Security (AReS 2011), pp. 627–632. IEEE (2011)
3. Gürses, S., Troncoso, C., Diaz, C.: Engineering privacy by design. In: Computers, Privacy and Data Protection, vol. 14 (2011)
4. Spiekermann, S.: The challenges of privacy by design. Commun. ACM **55**(7), 38–40 (2012)
5. Cavoukian, A.: Privacy by Design ... Take the Challenge. Office of the Information and Privacy Commissioner of Ontario (2009)
6. Cavoukian, A.: Privacy by Design: The 7 Foundational Principles Implementation and Mapping of Fair Information Practices (2010). https://www.ipc.on.ca/english/Resources/Discussion-Papers/Discussion-Papers-Summary/?id=953
7. Cavoukian, A., Shapiro, S., Cronk, R.J.: Privacy Engineering: Proactively Embedding Privacy by Design (2014). https://www.privacybydesign.ca/content/uploads/2014/01/pbd-priv-engineering.pdf
8. Antignac, T., Scandariato, R., Schneider, G.: A privacy-aware conceptual model for handling personal data. In: Margaria, T., Steffen, B. (eds.) ISoLA 2016. LNCS, vol. 9952, pp. 942–957. Springer, Cham (2016). https://doi.org/10.1007/978-3-319-47166-2_65
9. Möller, K.: Lifecycle models of data-centric systems and domains: the abstract data lifecycle model. Semant. Web **4**(1), 67–88 (2013)
10. United States Department of Health: Education and Welfare: Secretary's Advisory Committee on Automated Personal Data Systems: Records. Report. MIT Press, Computers and the Rights of Citizens (1973)

---

[6] http://www.omg.org/spec/UML/.

11. Cavoukian, A.: Creation of a Global Privacy Standard (2006). https://www.ipc. on.ca/images/Resources/gps.pdf
12. Spiekermann, S., Cranor, L.F.: Engineering privacy. IEEE Trans. Softw. Eng. **35**(1), 67–82 (2009)
13. Hoepman, J.-H.: Privacy design strategies. In: Cuppens-Boulahia, N., Cuppens, F., Jajodia, S., Abou El Kalam, A., Sans, T. (eds.) SEC 2014. IAICT, vol. 428, pp. 446–459. Springer, Heidelberg (2014). https://doi.org/10.1007/978-3-642-55415-5_38

Improvement of Security Costs Evaluation Process by Using Data Automatically Captured from BPMN and EPC Models

Dmitrij Olifer[(⊠)], Nikolaj Goranin, Justinas Janulevicius,
Arnas Kaceniauskas, and Antanas Cenys

Faculty of Fundamental Sciences, Vilnius Gediminas Technical University,
Saulėtekio al. 11, 10223 Vilnius, Lithuania
{dmitrij.olifer,nikolaj.goranin,
justinas.janulevicius,arnas.kaceniauskas,
antanas.cenys}@vgtu.lt

Abstract. Amount of security breaches and organizations' losses, related to them, is increasing every year. One of the key reasons is a high dependency of organization's key business processes on information and information technology. To decrease the risk of possible breaches, organizations have to ensure "due diligence" and "due care" principles. This means, organizations need to apply requirements or controls defined by existing security standards. One of the main issues in such approach is identification of critical areas and evaluation of cost for security requirements implementation.

In this paper we consider how our previously proposed method for information security requirements implementation cost evaluation could be linked with organizations' business processes. Our proposal could help us identify organization critical areas, which need to be protected and could let us to calculate security costs, related to the protected areas.

Keywords: Business process · Security standard · Security requirement
Security control

1 Introduction

According to the independent analysis performed by different organizations [1, 2], the amount of known/reported cyber security crimes is increasing each year. Losses related to the activities of these cyber security crimes are also increasing. Criminal organizations working in cyber area are well organized and are seeking to increase their profits.

Nowadays none of organizations could feel secure and protected against cyber-criminals attacks. Segregation by market sectors/areas, or organizations sizes is not valid anymore. As reported in 2016, the cyber attacks impact different sectors and areas, such as *government* (e.g., attack on DNC), *financial* (e.g., attack on Central Bank of Bangladesh, attack on Thailand state-run Government bank ATM) and international companies (e.g., fraud attack on Mattel CEO, and WADA data leakage). Following statistics given in [3, 4], cyber attacks do not depend on the organization size.

© Springer International Publishing AG 2018
E. Teniente and M. Weidlich (Eds.): BPM 2017 Workshops, LNBIP 308, pp. 698–709, 2018.
https://doi.org/10.1007/978-3-319-74030-0_56

Information security and data protection is becoming more and more important. However absolute information or data protection security could not be achieved. Each organization has to identify critical assets and to define needed level of information and data security. In other words organizations need to ensure "due diligence" and "due care" principles. From the security point of view "due care" is focused on taking reasonable ongoing care to protect the assets of organization. "Due diligence" is the background research related to security.

To ensure the "due diligence" and "due care" principles, organizations are implementing requirements defined in different IT Security standards. Some of these standards and acts are even mandatory to organizations working in the specific areas. For example Sarbane-Oxley Act 2002 [5] is mandatory for financial institutions; Payment Card Industry Data Security Standard 2016 [6] is mandatory for organization working with payment card holders' data; Health Insurance Portability and Accountability Act (HIPAA:2002) [7] is mandatory for organizations working in the US health assurance area and handling patient data.

From the security point of view, it is very important to identify organization's critical assets, which must be protected, and areas, where security requirements must be implemented. From the organization's point of view, it is vital to ensure that the cost of security solution implementation fits the purpose.

In our previous article [8] we have proposed a control-based method for organization security cost evaluation. However, during method verification, it was identified that initial security costs calculation requires manual actions, which are related to the gathering of information about organization. The goal of this paper is to improve the control-based security costs evaluation method. We propose a way to automate information gathering by extracting this information from the existing business processes models. In the following we will:

- explain the main principles of the control-based method for security cost evaluation (Sect. 2);
- explain what business processes models are the most suitable for our goal (Sect. 3) and provide what information needs to be gathered from the business processes models (Sect. 4);
- provide an approach which describes how information could be extracted from the business process models and imported to the control-based security cost evaluation method used for security cost calculation.

2   Information Security Requirements Costs Evaluation

In our previous work, we have verified existing evaluation methods for information security requirements implementation cost. These methods [9] were compared against 5 criteria: intelligibility for senior management, links with existing information security standards, calculation complexity, information security aspects coverage, and reusability.

The main goal defined of our information security implementation cost/benefits evaluation method is to calculate information security implementation costs/benefits, for organizations, which use two or more different security standards. The method steps and its calculation results should be understandable by senior management. It should be reusable and cover all security areas and controls types (such as administrative, technical, and physical).

The proposed method is based on two major areas for specific security requirements implementation: risk assessment process and security control implementation process. The main calculation formula (Eq. 1) is:

$$C_{Security} = \varphi \left(C_{Risk\_assessment} + \sum_{i=1}^{n} C_{Security\_control\_implementation_i} (standard) \right) \quad (1)$$

where φ – the complexity and maturity coefficient, $C_{Risk\_assessment}$ – risk assessment costs, which explanation will be defined and described below (Eq. 3) and $C_{Security\_control\_implementation_i}(standard)$ – security control implementation (Eq. 4).

φ – the complexity and maturity coefficient is calculated according to (Eq. 2):

$$\varphi = \frac{Complexity\_level}{Maturity\_level} \quad (2)$$

The risk assessment costs are calculated according to the following equation:

$$C_{Risk\_assessment} = C_{Asset\_analysis} + C_{Vulnerabilities\_analysis} + C_{Threat\_analysis} + C_{Impact}$$
$$+ C_{Penetration\_testing}(N) + C_{Gap\_analysis} \quad (3)$$

where $C_{Asset\_analysis}$ – costs related to critical asset analysis, $C_{Vulnerabilities\_analysis}$ – costs related to vulnerabilities analysis, $C_{Threat\_analysis}$ – costs related to threat analysis, $C_{Gap\_analysis}$ – costs related to gap analysis and $C_{Penetration\_testing}(N)$ – costs related to penetration testing needed for risk assessment, where N is amount of different organization systems, which have to be tested, C_{Impact} – costs related to impact evaluation.

Security control implementation costs are calculated according to (Eq. 4):

$$C_{Security\_control\_implementation} = \sum_{i=1}^{n} (m_i(Risk_i) * (C_{Mitigation\_strategy_i} + C_{Action_i})) \quad (4)$$

where $m_i(Risk_i)$ is control criticality coefficient

Mitigation strategy is defined as (risk acceptance; risk avoiding; risk remediation and risk transferring):

$$C_{Mitigation\_strategy} = \begin{cases} -C_{Action}, & where \ \frac{\Delta T(t_{in})}{T(l_j)} * \overline{W} \leq Risk\_apetite \ and \ C_{Action} \ is \ HIGH \\ 0, & where \ \frac{\Delta T(t_{in})}{T(l_j)} * \overline{W} \leq Risk\_apetite \ and \ C_{Action} \ is \ ACCEPTABLE \\ C_{Metrics\_control}, & where \ \frac{\Delta T(t_{in})}{T(l_j)} * \overline{W} > Risk\_apetite \ and \ C_{Action} \ is \ ACCEPTABLE \\ C_{insurance} + C_{Metrics_{control}} - C_{Action}, & where \ \frac{\Delta T(t_{in})}{T(l_j)} * \overline{W} > Risk\_apetite \ and \ C_{Action} \ is \ HIGH \end{cases} \quad (5)$$

where $\Delta T(t)$ – amount of security incidents during the defined time tin, $T(l_j)$ – amount of impacted systems, lj – asset impacted by security incident, j – asset number, $\overline{W}$ – impact average, $C_{Metrics\_control}$ – cost of metrics control operations, which could involve $C_{personal}$ and C_{Action} for additional specific tools, $C_{insurance}$ – cost of insurance, according to the signed off contract with the 3rd party.

Actions costs depends from implementation costs and operations costs (Eq. 6):

$$C_{Action} = C_{Implementation}(t) + C_{Operation} \qquad (6)$$

where $C_{Implementation}(t)$ is action implementation costs and $C_{Operation}$ is control operation costs.

$$C_{Implementation}(t) = C_{Environment\_purchase} + C_{deployment}(t) \qquad (7)$$

where $C_{Environment\_purchase}$ – are hardware and software procurement costs and $C_{deployment}(t)$ – are project deployment costs.

And operation costs could be calculated according to the equation (Eq. 8):

$$C_{Operation} = C_{Environment\_support} + \sum_{i=1}^{n} C_{Personal_i} + C_{Other\_services} \qquad (8)$$

where $C_{Environment\_support}$ is environment support costs, $C_{Personal_i}$ – organization employee costs, $C_{Other\_services}$ – cost of additional services needed for effective control functioning.

For more details on the method, detailed description, sample calculation and evaluation of application results please refer to [8].

3   Business Processes Model Techniques and Their Representation Tools

There exist a set of different business process definitions. However, they commonly state that business process is a collection/set of linked activities or tasks, that, once completed, will accomplish an organizational goal [10]. It is very important to have a clearly defined inputs and single output for the business process model. In our case, business process model would be a source to extract information about the main processes, stakeholders and related data of the organization. From the security point of view it is also very important to understand infrastructure, which will be used to handle these business processes.

Business processes could be presented in different ways. Johansson et al. [11] highlighted 4 graphical process oriented modeling techniques: Business Process Model and Notation (BPMN), UML-activity diagrams, Event-Driven process Chains (EPC) and flowchart/nodes maps. Aldin and de Cesare [12] presents a comparative analysis of business process modeling techniques. Their analysis involve flowchart, Petri Net, Data Flow Diagram (DFD), Role Activity Diagram (RAD), BPMN, business use case, and business object interaction diagram. These seven techniques for business process modelling were evaluated against flexibility, ease of use, understandability, simulation, and scope. It is important to note that Aldin and de Cesare [12] extract elements, which are commonly and generally accepted by the business modeling community. These elements are: process, activity, service and product, role, goal, event, and rule.

From the critical assets and environments identification point of view it is very important to identify the vital data, which will be involved in the business processes, and infrastructure/environment which will be handling this process. From the provided list of business process modelling techniques we have evaluated are common used notations, listed in Table 1.

Table 1. Business process components, which are able to provide information needed for security costs evaluation

Business process model techniques	Components, which could be used to present critical assets, stakeholders and infrastructure
BPMN	Artifacts (data object, groups and annotations)
EPC	Process owner, organization unit, information, material or resource object
Flowchart	Abstract or detailed description of units of work (rectangles), annotations
DFD	Entity and data store components

It is important to mention, that there exist a number of different tools for business processes representation (e.g., Microsoft Visio, SmartDraw, ConceptDraw, Luchicart, and other). These tools have a predefine list of objects, which later are used to present business process. Main disadvantages identified during the evaluation of these tools is a lack of libraries or classes for representation of infrastructure/environment components. This information can be presented in diagrams; however, it has to be entered manually by diagram creator through annotation, notes or other objects.

4 An Alignment of Security Requirements Cost Evaluation Method to Business Processes Models

At least 4 main calculations components (asset analysis; incident impact; control implementation and control operation) directly depend on organization business processes and elements, which are participating in them. If these elements are correctly defined in the business process diagrams, they will help in automating identification of organization's critical data assets. These assets (and their environment) could be integrated automatically in our security costs evaluation process.

To identify the most effective diagrams, we have to define the techniques evaluation criteria. We have chosen 2 criteria from set used by Aldin and de Cesare [12]. Other 3 criteria, proposed by author, are not so important for business process techniques integration to security cost evaluation process. 3rd criteria was proposed by us and used to evaluate data presented by business process techniques. In our evaluation we will use:

1. **Availability of details needed for security cost evaluation.** From the proposed security cost evaluation method point of view, very important *to identify hardware*

and software, which are participating in the critical business processes. Also it is *important to define the key stakeholders* or organization employees who are implementing these business processes and *data*, which is used in the processes. Thus, the business process diagram should have the possibility to present details about the components mentioned above.

2. **Ease-of-use.** This criterion helps us to understand the extent, to which the business stakeholders who do not have specialist knowledge of the technique *could be ready to apply* the *business process model technique*. It is important, because business process diagrams will be developed by different types of specialists and will need to have sufficient level of details about components used to ensure this process.

3. **Understandability.** This criterion helps us to understand the extent to which the business stakeholders who do not having specialist knowledge of the technique *could understand the business process model technique*. Argument for choosing this criterion is the same as for previous one.

The above identified business process modeling methods (i.e., BPMN, EPC, Flowchart, and Data Flow Chart) where evaluated against these 3 criteria. BPMN and EPC were identified as the most effective, because they provide more information needed for security cost evaluation.

In order to verify these models as the sources for initial input for the implementation costs evaluation method, we used the same simulation as it was done during verification of the security costs evaluation method itself. For the experiment an abstract organization ACME and implementation of logging and monitoring control (mandatory according to all IT security standards) processes were taken.

In Figs. 1 and 2 we can observe the kind of information they could provide and the kind of information is missing. It is necessary to be mentioned, that some diagram components are able to provide only part of needed information and some components are participating in the evaluation, however are not directly related to the control implementation costs.

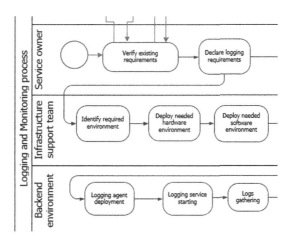

Fig. 1. Part of logging and monitoring process BPMN diagram

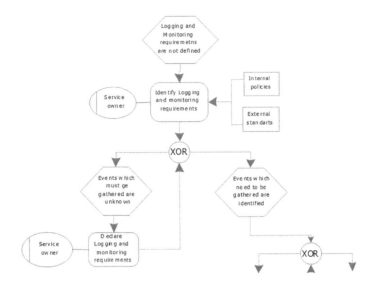

Fig. 2. Part of logging and monitoring process EPC diagram

To present a variety in our analysis we have used four classes: provide all needed information (Full), provide part of needed information (Partial), required information do not directly linked with control implementation (Not linked), and do not provide needed information (Missing). Table 2 summarizes our results.

The analysis shows that the possibility to use business process diagrams for security controls implementation directly depends on the level of details provided. Some parts of cost evaluation method calculation are out of the scope, since they are related to the organization itself and could not be identified from separate business processes diagrams. Other calculations are related to the risk assessment process, which is used to identify existing risk.

Table 2. Evaluation results

The cost evaluation method component	BPMN and EPC	Comments
Complexity and maturity coefficient	Not linked	This component is common for the whole organization and from the separate control evaluation point of view it could be ignored
Assets analysis	Partial	Figures 1 and 2 allows identification of main stakeholders and infrastructure participating in the process, but it does not provide enough information on software and hardware details
Vulnerability analysis	Partial	Lack of hardware and software details. Lack of information on organization procedures and policies. Automation is not possible

(*continued*)

Table 2. (*continued*)

The cost evaluation method component	BPMN and EPC	Comments
Threats analysis	Missing	This part of cost evaluation is performed by a security consultant; thus the diagram does not have this information
Impact analysis	Partial	Diagram provides details on assets, however could not provide information about losses
Penetration testing	Missing	This activity is optional and should not be present in business as a usual process, unless yearly test is planned
Gap analysis	Partial	Our diagrams can provide a part of needed details on infrastructure, however the list of applied security requirements is out of scope
Control criticality coefficient	Not linked	Risk identification is a result of risk assessment activities, which are out of scope for us
Mitigation strategy	Not linked	Mitigation strategy, from the business process point of view, is not directly related to business process flow
Environment purchase (Action implementation)	Partial	Business process flow allows identification of environment, which is needed to ensure process, however lack of details does not allow to define hardware or software
Implementation team (Deployment – Action implementation)	Full	Business process diagram allows us to identify teams, which will be participating in a new environment implementation process
Configuration tasks (Deployment – Action implementation)	Partial	Business process diagram does not provide enough details on hardware and software configurations, because of that it could be difficult to identify costs related to these activities
Training/Awareness (Deployment – Action implementation)	Partial	Information on infrastructure and main stakeholders could help in identifying missing trainings, but lack on details does not allow us to ensure, that all needed trainings are identified
Supporting team (Operation activities)	Full	Business process allows us to identify all main stakeholders and their functions in the defined business process flow
Environment support (Operation activities)	Partial	Business process allows us to identify infrastructure, which will be used, but lack of details does not allow to identify the level of support that will be needed and its price
Other services (Operation activities)	Partial	This component is optional and will not be presented in all business processes

However, it is necessary to be mentioned, that business process can provide a part of information for risk assessment, while other components, such as threat and vulnerabilities assessments, could not be linked to business processes.

5   Ways of Costs Evaluation Methods Automation

The analysis of existing business process modelling tools has shown that they all have a possibility to store data process in a portable format, which is close to XML or can be converted to the XML format. Such a possibility allows us to automate process. The high-level process design is presented in Fig. 3:

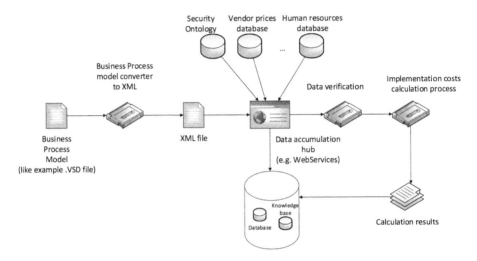

Fig. 3. Business process model data integration with the cost evaluation method

The initial file of business process diagram is generated by one of business process modeling tools (in our experiment Microsoft Vision was used). The generated file is transferred to a separate layer, where file data was converted to the XML format. Then the generated XML files are transferred to the middleware application. This application should help to extract and to collect data needed for the security costs evaluation. Data sources could be different, starting from business process models, security standards and finishing with ontologies [13]. As a middleware layer could be used Web-services solutions. In our experiment as middleware layer was used techniques, which allowed us to import data to Excel format. This data later was for security control cost calculations. Additional information required for calculations was entered into the XLSX spreadsheet manually.

It is necessary to be mentioned that in the principle diagram, the additional component "Data conversion verification" is used. In our experiment, this verification was done manually, by reviewing converted data. However from the automation point of view, this process should be independent from the manual review.

The last component is the calculation process. In our experiment for this purpose was used - Microsoft Excel application. Data from business process model was imported to Excel spreadsheet with help of open sources tools. After that the Microsoft Excel was used to calculate security cost for the Logging and Monitoring control.

6   Results and Discussion

The proposed approach for automatic data gathering from business processes and gathered data integration to security cost evaluation process was verified in the experiment, which was performed in the ACME organization. The cost evaluation was performed for the Logging and Monitoring control system. To simplify the experiment, the middleware components were replaced with the open source conversion and import tools while verification and additional data entry was performed manually.

The performed experiment has proved, that the approach could be used to gather initial information needed for security cost evaluation. However several gaps needs to be handled to make this approach more effective. The main issues spotted by the experiment are:

- a lack of details in the business process models. However this important information was not presented in the business process model developed by the business process owners by default;
- a lack of details on environment or activities costs. It is important to know the number of man-months and respective price to estimate the cost. Unfortunately the default business process model does not provide this data. This issue could be solved by performing additional mapping against vendors prices or data provided by the human resources (in case of internal resources used);
- a lack of details related to the organizational (policies, procedures, guidelines) controls, which are in place in the organization.
- a manual approach for a part of activities, that require further automation or application using more sophisticated tools.

During the experiment the following valuable features of the method proposed were approved:

- an approach allows us to collect initial information needed for security requirements cost implementation evaluation from sources, which are well understood by the business owners. Business process models help business owners to identify assets and stakeholders vital for their business effectiveness. The use of the XML format allows us to easily integrate the proposed model with other input systems needed for cost evaluation methods, if their data is presented in a compatible format. In future implementation WebServices could be used to integrate existing solution with the security ontology or other valuable sources of information.
- the approach can automate the cost evaluation method, because a huge amount of data and calculation could be done without operator interference. It is still critically important to automate the data verification process, which would allow us to ensure that data from different sources was imported without any errors.

7 Conclusion

The performed experiment has proved that the integration of the proposed business process modelling techniques and the cost evaluation method is possible. It can be automated and provides a set of advantages. However, we need to agree on several key principals, which should be implemented in the integration solution:

1. Data presented in business process must have a strict structure. This strict structure would help to minimize amount of errors during the integration and would provide a guidance for business owners on what data should be included and in which format.
2. Integration components should be implemented using the WebServices. As noted in [13], the approach could be integrated with at least security ontologies and could help us to identify security requirements according to security standards applicable in the organization.
3. One of the most important things, which should be implemented, is data import/integration verification. It should be an automatic process, which would let us identify import errors and to inform system owner about them. To increase the effectiveness of the proposed solution, the knowledge base can be created. Such a knowledge base would allow us to store information on controls and environment. In this way the approach would support the information reusability.

In the future we are going to implement the proposed approach on the base of WebServices, to ensure integration with security ontologies as a source base, and to implement data import verification.

References

1. PricewaterhouseCoopers: Information Security Breaches survey conducted by PwC (2016). http://www.pwc.be/en/news-publications/publications/2016/information-security-breaches-survey.html
2. Information Security Forum. Threat Horizon 2017 – Executive summary. https://www.securityforum.org/research/threat-horizon-2017-dangers-accelerate
3. Hackmageddon – Information Security timelines and statistics. http://www.hackmageddon.com/2017/01/19/2016-cyber-attacks-statistics
4. Symantec: 2016 Internet Security Threat Report. https://www.symantec.com/security-center/threat-report
5. Sarbane-Oxley act of 2002: US mandatory regulatory requirements
6. PCI DSS: Payment Card Industry Data Security Standard. International Information Security standard (2016)
7. HIPAA: Health Insurance Portability and Accountability Act. US mandatory regulatory requirements for Health Insurance sector (2002)
8. Olifer, D., Goranin, N., Kaceniauskas, A., Cenys, A.: Controls-based approach for evaluation of information security standards implementation costs. Technol. Econ. Dev. **23-1**, 196–219 (2017)

9. Brecht, M., Nowey, T.: A closer look at information security costs. In: Böhme, R. (ed.) The Economics of Information Security and Privacy. Springer, Heidelberg (2013). https://doi.org/10.1007/978-3-642-39498-0_1

10. Appian. About BPM – Business process definition. http://www.appian.com/about-bpm/definition-of-a-business-process

11. Johansson, L.O., Warja, M., Carlsson, S.: An evaluation of business process model techniques, using Moody's quality criterion for a good diagram. In: 11th International Conference on Perspectives in Business Informatics Research, pp 54–64. CEUR Workshop, Niznij Novgorod (2012)

12. Aldin. L., de Cesare, S.: A comparative analysis of business process modelling techniques. In: UK Academy for Information Systems Conference Proceedings 2009, Oxford, UK (2009). http://aisel.aisnet.org/ukais2009/2

13. Ramanauskaite, S., Olifer, D., Goranin, N., Cenys, A.: Security ontology for adaptive mapping of security standards. Int. J. Comput. Commun. Controls **8-6**, 878–890 (2013)

Joint International BPM 2017 Workshops on Theory and Application of Visualizations and Human-Centric Aspects in Processes (TAProViz 2017), Process Querying (PQ 2017), and Process Engineering (IWPE 2017)

Introduction to the Joint International BPM 2017 Workshops on Theory and Application of Visualizations and Human-Centric Aspects in Processes (TAProViz'17), Process Querying (PQ'17) and Process Engineering (IWPE'17)

Ross Brown[1], María Teresa Gómez-López[2], Simone Kriglstein[3], Henrik Leopold[4], Artem Polyvyanyy[1], Manfred Reichert[5], Stefanie Rinderle-Ma[6], Arthur H. M. ter Hofstede[1], Lucinéia Heloisa Thom[7], and Pablo David Villarreal[8]

[1] Queensland University of Technology, Brisbane, Australia
{r.brown,artem.polyvyanyy,a.terhofstede}@qut.edu.au
[2] Universidad de Sevilla, Seville, Spain
maytegomez@us.es
[3] Vienna University of Technology, Vienna, Austria
kriglstein@cvast.tuwien.ac.at
[4] VU University Amsterdam, Amsterdam, The Netherlands
h.leopold@vu.nl
[5] Ulm University, Ulm, Germany
manfred.reichert@uni-ulm.de
[6] University of Vienna, Vienna, Austria
stefanie.rinderle-ma@univie.ac.at
[7] Universidade Federal do Rio Grande do Sul, Porto Alegre, Brazil
lucineia@inf.ufrgs.br
[8] National Technological University, Santa Fe Faculty, Santa Fe, Argentina
pvillarr@frsf.utn.edu.ar

1 Introduction

BPM 2017 ran a number of joint workshops as part of the programme in Barcelona. Three workshops were merged to form a full day programme, viz. Theory and Application of Visualizations and Human-centric Aspects in Processes (TAProViz'17), Process Querying (PQ'17) and Process Engineering (IWPE'17). The next three sections introduce each of their programmes, organising committees and details of the papers published.

2   Sixth International Workshop on Theory and Application of Visualizations and Human-Centric Aspects in Processes (TAProViz'17)

Visualizations can make the structure and dependencies between elements in processes accessible in order to support users who need to analyze process models and their instances. However, effectively visualizing processes in a user-friendly way is often a big challenge, especially for complex process models, which can consist of hundreds of process components (e.g., process activities, data flows, and resources) and thousands of running process instances in different execution states. Many challenges remain to be addressed within the broad area of process visualization, human interaction and user led design such as: scalability, human-computer interaction, cognitive aspects, applicability of different approaches, collaboration, process evolution, run-time requirements of process instances and applications, user-engagement etc. The TAProViz workshop seeks to develop an international forum of quality to present and discuss research in this field. Three papers from four submissions were accepted for publication at the workshop. The presented papers fostered a lively discussion about the place of visualisation within the BPM research and practice community.

2.1 Introduction to TAProViz

This is the sixth TAProViz workshop being run at the 15th International Conference on Business Process Management (BPM). The intention this year is to consolidate on the results of the previous successful workshops by further developing this important topic, identifying the key research topics of interest to the BPM visualization community.

Submitted papers were evaluated by at least three program committee members, in a double blind manner, on the basis of significance, originality, technical quality and exposition. Three full papers were accepted for presentation at the workshop from four submissions.

The papers address a number of topics in the area of process model visualization, in particular:

- Aspect Oriented Business Process Modeling
- Human Physical Risks in Manufacturing Processes using BPMN
- Visual Analytics for Soundness Verification of Process Models

In their full paper, *AO-BPM 2.0: Aspect Oriented Business Process Modeling*, Luiz Paulo Silva, Flavia Santoro and Claudia Cappelli, described improvements in an existent aspect-oriented process modeling notation, enhancing readability, graphical scalability and simplicity. A case study in a real setting was presented to evaluate whether the notation produced a better visual representation.

Melanie Polderdijk, Irene Vanderfeesten, Jonnro Erasmus, Kostas Traganos, Tim Bosch, Gu van Rhijn and Dirk Fahland, presented their full paper *A Visualization of Human Physical Risks in Manufacturing Processes using BPMN* introducing a new extension that visualizes human physical risks (such as heavy lifting or repetitive work). Based on this graphical representation, users can see where in the process workers may encounter physical risks that should be mitigated through process redesign.

In the final full paper, *Visual Analytics for Soundness Verification of Process Models*, H. S. Garcia Caballero, M. A. Westenberg, H. M. W. Verbeek and Wil M. P. van der Aalst presented their solution to the problem of validating the soundness property of a process model via a novel visual approach and a new tool called PSVis (Petri net Soundness Visualization). The PSVis tool aims to guide expert users through the process models in order to obtain insights into the problems that cause the process to be unsound.

2.2 TAProViz Organizers

Ross Brown   Queensland University of Technology
Simone Kriglstein   Vienna University of Technology
Stefanie Rinderle-Ma University of Vienna

2.3 TAProViz Program Committee

Philip Abels, Germany
Massimiliano De Leoni, Netherlands
Phillip Effinger, Germany
Kathrin Figl, Austria
Hans-Georg Fill, Austria
Agnes Koschmider, Germany
Maya Lincoln, Israel
Cristiano Maciel, Brazil
Luciana Salgado, Brazil
Flavia Santoro, Brazil
Pnina Soffer, Israel
Irene Vanderfeesten, Netherlands
Eric Verbeek, Netherlands
Günter Wallner, Austria

3   Second International Workshop on Process Querying (PQ'17)

The objective of the second edition of the Process Querying (PQ 2017) workshop was to provide a forum to exchange findings and ideas on process querying research and practices. The program of the workshop included two accepted research papers (50% acceptance rate); each paper was refereed based on its full content by four program committee members. The two presentations of the accepted papers on the workshop day addressed the topics of multi-perspective process querying and methods for managing collaborative business processes. The workshop took place on September 11, 2017, in Barcelona, Spain.

3.1 PQ Aims and Background

The Second International Workshop on Process Querying (PQ 2017) aims to provide a high quality forum for researchers and practitioners to exchange research findings and ideas on technologies and practices in the area of process querying. *Process querying* studies (automated) methods for managing, e.g., filtering or manipulating, repositories of models of observed and envisioned processes, as well as their relationships, with the goal of converting process-related information into decision making capabilities.

Process-related information grows exponentially in organizations via workflows, guided procedures, business transactions, Internet applications, real-time device inter-actions, and other coordinative applications underpinning commercial operations. Event logs, application databases, process models, and business process repositories capture a wide range of process data, e.g., activity sequences, document exchanges, interactions with customers, resource collaborations, and records on product routing and service delivery. Process querying research spans a range of topics from theoretical studies of algorithms and the limits of computability of techniques for managing process-related information to practical issues of implementing process querying technologies in software.

Special Theme: Research combining process models and ontologies is increasingly gaining attention in recent years. One reason for this is that ontologies allow adding semantics to process models, which enables the automated inference of knowledge from business processes. This knowledge can be used to manage business processes at design and execution time. Hence, the goal of the special theme is to promote research on the application of ontologies to generate new or improve existing methods, tech-niques, tools, and process-aware systems that support the different phases of the business process management life cycle.

3.2 PQ Main Topics

The main topics of the Process Querying 2017 workshop include:

- Behavioral and structural methods for process querying
- Imperative and declarative process querying methods
- Exact and approximate process querying methods
- Expressiveness of process querying methods
- Decidability and complexity of process querying methods
- Process query languages and notations
- Indexing for fast process querying
- Empirical evaluation and validation of process querying methods
- Label management in process querying
- Information retrieval methods in process querying
- Process querying of big (process) data
- Automatic management of process models, e.g., process model repair
- Automatic management of process model collections
- Event log querying
- Event stream querying
- Process performance querying

- Multi-perspective process querying methods
- Process querying and rich ontology annotations
- Applications of process querying methods
- Experience reports from implementations of process querying tools
- Case studies in process querying

3.3 PQ Workshop Organizers

Artem Polyvyanyy	Queensland University of Technology, Australia
Arthur H. M. ter Hofstede	Queensland University of Technology, Australia
Henrik Leopold	VU University Amsterdam, The Netherlands
Lucinéia Heloisa Thom	Universidade Federal do Rio Grande do Sul, Brazil
Pablo David Villarreal	National Technological University, Argentina

3.4 PQ Program Committee

Agnes Koschmider	Karlsruhe Institute of Technology, Germany
Ahmed Awad	Cairo University, Egypt
Alistair Barros	Queensland University of Technology, Australia
Andreas Solti	Vienna University of Economics and Business, Austria
Artem Polyvyanyy	Queensland University of Technology, Australia
Arthur H. M. ter Hofstede	Queensland University of Technology, Australia
Avigdor Gal	Technion – Israel Institute of Technology, Israel
Boudewijn van Dongen	Eindhoven University of Technology, The Netherlands
Chun Ouyang	Queensland University of Technology, Australia
Claudio Di Ciccio	Vienna University of Economics and Business, Austria
David Knuplesch	Ulm University, Germany
Dirk Fahland	Eindhoven University of Technology, The Netherlands
Gero Decker	Signavio, Germany
Henrik Leopold	VU University Amsterdam, The Netherlands
Hyerim Bae	Pusan National University, South Korea
Jochen De Weerdt	Katholieke Universiteit Leuven, Belgium
Joos Buijs	Eindhoven University of Technology, The Netherlands
Jorge Munoz-Gama	Pontificia Universidad Católica de Chile, Chile
Luciano García-Bañuelos	University of Tartu, Estonia
Manfred Reichert	Ulm University, Germany
Lucinéia Heloisa Thom	Universidade Federal do Rio Grande do Sul, Brazil
Marcello La Rosa	Queensland University of Technology, Australia
Massimiliano de Leoni	Eindhoven University of Technology, The Netherlands
Matthias Weidlich	Humboldt University of Berlin, Germany
Minseok Song	Pohang University of Science and Technology, South Korea
Pablo David Villarreal	National Technological University, Argentina
Remco Dijkman	Eindhoven University of Technology, The Netherlands
Seppe vanden Broucke	Katholieke Universiteit Leuven, Belgium
Wil M. P. van der Aalst	Eindhoven University of Technology, The Netherlands

4 Third International Workshop on Process Engineering (IWPE'17)

The objective of 3rd International Workshop on Process Engineering (IWPE 2017) was to provide a forum for stimulating discussions of engineering approaches on the edge of software engineering and business process management. The program of the workshop included one accepted research paper (50% acceptance rate); each submitted paper was refereed by four program committee members. The presentation of the accepted paper on the workshop day carried on the necessity to integrate simulation services for process models in a feasible and flexible way. The workshop took place on September 11, 2017, in Barcelona, Spain.

5 IWPE 2017 Aims and Background

The 3rd International Workshop on Process Engineering (IWPE 2017) aims at bringing together researchers and practitioners interested in the engineering aspects of process-oriented information systems.

Business process management as a scientific discipline has been very successful in developing concepts, languages, algorithms, and techniques in different aspects of the domain. Many of those concepts and techniques, however, did not yet find their way to the operational business of companies. One of the reasons for the weak uptake of research results in business process management is the lack of research in engineering aspects of process-aware information systems.

Engineering focuses on the entire value chain of system development, from the elicitation of business requirements to the engineering of suitable architectures, components and user interfaces of process systems, as well as their testing, deployment, and maintenance. The proposed workshop aims at providing a forum for researchers and practitioners who are interested in all engineering aspects of the design and implementation of process systems.

6 IWPE 2017 Main Topics

The main topics of the International Workshop on Process Engineering 2017 workshop include:

- Methodologies for the design and engineering of process applications
- Requirements engineering of process systems
- Testing process systems
- Deployment aspects of process systems
- Novel architectures in process design and implementation
- Design of process engines
- Implementation concepts for large-scale process systems
- Empirical aspects of process systems
- Lessons learned from process implementation projects

- Service-based processes
- Cloud-based process enactment
- Smart processes in the physical world
- Event management in process systems

7   IWPE 2017 Workshop Organizers

Manfred Reichert	Ulm University, Germany
Stefanie Rinderle-Ma	Universität Wien, Austria
María Teresa Gómez-López	University of Seville, Spain

8   IWPE 2017 Program Committee

Leonardo Azevedo	Federal University of State of Rio de Janeiro, Brazil
Ruth Breu	Research Group Quality Engineering, Austria
Fabio Casati	University of Trento, Italy
Florian Daniel	Politecnico di Milano, Italy
Andrea Delgado	Instituto de Computación, Facultad de Ingeniería, Universidad de la República, Uruguay
Schahram Dustdar	TU Wien, Austria
Robert Heinrich	Karlsruher Institute of Technology, Germany
Angel Jesus	Varela Vaca, University of Seville, Spain
Dimka Karastoyanova	Kühne Logistics University, Germany
Ekkart Kindler	Technical University of Denmark, DTU Compute, Denmark
Akhil Kumar	Penn State University, USA
Marcello La Rosa	Queensland University of Technology, Australia
Frank Leymann	Institute of Architecture of Application Systems, Germany
Cesare Pautasso	University of Lugano, Switzerland
Manuel Resinas	University of Seville, Spain
Antonio Ruiz-Cortés	University of Seville, Spain
Andreas Solti	Vienna University of Economics and Business, Austria
Farouk Toumani	Limos, Blaise Pascal University, Clermont-Ferrand, France
Pablo Villarreal	CIDISI - Universidad Tecnológica Nacional - Facultad Regional Santa Fe, Argentina
Barbara Weber	Technical University of Denmark, Denmark
Matthias Weidlich	Humboldt-Universität zu Berlin, Germany
Mathias Weske	Hasso-Plattner-Institut, Germany

AO-BPM 2.0: Aspect Oriented Business Process Modeling

Luiz Paulo Carvalho[✉], Claudia Cappelli, and Flávia Maria Santoro

Federal University of the State of Rio de Janeiro, Rio de Janeiro, Brazil
{luiz.paulo.silva,claudia.cappelli,
flavia.santoro}@uniriotec.br

Abstract. Crosscutting concerns are process elements that, although not part of the core, permeate process models. Traditional methods of business process modeling do not usually address crosscutting concerns, causing them to remain scattered around the process model, hindering understanding and flexibility. The aspect orientation is a paradigm that provides mechanisms to modularize crosscutting concerns. The goal of this paper is to describe the improvements in an existent aspect-oriented process modeling notation, enhancing readability, graphical scalability and simplicity. A case study in a real setting is presented to evaluate whether the notation can produce a better visual representation.

Keywords: Graphical scalability · Aspects · Process visualization

1 Introduction

The growth of the Business Process Management (BPM) field has increased as much as the complexity and content of business processes models. All at once, the aspect orientation paradigm from software engineering has emerged to decrease complexity [6]. Primarily restricted to software development, aspects orientation has been also addressed in BPM by supporting the modeling of processes focusing on interests, i.e. more specifically on the crosscutting concerns [7], interlaced within the models. We can find different elements spread, entangled and repeated within process models, such as: activities, connectors, gateways, events; increasing the complexity and the possibility of errors [9]. Crosscutting concerns are not necessarily part of the core objective of the process and might harm the visualization of the model.

The main unit of modularity is the process itself, while a crosscutting concern is spanned in multiple processes [9]. Typical examples of these elements in process models include logging, error handling, and security (e.g., an activity which performs logging intertwines with other primary activities of a process). As the crosscutting behavior is scattered across some processes and tangled with other concerns, it is difficult to find and modify it. Therefore, process comprehension, evolution, and reuse are difficult to be practiced. Using aspect-oriented concepts, it is possible to modularize the crosscutting concerns and separate them from the main process flow.

There are some approaches to aspect-oriented business process modeling, such as AO4BPMN [8], AO-BPM [7] and AOBPMN [11]. Each notation successively refined concepts poorly addressed by the previous one, e.g. AO-BPM created an object to

© Springer International Publishing AG 2018
E. Teniente and M. Weidlich (Eds.): BPM 2017 Workshops, LNBIP 308, pp. 719–731, 2018.
https://doi.org/10.1007/978-3-319-74030-0_57

represent pointcuts, absent in AO4BPMN; AOBPMN presented the concept of precedence, allowing the ordering of aspects in the graphic model itself by the advice, which was absent in AO-BPM. Nevertheless, none of them provide an efficient visual representation to build a useful conceptual model. So, our work uses AO-BPM as the basis for designing a new notation, named AO-BPM 2.0, which aims to present improvements to AO-BPM in the conceptual modeling of business processes.

The goals of this paper are twofold: to propose a simple and of easy understanding aspect-oriented business process modeling notation, representing the crosscutting concerns and their respective actions; followed by a case study in a real setting to evaluate its capabilities and potential for representativeness.

This paper is structured as follows: Sect. 2 presents the background of this research; Sect. 3 presents the proposed notation, its characteristics, restrictions and potentials; Sect. 4 discusses the case study, Sect. 5 describes related work, and Sect. 6 concludes the paper with the final considerations and future work.

2   Research Background

2.1   Aspect Orientation in Programming and Business Process

This aspect paradigm comes from programming [6]. Requirements, such as reusability, performance and reliability, are examples that are not directly related to the system functionalities, but impact its overall quality and therefore cannot be neglected. Those concerns are scattered around the code, causing problems related to maintenance and reuse. The aspect orientation uses the orthogonal modularization [12], representing the aspects as separate entities. A concern is a functionality of a system which can be separated into two types: core and crosscutting. Core concerns are the module's main functionalities, i.e., the main task such module executes. Crosscutting concerns are peripheral functionalities that permeate various modules. Generally, they are non-functional requirements and do not relate to business rules.

The aspect orientation presents elements that enable the connection of transversal interests, which are the join point, the pointcut and the advice: (i) Aspects can only be invoked at well-defined points throughout program execution, called **join points**, which specify how or when aspects are invoked; (ii) A **pointcut** describes the set of junction points where aspects are invoked, using a series of predicates that detail every possible set of join points in a software application, thus determining the correct join point for each aspect; (iii) An **advice** is the implementation code of the aspect that is executed. In the context of processes, an advice is the description of the aspect itself, what it is and what it does [6].

The aspect orientation in BPM was started in the execution phase with AO4BPEL [13], followed by modeling, generating graphical representations. In business process, the aspects are connected to the main flow by the same elements, re-signified for conceptual modeling. Join points are points in the model where crosscutting concerns can be integrated; pointcuts represent the information about the cut in model where one or more aspects will be invoked; and the advice that indicates when the aspect will be invoked: before, around or after the join point [9]. Aspect-oriented business process modeling notations use orthogonal modularization [12] to unify and gather crosscutting concerns [7, 11]. When modeling complex processes, a hierarchical method is often an absolute necessity. This modularization strategy, based on the recognition of sub-processes, has another important advantage, since it allows reusing processes. Besides, users of process models need distinct types of visualization perspectives depending on their purposes: functional (process steps); data flow (data used in a process); operational (which operation is invoked to execute a process step); organizational (agents responsible to perform process steps); behavioral (causal dependencies among elements). One orthogonal perspective is the aspect representation, depending on the purpose of the visualization and domain modeling.

2.2   AO-BPM

Based on the aspect-oriented paradigm, Cappelli et al. proposed the Aspect Oriented Business Process Management (AO-BPM) [7] to improve the modularity and graphically represent aspects in process models. AO-BPM divides the process into two distinct elements: the core process, which contains the essence of the business; and the aspectual process that captures crosscutting information within the process core. Elements of a process model such as activities, events, and resources can be identified and represented as crosscutting interests [7] and these can be identified in the context of the same process (intra-process) or between different processes (inter-process).

Crosscutting interests are represented on a separate lane, orthogonal to the main elements, thus highlighting that they are interwoven with various activities. In this model, pointcuts are represented as ground elements, black arrows, and located near the core process element. The notation, its syntax and semantics are detailed in [7], not being addressed here in view of space limitations. Figures 1 and 2 show the same process modeled in BPMN and AO-BPM notation respectively. They both depict a software acquisition process. Non-functional activities are repeated many times throughout the process, making the model large and complex [16]. Crosscutting concerns are present in the vertical lane, and the crosscutting relationship connector, represented by the dashed line, indicates the elements that affect them.

In AO-BPM, a new connector, called crosscutting relationship, has the function to represent a transverse element (the origin) affecting another element (the target), in addition to representing the interaction between transversal interests and core elements. Although representing transversal interests and their influences on the process model, respecting the aspect orientation paradigm, it has some problems in conceptual modeling: (i) the loss of readability is proportional to the scalability increase of the scenario, making the model polluted [16]; (ii) to order the aspects, stating whether they will come before, after or around the grounded element, a textual description [7] is required in addition to the model; (iii) the size and complexity of the textual description is also proportional to the scalability of the scenario; (iv) model maintenance is hampered by overlapping crosscutting relationships; (v) there is no proposed solution to repeated aspects in a portfolio, e.g., if ten models have three equal aspects to each other, they will still be scattered, only in the portfolio and not in the models; (vi) even if the notation declares this possibility, there is no demonstration of the operation of resources as aspects [7], as observed in Fig. 2, if one aspect is performed by N actors, then a conjoint

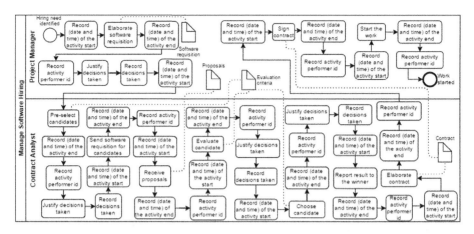

Fig. 1. Manage software acquisition business process model using BPMN (adapted from [1])

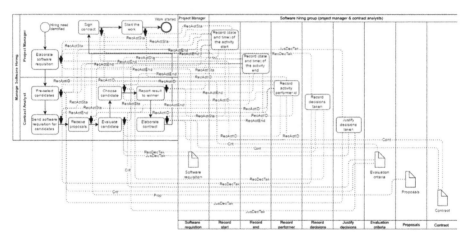

Fig. 2. Manage software acquisition process model using AO-BPM (adapted from [1])

lane will be created for the N (otherwise they would have to repeat each activity for each of the N actors); (vii) AO-BPM is not easy or intuitive to understand [15]. Thus, we propose AO-BPM 2.0 as a new graphical notation which is symmetrically compatible with the standard BPMN. We describe the notation in the next section.

3   AO-BPM 2.0

AO-BPM 2.0 is a symmetric extension to BPMN, which means that basic elements and crosscutting concerns are not distinguished in the specific abstraction. A specific composition mechanism composes the crosscutting elements with the basic elements [7]. The goal is to improve the graphic model of AO-BPM, aiming to solve the problems listed at the end of Sect. 2.2.

3.1   AO-BPM 2.0 Presentation

AO-BPM 2.0 uses the same elements for aspects operationalization of AO-BPM, i.e. pointcuts, join points and advices; however, they may have different syntax or semantics. The join point is represented as a small circle, attached to the process flow, having an identification number of the aspect within it. Several join points can occur in the same flow, so each join point needs its own circle with the identification of the aspect to which it refers. The advice is described by a text when the textual model is assembled to aid in the execution of the artifact connected to the join point. Finally, the pointcut should be presented by a textual description (which will not be modified), where the aspect has its pointcuts identified. This part should be next to the pool that presents the activities related to the aspect.

The aspect modeling should be done in a different pool from the core process. This change was proposed considering that, in small models, the way the aspect is presented in the AO-BPM, next to the model, works very well, but for large models, it can make it very difficult to read. The following pattern should be followed in the aspect pool: (i) A **lane** must be created for each aspect, and the identification must contain the element that will be used to identify the aspect in the business process model; (ii) The **resources** must be specified through "sub-lanes", since the aspect involves the actors that are defined in their model. They may have resources in the pool of aspects that will not be present in the main model; (iii) The **aspect** is a "part" of the process, so it must have its dependence and connection in the model, but it is not necessary to maintain the respective indicative flows. These flows must be implicit, on arrival at the first activity and at the exit of the last activity in the pool of aspects.

3.2   Aspect Objects and Scope Modeling

The identification of an aspect in the business process model is done through an identification number within a circle attached to the flow. The junction of the circle with the number will be the artifact denoting a join point. The purpose of the numbering is to label and track the aspects, but not to denote order of execution, e.g. in a process model, aspect 3 can be executed before aspect 1. There may be multiple join points in the same flow. To identify the execution sequence, the flow direction is used: the closest join point of the flow direction indication will be the last one to be executed, the farthest from this indication being the first one to be executed. Unlike AO-BPM, needing the textual description [7], it is possible to explicitly and graphically show the sequence of aspects. The advice meaning is graphically denoted in the model, showing if the join point is located before, around or after the point cut. A pool adjacent to the core process graphically represents the encapsulated aspects. Each aspect is contained in a lane of this pool, and the same lane can contain sub lanes for the specification of actors who carry out their respective activities. Aspects do not have initial or final events that delimit their internal flow. This behavior is observed in the BPMN[1] for sub-processes, where the presence of initial or final events it is not mandatory. Figure 3

---

[1] http://www.omg.org/spec/BPMN/2.0.

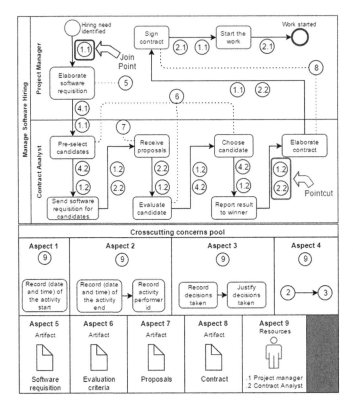

Fig. 3. Change management business process model built using AO-BPM 2.0

illustrates the same process shown in Fig. 1, but this time following the proposed notation, AO-BPM 2.0.

Figure 3 exemplifies the elements in AO-BPM 2.0. A join point is the invocation of a specific aspect of the aspect pool. A pointcut can have one or more join points, and the exact cutting point can call several aspects. Aspects can be instrumented by other aspects if, for example, aspect 4 uses aspect 2 and 3. One aspect, such as aspect 1, may have a more elaborate structure (respecting the rules already presented), while others, such as aspect 2, may only two tasks. Between task "Report result to winner" and "Elaborate contract", two join points (a pointcut) are called in sequence, and the aspect number does not determine its order of call, but its position between the two elements. In this case, aspect 2 will be executed first and aspect 1 secondly. The syntax for resource aspects consists of an index of suffixes representing them in redundant aspects: [aspect number] + [dot] + [resource suffix]. The suffix (number after dot) determines the actor performing the task, e.g., in a 1.1 join point aspect, the resource labeled as .1 performs the aspect 1.

3.3   Use of Flows, Events, Gateways and Data Objects

Some details concerning the notation elements need clarification and detailing, to preserve and to not harm the process model analyzability, comprehensibility, consistency [3], flow and readability [15, 16]. These items are: (i) Initial and final elements of the control-flow must not be aspects; (ii) Aspects must not contain the final element of a control flow; (iii) Join points should not represent the beginning or end of the process, whether intended or not; (iv) Gateways must be able to ensure the continuity of the logical sequence of activities defined by the process; (v) Gateways must be initiated and terminated inside the aspect so as not to adversely affect the interpretation of the model, since a gateway can be started within the aspect and its junction can be finished outside of it, leaving a gateway free from convergence element released by the model; (vi) Link events must not connect encapsulated elements with main flow elements, and vice-versa; Link events must also not connect two different encapsulated aspects; (vii) The control flow that takes place inside the aspect must not give margin to external structuring elements inside it; (viii) If the aspect encapsulates a control flow, then the quantity of sequence flows before the encapsulation must be the same as the sequence flows after the encapsulation; (ix) Boundary events must not be aspects; (x) Data objects and artifacts may be aspects and may be present and interact with other objects inside aspects.

3.4   Modularization of Resources as Aspects

When numerically analyzing the relationship between a resource and an aspect, if resource A is a resource involved in N aspects, then $(A \times N)$ aspects are required. Inductively, in case X amount of resources are involved in a Y number of aspects, $X \times Y$ aspect pools are seen to address all aspects relevant to their resources. Based on Fig. 3, if the project manager and contract analyst (who perform in various aspects: 2, 3, and 4) were not operationalized as aspects, then there would be redundancy of aspects based on resources performing them, e.g., Figure 2 with the "Software hiring group". There would be six aspects needed, three for the project manager, three for the contract analyst in the process of representing the necessary resources to perform their proper aspects. In this proposal, only four are needed, one aspect to sub-categorize the resources and one aspect to each sub-categorized resource. Six aspects reduced to four, in this simple scenario. In a scenario where, e.g., ten resources and thirty aspects take three hundred to represent all of it, if this approach is used, it can reduce to thirty-one, the amount of $(N + 1)$ sub categorizer resource aspect. Figure 5, using this concept, shows a reduction from fifteen to six aspects.

3.5   Pool of Crosscutting General Aspects

There were not only crosscutting concerns in one model, but crosscutting aspects permeating many models as well. AO-BPM 2.0 crosscutting general aspects pool seeks to solve this scenario, and Fig. 5 exemplifies its operationalization in the case study. Embedded tasks can only be invoked locally by its parent processes, from which they derive, and global tasks can be invoked by any process that requests it. Aspects that permeate many processes with the same concept will be a general or overall aspect; while specific aspects of those processes will be embedded. The more aspects eligible as global arise, the better the modularization of the crosscutting general aspects pool is. In turn, the use of a new pool exclusive for aspects culminates in the need for a new model as a repository exclusive for them as well. Thus, in scenarios where the general aspects appear in only a minimal number of processes and the macro process, the pool creation is not recommended. The use of the crosscutting general aspects pool in an unfavorable scenario, besides impairing the understanding by the excess of attention shift [17], will not reach any of the advantages of modularization proposed by the aspect-orientation (reuse, simplicity, flexibility, and so on).

As the general aspects pool serves as an index of aspects to the macro process, its presence recommended every time one analyses and evaluates the models that use this tool. It does not directly affect the base AO-BPM 2.0 complexity since the reading of the global aspects is accompanied by the reading of the embedded aspects. The reader does not need two steps or two different pieces of information, just the global aspect pool. To differentiate the embedded aspects of global aspects, a syntax change is proposed. The join points of the global aspects use numbers contoured in squares, and the embedded aspects use numbers contoured in circles.

4   Case Study: AO-BPM 2.0 Operationalization in a Real Setting

In this paper, we address the problem of improving an already existent modeling notation aiming to overcome some issues. In this sense, we should investigate the applicability of the proposal. So, our research question is "How complex is an AO-BPM 2.0 model in comparison with a BPMN model?" The model-building approach proposed herein was evaluated through a case study. The goal of this case study was to analyze the ability of AO-BPM 2.0 to be operationalized in a real setting to simplify the original model. We argue that the LOC Metric is suitable for that kind of analysis since it is a broadly form of measuring complexity [14] that simply counts objects in the model. The number of elements and arcs influences the understanding of a business process model [12, 15, 16], syntactically. The number of elements in the model is directly associated with the number of arcs and information. Understanding will be impaired as more arcs intersect, more bends the arcs has [5], more elements the model has [18] and more complex and unknown elements are, increasing the probability of error [12] and overload of information [10].

The scenario chosen to evaluate the AO-BPM 2.0 proposal was the processes from the Secretariat of the Information Systems School of a public university in Brazil, the Federal University of the State of Rio de Janeiro (UNIRIO). Those processes represent the administrative services provided to the students and they were modeled using the BPMN during a institutional initiative of the process group in 2014. Interactions between a student and the secretariat are based on a form called Secretariat Administrative Requirement (SAR). Through it, a student can require a service or product, in this case a document. The student selects one or more desired item(s) and the process begins. The process modeling was performed by the authors of this work. Afterwards, seven models were translated into AO-BPM and AO-BPM 2.0 in [2], where specific heuristics for finding aspects were used [7] and the main and transverse elements were counted by the before mentioned. The AO-BPM 2.0 models were built using the draw. io[2] online drawing tool. Due to space limitation, we discuss the results based on one model.

4.1   Requirements Breakdown Process

This process represents the claimant's application to attend a course which he or she does not meet the requirements. If the director and respective faculty member allows the breaking of these requirements, he or she can apply for attending the course. The core part of this process is composed of the following tasks: the analysis and the opinion of each evaluator of the application and consolidation of it in the system; the other tasks and elements indirectly support the core tasks being modeled as aspects (transport, authentication, bureaucracy and data objects).

Figure 4 depicts the process modeled using BPMN, and Fig. 5 in AO-BPM 2.0. The process begins when the secretariat receives the request form of requirement breakdown. It forwards the form to the faculty, who will read the request of the student and evaluate it. The faculty may reject or approve it. In the case of rejection, it is recorded on the form, stamped, signed and returned to the secretariat, finalizing the process; if it is approved, it is recorded on the form, stamped, signed and forwarded to the director. The director will also read the student's request and evaluate it. The director can reject or allow it. In any case, it is recorded on the form, stamped, signed and returned to the secretariat. In the secretariat, the requirement breakdown is registered in the Students Interaction System (SIS). In Fig. 4, twenty-seven elements were counted (excluding flow objects). In Fig. 5, elements that are eligible for aspects and non-functional elements were removed. Seven elements, apart from the join points, are enough for understanding the core process and the important tasks for delivering value to the client. There are thirteen

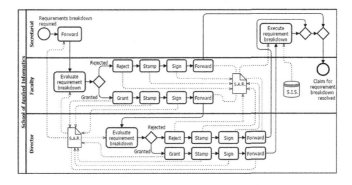

Fig. 4. Requirement breakdown process using BPMN.

---

[2] http://draw.io.

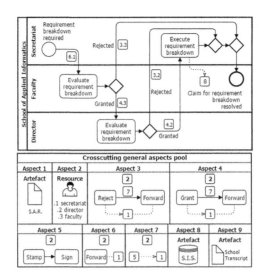

Fig. 5. Requirement breakdown process using BPMN and the general aspects pool

elements, of which six are responsible for the call to the aspect pool. The aspect pool elements also need to be read (twenty-two elements), without the join points, and twenty-eight with the join points (core process plus aspect pool).

4.2   Discussion

Table 1 shows the results of the LOC metric, an adaptation of Number of Activities, Joins and Splits (NOAJS), in the case study. The whole set of processes modeled is composed by: interaction with the secretariat (IS), selected items index (SII), requirements breakdown (RB), course completion statement (CCS), temporary course enrollment drop (TCED), test review (TR) and exemption or utilization of disciplines (EUD). They are presented respectively in the order presented in Carvalho [2].

Table 1. LOC metric results of the case study

	IS	SII	RB	CCS	TCED	TR	EUD	Total
BPMN	11	19	27	8	21	10	20	116
AO-BPM 2.0	11	19	13	6	11	7	12	79
Calls for crosscutting concern pool of aspects	5	1	6	1	8	4	6	31

The metric adapted from NOAJS included elements in the core process: activities, data objects, splits and joins. The reduction of the number of objects in the core process was clear. In the Requirement Breakdown, the number of elements is reduced by more than 74%, while in the Selected index, the number is maintained, by the process structure. The interaction with the secretariat process can be summed up to 27.2% of their activities if only the core activities are considered.

The low-level hierarchy processes (excluding interaction with the secretariat and selected index item) have more redundant aspects and crosscutting concerns related to each other and in themselves. The gain from the aspect orientation was higher in them. In total, there was a 63% reduction in the objects in the case study models. Removing the weakly functional or nonfunctional elements left over only 37% of the elements, which illustrates the objective processes directly and functionally. When counting artifacts that connect the core process in the crosscutting concern pools (pointcuts), the reduction is 32% in the amount of redundancy eliminated using the aspect orientation. The results of the LOC metric in this case study demonstrate a positive impact on simplicity using aspect orientation, by reducing redundancy, separating concerns and clarifying functional elements.

5   Related Work

Charfi et al. [8] propose the AO4BPMN, an extension of the BPMN, where aspects are detached from the main flow and, as sub processes, are allocated in their own lanes. There is a strong influence of execution elements in the notation, influenced by AO4BPEL [13]. Activities and events are join points, advices are detailed in annotations attached to the join points, and there is no graphical representation of the pointcut. Jalali et al. [11] propose the AOBPMN, an extension of BPMN, which uses the concept of precedence. Pointcuts are conditional boundary intermediate events, calling the aspects from the aspect pool to the main flow. Join points are annotations attached in the pointcut. When a process contains more than one aspect, it sets up an advice using the PROCEED concept of AO4BPMN. Precedence is configured by the ability to graphically represent the order of join points in pointcut and advice in the model itself, and is not addressed in the two notations mentioned above. Operationalization of data objects as aspects or aspect encapsulation (aspects in aspects) is not clear in AOBPMN. Jablonski et al. [4] state that users of process models need different types of visualization perspectives depending on their purposes.

6   Conclusion and Future Work

The main contribution of this paper is the AO-BPM 2.0 notation proposed as an improvement of AO-BPM regarding graphic pollution, without crosscutting relationships crossing the model, generating cross flows and a great deviation of heed; Simplicity and readability, where join points are ordered and there is no need for the textual model to interpret the order of execution in a pointcut with more than two join points; Linking the aspect with the core process, rather than flows that need to be connected to not hamper readability, using contoured numbers positioned with clarity.

The aspect orientation modularization has advantages in large scenarios, with a high number of non-functional requirements, soft goals, repeated non-essential elements or transversal interests. The simplification of the core process assists the understanding of the main objective. However, because the use of modularization using an aspect pool generates attention shift [17], it impairs the continuous reading of the model, as a limitation. Therefore, we conclude that AO-BPM 2.0 is a proposal which is graphically satisfactory for the representation of aspect orientation, but is restricted to the conditions of the scenario, as already mentioned. As future works, analysis and evaluation of AO-BPM 2.0 in complex cases and scenarios, with loops, laces etc.; graphical comparative study involving other modeling languages (e.g., AOBPMN).

References

1. Cappelli, C.: Uma Abordagem Para Transparência Em Processos Organizacionais Utilizando Aspectos. Doctorate's Thesis, PUC-Rio. Rio de Janeiro (2009). (in Portuguese)
2. Carvalho, L.P.: AO-BPM 2.0: improving aspect oriented business process modeling notation. Undergraduate Thesis, Federal University of the State of Rio de Janeiro, Rio de Janeiro (2016)
3. Russel, N., van der Aalst, W., ter Hofstede, A.H.M.: Workflow Patterns: The Definitive Guide. The MIT Press, Cambridge (2016)
4. Jablonski, S., Faerber, M., Jochaud, F., Götz, M., Igler, M.: Enabling flexible execution of business processes. In: Meersman, R., Tari, Z., Herrero, P. (eds.) OTM 2008. LNCS, vol. 5333, pp. 10–11. Springer, Heidelberg (2008). https://doi.org/10.1007/978-3-540-88875-8_5
5. Purchase, H.C., Cohen, R.F., James, M.: Validating graph drawing aesthetics. In: Brandenburg, Franz J. (ed.) GD 1995. LNCS, vol. 1027, pp. 435–446. Springer, Heidelberg (1996). https://doi.org/10.1007/BFb0021827
6. Kiczales, G., Lamping, J., Mendhekar, A., Maeda, C., Lopes, C., Loingtier, J.-M., Irwin, J.: Aspect-oriented programming. In: Akşit, M., Matsuoka, S. (eds.) ECOOP 1997. LNCS, vol. 1241, pp. 220–242. Springer, Heidelberg (1997). https://doi.org/10.1007/BFb0053381
7. Cappelli, C., Santoro, F.M., Leite, J.C.S.P., Batista, T., Medeiros, A.L., Romeiro, C.: Reflections on the modularity of business process models: the case for introducing the aspect-oriented paradigm. BPM J. **16**(4), 662–687 (2010)
8. Charfi, A., Müller, H., Mezini, M.: Aspect-oriented business process modeling with AO4BPMN. In: Kühne, T., Selic, B., Gervais, M.-P., Terrier, F. (eds.) ECMFA 2010. LNCS, vol. 6138, pp. 48–61. Springer, Heidelberg (2010). https://doi.org/10.1007/978-3-642-13595-8_6
9. Santos, F., Cappelli, C., Santoro, F.M., Leite, J.C.S.P., Batista, T.: Aspect-oriented business process modeling: analyzing open issues. BPM J. **18**(6), 964–991 (2012)
10. Ottensooser, A., Fekete, A., Reijers, H.A., Mendling, J., Menictas, C.: Making sense of business process descriptions: an experimental comparison of graphical and textual notations. J. Syst. Softw. **85**(3), 596–606 (2012)
11. Jalali, A., Wohed, P., Ouyang, C.: Aspect oriented business process modelling with precedence. In: Mendling, J., Weidlich, M. (eds.) BPMN 2012. LNBIP, vol. 125, pp. 23–37. Springer, Heidelberg (2012). https://doi.org/10.1007/978-3-642-33155-8_3
12. La Rosa, M., Wohed, P., Mendling, J., ter Hofstede, A.H.M., Reijers, H.A., van der Aalst, W.: Managing process model complexity via abstract syntax modifications. IEEE Trans. Industr. Inf. **7**(4), 614–629 (2011)
13. Charfi, A., Mezini, M.: Aspect-oriented web service composition with AO4BPEL. In: Zhang, L.-J., Jeckle, M. (eds.) ECOWS 2004. LNCS, vol. 3250, pp. 168–182. Springer, Heidelberg (2004). https://doi.org/10.1007/978-3-540-30209-4_13

14. Cardoso, J., Mendling, J., Neumann, G., Reijers, H.A.: A discourse on complexity of process models. In: Eder, J., Dustdar, S. (eds.) BPM 2006. LNCS, vol. 4103, pp. 117–128. Springer, Heidelberg (2006). https://doi.org/10.1007/11837862_13
15. Hipp, M., Strauss, A., Michelberger, B., Mutschler, B., Reichert, M.: Enabling a user-friendly visualization of business process models. In: Fournier, F., Mendling, J. (eds.) BPM 2014. LNBIP, vol. 202, pp. 395–407. Springer, Cham (2015). https://doi.org/10.1007/978-3-319-15895-2_33
16. Krogstie, J.: Quality in Business Process Modeling. Springer, Heidelberg (2016). https://doi.org/10.1007/978-3-319-42512-2
17. Turetken, O., Rompen, T., Vanderfeesten, I., Dikici, A., van Moll, J.: The effect of modularity representation and presentation medium on the understandability of business process models in BPMN. In: La Rosa, M., Loos, P., Pastor, O. (eds.) BPM 2016. LNCS, vol. 9850, pp. 289–307. Springer, Cham (2016). https://doi.org/10.1007/978-3-319-45348-4_17
18. Mendling, J., Reijers, H.A., van der Aalst, W.: Seven Process Modeling Guidelines (7PMG). J. Inf. Softw. Technol. **52**(2), 127–136 (2010)

A Visualization of Human Physical Risks in Manufacturing Processes Using BPMN

Melanie Polderdijk[1], Irene Vanderfeesten[1(✉)], Jonnro Erasmus[1],
Kostas Traganos[1], Tim Bosch[2], Gu van Rhijn[2], and Dirk Fahland[1]

[1] Eindhoven University of Technology, Eindhoven, The Netherlands
melaniepolderdijk@gmail.com,
{i.t.p.vanderfeesten,j.erasmus,k.traganos,d.fahland}@tue.nl
[2] TNO, Leiden, The Netherlands
{tim.bosch,gu.vanrhijn}@tno.nl

Abstract. Process models are schematic representations of business processes and support analysis for process redesign. A process model may be enhanced with additional information to further improve its analytical value (e.g. costs, throughput times, etc.). To support analysis of human factors in manufacturing processes, this paper introduces a new extension that visualizes human physical risks (such as heavy lifting or repetitive work). An existing human risk analysis method is integrated with the activity elements of BPMN. This integration facilitates a process wide risk analysis of occupational risk factors. Based on this graphical representation, users can see where in the process workers may encounter physical risks that should be mitigated through process redesign. This paper reports on the systematic design of a conceptual solution which is implemented in MS Visio and then applied and evaluated in practice.

Keywords: Visualization · Occupational risk factors
BPMN extension · Process modeling · Business Process Management

1 Introduction

A core aspect in Business Process Management (BPM) is process modeling [8,32]. Process models visualize how a particular business case should be handled and can be used to describe, analyze and enact a process [9,24]. A modeled business process is often a visual means of communication among process stakeholders to understand and analyze the process [32]. Based on the results of these analyses, redesign ideas may be generated on how to improve the process.

A process modeling language that is widely used for this is the Business Process Modeling and Notation (BPMN) from the Object Management Group (OMG) [5,23]. The origin of BPM and process modeling lies in the administrative domain. Therefore, BPMN is usually applied to model business and administrative processes [5]. However, increasingly often BPMN practices are translated to other domains such as manufacturing [34] and healthcare [4,19]. For basic modeling, this can easily be done since the abstract elements of the notation

© Springer International Publishing AG 2018
E. Teniente and M. Weidlich (Eds.): BPM 2017 Workshops, LNBIP 308, pp. 732–743, 2018.
https://doi.org/10.1007/978-3-319-74030-0_58

are widely applicable. However, sometimes it is required to model more domain specific information. In such cases, standard BPMN might not be able to capture the right amount of detail. To overcome this limitation, BPMN can be extended with customized elements [23]. To this date, already many extensions exists in multiple application domains [3]. Still, very few of these extensions focus on the manufacturing domain [28].

Only Zor et al. [34] have proposed a customized notation to represent explicit manufacturing-specific constructs such as assembly and material routes. However, there are more issues involved in the modeling and analysis of these processes. One of them is the well-being of workers. The issues related to this, such as health problems due to exposure to work-related risk factors, belong to the human factors domain [13]. In both decision-making and manufacturing processes, employees play an important role. The well-being of an operator can have a big influence on the throughput time or quality of a product [14]. Moreover, a company should take the responsibility in keeping its employees healthy and should comply to law and regulations on working conditions. It is already proven that applying human factors practices in the design of operational systems improves the well-being of workers and system performance [21]. When these practices are applied in early design stages, costs of potential changes to the process can be reduced and productivity can be improved [20,25]. Therefore, it is important to consider occupational risk factors during manufacturing process design and analysis. There is, however, no support available for a process wide analysis and visualization of occupational physical risk factors. This paper addresses this problem by systematically developing a BPMN extension that combines the two domains of process analysis and occupational risk analysis in the vizualization of a manufacturing process.

We proceed as follows. Section 2 discusses some background on human factors and related work on BPMN extensions and their visualizations. Next, in Sect. 3, the methodology to design our BPMN extension is explained. The actual design is introduced in Sect. 4 and Sect. 5 elaborates on the application of our BPMN extension in practice. We report on an evaluation of the proposed extension regarding usefulness and usability in industrial practice in Sect. 6. Finally, some conclusions and directions for future research are provided in Sect. 7.

2   Background and Related Work

Within the manufacturing industry, the occurrence of health related problems are present in various forms. Physical loads at a workplace, such as lifting and carrying too heavy objects or using vibrating tools for too long, are the most dominant causes of musculoskeletal disorders [10]. However, one can also consider the mental burden on workers who have to perform a repetitive task for several hours. These issues can reduce their well-being and may lead to stress-related disorders. Within the discipline of human factors engineering, it is aimed to identify these kind of risks and design solutions to reduce them [29]. The BPMN extension in this paper focuses on the physical risk factors, which are

Table 1. Physical risk factors [30]

Physical risk factor	Description
Lifting and carrying	The intensity of workers lifting and carrying objects
Pushing and pulling	The intensity of workers rolling, sliding, pushing or pulling objects in different ways
Hand-arm tasks	Duration and type of tasks workers perform by using their hands and/or arms
Working postures	Duration of different types of working postures
Computer-related work	Duration of work that requires a computer or laptop
Vibration	Intensity and duration of contact between workers and vibrating objects
Energetic overload	The energy a worker has to deliver beyond his limits
Energetic underload	The duration of sitting

summarized in Table 1, since these are the most prominent human factors risks in a manufacturing context.

A tool that can be used to assess the risk level of these physical risk factors is the Checklist Physical Load [30]. The Checklist Physical Load forms the base for the risk assessment enabled by the BPMN extension. It contains a questionnaire to assess the occurence and level of severity for each of the risk factors from Table 1. For instance to assess the carrying and lifting risk factor, the first question to be answered is: "Do workers have to carry or lift loads of more than 3 kg manually in the course of a working day?". If the answer to this question is yes, then a set of follow-up questions appears, such as: "Do they manually carry or lift loads of more than 15 kg?", "Do they manually carry or lift loads for more than 2 h?", "Do they manually carry or lift loads more than three times a minute?", etc. Based on the answers to these questions the risk level for each of the risks denoted in Table 1 is calculated. The objective of this work is to integrate the outcomes of this established human factors analysis tool into a BPMN model by developing a suitable visual extension in a systematic manner.

As discussed in the introduction, few studies have specifically extended BPMN for the manufacturing industry. Only Zor et al. [34] have proposed a BPMN extension for the manufacturing domain that is suitable to model processes on the factory floor. In their notation, they have adjusted some basic elements of BPMN to make it possible to model manufacturing-specific constructs such as assembly tasks and material routes, but they do not cover occupational risk related information.

In other domains, many BPMN extensions exist [3], e.g. Braun et al. [4] have extended BPMN for the healthcare domain and the extension of Brambilla et al. [2] focuses on social media interactions. General extensions such as those focusing on cost evaluation [15, 16] could be applied to model manufacturing processes, but the base of these extensions is standard BPMN and its original application domain.

The processes used in developing and demonstrating these extensions are usually high-level business processes. So, applying the extension directly to manufacturing processes may not always be possible or optimal.

Altogether, to the best of our knowledge, there is no prior work on incorporating occupational risks in a BPMN process model.

3   Methodology

To develop a BPMN extension that facilitates a human physical risk analysis on the process level, a design science approach was followed [11,26]. Along the lines of Peffers' design science research process [26], we briefly discuss our research methodology here. In the *problem identification and motivation* phase we reviewed literature on BPMN extensions and human factors. To gain some practical insight, we conducted exploratory interviews with experts from the manufacturing and human factors domain. Based on this information the *objectives* of the BPMN extension were elaborated, the scope of the solution was determined, and a number of general design principles were formulated. Next, in the *design* phase, three alternative conceptual designs were elaborated, taking into account the general design principles, and early feedback from human factors experts. The three alternative designs were then evaluated and compared through a theoretical evaluation using the physics of notations theory [17]. Additionally, a practical evaluation in which three user experience experts and three human factors experts were asked to give their opinion on the alternative designs. Based on these evaluations the best alternative was selected and refined addressing the experts' feedback. The resulting final conceptual design was implemented in a process model editing tool so that it can be used in practice. This tool was then used in two case studies in which several business processes were mapped and analyzed for physical risk factors together with the process stakeholders. Due to space limitations, in this paper, we will only report on the general design principles, the final conceptual design, its implementation in a tool, and an illustration with one of the two case studies, followed by a qualitative evaluation of the usefulness of the tool by the stakeholders from practice. Details are available in [28].

4   Design

The BPMN extension in this paper integrates the Checklist Physical Load [30] and its risk analysis results with BPMN. It facilitates a process analysis focused on occupational risk factors. The risk assessment is linked to each human activity individually. After filling in the questions of the risk assessment for a specific activity, the results are visualized on the activity element itself. When doing this for each human activity within a process model, the process model provides a process oriented overview of the results of the risk assessment. Below some insights in the design process are given by stating the general design principle made based on the problem analysis phase and by elaborating on a number of important design considerations. Next, the final conceptual design is presented and an implementation of this design as an add-on to an existing BPMN editor is discussed.

4.1   General Design Principles

After formulating the objectives, a number of constraints on the design were formulated. They are presented as general design principles here.

First, it is decided to extend the activity elements of standard BPMN rather than to introduce entirely new artifacts. The reason for this is that the physical risk is considered a property of an activity as it is determined by the characteristics of that activity. For instance, the duration of a computer-related task is crucial in determining whether the activity can reduce a worker's well-being: using a computer for an hour a day causes no harm but using it for many hours a day may lead to serious problems. To prevent separating the property (physical risk) from the object it belongs to (activity), we formulated the first general design principle:

GD_1: **Extend existing BPMN activity elements for displaying physical risks, instead of introducing new elements**

Secondly, it is decided that this extended activity element should replace the existing BPMN element of manual and user tasks, but that the human nature of the task should still be recognizable from the new design. Standard BPMN allows for defining types of activities. For instance, there are service tasks which are automated or message tasks that define the activity of sending or receiving a message. As the risk analysis is only relevant for activities involving humans, it would not be logical to perform this analysis for activities that are not related to people, such as service tasks. Generally, tasks involving humans are presented by a manual or user task in BPMN. Therefore, the second general design principle is:

GD_2: **The extended activity element is a replacement for the manual and user activity element in standard BPMN**

Based on the findings on visualization techniques used in existing BPMN extensions, it is decided to prevent a design that only uses color to express information. Colors limit the ability of making fast and simple sketches of process models by hand. Yet, such sketches are regularly used to improve communication about processes in practice. When only color is used, one should have pencils of different colors at hand to express the necessary information which is rarely the case. Another consequence of only using color is that it becomes difficult for color blind people to read the model correctly [33]. Therefore, the fourth general design principle is:

GD_3: **Do not use color as the only visualization technique to represent certain information**

Lastly, it is decided to preserve the representation of standard BPMN elements as much as possible. This makes the extension recognizable and easier to understand for users with experience in BPMN. Thus, in case a design uses icons on top of activity elements, i.e. markers, they are located in the upper left corner as this is consistent with the way information on activities is expressed in standard and other BPMN extensions as well [1,28]. This leads to the last general design principle:

GD_4: **Preserve the representation of standard BPMN elements as much as possible**

4.2 Final Design

Considering the general design principles, three conceptual designs for the BPMN extension were created, evaluated, and the best design was selected as the final design. Figure 1 shows the visual representation of this final conceptual design. It uses markers shaped as a human body to indicate the activity is a human activity. The markers are colored and contain a symbol to represent the different states of an activity (see Fig. 1):

– Green with checkmark symbol, if no physical risks are identified in this activity
– Orange with exclamation mark, if only few physical risks are identified (at most 2)
– Red with cross, if many physical risks are identified (more than 2)
– Grey with question mark, if the state is "undefined" since the questionnaire for this activity has not been filled yet.

This design thus indicates to the user, whether the physical risks of this activity still have to be assessed by filling in the risk analysis questions, whether the questions were filled and everything is OK, or whether some measures need to be taken due to risks present.

Based on the theory of occupational risk assessment, it was decided to distinguish between orange and red categories for few (1 or 2) and many (more than 2) risks present. It was also decided to explicitly indicate the "undefined" state. This design choice is mainly based on one of the usability principles of Norman [22], particularly that concerning user feedback. He argues that a system should always inform the user when his action has been completed, either

(a) Undefined   (b) No risk   (c) Few risks   (d) Many risks

Fig. 1. Final design: representation of risk level (Color figure online)

successfully or unsuccessfully. In this case it ensures that users have a visual feedback on whether the risk assessment for the activity is complete or still requires action.

It was decided to use a combination of visual dimensions for the design of the marker. First of all, the state is indicated by a symbol for their built-in mnemonics [27]. These mnemonics are based on previously-learned associations between objects and functions which speeds up recognition [18]. Additionally, color is used in order to make the symbols stand out a bit more. Studies of Christ [6] have shown that adding color as a redundant attribute improves visual search and identification accuracy in visual displays.

Fig. 2. Representation of risk type(s)

Furthermore, a textual notation is added close to the marker to express the amount of risks that are present relatively to the total amount of risks that could be present. And finally, the design has the property of hiding information that is less important. Figure 2 captures a screenshot of the visual representation of an activity when the user hovers over the shape with his mouse. Only then the type of risk is shown.

4.3   Implementation

The final conceptual design, as explained above, is implemented as a template in MS Visio. It comprises a new type of task (Human Task) that has built in a questionnaire of the Checklist Physical Load, and that visualizes the outcome of the occupational risk analysis when all questions are answered. By doing so,

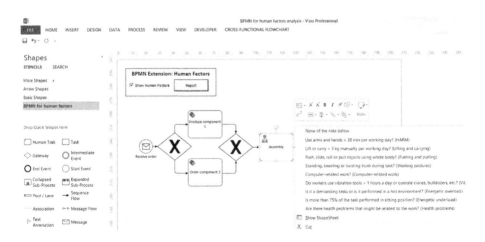

Fig. 3. A screenshot of the MS Visio tool. On the left side the Human Task element; on the right side the menu with questions from the physical load checklist that pops up when right-clicking the Human Task element in the model.

a practical tool is created in which the BPMN extension can be used to model and analyze processes in practice. Figure 3 shows a screenshot of the tool. A screencast demonstrating the usage of the tool can be found via https://tinyurl.com/BPMN-HF.

5   Application

The usefulness and usability of the BPMN extension was tested in practice. It was applied to model and analyze several processes of two real life manufacturing companies. Here, due to space limitations, only one application is discussed, particularly that of Thomas Regout International (TRI). TRI is a manufacturing company that develops and produces innovative telescopic slides and linear guides [31]. Their processes involve many interactions between humans and machines. Therefore, their business is very suitable to apply and test the BPMN extension at. In a workshop with the Director of Operations (process owner) process models for a subset of their manufacturing processes were created and analyzed on the occurence of physical risks.

As an illustration we focus on TRI's process of manually loading profiles on racks, which is a preparatory process for the surface treatment and final assembly of the profiles. During this process, workers have to carry heavy profiles and place them on a rack one by one. This is physically very demanding and a major contributor to the physical risk encountered in the company. Figure 4 represents this process after it is modeled and analyzed with our BPMN extension.

As can be seen in this figure, only the manual activity elements are replaced with the Human Task element from the BPMN extension. Since automatic activities (such as "crane moves filled rack back to baths") do not involve people, it is not necessary to analyze them on physical risks. Consequently, these activities do not need to be modeled with the Human Task.

Furthermore, the model clearly shows where in the process most of the physical problems occur. For instance, the activities *Pickup handful of profiles* and *Hang profiles one by one* show red icons due to a number of physical risks: (i) hand-arm tasks, (ii) lifting and carrying activities, (iii) prolonged standing, and (iv) energetic overload. The visualization in Fig. 4 provides the user with information on the presence of physical risks in the process and with that contributes to finding a suitable solution. In this case, the user may consider allocating these

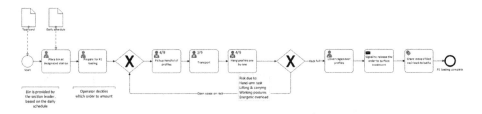

Fig. 4. Process model of PL2.1 manual loading of TRI

risky tasks to robots or find mechanical support for a worker to eliminate or alleviate the risk.

6   Evaluation

We modeled 6 processes from 2 different manufacturing companies using the tool described in Sect. 4.3. We evaluated both the tool and the visualization in semi-structured interviews with four process owners from the companies. Each process owner was interviewed separately based on the validated questionnaires of the Technology Acceptance Model [7] and Method Evaluation Model of Moody [17]. The aim of the interviews was to get insights on three aspects of the use of the tool [7], particularly:

- *Perceived Ease of Use (PEOU)*: the extent to which someone believes that using the tool is effortless.
- *Perceived Usefulness (PU)*: the extent to which someone believes that using the tool improves job performance.
- *Intention to Use (ITU)*: the extent to which someone is determined to use the tool.

From this evaluation, it is concluded that the tool is perceived as useful and usably by the interviewees [28]. It was mentioned that the BPMN extension presents relevant information with respect to the tasks of a process owner (e.g. managing the operations, streamlining the process, defining redesigns where needed). One of the interviewees thought it nice to receive some extra information on the practical implications of the results. For instance, he would prefer the results to entail some suggestions to improve the activities with an increased risk. It would help him in deciding which actions to take. Without these suggestions, he would need advice from, for instance, human factors experts or process engineers on how to reduce the risk levels.

In comparing the visualizations of the BPMN extension and the results of the Checklist Physical Load, the usefulness of representing the physical risk factors as a layer over the process model is definitely recognized, as compared to the format of the Checklist Physical Load in which the results are presented for one task at a time. The results presented in a process model provide a much better overview. According to one of the process owners, this makes it easier to identify smaller solutions for specific activities instead of changing an entire (sub)process.

With respect to the actual graphical notation of the BPMN extension, one interviewee noticed he would not mind having activities with increased risk stand out even more than they do in the current design. It was suggested to use mechanisms such as text-balloons or include more color.

Regarding the application of the tool in practice, some of the interviewees felt sufficiently competent to use the tool by themselves without the help of a process modeler or analyst. Others, as novices to process modeling and BPMN, however, would need some guidance in creating process models.

Finally, it is noticed that it would be nice to have some more information on the content of the risk assessment questions. In the Checklist Physical Load, information windows are used to provide the user with more knowledge on how to fill in the answer to the question. These are absent in the BPMN extension, but are perceived as a useful addition.

7   Conclusion

Concluding, this paper has introduced a BPMN extension that enables a process-oriented human physical risk assessment in the manufacturing domain. This is done by systematically designing a solution and delivering a practical tool which was shown to be useful in practice. Still, there are some limitations and areas for future research left.

Currently, the BPMN extension is limited to only one category of occupational risk factors: physical factors. However, there are other categories which are equally important, such as Organizational, Dangerous substances, Environmental and Safety factors [13]. Adding these risk categories would certainly enhance the usability and applicability of the visualization to other domains beyond manufacturing. But it also raises many new visualization challenges as there is more information to be shown on additional levels of aggregation. The user may for instance want to see the distinction between the different categories of risks. This can be done in many ways and a new design cycle should be started to see what visualization is most useful and usable.

Another direction for future work is the more prominent use of the process model in performing the physical risk analysis. In the current situation, the risk assessment is performed per activity and the result is visualized for each activity on its own. However, it could be imagined that a sequence of two activities with little risk can add up to a great risk when performed by the same worker. If the tool could accumulate or aggregate the assessed risks of individual activities and display them at higher process level, the results become more encompassing. Consequently, future research may also focus on adding other visual features to the BPMN extension. For instance, one could think of changing the appearance of a swimlane if it contains too many risky activities or of a subprocess element if the activities inside have an increased risk. Furthermore, the frequency a worker executes a certain activity or process during the course of a work day could be taken into account in the analysis.

Finally, more research should be conducted on customizing views in the visualization for the different stakeholders in the process. The current design of the BPMN extension is developed for and tested with Process Owners and is therefore limited to their preferences. However, it is considered that other stakeholders have different requirements. Some stakeholders might require more information and some might prefer a more abstract view. For instance, a Process Participant such as an operator will probably require different information on human factors than the Process Owner. Moreover, he might prefer a visualization that is more focused on the tasks he needs to perform rather than overseeing the whole process. On the other hand, Process Analysts with expertise in the area of human factors, also called OSH experts, probably need more detailed information on the results of the risk assessment compared to the Process Owner.

Acknowledgements. This research is conducted in the context of the HORSE project [12]. The HORSE project has received funding from the European Unions Horizon 2020 research and innovation programme under grant agreement No. 680734. The authors want to thank M. Douwes for her valuable feedback on early versions of the tool, and R. Keulen for the opportunity to apply this work in practice.

References

1. Allweyer, T.: BPMN 2.0: Introduction to the Standard for Business Process Modeling. BoD Books on Demand, Norderstedt (2016)
2. Brambilla, M., Fraternali, P., Vaca Ruiz, C.K.: Combining social web and BPM for improving enterprise performances: the BPM4 people approach to social BPM. In: Proceedings of the 21st International Conference on World Wide Web, pp. 223–226. ACM (2012)
3. Braun, R., Esswein, W.: Classification of domain-specific BPMN extensions. In: Frank, U., Loucopoulos, P., Pastor, Ó., Petrounias, I. (eds.) PoEM 2014. LNBIP, vol. 197, pp. 42–57. Springer, Heidelberg (2014). https://doi.org/10.1007/978-3-662-45501-2_4
4. Braun, R., Schlieter, H., Burwitz, M., Esswein, W.: BPMN4CP: design and implementation of a BPMN extension for clinical pathways. In: 2014 IEEE International Conference on Bioinformatics and Biomedicine (BIBM), pp. 9–16. IEEE (2014)
5. Chinosi, M., Trombetta, A.: BPMN: an introduction to the standard. Comput. Stand. Interfaces **34**(1), 124–134 (2012)
6. Christ, R.E.: Review and analysis of color coding research for visual displays. Hum. Factors **17**(6), 542–570 (1975)
7. Davis, F.D., Bagozzi, R.P., Warshaw, P.R.: User acceptance of computer technology: a comparison of two theoretical models. Manage. Sci. **35**(8), 982–1003 (1989)
8. Dumas, M., La Rosa, M., Mendling, J., Reijers, H.A.: Fundamentals of Business Process Management, vol. 1, p. 2. Springer, Heidelberg (2013). https://doi.org/10.1007/978-3-642-33143-5
9. Gordijn, J., Akkermans, H., van Vliet, H.: Business modelling is not process modelling. In: Liddle, S.W., Mayr, H.C., Thalheim, B. (eds.) ER 2000. LNCS, vol. 1921, pp. 40–51. Springer, Heidelberg (2000). https://doi.org/10.1007/3-540-45394-6_5
10. Griffith, L.E., Shannon, H.S., et al.: Individual participant data meta-analysis of mechanical workplace risk factors and low back pain. Am. J. Public Health **102**(2), 309–318 (2012)
11. Hevner, A.R., March, S.T., Park, J.: Design science in information systems research. MIS Quart. **28**, 75–105 (2004)
12. HORSE: HORSE Project (2016). http://www.horse-project.eu/. Accessed 07 Oct 2016
13. IEA: Denition and Domains of Ergonomics (2016). http://www.iea.cc/whats/index.html. Accessed 18 Oct 2016
14. Kahya, E.: The effects of job characteristics and working conditions on job performance. Int. J. Ind. Ergon. **37**(6), 515–523 (2007)
15. Lodhi, A., Küppen, V., Saake, G.: An extension of BPMN meta-model for evaluation of business processes. Sci. J. Riga Tech. Univ. Comput. Sci. **43**(1), 27–34 (2011)

16. Magnani, M., Montesi, D.: BPMN: how much does it cost? An incremental approach. In: Alonso, G., Dadam, P., Rosemann, M. (eds.) BPM 2007. LNCS, vol. 4714, pp. 80–87. Springer, Heidelberg (2007). https://doi.org/10.1007/978-3-540-75183-0_6

17. Moody, D.L.: The method evaluation model: a theoretical model for validating information systems design methods. In: ECIS 2003 Proceedings, p. 79 (2003)

18. Moody, D.: What makes a good diagram? Improving the cognitive effectiveness of diagrams in is development. In: Wojtkowski, W., Wojtkowski, W.G., Zupancic, J., Magyar, G., Knapp, G. (eds.) Advances in Information Systems Development, pp. 481–492. Springer, Boston (2007). https://doi.org/10.1007/978-0-387-70802-7_40

19. Müller, R., Rogge-Solti, A.: BPMN for healthcare processes. In: Proceedings of the 3rd Central-European Workshop on Services and their Composition (2011)

20. Neumann, W.P., Winkel, J., Medbo, L., Magneberg, R., Mathiassen, S.E.: Production system design elements influencing productivity and ergonomics: a case study of parallel and serial flow strategies. Int. J. Oper. Prod. Manag. **26**(8), 904–923 (2006)

21. Neumann, W.P., Dul, J.: Human factors: spanning the gap between OM and HRM. Int. J. Oper. Prod. Manag. **30**(9), 923–950 (2010)

22. Norman, D.A.: The Psychology of Everyday Things. Basic Books, New York (1988)

23. OMG: Notation (BPMN) version 2.0. Object Management Group (2011)

24. Ould, M.A., Ould, M.A.: Business Processes: Modelling and Analysis for Re-engineering and Improvement, vol. 598. Wiley, Chichester (1995)

25. Paquet, V., Lin, L.: An integrated methodology for manufacturing systems design using manual and computer simulation. Hum. Factors Ergon. Manuf. Serv. Ind. **13**(1), 19–40 (2003)

26. Peffers, K., Tuunanen, T., et al.: The design science research process: a model for producing and presenting information systems research. In: Proceedings of the First International Conference on Design Science Research in Information Systems and Technology (DESRIST 2006), pp. 83–106 (2006)

27. Petre, M.: Why looking isn't always seeing: readership skills and graphical programming. Commun. ACM **38**(6), 33–44 (1995)

28. Polderdijk, M.: Extending BPMN for analysis of human physical risk factors in manufacturing processes. Master thesis, Eindhoven University of Technology (2017). http://is.ieis.tue.nl/staff/ivanderfeesten/Papers/TAProViz2017/

29. Stanton, N.A., Hedge, A., et al. (eds.): Handbook of Human Factors and Ergonomics Methods. CRC Press, Boca Raton (2004)

30. TNO: Checklist Physical Load (2012). https://www.fysiekebelastingbeoordelen.tno.nl. Accessed 04 Oct 2016

31. Thomas Regout International (2016). http://www.thomasregout-telescopicslides.com/. Accessed: 06 Oct 2016

32. Van Der Aalst, W.M.P.: Business process management: a comprehensive survey. ISRN Software Eng. (2013)

33. Ware, C.: Information Visualization: Perception for Design. Elsevier, Amsterdam (2012)

34. Zor, S., Leymann, F., Schumm, D.: A proposal of BPMN extensions for the manufacturing domain. In: Proceedings of 44th CIRP International Conference on Manufacturing Systems, June 2011

Visual Analytics for Soundness Verification of Process Models

Humberto S. Garcia Caballero$^{(\boxtimes)}$, Michel A. Westenberg,
Henricus M. W. Verbeek, and Wil M. P. van der Aalst

Eindhoven University of Technology, Eindhoven, The Netherlands
`h.s.garcia.caballero@tue.nl`

Abstract. Soundness validation of process models is a complex task for process modelers due to all the factors that must be taken into account. Although there are tools to verify this property, they do not provide users with easy information on where soundness starts breaking and under which conditions. Providing insights such as states in which problems occur, involved activities, or paths leading to those states, is crucial for process modelers to better understand why the model is not sound. In this paper we address the problem of validating the soundness property of a process model by using a novel visual approach and a new tool called PSVis (Petri net Soundness Visualization) supporting this approach. The PSVis tool aims to guide expert users through the process models in order to get insights into the problems that cause the process to be unsound.

Keywords: Petri nets · Soundness · Verification · Process models
Process mining · Visualization · Visual analytics

1 Introduction

The use of process models for workflows has been studied for some decades now. It started back in 1979 with the work of Skip Ellis on office automation [9]. Still, it took two decades until notions such as workflow nets and soundness were defined [1] thus linking workflows to Petri nets. As a result of this link, a lot of existing Petri net theory (like [7,14,16,18]) became instantly applicable to the workflow process model domain. Nevertheless, some other approaches to the verification of these models also still emerged, in which was used [19] another graph-like notation for processes in combination with dedicated graph reduction rules.

The application of Petri nets to the workflow domain [2,3,12] triggered a new line of research focussing on the soundness verification tool *Woflan* [28], different variations on soundness [5,26], and extensions to the Petri net formalism (like EPCs [11,24,25], BPEL [15], and YAWL [29,31]). Verbeek et al. [28] also introduced the concepts of the problematic runs (called *sequences* in that paper), which are used in this paper.

None of these approaches offered a comprehensive visualization of the problems using the process model of choice (YAWL, EPC, Petri net, etc.).

© Springer International Publishing AG 2018
E. Teniente and M. Weidlich (Eds.): BPM 2017 Workshops, LNBIP 308, pp. 744–756, 2018.
https://doi.org/10.1007/978-3-319-74030-0_59

For instance, the Woflan tool did not visualize the Petri net it was checking soundness on, but just showed a series of messages that included the labels of the nodes in the net. Nevertheless, the Woflan tool was later included in the process mining framework ProM [21, 25], which did allow this net to be visualized. As a result of this inclusion, selected markings (like deadlock markings) could be visualized by projecting them onto the net, and other Petri-net related properties (like invariants) could also be visualized. However, such visualization means are still limited and users struggle to diagnose real problems.

This paper takes this initial and rudimentary visualization of soundness problems in ProM some steps further by focusing on the visualization. Also, whereas Woflan requires a unique final marking (which should be reached to achieve success in the workflow), the approach in this paper allows for any collection of such final markings. By visualizing any problem that prevents the workflow from reaching any of these final markings, the user is guided towards correcting the root cause of these problems.

The concept of runs as shown by the visualization is known in the Petri net field, and originates from Desel [6]. An example tool that supports these runs is VipTool [4, 8]. VipTool can provide the user with information whether a given *scenario* (say, a partial trace) fits the Petri net at hand. In this paper, we assume that such a partial trace fits the net. If not, we can use alignments to find the closest path in the model [22]. This is not supported by the VipTool, but is supported by ProM. We are more interested in visualizing the execution of the fitting partial trace in the net. Apart from this, VipTool can also synthesize a Petri net from a collection of scenarios. This connects VipTool to the field of process mining [21], which is the natural habitat of ProM. However, we do not use that feature of the tool.

PSVis (Petri net Soundness Visualization) is a tool to spot problems in Petri nets by means of visualization. The tool aims at guiding expert users through the process models to get insight into the problems that cause the process model to be unsound.

2   Problem Definition

In this section we give definitions of the core concepts that our visualization tool needs to handle. We define the problem that we address, and present a set of tasks. We designed our visualization tool accordingly.

2.1   Problem Analysis

Process modelers tend to ensure the soundness property of the process models. However, it is fairly easy to break this property with only a small number of changes on the model. In addition, models which are derived by discovery algorithms do not always ensure this property. Well-known miners like the Alpha-miner and the Heuristic-miner often produce models that are not sound (e.g., have deadlocks).

Because of the dynamic nature of the behavior of the Petri nets, understanding where the problems occur and the context in which they happen plays a key role for process modelers.

In order to address this problem, we define the following tasks, which form the design basis of our tool:

T1 Obtain an overview of all or a subset of final states of a Petri net.
T2 Compare disjoint sets of transitions and/or places belonging to a specific area.
T3 Find problematic states.
T4 Explore paths that lead to a problematic state.
T5 Determine when the problem occurs for a specific problematic state.
T6 Analyze the runs for a selection of states.
T7 Examine concurrency, loops and causal order in runs.

This set of tasks has been composed in collaboration with experts in the area of process mining and Petri nets in order to address the problem on soundness validation. To support these tasks, we chose appropriate visual encodings and made an interaction design.

2.2   Soundness Validation

We use an algorithm originally proposed by Verbeek [27] to compute the problematic states of a given Petri net. The output of the algorithm is used as input in our tool. The algorithm computes three sets of states, referred to as *Orange*, *Green* and *Red* areas. An abstraction of the resulting output of this algorithm is shown in Fig. 1.

In our tool, we focus on the border states, that is, the states which depict a transition from the *Orange* area to the either *Green* or *Red* area. In Fig. 1, the border states are linked by black-dotted arrows. Notice that there may be cases

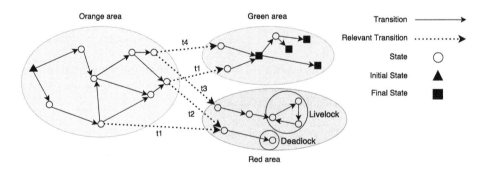

Fig. 1. Abstract representation of the concept of areas outlined by Verbeek et al. [27]. The state space is divided into three areas: *Orange*, *Green* and *Red*. Circles and arrows depict states of the state space and transitions, respectively. Two possible situations can occur in the *Red* area: deadlocks and livelocks, which are represented by circles without outgoing arrows and cycles. (Color figure online)

in which the same transition leads to different areas depending on the source state (e.g., *t1* leads to *Green* and *Red* areas). These states correspond to parts of the Petri net in which everything becomes right (henceforth, all reachable states are *Green*) or everything becomes wrong (henceforth, all reachable states are *Red*).

The *Red* area contains all those states from which is not possible to reach any final state. Clearly, all reachable states from a *Red* state are *Red* states too. Thus, states within this area are considered as wrong states. On the other hand, the *Green* area represents those states from which a final state can always be reached. Similar to the *Red* states, from a *Green* state only *Green* states can be reached. Lastly, the *Orange* area comprises states from which some *Red*, *Green* and *Orange* states can be reached.

Within the *Red* area, two major problems can be present: deadlocks and livelocks. In Fig. 1, two problems are indicated with circles. Deadlocks occur when a state from which no transition can be fired is reached, and it is not a final state. This can be seen in Fig. 1 as the state has no outgoing edges, that is, no transition can be fired at that state. Livelocks happen when a cycle is found, which means that it is possible to iterate between a subset of states forever.

3   PSVis

In this section we introduce PSVis (Petri net Soundness Visualization), its components and how they interact in order to execute the tasks depicted in Sect. 2.1. An overview of all components is given in Fig. 3. Every component enables to perform a specific task or a set of tasks.

Our tool assumes that the state space is computable in a reasonable amount of time. If the state space contains unbounded places, it is infinite in size and cannot be computed. But even if all places are bounded, the state space may still be too big to be constructed within reasonable time [28]. To avoid having to spend unreasonable time in constructing the state space, our tool uses a threshold that operates on the number of tokens in a state. If the threshold is set to b, then only states where every place contains less than b tokens are expanded. This is related to the notion of b-boundedness introduced earlier. Thus, if all places are b-bounded in a net, then setting the threshold to b or higher does not change the state space. If the threshold is reached at some state, we assume that state is a problematic one.

3.1   Glyphs on the Petri Net

In order to support T1 (Obtain an overview of all or a subset of final states of a Petri net), process modelers need to visualize the number of tokens of a specific place, and the states that place belongs to. As a result, we introduce *glyphs*, which decorate places in a Petri net. Glyphs are visual representations of a piece of the data where visual attributes are dictated by data attributes [30].

The number of tokens in a place is represented as dot shapes contained in the place, numbers, or a combination of both. The main problem of this representation arises when we want to visualize more than one final state at the same time. A final state according to the definition is a multiset of places. This implies that a specific place can belong to more than one final state. With the current way of presenting this information, expert users are not able to know the number of tokens contained inside each place for each state and whether a place belongs to more than one state. Therefore, we propose a new way to visually encode this information. This new encode is shown in Fig. 4(b).

In our approach, the state with the highest number of unique places determines the color of the places. Glyphs are then colored with the remaining states. The initial state constitutes an exception and the places that belong to it are always colored accordingly. Figure 2 shows an example of how our approach works. When the user selects *State 1*, the two places that belong to it are colored red. Next, the user selects *State 2* which has three unique places. Therefore, all the places are colored blue and two glyphs are created to present *State 1*. Lastly, when the user selects the initial state, two places are colored green and glyphs are created to show the remaining states.

If there is more than one token in a place, a label is attached to the glyph (or to the place) indicating the number of tokens. This label is colored dynamically depending on the brightness of the background color. Thus, labels can be colored black or white to make them readable.

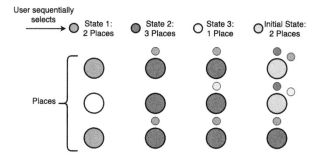

Fig. 2. Assignment of colors to places when selecting states. From left to right, the evolution of the coloring of the places when a user selects states is shown. Glyphs decorating the places of the Petri net are created on demand. (Color figure online)

3.2   Petri Net View

This is the main view of our tool. It presents the Petri net (see Fig. 3(1)) where circles and rectangles depict places and transitions. We implement a version of Sugiyama's approach [20] to layout the Petri net since it gives a good understanding of the flow of the process. Some parameters of the layout algorithm can be modified through the toolbar at the top of the view. In addition, users can perform zooming, panning and dragging of elements directly on the view.

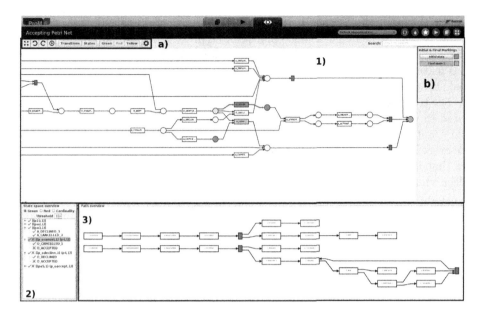

Fig. 3. Overview of our tool. (1) displays a Petri net in which four places are colored: two in red, indicating the last state reached in the *Red* area, one in blue, indicating that token was consumed in order to reach the *Red* state, and one in golden, indicating the final state of the Petri net, which was selected in (1b). (2) shows all the available problematic states, some of them are highlighted indicating that the user selected those ones. Lastly, 3) shows the corresponding runs. (Color figure online)

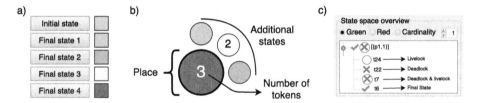

Fig. 4. Examples of the components of our tool: (a) component showing the initial and final states and their colors; (b) four different final states that involve the same place; (d) component showing a problematic state from which we can reach four different types of problems depending on the transition we fire. (Color figure online)

Last but not least, nodes can be hovered in order to show the label of such element by using a tooltip.

To the right of the Petri net view, there is a panel which shows the initial and final states of the model. Each state is presented in the tool by two components: a button and a colored rectangle, which are interactive. This component can be seen in Fig. 4(a). The example shows one single initial and five final states. Users can (de)select states with this component and change the assigned colors. When a state is selected, the Petri net view reacts by showing the current selection of states. Given the fact that a single place can be present in multiple states, we use a new approach to represent that a place belongs to several states (see Fig. 2). This feature directly relates to task T1 (Obtain an overview of all or a subset of final states of a Petri net).

The top part of this view (Fig. 3(1a)) shows two sets of buttons that are dedicated to perform a quick exploration of the problems that have been detected. These buttons enable users to explore places or transitions that belong to just one area. This is useful because it gives a quick overview of the parts of the Petri net that belong to the *Green/Red/Orange* area. In order to do this, users can select what they want to visualize (places/transitions) and which area they want to explore (*Green/Red/Orange*). Once the user selects one of these options, the Petri net view highlights those elements which belong to the selected area. This feature relates to task T2 (Compare disjoint sets of transitions and/or places belonging to a specific area).

3.3   States View

This component can be seen in more detail in Fig. 4(c) and it enables the exploration of the most important problematic states within the state space of the Petri net (T3 (Find problematic states)). Those states correspond to scenarios in which the process can lead from an *Orange* state to either a *Green* or *Red* state. This component only shows the border cases, which are derived from the relevant transitions of the state space (see Fig. 1).

Our approach uses a two-level tree to visualize the different states in which the net experiences a problem. The first indicates the states in which an *Orange-to-Red* or *Orange-to-Green* scenario was found, and the second level indicates the transition that is involved in the detected scenarios. For each state, the user can find a variable number of transitions that could trigger a step in the state space leading from an *Orange* state to a *Green* state, or a *Red* state in which two situations can ultimately happen: deadlock or livelock. This part relates to T5 (Determine when the problem occurs for a specific problematic state) since they show where (state and transition) problems occur.

In some cases, the algorithm that our tool uses detects a problem only because of the threshold that it uses. In those cases, this component indicates that by marking the problematic state with an asterisk. In this way, users can easily differentiate those states that are always problematic from those that are problematic because the algorithm did not continue exploring.

We use icons to give a quick overview of the scenarios found by the algorithm. They indicate the scenario for a specific state (aggregated view) and for a specific transition (specific view). Therefore, the first level of the tree might display multiple icons indicating all the possible scenarios that can be reached through that state. Seven scenarios are possible: (1) reaching a *Green* state; (2) a *Red* state that eventually leads to a deadlock is reached; (3) a *Red* state that eventually leads to a livelock is reached; (4) a *Red* state from which a deadlock and livelock can be reached; (5) either a *Green* state or a *Red* deadlocking state can be reached; (6) either a *Green* or a *Red* livelocking state can be reached; and (7) either a *Green* or *Red* dead/live-locking state can be reached.

The nodes of the tree can be sorted by three different criteria. By clicking on the corresponding radio buttons, the view sorts the states by the criterion chosen by the user. Thus, users can sort states by the type of scenario that they represent (either *Green* or *Red*) and the cardinality of the states.

Next to the sorting functions there is a spinner, which is used to set the threshold used by the algorithm that computes the state space. By default, this parameter is set to 1. When the user interacts with the spinner, the tool recomputes the state space, partitions the recomputed state space into *Green*, *Red* and *Orange*, and recomputes all the relevant information related to them, such as runs or disjoint sets of states and transitions.

Users can interact with the nodes of the tree to explore the different scenarios. This way, the nodes can be selected to be displayed in the Petri net view. When a node from the first level of the tree is selected, the main view shows all the available scenarios for that specific state by coloring the nodes of the Petri net that are involved in that specific state. In Fig. 3 state {*p_ocancel, p4*} is selected. The places that define the selected state are colored blue, while the transitions that can be triggered leading the process to a *Green* or *Red* state are colored green and red.

Once users select a (set of) state(s), it is possible to interact with the main view to explore the behavior of the net. This is done by enabling users to click on the transitions that have been colored to show the paths that lead to the selected state, and the final marking reached by triggering that transition. This feature connects this view and the runs view, which is described below.

3.4   Runs View

This view helps users perform tasks T4 (Explore paths that lead to a problematic state) and T6 (Analyze the runs for a selection of states). An example is shown in Fig. 3(3). Runs are displayed as disconnected graphs, which can be projected as paths in the Petri net view. When users select a state from the *states view*, this component shows the runs that lead to the chosen states. Then, two major interactions are provided: nodes hovering, and path selection. On the one hand, the first interaction aids users in linking nodes of the runs to nodes of the Petri net view. On the other hand, the second interaction assists in visualizing the path that goes from the initial state to the selected state directly on the Petri net.

Through these two interactions, users can detect states that share similar segments of path, helping users get insights into the problematic scenarios of the model.

There is a third interaction which links directly with the Petri net view. When the user clicks on a transition that is involved in the problematic scenario that the user is exploring, the Petri net view takes the run that leads to the ultimate state reachable from the current state, that is, a deadlock or livelock, and shows the path. Providing the context in which the process ends up in a *Red* state helps the user to understand how the process led to that problem.

3.5 Design Decisions

The design decisions made in this work support some basic notions on human perception [13]. The usage of glyphs to represent the belonging of a place to different states is a natural way to show that type of information. They are located next to the elements for which they provide information, and they use a basic color code to show the state that they represent. We use this notation since it is known that the color is a cognitively effective visual variable [13]. Also, we use colors to display useful information on top of the Petri net. We consider this approach to be acceptable since the usage of other ones (e.g., shapes) would probably interfere with the Petri net notation itself. Even though color representations are a limitation for our tool, we consider satisfactory for the purposes of this work.

3.6 Implementation Details

Our tool is implemented as a plug-in within the ProM [23] framework using Java v6. Prefuse [10] is used to manage the graph structures and the visual properties of the visualization of the Petri net. The jBPT library [17] for Petri nets is used to compute the runs.

4 Use Cases

In this section we demonstrate how our tool can be used to assess Petri nets. We focus on two nets, which were designed by students who participated in the study developed in [28]. These nets include information on initial and final states.

The first case exposes an example in which no final state can be reached. Figure 5 summarizes this use case. It shows an overview of the Petri net and an area of interest in more detail. As can be seen, there are no problematic states in the states view. However, we observe that some of the places are colored red, which means that this place can contain tokens but that no final state can be reached from any state in which this place contains tokens, and some are not colored at all. The latter places do not appear in the state space, which means they are not reachable since if they were reachable, they would be either colored or present in the problematic states view. The final state is also not colored, and therefore the process can never reach the final state. Observe that all three input places of the *book_hotel...* transition can contain tokens, but the single output place cannot. Apparently, not all input places of this transition can contain tokens at the same time. The source of this problem can be found in the second place from the left in the overall net, which corresponds to a three-way choice. If the highlighted path is chosen, then only one of the input places can contain a token, otherwise only the other two can contain tokens.

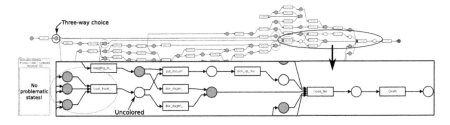

Fig. 5. Screenshots of the usage of the tool for a dataset. (Color figure online)

The second use case (Fig. 6) depicts a Petri net in which several problematic scenarios have been found. Initially, we proceed by exploring the disjoint sets of places and transitions. We can easily spot some straightforward paths that lead to the final state as well as some paths that eventually finish in the *Red* area (Fig. 6(a)). Furthermore, we see some places and transitions that have not been colored. We now focus on those elements in order to explore what occurs there. By hovering on some of the places that have not been colored, we can see their labels. One interesting place is the one labeled as *Status7* since it is present in three of the problematic states shown in the states view. We pick one of those states to explore the runs that lead to that problematic state. The tool shows two different runs (see Fig. 6(a)). By looking at these, we can see

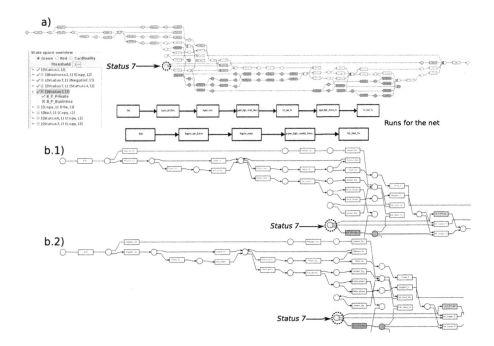

Fig. 6. Screenshots of the usage of the tool for a dataset. (Color figure online)

that they share some of the initial steps in the process, but that they divert at some point. Clicking a run visualizes the paths that those runs represent in the Petri net. If we compare the two runs in this way, we immediately see that they finish in the same problematic state, although they followed different paths. We can also see where the deadlock occurs by clicking on the transition colored red. Figure 6(b) shows the two paths (edges colored blue) for the two runs, as well as the deadlock that is reached (place colored red). The place colored blue indicates that the token must be consumed by firing the red transition.

5   Conclusion

We have presented PSVis, a tool to visually assess the soundness of a Petri net. We have formulated the most important analysis tasks, and have demonstrated the usage of our tool through the exploration of two Petri nets. The first use case showed a simple scenario in which we discovered why the final state was not reachable. The second use case represented a more complex example in which more actions were performed. Through different actions, we observed different aspects of the Petri net such as deadlocks and common paths that lead to them.

One of the main limitations of our tool is that it relies on the state space, which cannot not always be computed in reasonable time. Even though we have some workarounds (setting a threshold to limit the computation of branches), it may take a considerable amount of time to finish. We plan to study alternatives to compute the parts of the state space that are used in the analysis. One option might be to explore the state space incrementally by computing just portions of it.

In future work we aim to perform experiments to compare different approaches for displaying the runs since the current representation lacks of an explicit way to display loops and concurrency, which would help to perform task T7 (Examine concurrency, loops and causal order in runs).

Acknowledgment. This research was performed within the framework of the strategic joint research program on Data Science between TU/e and Philips Electronics Nederland B.V.

References

1. van der Aalst, W.M.P.: Verification of workflow nets. In: Azéma, P., Balbo, G. (eds.) ICATPN 1997. LNCS, vol. 1248, pp. 407–426. Springer, Heidelberg (1997). https://doi.org/10.1007/3-540-63139-9_48
2. van der Aalst, W.M.P.: The application of Petri nets to workflow management. J. Circ. Syst. Comput. **8**(1), 21–66 (1998)
3. Adam, N.R., Atluri, V., Huang, W.K.: Modeling and analysis of workflows using Petri nets. J. Intell. Inf. Syst. **10**(2), 131–158 (1998)
4. Bergenthum, R., Desel, J., Juhás, G., Lorenz, R.: Can i execute my scenario in your net? viptool tells you!. In: Donatelli, S., Thiagarajan, P.S. (eds.) ICATPN 2006. LNCS, vol. 4024, pp. 381–390. Springer, Heidelberg (2006). https://doi.org/10.1007/11767589_21

5. Dehnert, J., Rittgen, P.: Relaxed soundness of business processes. In: Dittrich, K.R., Geppert, A., Norrie, M.C. (eds.) CAiSE 2001. LNCS, vol. 2068, pp. 157–170. Springer, Heidelberg (2001). https://doi.org/10.1007/3-540-45341-5_11

6. Desel, J.: Validation of process models by construction of process nets. In: van der Aalst, W., Desel, J., Oberweis, A. (eds.) Business Process Management. LNCS, vol. 1806, pp. 110–128. Springer, Heidelberg (2000). https://doi.org/10.1007/3-540-45594-9_8

7. Desel, J., Esparza, J.: Free Choice Petri Nets. Cambridge Tracts in Theoretical Computer Science, vol. 40. Cambridge University Press, Cambridge (1995)

8. Desel, J., Juhás, G., Lorenz, R., Neumair, C.: Modelling and validation with vip-tool. In: van der Aalst, W.M.P., Weske, M. (eds.) BPM 2003. LNCS, vol. 2678, pp. 380–389. Springer, Heidelberg (2003). https://doi.org/10.1007/3-540-44895-0_26

9. Ellis, C.A.: Information control nets: a mathematical model of office information flow. In: Proceedings of the Conference on Simulation, Measurement and Modeling of Computer Systems, Boulder, Colorado, USA, pp. 225–240. ACM Press (1979)

10. Heer, J., Card, S.K., Landay, J.A.: Prefuse: a toolkit for interactive information visualization. In: Proceedings of the SIGCHI Conference on Human Factors in Computing Systems, pp. 421–430. ACM (2005)

11. Mendling, J., Verbeek, H.M.W., van Dongen, B.F., van der Aalst, W.M.P., Neumann, G.: Detection and prediction of errors in EPCs of the SAP reference model. Data Knowl. Eng. **64**(1), 312–329 (2008). Fourth International Conference on Business Process Management (BPM 2006), 8th International Conference on Enterprise Information Systems (ICEIS 2006), Four Selected and Extended Papers, Three Selected and Extended Papers

12. De Michelis, G., Ellis, C., Memmi, G., (eds.): Modelling workflow management systems with high-level Petri nets. In: Proceedings of the Second Workshop on Computer-Supported Cooperative Work, Petri nets and Related Formalisms, Zaragoza, Spain, June 1994

13. Moody, D.: The physics of notations: toward a scientific basis for constructing visual notations in software engineering. IEEE Trans. Softw. Eng. **35**(6), 756–779 (2009)

14. Murata, T.: Petri nets: properties, analysis and applications. Proc. IEEE **77**(4), 541–580 (1989)

15. Ouyang, C., Verbeek, H.M.W., van der Aalst, W.M.P., Breutel, S., Dumas, M., ter Hofstede, A.H.M.: Formal semantics and analysis of control flow in WS-BPEL. Sci. Comput. Program. **67**(2), 162–198 (2007)

16. Peterson, J.L.: Petri Net Theory and the Modeling of Systems. Prentice-Hall, Englewood Cliffs (1981)

17. Polyvyanyy, A., Weidlich, M.: Towards a compendium of process technologies: the jBPT library for process model analysis. In: Proceedings of the CAiSE 2013 Forum at the 25th International Conference on Advanced Information Systems Engineering (CAiSE), pp. 106–113. Sun SITE Central Europe (2013)

18. Reisig, W.: Petri Nets: An Introduction. EATCS Monographs on Theoretical Computer Science, vol. 4. Springer, Heidelberg (1985)

19. Sadiq, W., Orlowska, M.E.: Analyzing process models using graph reduction techniques. Inf. Syst. **25**(2), 117–134 (2000)

20. Sugiyama, K., Tagawa, S., Toda, M.: Methods for visual understanding of hierarchical system structures. IEEE Trans. Syst. Man Cybern. **11**(2), 109–125 (1981)

21. van der Aalst, W.M.P.: Process Mining: Data Science in Action. Springer, Berlin (2016)

22. van der Aalst, W.M.P., Adriansyah, A., van Dongen, B.F.: Replaying history on process models for conformance checking and performance analysis. Wiley Interdiscip. Rev. Data Min. Knowl. Discov. **2**(2), 182–192 (2012)

23. van Dongen, B.F., de Medeiros, A.K.A., Verbeek, H.M.W., Weijters, A.J.M.M., van der Aalst, W.M.P.: The ProM framework: a new era in process mining tool support. In: Ciardo, G., Darondeau, P. (eds.) ICATPN 2005. LNCS, vol. 3536, pp. 444–454. Springer, Heidelberg (2005). https://doi.org/10.1007/11494744_25

24. van Dongen, B.F., Jansen-Vullers, M.H., Verbeek, H.M.W., van der Aalst, W.M.P.: Verification of the SAP reference models using EPC reduction, state-space analysis, and invariants. Comput. Ind. **58**(6), 578–601 (2007)

25. van Dongen, B.F., van der Aalst, W.M.P., Verbeek, H.M.W.: Verification of EPCs: using reduction rules and Petri nets. In: Pastor, O., Falcão e Cunha, J. (eds.) CAiSE 2005. LNCS, vol. 3520, pp. 372–386. Springer, Heidelberg (2005). https://doi.org/10.1007/11431855_26

26. van Hee, K., Sidorova, N., Voorhoeve, M.: Generalised soundness of workflow nets is decidable. In: Cortadella, J., Reisig, W. (eds.) ICATPN 2004. LNCS, vol. 3099, pp. 197–215. Springer, Heidelberg (2004). https://doi.org/10.1007/978-3-540-27793-4_12

27. Verbeek, H.M.W.: Verification of WF-nets. Eindhoven University of Technology Eindhoven, The Netherlands (2004)

28. Verbeek, H.M.W., Basten, T., van der Aalst, W.M.P.: Diagnosing workflow processes using Woflan. Comput. J. **44**(4), 246–279 (2001)

29. Verbeek, H.M.W., Wynn, M.T.: Verification. In: ter Hofstede, A.H.M., van der Aalst, W.M.P., Adams, M., Russell, N. (eds.) Modern Business Process Automation: YAWL and its Support Environment, pp. 513–539. Springer, Heidelberg (2010). https://doi.org/10.1007/978-3-642-03121-2_20. Database Management & Info Retrieval, Chap. 20

30. Ward, M.O.: Multivariate data glyphs: principles and practice. In: Chen, C.-H., Härdle, W., Unwin, A. (eds.) Handbook of Data Visualization. SHCS, pp. 179–198. Springer, Heidelberg (2008). https://doi.org/10.1007/978-3-540-33037-0_8

31. Wynn, M.T., Verbeek, H.M.W., van der Aalst, W.M.P., ter Hofstede, A.H.M., Edmond, D.: Business process verification - finally a reality!. Bus. Process Manag. J. **15**(1), 74–92 (2009)

An Architecture for Querying Business Process, Business Process Instances, and Business Data Models

María Teresa Gómez-López[1]([✉])(iD), Antonia M. Reina Quintero[1](iD),
Luisa Parody[1](iD), José Miguel Pérez Álvarez[1](iD), and Manfred Reichert[2]

[1] Departamento de Lenguajes y Sistemas Informáticos,
Universidad de Sevilla, Seville, Spain
{maytegomez,reinaqu,lparody,josemi}@us.es
[2] Institute of Databases and Information Systems, Ulm University, Ulm, Germany
manfred.reichert@uni-ulm.de

Abstract. Business data are usually managed by means of business processes during process instances. These viewpoints (business, instances and data) are strongly related because the life-cycle of business data objects need to be aligned with the business process and process instance models. However, current approaches do not provide a mechanism to integrate these three viewpoints nor to query them all together while maintaining the information in the distributed, heterogeneous systems where they have been created. In this paper, we propose the integration of the business process, business process instance, and business data models by using their metamodels and also an architecture to support this integration. The goal of this integration is to make the most of the three models and the technologies that support them in an isolated way. In our approach, it is not necessary to change the source data formats nor transforming them into a common one. Furthermore, the proposed architecture allows us to query the three models even though they come from three different technologies.

Keywords: Business Data Model · Business Process Model
Business process instance model · Model integration
Heterogeneous data sources

1 Introduction

The volume, variety and velocity of process-related data are growing drastically. Some of these data are created and managed by Business Process Management Systems (BPMS). A business process (BP for short) consists of a set of activities whose execution needs to be coordinated in an organisational and technical environment in order to achieve a particular business goal [24]. The coordinated execution of activities is a fundamental principle from the viewpoint of Business Process Management, and it provokes the change of data objects during

E. Teniente and M. Weidlich (Eds.): BPM 2017 Workshops, LNBIP 308, pp. 757–769, 2018.
https://doi.org/10.1007/978-3-319-74030-0_60

process instantiation. In general, a BP can be implemented and operationalized by a BPMS. To make the most of the heterogeneous information provided by the different viewpoints (i.e., business process, business process instance and business data), it is necessary to integrate the different models and provide a mechanism to query them all together, since they are frequently supported by different technologies. Previous solutions based on semantic models [10] have provided mechanisms to create homogeneous data stores, using the information from heterogeneous sources. The problem of this type of approaches is that they go+ against the requirements established when there are a high volume, variety and velocity of data, as in Big Data scenarios, where the changeability and quantity of data make not possible transform the sources. For this reason, this paper proposes an architecture where process querying mechanisms are inspired in the Map-Reduce paradigm. Map-Reduce is a programming paradigm that allows for massive scalability across hundreds or thousands data nodes, and it lets: (1) divide the query into the subsystems that contain the data, and (2) merge the outputs of the distributed databases.

In order to define how to divide the query and combine the obtained information, it is necessary to determine the relation between subsystems. We propose the use of metamodels and the combination of them. A metamodel defines a frame and a set of rules for creating models for a specific application domain. A viewpoint is defined in relation to one or more metamodels [9]. In our scenario, each viewpoint is supported by a metamodel that defines the main elements managed in each case and their relationships. In general, there exists a fundamental relation between the three models: Business Process Model (BPM), Business Process Instances Model (BPIM), and Business Data Model (BDM). Accordingly, it is not sufficient to query each of these models separately, but to provide integrated access to the process, data and instance perspectives. In particular, process-centric queries across these different viewpoints need to be enabled [4].

The insufficient understanding of the inherent relationships existing between business processes and business data is deficient [21]. As far as we know, existing querying approaches focus on one specific model type, but they do not exploit the information resulting from that integration. This paper shows how to integrate the BPM, BPIM, and BDM by means of a weaving model that makes explicit the relationships between the elements of the three models.

As an example of a BP considers the organisation of a conference. The corresponding BP model is shown in Fig. 1. Three different pools cover the main functions needed to organize the conference, to submit a paper, and to register for the conference: (1) the first pool describes how to manage the conference organisation by the conference chair who is in charge of the conference requirements; (2) the second pool shows the submission process from the viewpoint of an author; (3) the third pool deals with the conference registration and related payment. Usually, the execution of a process instance is persisted in the database of the BPMS to which the BP is deployed. According to the framework proposed in [20], this situation corresponds to a process repository composed of simulation

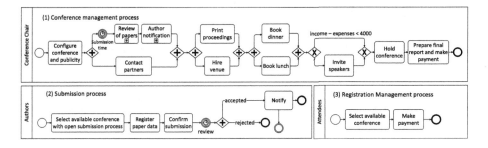

Fig. 1. Business processes for conference organisations

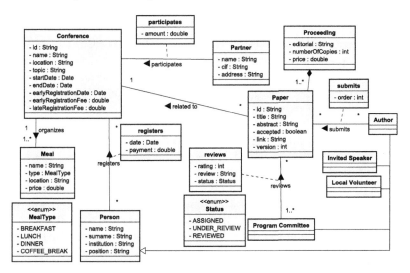

Fig. 2. Conceptual model of the conference process

models (see Def. 3.4 in [20]) This kind of information must follow the BP instance according to the used BPMS. Moreover, business data that flows through the process is persisted in a database. For the conference example, the corresponding conceptual data model is depicted in Fig. 2. It presents the entities involved in the conference process. A *Conference* is defined by its *id, name, location, scope* (e.g., "Software Engineering", "Business Process Management", or "Big Data"), *start* and *end dates*, and the date until which the *early registration fee* is valid. In turn, an author may submit several papers, which are going to be reviewed by the *Program Committee* that decides on the *rating* and *status* of the paper.

Regarding the conference scenario, the three models (i.e., BPM, BPIM and BDM) play an important role in extracting relevant information about the business process. Examples of queries making use of the integration of these viewpoints are "Average of execution time of the conferences with scope *Business Process*", and "Evolution of the *number of registrations* variable for the process named 'Conference Management Process' of those conferences taking place in Barcelona". In particular, this paper shows how the three models may be integrated to process such queries.

The paper is organized as follows: Sect. 2 introduces the metamodels related to the business process, business process instance and business data viewpoints as well as the metamodel that supports the integration of them all. After that, Sect. 3 presents our proof-of-concept prototype. Section 4 discusses related work. Finally, Sect. 5 concludes the paper.

2   Business Viewpoints

This section describes the metamodels covering the three viewpoints on a BP, i.e., BPM, BPIM, BDM as well as the metamodel used to integrate them all. Then, these metamodels are used to enable an integration of the viewpoints.

A BP corresponds to a set of coordinated activities, carried out manually or automatically, to achieve a specific business goal. The metamodel for creating BP models is summarised in Sect. 2.1. In turn, the metamodel related to the process instance viewpoint is presented in Sect. 2.2. Finally, concerning to the data viewpoint, a simplified metamodel of Unified Modeling Language (UML) [15] is introduced in Sect. 2.3. In particular, a BP needs to control its data flow, which reflects the business data managed in the context of the BP and the way this data is transferred through the activities. In general, this process-related data can be described by a conceptual model which usually is defined using UML. In practice, the three models must be considered in an integrated way due to their many interdependencies; e.g., BP instances are related to a BP model, data entities are used to define the data flow of a BP, and the various BP instances create data objects or update the values of the data.

2.1   Viewpoint 1: Business Process Metamodel

The main standard used to model business processes is the Business Process Model and Notation (BPMN) as proposed by OMG [14]. Besides other elements, BPMN 2.0 includes data objects, events of various types, and artefacts. Figure 3 introduces a simplified version of the BP metamodel (BPMM for short), which includes the main metaclasses used in this paper to model the BPs as well as their relationships and attributes.

The root of the metamodel is `BPMNProcess`, which is composed of a set of `BPMNElement`. A `BPMNElement`, in turn, may be a `Swimlane`, a `ConnectingObject`, a `FlowObject`, and an `Artifact`. Note that these metaclasses are abstract. The concrete metaclasses are subtypes of these ones. Thus, a `ConnectingObject` may be a `MessageFlow`, a `SequenceFlow`, and an `Association`. Furthermore, there are three types of `FlowObject`: `Event`, `Gateway` and `Activity`. The three of them are also abstract metaclasses. An `Event` may be an `InitialEvent`, an `IntermediateEvent`, and a `FinalEvent`. A `Gateway` may be a `XOR`, `OR`, and `AND` gateway. An `Activity`

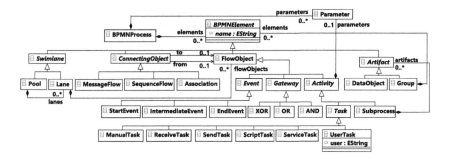

Fig. 3. Simplified business process metamodel

may be a `Task` (also an abstract metaclass) or a `Subprocess`. There are different types of `Task`: `ManualTask`, `ReceivedTask`, `SendTask`, `ScriptTask`,
`ServiceTask`, and `UserTask`. Note that the semantics of these metaclasses
is exactly the specified by the BPMN2.0 standard and also that this metamodel
is a simplified version of it. We refer interested readers to [14] for knowing the
semantics of each metaclass.

2.2 Viewpoint 2: Business Process Instance Metamodel

A business process instance represents a concrete case in the operational
business of a company [24]. Figure 4 shows our proposal of Business Process Instance Metamodel (BPIMM for short). `ProcessEngine` is the root
of the metamodel and it represents the Business Process Management System (BPMS) that executes the BP. A BPMS may deploy various processes.
Each process is described by means of the `ProcessDefinition` metaclass. Note that `ProcessDefinition` is related to a Business Process (cf.
Sect. 2.1). `ProcessInstance` represents an execution of a specific process.
Thus, a `ProcessDefinition` is related to a set of `ProcessInstance`.
A `ProcessDefinition` is composed of a sequence of `Activity`. Accordingly, a `ProcessInstance` is composed of a set of `ActivityInstance`
(`childActivities` relation). There is also a parent-child relation between
activities instances. The first activity that is instantiated during a process execution is an activity instance that has no parent. Finally, a set of `Variable`
can be associated to a `Process Instance`.

Note that BPIMM deals with concepts related to process execution and that
the definition of processes included in many BPMSs usually has more properties of processes and activities than the standard BPMN 2.0. Having this into
account, the ProcessDefinition and Activity metaclasses model these properties
that are present in the execution environments.

2.3 Viewpoint 3: Business Data Metamodel

In order to be able to create BDM, we use the OMG Metamodel to represent Conceptual Models introduced in Fig. 5, which includes the main entities of the Business Data metamodel (BDMM) presumed in this paper.

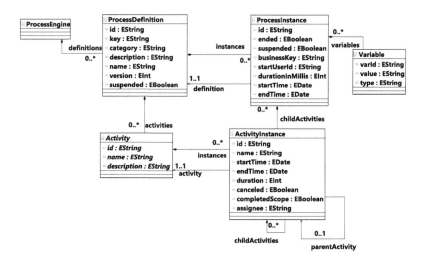

Fig. 4. Business Process Instance Metamodel (BPIMM)

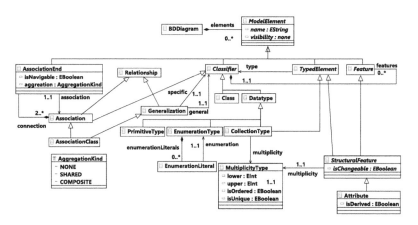

Fig. 5. Simplified Business Data Metamodel (BDMM)

The root entity is BMDiagram, which is composed of a set of ModelElement. A ModelElement may be an AssociationEnd, a Relationship, a Classifier, a TypedElement, or a Feature. The entities of the system are mainly represented by a Classifier, either an Association, Class, or DataType. Furthermore, through the AssociationEnd and Relationship, Classifier entities are related. Note that an instance of this metamodel which depicts entities related to the organisation of a conference is presented in Fig. 2.

2.4   Integrating BPM, BPIM and BDM

Business process, business process instance and business data metamodels represent different, but complementary, viewpoints of a business process. Linking entities across these metamodels is a must in order to exploit the information provided by their specific instances, i.e., their models. In this context, model weaving is a generic operation that establishes correspondences between model elements. The resulting weaving models are a special kind of models that link together other models. In general, weaving models contain a set of links between the elements of two different models [8]. A weaving model conforms to a weaving metamodel, which defines the kind of links that may be established between the elements of the woven models.

Atlas Model Weaver (AMW) [8] is a tool that allows creating and handle weaving models and metamodels. New weaving metamodels should be implemented by extending a Core Weaving Metamodel that it is included and implemented in AMW. Thus, to define our integration metamodel we extend this Core Weaving Metamodel. Figure 6 shows our extension (see the metaclasses with grey background colour), which specifies the semantics of the links between the elements of Business Process, Business Process Instance and Business Data models. These elements are specific instances of the corresponding metamodels (i.e., BPMM, BDMM, and BPIMM). Note that classes belonging to the core weaving metamodel are abstract, and are extended by metaclasses (the ones coloured in grey) that refine them by adding the semantics needed in the context of our scenario.

An Integration Model can be seen as a kind of weaving model that let us integrate the three kinds of models (BPIM, BDM, and BPM). This integration model defines the relationships between the elements in the three models. Match is a kind of link representing the relation between two elements from different models.

Figure 7 shows how three instances (i.e., models) of BPIMM, BPMM and BDMM may be related by means of an instance of the Integration Metamodel. Classes with <<BPIM>> stereotype represent instances of Business Process Instance metaclasses, classes with <<BPM>> represent instances of Business

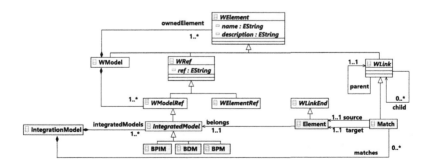

Fig. 6. Integration metamodel

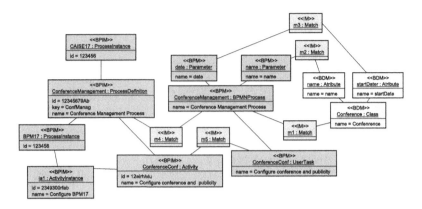

Fig. 7. Integrated BPIM, BPM and BDM

Process metaclasses, classes with <<BDM>> stereotype represent instances of Business Data metaclasses, and classes with <<IM>> represent instances of Integration metaclasses. Thus, the `ConferenceManagement::BPMNProcess` instance represents the process named Conference Management Process, as depicted in Fig. 1, whereas the instance `ia1::ActivityInstance` represents one execution of the activity that configures the Conference BPM17.

3 Architecture for Combined Queries

Data combination is an important open problem in business intelligence, as explained in [20], where the authors propose a general, abstract framework for devising process querying methods. Our architecture, based on Map-Reduce paradigm, defines how the query is divided (Map), and the partial queries are combined (Reduce). Our proposal can be seen as an instance of the abstract framework presented in [20], in such a way that it includes a real combination of technologies that can be thought as the implementation of some of the generic functionalities described in the framework. Assume one wishes to know how many institutions are involved in the execution of each task, in a context where the BPMS Bonita^{TM}, data are persisted in an Oracle database and the BPM is persisted by using Neo4J a query similar to the one shown in Listing 1 should be executed:

```
SELECT Task.id, COUNT(institution)
FROM (Task FULL JOIN ActivityInstance FULL JOIN Person)
GROUP BY Task.id WHERE Task.idProcess = idP
```

Listing 1. Query example

Note that obtain the query results, you need to query data related to the three viewpoints. In other words, you need information about Tasks from the BP viewpoint, information about Activity Instances from the BPI viewpoint, and information about Persons from the BD viewpoint. Note that to obtain this information we have to deal with three different technologies (Neo4J, Bonita, and Oracle, respectively). We propose a two-step process to execute the query aligned with Map-Reduce methods. The first step, data extraction, is the responsible for transforming the original query into a set of queries that are specific for each viewpoint. The second step, data integration, has as input the results of the previous queries and is the responsible for combining them. Figure 8 shows the architecture that supports the two-steps query execution. The following sections give some details about these two steps.

3.1 Data Extraction

In the data extraction step, the original query is transformed into three different queries, in order to query the three different viewpoints in an isolated way (see (1) in Fig. 8). The result of these specific queries (step 2) is a set of JSON files that will be combined in step (3). The interaction with the specific technology that supports each viewpoint (i.e., Neo4J, Bonita and Oracle) is in charge of drivers. Drivers are the components that deal with the platform specific features of each technology. In other words, drivers transform a generic query to a specific query on the respective data repository. Note that these three queries can be executed in parallel since each result is obtained from isolated databases.

All drivers should implement the generic operation select. This select operation extracts a set with all of the objects of a given class that fulfil certain predicate. That is, every driver should implement an operation with a prototype similar to the following one: Set<T> select (Class<?> name, Predicate<T> pred). Thus, our original query (see Listing 1) is transformed into three select calls (one to obtain a set of tasks, another one to obtain a set of Activity Instances, and, finally, one to obtain a set of Persons) that will be executed by one driver each.

The Neo4J Driver transforms the select operation into a Neo4J query to obtain the set of tasks of the process whose id is idP. The driver uses Spring Data and neo4j-ogm-bolt-driver to map the graph database into Java Objects. The Bonita Driver transforms the select operation into a set of calls to the

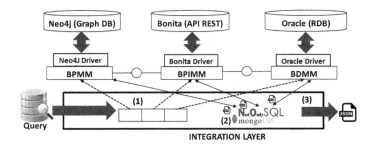

Fig. 8. Combination of technologies for the architecture

BONITA REST API to get the set of Activity Instances. The Oracle Driver transforms the select operation into an Oracle query to obtain the set of Persons stored in the database.

3.2 Data Integration

In the data integration step, the results are integrated to obtain the result of the generic query. Our approach uses *JSON* as data interchange mechanism, in such a way that the *JSON* files can be uploaded in the end-point and then be combined using the MongoDB query language. Thus, full join and aggregation operations allow us to combine the partial solutions in a single one.

The service that manages the specific queries to the various drivers, stores the results of each specific query in a centralised MongoDB. The query process in the MongoDB is a Spring application. This service further uses Spring Data MongoDB in order the manage data in the database.

Although the proposed architecture is a step beyond BI data extraction [20], some limitations are still a challenge: do the current technologies support the attributes defined in each metamodel, how can affect the query operations supported by subsystems to the development of the drivers, and can the integration layer apply any query operation to combine subsets of data.

4   Related Work

The approaches related to our work can be classified into two different groups:

Approaches that specify a viewpoint or a combination of viewpoints.
Our contribution in this area is the Business Process Instance metamodel, as our business process and business data metamodels are a simplification of the corresponding standards proposed by OMG, BPMN2.0 [14] and UML2.0 [15]. As a consequence, this group of approaches focuses only on those ones that deal with the modelling of process instances. In relation to this, the necessity to store the history of process instances has been analysed in [13] and in [17]. However, there is no standard definition of the process instance metamodel. Frequently, ad-hoc models are used according to the BPMS that supports the solution. In [18] a multi-view metamodel is proposed for business process instance model. The metamodel deals with three different viewpoints: process execution path, process instance data and process instance metadata.

The relation between data stored and process activities have been studied in [6,11,12], but not having as a goal to query both at the same time, only to have a way to create a model where both aspects are combined.

Approaches for querying the viewpoints. Approaches that query the different viewpoints can be classified according to their capacities to query data into three groups: those that can query only a viewpoint in an isolated way, those that can query a combination of two of the viewpoints, and, finally, those that can query the combination of all of them.

In the first group, there are approaches such as BP-QL [1], BPMN-Q [22], APQL [23] that only query the business process viewpoint or IPMPQL [5], BeehiveZ [16], BP-Mon [2] and BP-SPARQL [3] only query the business process instance viewpoint. In the second group, is PIQL [19] that queries, on the one hand, the business process viewpoint, and, on the other hand, the combination of the business process instances and data viewpoints. Data-Aware POQL [7] also belongs to the second group by querying the combination of Business Process Instance and Business Data viewpoints. Finally, note that none of the proposals allows us to combine the three models in the same query by using the same language, while our proposal allows any combination of metamodels in a query.

5   Conclusions

Business process models are usually executed in a Business BPMS, which registers each execution of the model, the so-called business process instance. Furthermore, the data that flows through the process is persisted in an external database belonging to the company. Thus, business processes can be viewed from three different perspectives or viewpoints, namely, business process, business process instance and business data. In this paper, we specify the elements that are involved in each viewpoint by means of a metamodel. The relations between the models that represent the different viewpoints are made explicit by means of an integration model that also conforms to an integration metamodel. The integration model is a kind of weaving model that specifies the semantics of the relationships between the elements that belong to the different viewpoints.

This paper presents an architecture that allows us to extract in a combined way the different parts of the business process knowledge, providing two main benefits: (1) no data source modifications must be done, evaluating the subqueries into each data repository, reducing the time and complexity of data transformation; and, (2) an architecture for a possible combination of concrete technologies has been proposed.

Acknowledgment. This work has been partially funded by the Ministry of Science and Technology of Spain (TIN2015-63502-C3-2-R, TIN2013-40848-R and TIN2016-75394-R) and the European Regional Development Fund (ERDF/FEDER).

References

1. Beeri, C., Eyal, A., Kamenkovich, S., Milo, T.: Querying business processes with BP-QL. Inf. Syst. **33**(6), 477–507 (2008)
2. Beeri, C., Eyal, A., Milo, T., Pilberg, A.: BP-Mon: query-based monitoring of BPEL business processes. SIGMOD Rec. **37**(1), 21–24 (2008)
3. Beheshti, S.-M.-R., Benatallah, B., Motahari-Nezhad, H.R.: Enabling the analysis of cross-cutting aspects in ad-hoc processes. In: Salinesi, C., Norrie, M.C., Pastor, Ó. (eds.) CAiSE 2013. LNCS, vol. 7908, pp. 51–67. Springer, Heidelberg (2013). https://doi.org/10.1007/978-3-642-38709-8_4

4. Beheshti, S.-M.-R., Benatallah, B., Sakr, S., Grigori, D., Motahari-Nezhad, H.R., Barukh, M.C., Gater, A., Ryu, S.H.: Concepts and techniques for querying and analyzing process data. Process Analytics. Springer, Cham (2016). https://doi.org/10.1007/978-3-319-25037-3

5. Choi, I., Kim, K., Jang, M.: An XML-based process repository and process query language for integrated process management. Knowl. Process Manag. **14**(4), 303–316 (2007)

6. González López de Murillas, E., Reijers, H.A., van der Aalst, W.M.P.: Connecting databases with process mining: a meta model and toolset. In: Schmidt, R., Guédria, W., Bider, I., Guerreiro, S. (eds.) BPMDS/EMMSAD-2016. LNBIP, vol. 248, pp. 231–249. Springer, Cham (2016). https://doi.org/10.1007/978-3-319-39429-9_15

7. González López de Murillas, E., Reijers, H.A., van der Aalst, W.M.P.: Everything you always wanted to know about your process, but did not know how to ask. In: Dumas, M., Fantinato, M. (eds.) BPM 2016. LNBIP, vol. 281, pp. 296–309. Springer, Cham (2017). https://doi.org/10.1007/978-3-319-58457-7_22

8. Fabro, M.D.D., Valduriez, P.: Towards the efficient development of model transformations using model weaving and matching transformations. Softw. Syst. Model. **8**(3), 305–324 (2009)

9. Fischer, K., Panfilenko, D., Krumeich, J., Born, M., Desfray, P.: Viewpoint-based modeling-towards defining the viewpoint concept and implications for supporting modeling tools. In: EMISA, pp. 123–136 (2012)

10. Di Francescomarino, C., Corcoglioniti, F., Dragoni, M., Bertoli, P., Tiella, R., Ghidini, C., Nori, M., Pistore, M.: Semantic-based process analysis. In: Mika, P., et al. (eds.) ISWC 2014. LNCS, vol. 8797, pp. 228–243. Springer, Cham (2014). https://doi.org/10.1007/978-3-319-11915-1_15

11. Gómez-López, M.T., Borrego, D., Gasca, R.M.: Data state description for the migration to activity-centric business process model maintaining legacy databases. In: Abramowicz, W., Kokkinaki, A. (eds.) BIS 2014. LNBIP, vol. 176, pp. 86–97. Springer, Cham (2014). https://doi.org/10.1007/978-3-319-06695-0_8

12. Gómez-López, M.T., Pérez-Álvarez, J.M., Varela-Vaca, A.J., Gasca, R.M.: Guiding the creation of choreographed processes with multiple instances based on data models. In: Dumas, M., Fantinato, M. (eds.) BPM 2016. LNBIP, vol. 281, pp. 239–251. Springer, Cham (2017). https://doi.org/10.1007/978-3-319-58457-7_18

13. Grigorova, K., Kamenarov, I.: Object relational business process repository. In: CompSysTech, pp. 72–78 (2012)

14. O.M. Group. Business Process Model and Notation (BPMN) Version 2.0. OMG Standard (2011)

15. O. M. Group. Unified Modeling Language Reference Manual, Version 2.5. OMG Standard (2015)

16. Jin, T., Wang, J., Rosa, M.L., ter Hofstede, A.H.M., Wen, L.: Efficient querying of large process model repositories. Comput. Ind. **64**(1), 41–49 (2013)

17. Ma, Z., Wetzstein, B., Anicic, D., Heymans, S., Leymann, F.: Semantic business process repository. In: SBPM (2007)

18. Moghadam, N.N., Paik, H.-Y.: BPIM: a multi-view model for business process instances. In: APCCM, pp. 23–32 (2015)

19. Pérez-Álvarez, J.M., Gómez-López, M.T., Parody, L., Gasca, R.M.: Process instance query language to include process performance indicators in DMN. In: EDOC Workshops 2016, pp. 1–8 (2016)

20. Polyvyanyy, A., Ouyang, C., Barros, A., van der Aalst, W.M.P.: Process querying: enabling business intelligence through query-based process analytics. Decis. Support Syst. **100**, 41–56 (2017)

21. Reichert, M.: Process and data: two sides of the same coin? In: Meersman, R., et al. (eds.) OTM 2012. LNCS, vol. 7565, pp. 2–19. Springer, Heidelberg (2012). https://doi.org/10.1007/978-3-642-33606-5_2
22. Sakr, S., Awad, A.: A framework for querying graph-based business process models. In: WWW, pp. 1297–1300 (2010)
23. ter Hofstede, A.H.M., Ouyang, C., La Rosa, M., Song, L., Wang, J., Polyvyanyy, A.: APQL: a process-model query language. In: Song, M., Wynn, M.T., Liu, J. (eds.) AP-BPM 2013. LNBIP, vol. 159, pp. 23–38. Springer, Cham (2013). https://doi.org/10.1007/978-3-319-02922-1_2
24. Weske, M.: Business Process Management: Concepts, Languages, Architectures. Springer, Heidelberg (2007). https://doi.org/10.1007/978-3-642-28616-2

Formal Semantics for Modeling Collaborative Business Processes Based on Interaction Protocols

Emiliano Reynares[1,2(✉)] ⓘ, Jorge Roa[1], María Laura Caliusco[1,2], and Pablo David Villarreal[1,2]

[1] CIDISI, Universidad Tecnológica Nacional Facultad Regional Santa Fe, Santa Fe, Argentina
{ereynares,jroa,mcaliusc,pvillarr}@frsf.utn.edu.ar
[2] CONICET, Santa Fe, Argentina

Abstract. Collaborative business processes (CBPs) are expected to conduct the behavior among organizations that participate in collaborative networks. Although languages for CBPs emerged from the industry and academic sides, not much effort has been put to add formal semantics to the constructs of the languages to reason on structural aspects of CBP models. In particular, the UP-ColBPIP language supports the modeling of collaborative business processes in terms of interaction protocols, which describes a choreography of business messages based on speech acts. This paper presents an approach to add formal semantics to the UP-ColBPIP language, by defining an OntoUML conceptual model of the constructs that allows modeling CBPs as interaction protocols. The formal semantics of the constructs of UP-ColBPIP enables the definition of design guidelines and the development of techniques for the structural analysis of CBP models represented in terms of interaction protocols. Finally, the work depicts an OWL ontology implementing the proposed conceptual model, with the purpose of answering queries about the messages based on speech acts as well as other structural aspects of CBP models.

Keywords: Cross-organizational collaborations · Speech act · Ontology
OntoUML · Conceptual model

1 Introduction

Collaborative networks consisting of tightly collaborating organizations are becoming an increasingly important issue of contemporary Business Process Management (BPM). In a collaborative network, collaborative business processes (CBPs) are expected to conduct the collaboration among organizations to coordinate the flow of information among them and integrate their business processes. A collaborative business process specifies the global view of interactions between organizations to achieve common business goals and serves as a contractual basis for the collaboration [5, 6]. Thus, collaborative networks require that organizations can define and represent the agreed CBPs.

BPMN supports the modeling of CBPs by means of choreography diagrams. This language provides a foundation for basic collaboration functionalities, such as

© Springer International Publishing AG 2018
E. Teniente and M. Weidlich (Eds.): BPM 2017 Workshops, LNBIP 308, pp. 770–781, 2018.
https://doi.org/10.1007/978-3-319-74030-0_61

partnership representation, global view of interactions between participants and enterprise autonomy. However, there are other domain-specific languages proposed to meet specific requirements of the collaborative networks domain that add new concepts and semantics to model CBPs, such as UP-ColBPIP [5] or UMM [13].

In particular, UP-ColBPIP (UML Profile for Collaborative Business Processes based on Interaction Protocols) is a language that supports the modeling of collaborative business processes in terms of interaction protocols, to represent the global view of interactions through a choreography of business messages between organizations playing different roles. In cross-organizational collaborations, the focus should be not only on the representation of the information exchange but also on the representation of the communication aspects between organizations. These aspects are expressed in interaction protocols by using *speech acts*. In an interaction protocol, a business message is associated with a speech act, which represents the intention the sender has with respect to the business document (information) exchanged in the message. Thus, decisions and commitments between organizations can be known from the semantics of the speech acts. This enables the definition of complex negotiations and avoids the ambiguity in the understanding of the business messages of CBPs.

Formal semantics based on Petri nets were defined for the UP-ColBPIP language for checking properties of the behavior of CBP models expressed as interaction protocols, as well as to derive and guarantee alignment between a collaborative process model and its implementation [7]. Anti-Patterns for the UP-ColBPIP language were also defined by using formal semantics of the behavior of the constructs of the language [14]. However, there is not yet a formal representation for the static (not behavioral) aspects of CBPs, e.g. to detect structural issues regarding how the elements of this language can be combined. An important issue in the modeling of CBPs through interaction protocols is to know the semantics of speech acts and their appropriate combinations to represent suitable negotiations and commitments.

In this paper we defined a formal semantics for the static structure of UP-ColBPIP by means of an OntoUML conceptual model that represents its Interaction Protocols View. The formal semantics considers both UP-ColBPIP constructs as well as speech acts related to business messages. From this conceptual model, we generated an OWL ontology that implements such a model. This ontology provides a basis for defining queries about the structural aspects of CBP models.

This paper is organized as follows. Section 2 describes UP-ColBPIP. Section 3 presents an OntoUML model that describes the conceptual model of UP-ColBPIP. Section 4 presents a case study used to depict an OWL implementation of the proposed conceptual model for testing structural issues. Section 5 discusses related work. Section 6 concludes with final remarks and future work.

2   Collaborative Business Processes Based on Interaction Protocols

UP-ColBPIP is a language to model technology-independent collaborative processes [5, 6]. One of the main contributions of UP-ColBPIP is the use of a modeling approach based on interaction protocols to represent the behavior of CBPs. An *interaction protocol* defines a high-level communication pattern through a choreography of business messages between roles performed by participants (organizations).

The modeling of interaction protocols focusses on representing the global view and dependencies of interactions between participants and the responsibilities of the roles they fulfill. The coordination and communication aspects of the interactions are represented by enhancing the semantics of business messages using the speech act theory [5]. A *speech act* represents the intention of a speaker with respect to some propositional content. In UP-ColBPIP, speech acts are used to provide a common understanding of the message-based interactions by representing the intention of the participants.

UP-ColBPIP extends the UML2 Interactions to model interaction protocols through UML2 Sequence Diagrams. *Partners* and the *Role* they fulfill are represented through lifelines (Table 1a). An interaction protocol is described by a temporal and ordered sequence of elements that cross the lifelines: business messages, terminations, protocol references, control flow segments and interaction paths.

Table 1. Graphical notations of the main interaction protocol elements

Lifeline	Messages and Time Constraint	Protocol Reference	Control Flow Segment
Participant :Role	SpeechAct(BusinessDocument) {t=now} SpeechAct(BusinessDocument) tc1: {t..t+2d}	ref Subprotocol name c)	Operator [condition 1] Interaction path 1 [condition N] Interaction path N
		Termination ⬤ [Failure] ⬤ [Success] d)	
a)	b)	d)	e)

A *business message* (Table 1b) is the atomic building block of an interaction protocol and defines a one-way asynchronous interaction between two roles, a sender and a receiver. Its semantics is defined by the associated *speech act* that represents the intention the sender has with respect to the *business document* (information) exchanged in the message. Also, the message indicates that the sender's expectation is that the receiver should act according to the semantics of the speech act.

A *protocol reference* (Table 1c) represents a sub-protocol. Protocols have an implicit termination. A *termination* (see Table 1d) defines an explicit end event of a protocol, which can be *success* or *failure*, to provide a logical indication of the end.

A *time constraint* (Table 1b) denotes a deadline or time-out on messages, control flow segments or protocols; i.e. the available time limit for the execution of such elements. A time constraint can be defined using relative or absolute date and time.

A *control flow segment (CFS)* represents complex message sequences (Table 1e). It contains a *control flow operator* and one or more interaction paths. An *interaction path* contains an ordered sequence of elements: messages, protocol references, terminations and nested CFSs. An interaction path can contain a *condition* that constrains its execution. The condition expression is based on the evaluation of business documents' attributes and written in natural or formal language such as OCL. The semantics of a CFS depends on the operator used, such as And (parallelism), Xor (exclusive choice), Or (multiple choice), Loop (repetition), MI (Multiple instances), Exception or Cancel (exception handling). Hence, UP-ColBPIP is a block-structured language, a CFS (block) is a construct that represents the divergence of parallel/alternative paths and its end represents the convergence of such paths.

2.1 Speech Acts

A business message defines an interaction or communication between two roles, the message's sender and the receiver. The real semantics of the message is given by the speech act associated with the message. A business message expresses that the role sender has made an action that generated the sending of a speech act that represents sender's intent in the interaction, and also implies the expectation that the receiver reacts according to the semantics of the speech act.

A speech act represents the intent that a given role has (such as accept, propose, reject, inform, etc.) with respect to the information that is being sent in a business message. Thus, a business message is not only understood by the information that is being sent, but also by the intention the sending role has with respect to the information sent in the message. In addition, the act of communication associated with a message represents the action that the sender of the message has made, which is communicated to the receiver and through which a compromise between the parties is created. For example, a message with the speech act "propose" represents a proposal with the commitment of the sender to perform a certain action, which, in order to be effective, must be accepted by the recipient. For this, the receiver can send a speech act "accept-proposal" indicating the acceptance of the proposal. This finally represents the commitment established between the two roles to carry out the initially proposed action.

Speech acts meet specific requirements pursued when modeling CBPs through interaction protocols: modeling interactions between organizations focusing on the public communication aspects of CBPs; representation of complex negotiations; and creation, modification, cancelation, or fulfillment of commitments. The use of speech acts simplifies the design of collaborative processes. Designers can use intuitive concepts closer to the natural language and social interactions between human beings.

However, for a common understanding of collaborative processes based on interaction protocols, speech acts must be known and understood in the same way by all the parties. Consequently, it is necessary to use a rich set of speech acts suitable for expressing any type of intention in collaborative processes. Currently, there is no organization or consortium providing a standard for speech acts that satisfy this requirement for business processes. One solution is to use agent theory, in which there exists a set of properties or attributes that characterize the agents, such as the sociability (communication) due to their interaction with other agents, using a language of communication between agents. Therefore, a well-defined speech acts library to be used in interaction protocols is FIPA ACL [2]. This library proposes a suitable set of speech acts that can be used for collaborative processes.

3   OntoUML Conceptual Model for CBPs Based on Interaction Protocols

In order to provide a conceptual model for CBPs based on interaction protocols we follow the proposal of Guizzardi [10]. This way, the definition of an OntoUML conceptual model of UP-ColBPIP is followed by the definition of an OWL ontology derived from such ontological conceptual model [1]. While OntoUML is a highly-expressive language able to define strongly axiomatized ontologies, the derived OWL ontology guarantees desirable computational properties to execute queries.

OntoUML is a language for ontology-driven conceptual modeling whose constructs represents the ontological distinctions put forth by UFO as well as constraints on how these constructs can be combined. The resulting conceptual models consist of a collection of types (classes) of individuals in the subject domain (e.g. the "Person" kind, the "Child" phase, the "Student" role). Each of these domain types instantiate types in the foundational ontology (e.g. kind, subkind, role, phase, etc.). For a complete presentation and formal characterization of the language, the reader is referred to [11]. In OntoUML, the distinctions of the foundational ontology can be used to provide useful constraints and modeling guidelines, ultimately leading to ontologically well-founded conceptual models.

In this paper, we used the Menthor (available at http://www.menthor.net/) which is a model-driven engineering platform that supports OntoUML modelling. It provides support for model specification (through ontological patterns), automatic syntax verification, validation (through visual simulation) and code generation for the semantic web (RDF/OWL) and software development (information models).

Since the aim of this work is to describe the semantics of CBPs by means of interaction protocols, and hence, by the exchange of messages between organizations, the domain of the models proposed in this work is bounded to the Interaction Protocols View of UP-ColBPIP and to the standard FIPA-ACL. The use of both models in conjunction enables a semantic description of messages exchanged between organizations in a cross-organizational collaboration.

We defined a model for the Interaction Protocols view of UP-ColBPIP that allows the specification of the behavior of CBPs using interaction protocols (Fig. 1). An interaction protocol is conceived as a flow of communications among the sender and receiver roles. Such flow is depicted as a set of associations among an interaction path and some protocol element, i.e., a reference to a business message, a termination, a control flow, and/or a reference to another interaction protocol (Fig. 2). Both, control flow and protocol reference gets its own interaction path as a way to represent a complex structure that allows nested communication flows. The termination elements make it possible to represent a successful or a failed communication. The core element of this view is the business message, which defines an interaction or communication among the message's sender and the receiver. The semantics and information content of each business message is given by the associated speech act and business document associated to the message. It is important to highlight that the business messages are referenced in the context of a given interaction protocol, that is, a business message does not "belong" to any interaction protocol.

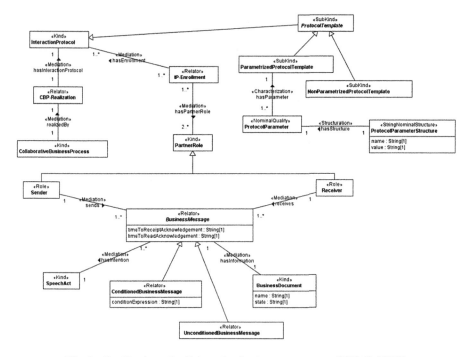

Fig. 1. Realization of collaborative business processes of UP-ColBPIP

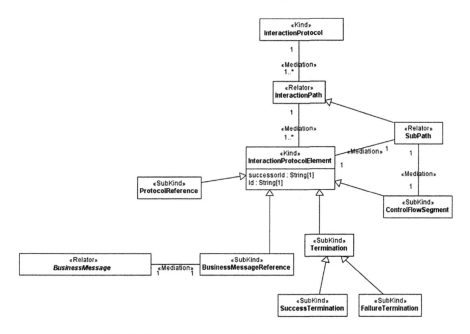

Fig. 2. Interaction protocol elements of UP-ColBPIP

Regarding the speech acts of the business messages, we defined a taxonomy for the standard FIPA-ACL which represents the "intention" of a given communication act (a business message) among the partners. Such taxonomy is based on the work proposed in Villarreal et al. [5] to define speech acts that can be used in CBPs.

Figure 3 shows an OntoUML model with the taxonomy for FIPA-ACL speech acts. In this model, kind SpeechAct is the root of the speech acts taxonomy. The intention relationship between BusinessMessage and SpeechAct refers to the speech act associated with the business message, which represents the intention of the sender with respect to the business document exchanged in the message. See that this class BusinessMessage refers to the same class in Fig. 1. ErrorHandling, InformationExchange, Information-Request and Negotiation classify speech acts according to the type of interactions that could be carried out in a protocol, such as information exchange, request for information, negotiations, commitments to carry out an action, and the handling of errors. This taxonomy could be used as a starting point to define how speech acts could be related to each other in interaction protocols.

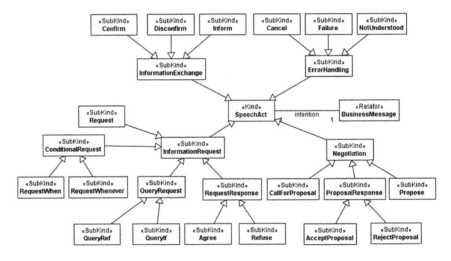

Fig. 3. Taxonomy of speech acts of FIPA-ACL

4 Case Study: A CBP for Collaborative Demand Forecast

This section presents a case study of a CBP [7]. The case study refers to a CBP for the management of a collaborative demand forecast, which occurs in the context of a Vendor Managed Inventory collaborative network to manage the supply of computers and note-books from the supplier company "TK Computers" to the customer company "Computer's Market". Figure 4 shows this CBP modeled with UP-ColBPIP. This process begins with the customer who requests the supplier to make a demand forecast for a product, detailing the data to be considered in the forecast. The supplier may respond in two ways: either accepting the request and agreeing to make the demand forecast requested, or refusing the request and ending the process with a failure. If the supplier agrees the

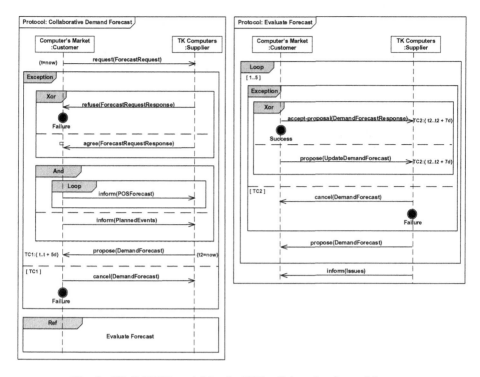

Fig. 4. UP-ColBPIP model for the CBP collaborative demand forecast

request, the customer then sends in parallel (represented by the *And* segment): (i) the sales forecast of each point of sale for the concerned product (represented by a loop together with the message *inform(POSForecast)*), and a program of planned events (message *inform(PlannedEvents)*, e.g., promotions and customer-defined sales strategies. The supplier uses this information to generate and sent to the customer a demand forecast (message *propose(DemandForecast)*). The supplier must respond to the customer within five days from the request of forecast made by the customer. Otherwise, the exception must be handled (represented by the construct *Exception* and the label *TC1: {t..t+5d}*), where the customer must send a cancellation and the process ends with a failure (message *cancel(DemandForecast)*). Then, the customer evaluates the supplier's forecast, which is represented by the subprotocol *Evaluate Forecast*, by comparing it to his sales forecasts and his demand plan, and can respond in two ways: either sending an acceptance response (message *accept-proposal(DemandForecastResponse)*) and ending successfully the process, or sending an updated forecast representing a counterproposal (message *propose(UpdateDemandForecast)*). The customer has a period of one week to respond to the provider if he accepts or updates the forecast from the moment he received a forecast from the provider. If the customer does not respond within that period, an exception must be handled, where the provider must send a cancellation (message *cancel(DemandForecast)*) and the process ends with a failure. A new negotiation cycle starts when the customer sends to the supplier an updated forecast, i.e. the supplier must make a new proposal and the customer can accept or re-update. As restriction to the negotiation, it is assumed that more than five cycles of negotiation cannot occur (represented by the Loop). If the negotiation reaches the maximum number of iterations and does not end with an acceptance, a message must be sent from the supplier to the customer indicating the problems (message *inform(Issues)*) and the protocol ends.

The OntoUML conceptual models of Sect. 3 were implemented as an OWL ontology, according to the automated transformations proposed in [1]. Such ontology was instantiated according to the UP-ColBPIP model of Fig. 4, with the aim to check some structural aspects by means of SPARQL queries. Instantiation was performed using the free ontology editor called Protégé and the Pellet inference engine that provides sound-and-complete OWL-DL reasoning services. The ontology was written in the OWL-DL 2.0 ontology language and serialized in OWL/RDF format[1]. Some of these queries are described below.

Query 1. Documents sent through business messages with speech act *inform*:
        PREFIX up-colbpip: <http://cidisi.frsf.utn.edu.ar/ontologies/up-colbpip#>
        SELECT ?businessDocument
        WHERE {
        ?businessMessage up-colbpip:has_business_document ?businessDocument.
        ?businessMessage up-colbpip:has_speech_act ?speechAct.
        ?speechAct a up-colbpip:Inform.}

This query returns the following elements: POSForecast, PlannedEvents, Issues.

Query 2. Business messages with speech act *negotiation*, classified by partner role :
        PREFIX up-colbpip: <http://cidisi.frsf.utn.edu.ar/ontologies/up-colbpip#>
        SELECT DISTINCT ?businessMessage ?speechAct ?sender ?receiver
        WHERE {
        ?businessMessage up-colbpip:has_speech_act ?speechAct.
        ?speechAct a up-colbpip:Negotiation.
        ?businessMessage up-colbpip:is_sent_by ?sender.
        ?businessMessage up-colbpip:is_received_by ?receiver.}

This query returns the following business messages classified by sender and receiver: *propose(DemandForecast)*,   *accept-proposal(DemandForecastResponse)*   and *propose(UpdateDemandForecast)*.

The classification of speech acts defined in the OntoUML conceptual model of UP-ColBPIP is useful to determine the speech acts related to a *propose* speech act. From the results of this query, it is possible to conclude that interactions that follow the *propose* message in this process model are semantically correct, since the messages contain the speech act *accept-proposal* (which indicates that a part accepts the action proposed by other part) and the speech act *propose* (to represent a counter-proposal), which are part of the speech acts for negotiations according to the model of Fig. 3. However, according

---

[1] Ontology can be founded in: http://dx.doi.org/10.17632/rdkmdr6xfn.2.

to the classification of these speech acts, there is no message with the speech act *reject-proposal* after the propose message. This may suggest that there is no way for partners to finish the negotiation, other than by finishing the five iterations of the negotiation cycle. This issue may indicate a warning to the designer of the CBP to improve the semantics of the negotiation.

Query 3. To know the elements of the protocol that follow an element termination with failure:

        PREFIX up-colbpip: <http://cidisi.frsf.utn.edu.ar/ontologies/up-colbpip#>
        SELECT ?element
        WHERE {
        ?termination a up-colbpip:FailureTermination.
        ?termination up-colbpip:ipe_successor_id ?id.
        ?element up-colbpip:ipe_id ?id.}

This query returns no elements for this process model, since interactions between roles cannot occur once the interaction protocol reached a termination state.

5 Related Work

Different proposals for ontological formalization of business processes are available in the state-of-the-art, remarking the importance of having such an artefact. The first BPMN ontology was called the sBPMN ontology [11] developed within the SUPER project and based on the BPMN 1.0. In [12], a BPMN ontology is defined both to provide a knowledge base useful during the analysis of the process and useful to help in the understanding of BPMN notation, as well as to provide a syntax checker to validate the BPMN model. More recently, BPMN 2.0 was formalized in [3, 4]. Authors in [8] used such ontology to formalize and reason on the behavior of business process models by means of anti-patterns. All these works focus on the ontological representation of the BPMN elements that allow modeling private processes in Business Process Diagrams. However, they do not provide semantics for the elements related with the modeling of CBPs in terms of process choreographies, which can be used from the BPMN version 2.0.

Many efforts have been made to define a best fit methodology for collaborative business process modeling. These approaches lack of formalization and simplicity in terms of their implementation and understanding of the terminologies by non-technical persons [9]. Most of them are based on traditional modeling approaches and no effort has been done in adding semantics for solving those problems.

6 Conclusion and Future Work

This work presented an approach to add semantics to the constructs of the UP-ColBPIP language to model collaborative business processes (CBP) in terms of interaction protocols. The main challenge was to get a strongly axiomatized formal representation of UP-ColBPIP allowing the derivation of a model with proved computational properties and able to support the execution of queries for checking structural properties of interaction protocols.

The defined OntoUML model addresses the requirement of formally representing the semantical structural aspects of the constructs of the UP-ColBPIP language, as well as the definition of a taxonomy of speech acts to indicate all possible intentions that can be associated in business messages to build complex negotiations and commitments in collaborative process models. The generated OWL ontology for the UP-ColBPIP language enables querying and reasoning services over a given instance of the ontology representing a CBP model. This way, it is possible to define queries to investigate structural aspects of a CBP model, as well as queries and rules to provide design guidelines such as suitable combinations of speech acts. This was demonstrated through a case study on the domain of collaborative supply chain management.

A benefit of using OntoUML is a separation of concerns between the conceptualization of the domain and its implementation in OWL. Although OWL and SPARQL are useful for checking properties of a model, we are evaluating other languages and formalisms. As future work, we will explore the capabilities of a mix-in of property graph processing and structured queries. In addition, we will revise and update the defined OntoUML model and the OWL ontology of UP-ColBPIP to add explicit representation of the combinations of speech acts that conduct to appropriate negotiations and commitments in CBPs. The purpose is to build a semantic verification tool that enables the semantic analysis of speech act-based messages in interaction protocols to provide indications about how well it will be conducted the commitments and negotiations in a CBP model. It is also possible to apply this approach on the generation of the private business processes of the organizations from a collaborative business process to keep consistence and conformance about the agreed behavior of the collaboration; and the semantical annotation of business documents indicated in messages of the protocols based on a domain ontology. Another research line is the application of the conceptual models proposed in this work to BPMN by extending choreography diagrams with interaction protocols and speech acts. This would benefit BPMN by enabling it to express the communication aspects of the messages exchanged between organizations and to reason on structural issues of models.

References

1. Barcelos, P., dos Santos, V.A., Silva, F.B., Monteiro, M., Garcia, A.S.: An automated transformation from OntoUML to OWL and SWRL. In: Ontobras, pp. 130–141 (2013)
2. FIPA-ACL: FIPA Communicative Act Library Specification (2002). http://www.fipa.org/specs/fipa00037/SC00037J.html
3. Guido, A., Pandurino, A., Paiano, R.: An ontological meta-model for business process model and notation (BPMN). Int. J. Bus. Res. Manage. 7(3), 40–52 (2016)
4. Rospocher, M., Ghidini, C., Serafini, L.: An ontology for the business process modelling notation. Front. Artif. Intell. Appl. 267, 133–146 (2014)
5. Villarreal, P., Salomone, H.E., Chiotti, O.: Modeling and specifications of collaborative business processes using a MDA approach and a UML profile. In: Rittgen, P. (eds): Enterprise Modeling and Computing with UML. Idea Group Inc., Argentina (2007)

6. Villarreal, P.D., Lazarte, I., Roa, J., Chiotti, O.: A modeling approach for collaborative business processes based on the UP-ColBPIP language. In: Rinderle-Ma, S., Sadiq, S., Leymann, F. (eds.) BPM 2009. LNBIP, vol. 43, pp. 318–329. Springer, Heidelberg (2010). https://doi.org/10.1007/978-3-642-12186-9_30

7. Roa, J., Chiotti, O., Villarreal, P.: Behavior alignment and control flow verification of process and service choreographies. J. Univers. Comput. Sci. **18**(17), 2383–2406 (2012)

8. Roa, J., Reynares, E., Caliusco, M.L., Villarreal, P.: Ontology-based heuristics for process behavior: formalizing false positive scenarios. In: Dumas, M., Fantinato, M. (eds.) BPM 2016. LNBIP, vol. 281, pp. 106–117. Springer, Cham (2017). https://doi.org/10.1007/978-3-319-58457-7_8

9. Aleem, S., Lazarova-Molnar, S., Mohamed, N.: Collaborative business process modeling approaches: a review. In: Proceedings of the 2012 IEEE 21st International workshop on Enabling Technologies: Infrastructure for Collaborative Enterprises, pp. 274–279 (2012)

10. Guizzardi, G.: On ontology, ontologies, conceptualizations, modeling languages, and (meta) models. Front. Artif. Intell. Appl. **155**, 18–39 (2007)

11. Guizzardi, G.: Ontological foundations for structural conceptual models. Ph.D. Thesis. Telematica Instituut Fundamental Research Series No. 15, The Netherlands (2005)

12. Abramowicz, W., Filipowska, A., Kaczmarek, M., Kaczmarek, T.: Semantically enhanced business process modelling notation. In: CEUR Workshop Proceedings, pp. 1–4 (2007)

13. Huemer, C., Liegl, P., Motal, T., Schuster, R., Zapletal, M.: The development process of the un/cefact modeling methodology. In: Proceedings of the 10th International Conference on Electronic Commerce, ICEC 2008 pp. 36:1–36:10. ACM, New York (2008)

14. Roa, J., Chiotti, O., Villarreal, P.: Specification of behavioral anti-patterns for the verification of block-structured collaborative business processes. Inf. Soft. Technol. **75**, 148–170 (2016)

Design of an Extensible BPMN Process Simulator

Luise Pufahl[✉], Tsun Yin Wong, and Mathias Weske

Hasso Plattner Institute, University of Potsdam, Potsdam, Germany
{Luise.Pufahl,Mathias.Weske}@hpi.de, TsunYin.Wong@student.hpi.de

Abstract. Business process simulation is an important means for quantitative analysis of a business process and to compare different process alternatives. With the Business Process Model and Notation (BPMN) being the state-of-the-art language for the graphical representation of business processes, many existing process simulators support already the simulation of BPMN diagrams. However, they do not provide well-defined interfaces to integrate new concepts in the simulation environment. In this work, we present the design and architecture of a proof-of-concept implementation of an open and extensible BPMN process simulator. It also supports the simulation of multiple BPMN processes at a time and relies on the building blocks of the well-founded discrete event simulation. The extensibility is assured by a plug-in concept. Its feasibility is demonstrated by extensions supporting new BPMN concepts, such as the simulation of business rule activities referencing decision models and batch activities.

Keywords: Business process simulation · Extensibility · BPMN

1 Introduction

Business process management (BPM) has been introduced by organizations to make their daily business operations efficient and flexible with respect to their execution environment. For this purpose, business processes are captured by process models with a process modeling language. The resulting process models are used as blueprint for process execution [19]. Once a process is documented, it can be used for simulation to identify its most optimized form [5]. Business process simulation (BPS) is a quantitative analysis technique which provides insights into throughput times, resource utilization and process costs of different process alternatives [4]. It is a cost-effective way to gain insights into an existing or a future situation of the business process execution [1].

The most common process modeling language in BPM practice as well as research is the Business Process Model and Notation (BPMN) which also includes semantics for an automated process execution [11]. Some BPM tool vendors already provide BPMN simulators, e.g., Bizagi Modeler, BonitaSoft, Visual Paradigm, and Trisotech Modeler [5], for which a translation of the BPMN

© Springer International Publishing AG 2018
E. Teniente and M. Weidlich (Eds.): BPM 2017 Workshops, LNBIP 308, pp. 782–795, 2018.
https://doi.org/10.1007/978-3-319-74030-0_62

process diagram in a specific simulation language is not necessary anymore. This avoids errors in the pre- and post-phase of process simulation and eases the usability.

BPS is also used by researchers to evaluate new process modeling artifacts. It eases the communication of the value of new ideas to the BPM community. However, commercial BPMN simulators of tool vendors and academic products, such as BIMP [2], are mainly proprietary and do not support the extensibility by new BPMN constructs. Thus, CPN Tools [15] relying on colored petri nets (CPNs) is mainly applied for process simulation in research [8]. However, it requires a manual translation of the BPMN process diagrams into CPNs and expert knowledge to work with the tool. Further, existing simulators do not support the simulation of multiple processes which run concurrently and compete for the same resources. This is relevant in order to observe the resource utilization in case of several business processes.

In this paper, we present the design and architecture of a proof-of-concept implementation for an extensible BPMN process simulator which also considers the simulation of multiple concurrent business process models. It relies on the building blocks of computer simulation as well as on aspects of existing BPS software. The architecture requires a framework for discrete event simulation (DES). With DES, a real-world process is captured as a finite set of events in time, i.e., each event occurs at a certain point in time and marks the change of the process state [3]. As no changes occur between events, the simulation jumps from one event to another, allowing DES to run fast and independently from real process time in contrast to continuous systems. Tumay [17] describes DES as the "most powerful and realistic tool for analyzing the performance of business processes", which provides "statistical input and output capabilities and advanced modeling elements [...]." The extensibility mechanism of the simulator is based on a designed plug-in structure which provides well-defined entry points into the simulation environment. Additionally, the extensibility has the advantage that BPMN features can be separated in mandatory and optional ones. By outsourcing certain BPMN features in a plug-in, the performance of simulator can be increased, because only those BPMN features are used which are necessary for the respective simulation use case. The feasibility of the plug-in structure is demonstrated based on two recently introduced BPM constructs: (1) business rule activities referencing decision models [12], and (2) batch activities [14].

The remainder of the paper is structured as follows: in Sect. 2 the related work, specifically other academic simulation prototypes and architectures are discussed. Based on the background in Sect. 3 about BPS and DES, the design and architecture of the extensible BPMN process simulator is presented in Sect. 4. Thereby, we discuss the functionality of the different components, the mapping of BPMN constructs into DES, the selection of mandatory BPMN constructs from a simulation perspective, and which are provided by plug-ins, and, finally, the plug-in structure. The feasibility of plug-in structure is demonstrated in Sect. 5 by extending the simulator with features of current BPM research. The paper concludes in Sect. 6.

2   Related Work

Business process simulation (BPS) software can be differentiated into various categories: (1) BPM tools extended with a simulation functionality (e.g., Bizagi Modeler, Trisotech Modeler), (2) stand-alone BPMN simulation tools (e.g., BIMP), and (3) general purpose simulation tools (e.g., CPN, Arena being a discrete event simulator). Several survey exists on the evaluation of different BPS tools, for instance, by Jansen-Vullers and Netjes [8], or by Freitas and Pereira [5] which is specifically on BPMN simulators. However, they do not discuss the general extensibility of the BPS software tools. Existing commercial tools are proprietary and do not provide any extension mechanism; therefore, we focus in the following on architectures and prototypes developed in the academic context:

Wynn et al. [21] present a procedure for BPS with the input and output data of each step. Based on observation of historical data from a process execution engine, the simulation model is instantiated with values referring to a certain point of time in execution, allowing short-term simulation of business processes. The approach is evaluated by using the YAWL workflow engine and CPN tools. This work focuses on the connection between a process execution engine and a simulator; extensibility is not discussed and the BPS is done based on YAWL models.

In [6], García-Bañuelos and Dumas describe an open and extensible business process simulator which transforms a BPMN process diagram into a hierarchical colored petri net (CPN) which is then simulated by CPN tools. The transformation is based on templates describing for each BPMN construct how to map it into CPN fragments. The templates can be adapted or extended for new BPMN constructs. However, the transformation requires very detailed and technical CPNs in order to represent the resource handling and corresponding timing aspects [8]. This leads to a high overhead for developers and needs a profound understanding of CPNs. A similar approach can be found by Krumnow et al. [10] presenting an architectural blueprint for a BPMN simulator in which BPMN concepts are first mapped into a formal language, petri nets, which are easily translatable into DES concepts. In this work, extensibility of the simulator is not discussed. Petri nets can replay the control flow of a business process, but they are constrained in executing specific activity behavior, for instance, simulating a service call.

In contrast, Rücker developed a BPS solution [16] based on the DES software DESMO-J [7]. At the time of development, it was integrated in the open source business process engine JBoss jBPM[1] and supported models in the proprietary language jPDL. Today, it allows the simulation of BPMN models, but it is not open source. It is integrated in the commercial JBoss BPM Suite[2]. In contrast, Wagner et al. [18] suggest to modify DES specification itself to represent

---

[1] https://www.jbpm.org/ [accessed 23 May 2017].
[2] http://developers.redhat.com/products/bpmsuite/overview/ [accessed 23 May 2017].

process activities. The authors present a conceptual model in which activities are represented as a complex DES event having a start event and an end event, and optionally a resource assigned. However, its detail level is not sufficient in comparison to the BPMN specification [11]. For example, the enablement of an activity, which can be different to its start, is missing.

The BIMP simulator developed in the work of Abel [2] is a academic process simulator which has been optimized for performance. This simulator does not base on any existing simulation framework and is not open for extensions. In summary, most of the presented simulation tools lack to support extensibility and multiple concurrently running business processes using the same set of resources. We address these aspects after having introduced some basics on BPS and DES in the next section.

3   Background

This section serves as brief introduction into business process simulation in Sect. 3.1 and discrete event simulation in Sect. 3.2, which are the conceptual basis for the extensible BPMN process simulator.

3.1   Business Process Simulation

To explain the functionality of business process simulators, Fig. 1 provides a generalized architecture. The goal of a business process simulator is to imitate the execution of a number of process instances – different process executions – based on the given simulation input to generate an artificial history [4]. This artificial history delivers information about the probable throughput time, process costs and resource utilization of the current process design. The simulation input consists of the process model, the resource information (resource types, number of resources, timetable etc.) and the stochastic information about the process, such as the arrival rate with its associated distribution, the distribution of the activity duration, and the branch probability [21]. Stochastic information about the process can be either estimated by process experts or identified based on the analysis of process execution logs. The *simulation input* is provided by the *process analyst* with the help of a *BPS modeler* providing means

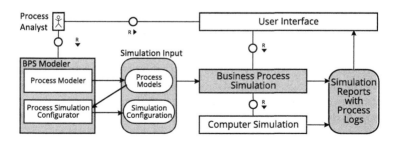

Fig. 1. Architecture of a business process simulator (as FMC block diagram [9]).

to model the process and to provide all relevant information for the simulation. The process modeling functionality is not always part of a business process simulator. When the process analysts starts a simulation run, the *Business Process Simulation* parses the simulation input, sets up the simulation experiment, and runs it for the defined number of instances [21]. In case a general computer simulation tool is used for the simulation, e.g., CPN tools or a DES simulator, the BPMN constructs of the process diagram have to be translated into the concepts of the *Computer Simulation*. When a simulation run is finished, the artificial process logs are generated. They are enriched with resource and cost information. Based on them, simulation reports with statistics on the flow time, process costs, resource utilization etc. can be generated which are presented to the process analyst.

3.2   Discrete Event Simulation

In contrast to continuous simulation in which the system dynamics are tracked over time, in a discrete event simulation (DES) the behavior of a complex system is mapped to a discrete sequence of well-defined events [3]. An event marks a specific change of the system and it is happening at specific instants in time. The simulation can jump from event to event as no state change occurs between two consecutive events which makes the simulation time efficient.

Different frameworks exist that guide the creation of simulation models in terms of time and state, the so-called world views, e.g., the event scheduling approach, where each event contains a routine which decides on the next events to be scheduled, or the activity scanning approach, where activities are due by condition. The activities define a fixed incrementation of simulation time. When an activity finishes, the conditions of all other activities are checked. In addition to the primary world views, several modifications exist, such as the three-phase simulation approach [13]. It is based on the activity scanning approach where each point of time is simulated, leading to slow execution. This drawback is addressed by complementing it with features of the event scheduling.

For our business process simulator, we will use the DES framework DESMO-J[3] which natively supports the process interaction approach and the event scheduling approach. The framework is open for extensions, thus enabling the modification of the event scheduling approach to use the three-phase simulation approach. DESMO-J has not only gained wide acceptance in the academic community; it is also the foundation of many commercial simulation software. It provides blueprints for simulation models in which discrete events and entities can be defined. A simulation model collects and references resources which are required throughout the simulation, defines the built-in logging configuration, and defines the first steps of a simulation, for instance, the initialization of discrete events and the scheduling of the first event. A discrete event is defined by an event routine. It is optionally assigned to an entity which can be modified by the event. To start a simulation, the simulation model is connected to a DESMO-J

---

[3] http://desmoj.sourceforge.net/.

experimentation component, optionally with a temporal end condition. This "black-box" component executes the first steps defined in the simulation model and is responsible to monitor, log, and terminate the simulation.

4   Design and Architecture of the BPMN Process Simulator

After having introduced the basics of BPS and DES, this section introduces the design and architecture of an extensible BPMN process simulator. First, the architecture is sketched in Sect. 4.1 and the functionality of the different components is described. Next, the mapping of the BPMN constructs into DES is presented in Sect. 4.2. Thereby, the coverage of the BPMN constructs by the simulator is discussed. Here, we differentiate between BPMN constructs being covered by the basic simulator and those being captured as plug-in to ease the performance of the simulator. Finally, the plug-in concept is presented in Sect. 4.3 to allow a flexible extensibility of the simulator.

4.1   Architecture

The architecture of the extensible simulator for business process simulation is shown in Fig. 2. To focus on the internal behavior of the process simulator, we limited the architecture regarding the interaction of the simulator with its environment to the mandatory aspects.

Simulation Manager. The *Process Analyst* accesses the *Business Process Simulator* via the *Simulation Controller* and selects the process models and configurations for a simulation run. The selected simulation input is loaded by the

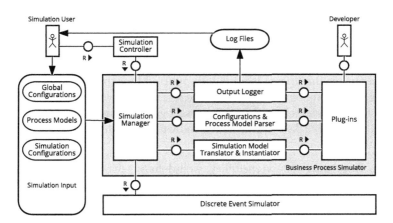

Fig. 2. Architecture of prototypical implementation of the extensible BPMN process simulator (as FMC block diagram [9]).

Simulation Manager which accesses all other components of the process simulator. The *Configurations & Process Model Parser* validates the input and prepares it for translation by the *Simulation Model Translator & Instantiator*. The latter builds the process-specific DES simulation model which is then called by the DES experiment running the simulation. From the insights collected by the Simulation Manager, the *Output Logger* creates *Log Files* which are the output of the simulation.

Simulation Input. As resources can be involved in multiple processes of their organization, the resources are defined in the *Global Configuration*. In this, the type of resources, the number of resources, and their time tables are described; concrete resources can be defined, too. Summarized, the input for the process simulator consists of an arbitrary number of *BPMN Process Diagrams*, a *Simulation Configuration* for each process diagram, in which the arrival rate distribution, the activity duration distribution, the branching probability etc. is given, and one *Global Configuration* file for the organizational details. To ease extensibility, all configuration files are provided in an open format, such as XML.

Configurations and Process Model Parser. This component consists of several parsers to convert each BPMN process diagram and its simulation configuration as well as the global configuration into structures which can be efficiently queried during simulation. These structures are designed in a way, such that they are able to cover the openness of the format used for the configuration files.

Simulation Model Translator and Instantiator. This component is responsible to translate the simulation input into a DES simulation model and to initialize the DES experiment executing the simulation model. Therefore, first, the parsed input is translated into static model components like queues, distributions and structures for data collection. Second, the initial state of the simulation model is instantiated. The initial events, e.g., a event representing the generation of process instance, for the event list of the *Discrete Event Simulator*, are created and activated.

Output Logger. The simulator supports low-level DES logging as provided by the *Discrete Event Simulator* to debug its implementation. For providing a simulation output, high-level logging is supported by capturing process-relevant events (e.g., the enabling of an activity, the start of a new process instance). This is used to finally create the artificial process logs and a simulation report including several calculated key performance indicators (KPIs), such as the average throughput time.

Plug-Ins. Regarding the extensibility, well-defined entry points are desired for the architecture of the process simulator. We provide the concept of *Plug-Ins*, allowing the development of new functionality for the process simulator. A plug-in refers to one single feature and can be switched on and off before simulation, thus supporting modularity on simulation usage level. The plug-in concept will be presented in Sect. 4.3.

Details on the actual prototypical implementation of the extensible BPMN process simulator called *Scylla*[4] can be found in [20].

4.2   Mapping of BPMN Constructs into a Discrete-Event Simulation

After having parsed the simulation inputs and represented them in an internal structure, each identified BPMN construct is translated into a DES construct in order to design the DES simulation model. BPMN constructs which represent an occurrence at a point in time, i.e., start, intermediate and end events, and gateways are mapped directly to discrete events. The routines of the discrete events are responsible for selection and scheduling of the next discrete event based on the configurations of the simulation model. As BPMN tasks are executed over a period of time, their states are mapped to a set discrete events according to the BPMN activity lifecycle [11] with each routine of the event describing the transition from the state. The specific type of each of these constructs is represented in the discrete event by attributes. The events process an entity which represents a process instance. During simulation, the entity stores information on the resources used and events to be scheduled at a later point in time.

If a simulation experiment is started, a discrete event generates a process instance for each process model by creating and scheduling the event which represents the BPMN start event of the process model. To simulate more process instances, it reschedules itself for a later point in time according to the arrival rate, which is defined in the simulation model configurations. The discrete event representing the BPMN start event immediately executes its event routine as described in the previous paragraph. The simulation experiment terminates by either a time constraint defined in the configuration or if all process instances have been completed.

After presenting the mapping, we want to continue with discussing the coverage of BPMN constructs. A plug-in concept does not only support new developers to implement features without modifying the system, it also allows to reduce the size of the simulator by moving optional parts to plug-ins; hence easing the performance of the system. Therefore, it is necessary to define a minimum set of BPMN elements which we consider as sufficient to conduct a BPS. In case of BPMN constructs which are not part of the minimum set, the core process simulator is designed in a way that it will not stop its simulation; instead, the simulator will skip or replaces them with supported constructs of similar type. Further, BPMN constructs have to be identified which are very relevant for process simulations and should be supported by plug-ins.

---

[4] Its source code is available at https://github.com/bptlab/scylla.

As a result, the standard representations of the BPMN constructs *Flow, Task, Start Event, End Event* are part of the minimum set. Since these constructs are located in *Pools* and *Lanes*, these two are supported as well. They refer to task roles and to the idea of resource allocation, which is essential in BPS. Furthermore, gateways are part of the minimum set, because they are

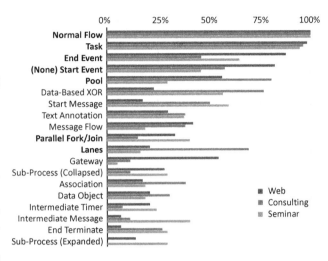

Fig. 3. List of frequently used BPMN constructs and their support in the business process simulator, cf. [23].

responsible for creating, collecting, and directing tokens in the execution of a process model. An exception is the exclusive gateway which should be available as a plug-in to allow different versions of the branch selection. It can be either driven by stochastic distributions or by the values of data objects.

Additional BPMN constructs to be supported by plug-ins are selected based on their usage frequency by practitioners which has been thoroughly investigated by zur Muehlen and Recker [23]. Figure 3 shows a list of the most frequent BPMN constructs, descending order sorted by their coverage in the diagrams [23]. Constructs considered for the core process simulator are written in bold, whereas constructs to be supported in plug-ins are not emphasized. Here, we want to highlight that, in contrast to existing BPS, also a data object plug-in is provided to simulate data in business processes. Additionally, constructs which are not supported are displayed in gray, e.g., *Text Annotations* having no influence on the process execution or *Message Flow* which requires an interaction with another organization.

4.3 Plug-In Structure

In software engineering, extensibility of a system is crucial for third-party developers aiming at implementing new features. Different extensibility mechanisms can be differentiated [22]: Whereas an *open-box* system allows the direct modification of the source code, a *glass-box* system allows the extension of the open source code without changing it. A system is considered as *black-box*, if system details are completely hidden and extensions can only be added by a finite set of mechanism. Extensibility of the BPMN process simulator is allowed by a *grey-box* approach where a list of abstractions is provided which can be refined by plug-in developers like in the black-box approach, but it also provides access to the source code like in the glass-box approach. Thereby, we want to provide a

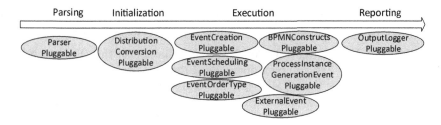

Fig. 4. Abstract classes as entry points for writing plug-ins categorized into the different steps of a simulation.

plug-in structure with a high degree of extensibility which cover the complete simulation flow from input parsing to output logging.

As result, several abstract classes are defined as entry points for developers to write plug-ins shown in Fig. 4 in which they are categorized into the different stages of simulation: parsing, initialization, execution, and reporting. The *ParserPluggable* offers an entry point to the input parsers, such that new simulation input, e.g., a new BPMN element, can be parsed. During initialization of the DES simulation model, the DESMO-J distribution for the arrival rate and activity distributions are set. The entry point *DistributionConversionPluggable* allows the initialization of additional distributions. While executing a simulation experiment, events are generated, stored in queues, and if an event occurs, their event routines are executed which usually results in new events. Different entry points are available to influence DES events: *EventCreationPluggable* to generate new type of events, *EventSchedulingPluggable* to influence the scheduling of events, and *EventOrderTypePluggable* to adapt the priorities of events and changing their order in the queues. For influencing the implemented BPMN behavior of the simulator, two entry points exists: one on the process instance level – *ProcessInstanceGenerationEventPluggable*– and one on the BPMN events level – *BPMNConstructsPluggable*. The latter ones includes several sub-classes to influence the behavior of the minimum set of BPMN elements supported by the basic simulator (cf. the BPMN constructs written in bold in Fig. 3), for instance, *BPMNStartEventPluggable*, *TaskEnableEventPluggable*. The *ExternalEventPluggable* offers the opportunity to add behavior which is not strictly related to a single process instance, but to the general behavior of business process simulation. Finally, a plug-in entry point, the *OutputLoggerPluggable*, is available to extend the simulator logs and output reports.

Our implementation [20] contains plug-ins for several advanced BPMN constructs, for the logging of simulated process executions and for the calculation of their KPIs.

5   Demonstration of Feasibility

In this section, we discuss the feasibility of the plug-in structure by integrating new plug-ins for two new types of BPMN activities: One the hand to enable the simulation of decision-aware activities (i.e., business rule tasks) based on DMN

(Decision Model and Notation) models [12] and on the other hand the simulation of batch activities which allow batch processing in business processes:

Simulation of DMN Models: With the introduction of the DMN standard, a possibility of modeling complex operational decisions was created. The DMN standard complements the BPMN standard and leads to decision-aware business processes. In current BPS, decision models are ignored. Therefore, we extended the simulator to allow the simulation of business rule tasks referencing DMN decision models. For this, two plug-ins were necessary: On the one hand, the *DMNTaskBeginPlugIn* was created which extends the *TaskBeginPluggable* being one of the *BPMNConstructsPluggables*. In case of a business rule task, it uses the input data of the currently simulated process instance and the referenced decision rules to call a rule engine. The output of the rule engine is then written into a output data object of the business rule task. This output can be later used, e.g., for the branching at the exclusive gateway, then being driven by data. On the other hand, a *DMNStatisticLoggerPlugIn* extending the *OutputLoggerPluggable* was created which documents how often which output was returned and which branch was then selected with its associated throughput time and connected costs. The plug-in structure allows a quick way to integrate a new activity behavior by adapting the event routine of the task-begin-event. Further, the access to the output logger allows to generate specific simulation reports.

Simulation of Batch Processing: Recent works, e.g., in [14], have integrated so-called *batch activities* into business processes to allow batch processing. Batch processing enables a business process, which usually act on a single item, to bundle the execution of groups of process instances for particular activities in order to improve its performance. Incorporating batch activities in the BPMN simulator requires an extension with regards to all steps of a simulation: parsing of a new type of BPMN element, changing the execution to allow the collection of process instances and their synchronized execution, and logging of batch-specific KPIs (e.g., costs reduction or waiting time due to batch processing). For realizing a batch activity, five entry points were used. First of all, the *BatchParserPlugIn* extending the *ParserPluggable* was created to parse the extension elements used for a batch activity consisting of different batch configuration parameters, such as the batch activation rule. For a batch activity, the challenge is that the normal activity behavior has to be adapted. Activity instances are interrupted in their execution and only if a certain condition is fulfilled, a batch of instances is executed. Therefore, the *BatchTaskEnablePlugIn* extending the *TaskEnablePluggable* assures that enabled batch activity instances are assigned to a batch cluster after activity enablement. In case a new batch cluster has to be created, a new type of event – a *BatchClusterStartEvent* – is prepared. It implements the blueprint of a DESMO-J discrete event with the batch cluster as its entity. Its event routine schedules all task-begin-event objects of the instances in the batch cluster and selects one process instance which represents the batch execution of the others and is executed as usual. For the remaining instances, the *BatchTaskBeginPlugIn* ensures that the task-begin-event routine is not executed, instead their task-terminate-events are scheduled. After the representative instance was executed, the *BatchTaskTerminatePlugIn* allows that the task-terminate-events of the others are activated and that resulted logging data of the representative ones is taken over by the others. Finally, the *BatchLogger* extending the *OutputLoggerPluggable* creates a report with batch-specific KPIs, such as reduced costs, the maximum and average waiting time of the batch clusters. With the existing plug-in structure, it was possible to integrate a new activity type which also changed the execution semantics. Thereby, the main challenge was after interrupting the process instance execution to return to the normal execution. This could be solved by adapting the scheduling of events using the *TaskEnable-*, *TaskBegin-*, and *TaskTerminatePluggable*.

6   Conclusion

In business process management, simulation has gained acceptance as a tool for quantitative analysis of business processes. This paper showed that existing BPMN simulators are limited in their extensibility. Therefore, the design and architecture was presented for an extensible BPMN process simulator to support also academic researchers in estimating the impact of new concepts on the process execution. From a set of well-founded computer simulation paradigms, discrete event simulation was chosen as the basis of our simulator since the event-based behavior maps directly to the tokenized behavior in process execution. As opposed to existing simulation software, it also supports simulation models consisting of multiple business processes which may be run independently from each other.

The presented simulator enables researchers with programming knowledge to create plug-ins. Plug-ins extend the behavior of the simulator at well-defined interface points which are distributed over all core components, i.e. input parsing and instantiation, simulation experimentation and output logging. Furthermore, they provide modularity on development and usage level and offer separation between mandatory and optional features. The feasibility of the architecture was evaluated by extending the simulator with new BPMN concepts. The extensions cover the full range of pluggables. It showed that the plug-in structure allowed an easy and structured means for integrating new types of BPMN activities including parsing and generation of special simulation output reports. However, further extensions will show whether the current design of plug-in interfaces is complete or need further extensions. We plan, for example, to integrate a resource pluggable with which other resource allocation mechanism (currently the direct and role allocation is supported) can be integrated. Additionally, a plug-in may work properly when used alone, but the interaction between certain plug-ins may cause problems. In future, we plan to evaluate the performance of the simulator with its plug-in concept.

References

1. Van der Aalst, W.M., Nakatumba, J., Rozinat, A., Russell, N.: Business process simulation. Handbook on Business Process Management 1, pp. 313–338. Springer, Heidelberg (2010). https://doi.org/10.1007/978-3-642-00416-2_15
2. Abel, M.: Lightning fast business process simulator. Master's thesis, Institute of Computer Science, University of Tartu (2011)
3. Banks, J.: Discrete-Event System Simulation. Pearson Education, India (1984)
4. Dumas, M., La Rosa, M., Mendling, J., Reijers, H.A., et al.: Fundamentals of Business Process Management, vol. 1. Springer, Heidelberg (2013). https://doi.org/10.1007/978-3-642-33143-5
5. Freitas, A.P., Pereira, J.L.M.: Process simulation support in BPM tools: The case of BPMN. In: 5th International Conference on Business Sustainability. 2100 Projects Association (2015)
6. Garcıa-Banuelos, L., Dumas, M.: Towards an open and extensible business process simulation engine. In: CPN Workshop 2009 (2009)
7. Göbel, J., Joschko, P., Koors, A., Page, B.: The discrete event simulation framework DESMO-J. In: ECMS. pp. 100–109 (2013)
8. Jansen-Vullers, M., Netjes, M.: Business process simulation-a tool survey. In: 7th Workshop on the Practical Use of Coloured Petri Nets and CPN Tools (2006)
9. Knöpfel, A., Gröne, B., Tabeling, P.: Fundamental Modeling Concepts: Effective Communication of IT Systems. Wiley, San Francisco (2005)
10. Krumnow, S., Weidlich, M., Molle, R.: Architecture blueprint for a business process simulation engine. EMISA. vol. 172, pp. 9–23 (2010)
11. OMG: Business Process Model and Notation (BPMN), Version 2.0 (January 2011)
12. OMG: Decision Model and Notation (DMN), Version 1.1 (June 2016)
13. Pidd, M., Cassel, R.A.: Three phase simulation in java. In: Proceedings of the 30th conference on Winter simulation, pp. 367–372. IEEE Computer Society Press (1998)
14. Pufahl, L., Meyer, A., Weske, M.: Batch regions: process instance synchronization based on data. In: 18th International Enterprise Distributed Object Computing Conference (EDOC), pp. 150–159. IEEE (2014)
15. Ratzer, A.V., Wells, L., Lassen, H.M., Laursen, M., Qvortrup, J.F., Stissing, M.S., Westergaard, M., Christensen, S., Jensen, K.: CPN Tools for editing, simulating, and analysing coloured Petri Nets. In: van der Aalst, W.M.P., Best, E. (eds.) ICATPN 2003. LNCS, vol. 2679, pp. 450–462. Springer, Heidelberg (2003). https://doi.org/10.1007/3-540-44919-1_28
16. Rücker, B.: Building an Open Source Business Process Simulation tool with JBoss jBPM. BA Thesis, Stuttgart University of Applied Science (2008)
17. Tumay, K.: Business process simulation. In: Proceedings of the 27th conference on Winter simulation, pp. 55–60. IEEE Computer Society (1995)
18. Wagner, G., Nicolae, O., Werner, J.: Extending discrete event simulation by adding an activity concept for business process modeling and simulation. In: 2009 Winter Simulation Conference (WSC), pp. 2951–2962. Winter Simulation Conference (2009)
19. Weske, M.: Business Process Management: Concepts, Languages, Architectures. Springer, Heidelberg (2010)
20. Wong, T.Y.: Extensible BPMN Process Simulator. Master's thesis, Hasso Plattner Institute, University of Potsdam (2017)

21. Wynn, M.T., Dumas, M., Fidge, C.J., ter Hofstede, A.H.M., van der Aalst, W.M.P.: Business process simulation for operational decision support. In: ter Hofstede, A., Benatallah, B., Paik, H.-Y. (eds.) BPM 2007. LNCS, vol. 4928, pp. 66–77. Springer, Heidelberg (2008). https://doi.org/10.1007/978-3-540-78238-4_8
22. Zenger, M.: Programming Language Abstractions For Extensible Software Components. Ph.D. thesis, École polytechnique fédérale de Lausanne (2004)
23. Muehlen, M., Recker, J.: How much language is enough? theoretical and practical use of the business process modeling notation. In: Bellahsène, Z., Léonard, M. (eds.) CAiSE 2008. LNCS, vol. 5074, pp. 465–479. Springer, Heidelberg (2008). https://doi.org/10.1007/978-3-540-69534-9_35

Author Index

Printed in the United States
By Bookmasters